TABLE 1–2 *SI Derived Units*

Quantity	Unit	SI Symbol	Formula
Acceleration	meter per second squared	—	m/s^2
Area	square meter	—	m^2
Density (mass per unit volume)	kilogram per cubic meter	—	kg/m^3
Force	newton	N	$\dfrac{kg \cdot m}{s^2}$
Pressure or stress	pascal	Pa	N/m^2
Volume	cubic meter	—	m^3
Section modulus	meter to third power	—	m^3
Moment of inertia	meter to fourth power	—	m^4
Moment of force (torque)	newton meter	—	$N \cdot m$
Force per unit length	newton per meter	—	N/m
Mass per unit length	kilogram per meter	—	kg/m
Mass per unit area	kilogram per square meter	—	kg/m^2
Energy or work	joule	J	$N \cdot m$
Power	watt	W	J/s or $\dfrac{N \cdot m}{s}$
Rotational speed	radian per second	—	rad/s

Applied Statics and Strength of Materials

Second Edition

Leonard Spiegel, P.E.
Consulting Engineer

George F. Limbrunner, P.E.
Hudson Valley Community College

Merrill, an imprint of
Macmillan Publishing Company
New York

Maxwell Macmillan Canada, Inc.
Toronto

Maxwell Macmillan International Publishing Group
New York   Oxford   Singapore   Sydney

Cover photo: Thomas Leighton
Editor: Stephen Helba
Developmental Editor: Monica Ohlinger
Production Editor: Christine M. Harrington
Art Coordinator: Lorraine Woost
Text Designer: Anne Flanagan
Cover Designer: Thomas Mack
Production Buyer: Patricia A. Tonneman

This book was set in Times Roman by Bi-Comp, Inc. and was printed and bound by R.R. Donnelley & Sons Company. The cover was printed by Phoenix Color Corp.

The Publisher offers discounts on this book when ordered in bulk quantities. For more information, write to: Special Sales Department, Macmillan Publishing Company, 445 Hutchinson Avenue, Columbus, OH 43235, or call 1-800-228-7854.

Macmillan Publishing Company
866 Third Avenue
New York, NY 10022

Macmillan Publishing Company is part of the
Maxwell Communication Group of Companies.

Maxwell Macmillan Canada, Inc.
1200 Eglinton Avenue East, Suite 200
Don Mills, Ontario M3C 3N1

Library of Congress Cataloging-in-Publication Data

Spiegel, Leonard.
    Applied statics and strength of materials / Leonard Spiegel,
  George F. Limbrunner.—2nd ed.
      p.   cm.
    Includes index.
    ISBN 0-02-414961-6
    1. Statics. 2. Strength of materials. 3. Structural engineering.
  I. Limbrunner, George F. II. Title.
  TA351.S64 1993
  620.1′123—dc20   93-7468
                                                                CIP

Printing: 1 2 3 4 5 6 7 8 9 Year: 4 5 6 7

Applied Statics and
Strength of Materials

Second Edition

MERRILL'S INTERNATIONAL SERIES IN ENGINEERING TECHNOLOGY

Zanger & Zanger, *Fiber Optics: Communication and Other Applications*, 0-675-20944-7

Microcomputer Servicing

Adamson, *Microcomputer Repair*, 0-02-300825-3

Asser, Stigliano, & Bahrenburg, *Microcomputer Servicing: Practical Systems and Troubleshooting*, 2nd Edition, 0-02-304241-9

Asser, Stigliano, & Bahrenburg, *Microcomputer Theory and Servicing*, 2nd Edition, 0-02-304231-1

Programming

Adamson, *Applied Pascal for Technology*, 0-675-20771-1

Adamson, *Structured BASIC Applied to Technology*, 2nd Edition, 0-02-300827-X

Adamson, *Structured C for Technology*, 0-675-20993-5

Adamson, *Structured C for Technology (with disk)*, 0-675-21289-8

Nashelsky & Boylestad, *BASIC Applied to Circuit Analysis*, 0-675-20161-6

Instrumentation and Measurement

Berlin & Getz, *Principles of Electronic Instrumentation and Measurement*, 0-675-20449-6

Buchla & McLachlan, *Applied Electronic Instrumentation and Measurement*, 0-675-21162-X

Gillies, *Instrumentation and Measurements for Electronic Technicians*, 2nd Edition, 0-02-343051-6

Transform Analysis

Kulathinal, *Transform Analysis and Electronic Networks with Applications*, 0-675-20765-7

Biomedical Equipment Technology

Aston, *Principles of Biomedical Instrumentation and Measurement*, 0-675-20943-9

Mathematics

Monaco, *Essential Mathematics for Electronics Technicians*, 0-675-21172-7

Davis, *Technical Mathematics*, 0-675-20338-4

Davis, *Technical Mathematics with Calculus*, 0-675-20965-X

INDUSTRIAL ELECTRONICS/INDUSTRIAL TECHNOLOGY

Bateson, *Introduction to Control System Technology*, 4th Edition, 0-02-306463-3

Fuller, *Robotics: Introduction, Programming, and Projects*, 0-675-21078-X

Goetsch, *Industrial Safety and Health: In the Age of High Technology*, 0-02-344207-7

Goetsch, *Industrial Supervision: In the Age of High Technology*, 0-675-22137-4

Geotsch, *Introduction to Total Quality: Quality, Productivity, and Competitiveness*, 0-02-344221-2

Horath, *Computer Numerical Control Programming of Machines*, 0-02-357201-9

Hubert, *Electric Machines: Theory, Operation, Applications, Adjustment, and Control*, 0-675-20765-7

Humphries, *Motors and Controls*, 0-675-20235-3

Hutchins, *Introduction to Quality: Management, Assurance, and Control*, 0-675-20896-3

Laviana, *Basic Computer Numerical Control Programming*, 0-675-21298-7

Pond, *Fundamentals of Statistical Quality Control*

Reis, *Electronic Project Design and Fabrication*, 2nd Edition, 0-02-399230-1

Rosenblatt & Friedman, *Direct and Alternating Current Machinery*, 2nd Edition, 0-675-20160-8

Smith, *Statistical Process Control and Quality Improvement*, 0-675-21160-3

Webb, *Programmable Logic Controllers: Principles and Applications*, 2nd Edition, 0-02-424970-X

Webb & Greshock, *Industrial Control Electronics*, 2nd Edition, 0-02-424864-9

MECHANICAL/CIVIL TECHNOLOGY

Dalton, *The Technology of Metallurgy*, 0-02-326900-6

Keyser, *Materials Science in Engineering*, 4th Edition, 0-675-20401-1

Kokernak, *Fluid Power Technology*, 0-02-305705-X

Kraut, *Fluid Mechanics for Technicians*, 0-675-21330-4

Mott, *Applied Fluid Mechanics*, 4th Edition, 0-02-384231-8

Mott, *Machine Elements in Mechanical Design*, 2nd Edition, 0-675-22289-3

Rolle, *Thermodynamics and Heat Power*, 4th Edition, 0-02-403201-8

Spiegel & Limbrunner, *Applied Statics and Strength of Materials*, 2nd Edition, 0-02-414961-6

Spiegel & Limbrunner, *Applied Strength of Materials*, 0-02-414970-5

Wolansky & Akers, *Modern Hydraulics: The Basics at Work*, 0-675-20987-0

Wolf, *Statics and Strength of Materials: A Parallel Approach to Understanding Structures*, 0-675-20622-7

DRAFTING TECHNOLOGY

Cooper, *Introduction to VersaCAD*, 0-675-21164-6

Ethier, *AutoCAD in 3 Dimensions*, 0-02-334232-3

Goetsch & Rickman, *Computer-Aided Drafting with AutoCAD*, 0-675-20915-3

Kirkpatrick & Kirkpatrick, *AutoCAD for Interior Design and Space Planning*, 0-02-364455-9

Kirkpatrick, *The AutoCAD Book: Drawing, Modeling, and Applications*, 2nd Edition, 0-675-22288-5

Kirkpatrick, *The AutoCAD Book: Drawing, Modeling, and Applications, Including Release 12*, 3rd Edition, 0-02-364440-0

Lamit & Lloyd, *Drafting for Electronics*, 2nd Edition, 0-02-367342-7

Lamit & Paige, *Computer-Aided Design and Drafting*, 0-675-20475-5

Maruggi, *Technical Graphics: Electronics Worktext*, 2nd Edition, 0-675-21378-9

Maruggi, *The Technology of Drafting*, 0-675-20762-2

Sell, *Basic Technical Drawing*, 0-675-21001-1

TECHNICAL WRITING

Croft, *Getting a Job: Resume Writing, Job Application Letters, and Interview Strategies*, 0-675-20917-X

Panares, *A Handbook of English for Technical Students*, 0-675-20650-2

Pfeiffer, *Proposal Writing: The Art of Friendly Persuausion*, 0-675-20988-9

Pfeiffer, *Technical Writing: A Practical Approach*, 2nd Edition, 0-02-395111-7

Roze, *Technical Communications: The Practical Craft*, 2nd Edition, 0-02-404171-8

Weisman, *Basic Technical Writing*, 6th Edition, 0-675-21256-1

Preface

to the Second Edition

Applied Statics and Strength of Materials presents an elementary, analytical, and practical approach to the principles and physical concepts of statics and strength of materials. It is written at an appropriate mathematics level for engineering technology students, utilizing algebra, trigonometry, and analytic geometry. A knowledge of calculus is not required for understanding the text or for working the homework problems.

The book is intended primarily for use in two-year or four-year technology programs in engineering, construction, or architecture. Much of the material has been classroom tested in our ABET (Accreditation Board for Engineering and Technology) accredited engineering technology programs as well as in our non-ABET accredited technology programs. The text could also serve as a concise reference guide for undergraduates in a first Engineering Mechanics (Statics) and/or Strength of Materials course in engineering programs. Although it is written primarily for the technology student, it could also serve as a valuable guide for practicing technologists and technicians, as well as for those preparing for state licensing exams for professional registration in engineering, architecture, or construction.

The emphasis of the book is on the mastery of basic principles, since it is this mastery that leads to successful solutions of real-life problems. This emphasis is achieved through abundant worked-out example problems, a logical and methodical presentation, and a topical selection geared to student needs. The problem-solving method that we emphasize is a consistent, comprehensive, step-by-step approach. The principles and applications (both example problems and homework problems) presented are applicable to many fields of engineering technology, among them civil, mechanical, construction, architectural, industrial, and manufacturing.

This second edition has been prepared with the objective of including some new topics that will help to prepare the student for more advanced topics and for impending changes in some design approaches and methodology. The authors feel that the trend toward ultimate strength design methods for increasing numbers of materials requires that these topics be addressed in introductory strength of materials courses.

New sections dealing with elastic-inelastic behavior and inelastic bending of beams have been added along with an introduction to ultimate strength design based on the AISC Load and Resistance Factor Design (LRFD) method. A new chapter is included that introduces statically indeterminate beams. New sections on engineering materials and square-threaded screws have been added.

Many sections have been expanded or rewritten. Homework problems have been added and many of the existing homework problems have been revised with the intent of having them better reflect the practical aspects of a range of engineering technology fields. A listing of notation is now included.

The book includes the following features:

- Each chapter is written to introduce gradually the more complex material.
- Homework problems are furnished at the end of each chapter grouped and referenced to a specific section. These are then followed by a group of supplemental problems provided for review purposes. All problems are arranged in order of increasing difficulty.
- Most chapters contain computer homework problems following the section problems. These problems require students to develop computer programs to solve problems pertinent to the topics of the chapter. Any appropriate computer language may be used. The computer problems are another tool with which to reinforce students' understanding of the concepts under consideration.
- Answers to selected problems are provided at the back of the text.
- The primary unit system in this book is the U.S. Customary System. We recognize, however, that the introduction of, and total conversion to, the metric (SI) system in the technology field in the United States will undoubtedly occur in the near future. Therefore, as a means of introducing the SI System, we have written a section in each chapter entitled "SI System Examples." These sections contain additional example problems in which the SI System is used exclusively. SI homework problems are also provided.
- To make the book self-contained, design and analysis aids are furnished in an extensive appendix section. Both U.S. Customary and SI data are presented.
- Calculus-based proofs are introduced in the appendixes.
- The Instructor's Manual includes 300 supplementary problems that can be used for additional assignments or test problems. It also includes complete solutions for all the homework problems in the text and the supplementary problems.

There is sufficient material in this book for two semesters of work in Statics and Strength of Materials. In addition, by selecting certain chapters, topics, and problems, the instructor can adapt the book to other situations, such as separate courses in statics (or mechanics) and strength of materials.

We wish to extend our thanks to our many colleagues, associates and students who with their enthusiastic encouragement, insightful comments and constructive criticisms have helped with the input for this edition.

A special word of appreciation goes to the editorial and production staffs at Macmillan, with particular thanks to Monica S. Ohlinger, our developmental editor, who has been steadfastly patient and helpful. Thanks is extended also to the reviewers for this edition for their help and constructive suggestions: John H. Erion, Jr., Bowling Green State University; Ross C. Lyman, College of Lake County; Neil W. Morgan, Ricks College; George Pillainayagam, Lorain County Community College; Jack K. Poplin, Louisiana State University.

We continue to be indebted to our wives and families for their enduring support, patience and understanding during the term of this project. We affectionately dedicate this edition to them.

Contents

4   **Equilibrium of Coplanar Force Systems**   77

5   **Analysis of Structures**   109

6   **Friction**   143

11 Stress Considerations   295

12 Torsion in Circular Sections   333

13 Shear and Bending Moment in Beams   359

20 Pressure Vessels 657

21 Statically Indeterminate Beams 671

Appendixes 697

1 Introduction

1–1
MECHANICS
OVERVIEW

Mechanics is the oldest and most fundamental of the physical sciences. Its laws and principles underlie all branches of engineering. In fact, so universal are the applications of mechanics that they often seem nonspectacular—of no compelling interest. Like mathematics, mechanics is sometimes thought of as a necessary evil—a means to perhaps more interesting ends, such as design and analysis or research and development of space vehicles, buildings, bridges, automobiles, aircraft, and the like. While mechanics is essential to understand and participate in these endeavors, it also deserves attention in its own right.

Mechanics essentially deals with the study of forces and their effects on bodies that are at rest or in motion. Figure 1–1 illustrates the broad categories encompassed by the field of mechanics.

The first eight chapters of this text are concerned solely with the mechanics of solids, and only with that specialized area designated as *statics*. In statics, we consider forces and force systems acting on rigid bodies which are, and which remain, at rest. In the study of statics, it is assumed that all solid bodies (parts of the structure or machine being considered) are perfectly rigid and do not deform, even under large forces. Statics is basic to the understanding of how structural components and complex systems of buildings, bridges, machines, and equipment perform their function. The spectrum of applications ranges from the very simple (e.g., a child's backyard playgym) to the highly complex (e.g., aircraft/spacecraft structural systems).

Statics also provides a basic foundation for the study of strength of materials, which may be described as a study of the relationships between external forces acting on solid bodies and the internal responses generated by these forces. Here, solid bodies are assumed to be deformable, not rigid.

To describe the scope and use of statics and strength of materials, consider an example: the design of a structure or machine to serve some definite purpose. Such design almost always involves consideration of the following questions: (a) What are the loads that come upon the structure and its parts? (b) How large, in what form, and of what material should these parts be made so that they may sustain these loads safely and economically?

Some of the loads to be supported may be known at the outset; others may be assumed in accordance with engineering experience and judgment. Some loads are prescribed by codes; others must be computed or solved for.

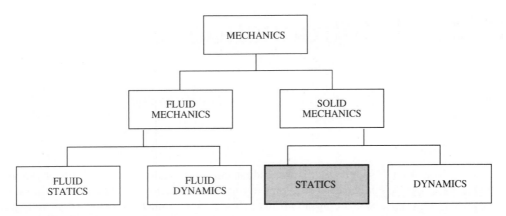

FIGURE 1–1   The field of mechanics.

When all the forces that act on a given part are known, their effect with respect to the physical integrity of the part (that is, their effect in elongating, bending, twisting, compressing, or breaking it) still must be determined. The study of the relations between the forces that act on a body and the changes they produce in its size and form, or the tendency they have to break it, is the province of strength of materials.

Thus, the sequence of statics and strength of materials allows for a beginning understanding of the basic laws and principles involved in both the design and investigation of machine and structural elements.

**1–2
APPLICATIONS
OF STATICS**

Perhaps some of us remember the mistake of teeter-tottering with a bigger kid who laughingly held us aloft at the high end of the plank (see Fig. 1–2). Was there any thought, at the time, that we were at the mercy of the principles of statics (rather than at the mercy of the bigger kid)? Probably not. However, we quickly learned to move the plank so that pivot point would be farther away and more of the plank would be on our side. We couldn't explain why it worked, but it did, and we added that experience to our accumulated knowledge. What we were actually doing was applying one of the principles of statics.

Many everyday examples may be cited in which the understanding of an application is made possible through the science of what we call statics. A few of these are illustrated in Fig. 1–3. All involve the analysis of forces and

FIGURE 1–2   Teeter-totter example.

FIGURE 1–3   Everyday applications of statics.

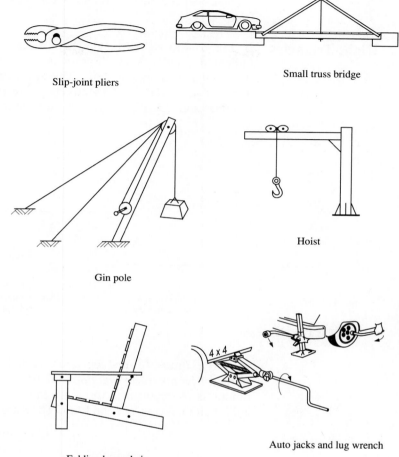

Slip-joint pliers

Small truss bridge

Gin pole

Hoist

Folding lawn chair

Auto jacks and lug wrench

force systems. The basic understanding of forces in many structures and machines is intuitive, or perhaps based on experience, but a detailed analysis can be made only through the rigorous application of the principles of statics. One such application is the analysis of roof and bridge trusses. Trusses are structures whose individual members are so connected as to form a series of triangles.

1–3 THE MATHEMATICS OF STATICS

Statics is an analytical subject that usually requires the physical conceptualization, as well as the mathematical modeling, of a problem. Complicated mathematics is not required in our treatment of the subject. A knowledge of basic arithmetic, algebra, geometry, and trigonometry is sufficient. Because of the importance of directions of forces, and the geometric layout of machines, trusses, frames, and the like, familiarity with the trigonometry of triangles is necessary. A brief review of essential trigonometric relationships follows.

Right Triangles   In Fig. 1–4(a) right triangle *ABC* is shown. The right angle (90°) at *C* is indicated. Angles *A* and *B* are acute (less than 90°) angles. The sum of the three interior angles is 180°. The sides opposite angles *A*, *B*, and *C* are denoted *a*, *b*, and *c*, respectively. Side *c* is the hypotenuse of the right triangle, and the other two sides are the legs (or simply, the sides).

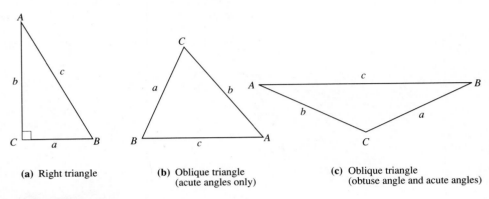

(a) Right triangle   (b) Oblique triangle (acute angles only)   (c) Oblique triangle (obtuse angle and acute angles)

FIGURE 1–4   Types of triangles.

The ratios formed between various sides of the right triangle are termed *trigonometric functions* (or *trig functions*) of the acute angles. The functions of importance to us are the sine, cosine, and tangent. These are abbreviated as sin, cos, and tan. They are defined as follows:

$$\sin = \frac{\text{opposite side}}{\text{hypotenuse}}$$

$$\cos = \frac{\text{adjacent side}}{\text{hypotenuse}}$$

$$\tan = \frac{\text{opposite side}}{\text{adjacent side}}$$

From the preceding definitions, and with reference to the right triangle of Fig. 1–4(a), the following may be written:

$$\sin B = \frac{b}{c} \qquad \cos B = \frac{a}{c} \qquad \tan B = \frac{b}{a}$$

$$\sin A = \frac{a}{c} \qquad \cos A = \frac{b}{c} \qquad \tan A = \frac{a}{b}$$

These values are constant for a given angle, regardless of the size of the triangle; they may be obtained from reference books and scientific hand-held calculators.

A relationship formulated by Pythagoras, a Greek philosopher and mathematician, gives us another tool for use with right triangles. The Pythagorean theorem states that in a right triangle, the square of the hypotenuse equals the sum of the squares of the other two sides. With reference to Fig. 1–4(a),

$$c^2 = a^2 + b^2$$

Knowing two sides of a right triangle, or one side and one of the acute angles, the unknown sides and angles can be computed using the Pythagorean theorem and/or the trig functions.

Oblique Triangles   An *oblique triangle* is one in which no interior angle is equal to 90°. It may have three acute (less than 90°) angles, or two acute angles and one obtuse (greater than 90°) angle, as shown in Fig. 1–4(b) and Fig. 1–4(c). As with the right triangle, the sum of the three interior angles is 180°.

Knowing three sides, or two sides and the included angle, or two angles and the included side, the unknown sides and angles can be computed using the following laws:

1. The law of cosines:

$$a^2 = b^2 + c^2 - 2bc(\cos A)$$
$$b^2 = a^2 + c^2 - 2ac(\cos B)$$
$$c^2 = a^2 + b^2 - 2ab(\cos C)$$

2. The law of sines:

$$\frac{a}{\sin A} = \frac{b}{\sin B} = \frac{c}{\sin C}$$

The letter designations are shown in Fig. 1–4.

The following examples illustrate solutions of both the right triangle and oblique triangle. (Refer to Fig. 1–5 for Examples 1–1 and 1–2.)

☐ **EXAMPLE 1–1**   A 14 ft long ladder leans against a wall with the bottom of the ladder placed 4 ft from the base of the wall as shown in Fig. 1–5(a). How high on the wall will the ladder reach?

FIGURE 1–5   Mathematics of statics examples.

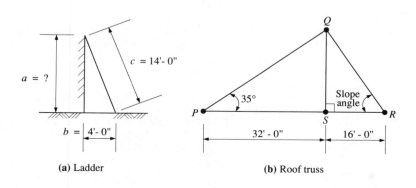

(a) Ladder   (b) Roof truss

Solution   The Pythagorean theorem is used:

$$c^2 = a^2 + b^2$$

Rewriting, substituting, and solving for a,

$$a^2 = c^2 - b^2 = 14^2 - 4^2 = 180 \text{ ft}^2$$

from which

$$a = 13.42 \text{ ft}$$

□ **EXAMPLE 1–2**   For the roof truss shown in Fig. 1–5(b), determine the height QS, the length of the steep slope QR, and the slope angle at R.

Solution   To determine QS, use the tangent of the 35° slope and the 32 ft base, PS:

$$\tan 35° = \frac{\text{opposite}}{\text{adjacent}} = \frac{QS}{32}$$

$$QS = 32(\tan 35°) = 22.4 \text{ ft}$$

To determine QR, use the Pythagorean theorem for triangle QRS:

$$(QR)^2 = (QS)^2 + (SR)^2$$
$$= (22.4)^2 + (16)^2 = 758 \text{ ft}^2$$
$$QR = 27.5 \text{ ft}$$

Now find the angle at R using any of the three trig functions (since all three sides of the triangle QRS are known):

$$\tan R = \frac{\text{opposite}}{\text{adjacent}} = \frac{22.4}{16} = 1.40$$

Determine the angle that has a tangent of 1.40. This is called the *arc tangent* of 1.40 and is written

$$R = \tan^{-1}(1.40)$$
$$R = 54.5°$$

□ **EXAMPLE 1–3**   Compute the angle B between cables AB and BC if a force (load) is applied as shown in Fig. 1–6.

Solution   The sides of triangle ABC (which is *not* a right triangle) have been designated a, b, and c, as shown. Compute distance b using triangle ACD and the Pythagorean theorem:

$$b^2 = 12^2 + 4^2$$
$$b = 12.65 \text{ ft}$$

Compute angle B using triangle ABC and the law of cosines:

$$b^2 = a^2 + c^2 - 2ac(\cos B)$$
$$12.65^2 = 6^2 + 8^2 - 2(6)(8)\cos B$$
$$\cos B = -0.625$$

Therefore,

$$B = \cos^{-1}(-0.625) = 128.7°$$

Note that since cos B is negative, the angle B must lie in the second or third quadrant (where the cosine is negative) and must have a value between 90° and 270°. We select 128.7° because it is apparent that angle B cannot exceed 180°.

FIGURE 1–6 Cable structure.

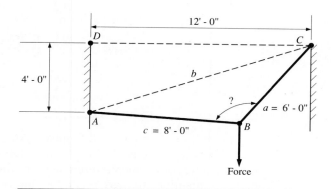

□ **EXAMPLE 1–4**   A rigging boom is supported by means of a boom cable BC as shown in Fig. 1–7. Compute the length of the cable and the angle it makes with the boom (angle C).

FIGURE 1–7 Cable-supported boom.

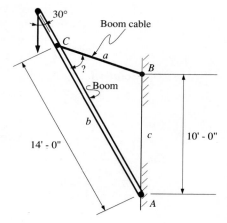

Solution   The sides of triangle ABC are designated a, b, and c, as shown. Compute the length of the boom cable a using the law of cosines. Note that the data needed for the law of cosines are two sides and the included angle, and that the side to be found is opposite the known angle.

$$a^2 = b^2 + c^2 - 2bc(\cos A)$$
$$= 14^2 + 10^2 - 2(14)(10)\cos 30°$$
$$= 53.51 \text{ ft}^2$$

from which

$$a = 7.32 \text{ ft}$$

Then compute the angle that the cable makes with the boom (angle C) using the law of sines:

$$\frac{a}{\sin A} = \frac{c}{\sin C}$$

$$\frac{7.32}{\sin 30°} = \frac{10.0}{\sin C}$$

$$\sin C = \frac{10.0(\sin 30°)}{7.32} = 0.683$$

from which

$$C = \sin^{-1}(0.683) = 43.1°$$

In addition to a required familiarity with trigonometry, one must also be familiar with various algebraic manipulations and equations. One type of problem that is often encountered involves the need to solve for two or more unknown quantities that are related by linear equations. Such equations are called *simultaneous equations*.

The following examples will illustrate two solution methods for this type of problem.

□ **EXAMPLE 1–5**   An engineer lives 5 mi from his office. In an attempt to get some regular exercise, he decides to ride his bicycle for part of the distance and jog the rest. He knows that he can average 18 mph on the bike and 6.0 mph jogging.

He must get to work in one-half hour. How long should he ride and how long should he jog?

Solution   Let

$$x = \text{the length of time to ride (hr)}$$
$$y = \text{the length of time to jog (hr)}$$

The two equations may be expressed as follows:

$$x + y = 0.5 \text{ hr} \qquad \text{(Eq. 1)}$$

and

$$18 \text{ mph } (x) + 6 \text{ mph } (y) = 5.0 \text{ mi}$$

from which

$$18x + 6y = 5.0 \text{ mi} \qquad \text{(Eq. 2)}$$

Algebraic solution method: In order to eliminate one unknown, multiply Eq. 1 by -6 and then add the two equations:

$$
\begin{array}{ll}
-6x - 6y = -3.0 & \text{(Eq. 1)} \\
+18x + 6y = +5.0 & \text{(Eq. 2)} \\
\hline
+12x \qquad\ = +2.0 &
\end{array}
$$

from which

$$x = 0.1667 \text{ hr}$$

Since $x + y = 0.5$, substitute for x:

$$0.1667 + y = 0.5$$
$$y = 0.5 - 0.1667$$
$$y = 0.333 \text{ hr}$$

Substitution method: Solve one of the equations for one of the variables and substitute this expression into the other equation.

Solve Eq. 1 for y:

$$x + y = 0.5 \text{ hr} \qquad\qquad \text{(Eq. 1)}$$
$$y = 0.5 - x$$

Substitute this expression into Eq. 2:

$$18x + 6y = 5.0 \text{ mi} \qquad\qquad \text{(Eq. 2)}$$
$$18x + 6(0.5 - x) = 5.0$$
$$18x + 3.0 - 6x = 5.0$$
$$+12x = 2.0$$
$$x = 0.1667 \text{ hr}$$

Then

$$y = 0.5 - x$$
$$y = 0.5 - 0.1667 = 0.333 \text{ hr} = 20 \text{ min}$$

Another type of algebraic problem that is often encountered is the solution of a quadratic equation. Recall that the general form of a quadratic equation is

$$ax^2 + bx + c = 0$$

Solving for x, or finding the roots of the equation, can be accomplished algebraically by either "completing the square" or by using the quadratic formula. A programmable calculator can also be used for this calculation.

The method of completing the square is accomplished as follows. Use the general form of a quadratic equation, subtract c from each side, and divide through by a:

$$x^2 + \frac{b}{a}x = -\frac{c}{a}$$

Complete the square on the left side of the equation by adding the square of one-half of the coefficient of the linear term. (Add this value to the right side, also.)

$$x^2 + \frac{b}{a}x + \left(\frac{b}{2a}\right)^2 = \frac{b^2}{4a^2} - \frac{c}{a}$$

Rewrite this equation:

$$\left(x + \frac{b}{2a}\right)^2 = \frac{b^2 - 4ac}{4a^2}$$

Take the square root of each side:

$$x + \frac{b}{2a} = \frac{\pm\sqrt{b^2 - 4ac}}{2a}$$

from which

$$x = \frac{-b \pm \sqrt{b^2 - 4ac}}{2a} \qquad\qquad \textbf{(1–1)}$$

Equation (1–1) is called the *quadratic formula*. The use of the quadratic formula or the method of completing the square will lead to the same result.

□ **EXAMPLE 1–6** Determine the roots of the following equation using the method of completing the square:

$$1.230x^2 - 4.71x - 11.30 = 0$$

Solution The coefficients are $a = 1.230$, $b = -4.71$, and $c = -11.30$. First, subtract c from each side (note that the sign of c is negative) and divide the result by a. This result is

$$x^2 - 3.83x = +9.19$$

Add the square of $(b/2)$ to each side of the equation:

$$x^2 - 3.83x + \left(\frac{-3.83}{2}\right)^2 = \left(\frac{-3.83}{2}\right)^2 + 9.19$$

Rewrite as

$$(x - 1.915)^2 = 12.86$$

Take the square root of each side of the equation:

$$x - 1.915 = \pm 3.59$$

from which $x = +5.51$ and -1.675. You may wish to verify the result using the quadratic formula.

**1–4
CALCULATIONS
AND NUMERICAL
ACCURACY**

Solutions to problems cannot be more accurate than the engineering data that are used. When dealing with statics problems, we must consider that the dimensions of structural and machine parts and the loads used in the analysis are accurate only to a certain degree. While calculation methods and tools are capable of handling numbers having many digits, this is usually not warranted.

A *significant digit* is a meaningful digit, one that reflects a quantity that has been measured and is thought to be accurate. The accuracy of a number is implied by the significant digits shown. The number 58 has two significant digits, while the number 7 has only one significant digit.

An ordinary electronic calculator will yield the following:

$$\frac{58}{7} = 8.285714286$$

However, the ten significant digits of the result indicate an accuracy far greater than that of the numbers going into the calculation (one significant digit in the denominator). Logically the result of a calculation should not reflect an accuracy greater than that of the data from which that result is obtained. Therefore, one should guard against implying unwarranted accuracy in this manner and simply round off the result of the preceding calculation to 8 (not 8.0 or 8.00, since these contain two and three significant digits, respectively).

To determine the number of significant digits in a number, begin at the left with the first nonzero digit and count left-to-right across the number. Stop counting at the last nonzero digit unless any trailing zeros are to the right of the decimal point, in which case they are considered significant. Note that the location of the decimal point does not establish the number of significant digits. For example, each of the following numbers has three significant digits:

4.78   47.8   0.478   0.00478 4.78×10^6 0.470

A predicament may exist with a number such as 47,800. The usual assumption is that the two zeros exist to place the decimal point only and are not significant. However, the possibility exists that the two zeros were measured and that there is actually five significant digits. One way around this problem is to use exponential notation:

478×10^2 (three significant digits)
478.00×10^2 (five significant digits)

While it may be difficult to attest to the accuracy of the data available, it is generally agreed that engineering data are rarely known to an accuracy of greater than 0.2 percent. This would be equivalent to a possible error of 100 lb in a 50,000 lb load:

$$\frac{100}{50,000} \, (100) = 0.2 \text{ percent}$$

Rather than tediously determining 0.2 percent of each numerical solution, a general rule of thumb for engineering calculations has evolved: *Represent solutions numerically to an accuracy of three significant digits. If the number begins with 1, then use four significant digits*. This rule keeps us true to the spirit (if not within the letter) of the 0.2 percent guideline.

Adhering to the preceding rule of thumb would result in these numerical representations:

4.78	728	1.742	0.1781
32.1	88,300	0.00968	1056

We have attempted to maintain consistency in our problem solutions by rounding off intermediate and final numerical solutions in accordance with the rule of thumb. For the text presentation, we have used the rounded intermediate solution in the subsequent calculations. When working on a calculator, however, one would normally maintain all digits and round only the final answer. For this reason, the reader may frequently obtain numerical results that differ slightly from those given in the text. This should not cause undue concern.

In rounding off numbers, the following method is used:

1. If the digit to be dropped is 5 or greater, the digit to the left is increased by 1. Example: 47.68 becomes 47.7.
2. If the digit to be dropped is less than 5, the digit to the left remains unchanged. Example: 47.62 becomes 47.6.

1–5
CALCULATIONS AND DIMENSIONAL ANALYSIS

An integral part of the calculation process in mechanics deals with the proper handling of units. In most cases (not all), a unit must be included with a numerical result in order to accurately describe the quantity in question. If the result is to be a calculated distance, for example, the associated unit must be a length unit (ft, in., or the like).

Thoroughness in the calculation process, particularly for the beginner in the mechanics field, should incorporate inclusion of all units in the calculation. For instance, for the simple conversion of 87.3 ft to miles (mi), the calculation would be

$$87.3 \text{ ft}\left(\frac{1 \text{ mi}}{5280 \text{ ft}}\right) = 0.01653 \text{ mi}$$

Notice that the ft units of the original quantity will be cancelled by the ft units in the denominator of the conversion factor (within the parentheses), since ft/ft = 1. Therefore, the resulting unit will be the mile. The cancelling process is indicated by strike marks through the units.

In a similar example, when converting 185 yards to miles, using familiar conversions,

$$185 \text{ yd}\left(\frac{3 \text{ ft}}{1 \text{ yd}}\right)\left(\frac{1 \text{ mi}}{5280 \text{ ft}}\right) = 0.1051 \text{ mi}$$

The inclusion of units is also important (and not only for the beginner) when formulas and calculations become long and complex. When all units are included, they point out any necessary conversions and occasionally also provide clues as to substitution errors. For our purposes, we followed this practice only when it was thought to be particularly instructive (such as in calculations dealing with the metric (SI) system).

An example of a more complex calculation is the determination of the deflection (the sag) of a beam. The terms in this equation are fully discussed in Chapter 16 (see Example 16–3 and Appendix H). Because of the magnitude of the deflection, the result of this calculation is usually expressed in inches (in.). The equation, and the appropriate substitutions, would appear as follows:

$$\Delta = \frac{5wL^4}{384EI}$$

$$= \frac{5(1.2 \text{ kips/ft})(20 \text{ ft})^4(12 \text{ in./ft})^3}{384(30{,}000 \text{ kips/in}^2)(97 \text{ in.}^4)}$$

$$= 1.48 \text{ in.}$$

Since the desired unit for the result is inches, any length units of feet in the equation are converted to inches through the use of the (12 in./ft) conversion factor. The fourth term in the numerator is cubed, since the product of the second and third terms in the numerator yields length units of cubic feet (ft³). The force units of kips (which is short for kilopounds or 1000 lb) will cancel, since *kips* appears in both the numerator and the denominator. In an equation such as this one, it is sometimes helpful to isolate the units for the purposes of checking:

$$\frac{\left(\dfrac{\text{kips}}{\text{ft}}\right)(\text{ft}^4)\left(\dfrac{\text{in.}^3}{\text{ft}^3}\right)}{\left(\dfrac{\text{kips}}{\text{in.}^2}\right)(\text{in.}^4)} = \text{in.}$$

Therefore, the resulting unit is the inch unit, as expected. That the units cancel out, leaving only inches, is a requirement (but does not guarantee a correct numerical result). This procedure is sometimes called *dimensional analysis*.

The preceding discussion applies when one is dealing with basic physical equations that derive from physical laws in which any system of units is valid. For instance, $V = LWH$ will yield the volume of a rectangular prism in cubic units of whatever unit is used for length (L), height (H), and width (W). The units may be in., ft, mm, rods, etc. As in the previously shown calculation for beam deflection Δ, any units may be used providing that proper conversions are made to achieve compatibility. Equations of this type are used, for the most part, in this book.

There is another category of equations, however, in which the equations are set up to yield specific results. These equations are dedicated to use in certain situations. One must ensure that all terms substituted in these equations are in accordance with requirements. An example of such a *dedicated equation* is found in Chapter 12, Eq. (12–7):

$$\text{HP} = \frac{Tn_r}{63{,}025}$$

This equation will yield the power (in arbitrary units of horsepower, hp) transmitted by a rotating shaft as a function of torque T, in inch-pounds, and the speed of rotation of the shaft n_r, in revolutions per minute. The substitutions must be made in those units. A dimensional analysis will lead nowhere. All conversions have been combined in the denominator.

Dedicated equations are extremely useful when many repetitions of the same calculation are called for. They also emphasize the need to understand the equation being used so that it can be used correctly.

1–6
SI UNITS FOR STATICS AND STRENGTH OF MATERIALS

The United States Customary System of weights and measures is used as the primary unit system in this book. Despite the fact that this system eventually will be replaced by the metric system, it is probable that total conversion will not occur soon.

The metric system was legalized for use in the United States by an act of Congress in 1866; however, since its usage was not made mandatory, it was never adopted by the general public. The issue of switching to the metric system in this country has surfaced regularly. Eventually, Congress responded again by enacting the Metric Conversion Act of 1975, which established a U.S. Metric Board to carry out the planning, coordination, and public education to facilitate a conversion to a modern metric system. The Act, however, still did not mandate conversion; it merely encouraged voluntary conversion by those who so desired. Today, the United States stands alone among all the industrialized nations of the world in not having adopted the metric system.

The confusion resulting from the many versions of the metric system that had evolved in different countries over the years prompted the adoption of a modernized metric system by the International General Conference on Weights and Measures in 1960. It was named Le Système International d'Unités (International System of Units), with the international abbreviation SI. Despite the slow progress of adopting the SI system in the United States, a commitment to convert to this system has been generally accepted.

As a means of introducing the SI system into the areas of statics and strength of materials, numerous example problems and problems for solution are furnished at the end of each chapter in this text. The SI applications stand alone as a separate entity as opposed to a mixture of the two systems. In addition, Tables 1–1 through 1–4 provide handy reference information. These tables have been placed just inside the covers of the book for easy access by the reader.

The SI system consists of a limited number of base units that correspond to fundamental quantities, and a large number of derived units (derived from the base units) to describe other quantities. The SI base units and derived units pertinent to statics and strength of materials are listed in Tables 1–1 and 1–2, respectively.

Since the orders of magnitude of many quantities in the SI system cover wide ranges of numerical values, prefixes are used to deal with decimal point placement. SI prefixes representing steps of 1000 are recommended to indicate orders of magnitude. Those recommended for use in statics and strength of materials are listed in Table 1–3. Referring to Table 1–3, it is clear that one has a choice of ways to represent numbers. It is preferable to use numbers between 0.1 and 1000 by selecting the appropriate prefix. For example, 18 m (meters) is preferred to 0.018 km (kilometers) or 18 000 mm (millimeters.)

In the presentation of numbers, the recommended international practice is to set off groups of three digits with a gap as shown in Table 1–3. Note that this method is used on both the left and right sides of the decimal marker for any string of five or more digits. A group of four digits on either side of the decimal marker need not be separated.

It should be noted that in the U.S. Customary System the terms *force* and *weight* are used interchangeably and are expressed in the same units of measure (e.g., pounds, kips, tons). In the SI system, force is derived from the base unit of mass. Mass represents a quantity of matter in an object and is expressed in kilograms, whereas force is expressed in newtons. When the term *weight* is used, it should not be confused with mass. Weight is defined as the force of gravity on a body. Since gravity may vary, the weight of a body may vary. Weight may be computed using Newton's second law of motion, which may be stated mathematically as

$$F = ma$$

where F = force

m = mass

a = acceleration

For the determination of weight, which has been defined as the force of gravity,

Weight = force of gravity on a body

= (mass of the body) × (acceleration of gravity)

or

$$\text{Weight} = mg$$

where g is the acceleration of gravity.

For the determination of the weight of a given mass for purposes of analysis, the recommended value for the acceleration of gravity in the United States may be taken as 9.81 m/s^2.

As an example, for a truck carrying 500 kg (kilograms) of stone, the weight of the stone can be obtained from

$$\begin{aligned}
\text{Weight} = mg &= (500 \text{ kg})(9.81 \text{ m/s}^2) \\
&= 4910 \text{ kg·m/s}^2 \\
&= 4910 \text{ N} \\
&= 4.91 \text{ kN}
\end{aligned}$$

Note that the preceding reflects the fact that 1 kg·m/s^2 is defined as one newton (N).

Table 1–4 is provided to enable rapid conversion from the U.S. Customary System to SI units. This table includes those quantities frequently used in statics and strength of materials.

□ EXAMPLE 1–7   A steel beam weighs 48 lb/ft. Determine the weight per unit length in the SI system.

Solution   The weight of lb/ft in the U.S. Customary System will be converted to N/m (newtons per meter) in the SI system. In the following expression, the second and third terms will convert lb to kg and ft to m, which will yield mass per unit length. The fourth term will convert the mass to force (weight) using $F = ma$. All conversion factors are from Table 1–4.

$$48 \frac{\text{lb}}{\text{ft}} \left(\frac{0.454 \text{ kg}}{1 \text{ lb}}\right)\left(\frac{1 \text{ ft}}{0.3048 \text{ m}}\right)\left(9.81 \frac{\text{m}}{\text{s}^2}\right) = 701 \frac{\text{kg·m}}{\text{m·s}^2} = 701 \text{ N/m}$$

or, using the direct conversion factor,

$$48 \times 14.594 = 701 \text{ N/m}$$

□ EXAMPLE 1–8   In the design of a reinforced concrete beam (bending member), the weight of the beam, itself, is always considered as a load, or force, per unit length of span. In the U.S. Customary System, reinforced concrete is usually assumed to weigh 150 lb/ft³ (pcf).

Calculate the load per unit length of beam in the SI system if the width of the beam is 500 mm and the depth of the beam is 1000 mm.

Solution   Use the conversion factors from Table 1–4 to determine mass per unit volume:

$$150 \frac{\text{lb}}{\text{ft}^3} \left(\frac{1 \text{ ft}}{0.3048 \text{ m}}\right)^3 \left(\frac{0.454 \text{ kg}}{1 \text{ lb}}\right) = 2405 \text{ kg/m}^3$$

The value of 2405 kg/m³ represents a mass per unit volume where the unit volume is one cubic meter. To obtain a force or weight per cubic meter, Newton's law must be applied:

$$\text{Weight} = mg = \left(2405 \frac{\text{kg}}{\text{m}^3}\right)\left(9.81 \frac{\text{m}}{\text{s}^2}\right)$$

$$= 23\,590 \frac{\text{kg·m}}{\text{m}^3 \cdot \text{s}^2} = 23.59 \frac{\text{kN}}{\text{m}^3}$$

Since the beam dimensions are 500 mm by 1000 mm (or 0.500 m by 1.000 m), the weight or load per unit length can be calculated from

$$(0.500 \text{ m})(1.000 \text{ m})\left(23.59 \frac{\text{kN}}{\text{m}^3}\right) = 11.80 \text{ kN/m}$$

Thus, the load per unit length of one meter, due to the weight of the beam, is 11.80 kN.

SUMMARY—BY SECTION NUMBER

1–1 Statics deals with forces and force systems acting on rigid bodies at rest.

1–3 A right triangle may be solved using the Pythagorean theorem and/or trig functions (sin, cos, tan). An oblique triangle may be solved using the law of cosines or the law of sines.

1–4 Numerical accuracy of calculations will be considered sufficient, for our purposes, if solutions are rounded to three significant digits. If the number begins with 1, then use four significant digits. When working on a calculator, maintain all digits and round only the final answer.

1–5 Dimensional analysis is a process used in computations to establish that units on each side of an equation are of the same dimensional form.

PROBLEMS

Section 1–3   The Mathematics of Statics

1. A 36 ft long ladder leaning against a wall makes an angle of 72° with the ground. Determine the vertical height to which the ladder will reach.

2. Two legs of a right triangle are 6 ft and 10 ft. Determine the length of the hypotenuse and the angle opposite the short side.

3. In the roof truss shown in Fig. 1–8, the bottom chord members *AD* and *DC* have lengths of 16 ft and 32 ft, respectively. The height *BD* is 12 ft. Determine the lengths of the top chords *AB* and *BC* and find the angles at *A* and *C*.

5. Freehand sketch the following triangles and solve using the law of cosines: (a) $a = 10$ ft, $b = 12$ ft, $C = 80°$; (b) $b = 78$ ft, $c = 83$ ft, $A = 72°$.

6. One side of a triangular lot is 100 ft and the angle opposite this side is 55°. Another angle is 63°. Sketch the shape of the lot and determine how much fencing is needed to enclose it.

7. A box containing 8 shock absorbers and 10 brake pad sets weighs 101.6 lb. Another box containing 10 shock absorbers and 6 brake pad sets weighs 106.2 lb. Determine the weight of one shock absorber and one set of brake pads. Neglect the weight of the boxes.

8. Calculate the length of AB in Fig. 1–9.

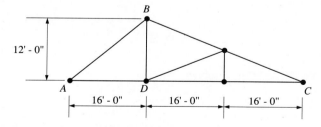

FIGURE 1–8   Problem 3.

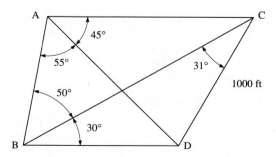

FIGURE 1–9   Problem 8.

4. A plane flies due north for 50 mi, then due west for 15 mi, and then southwest for 35 mi, at which time it lands. At the end of its flight, how far is the plane from its starting point?

9. An Egyptian pyramid has a square base and symmetrical sloping faces. The inclination of a sloping face is 53°07′. At a distance of 600 ft from the base, on level

ground, the angle of inclination to the apex is 27°09'. Find the vertical height and the slant height of the pyramid and the width of the pyramid at its base.

Section 1–5   Calculations and Dimensional Analysis

10. Express 0.015 tons (a) in pounds (lb) and (b) in ounces (oz).

11. Express 5 mi (a) in yards (yd) and (b) in feet (ft).

12. Express 60 miles per hour (mph) in units of feet per second (fps).

13. The soil pressure under a concrete footing is calculated to be three tons per square foot. Express this value in pounds per square foot (psf) and pounds per square inch (psi).

SI System Problems

14. Using Table 1–4, establish your height in mm, your weight in N, and your mass in kg.

15. The dimensions on a baseball field are as follows: pitcher's mound to homeplate, 60.5 ft; first base to second base, 90.0 ft; and homeplate to center field fence, 413.0 ft. Calculate the SI equivalents in meters.

16. Convert the following quantities from U.S. Customary System units to designated SI units: (a) 2212 lb to kg, (b) 87 ft² to mm² and m², (c) 18,460 psi to MPa and kPa.

17. Convert the following quantities to SI units of millimeters and meters: (a) 18.6 ft, (b) 8 ft–10 in., (c) $27\frac{1}{2}$ in.

18. Convert the following quantities to SI units of newtons and kilonewtons: (a) 248 lb, (b) 3.65 kips, (c) 8.7 tons.

19. Convert the following quantities to the designated SI units: (a) 627 in.² to mm² and m², (b) 14 yd² to mm² and m², (c) 3.5 kips/ft to kN/m, (d) 8470 psi to MPa and kPa, (e) 1740 psf to Pa and kPa, (f) 2.8 kips/ft² to kPa, (g) 247 pcf to kN/m³.

Supplemental Problems

20. A ladder rests against a vertical wall at a point 12 ft from the floor. The angle formed by the ladder and the floor is 63°. Calculate (a) the length of the ladder, (b) the distance from the base of the wall to the foot of the ladder, and (c) the angle formed by the ladder and the wall.

21. A surveyor stands on a cliff 25 ft above the water at the edge of the river directly below. The angle of depression from the edge of the cliff to the water's edge on the opposite bank is 12°. How wide is the river at this point?

22. From the top of a 30 ft building, the angle of depression to the foot of a building across the street is 70° and the angle of elevation to the top of the same building is 71°. How tall is the building across the street?

23. Two planes, starting from airport X, fly for two hours, one at 500 mph and the other at 650 mph. The slower plane flies in a direction of 60° east of north and the faster plane flies in a direction of 10° south of east, with reference to X. How far apart are the planes at the end of the two hours?

24. Two observation towers A and B are located 300 ft apart, as shown in Fig. 1–10. An object on the ground between the towers is observed at point C, and the observer on tower A notes that the angle CAB is 68°10'. At the same time, the observer on tower B notes that the angle CBA is 72°30'. Assuming a horizontal surface and towers of equal height, how far is the object from each tower and how high are the towers?

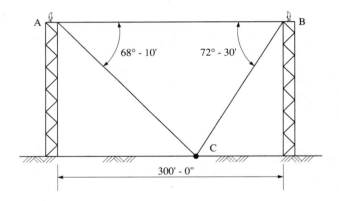

FIGURE 1–10   Problem 24.

25. Solve the following oblique triangles. Note that the sides may have any units of length. (a) $C = 15°$, $b = 60$, $a = 15$; (b) $C = 30°$, $a = 20$, $b = 20$; (c) $C = 74°$, $a = 10$, $b = 20$.

26. What is the average speed in miles per hour (mph) of a runner who runs a mile in 4 min? What is the average speed (mph) of a runner who runs a 26 mi race in 2 hr and 47 min?

27. The volume flow of a river is expressed as 5000 cubic feet per second (cfs). Express the flow in gallons per minute and cubic miles per year. (*Note*: 7.481 gal = 1 ft³.)

28. A 55 gal drum filled with sand weighs 816 lb. The empty drum weighs 38 lb. Find the unit weight of the sand in pounds per cubic foot (pcf).

29. In Fig. 1–11, ABC is a triangular tract of land. The one acre tract DEFG is to be subdivided. AE is 400 ft and DC is 200 ft. Determine the lengths of DG and CB.

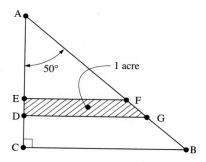

FIGURE 1–11 Problem 29.

30. The total area to be occupied by a building and a parking lot is 90,000 ft². If the area occupied by the building is to be 5000 ft² more than that occupied by the parking lot, what is the area occupied by each?

31. A surveying problem that is sometimes encountered is to determine the distance between two inaccessible points. A and B are the points (both inaccessible) as shown in Fig. 1–12. One solution is to select two other points C and D, from each of which both A and B are visible. Then measure line CD (which is called a *baseline*) and the four angles shown. Assume the following: CD = 456.3 ft, α = 30°41′, β = 40°15′, ϕ = 35°16′, and θ = 56°47′. Find the distance AB.

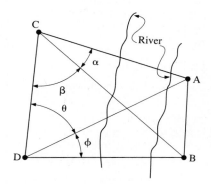

FIGURE 1–12 Problem 31.

2 Principles of Statics

2–1
FORCES AND THE EFFECTS OF FORCES

Statics has been previously described as the science that treats the relationships between forces acting on rigid bodies at rest. With this description in mind, it is sufficient to define a *force* as a push or a pull exerted on one body by another. A more precise definition of a force (although outside of the realm of statics) would be a push or pull tending to produce a change in the motion of the body acted upon. This definition applies to the external effects of a force.

In general, forces have two effects on a body: (a) to cause it to move if it is at rest or to change its motion if it was already moving, and (b) to deform it. In statics we are not concerned with the deformation of a body. The deformation and the internal behavior of a body when subjected to a force lies within the province of strength of materials. Also, the study of the relationships between forces and motion lies within the province of dynamics, rather than statics. In statics our primary concern is with bodies at rest (or moving with zero acceleration), which we will hereafter refer to as bodies in equilibrium. In statics, then, the effect of a force may be described as a tendency to preserve or to upset equilibrium. Also, since supporting or reacting forces are produced by the application of forces to a body not free to move, we may also say that one of the effects of a force is to produce, or bring into action, other forces.

2–2
CHARACTERISTICS OF A FORCE

For practical applications, a force must be completely described. A complete description includes the following information:

1. **Magnitude**—Refers to the size or amount of the force in acceptable units. A 1000 lb force has a larger magnitude than a 500 lb force.
2. **Direction**—Refers to the path of the line along which the force acts and is commonly called the *line of action*. The force may act vertically, horizontally, or at some angle with the vertical or horizontal.
3. **Sense**—Refers to the direction in which a force acts along its line of action. The direction of a force may be vertical, but the sense could be up or down. Similarly, the direction may be horizontal, but the sense could be to the left or to the right. Sense is generally denoted by an arrowhead in a diagram.
4. **Point of application**—Refers to the point on, or in, the object at which the force is applied.

Figure 2–1 portrays the characteristics of a force.

FIGURE 2–1 Characteristics of a force.

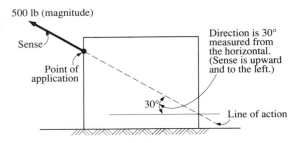

2–3
UNITS OF A FORCE

The unit generally used for expressing the magnitude of a force in the U.S. Customary System is the pound (lb). Multiples of the pound, also commonly used, are the kip, which is equal to 1000 lb, and the ton, which is equal to 2000 lb.

In the SI system, the unit of force is the newton (N). The various prefixes discussed in Chapter 1 are also used, with the kilonewton (kN), which is equal to 1000 N, being very common.

2–4
TYPES AND OCCURRENCE OF FORCES

Forces may be broadly classified according to the way in which they are exerted. Forces may be exerted by contact of one solid body on another solid body (e.g., through a push or pull). Forces may also be exerted directly against a solid body by wind or by water. In addition, forces may be exerted without contact, as in the gravitational attraction of one body for another. Forces in engineering applications that must be dealt with using the principles of statics may encompass any of the foregoing forces acting singly or in combination.

The place of application of most forces with which we are concerned is some portion of the surface area of the body to which the force is applied. If this area of application is large in comparison with the total surface area, the force may be categorized as a *distributed force*. Examples of distributed forces are water pressure against the side of a tank or dam, wind pressure against the side of a building, or the weight of a structural member itself (a beam, for example), which is actually a gravitational force. If the area of application is small in comparison with the total surface area, the force may be regarded as acting at a point. It is then categorized as a *concentrated force*. While the force is actually spread over a small area, and not concentrated at a single point, we consider it to be so concentrated. Moreover, this convention is satisfactory for most engineering solutions. Examples of concentrated forces are the weight of a table and its contents, transmitted to the floor through the leg of the table, and a weight supported by a cable fastened to a beam.

Forces may also be categorized as external or internal. A force is said to be external to a body if it is exerted on that body by some other body. It is said to be internal if it is exerted on a part of that body by some other part of the same body. For example, the external forces acting on the truss of Fig.

FIGURE 2–2   Forces acting on a truss.

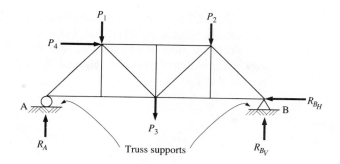

2–2 produce internal forces in the various truss members. The externally applied forces are P_1, P_2, P_3, and P_4. These forces are, in turn, transmitted from their points of application through the various members of the truss to the external supports at A and B. Internal forces are developed in the truss members as a result of the members pushing or pulling on other members at points of intersection.

Note that the externally applied forces are all concentrated forces and are exerted by other bodies. They are said to be acting on the truss. External resisting forces R_A, R_{B_V}, and R_{B_H} are shown at the external supports. These are commonly called *reacting forces*, or *reactions*.

2–5
SCALAR AND
VECTOR QUANTITIES

A *scalar quantity* is a quantity that has magnitude only. For example, the population of a city, a volume of water, a duration of time, and an amount of money are all scalar quantities. Scalar quantities may be added algebraically, with the result still having magnitude only.

A *vector quantity* is any quantity that has magnitude, direction, and sense. These attributes can be represented by a *vector*, which is a segment of a straight line, with the sense expressed by an arrowhead at one end. If drawn to scale, the vector will be of a definite length, representing its magnitude. A force is vector quantity because a push or pull is exerted in a certain direction and sense, as well as with a certain magnitude. Other quantities commonly represented by vectors are displacements, velocities, and accelerations.

Vector quantities cannot be added algebraically, since the direction of the vector is as significant as the magnitude. One exception is illustrated in Fig. 2–3 where two vectors (forces) have the same line of action. In such a case, the vectors may be added algebraically, in the manner of scalar quantities.

FIGURE 2–3   Vectors with same line of action.

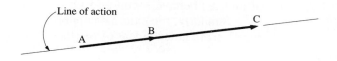

With forces having different lines of action, the vectors must be added vectorially (geometrically). For example, in Fig. 2–4, vector AB represents the vectorial sum of vectors AC and BC. In effect, the single force vector AB is equivalent to vectors AC and BC and produces the same effect as the forces it replaces.

FIGURE 2–4   Vectors with different lines of action.

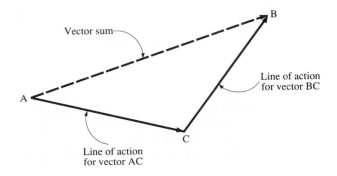

2–6
THE PRINCIPLE OF TRANSMISSIBILITY

In Fig. 2–5, a block is being pulled along a horizontal surface by a horizontal force P. The motion of the block is independent of whether it is pulled or pushed by the force, as long as P has the same magnitude, direction, and sense and acts along the same line of action. This is termed the *principle of transmissibility*. It states, in effect, that the external effect of a force on a body is the same for all points of application along its line of action. The effect is independent of the point of application.

FIGURE 2–5   Principle of transmissibility.

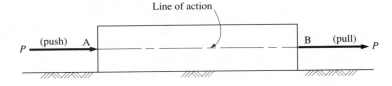

When applying the principle of transmissibility, one must be aware that only external effects should be considered. The internal effect of a force is dependent on its point of application. Consider the forces within the block of Fig. 2–5. If the force P were applied at A, the block would tend to be crushed; on the other hand, if the force were applied at B, the block would tend to be elongated. Therefore, we see that two completely different types of internal behavior would result.

Similarly, the external support forces at A and B of the truss shown in Fig. 2–6 are independent of whether P is applied at C or at D. However, the

FIGURE 2–6   Principle of transmissibility.

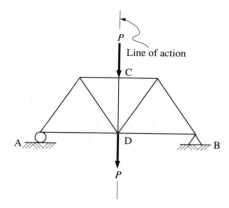

internal forces in some of the truss members will be appreciably different depending on whether the point of application is at C or D.

**2–7
TYPES OF FORCE
SYSTEMS**

A *force system* may be defined as any number of forces that are collectively considered. There are two broad classifications of force systems: coplanar and noncoplanar. A coplanar force system is one in which the lines of action of all the forces lie in the same plane. A noncoplanar system of forces is one in which the lines of action of all the forces do not lie in the same plane.

The coplanar and noncoplanar force systems may be further classified. When the lines of action of all the forces intersect at a common point, the system is said to be concurrent. If the lines of action of all the forces are parallel, the system is said to be a parallel force system. If the action lines do not intersect at a common point and are not parallel, the system is said to be a noncurrent system of forces.

If all the forces in a parallel system act along a single line of action, the system is said to be collinear. Actually, this is considered a special case of the coplanar concurrent force system.

The classification of force systems may be summarized as follows:

1. coplanar concurrent
2. coplanar nonconcurrent
3. coplanar parallel
4. noncoplanar concurrent
5. noncoplanar nonconcurrent
6. noncoplanar parallel

Since most structural systems can be reduced to coplanar force systems, from a design and analysis standpoint, only coplanar force systems will be considered in this text. These are summarized graphically in Fig. 2–7.

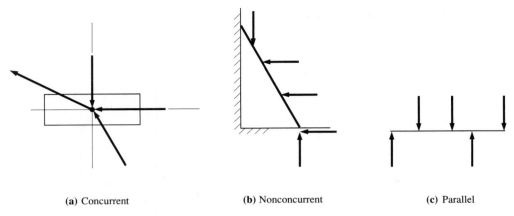

(a) Concurrent   (b) Nonconcurrent   (c) Parallel

FIGURE 2–7 Types of coplanar force systems.

2–8
COMPONENTS OF A FORCE

It is frequently convenient to replace a given force in order to simplify an analysis. Any given force may be replaced by an alternate set of forces provided that the replacement forces have the same effect as that of the original force.

As an example, consider the force due to wind against the roof of a building. The direction of the wind force is perpendicular to the sloping surface, as shown in Fig. 2–8(a). To simplify certain aspects of this problem, we may choose to replace the inclined force with the vertical and horizontal forces shown in Fig. 2–8(b). The vectorial sum of these two forces would be equivalent to the inclined force. In other words, the inclined force would be replaced by an equivalent rectangular force system, with the two forces acting at right angles to each other and having the same effect as the applied inclined force.

The process of replacing a force with a rectangular system in which the two forces are at right angles to each other, or the special case in which the forces are vertical and horizontal, is called resolving a force into its components. When the two components are at right angles to each other, they are called *rectangular components* and are usually designated as X and Y when the conventional X–Y coordinate axes are used.

FIGURE 2–8 Inclined wind force.

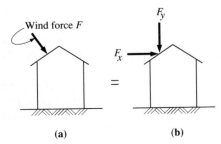

(a)   (b)

There are an infinite number of possible sets of components for a given force, but usually only one or two specific sets are of interest. In most engineering problems, the components of interest are the rectangular components with forces at right angles to each other.

Rectangular components may be computed analytically using the conventional X–Y (horizontal–vertical) coordinate axes system as shown in Fig. 2–9. An arbitrary vector representing a force F is shown applied to a hypothetical body at point O. The force acts at angle θ_x with the horizontal X axis. As an arbitrary sign convention, horizontal forces acting to the right and vertical forces acting upward will be taken as positive, while horizontal forces to the left and vertical forces acting downward will be taken as negative. The sign convention is indicated in the figure.

FIGURE 2–9 Rectangular components of a force.

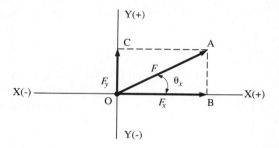

If the vector F is projected upon the X and Y axes as shown, the rectangular components F_x and F_y are formed and can then be determined. Note that the components form a concurrent force system.

Since triangle OAB is a right triangle, the relationship between the components and the force F can be determined by the sine and cosine functions of the angle θ_x. Since BA = F_y and CA = F_x,

$$\cos \theta_x = \frac{F_x}{F} \quad \text{and} \quad \sin \theta_x = \frac{F_y}{F}$$

which can be rewritten as

$$F_x = F \cos \theta_x \tag{2–1}$$
$$F_y = F \sin \theta_x \tag{2–2}$$

Note that the choice of the orientation of the X and Y axes is arbitrary. Any orientation may be chosen, not necessarily vertical and horizontal (although this is usually the most convenient).

It is evident from Fig. 2–9 that if the two rectangular component vectors are known, the single vector F, which would have the same effect on the body, can be computed. This single vector is called the *resultant vector*, the *resultant force* or, simply, the *resultant*. (Resultants will be discussed in detail in Chapter 3.) Using the right triangle OAB and the Pythagorean theorem,

$$F^2 = F_x^2 + F_y^2$$

from which

$$F = \sqrt{F_x^2 + F_y^2} \tag{2-3}$$

Additionally, the direction (inclination with the horizontal X axis) of the vector F would be

$$\theta_x = \tan^{-1} \frac{F_y}{F_x} \tag{2-4}$$

The sense of F would be determined by the signs of the component vectors, as previously discussed, in combination with a diagram such as that of Fig. 2–9. In this case, the sense would be upward and to the right.

It should be noted, at this point, that the preceding discussion has been developed using the notation of F for vector, or force. It must be recognized that this is only a general notation, and that a force may be denoted in many ways, either subscripted or not. Resultant forces are frequently, but not always, denoted R. Equation (2–3), for example, may be referenced but the expression written as

$$R = \sqrt{Q^2 + S^2}$$

where the forces involved are denoted R, Q, and S. This is an equally valid use of the principle of the Pythagorean theorem.

□ **EXAMPLE 2–1**   Compute the rectangular components (vertical and horizontal) for the 400 lb force shown in Fig. 2–10.

FIGURE 2–10   Rectangular components of a force.

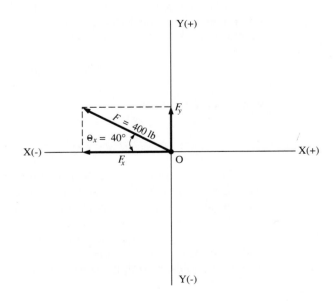

Solution   The force (labeled F) is shown acting through the origin of X–Y coordinate axes. Its inclination with the horizontal X axis is 40°. Note that the diagram does not have to be drawn to scale; however, to the extent possible, the various parts should be drawn in proportion to each other.

Projecting force F upon the X and Y axes reveals that the sign of F_y is positive (acting upward) and that the sign of F_x is negative (acting to the left). Therefore, assign a negative sign to the F_x component. Using Eqs. (2–1) and (2–2),

$$F_x = -F \cos \theta_x = -400(\cos 40°) = -306 \text{ lb}$$
$$F_y = +F \sin \theta_x = +400(\sin 40°) = +257 \text{ lb}$$

☐ **EXAMPLE 2–2**   Figure 2–11 shows a body on a 20° inclined plane. The weight of the body represented by the force W is 50 lb. Resolve this force at point O into components parallel and perpendicular to the inclined plane.

FIGURE 2–11   Rectangular components of a force acting on an inclined plane.

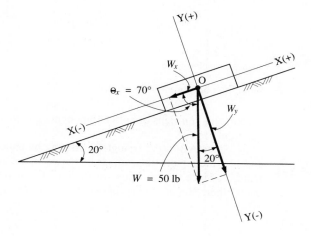

Solution   The X–Y coordinate axes are sketched with the X axis parallel to the incline and the Y axis perpendicular to the incline. The 50 lb force is shown acting through the origin. Note that since the force W represents the weight of the body, it is a gravity force and acts vertically downward.

Projecting the force W upon the X and Y axes reveals that the signs of both components are negative. W_x acts to the left and W_y acts downward. Using Eqs. (2–1) and (2–2),

$$W_x = -W \cos \theta_x = -50(\cos 70°) = -17.1 \text{ lb}$$
$$W_y = -W \sin \theta_x = -50(\sin 70°) = -47.0 \text{ lb}$$

☐ **EXAMPLE 2–3**   In Fig. 2–12, a force P of 25 lb is acting at a slope as shown. Resolve the force into its rectangular components (vertical and horizontal).

Solution   Note that the direction of the force is defined by a slope triangle, as shown in Fig. 2–12(b), rather than by an angular value. Projecting force P upon the X and Y axes

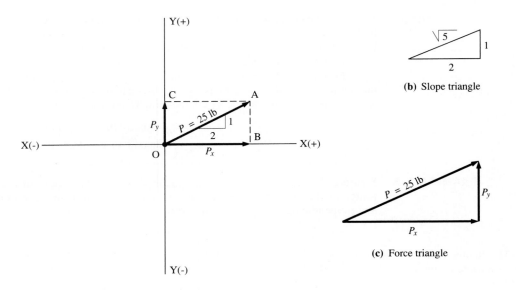

(b) Slope triangle

(c) Force triangle

(a) X-Y axes with applied force

FIGURE 2–12   Rectangular components of a force.

reveals that the signs of both components are positive. P_y acts upward and P_x acts to the right.

Since AB $= P_y$, triangle OAB may be called a *force triangle* and is shown in Fig. 2–12(c.) Note the similarity between the force triangle and the given slope triangle. Since the triangles are similar, the corresponding sides are proportional to each other. Hence, the components P_x and P_y can be computed as follows:

$$\frac{P_y}{1} = \frac{P_x}{2} = \frac{P}{\sqrt{5}}$$

from which

$$P_x = +\frac{2P}{\sqrt{5}} = +\frac{2(25)}{\sqrt{5}} = +22.4 \text{ lb}$$

$$P_y = +\frac{1P}{\sqrt{5}} = +\frac{1(25)}{\sqrt{5}} = +11.2 \text{ lb}$$

The force triangle concept is a convenient means for visualizing the relationship between resultant and component forces. Note that component P_y still acts through point O. The construction of the force triangle is merely a graphical representation of the vector relationships.

☐ **EXAMPLE 2–4**   The rectangular components of a force are +100 lb in the Y direction (F_y) and −200 lb in the X direction (F_x). Both forces act through point O, as shown in Fig. 2–13(a). Calculate the magnitude, inclination with the X axis, and the sense of the resultant force F.

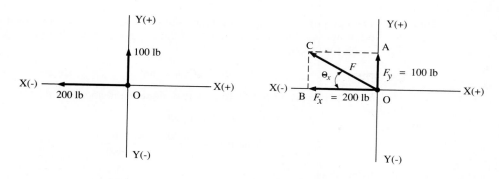

(a) Component forces

(b) Resultant calculation

FIGURE 2–13 Force relationships.

Solution Using an X–Y coordinate axis system, construct a rectangle by drawing lines AC parallel and equal to F_x, and BC parallel and equal to F_y, as shown in Fig. 2–13(b). The diagonal of the rectangle represents the unknown force F.

Using the right triangle OBC where BC = F_y along with Eq. (2–3),

$$F = \sqrt{F_x^2 + F_y^2} = \sqrt{(-200)^2 + 100^2} = 224 \text{ lb}$$

The inclination of the resultant with the X axis can be obtained using Eq. (2–4):

$$\theta_x = \tan^{-1}\frac{F_y}{F_x} = \tan^{-1}\frac{100}{200} = 26.6°$$

The sense of force F is upward and to the left and is shown in Fig. 2–13(b).

□ **EXAMPLE 2–5** A heavy weight is being dragged by two tractors B and C, as shown in Fig. 2–14. Tractor B exerts a pull of 800 lb at an angle of 20°. Tractor C exerts a pull of 400 lb. Determine the angle θ required to move the weight along the path of the X axis.

FIGURE 2–14 Plan view—
Example 2–5.

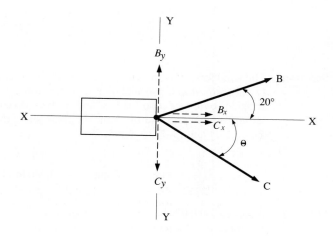

Solution   The components of each force are shown. The Y components must balance each other out, so that only the effects of the components parallel to the X axis remain. From Eq. (2–2),

$$B_y = B \sin 20° \quad \text{and} \quad C_y = C \sin \theta$$

Since you want B_y and C_y to be equal, equate and solve for θ:

$$C \sin \theta = B \sin 20°$$

$$\sin \theta = \frac{B \sin 20°}{C} = \frac{800(\sin 20°)}{400} = 0.684$$

from which

$$\text{Required } \theta = 43.2°$$

2–9
SI SYSTEM
EXAMPLES

□ **EXAMPLE 2–6**

The roof of a building is subjected to a wind force F of 600 N, as shown in Fig. 2–15. Resolve this force into its vertical and horizontal rectangular components.

FIGURE 2–15   Wind force on inclined surface.

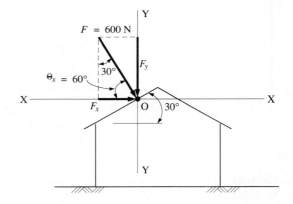

Solution   The wind force F is shown acting through the origin of the X–Y coordinate axes. Due to the roof slope, the inclination of the force with the horizontal X axis is 60°. Projecting force F upon the X and Y axes reveals that the sign of F_y is negative (acting downward) and the sign of F_x is positive (acting to the right). Therefore, use Eqs. (2–1) and (2–2) to obtain

$$F_x = +F \cos \theta_x = +600(\cos 60°) = +300 \text{ N}$$
$$F_y = -F \sin \theta_x = -600(\sin 60°) = -520 \text{ N}$$

□ **EXAMPLE 2–7**   A force F of 950 N is applied to a steel tank at a slope, as shown in Fig. 2–16. Resolve the force into its rectangular components (vertical and horizontal).

FIGURE 2–16   Inclined force
acting on a tank.

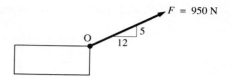

Solution   The direction of the force is defined by a slope triangle, rather than by an angle. Projecting force F upon the X and Y axes reveals that the signs of both components are positive. F_y acts upward and F_x acts to the right.

In Fig. 2–17(a), F is shown as diagonal OA of rectangle OCAB. Note that in OCAB, AB is equal to F_y. Triangle OAB may be considered to be a force triangle, as shown in Fig. 2–17(c). Note that the force triangle and the slope triangle are similar, with the corresponding sides proportional to each other. Express this mathematically as

$$\frac{F_x}{12} = \frac{F}{13} = \frac{F_y}{5}$$

from which

$$F_x = \frac{12F}{13} = \frac{12(950)}{13} = 877 \text{ N}$$

and

$$F_y = \frac{5F}{13} = \frac{5(950)}{13} = 365 \text{ N}$$

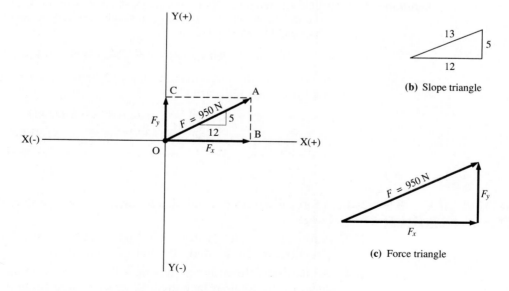

(b) Slope triangle

(a) X-Y axes with applied force

(c) Force triangle

FIGURE 2–17   Rectangular components of a force.

☐ **EXAMPLE 2–8**   A block having a mass of 400 kg rests on an inclined surface, as shown in Fig. 2–18. Determine the components of gravitational force exerted on the block perpendicular and parallel to the inclined surface.

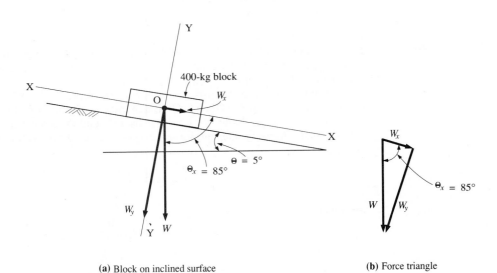

(a) Block on inclined surface   **(b)** Force triangle

FIGURE 2–18   Rectangular components of a force.

Solution   For convenience, reference X–Y axes are established with the X axis parallel to the incline. The weight W (which is the force of gravity exerted on the block), shown in Fig. 2–18(a), can be calculated from

$$W = mg = 400 \text{ kg}(9.81 \text{ m/sec}^2) = 3924 \text{ N} = 3.924 \text{ kN}$$

Noting θ_x and the force triangle of Fig. 2–18(b), calculate the rectangular components from Eqs. (2–1) and (2–2):

$$W_x = W \cos \theta_x = 3.924(\cos 85°) = 0.342 \text{ kN}$$
$$W_y = -W \sin \theta_x = -3.924(\sin 85°) = -3.91 \text{ kN}$$

SUMMARY—BY SECTION NUMBER

2–1   A force is a push or a pull that tends to change the state of motion of a body.

2–2   A force is completely described by four quantities: (a) magnitude, (b) direction, (c) sense, and (d) point of application.

2–4   A distributed force is one having a large area of application with respect to the total surface area. A concentrated force is one having a small area of application with respect to the total surface area.

2–5   A force is a vector quantity. A vector is a graphical representation of a vector quantity.

2–6 The principle of transmissibility states that the external effect of a force on a body is the same for all points of application along its line of action.

2–7 Coplanar forces lie in the same plane. Noncoplanar forces do not lie in the same plane. The lines of action of concurrent forces intersect at a common point; those of nonconcurrent forces do not. Force systems (coplanar and noncoplanar) may be categorized as (a) concurrent, (b) nonconcurrent, (c) parallel. A force system is collinear if all forces act along a single line of action.

2–8 A single force may be replaced by two or more forces called components that will produce the same effect as the single force they replace. The rectangular components of a single force F, in terms of the angle θ_x at which it is inclined with the X axis, are expressed by

$$F_x = F \cos \theta_x \tag{2–1}$$

$$F_y = F \sin \theta_x \tag{2–2}$$

Conversely, if the components are known, the single force, called a resultant, can be obtained using the Pythagorean theorem:

$$F = \sqrt{F_x^2 + F_y^2} \tag{2–3}$$

and the inclination (θ_x) of the force F with the X axis would be expressed as

$$\theta_x = \tan^{-1} \frac{F_y}{F_x} \tag{2–4}$$

PROBLEMS

Section 2–8   Components of a Force

1. Compute the vertical and horizontal components for each of the forces shown in Fig. 2–19.

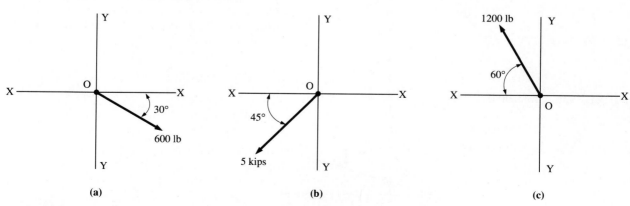

(a)   (b)   (c)

FIGURE 2–19   Problem 1.

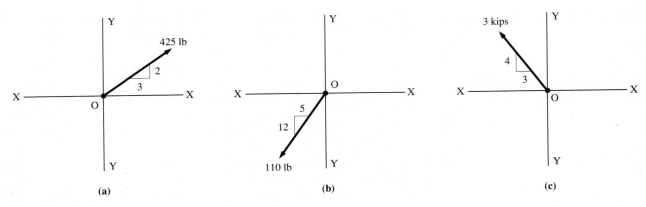

(a)   **(b)**   **(c)**

FIGURE 2–20 Problem 2.

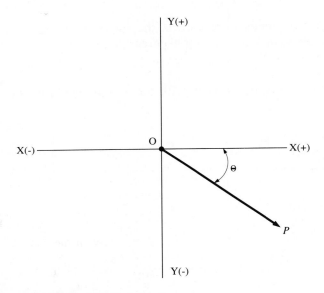

FIGURE 2–21 Problem 3.

2. Compute the vertical and horizontal components for each of the forces shown in Fig. 2–20.

3. Compute the vertical and horizontal components for the given values of P and θ. The angle is measured clockwise from the positive X axis to the force P, as shown in Fig. 2–21. (a) 320 lb, 20°; (b) 640 lb, 30°; (c) 320 lb, 40°; (d) 320 lb, 88°.

4. Compute the rectangular components parallel and perpendicular to the inclined planes shown in Fig. 2–22.

SI System Problems

5. Compute the vertical and horizontal components of the given values of P and θ. The angle is measured clockwise from the positive X axis to the force P (refer to Fig. 2–21). (a) 120 kN, 30°; (b) 120 kN, 75°; (c) 120 kN, 105°.

6. The rectangular components of a force are $F_y = +1.3$ kN and $F_x = +103$ N. Calculate the magnitude, inclination with the horizontal axis, and the sense of the resultant force F.

FIGURE 2–22 Problem 4.

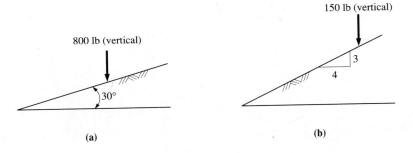

(a)   **(b)**

FIGURE 2–23   Problem 13.

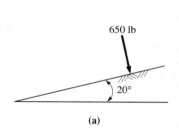

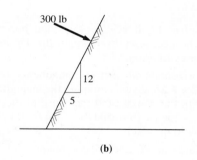

(a)

(b)

7. A rope is tied to the top of a 10 m flagpole. The rope is tensioned to 120 N and then tied to a stake in the ground 20 m from the bottom of the pole. The ground is level. Find the vertical and horizontal components of force applied at the top of the flagpole.

8. A body having a mass of 40 kg rests on a frictionless surface inclined at 20° with the horizontal. The body is kept in position by a force acting parallel to the inclined surface. Compute the required magnitude of the force.

9. Rework Problem 8 changing the force keeping the body in position to one that acts horizontally.

Computer Problems

For the following computer problems, any appropriate programming language may be used. Input prompts should fully explain what is required of the user (the program should be "user friendly"). The resulting output should be well labeled and self-explanatory.

10. Write a program that will calculate the rectangular components of a force. User input is to be the magnitude of the force (lb) and its inclination with the positive X axis, as defined in Fig. 2–9.

11. Write a program that will calculate the magnitude of a resultant force, given the X and Y rectangular components of the force. Also include the calculation of the inclination of the resultant force with the positive X axis.

12. Assume that the horizontal force of Problem 17 will range from 100 lb to 250 lb in 10 lb steps. Solve for the associated values of θ. Display the results in tabular format.

Supplemental Problems

13. Compute the vertical and horizontal components for each of the forces shown in Fig. 2–23. In each case the force is perpendicular to the incline.

14. As shown in Fig. 2–24, a telephone pole is braced by a guy wire that exerts a 300 lb pull on the top of the pole.

The guy wire forms an angle of 42° with the pole. Compute the vertical and horizontal components of the pull on the pole.

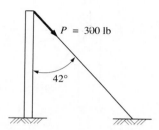

FIGURE 2–24   Problem 14.

15. A person is walking along the bank of a stream pulling a small boat by a rope. The rope is inclined at 20° to the direction of the boat's motion. If the person exerts a force of 30 lb on the rope, what is the effective force in the direction of the boat's motion?

16. A child pulls a rope attached to a sled with a force of 20 lb. The rope forms an angle of 25° with the ground. Compute the effective value of the pull tending to move the sled along the ground and the effective value tending to lift the sled vertically.

17. A force of 225 lb, acting horizontally, maintains a 450 lb cylindrical roller at rest on a smooth, inclined surface (assumed to be frictionless). The surface is inclined at an angle of θ to the horizontal, as shown in Fig. 2–25. Calculate the value of θ.

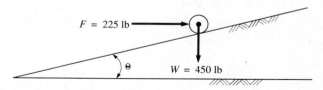

FIGURE 2–25   Problem 17.

18. In Problem 17, if θ were 45°, what horizontal force would be necessary to maintain the 450 lb roller in a state of equilibrium?

19. The horizontal and vertical components, respectively, of a force F are given. Compute the magnitude, inclination with the X axis, and the sense of the force F. (a) 300 lb, −200 lb; (b) −500 lb, −300 lb; (c) −240 lb, 360 lb; (d) 250 lb, 460 lb.

20. Calculate the X and Y components of each force shown in Fig. 2–26.

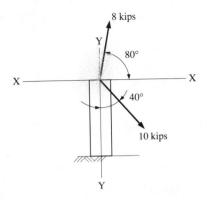

FIGURE 2–26   Problem 20.

21. A drawbar support assembly is shown in Fig. 2–27. The force in the lifting-link is 2000 lb. Determine the vertical and horizontal components of the force.

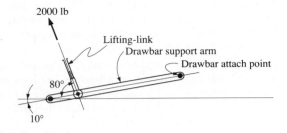

FIGURE 2–27   Problem 21.

22. The Howe roof truss shown in Fig. 2–28 is subjected to wind loads as shown. The wind loads are perpendicular to the inclined top chord. Resolve each force into vertical and horizontal components.

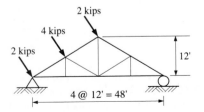

FIGURE 2–28   Problem 22.

23. The release cam mechanism is subjected to the forces shown in Fig. 2–29. Resolve each force into components parallel and perpendicular to a line connecting points A and B.

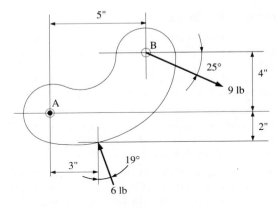

FIGURE 2–29   Problem 23.

3 Resultants of Coplanar Force Systems

3–1
RESULTANT OF TWO
CONCURRENT
FORCES

In Chapter 2 we established a rationale and method for resolving a force into its rectangular components. It was shown that when the components act simultaneously, they are, in effect, equivalent to the original force. A reverse process was also described, in which two concurrent forces at right angles to each other were replaced by a single force that had the same effect on the body as the two original forces. This single force, called a *resultant*, is denoted for convenience as R in Fig. 3–1. The resultant of two forces acting on a body may be defined as the single force that, if acting alone, would produce the same effect as the two forces combined.

Two forces (F_x and F_y) acting at right angles to each other, as shown in Fig. 3–1, is a special case frequently encountered in solutions of concurrent force system problems. Equations (2–3) and (2–4), which stated that

$$R = \sqrt{F_x^2 + F_y^2} \qquad \text{and} \qquad \tan \theta_x = \frac{F_y}{F_x}$$

are applicable in this case.

A more general case is one in which the coplanar concurrent forces are not acting at right angles to each other, as shown in Fig. 3–2. Forces F_1 and F_2 represent any two concurrent forces acting in any direction at any angle with each other. Point O is the point of concurrence of their lines of action. The resultant of these two nonrectangular forces can be computed using the parallelogram law. The principle of the parallelogram law is that two concurrent forces can be replaced by their resultant, which is represented by the diagonal of a parallelogram, the sides of which are equal and parallel to the two forces.

Note that the parallelogram OABC is constructed by drawing AB and BC parallel to OC and OA (which are, respectively, forces F_1 and F_2). The parallelogram composed of forces (vectors) and angles may be drawn to some selected scale and the resultant determined using a graphical method of solution. However, using an electronic calculator makes it simpler and more accurate to use a freehand sketch of the parallelogram along with geometric and trigonometric relationships to obtain the resultant. Our discussion will be limited to mathematical solutions rather than to graphical solutions. You may wish to verify the solutions graphically.

In order to compute the magnitude and direction of the resultant, the magnitudes and directions of the two forces F_1 and F_2 must be known. It may

FIGURE 3–1   Resultant and rectangular components.

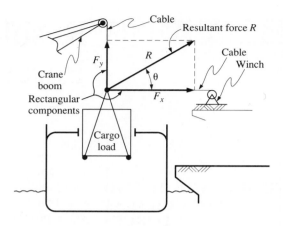

FIGURE 3–2   Parallelogram method.

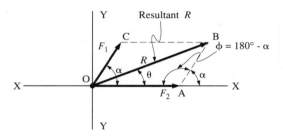

be recognized in the parallelogram of Fig. 3–2 that triangle OAB has been created in which two sides and the included angle are known. The resultant R can then be calculated using the law of cosines, which, with reference to Fig. 3–2, can be written as

$$R^2 = F_1^2 + F_2^2 - 2F_1F_2 \cos \phi$$

The direction of the resultant can be obtained using the law of sines, which, again with reference to Fig. 3–2, can be written

$$\frac{R}{\sin \phi} = \frac{F_1}{\sin \theta}$$

from which

$$\sin \theta = \frac{F_1 \sin \phi}{R}$$

Note that in a concurrent force system the line of action of the resultant will always pass through the point of concurrence (point O). Therefore, the location of the resultant is always known, and it is necessary to determine only its magnitude, direction, and sense.

An examination of Fig. 3–2 reveals that it is necessary to sketch and make computations for only one of the triangles of the parallelogram. Both triangles will yield the same result. The triangle is simply a force triangle, similar to that shown in Fig. 2–12, except that this force triangle is not a right triangle. The relationship shown by the force triangle is sometimes set forth as a principle itself and is given the name the *triangle law*. The basis for this principle is that if the tail end of either force vector is placed at the arrow end of the other, the resultant force vector is the third side of the triangle, and it has a direction from the tail end of the first vector to the arrow end of the other. Note that the computations to determine the resultant are identical, based on trigonometric relationships, whether a parallelogram or a triangle is considered. A graphical solution may be used in either case.

☐ **EXAMPLE 3–1**   Determine the magnitude, direction, and sense of the resultant of the two concurrent forces F_1 and F_2 shown in Fig. 3–3. The forces have magnitudes of 100 lb and 140 lb, respectively, and are concurrent at point O. F_1 acts at an angle of 30° above the X axis and F_2 acts at an angle of 45° below the X axis.

FIGURE 3–3 Resultant of concurrent force system.

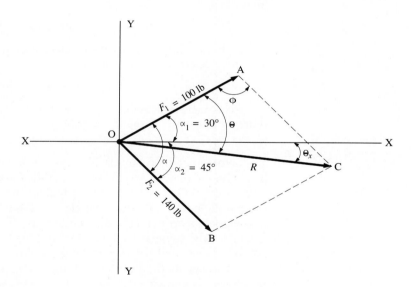

Solution   The parallelogram law will be used. First, construct parallelogram OACB as shown, with AC equal and parallel to F_2 and BC equal and parallel to F_1. Then from Fig. 3–3, calculate angle α (the angle at O between line segments OA and OB, which represent the forces):

$$\alpha = \alpha_1 + \alpha_2 = 30° + 45° = 75°$$

Since OB and AC are parallel, angle ϕ can be calculated:

$$\phi = 180° - \alpha = 180° - 75° = 105°$$

Using triangle OAC, now calculate the magnitude of R, recognizing that side AC is equal to F_2 (or 140 lb):

$$R^2 = F_1^2 + F_2^2 - 2F_1F_2 \cos \phi$$
$$= 100^2 + 140^2 - 2(100)(140)(-0.259)$$

from which

$$R = 192 \text{ lb}$$

Note that the cosine of 105° is a negative value. Recall that the cosine of an angle between 0° and 90° is positive, and that it is negative from 90° to 180°.

The direction θ of the resultant with respect to force F_1 can be calculated using the law of sines. Again, using triangle OAC,

$$\frac{R}{\sin \phi} = \frac{AC}{\sin \theta}$$

$$\sin \theta = \frac{AC \sin \phi}{R} = \frac{140(\sin 105°)}{192} = 0.7043$$

from which

$$\theta = 44.8°$$

The direction of the resultant with respect to the horizontal X axis, designated θ_x, can be determined from Fig. 3–3:

$$\theta_x = \theta - \alpha_1 = 44.8° - 30° = 14.8°$$

This angle is measured clockwise from the X axis. Therefore, the sense of the resultant is downward and to the right.

Note that in Example 3–1 the geometric diagram of the forces was based on the parallelogram law. A parallelogram was sketched and the triangular portion OAC was used to calculate the unknown resultant. Instead of the parallelogram, a force triangle could have been sketched based on the triangle law as previously defined and as shown in Fig. 3–4. Note that this triangle is the same as triangle OAC in Fig. 3–3. The computations using the law of cosines and the law of sines would be identical.

Another method of determining the resultant of two coplanar concurrent forces is known as the *method of components*. This is probably the most

FIGURE 3–4 Force triangle.

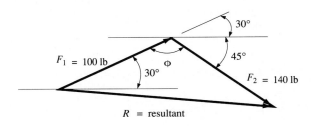

general method, and, therefore, the most common method used, with applications in other types of force problems as we will see in subsequent sections.

The method of components again makes use of a selected X–Y rectangular coordinate axis system. The sequence of steps in the application of the method is as follows:

1. Calculate and algebraically sum the X components for each force (ΣF_x).
2. Calculate and algebraically sum the Y components for each force (ΣF_y).
3. The results from steps 1 and 2 represent the X and Y rectangular components of the resultant force. Designating ΣF_x as R_x and ΣF_y as R_y, the resultant R can be calculated using Eq. (2–3):

$$R = \sqrt{R_x^2 + R_y^2}$$

This represents the magnitude of the resultant. In order to determine the sense of the resultant, a sketch should be drawn indicating the two rectangular components, with the resultant superimposed on an X–Y coordinate set of axes.

4. The angle of inclination between the resultant and the X axis can be observed in the sketch of step 3. It can be calculated using Eq. (2–4):

$$\tan \theta_x = \frac{R_y}{R_x} = \frac{\Sigma F_y}{\Sigma F_x}$$

☐ **EXAMPLE 3–2**   Determine the resultant R of the two-force coplanar system of Fig. 3–5. Compute its magnitude, sense, and angle of inclination with the horizontal X axis.

Solution   Use the method of components. The X–Y coordinate axes with forces F_1 and F_2 are shown in Fig. 3–5. Positive directions are upward and to the right.

FIGURE 3–5 Two-force concurrent force system.

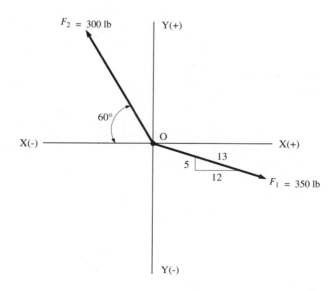

1. The X component R_x of the resultant force is equal to the algebraic summation of the X components of F_1 and F_2:

$$R_x = \Sigma F_x = F_{1_x} + F_{2_x}$$
$$= 350\left(\frac{12}{13}\right) - 300(\cos 60°)$$
$$= +173.1 \text{ lb} \rightarrow$$

The positive sign (and the arrow) indicates that the sum of the force components acts to the right.

2. The Y component R_y of the resultant force is equal to the algebraic summation of the Y components of F_1 and F_2:

$$R_y = \Sigma F_y = F_{1_y} + F_{2_y}$$
$$= -350\left(\frac{5}{13}\right) + 300(\sin 60°)$$
$$= +125.2 \text{ lb} \uparrow$$

The positive sign (and the arrow) indicates that the sum of the force components acts upward.

3. The results obtained in steps 1 and 2 are shown in Fig. 3–6. Knowing R_x and R_y, the parallelogram (actually a rectangle) OCBA is sketched along with the resultant shown as diagonal R. Using triangle OAB (in which AB = OC),

$$R = \sqrt{R_x^2 + R_y^2}$$
$$= \sqrt{173.1^2 + 125.2^2} = 214 \text{ lb}$$

The sense of the resultant is upward and to the right.

4. Using triangle OAB, compute the angle of inclination θ_x:

$$\theta_x = \tan^{-1}\frac{R_y}{R_x} = \tan^{-1}\frac{125.2}{173.1} = 35.9°$$

FIGURE 3–6 Parallelogram of forces.

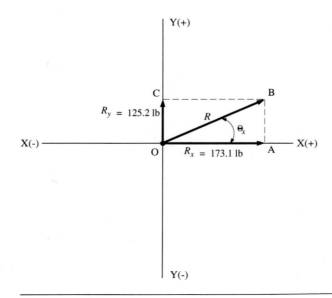

**3–2
RESULTANT OF
THREE OR MORE
CONCURRENT
FORCES**

The method of components, as used in Section 3–1, may not appear to offer much advantage when only two forces are involved. However, for three or more concurrent coplanar forces, it is the recommended method.

 The sequence of steps described in Section 3–1 is followed, with the only difference being the number of forces involved. The following example is typical of cases involving three or more concurrent forces.

☐ **EXAMPLE 3–3**   Determine the resultant R of the four-force coplanar force system of Fig. 3–7. Compute its magnitude, sense, and angle of inclination with the horizontal X axis. Use the method of components.

FIGURE 3–7   Four-force concurrent force system.

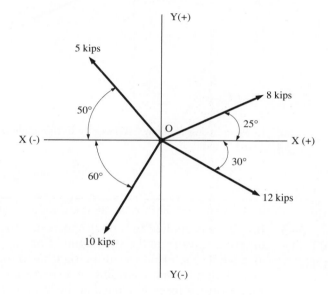

Solution   Force components upward and to the right are considered positive. Use the following steps:

1. Calculate the X component:

$$R_x = \Sigma F_x$$
$$= -5(\cos 50°) - 10(\cos 60°) + 8(\cos 25°) + 12(\cos 30°)$$
$$= +9.43 \text{ kips} \rightarrow$$

2. Calculate the Y component:

$$R_y = \Sigma F_y$$
$$= +5(\sin 50°) - 10(\sin 60°) + 8(\sin 25°) - 12(\sin 30°)$$
$$= -7.45 \text{ kips} \downarrow$$

3. The results of steps 1 and 2 represent the X and Y components of the resultant force and are shown in Fig. 3–8. Parallelogram OABC is drawn, with diagonal R indicating the resultant. Using triangle OAB (where AB = OC by construction),

$$R = \sqrt{R_x^2 + R_y^2} = \sqrt{9.43^2 + (-7.45)^2} = 12.02 \text{ kips}$$

FIGURE 3–8   Parallelogram of forces.

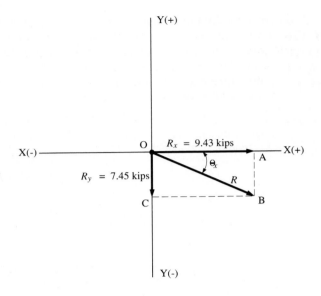

4. Again using triangle OAB compute the angle of inclination of the resultant with the X axis:

$$\theta_x = \tan^{-1}\frac{R_y}{R_x} = \tan^{-1}\frac{7.45}{9.43} = 38.3°$$

**3–3
MOMENT OF
A FORCE**

If a force is applied to a body "at rest," the body can be disturbed in two different ways from the standpoint of planar motion. Either it can be moved as a whole, up or down, to the right or to the left (translation), or it can be turned about some fixed line or axis (rotation). For example, as shown in Fig. 3–9(a), a force P applied at the midpoint of a free, rigid, uniform object will slide the object such that every point moves an equal distance. The object is said to *translate*. If the same force is applied at some other point as in Fig. 3–9(b), then the object will both translate and rotate. The amount of rotation depends on the point of application of the force. If a point on the object is fixed against translation, as at point A in Fig. 3–9(c), then the applied force P causes the object to rotate only. Such rotation occurs when one pushes on a door or pulls on the handle of a wrench to turn a nut. This

FIGURE 3–9   Types of motion.

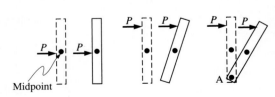

(a) Translation   (b) Translation   (c) Rotation
                            and rotation

tendency of a force to produce rotation about some axis is called the *moment of a force*. The magnitude of this tendency is directly proportional to (a) the magnitude of the force and (b) the perpendicular distance between the axis and the line of action of the force.

The perpendicular distance from the line of action of the force to the axis about which rotation is assumed to take place is called the *moment arm* (or lever arm). Since the tendency to cause rotation depends on both the magnitude of the force and the length of the moment arm, the moment of a force (or simply, the moment) is defined as the product of the force and its moment arm.

With reference to Fig. 3–10, the moment about axis O may be expressed as the product *Fd*. *F* represents a force. Point O, called the *moment center*, is a point on an axis of rotation that is perpendicular to the plane of the page. OG, also designated *d*, is the perpendicular distance between the moment center and force *F* and represents the moment arm. Note that *d* is perpendicular to the force *F* as well as perpendicular to the axis of rotation. Therefore, as previously stated, the moment of *F* with respect to point O is equal to *Fd*.

FIGURE 3–10 Moment of a force.

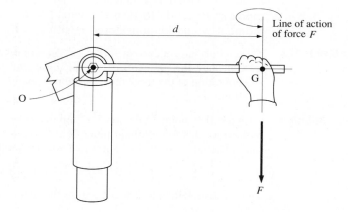

Since moment is the product of force and distance, the units of moment are the product of the applicable units generally used for force and distance. In the U.S. Customary System, units for moment are in.-lb, ft-lb, in.-kips, and ft-kips. In the SI system, the units are newton-meter (N·m).

Since a moment tends to produce rotation about an axis or point, a sign convention is generally used to identify the direction of the rotation. An acceptable convention, and the one used throughout this text, is to identify a clockwise rotation as negative and a counterclockwise rotation as positive. For example, in Fig. 3–11(a), the force *F* acts along the line of action shown. Its moment arm is the distance *d*, and the moment of the force about the point (or axis) O is *Fd*. Note that *d* is at right angles to the line of action. The moment has a tendency to produce rotation about point O, the moment

FIGURE 3–11   Moment examples.

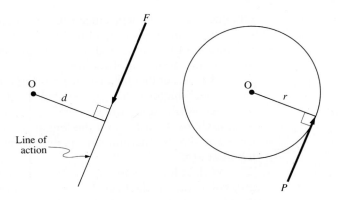

(a) Clockwise rotation of a
moment—negative (-)

(b) Counterclockwise rotation of a
moment—positive (+)

center. It is seen that the rotation is in a clockwise direction. This moment, according to our sign convention, would be negative (−).

In Fig. 3–11(b), the force P, acting tangent to the wheel, has a moment about O, which is equal to Pr, where r is the radius of the wheel. The moment Pr tends to turn the wheel in a counterclockwise direction. This moment would be considered positive (+).

☐ **EXAMPLE 3–4**   Three coplanar concurrent forces act on a body at point A as shown in Fig. 3–12. (a) Calculate the moment of each of these forces about point O. (b) Calculate the algebraic sum of the three moments about point O and determine the direction of the rotation.

Solution   (a) Note that point O lies on the line of action of the 40 lb force. Therefore, the moment of this force (about point O) is zero.

The moment of the 75 lb force is calculated from

$$M = Fd = 75(5) = +375 \text{ ft-lb}$$

The positive sign is assigned because the moment is counterclockwise.

FIGURE 3–12   Moment of concurrent forces.

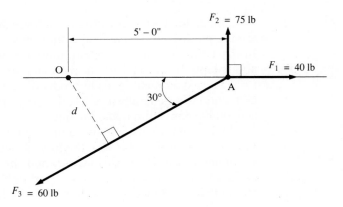

The moment of the 60 lb force is calculated from

$$M = Fd = 60[5.0(\sin 30°)] = -150 \text{ ft-lb}$$

Since the rotation is clockwise, the moment is assigned a negative sign.

(b) The algebraic summation of the three moments is

$$\Sigma M = 0 + 375 - 150 = +225 \text{ ft-lb}$$

which indicates that the effect of the moments of the three forces about point O is to produce a counterclockwise rotation.

3–4
THE PRINCIPLE OF MOMENTS— VARIGNON'S THEOREM

In Chapter 2 it was shown that the components of a force will produce the same effect on a body as the original force. Therefore, with respect to any point, the algebraic summation of the moments of the components of a force must equal the moment of the original force. This principle is known as *Varignon's theorem*. A specific numerical example will illustrate the principle.

☐ **EXAMPLE 3–5**

Calculate the moment about point O of the 200 lb force that lies in the X–Y plane of Fig. 3–13. Use the following techniques: (a) Solve directly using the perpendicular distance from the line of action to point O. (b) Resolve the force into rectangular components at point M. (c) Resolve the force into rectangular components at point N.

FIGURE 3–13 Example of Varignon's theorem.

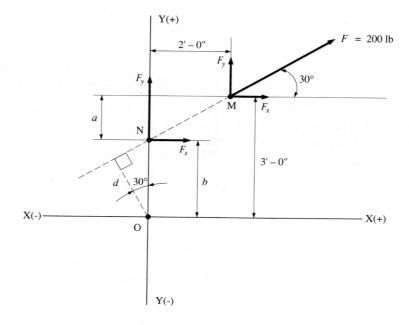

Solution   (a) Compute distance d shown in Fig. 3–13:

$$a = 2.0(\tan 30°) = 1.15 \text{ ft}$$
$$b = 3.0 - a = 3.0 - 1.15 = 1.85 \text{ ft}$$
$$d = b \cos 30° = 1.85(0.866) = 1.60 \text{ ft}$$

Therefore, using $M = Fd$, and assigning a negative sign since the rotation is clockwise,

$$M = -200(1.60) = -320 \text{ ft-lb}$$

(b) Resolving the force F into X and Y components at point M yields

$$F_x = 200 \cos 30° = 173.2 \text{ lb}$$
$$F_y = 200 \sin 30° = 100.0 \text{ lb}$$

The algebraic summation of moments about point O is

$$\Sigma M_O = -F_x(3.0) + F_y(2.0)$$
$$= -173.2(3.0) + 100(2.0)$$
$$= -320 \text{ ft-lb}$$

where the negative sign indicates a clockwise rotation.

(c) Using the principle of transmissibility, the force F can be resolved into X and Y components at point N instead of point M, and the moment about point O should be the same. Note that the moment of F_y about the moment center O is equal to zero since the line of action of this component passes through point O, making the moment arm equal to zero.

$$\Sigma M_O = -F_x b = -173.2(1.85) = -320 \text{ ft-lb}$$

Again, the negative sign indicates a clockwise rotation.

In Example 3–5, the moment about point O was the same in all three cases. It is apparent, then, that the algebraic summation of moments of the components of a force is equal to the moment of the force itself.

3–5
RESULTANTS OF PARALLEL FORCE SYSTEMS

A parallel force system is one in which the lines of action of all the forces are parallel. The resultant of such a system will be parallel to the lines of action of the forces in the system. The magnitude, direction, and sense of the resultant can be determined by an algebraic summation of the forces. The position (location) of the line of action of the resultant may be determined by the principle of moments (Varignon's theorem); namely, that the moment of the resultant force is equal to the algebraic sum of the moments of the given forces.

As an example, consider the system of vertical parallel forces applied to a horizontal member, as shown in Fig. 3–14. Assume that the member is weightless. The magnitude and direction of the resultant R is computed from an algebraic summation of the vertical forces, which can be written as

$$R = \Sigma F_y = F_1 + F_2 + F_3$$

FIGURE 3–14 Parallel force system.

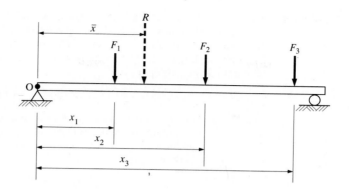

The resultant is shown as a dashed arrow. The position at which the resultant is indicated is an assumed position and is dimensioned as distance $\bar{x}$ from a moment center at O. Applying Varignon's theorem to solve for the distance $\bar{x}$,

$$\Sigma M_O = R\bar{x} = F_1 x_1 + F_2 x_2 + F_3 x_3$$

from which

$$\bar{x} = \frac{F_1 x_1 + F_2 x_2 + F_3 x_3}{R}$$

In more general terms, for n forces,

$$R = \Sigma F_y = F_1 + F_2 + \cdots + F_n \tag{3–1}$$

$$\bar{x} = \frac{\Sigma M}{R} \tag{3–2}$$

☐ **EXAMPLE 3–6** Determine the resultant of the parallel force system of Fig. 3–15 acting on the horizontal beam AB. All forces are vertical. Neglect the weight of the beam.

Solution The resultant force is equal in magnitude to the algebraic summation of the vertical forces it replaces. Assuming upward acting forces to be positive,

$$R = \Sigma F_y = -25 - 15 + 10 - 20 = -50 \text{ lb} \downarrow$$

FIGURE 3–15 Parallel force system.

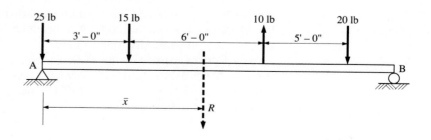

The resultant acts vertically downward. (Recall that it must be parallel to the forces in the parallel force system.) The resultant (dashed arrow) is shown in the diagram at an arbitrary location $\bar{x}$ from point A. The location of the resultant can be determined by applying Eq. (3–2). Using point A as the moment center and taking clockwise moments as negative,

$$\Sigma M_A = -15(3) + 10(9) - 20(14) = -235 \text{ ft-lb}$$

This summation of moments yields a clockwise moment about point A. Further, note that since the resultant force must also produce a clockwise moment about point A, the relative position of R with respect to point A has been correctly assumed. Equation (3–2) then yields

$$\bar{x} = \frac{\Sigma M_A}{R} = \frac{235 \text{ ft-lb}}{50 \text{ lb}} = 4.70 \text{ ft}$$

Again, note that the downward-acting resultant produces a clockwise moment, as did the original forces. Always verify the result, at this point, to ensure that it is logical. A force acting with a negative sense does not necessarily produce a negative moment. The sign conventions for forces and moments should be considered as separate entities.

Up to this point problems and discussions have been limited to concentrated forces (loads). However, be aware that the principles and methods used in determining the resultant of a force system are also applicable when the forces are of a distributed type or when the forces are combinations of concentrated and distributed forces.

Distributed forces were defined and briefly discussed in Section 2–4. Recall that a distributed force is one that acts over an area or length rather than at a point. The distributed force may be considered to be spread out over some length of a member.

For practical purposes, we will discuss distributed forces in terms of distributed loads applied to members. Two common types of distributed loads are usually encountered: uniformly distributed and nonuniformly distributed loads. If the distributed load is of equal magnitude for each unit of length, it is said to be *uniformly distributed*. It may exist over the entire length of a member or over a portion of it.

A concentrated force is pictorially represented by a single arrow as shown in Fig. 3–16(a). A uniformly distributed load is represented by a rectangular block, as shown in Fig. 3–16(b) and (d). This representation defines both the intensity of the uniformly distributed load and the length of the member over which it acts. The notation commonly used for the intensity of the uniformly distributed load is w. Units for the load intensity are usually lb/ft or kips/ft in the U.S. Customary System and N/m in the SI system. As an example, if we consider a beam with a uniform cross section throughout its length, the weight of the beam (lb/ft) would be a uniformly distributed load.

FIGURE 3–16 Types of loads.

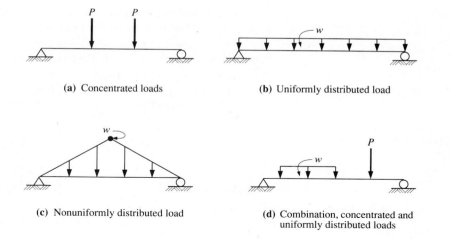

(a) Concentrated loads   **(b)** Uniformly distributed load

(c) Nonuniformly distributed load   **(d)** Combination, concentrated and uniformly distributed loads

The distributed load may also have a varying intensity. In this case, it is said to be *nonuniformly distributed*. Generally, a nonuniformly distributed load increases or decreases at a definite and known rate along the length of the member. The pictorial representation for this type of load usually results in a triangular or trapezoidal block, as shown in Fig. 3–16(c). Note that in this particular case, the maximum intensity of the distributed load is w at the midpoint and decreases to zero at the ends of the member.

For the purpose of determining the resultant of a force system, each distributed load may be replaced by its equivalent concentrated resultant load. This equivalent load acts through the centroid of the geometric shape of the distributed load. The *centroid* is that point at which all of the area may be considered to be concentrated; it will be discussed in detail in Chapter 7. The shape is a rectangle if the load is a uniformly distributed load. The shape is a triangle or a trapezoid if the load is a nonuniformly distributed load.

As shown in Fig. 3–17, distributed loads may be considered to be parallel force systems. The uniformly distributed load has an intensity of 2 kips/ft and is 6 ft long. Consider the distributed load to be divided up into six 1-ft-long segments. Each segment will have a load of 2 kips. The resultant will have a magnitude equal to 12 kips, or (6 ft × 2 kips/ft). Its location will be at the center of the 6 ft length, since the load is uniformly distributed (a

FIGURE 3–17 Resultant of uniformly distributed load.

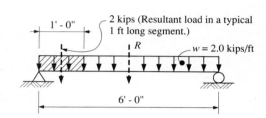

rectangular load area). Apply the same approach to nonuniformly distributed loads (triangular or trapezoidal load areas).

The centroid of a rectangle is at the geometric center of the figure. The centroid of the triangular areas may be obtained from Table 7–1. When dealing with trapezoidal shapes, it is recommended that the trapezoid be replaced with a rectangle and a triangle as shown in Fig. 3–18. (Alternatively, the trapezoid may be replaced with two triangles.)

FIGURE 3–18 Centroid of load areas.

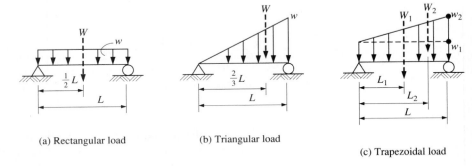

(a) Rectangular load   (b) Triangular load

(c) Trapezoidal load

□ **EXAMPLE 3–7**   Determine the magnitude and location of the resultant R of the parallel force system of Fig. 3–19. The force system acts on a horizontal beam AB. All forces are vertical. Neglect the weight of the beam.

Solution   Note that this is a combined force system consisting of concentrated loads and a uniformly distributed load. The uniformly distributed load may be replaced by its equivalent concentrated resultant force. This equivalent concentrated force is often denoted as W. Therefore,

$$W = 2 \text{ kips/ft} \times 14 \text{ ft} = 28 \text{ kips}$$

Note also that the resultant W of the uniformly distributed load is indicated by a dashed arrow to distinguish it from the given concentrated loads, and the location of the dashed arrow is shown at the center of the distributed-load portion. (The equivalent concentrated resultant force of the uniformly distributed load always acts through the centroid of the distributed-load portion.)

FIGURE 3–19 Parallel force system.

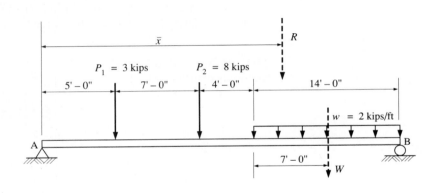

The magnitude of the resultant force of the total force system is equal to the algebraic summation of the vertical forces it replaces. Assume upward acting forces are positive:

$$R = \Sigma F_y = -3 - 8 - 28 = -39 \text{ kips} \downarrow$$

The resultant force acts vertically downward. It is indicated in Fig. 3–19 by a dashed arrow R. The location of the resultant is unknown at this time, but it is shown acting a distance $\bar{x}$ from a moment center at point A. Determine $\bar{x}$ by applying Eq. (3–2). Use point A as the moment center and assume clockwise moments negative:

$$\Sigma M_A = -3(5.0) - 8(12) - 28(23) = -755 \text{ ft-kips}$$

The negative sign indicates that the moment about point A is clockwise. Resultant R must also produce an equal clockwise moment about point A. Equation (3–2) yields

$$\bar{x} = \frac{\Sigma M_A}{R} = \frac{755 \text{ ft-kips}}{39 \text{ kips}} = 19.4 \text{ ft}$$

Since the moment of the given forces was clockwise, the downward acting R must be located to the right of point A, as shown, in order to also produce a clockwise moment.

**3–6
COUPLES**

A coplanar parallel force system composed of two parallel forces having different lines of action, equal in magnitude, but opposite in sense, constitutes a special case. This force system is called a *couple* and is illustrated in Fig. 3–20.

FIGURE 3–20 Couple.

Equal parallel forces

Plane of the couple   Arm of the couple

The perpendicular distance between the lines of action of the two parallel forces is called the *arm of the couple*, and the plane in which the lines of action lie is called the *plane of the couple*. A couple either causes or tends to cause rotation about an axis perpendicular to its plane. When the driver of an automobile grasps opposite sides of the steering wheel and turns it, a couple is being applied to the wheel.

The following characteristics of couples should be noted:

1. The moment of a couple is the product of one of the forces and the arm of the couple.

2. The moment of a couple is independent of the choice of the axis of moments (moment center). The moment of a couple is the same with respect to any axis perpendicular to the plane of the couple (or any point in the plane of the couple.)

3. The magnitude of a resultant force of a couple is zero. Therefore, a couple cannot be replaced with a single equivalent resultant force.
4. If a given force system is composed entirely of couples in the same plane, the resultant moment will consist of another couple equal to the algebraic summation of the original couples.
5. A couple can be balanced only by an equal and opposite couple in the same plane.
6. A couple is fully defined by its magnitude and sense of rotation. If the sense of rotation is counterclockwise, it will be considered positive.
7. A couple may be transferred to any location in its plane and still have the same effect.

☐ **EXAMPLE 3–8**   Calculate the magnitude of the moment of the two parallel forces shown in Fig. 3–21. (a) Use point O as the moment center. (b) Use point A as the moment center.

FIGURE 3–21   Special parallel force system.

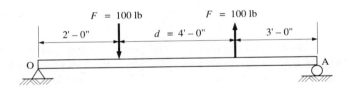

Solution   Since the two forces are parallel with different lines of action, equal in magnitude and opposite in sense, they constitute a couple. The magnitude of a couple is equal to the product of one of the forces and the perpendicular distance between the forces. This may be expressed as

$$M = Fd = 100(4.0) = +400 \text{ ft-lb}$$

Note that the result is a positive value since the couple is acting counterclockwise.

One of the characteristics of a couple is that its magnitude is the same with respect to any point in the plane of the couple. Next compute the moment of the couple with respect to points O and A:

$$\Sigma M_O = -100(2) + 100(6) = +400 \text{ ft-lb}$$
$$\Sigma M_A = -100(3) + 100(7) = +400 \text{ ft-lb}$$

Note that the couple has the same counterclockwise moment effect of 400 ft-lb, irrespective of where the moment center is chosen.

☐ **EXAMPLE 3–9**   Four vertical forces with parallel lines of action acting on a horizontal beam are shown in Fig. 3–22. Calculate the resultant moment with respect to (a) point O, (b) point B, and (c) point C.

FIGURE 3–22 Parallel force systems.

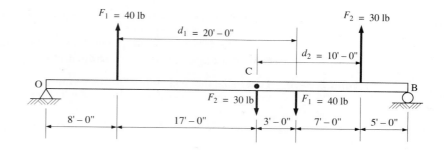

Solution   Moment summations at the individual points are taken as follows:

$$\Sigma M_O = +40(8) - 30(25) - 40(28) + 30(35) = -500 \text{ ft-lb}$$
$$\Sigma M_B = -40(32) + 30(15) + 40(12) - 30(5) = -500 \text{ ft-lb}$$
$$\Sigma M_C = -40(17) + 30(10) - 40(3) = -500 \text{ ft-lb}$$

Note that the moment effect is the same in each case and is irrespective of the moment center. Also note that the member is subjected to two couples. You can determine a resultant couple equal to the algebraic summation of the two couples acting on the member as follows:

$$M = M_1 + M_2 = F_1 d_1 + F_2 d_2$$
$$= -40(20) + 30(10) = -500 \text{ ft-lb}$$

The result—a clockwise moment of 500 ft-lb—is identical to the moment summations at the individual points.

3–7 RESULTANTS OF NONCONCURRENT FORCE SYSTEMS

In a concurrent force system, the lines of action of the forces meet at a common point, while in a nonconcurrent force system, they do not. Hence, in the latter system, in addition to the unknown magnitude, sense, and direction, the position of the line of the action of the resultant is also unknown. As with every force, the resultant of a coplanar nonconcurrent force system is defined by its magnitude, direction, sense, and position (or location) of its line of action.

The magnitude, direction, and sense can be calculated in a way similar to that for the concurrent force system, that is, by utilizing an X–Y coordinate axis system and taking an algebraic summation of the X and Y components of each force. The location of the line of action of the resultant can then be determined by computing its moment arm with respect to some convenient moment center. This latter step is similar to the procedure for locating the resultant of a coplanar parallel force system and is based on Varignon's theorem.

A common error in the analysis of this type of force system is to place the resultant on the wrong side of the moment center. However, as with the parallel force system, the location of the resultant with respect to a moment center is based on the fact that the resultant must produce the same effect as the original force system.

The following example will illustrate a method of determining the resultant of a coplanar nonconcurrent force system.

☐ **EXAMPLE 3–10**   Determine the magnitude, direction, and sense of the resultant of the coplanar nonconcurrent force system of Fig. 3–23. Locate the resultant with respect to point O.

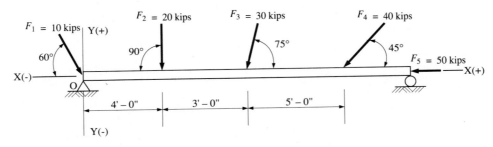

FIGURE 3–23   Nonconcurrent force system.

Solution   Forces upward and to the right are positive.

1. First compute the X component of the resultant force:

$$R_x = \Sigma F_x$$
$$= +10 \cos 60° - 30 \cos 75° - 40 \cos 45° - 50$$
$$= -81.0 \text{ kips} \leftarrow$$

2. Next compute the Y component of the resultant force:

$$R_y = \Sigma F_y$$
$$= -10 \sin 60° - 20 - 30 \sin 75° - 40 \sin 45°$$
$$= -85.9 \text{ kips} \downarrow$$

3. Using R_x and R_y, sketch a parallelogram with the resultant shown as the diagonal (see Fig. 3–24). Using triangle OAB (with AB equal to OC, by construction), compute the magnitude of the resultant from Eq. (2–3):

$$R = \sqrt{R_x^2 + R_y^2} = \sqrt{(-81.0)^2 + (-85.9)^2} = 118.1 \text{ kips}$$

Note that the sense of the resultant is downward and to the left.

4. Again using triangle OAB, compute the angle of inclination θ_x:

$$\theta_x = \tan^{-1} \frac{R_y}{R_x} = \tan^{-1} \frac{85.9}{81.0} = 46.7°$$

5. Note that the magnitude, direction, and sense of the resultant have been determined. However, you must determine its location relative to some moment center. The point O will be used as the moment center, and the location of the resultant ($\bar{x}$ from point O) will be determined (see Fig. 3–25).

Use Varignon's theorem, taking moments about point O in Fig. 3–23. The most convenient solution is to consider the moments due to the Y components of the applied forces. The moment arms for these components are measured along the X axis. Since the lines of action of the X components pass through the moment center at O, their effects disappear.

FIGURE 3–24 Parallelogram of forces.

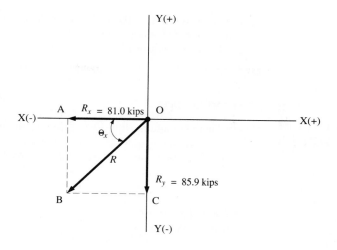

FIGURE 3–25 Location of resultant.

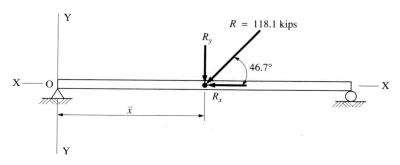

The summation of moments of the Y components of the applied forces with respect to point O (clockwise negative) is

$$\Sigma M_O = -20(4.0) - 30(\sin 75°)(7.0) - 40(\sin 45°)(12.0)$$
$$= -622 \text{ ft-kips}$$

Since the resultant must have the same moment effect about point O as all the other given forces, it is evident from Fig. 3–25 that the location of the resultant must be to the right of point O, as shown. This location will provide a clockwise moment, which is in agreement with the moment effect of the given forces.

Resolving the resultant R into its components (where it intersects the X axis) and taking moments about point O,

$$\Sigma M_O = R_y \bar{x}$$

from which

$$\bar{x} = \frac{\Sigma M_O}{R_y} = \frac{622}{85.9} = 7.24 \text{ ft}$$

Note that the signs of the moment and the force in the preceding expression are neglected. The signs for moment and force are separate entities. Using this approach, as discussed, the $\bar{x}$ distance will always be positive, and you establish the necessary correlation of the senses of the force and moment in the final calculation by inspection of Figs. 3–23, 3–24, and 3–25.

3–8

**SI SYSTEM
EXAMPLES**

☐ **EXAMPLE 3–11**

Determine the resultant R of the two-force concurrent coplanar force system of Fig. 3–26. Compute its magnitude, sense, and angle of inclination with the X axis. Use the method of components. Positive directions are upward and to the right.

FIGURE 3–26 Two-force concurrent force system.

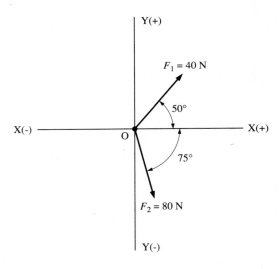

Solution   The X and Y components of the resultant force are found by summing, respectively, the X and Y components of the given forces:

$$R_x = \Sigma F_x = F_{1x} + F_{2x}$$
$$= F_1(\cos 50°) + F_2(\cos 75°)$$
$$= 40(0.6428) + 80(0.2588) = +46.42 \text{ N} \rightarrow$$

$$R_y = \Sigma F_y = F_{1y} + F_{2y}$$
$$= F_1(\sin 50°) - F_2(\sin 75°)$$
$$= 40(0.7660) - 80(0.9659) = -46.63 \text{ N} \downarrow$$

The components of R are shown in Fig. 3–27. The magnitude of the resultant is

$$R = \sqrt{R_x^2 + R_y^2} = \sqrt{(46.42)^2 + (-46.63)^2} = 65.8 \text{ N}$$

Note that the sense of the resultant is downward and to the right.
The angle of inclination θ_x is calculated from

$$\theta_x = \tan^{-1} \frac{R_y}{R_x} = \tan^{-1} \frac{46.63}{46.42} = 45.1°$$

FIGURE 3–27   Components and resultants.

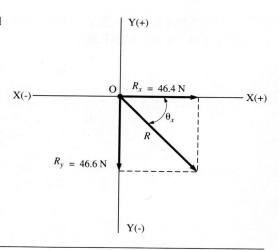

☐ **EXAMPLE 3–12**   Determine the magnitude and location of the resultant R of the parallel force system of Fig. 3–28. The force system acts on a horizontal beam BC. Neglect the weight of the beam.

Solution   The resultant R is shown acting down at an arbitrary location $\bar{x}$ from point B. First replace the uniformly distributed load with its equivalent concentrated resultant force. Its magnitude is calculated from

$$W = 70 \text{ N/m} \times 12 \text{ m} = 840 \text{ N} = 0.840 \text{ kN}$$

The magnitude of the resultant is obtained by algebraically summing the vertical forces. Assuming upward acting forces are positive,

$$R = \Sigma F_y = -10 - 0.84 - 6 = -16.84 \text{ kN} \downarrow$$

The location is determined by equating the moment of the resultant about an arbitrary point (moment center) to the sum of the moments of the applied forces about the same point. Select point B as the moment center. Counterclockwise moments are assumed positive. Note that the moment of the 10 kN force about point B is zero; it does not appear in the calculation.

$$\Sigma M_B = -W(15 \text{ m}) - (6 \text{ kN})(21 \text{ m})$$
$$\bar{x} = \frac{\Sigma M_B}{R} = \frac{-(0.84 \text{ kN})(15 \text{ m}) - (6 \text{ kN})(21 \text{ m})}{-16.84 \text{ kN}} = 8.23 \text{ m}$$

FIGURE 3–28   Parallel force system.

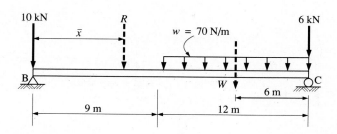

**SUMMARY—BY
SECTION NUMBER**

3–1   and **3–2** In these sections, the resultant is defined. Resultants of concurrent coplanar force systems may be found using the parallelogram law, the triangle law, or the method of components.

3–3   The magnitude of the moment of a force equals the product of the force and moment arm. The moment arm is the perpendicular distance from the line of action of the force to the moment center. In the sign convention for this text, counterclockwise is positive.

3–4   Varignon's theorem: the algebraic sum of the moments of the components of a force system is equal to the moment of the resultant force of the system.

3–5   Resultant of a parallel force systems:

$$R = \Sigma F_y = F_1 + F_2 + \cdots + F_n \qquad (3\text{–}1)$$

$$\bar{x} = \frac{\Sigma M}{R} \qquad (3\text{–}2)$$

Resultants of distributed loads act at the centroids of the load areas.

3–6   A couple is a special case of a coplanar parallel force system. It is composed of two equal, opposite, and parallel forces that are noncollinear.

3–7   Resultants of nonconcurrent coplanar force systems are defined by magnitude, direction, sense, and position of the line of action. The resultant must produce the same effect as the original force system.

PROBLEMS

Section 3–1   Resultant of Two Concurrent Forces

1–3.   Determine the magnitude, direction, and sense of the resultant for the coplanar concurrent force systems of Figs. 3–29 through 3–31. Use the parallelogram law. Also sketch the force triangle.

4–6.   Solve Problems 1 through 3 using the method of components.

7.   The 150 lb force shown in Fig. 3–32 is the resultant of two forces, one of which is shown. Determine the other force.

8.   A rope tow is used to pull skiers up a snow-covered slope which is 1000 ft long and has a slope of 21° measured from the horizontal. On a recent good skiing day it was noted that the maximum number of people on the tow never exceeded 44. Calculate the maximum tension in the rope under these conditions. Assume that the total weight of each skier is 175 lb. Assume that the slope is frictionless.

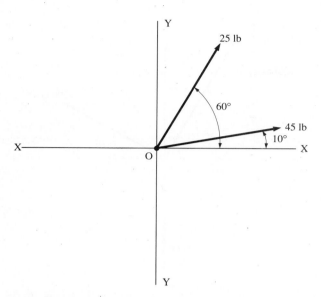

FIGURE 3–29   Problem 1.

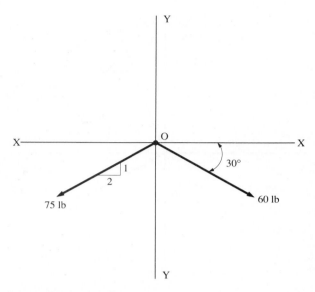

FIGURE 3–31   Problem 3.

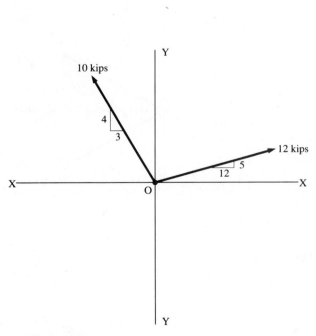

FIGURE 3–30   Problem 2.

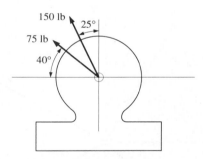

FIGURE 3–32   Problem 7.

Section 3–2   Resultant of Three or More Concurrent Forces

9–11. Determine the resultant of the coplanar concurrent force systems of Figs. 3–33 through 3–35. Compute the magnitude, sense, and angle of inclination with the X axis. Use the method of components.

12. The resultant of the concurrent force system of Fig. 3–36 has a magnitude of 300 lb acting vertically upward along the Y axis. Compute the magnitude of the force F_1 and its required angle of inclination θ for the resultant to act as described.

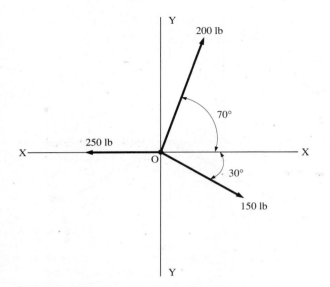

FIGURE 3–33 Problem 9.

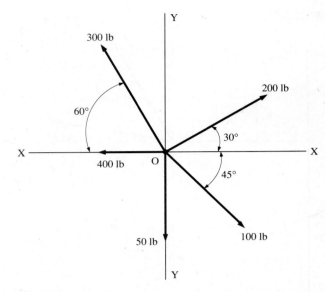

FIGURE 3–35 Problem 11.

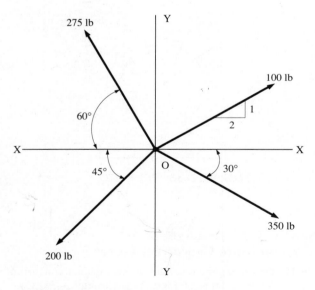

FIGURE 3–34 Problem 10.

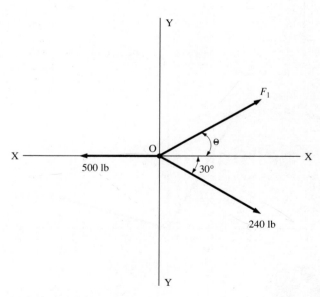

FIGURE 3–36 Problem 12.

FIGURE 3–37   Problem 14.

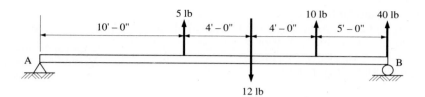

13. Three forces of 800 lb, 1000 lb, and 600 lb are acting on a boat. The first force acts due north, the second acts due east, and the third acts 30° east of south. Find the magnitude and direction of the resultant force on the boat.

Section 3–3   Moment of a Force

14. The four forces shown in Fig. 3–37 have parallel lines of action. Member AB is perpendicular to the lines of action of these forces. (a) Determine the moment of each force with respect to point A. (b) Compute the algebraic summation of the moments with respect to point A and determine the direction of rotation of the member.

15. Three coplanar concurrent forces act as shown in Fig. 3–38. (a) Calculate the moment of each force about point O that lies in the line of action of the F_1 force. (b) Calculate the algebraic summation of the

three moments about point O and determine the direction of rotation. (c) Calculate the magnitude of the resultant and angle of inclination with the X axis. (d) Compute the moment of the resultant about point O and compare with the result of part (b).

16. Four coplanar concurrent forces act as shown in Fig. 3–39. (a) Calculate the moment of each force about point O that lies in the line of action of the F_4 force. (b) Calculate the algebraic summation of the four moments and determine the direction of rotation.

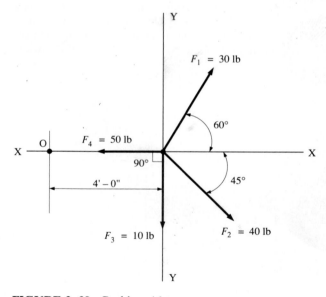

FIGURE 3–39   Problem 16.

17. Determine the resultant of the four forces of Problem 16 (magnitude and angle of inclination with respect to the X axis). Compute the moment of the resultant with respect to point O and compare with the results of Problem 16.

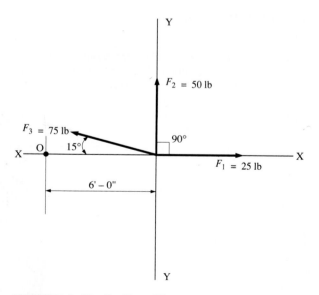

FIGURE 3–38   Problem 15.

18. The rectangular body in Fig. 3–40 measures 6 ft by 15 ft. (a) Calculate the algebraic summation of the moments of the forces shown about point A. (b) Calculate the algebraic summation of the moments of the forces about point B.

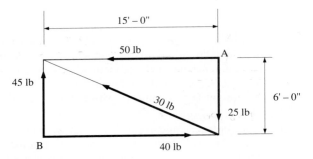

FIGURE 3–40   Problem 18.

19. The transmission tower shown in Fig. 3–41 is subjected to a horizontal wind force of 400 lb acting at point B. Supported cables transmit a force of 900 lb at two different levels acting as shown. Calculate the moment about point A at the base of the tower.

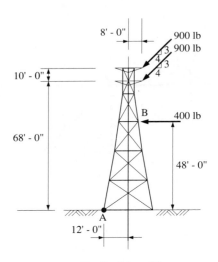

FIGURE 3–41   Problem 19.

Section 3–4 The Principle of Moments—Varignon's Theorem

20. Calculate the moment of the 550 lb force about point O shown in Fig. 3–42 without using Varignon's theorem. Make a similar calculation using the theorem, and resolving the force into its X and Y components at point A.

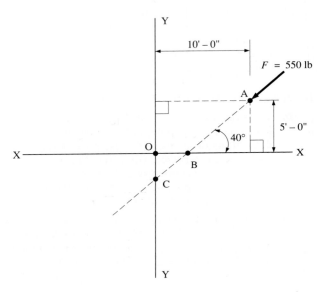

FIGURE 3–42   Problem 20.

21. In Problem 20, calculate the moment about point O, using Varignon's theorem, resolving the force into its X and Y components at point B and at point C.

22. Compute the moment about point A for the linkage shown in Fig. 3–43.

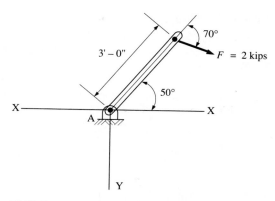

FIGURE 3–43 Problem 22.

23. Compute the moment of the force F about point A for the conditions indicated in Fig. 3–44.

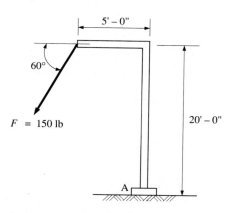

FIGURE 3–44 Problem 23.

Section 3–5 Resultants of Parallel Force Systems

24–27. Determine the magnitude of the resultant of the parallel force systems of Figs. 3–45 through 3–48. Locate the resultant with respect to point A. Assume that all forces are vertical and that the members are horizontal.

28. Determine the resultant and its location for the standard truck loading used in highway bridge design. The loads shown in Fig. 3–49 represent axle loads.

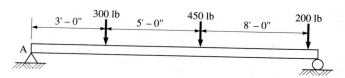

FIGURE 3–45 Problem 24.

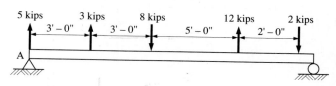

FIGURE 3–46 Problem 25.

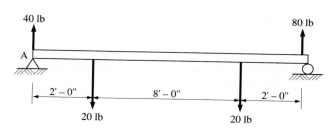

FIGURE 3–47 Problem 26.

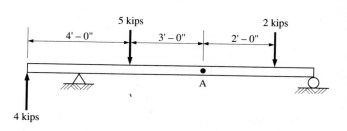

FIGURE 3–48 Problem 27.

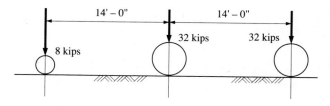

FIGURE 3–49 Problem 28.

29. Compute the magnitude, direction, and location of a load F_1 if the other forces of the parallel force system and the resultant are as shown in Fig. 3–50.

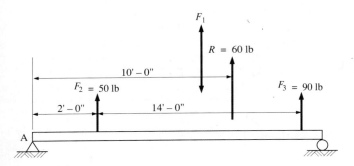

FIGURE 3–50   Problem 29.

30. Compute the magnitude and location of the resultant for the load system shown in Fig. 3–51.

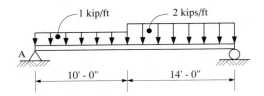

FIGURE 3–51   Problem 30.

31. Determine the magnitude and location of the resultant force for the load system shown in Fig. 3–52.

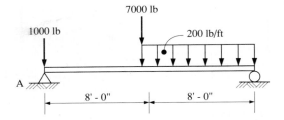

FIGURE 3–52   Problem 31.

Section 3–6   Couples

32–34. Compute the magnitude and direction of the resultant couples acting on the bodies shown in Figs. 3–53 through 3–55.

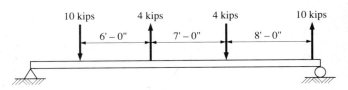

FIGURE 3–53   Problem 32.

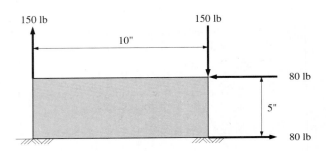

FIGURE 3–54   Problem 33.

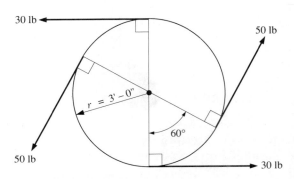

FIGURE 3–55   Problem 34.

35. A body is subjected to the following three couples: (a) 30 lb forces, 3 in. arm, counterclockwise; (b) 20 lb forces, 6 in. arm, counterclockwise; (c) 10 lb forces, 5 in. arm, clockwise. Determine the required magnitude of the forces of a single resultant couple, equivalent to the three given couples, and having a 2.5 in. arm.

Section 3–7 Resultants of Nonconcurrent Force Systems

36. Determine the magnitude, direction, and sense of the resultant force of the nonconcurrent force system of Fig. 3–56. Locate the resultant with respect to point O.

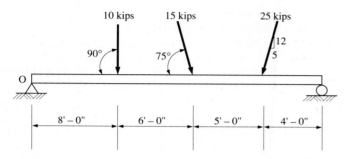

FIGURE 3–56 Problem 36.

37 and **38.** Determine the magnitude, direction, and sense of the resultant forces shown in Figs. 3–57 and 3–58. Determine where the resultant intersects the bottom of the body with respect to point O in each case.

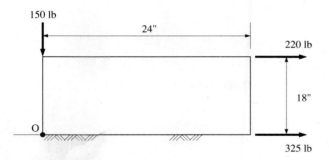

FIGURE 3–57 Problem 37.

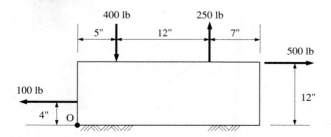

FIGURE 3–58 Problem 38.

39. For the concrete structure of Fig. 3–59, determine the magnitude, direction, and sense of the resultant force. Determine where the resultant intersects the bottom of the footing with respect to point A.

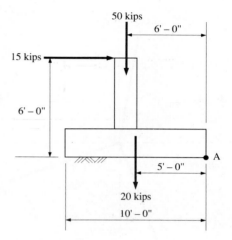

FIGURE 3–59 Problem 39.

40. A gravity-type masonry dam, shown in Fig. 3–60, depends on its own weight for stability. The horizontal force represents the total pressure of the water acting on a section of the dam having a width of one foot (into the page). The vertical force represents the weight of the section. Find the resultant of these two forces and locate the point where its line of action intersects the base AB of the dam. This point should fall within the middle 1/3 of the length AB. Does it?

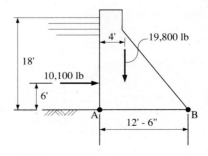

FIGURE 3–60 Problem 40.

SI System Problems

41. Determine the magnitude, direction, and sense of the resultant force of the coplanar concurrent force system of Fig. 3–61.

42. Determine the magnitude and location of the resultant of the parallel force system acting on the horizontal member AB of Fig. 3–62. Use point A as the reference point. Neglect the weight of the member.

43. (a) Calculate the moments about points A and B due to the nonconcurrent force system of Fig. 3–63. (b) Determine the location and direction of the resultant force.

44. Determine the magnitude, direction, and sense of the resultant of the three forces acting on a wall, the cross section of which is shown in Fig. 3–64. Also determine, with respect to point O, where the resultant intersects the base of the wall.

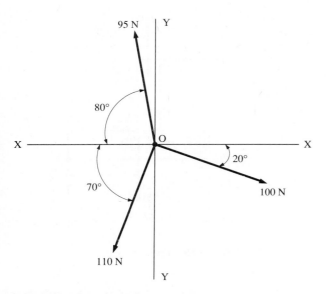

FIGURE 3–61   Problem 41.

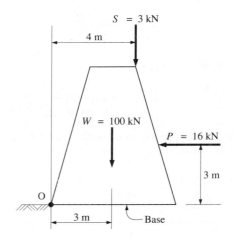

FIGURE 3–64   Problem 44.

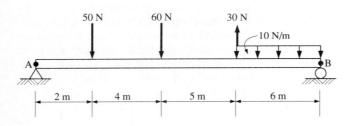

FIGURE 3–62   Problem 42.

FIGURE 3–63   Problem 43.

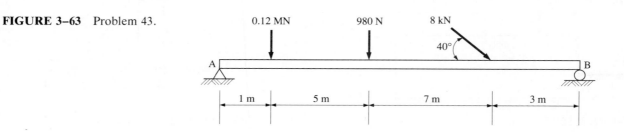

Computer Problems

For the following computer problems, any appropriate programming language may be used. Input prompts should fully explain what is required of the user (the program should be "user friendly"). The resulting output should be well labeled and self-explanatory.

45. Write a program that will calculate the magnitude and direction of the resultant of two concurrent forces that lie in the first quadrant, as shown in Fig. 3–29. User input is to be the magnitude of each force and its angle of inclination with the positive X axis.

46. Write a program that will determine the resultant of *n* concurrent forces. User input is to be the number of forces (*n*) and the magnitude and direction of each. For convenience, assume that the layout is similar to that shown in Fig. 3–7. (**Hint:** you may wish to consider azimuth angles from the positive X axis, or designation of quadrants.)

47. Write a program to solve Problem 28. The distance between the two 32 kip loads may range from 14 ft to 30 ft. The user should be allowed to input this axle spacing.

Supplemental Problems

48. The resultant and one component force of a two-force concurrent force system are shown in Fig. 3–65. Compute the other component force F_2 (not shown) that would be required. Determine magnitude, direction, and sense.

49. The resultant force of a concurrent force system is given in Fig. 3–66. Determine the magnitude of the two concurrent components if their angles of inclination are known.

50. The resultant force of a concurrent force system is given in Fig. 3–67. Determine the magnitudes of the components F_1 and F_2 if their directions and senses are as shown.

51 and **52.** Determine the resultant force for each of the coplanar concurrent force systems of Figs. 3–68 and 3–69. Compute the magnitude, sense, and angle of inclination with the horizontal X axis. Use the method of components.

53. As shown in Fig. 3–70, the resultant of the three concurrent forces is 100 lb with an angle of inclination of 20° to the X axis. Determine the magnitude of F_1 and F_2.

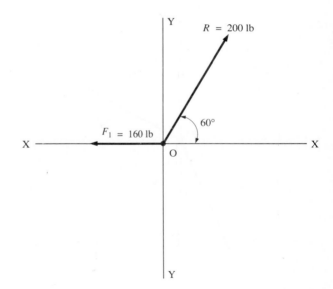

FIGURE 3–65 Problem 48.

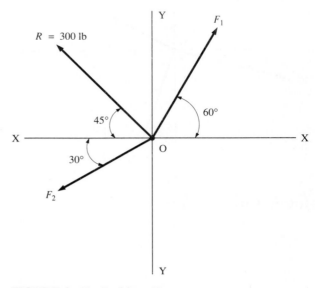

FIGURE 3–66 Problem 49.

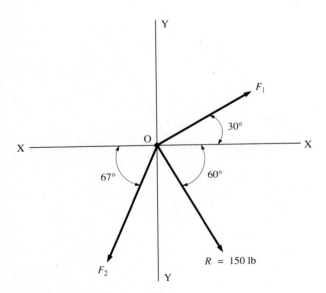

FIGURE 3–67 Problem 50.

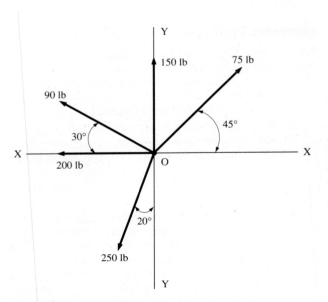

FIGURE 3–69 Problem 52.

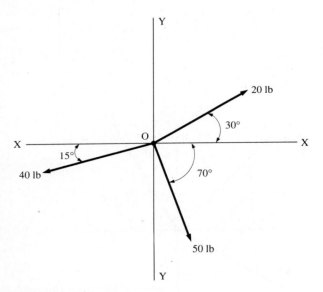

FIGURE 3–68 Problem 51.

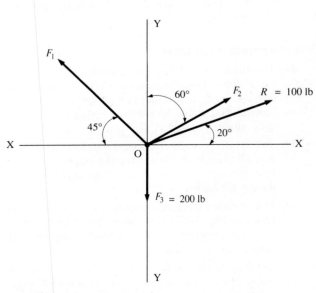

FIGURE 3–70 Problem 53.

72

54. Calculate the moment of the forces shown in Fig. 3–71 with respect to point B. The forces are vertical and the member is horizontal.

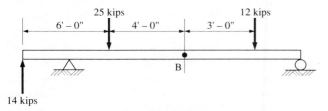

FIGURE 3–71 Problem 54.

55. Determine the moment (about point A) of the applied loads shown in Fig. 3–72.

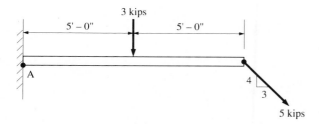

FIGURE 3–72 Problem 55.

56. The lift force on the wing of an aircraft is approximately represented in Fig. 3–73. Calculate the magnitude and location of the resultant of the distributed forces.

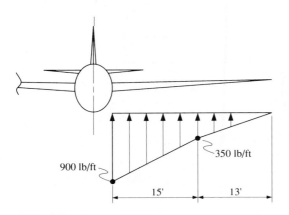

FIGURE 3–73 Problem 56.

57. A beam is subjected to distributed loads as shown in Fig. 3–74. Determine the magnitude and location of the resultant of the distributed forces.

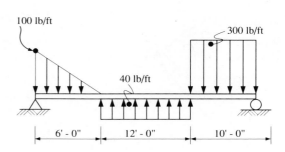

FIGURE 3–74 Problem 57.

58. The force F shown in Fig. 3–75 produces a clockwise moment of 400 ft-lb about point O. Determine the magnitude of the force.

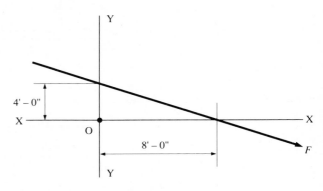

FIGURE 3–75 Problem 58.

59. (a) Compute the moment (about point A) of the forces shown in Fig. 3–76. (b) Find the resultant of the forces. Determine where it intersects a vertical line through point A.

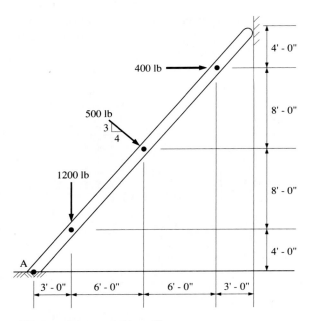

FIGURE 3–76   Problem 59.

60. Determine the resultant of the three forces acting on the horizontal beam of Fig. 3–77.

61. Determine the magnitude and location of the resultant of the parallel force system acting on a horizontal member, as shown in Fig. 3–78. Use point A as the reference point.

62. Compute the magnitude and direction of the resultant couple acting on the body in Fig. 3–79.

63. Determine the magnitude of F_1 and F_2 in Fig. 3–80 such that the resultant will be a counterclockwise couple with a moment of 220 ft-lb.

64. Calculate the magnitude, direction, and sense of the resultant force of the noncurrent force system of Fig. 3–81. Determine where the resultant intersects the bottom of the shape with respect to point A.

FIGURE 3–77   Problem 60.

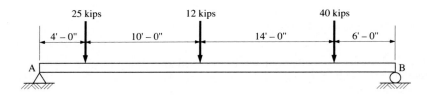

FIGURE 3–78   Problem 61.

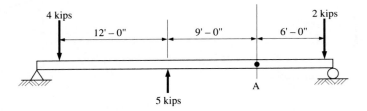

FIGURE 3–79 Problem 62.

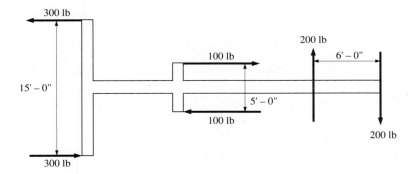

FIGURE 3–80 Problem 63.

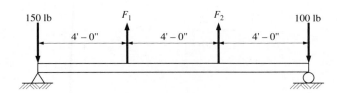

FIGURE 3–81 Problem 64.

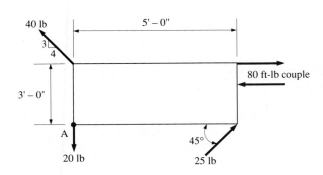

4 Equilibrium of Coplanar Force Systems

4–1
INTRODUCTION

As we have learned, statics deals essentially with the action of forces on rigid bodies at rest—a state also defined as one of equilibrium, or of zero motion. We may therefore conclude that, with zero motion, both the body and the entire system of external forces and moments acting on the body, no matter how complex, are in equilibrium.

If a system of forces is in equilibrium, the resultant of the force system is equal to zero. When a force system that is in equilibrium acts upon a body, the body is also said to be in equilibrium.

In this chapter we will establish the conditions a force system must satisfy in order for it to be in equilibrium.

4–2
CONDITIONS OF
EQUILIBRIUM

The resultant of a force system must be zero if the force system is to be in equilibrium. Two general conditions, based on fundamental laws, must be satisfied if the resultant is to be zero:

1. For a force system to be in equilibrium, the algebraic sum of all forces (or components of forces) along any axis, in any direction, must be equal to zero.
2. For a force system to be in equilibrium, the algebraic sum of the moments of the forces about any axis or point must be equal to zero.

These two conditions for equilibrium can be expressed mathematically as

$$(1)\ \Sigma F = 0 \quad \text{and} \quad (2)\ \Sigma M = 0$$

Considering only the two-dimensional case and the usual X–Y coordinate axes system, the algebraic summation of forces must be zero in both the X and the Y directions. The $\Sigma F = 0$ condition can then be rewritten as

$$\Sigma F_x = 0 \quad \text{and} \quad \Sigma F_y = 0$$

which states that, for equilibrium, the algebraic sums, respectively, of the X and Y components of the force system must equal zero. Frequently, the X–Y coordinate axes system is oriented with a vertical Y axis (rather than inclined), as defined by the direction of gravity. In this situation, the horizontal and vertical directions are, for convenience, denoted as such. Therefore, it is common to use $\Sigma F_H = 0$ and $\Sigma F_V = 0$ in place of the preceding expressions.

The moment condition for equilibrium can still be expressed as

$$\Sigma M = 0$$

which states that, for equilibrium, the algebraic sum of the moments of the forces of the system about any point in the plane must equal zero. This, in essence, states that the sum of the clockwise moments must equal the sum of the counterclockwise moments.

It should be noted that $\Sigma F_x = 0$, $\Sigma F_y = 0$, and $\Sigma M = 0$ represent what are generally termed the *three laws of equilibrium*. They are fundamental laws for bodies at rest and cannot be proven mathematically. They are based on the results of observations and were first advanced by Sir Isaac Newton (1642–1727) in his statement on the laws of motion. They stem from Newton's first law which states, in part, that when a body is at rest, the resultant of all the forces acting on the body is zero.

4–3 THE FREE-BODY DIAGRAM

The following sections of this chapter will deal with the applications of the conditions of equilibrium. Our primary purpose will be to determine desired information about certain forces that result from the effects of other forces acting on bodies.

Most problems encountered in statics result from the interaction of bodies. To solve these problems generally requires that a body be isolated, and that the given force system acting on the body be identified and analyzed so that unknown forces can be determined. The method most useful in meeting these requirements for solution is known as the *free-body diagram method*.

The free-body diagram is a sketch or pictorial representation (not necessarily to scale) indicating (a) the body in question, by itself, entirely isolated from the other bodies, and (b) all the external forces exerted on that body as a result of the interaction between the free body and other bodies. The external interactive forces acting on the free body may be direct forces due to contact between the free body and other bodies external to it (which could be solid, liquid, or gaseous) or indirect forces, such as gravitational or magnetic forces, which act without bodily contact. The free-body diagram is one of the most important and useful analytical tools in engineering mechanics.

A summary of the procedure for sketching a free body is as follows:

1. Sketch the subject body isolated from all other bodies. This body may be an entire structure consisting of many component parts but treated as a unit, or it may be any individual part of the structure.
2. Show all the known forces acting on the body, indicating a magnitude, line of action, and sense. These forces will represent the effect of the contact bodies removed and/or the noncontact gravitational or magnetic effects.
3. Indicate all desired unknown forces with a symbol and include as much known information as possible. Point of application is usually known;

direction *may* be known. Sense and/or components may have to be assumed.

Figure 4–1(a) shows a weight suspended by a rope. Free-body diagrams of the rope and the weight are shown in Fig. 4–1(b) and (c). Note how the rope and the weight are isolated and how the external forces acting on them are represented. In each case, the line of action of the force must coincide with the vertical centerline of the rope.

FIGURE 4–1 The free-body diagram.

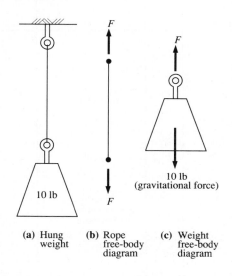

(a) Hung weight   **(b)** Rope free-body diagram   **(c)** Weight free-body diagram

Since the free-body diagram is a sketch showing known and unknown external forces acting on the free body, conventional symbols should be used to represent the action of the forces. Figure 4–2 shows some types of bodily contact or support and illustrates how the specific conditions should be represented in a free-body diagram. If the supporting member of a body is flexible, as in the case of a cable or rope (Fig. 4–2(a)), the force, in the nature of a pull (tension) is directed along the centerline of the member. A roller-type contact or support (Fig. 4–2(b)) permits motion parallel to the supporting surface. Therefore, the roller provides a reaction that is perpendicular to the supporting surface. The action of the smooth surface (Fig. 4–2(c)) on the free body is similar to the roller-type support in that it provides a reaction perpendicular to the supporting surface. A pinned-type contact or support (also called a hinged, or knife-edged, support) (Fig. 4–2(d)) does not permit any linear motion, but does permit rotational motion. The direction of the reaction at a pinned support is unknown; therefore, the reaction is generally indicated as two independent components. This is similar to the case in which a body is supported by a rough surface. A fixed-type contact or support (Fig. 4–2(e)) allows no linear or rotational motion. Therefore, the reaction is indicated as two independent components and a resisting couple (moment).

Description of Support	Sketch of Idealized Support	Effect on Free-Body (How Represented)
(a) Flexible cable, rope, chain, or wire	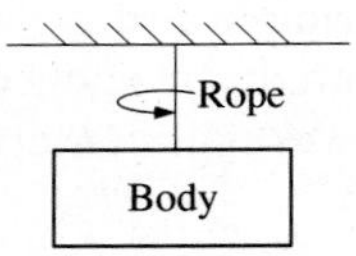	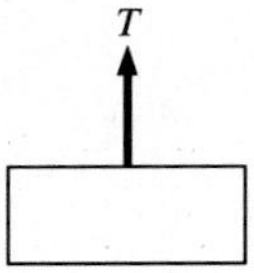
(b) Roller (zero friction)	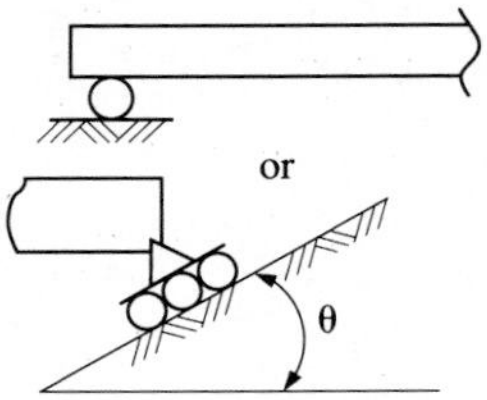	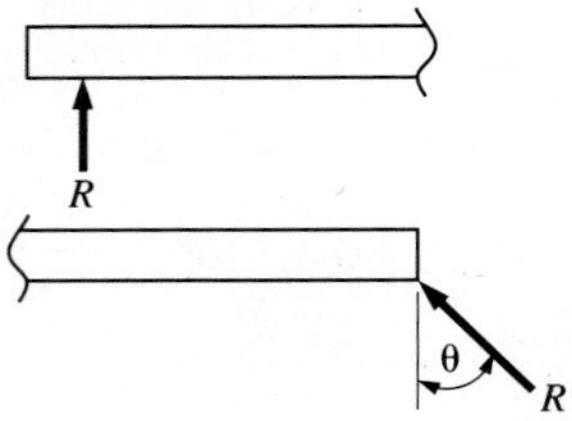
(c) Smooth surface (zero friction)	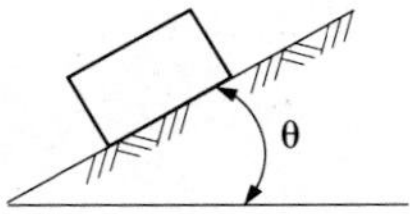	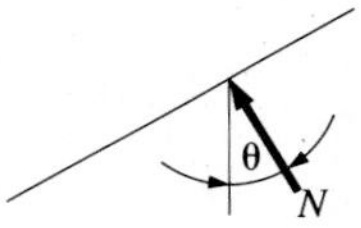
(d) Pinned, hinged, or knife-edged (rough surface)	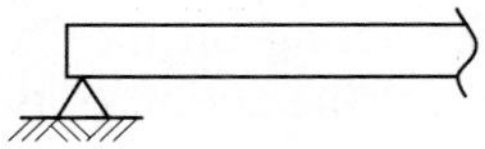	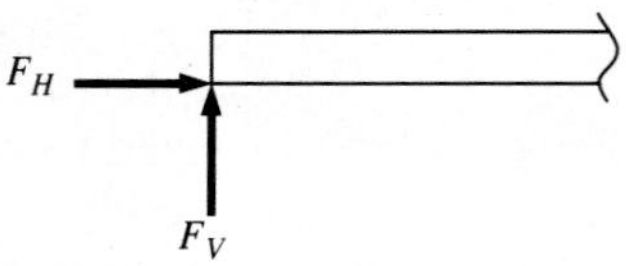
(e) Fixed	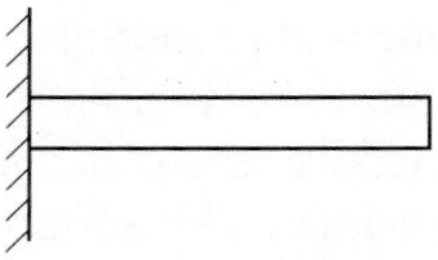	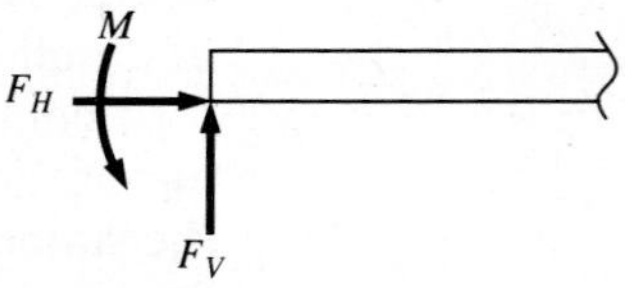

FIGURE 4–2 Supports and their representation in free-body diagrams.

☐ **EXAMPLE 4–1**   The beam shown in Fig. 4–3(a) is supported by a pinned support at A and a roller on a horizontal surface at B. Sketch the free-body diagram for the beam.

Solution   The free-body diagram for the beam is shown in Fig. 4–3(b). The weight of the beam, itself, which acts vertically downward through the center of the beam, is represented by the force W. The inclined load P is indicated on the free-body diagram exactly as on the beam diagram. Since the beam is supported by a pinned connection at A, the direction of the reaction at this point is unkown. The reaction is represented by its rectangular components A_V and A_H with assumed senses as shown. The reaction supplied at B is represented by a single vertical force B_V, which is perpendicular to the horizontal surface on which the roller is supported.

FIGURE 4–3 Beam for Example 4–1.

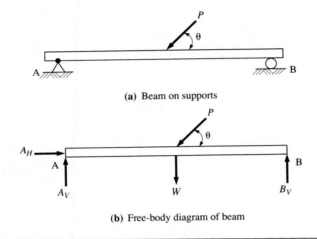

(a) Beam on supports

(b) Free-body diagram of beam

☐ **EXAMPLE 4–2**   The horizontal beam shown in Fig. 4–4(a) is supported by a pinned support at A and a roller on an inclined plane at B. Sketch the free-body diagram of the beam.

FIGURE 4–4 Beam for Example 4–2.

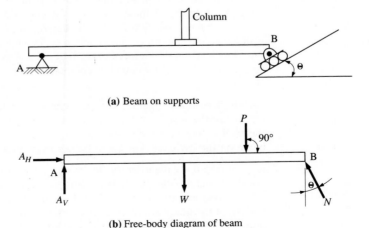

(a) Beam on supports

(b) Free-body diagram of beam

Solution   The free-body diagram for the beam is shown in Fig. 4–4(b). The weight of the beam and the reaction at A are similar to that in Example 4–1 and are indicated in the same manner. The external force from the column, which is supported by the beam, is represented by P and is acting vertically downward. The beam is supported by a roller at point B. The reaction N acting on the beam must be perpendicular to the supporting surface (since the roller permits motion parallel to the inclined plane). Therefore, N is known to be acting at an angle of inclination of θ with the vertical, as shown.

☐ **EXAMPLE 4–3**   A cylinder, shown in Fig. 4–5, is supported by a vertical wall, a beam that is pin connected to the wall at point A, and a flexible cable. The two contact surfaces with the cylinder are smooth.

FIGURE 4–5 Support system for cylinder.

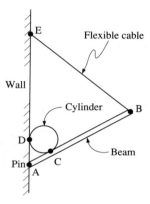

        Sketch the free-body diagram for (a) the entire system, considering the cylinder, beam, and cable as a single body; (b) the cylinder alone; and (c) the beam alone.

Solution   (a) The free-body diagram for the entire system is shown in Fig. 4–6(a). The external forces are the weight of the cylinder W_C, the weight of the beam W_B, the pull T of the wall on the cable, the push P of the wall at point D, and the force exerted by the pin at point A. Note that T must act along the centerline (or axis) of the cable, since the cable is flexible, and that P must be horizontal, acting perpendicular to the vertical wall, since the smooth contact surface of the wall permits vertical motion. Also note that the direction of the force exerted at point A is unknown. Therefore, this force is represented by its rectangular components A_V and A_H with assumed senses as shown.

(b) The free-body diagram for the cylinder is shown in Fig. 4–6(b). The external forces are the weight of the cylinder W_C acting vertically downward, the push P of the wall acting perpendicularly to the surface of the wall, and the reaction of the beam on the cylinder (known to be perpendicular to the beam).

(c) The free-body diagram for the beam is shown in Fig. 4–6(c). The external forces are the weight of the beam W_B acting vertically downward, the pull T of the cable acting along its axis, the force F of the cylinder, and the rectangular components A_V and A_H of the unknown force exerted by the pin connection at point A. In particular, note the equal and opposite relationship of force F acting on the beam and the cylinder.

FIGURE 4–6 Free-body diagrams.

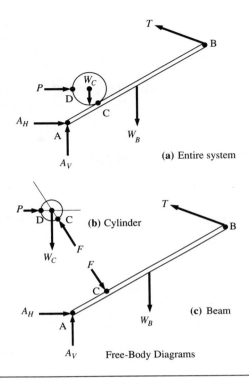

(a) Entire system

(b) Cylinder

(c) Beam

Free-Body Diagrams

4–4
EQUILIBRIUM OF CONCURRENT FORCE SYSTEMS

When a coplanar concurrent force system is in equilibrium, the algebraic sum of the vertical and horizontal components of all the forces must, respectively, equal zero. This has been expressed in Section 4–2 as

$$\Sigma F_y = 0 \quad \text{and} \quad \Sigma F_x = 0$$

Conversely, if it can be demonstrated that $\Sigma F_y = 0$ and $\Sigma F_x = 0$ in a concurrent force system, then we can say that the system is in equilibrium and that the resultant is equal to zero.

Recall from previous chapters that a concurrent force system is one in which the action lines of all the forces intersect at a common point. This system cannot cause rotation of the body on which it acts, thereby implying that only two equations of equilibrium are sufficient for analyzing this type of force system.

Problems involving concurrent force systems frequently involve special members that are in equilibrium under the action of two equal and opposite forces. These are called *two-force members*. The forces are normally applied at the ends. In this text, we will consider only straight two-force members.

A study of the free-body diagram for a two-force member will show that, for equilibrium to exist, the lines of action of the forces must be collinear with the axis of the member. This will usually be an important key in the solution of the problem since the force effect of a two-force member in contact with any other member must act in the direction established by the axis of the two-force member. The rope shown in Fig. 4–1(b) and the boom and the cable of Example 4–4 are two-force members.

☐ **EXAMPLE 4–4**   A 100 lb weight is supported by a tied boom, as shown in Figure 4–7(a). Determine the magnitude of the force C in the boom and the force T in the cable.

FIGURE 4–7   Concurrent force system.

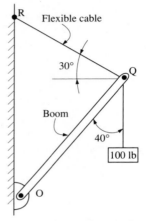

(a) Tied boom-supported weight

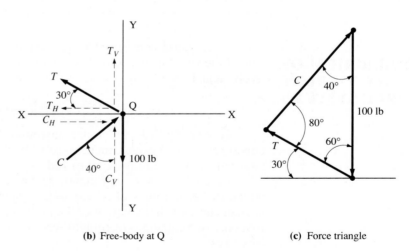

(b) Free-body at Q   **(c)** Force triangle

Solution   The free-body diagram of the joint at Q is shown in Fig. 4–7(b). The boom and the cable are two-force members. Therefore, the lines of action of unknown forces C and T, respectively, are known. The two forces are shown acting with assumed senses, which are readily apparent in this case. At times, however, the situation will not be as clear. In those cases, a sense should be assumed for each unknown force and verified by calculation.

This force system is categorized as a coplanar concurrent force system. The two unknown forces may be found either by the force triangle method or by the method of components, applying the two laws of equilibrium. For illustrative purposes, we will show both methods.

The Force Triangle Method

A concurrent coplanar force system in equilibrium must have a zero resultant. With the system comprising three forces, the forces must form a closed triangle. The tip of each force vector must touch the tail of another force vector, resulting in a closed triangle (see Fig. 4–7(c)).

It is not necessary to assume the sense of the unknown force in either the cable or the boom; they can be determined from the force triangle. As shown in Fig. 4–7(c), begin the sketch of the force triangle with the 100 lb force, which is known in magnitude, direction, and sense. Then draw lines parallel to the lines of action of the two unknown forces, one through the tail and one through the tip of the known force vector. The lines must intersect to form a closed triangle. Since the force vectors must lie tip to tail, the senses of the unknown forces are established in the force triangle. These senses are then related to the forces acting at the point of concurrency Q in Fig. 4–7(b). Force C acts upward and to the right; force T acts upward and to the left.

Using the law of sines to solve for the unknown forces,

$$\frac{100}{\sin 80°} = \frac{T}{\sin 40°} = \frac{C}{\sin 60°}$$

from which

$$T = \frac{\sin 40°}{\sin 80°} (100) = 65.3 \text{ lb}$$

$$C = \frac{\sin 60°}{\sin 80°} (100) = 87.9 \text{ lb}$$

The Method of Components

Applying the two laws of force equilibrium ($\Sigma F_H = 0$ and $\Sigma F_V = 0$), and with reference to Fig. 4–7(b), the two unknown forces may be determined. Assume positive senses to be upward and to the right. The assumed senses of T and C are as shown.

Summing forces in the horizontal direction,

$$\Sigma F_H = -T_H + C_H = 0$$
$$= -T \cos 30° + C \sin 40° = 0$$

from which

$$T = \frac{\sin 40°}{\cos 30°} C = 0.7422\,C \qquad \text{(Eq. 1)}$$

Summing forces in the vertical direction,

$$\Sigma F_V = +T_V + C_V - 100 = 0$$
$$= +T \sin 30° + C \cos 40° - 100 = 0$$

Substituting from Eq. 1,

$$0.7422C(\sin 30°) + C \cos 40° - 100 = 0$$

from which

$$C = +87.9 \text{ lb}$$

From Eq. 1,

$$T = 0.7422C = +65.3 \text{ lb}$$

The positive signs indicate that the senses of the forces are, in fact, as they were assumed. This is an important concept. A positive sign resulting in the final calculation does not necessarily mean that the force is acting with a positive sense. It means that the sense of the force is as it was assumed.

Note that a problem involving three coplanar concurrent forces is very conveniently solved using the force triangle. For more than three forces, the use of force components and the two laws of equilibrium is generally more efficient.

□ **EXAMPLE 4–5**   Two straight, rigid bars, AB and BC, are pin connected to a horizontal supporting floor at their lower ends and to each other at their upper ends, as shown in Fig. 4–8(a). Applied loads at point B are 2000 lb vertically and 1800 lb horizontally. Assume that the weights of the bars are negligible and that the system is coplanar. Compute the magnitude and sense of the forces in the two bars using the method of components.

FIGURE 4–8   Structure for Example 4–5.

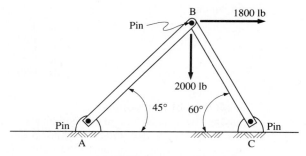

(a) Rigid pin-connected structure

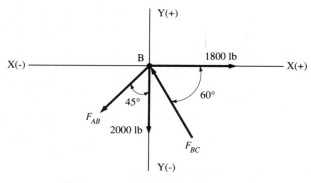

(b) Free-body diagram of pin at point B

Solution   The free-body diagram for the pin at point B is shown in Fig. 4–8(b). The forces exerted on the pin are the horizontal pull of 1800 lb, the vertically downward pull of 2000 lb, the force exerted by AB, and the force exerted by BC. Note that AB and BC are two-force members. Therefore, the lines of action of forces F_{AB} and F_{BC} will lie, respectively, along the axes of members AB and BC. That is, the directions of the forces are known. Only the magnitudes and senses must be determined.

The senses of the forces are assumed, as shown in Fig. 4–8(b). The correct sense will be determined by the sign of the computed magnitude of the force: a positive sign will indicate that the sense of the force is as assumed; a negative sign will indicate that the sense of the force is opposite to that assumed.

Since the force system is coplanar and concurrent, the two equations of equilibrium may be applied using the conventional X–Y coordinate axes. Forces upward and to the right will be assumed positive. The assumed senses of F_{AB} and F_{BC} are as shown.

$$\Sigma F_x = +1800 - F_{BC} \cos 60° - F_{AB} \cos 45° = 0$$
$$= +1800 - 0.5F_{BC} - 0.707F_{AB} = 0 \qquad \text{(Eq. 1)}$$
$$\Sigma F_y = -2000 + F_{BC} \sin 60° - F_{AB} \sin 45° = 0$$
$$= -2000 + 0.866F_{BC} - 0.707F_{AB} = 0 \qquad \text{(Eq. 2)}$$

Solving Eq. 2 for F_{AB} in terms of F_{BC},

$$0.707F_{AB} = 0.866F_{BC} - 2000$$

Substituting the preceding in Eq. 1 yields

$$+1800 - 0.5F_{BC} - (0.866F_{BC} - 2000) = 0$$
$$+1800 - 1.366F_{BC} + 2000 = 0$$

from which

$$F_{BC} = +2782 \text{ lb}$$

Substituting this in Eq. 2 and solving yields

$$-2000 + 0.866(2782) - 0.707F_{AB} = 0$$
$$F_{AB} = +579 \text{ lb}$$

The positive signs indicate that the senses of the forces are as assumed. Member BC acts toward the pin (at B) and member AB pulls away from the pin.

**4–5
EQUILIBRIUM
OF PARALLEL
FORCE SYSTEMS**   When a coplanar parallel force system is in equilibrium, the algebraic sum of the forces of the system must equal zero. In addition, the algebraic sum of the moments of the forces of the system about any point in the plane must equal zero. These requirements have been expressed in Section 4–2 as

$$\Sigma F = 0 \quad \text{and} \quad \Sigma M = 0$$

Conversely, if it can be demonstrated that the two preceding requirements are satisfied (where the moment *M* may be about any point in the plane), then we can say that the parallel force system is in equilibrium, and that the force and moment resultants are equal to zero.

It is important to note that equilibrium of parallel force systems cannot be verified through the use of the force summation equations only. In all cases, at least one moment summation equation must be considered.

A common type of problem associated with parallel force systems is determining two unknown support reactions for a beam or member. These are external reactions and, as will be shown in later chapters, must be calculated prior to the internal behavior investigation of the member.

In computing reactions of parallel force systems, care must be taken to adhere to a sign convention. Whether a clockwise moment (rotation) is taken as positive or negative is inconsequential, but the adherence to one or the other convention throughout the solution of any particular problem is of utmost importance. We will assume a counterclockwise moment (rotation) about a moment center to be positive and a clockwise moment to be negative. Further discussion on computing reactions with examples using more complex members and loadings is provided in Chapter 13.

☐ **EXAMPLE 4–6**   A beam carries vertical concentrated loads as shown in Fig. 4–9(a). The beam is pin supported at A and supported by a roller on a horizontal surface at B. A beam of this type with the indicated supports is called a *simple beam*. The supports are called *simple supports*. The reactions at the supports are assumed to be parallel to the loads. Calculate the reactions at each support. Neglect the weight of the beam.

FIGURE 4–9   Beam for Example 4–6.

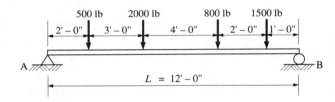

(a) Parallel force system on beam

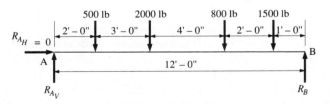

(b) Free-body diagram

Solution   The free-body diagram is shown in Fig. 4–9(b). Note that the supports at A and B have been replaced with forces (reactions) with a known line of action and assumed sense. The pin support at A could provide a horizontal reaction, but there are no horizontally applied forces or components. Therefore, this reaction would be zero and may be neglected. As will be discussed further in Chapter 13, the free-body diagram of the beam is commonly called a *load diagram*. The load diagram has the same geometry as the original structure (beam), the same loads are shown, and the supports have been replaced with the reactions expected from those supports.

Assuming counterclockwise rotation is positive, the reaction at B may be computed by taking an algebraic summation of the moments of the forces about point A:

$$\Sigma M_A = +R_B(12.0) - 500(2.0) - 2000(5.0) - 800(9.0) - 1500(11.0) = 0$$

from which

$$R_B = +2892 \text{ lb}$$

The positive sign indicates that the sense (upward) is as assumed for the reaction at point B. Had this result turned out to be negative, it would have meant that the sense of the reaction was opposite to that assumed, that is, that the reaction was actually downward.

The reaction at point A may be computed by taking an algebraic summation of the moments of the forces about point B:

$$\Sigma M_B = -R_{A_V}(12.0) + 1500(1.0) + 800(3.0) + 2000(7.0) + 500(10.0) = 0$$

from which

$$R_{A_V} = +1908 \text{ lb}$$

As an independent check of the computations, an algebraic summation of all the vertical forces should be made. In order that the beam be in equilibrium, it is required that the sum be equal to zero. Assuming upward acting forces to be positive and downward to be negative,

$$\Sigma F_V = +1908 + 2892 - 500 - 2000 - 800 - 1500 = 0 \quad \textbf{OK}$$

☐ **EXAMPLE 4–7** A simple beam is subjected to vertical concentrated and uniformly distributed loads as shown in Fig. 4–10(a). The beam is pin supported at A and roller supported (on a horizontal surface) at B. Calculate the reactions at each support. Neglect the weight of the beam.

Solution The free-body diagram (the load diagram) is shown in Fig. 4–10(b). The reactions are assumed to be parallel to the loads. The supports at A and B have been replaced with forces (reactions) with known lines of action and assumed senses. The pin support at A could provide a horizontal reaction, but there are no horizontally applied loads or components. Therefore, this reaction would be zero and may be neglected.

The equivalent concentrated resultant force for the uniformly distributed load is

$$W = 2 \text{ kips/ft} \times 18 \text{ ft} = 36 \text{ kips}$$

This is indicated by a dashed arrow acting at the center of the distributed load.

Next compute the reaction at B by taking an algebraic summation of the moments of the forces about point A. Assume counterclockwise rotation is positive and also that both reactions are acting upward.

$$\Sigma M_A = +R_B(36) - 20(32) - 10(24) - 36(9) = 0$$

from which

$$R_B = +33.4 \text{ kips}$$

Since the result is positive, the sense of the reaction is as assumed (upward).

The reaction at A may be computed by summing the moments of the forces about point B:

$$\Sigma M_B = -R_{A_V}(36.0) + 36(27.0) + 10(12.0) + 20(4.0) = 0$$

from which

$$R_{A_V} = +32.6 \text{ kips}$$

Finally, check the calculations with a vertical summation of forces (upward is positive):

$$\Sigma F_V = +33.4 + 32.6 - 36 - 10 - 20 = 0 \quad \textbf{OK}$$

FIGURE 4–10   Beam for Example 4–7.

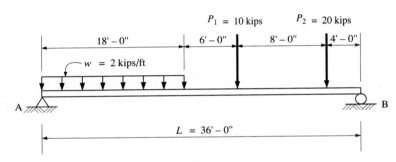

(a) Parallel force system on simple beam

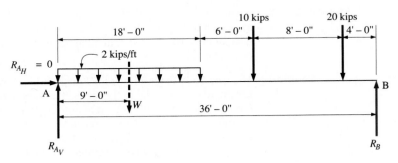

(b) Free-body diagram

4–6
EQUILIBRIUM OF NONCONCURRENT FORCE SYSTEMS

When a coplanar, nonconcurrent, nonparallel force system is in equilibrium, the algebraic sum of the vertical and horizontal components of all the forces must, respectively, equal zero. Additionally, the algebraic sum of the moments of the forces about any point in the plane must equal zero.

Conversely, if $\Sigma F_V = 0$, $\Sigma F_H = 0$, and $\Sigma M = 0$, then we can say that the force system is in equilibrium and that the force and moment resultants are equal to zero.

Equilibrium of this system cannot be verified through the use of the force summation equations only. In all cases, at least one moment summation equation must be considered. In choosing the moment center about which to sum moments, it must be remembered that forces having lines of action passing through the moment center have zero moment about that moment center. Therefore, a moment center should be selected that will eliminate as many forces as possible (or selected forces) from the moment equation.

☐ **EXAMPLE 4–8** Compute the reactions at A and B on the truss shown in Fig. 4–11(a). There is a roller support at A and a pin support at B.

FIGURE 4–11 Truss for Example 4–8.

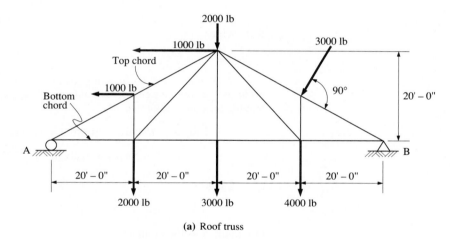

(a) Roof truss

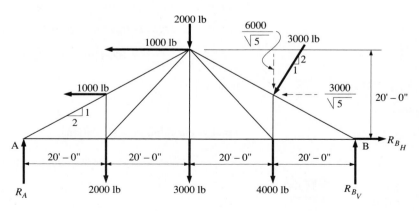

(b) Free-body diagram

Solution   The free-body diagram for the truss is shown in Fig. 4–11(b). The external forces are as follows:

1. On the bottom chord: three vertical concentrated loads.
2. On the top chord: two horizontal concentrated loads, one vertical concentrated load, and one inclined load acting perpendicularly to the top chord. (For convenience, the inclined load is resolved into its components, shown dashed.)
3. The reaction at A, known to be vertical since the support is furnished by rollers. The sense is assumed upward.
4. The reaction at B. Since the support is pinned, the direction is unknown; therefore, the reaction is represented by its vertical and horizontal components. The senses are assumed upward and to the right, respectively.

This external force system is coplanar, nonconcurrent, and nonparallel. Therefore, utilize the three laws of equilibrium: $\Sigma F_V = 0$, $\Sigma F_H = 0$, and $\Sigma M = 0$. Assume forces upward and to the right and counterclockwise moments are positive.

$$\Sigma F_H = R_{B_H} - 1000 - 1000 - 3000\left(\frac{1}{\sqrt{5}}\right) = 0$$

from which

$$R_{B_H} = +3342 \text{ lb}$$

Next determine R_A by summing moments about point B. The selection of this point as the moment center will eliminate R_{B_V} and R_{B_H} from the moment equation, leaving only one unknown: R_A.

$$\Sigma M_B = -R_A(80) + 2000(60) + 3000(40) + 4000(20) + 2000(40)$$
$$+ 1000(10) + 1000(20) + \left(\frac{6000}{\sqrt{5}}\right)(20) + \left(\frac{3000}{\sqrt{5}}\right)(10) = 0$$

from which

$$R_A = +6214 \text{ lb}$$

Use point A as the moment center to compute R_{B_V}, after which an algebraic summation of the vertical forces can be done as a check on your calculations.

$$\Sigma M_A = +R_{B_V}(80) - 2000(20) - 3000(40) - 4000(60) - 2000(40)$$
$$+ 1000(10) + 1000(20) - \left(\frac{6000}{\sqrt{5}}\right)(60) + \left(\frac{3000}{\sqrt{5}}\right)(10) = 0$$

from which

$$R_{B_V} = +7470 \text{ lb}$$

Finally, checking the calculations,

$$\Sigma F_V = +6214 + 7470 - 2000 - 3000 - 4000 - 2000 - \left(\frac{6000}{\sqrt{5}}\right) = 0$$

(very close)   **OK**

□ **EXAMPLE 4–9**   A cylinder is supported by an inclined wall and a vertical bar, as shown in Fig. 4–12. The bar is pin connected to the wall at point A and connected to a flexible cable at point B. All surfaces are smooth. The cylinder weight W is 2500 lb. Determine the forces acting on the bar; namely, the reaction at point A and the pull in the cable at point B. Neglect the weight of the bar.

FIGURE 4–12   Cylinder support system.

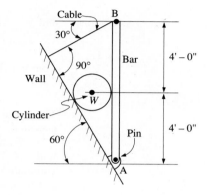

Solution   The free-body diagram for the bar is shown in Fig. 4–13(b). The force system acting on the bar is coplanar, nonconcurrent, and nonparallel. Note that four unknown forces are acting on the bar. Since there are only three equations of equilibrium, you will not be able to determine all of these forces using this free body alone. One of the forces must be determined elsewhere.

The cylinder is in contact with the bar and creates the force N on the bar. Investigate the cylinder as a free body and, utilizing the laws of equilibrium, determine the contact force N. As shown in Fig. 4–13(a), the force system acting on the

FIGURE 4–13   Free-body diagrams.

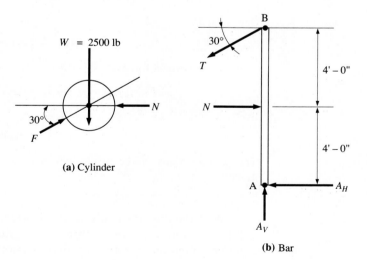

(a) Cylinder

(b) Bar

cylinder is coplanar and concurrent. For the cylinder, assuming forces upward and to the right to be positive, apply $\Sigma F_V = 0$ and $\Sigma F_H = 0$ as follows:

$$\Sigma F_V = -W + F \sin 30° = -2500 + F(0.5) = 0$$

from which

$$F = +5000 \text{ lb}$$

and

$$\Sigma F_H = -N + F \cos 30° = -N + 5000(0.866) = 0$$

from which

$$N = +4330 \text{ lb}$$

The positive signs for F and N indicate that the senses for these forces are as assumed.

With force N known, the laws of equilibrium may now be applied to the free body of the bar shown in Fig. 4–13(b). Using point A as a moment center will eliminate A_V and A_H from the summation. Assuming counterclockwise moment to be positive, and assuming T to be acting as shown (where only the horizontal component of T will create moment about point A), the moment summation yields

$$\Sigma M_A = -N(4.0) + T \cos 30°(8.0) = -4330(4.0) + T(0.866)(8.0) = 0$$

from which

$$T = +2500 \text{ lb}$$

Since the result is positive, the sense of T is as assumed. Next, with T and N known, the two components of the reaction at point A may be determined. Assuming the forces at A to be acting as shown,

$$\Sigma F_H = +N - T \cos 30° - A_H = +4330 - 2500(0.866) - A_H = 0$$

from which

$$A_H = +2165 \text{ lb}$$

And for the vertical forces,

$$\Sigma F_V = +A_V - T \sin 30° = +A_V - 2500(0.5) = 0$$

from which

$$A_V = +1250 \text{ lb}$$

The positive signs of the results indicate that the senses for A_H and A_V are as assumed.

We have previously discussed two-force members. Bar AB in Example 4–9 is acted on by forces at three points along its length. Members such as this one, acted on by forces at three or more points along their lengths, are sometimes called *multiple-force members*. They differ radically from two-force members in that some, or all, of the forces are applied transversely to the axis of the member, causing it to bend. Recall that in the two-force member the forces, of necessity, were collinear with (acted along) the axis of the member. The direction of a force applied to a multiple-force member (such as at A on bar AB) will frequently be unknown. Multiple-force members in frames will be discussed in Section 5–6.

4–7
SI SYSTEM
EXAMPLES

□ **EXAMPLE 4–10**

A block of 200 kg mass is supported by two cables, as shown in Fig. 4–14(a). Find the magnitude of the forces in the cables.

Solution   The load due to the block is calculated from

$$W = mg = 200 \text{ kg}(9.81 \text{ m/s}^2) = 1962 \text{ N} = 1.962 \text{ kN}$$

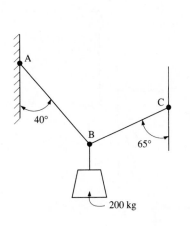

(a) Suspended weight

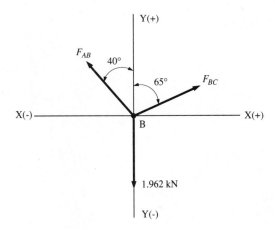

(b) Free-body diagram of joint B

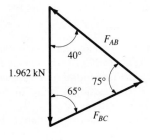

(c) Force triangle

FIGURE 4–14   Equilibrium of concurrent forces

The force system is coplanar and concurrent. Both cables must be in tension. The force system may be solved by using the method of components and applying the two equations of equilibrium to the free body of joint B, shown in Fig. 4–14(b), or by solution of the force triangle, shown in Fig. 4–14(c). The latter is the more direct solution.

$$\frac{1.962 \text{ kN}}{\sin 75°} = \frac{F_{BC}}{\sin 40°} = \frac{F_{AB}}{\sin 65°}$$

$$F_{BC} = \frac{\sin 40°}{\sin 75°} (1.962) = 1.306 \text{ kN}$$

$$F_{AB} = \frac{\sin 65°}{\sin 75°} (1.962) = 1.842 \text{ kN}$$

☐ **EXAMPLE 4–11** The tied boom in Fig. 4–15(a) supports a gravity load of 100 N. The boom is pinned at A. Determine the force in the tie and the reactions at A.

FIGURE 4–15 Nonconcurrent force system.

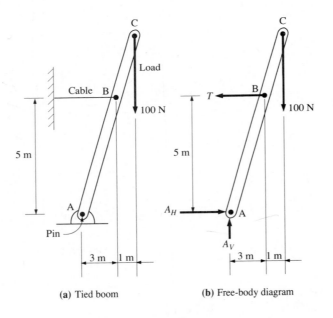

(a) Tied boom   (b) Free-body diagram

Solution This force system is coplanar and nonconcurrent. The free-body diagram of the boom is shown in Fig. 4–15(b). T is the tensile force in the cable. The pin support at A has been replaced with horizontal and vertical reactions A_H and A_V, with assumed senses as shown. Forces acting upward or to the right and counterclockwise moments are considered positive.

T is calculated by summing moments about point A:

$$\sum M_A = +T(5 \text{ m}) - 100 \text{ N}(4 \text{ m}) = 0$$

$$T = \frac{+400 \text{ N·m}}{+5 \text{ m}} = +80 \text{ N} \leftarrow$$

T is a tensile force, acting to the left, as assumed.

Next determine A_H and A_V by summing horizontal forces ($\Sigma F_H = 0$) and vertical forces ($\Sigma F_V = 0$), respectively:

$$\Sigma F_H = -T + A_H = 0$$
$$A_H = T = +80 \text{ N} \rightarrow$$
$$\Sigma F_V = -100 \text{ N} + A_V = 0$$
$$A_V = +100 \text{ N} \uparrow$$

The positive signs indicate that the senses are as assumed.

SUMMARY—BY SECTION NUMBER

4–1   Equilibrium is an "at rest" state, or a state of zero motion.

4–2   The resultant of any force system acting on a body in equilibrium must be zero. For the two-dimensional case, the three laws of equilibrium are

$$\Sigma F_x = 0 \qquad \Sigma F_y = 0 \qquad \Sigma M = 0$$

4–3   The free-body diagram is a sketch of an isolated body indicating all forces that act on the body.

4–4   Coplanar concurrent force systems are in equilibrium if $\Sigma F_x = 0$ and $\Sigma F_y = 0$. Generally, such systems involve two-force members. In a two-force member, forces are collinear with the axis of the member and act at two points only, usually at the ends.

4–5   Coplanar parallel force systems are in equilibrium if $\Sigma F = 0$ and $\Sigma M = 0$. A common parallel force system problem involves determining two unknown support reactions for a beam.

4–6   Coplanar nonconcurrent force systems are in equilibrium if $\Sigma F_x = 0$, $\Sigma F_y = 0$, and $\Sigma M = 0$. Such systems may involve multiple-force members, which are members acted on by forces at three or more points along their length. The reactions for multiple-force members are not collinear with the member.

PROBLEMS

Section 4–3   The Free-Body Diagram

1 and **2.** Sketch free-body diagrams for the members shown in Figs. 4–16 and 4–17.

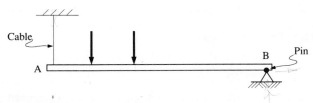

FIGURE 4–16   Problem 1.

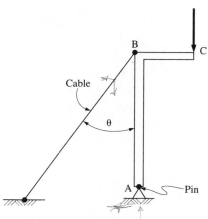

FIGURE 4–17   Problem 2.

3. A cylinder is supported by two smooth inclined planes, as shown in Fig. 4–18. Sketch the free-body diagram for the cylinder.

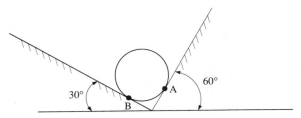

FIGURE 4–18   Problem 3.

4. A cylinder is supported on an inclined plane by a cable, as shown in Fig. 4–19. Sketch the free-body diagram for the cylinder.

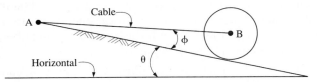

FIGURE 4–19   Problem 4.

5. A weight W is supported by a flexible cable and an inclined bar, as shown in Fig. 4–20. The bar is pin connected at a vertical wall. Sketch the free-body diagram for the bar.

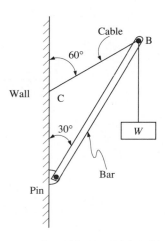

FIGURE 4–20   Problem 5.

6. The ladder in Fig. 4–21 is supported by a smooth frictionless vertical wall and is pin connected at point A. The ladder supports a weight W at point C. Assuming the weight of the ladder is acting at its center of gravity, sketch the free-body diagram for the ladder.

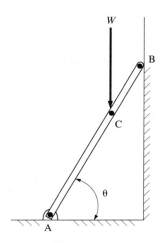

FIGURE 4–21   Problem 6.

Section 4–4   Equilibrium of Concurrent Force Systems

7. Calculate the reactions of the two smooth inclined planes against the cylinder in Fig. 4–22. The cylinder weighs 100 lb.

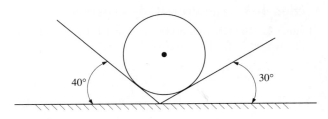

FIGURE 4–22   Problem 7.

8. Calculate the force in each cable for the suspended weight in Fig. 4–23.

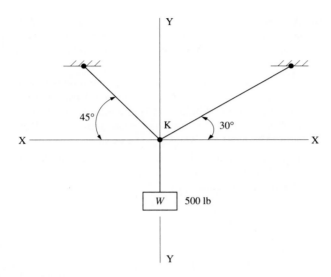

FIGURE 4–23   Problem 8.

9. What horizontal force *F* applied at the center of the cylinder in Fig. 4–24 is required to start the cylinder to roll over the 5 in. curb? What is the reaction of the curb? The cylinder weighs 250 lb.

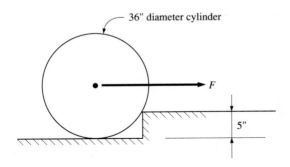

FIGURE 4–24   Problem 9.

10. Calculate the force in cable AB and the angle θ for the support system of Fig. 4–25.

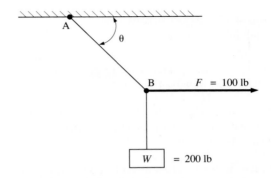

FIGURE 4–25   Problem 10.

11. Calculate the horizontal force *F* that should be applied to the 100 lb weight in Fig. 4–26 in order that the cable AB be inclined at an angle of 30° with the vertical.

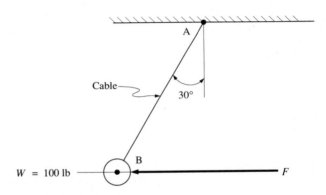

FIGURE 4–26   Problem 11.

12. The bell crank shown in Fig. 4–27 is supported by a pin bearing at B. A force of 500 lb is applied vertically at C. Rotation is prevented by the force *F* acting at A. Calculate the value of *F* and the bearing reaction at B.

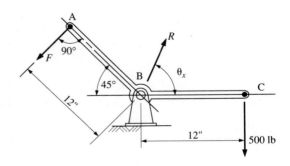

FIGURE 4–27   Problem 12.

Section 4–5   Equilibrium of Parallel Force Systems

13. The beam in Fig. 4–28 carries vertical concentrated loads as shown. Calculate the reaction at each support. Neglect the weight of the beam.

14. The beam in Fig. 4–29 carries vertical loads as shown. Calculate the reaction at each support. Neglect the weight of the beam.

15. Calculate the reaction at each support for the truss in Fig. 4–30. Neglect the weight of the truss.

16. A 12 ft simple beam weighing 30 lb per linear ft is supported at each end. How far from the left support should a concentrated load of 150 lb be applied in order that the left reaction equal 225 lb?

17. A 12 ft simple beam is supported at each end. It supports a concentrated load of 800 lb at 3 ft from the left support. Where should a second concentrated load of 1500 lb be placed so that the beam reactions will be equal? Neglect the weight of the beam.

18. The beam in Fig. 4–31 carries vertical loads as shown. Calculate the reaction at each support. Neglect the weight of the beam.

FIGURE 4–28   Problem 13.

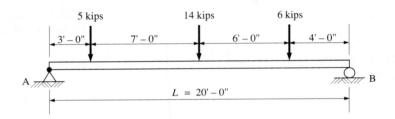

FIGURE 4–29   Problem 14.

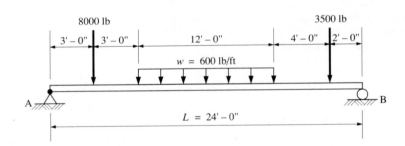

FIGURE 4–30   Problem 15.

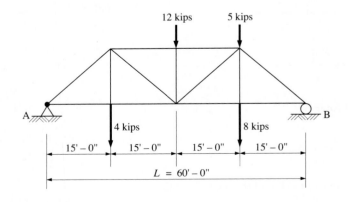

FIGURE 4–31   Problem 18.

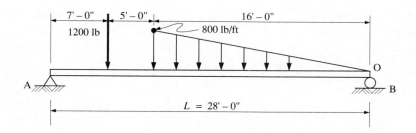

Section 4–6   Equilibrium of Nonconcurrent Force Systems

19. Determine the reactions for the beam in Fig. 4–32. The beam has a pinned support at one end and a roller support at the other end. Neglect the weight of the beam.

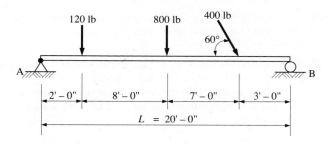

FIGURE 4–32   Problem 19.

20. Calculate the reaction at each support for the truss in Fig. 4–33. Neglect the weight of the truss.

21. Calculate the wall reactions for the cantilever truss in Fig. 4–34. The upper support is pinned and the lower support is a roller. Neglect the weight of the truss.

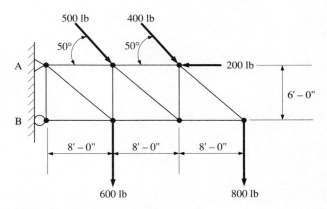

FIGURE 4–34   Problem 21.

FIGURE 4–33   Problem 20.

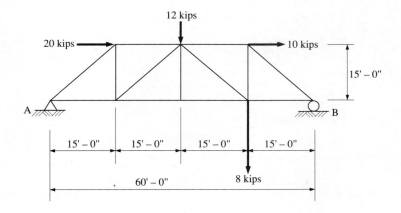

FIGURE 4–35   Problem 22.

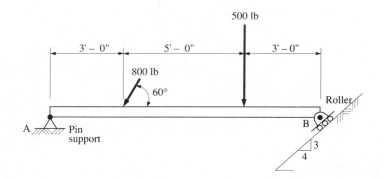

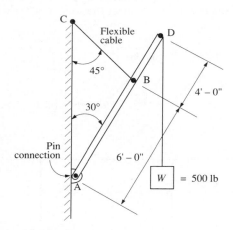

FIGURE 4–36   Problem 23.

22. Determine the reactions at supports A and B of the beam in Fig. 4–35. Neglect the weight of the beam.

23. A 500 lb weight is carried by a boom-and-cable arrangement, as shown in Fig. 4–36. Determine the force in the cable and the reactions at point A.

24. Calculate the force in the tie rod BC and the reaction at the pinned support at point A for the rigid frame shown in Fig. 4–37.

25. The davit shown in Fig. 4–38 is used in pairs for supporting lifeboats onboard ships. The load of 3500 lb represents that portion of the weight of the boat and its occupants supported by one davit. CG represents the center of gravity of the davit. The weight of the davit

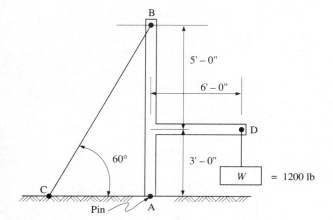

FIGURE 4–37   Problem 24.

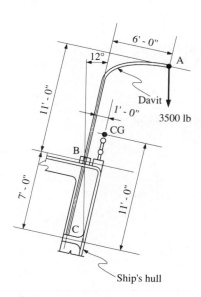

FIGURE 4–38   Problem 25.

itself is 900 lb. The ship has a "list" of 12°. Calculate the reactions at supports B and C. Assume that the reaction at B is at 90° to the axis of the davit and that the support at C is a pocket.

SI System Problems

26. Rework Problem 7. Assume that the mass of the cylinder is 50 Mg.

27. A strut having a mass of 40 kg/m is supported by a cable, as shown in Fig. 4–39. The structure supports a block having a mass of 500 kg. Find the force in the cable and the horizontal and vertical reactions at the pin connection at B. Neglect the mass of the cable.

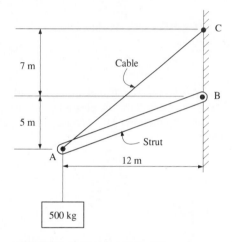

FIGURE 4–39   Problem 27.

28. A beam supports a distributed mass, as shown in Fig. 4–40. Calculate the reaction at each support. Neglect the mass of the beam.

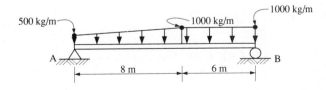

FIGURE 4–40   Problem 28.

29. Compute the reactions at each support for the beam in Fig. 4–41. Neglect the mass of the beam.

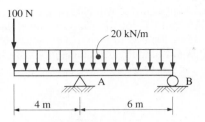

FIGURE 4–41   Problem 29.

30. The truss in Fig. 4–42 is supported by a pin at A and a roller at B. Determine the reactions at these points.

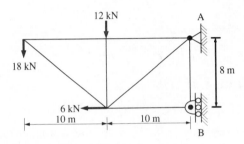

FIGURE 4–42   Problem 30.

Computer Problems

For the following computer problems, any appropriate programming language may be used. Input prompts should fully explain what is required of the user (the program should be "user friendly"). The resulting output should be well labeled and self-explanatory.

31. Write a program that will calculate the reactions for the beam shown in Fig. 4–43. User input is to be L, L_1, L_2, P, and w. Neglect the weight of the beam.

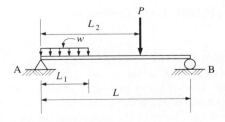

FIGURE 4–43   Problem 31.

32. Write a program that will calculate the forces in members AB and CB for the support frame shown in Fig. 4–44. User input is to be θ and P (where $0° \le \theta \le 90°$ and P is not limited). All connections are pinned.

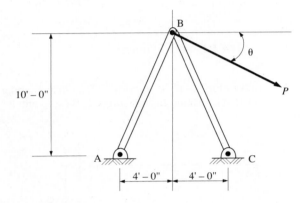

FIGURE 4–44   Problem 32.

33. With reference to Problem 48, write a program that will calculate the force in the cable DC. User input is to be the weight W, as well as lengths AD, AC, and CB.

Supplemental Problems

34. The upper beam in Fig. 4–45 is supported by a pin connection at point A and a roller at point B, supported, in turn, by a lower beam as shown. Sketch a free-body diagram for each member.

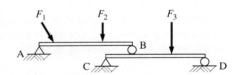

FIGURE 4–45   Problem 34.

35. A 1200 lb load is supported by a cable that runs over a small pulley at E and is anchored to a bar DA as shown in Fig. 4–46. Sketch free-body diagrams of bars EB and DA and of the pulley. The bars are pin connected at each end. Neglect the weights of the members and the pulley.

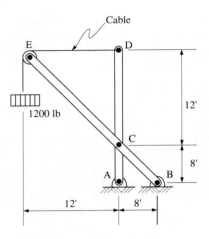

FIGURE 4–46   Problem 35.

36. For the pin-connected frame in Fig. 4–47, sketch a free-body diagram of (a) the entire frame, (b) the member AC, (c) the member DF. Neglect the weight of the members.

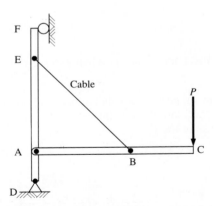

FIGURE 4–47   Problem 36.

37. For the concurrent force system of Fig. 4–48, calculate the maximum load W that could be supported if the maximum allowable force in each cable was 1200 lb.

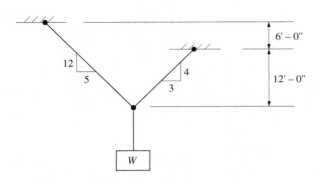

FIGURE 4–48   Problem 37.

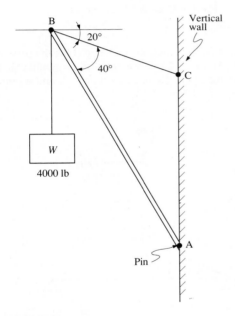

FIGURE 4–49   Problem 38.

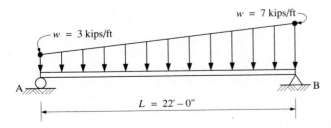

FIGURE 4–50   Problem 39.

38. For the system of Fig. 4–49, what are the forces in AB and BC when a weight of 4000 lb is applied?

39. A beam supports a nonuniformly distributed load as shown in Fig. 4–50. Calculate the reaction at each support. Neglect the weight of the beam.

40. Compute the reaction at each support for the beam in Fig. 4–51. Neglect the weight of the beam.

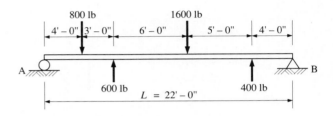

FIGURE 4–51   Problem 40.

41. Compute the reaction at each support for the beam in Fig. 4–52. Neglect the weight of the beam.

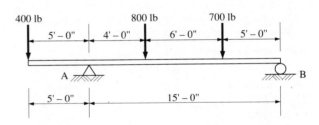

FIGURE 4–52   Problem 41.

42. Compute the reaction at each support for the beam in Fig. 4–53. Neglect the weight of the beam.

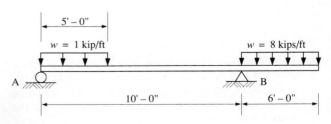

FIGURE 4–53   Problem 42.

43. A rod of uniform cross section weighs 4 lb/ft and is pin connected at point A, as shown in Fig. 4–54. The rod supports a load of 48 lb at point B and is held horizontal by a vertical wire attached 2 ft from point B. With a force of 85 lb in the wire, determine the length of the rod.

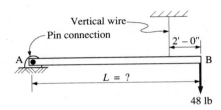

FIGURE 4–54   Problem 43.

44. A 12 ft member weighs 35 lb/ft and supports two loads, as shown in Fig. 4–55. The member is held horizontal by two flexible cables attached to its ends. Determine the force in each cable and the angle θ required for the system to be in equilibrium.

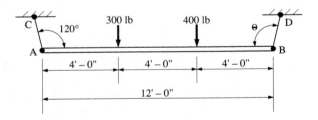

FIGURE 4–55   Problem 44.

45. A uniform rod AB, having a weight of 5.00 lb and a length of 20.0 in., is free to slide within a vertical slot at A. A rope supports the rod at point B. For the situation shown in Fig. 4–56, determine the angle θ and the tension in the rope.

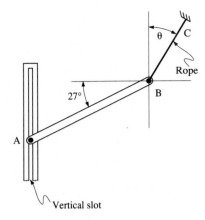

FIGURE 4–56   Problem 45.

46. Determine the reactions at A and B for the truss in Fig. 4–57. The two 2.6 kip loads are perpendicular to the upper member. There is a roller at A and a pin at B.

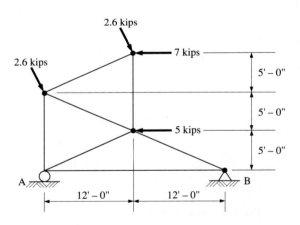

FIGURE 4–57   Problem 46.

47. A 40 ft ladder weighing 130 lb is pin connected to the floor at point A and rests against a smooth, frictionless wall at point B, as shown in Fig. 4–58. The ladder forms an angle of 60° with the horizontal floor and supports a load of 200 lb located 5 ft from the top end. Calculate the reactions at the top and bottom of the ladder.

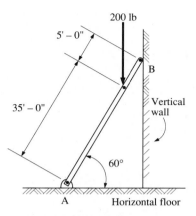

FIGURE 4–58   Problem 47.

48. The frame in Fig. 4–59 is pin connected at point A and held horizontal by the cable CD. The frame is subjected to a vertical load of 16 kips applied at point B. Calculate the force in the cable and the vertical and horizontal components of the reaction at point A.

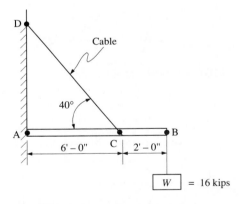

FIGURE 4–59   Problem 48.

49. A crane consists of a post AB, a boom CD, and a brace EF, weighing, respectively, 400 lb, 600 lb, and 300 lb. The crane supports a load of 3000 lb as shown in Fig. 4–60. The post is pin connected at point A and has only lateral support at point B. Thus, the reaction at point B will be horizontal. Determine the reactions at points A and B.

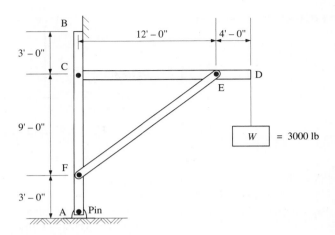

FIGURE 4–60   Problem 49.

50. A horizontal beam is pin connected to a wall at one end and braced diagonally at point D, as shown in Fig. 4–61. The beam carries a uniformly distributed load of 800 lb/ft and a concentrated load at its free end of 500 lb. Determine the horizontal and vertical components of the reaction at point A and the force in member BD. Neglect the weight of the members.

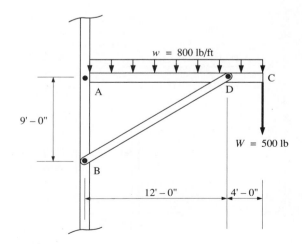

FIGURE 4–61   Problem 50.

FIGURE 4–62   Problem 51.

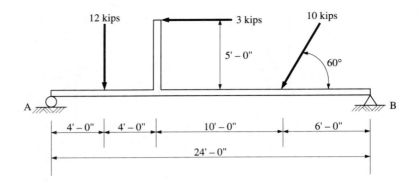

51. Compute the reactions at points A and B for the member shown in Fig. 4–62. Neglect the weight of the member.

52. Calculate the force in the cable for the structure shown in Fig. 4–63.

53. A Thénard shutter dam, originally developed and used in 1831, is shown in Fig. 4–64. The resultant pressure of the water on the dam is a force of 4000 lb, acting perpendicular to the face of the dam. Assume frictionless hinges at B and C and neglect the weight of the dam. Calculate the reactions at C and D.

54. An inclined railway can be used to lift heavy loads up steep inclines. The situation is as shown in Fig. 4–65. Determine the tension in the tow cable and the reactions at each wheel. Neglect the weight of the car.

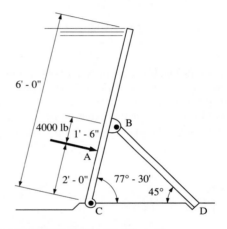

FIGURE 4–64   Problem 53.

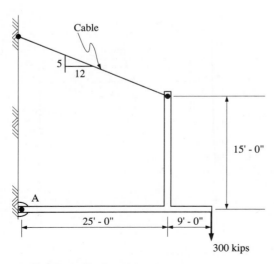

FIGURE 4–63   Problem 52.

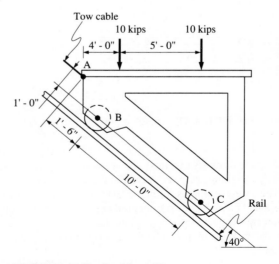

FIGURE 4–65   Problem 54.

5 Analysis of Structures

**5–1
INTRODUCTION**

The analysis of structures involves determining how an external load, from its point of application, is transmitted through the various members of a structure to its external supports. This occurrence is often termed the *flow of load*.

The two types of structures considered in this chapter are pin-connected trusses and pin-connected frames. The difference between them is as follows: In trusses, all the members are two-force members, meaning that the member is subjected to equal, opposite, and collinear forces (either a push or a pull), with lines of action coinciding with the longitudinal axis of the member. In frames, all or some of the members are multiple-force members, in which there is a bending of the member in combination with a longitudinal push or pull.

The analysis process, as described in this chapter, involves the analytical application of the conditions and laws of equilibrium of coplanar force systems. These were presented in Chapters 3 and 4. The object of our analyses will be to determine the forces that are developed in, or on, the various members of the structures.

**5–2
TRUSSES**

A *truss* may be described as a structural framework consisting of straight individual members, all lying in the same plane and so connected as to form a triangle or a series of triangles. The triangle is the basic stable element of the truss. It may be readily observed that practically all trusses are composed of members placed in triangular arrangements, although some specialized trusses do not have this configuration. The trusses considered in this text are planar trusses; that is, all of the truss elements and all of the applied loads or forces lie in the same plane. Further, all loads are applied at points of intersection of the members.

In the common types of trusses shown in Fig. 5–1, the truss members are assumed to be connected at their points of intersection with frictionless hinges or pins; in effect, they permit the ends of the member the freedom to rotate. Because the ends of the members are assumed to be pin connected, the members must be arranged in the triangular shape if they are to form a stable structure. As shown in Fig. 5–2, structures of four or more sides that are connected with frictionless pins at their points of intersection are not stable and will collapse under load.

109

FIGURE 5–1   Types of trusses.

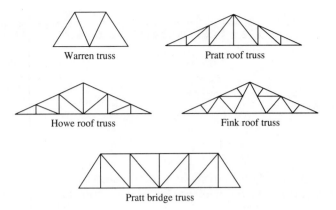

Warren truss   Pratt roof truss

Howe roof truss   Fink roof truss

Pratt bridge truss

FIGURE 5–2   Member relationships.

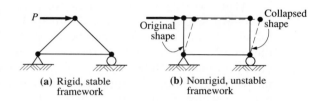

(a) Rigid, stable
framework

(b) Nonrigid, unstable
framework

Trusses are fabricated units and may be considered to be very large beams. When loads are extremely heavy and/or spans are very long, normal beam sections will not be adequate and trusses will be used. Most trusses are constructed of either metal or wood. As compared to a solid bending member, trusses are generally economical with respect to material; however, fabrication costs are high.

Trusses, being somewhat specialized, have an associated terminology descriptive of the various component parts. As indicated in Fig. 5–3, members that form the upper and lower outline of the truss are generally termed the *upper chord* and *lower chord* respectively.

FIGURE 5–3   Truss terminology.

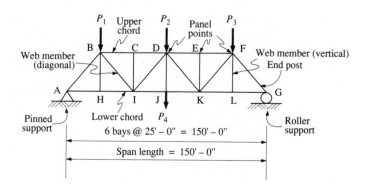

Points A, B, C, D, etc. are called *joints* or *panel points*. They are points of intersection of the longitudinal axes of the truss members. The interior members connecting the joints of the chords are called *web members* (either *vertical* or *diagonal*, depending on their direction in the web system). The distance between panel points (e.g., between B and C, or D and E) represents a *panel length*. The general area between panel points is commonly called a *bay*. Assuming equal panel lengths for all bays, the total span length of a truss would be the number of bays multiplied by the panel length.

In the ideal case, trusses are loaded at their panel points (joints), and each member plays a role in transmitting the applied loads to the supports. The members are two-force members and the applied forces are collinear with the longitudinal axes of the members. The applied forces tend to either stretch or shorten the members. They are called *axial forces* or *direct forces*, as distinguished from forces that produce bending (see Fig. 5–4). A member that is stretched is said to be in *tension* and a member that is shortened is said to be in *compression*. The analysis of a truss involves the determination of the magnitude of force in the members as well as the determination of whether the member is in tension or compression.

FIGURE 5–4 Member behavior.

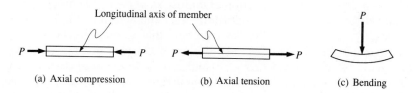

Longitudinal axis of member

(a) Axial compression   (b) Axial tension   (c) Bending

5–3
FORCES IN MEMBERS OF TRUSSES

In order to simplify the analysis of a truss, the following assumptions are made:

1. All members of the truss lie in the same plane.
2. Loads and reactions are applied only at the panel points (joints) of the truss.
3. The truss members are connected with frictionless pins.
4. All members are straight and are two-force members; therefore, the forces at each end of the member are equal, opposite, and collinear.
5. The line of action of the internal force within each member is axial.
6. The change in length of any member due to tension or compression is not of sufficient magnitude to cause an appreciable change in the overall geometry of the truss.
7. The weight of each member is very small in comparison with the loads supported and is therefore neglected. (If the weight is to be considered, it may be assumed a concentrated load acting partially at each end of the member.)

Based on these assumptions, and using the principles and laws of static equilibrium as discussed in Chapter 4, the force in each member (tension or compression) may be determined by means of either of two analytical techniques. One is called the *method of joints*; the other is called the *method of sections*.

At this point, it should be noted that prior to determining the internal force in each truss member, the external equilibrium of the truss must be considered. This means that the truss reactions at the external supports should be determined in a manner similar to that described in Chapter 4 for beams and/or frames.

5–4 THE METHOD OF JOINTS

The method of joints consists of removing each joint in a truss and considering it as if it were isolated from the remainder of the truss. A free-body diagram of the pin is the basis of the approach. The free-body is prepared by cutting through all the members framing into the joint being considered. Since all members of a truss are two-force members carrying axial loads, the free-body diagram of each joint will represent a coplanar concurrent force system.

Since the truss as a whole is in external equilibrium, any isolated portion of it must likewise be in equilibrium. Therefore, each joint must be in equilibrium under the action of the external loads and the internal forces of the cut members that frame into the joint.

As shown in Chapter 4, only the two equations of force equilibrium ($\Sigma F_x = 0$ and $\Sigma F_y = 0$) are necessary to determine the unknown forces in a coplanar concurrent force system. Again, for convenience, we will use the subscripts H and V to indicate horizontal and vertical directions, since, in our applications here, the Y axis is assumed to be vertical (coincident with the direction of gravity). Therefore, our two equations of force equilibrium may be written as $\Sigma F_H = 0$ and $\Sigma F_V = 0$. The third equation of equilibrium ($\Sigma M = 0$) is not applicable, since all the forces intersect at a common point.

The use of force components and the two equations of force equilibrium is usually the method of choice at each joint. However, in situations where only three members are connected at a joint, and particularly when two unknown force members are sloped at that joint, analysis of the force triangle will provide a quick solution, saving time and effort.

When using the method of joints, no more than two unknown member forces can be determined at any one joint. Once these unknown forces have been calculated for one joint, their effects on adjacent joints are known. Successive joints may be then considered until the unknown forces in all the members have been determined. This procedure is demonstrated in Example 5–1.

☐ **EXAMPLE 5–1**   A simply supported Warren truss is loaded as shown in Fig. 5–5(a). Determine the truss reactions and the force in each member. Use the method of joints. The truss is supported by a pin at point A and a roller at point E.

FIGURE 5–5   Truss for Example 5–1.

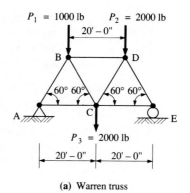

(a) Warren truss

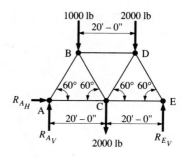

(b) Free-body diagram

Solution   The truss must be in equilibrium. Therefore, the reactions at A and E will be computed first. Figure 5–5(b) is a free-body diagram of the entire truss. Forces upward and to the right and counterclockwise moments are considered positive. Summing moments about point E,

$$\Sigma M_E = -R_{A_V}(40) + 2000(10) + 1000(30) + 2000(20) = 0$$

from which

$$R_{A_V} = +2250 \text{ lb}$$

Summing moments about point A,

$$\Sigma M_A = +R_{E_V}(40) - 2000(20) - 2000(30) - 1000(10) = 0$$

from which

$$R_{E_V} = +2750 \text{ lb}$$

Checking the calculations by a summation of vertical forces,

$$\Sigma F_V = +2250 + 2750 - 2000 - 1000 - 2000 = 0 \quad \textbf{OK}$$

Note that there are no horizontal forces or horizontal components of diagonal forces; therefore, $R_{A_H} = 0$.

Now, with the truss reactions known, the internal forces in all the truss members can be calculated. Each joint will be isolated in sequence as a free body. Since a joint with more than two unknown forces cannot be solved completely, a logical joint to start with would be joint A or joint E.

The free-body diagram for joint A is shown in Fig. 5–6(a). Note that the force system is a coplanar concurrent force system consisting of one known force, R_{A_V}, and two unknown forces. The unknown forces are designated AB and AC and should be interpreted as the *forces* in members AB and AC. The lines of action of the two unknown forces are known, but the sense and magnitude of each are unknown. If the sense of an unknown force is not obvious, it must be assumed. A negative result would indicate that the sense was opposite to that assumed.

FIGURE 5–6 Joint free-body diagrams.

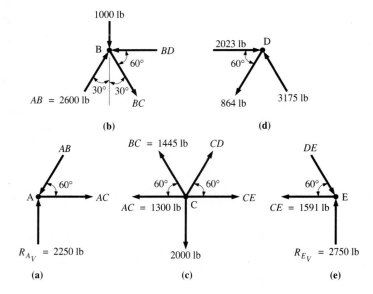

In Fig. 5–6(a), force AB is shown pushing into the joint (acting toward the joint). This means that member AB is assumed to be in compression and is pushing into the joint. Force AC is shown acting away from the joint. Member AC, therefore, is assumed to be in tension, pulling away from the joint.

Using a vertical and horizontal coordinate axis system with upward-acting forces and forces acting to the right as positive, first sum the vertical forces:

$$\Sigma F_V = +R_{A_V} - AB \sin 60° = 0$$
$$= +2250 - AB(0.866) = 0$$

from which

$$AB = +2600 \text{ lb (compression)}$$

Next, sum the horizontal forces:

$$\Sigma F_H = +AC - AB \cos 60° = 0$$
$$= +AC - 2600(0.50) = 0$$

from which

$$AC = +1300 \text{ lb (tension)}$$

Since the results are positive values, the senses for forces AB and AC are as assumed. (Member AB is in compression and member AC is in tension.)

Alternatively, it is convenient to use a force triangle solution at joint A. From the force triangle shown in Fig. 5–7,

$$AC = 2250 \tan 30° = 1300 \text{ lb (tension)}$$

$$AB = \frac{2250}{\cos 30°} = 2600 \text{ lb (compression)}$$

That members AC and AB are in tension and compression, respectively, is determined by observing their senses in the force triangle and comparing their positions in the free-body diagram. If the sense of a force is incorrectly assumed in the free-body diagram, it will become apparent when the force triangle is drawn.

FIGURE 5–7 Force triangle at joint A.

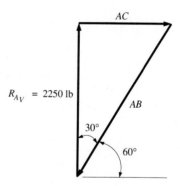

The next joint to be considered is joint B. The free-body diagram is shown in Fig. 5–6(b). There is one known force (AB), two unknown forces (BD and BC), as well as one external load of 1000 lb. In this free body, force AB is known to be compressive (as was determined at joint A), and it is shown acting into the joint. The 1000 lb load is shown acting directly on the joint. The lines of action for forces BC and BD are known, but their senses and magnitudes are unknown. Assume member BC to be in tension and member BD to be in compression, as shown in Fig. 5–6(b).

First sum the vertical forces:

$$\Sigma F_V = -1000 - BC \cos 30° + (2600)\cos 30° = 0$$

from which

$$BC = +1445 \text{ lb (tension)}$$

Next, sum the horizontal forces:

$$\Sigma F_H = -BD + (2600)\sin 30° + BC \sin 30° = 0$$
$$= -BD + (2600)(0.500) + 1445(0.500) = 0$$

from which

$$BD = +2023 \text{ lb (compression)}$$

The positive results indicate that the senses for forces BC and BD are as assumed. (Member BC is in tension and member BD is in compression.)

The next joint to be considered is joint C. The free-body diagram is shown in Fig. 5–6(c). There are two known forces (AC and BC), two unknown forces (CD and CE), and one external load of 2000 lb. The two known forces, AC and BC, are shown pulling on the joint, since from our previous calculations at joints A and B, both were found to be tensile forces. Members CD and CE are assumed to be in tension as shown.

First sum the vertical forces:

$$\Sigma F_V = -2000 + 1445(\sin 60°) + CD(\sin 60°) = 0$$

from which

$$CD = +864 \text{ lb (tension)}$$

Next, sum the horizontal forces:

$$\begin{aligned} \Sigma F_H &= +CE - 1300 + CD(\cos 60°) - 1445(\cos 60°) = 0 \\ &= +CE - 1300 + 864(0.500) - 1445(0.500) = 0 \end{aligned}$$

from which

$$CE = +1591 \text{ lb (tension)}$$

The results are positive values; therefore, the senses for forces CD and CE are as assumed. (Both members are in tension.)

The next joint to be considered may be either joint D or joint E. The last remaining member force to be determined is that of member DE. Select joint E. The free-body diagram is shown in Fig. 5–6(e). Note that only force DE is unknown. One equation of equilibrium will be used to solve for force DE. The second equation of equilibrium can be used as a check. Member DE is assumed to be in compression (force DE is acting into joint E).

$$\Sigma F_V = +2750 - DE(\sin 60°) = 0$$

from which

$$DE = +3175 \text{ lb (compression)}$$

The positive sign indicates that the sense of force DE is as assumed; therefore, member DE is in compression.

Now use a summation of horizontal forces as a check on the calculations:

$$\begin{aligned} \Sigma F_H &= -1591 + DE(\cos 60°) = 0 \\ &= -1591 + 3175(0.500) = 0 \qquad \text{(very close) } \mathbf{OK} \end{aligned}$$

The small discrepancy results from the rounding of numbers; it should not cause undue concern. Note that a check on the calculations could also be done by applying the equations of equilibrium at joint D, where all of the forces are now known. Joint

D has not been used in the calculation process. However, the free-body diagram of joint D is shown in Fig. 5–6(d).

As a means of summarizing the forces in the truss members, a force summary diagram, as shown in Fig. 5–8, should be sketched. Note that the member forces are designated T and C for tension and compression, respectively.

FIGURE 5–8 Force summary diagram.

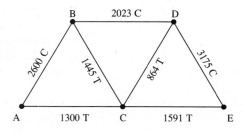

Summary of Procedure—Method of Joints

1. Calculate the reactions for the truss using a free-body diagram for the entire truss. Replace sloping forces with their vertical and horizontal components.
2. Isolate each joint in sequence as a free body and sketch the free-body diagram showing all forces acting on the joint. Each joint represents a coplanar concurrent force system and can involve no more than two members in which the forces are unknown.
3. By applying the two equations of force equilibrium ($\Sigma F_V = 0$ and $\Sigma F_H = 0$), the two unknown member forces for each joint may be determined. These forces are then carried over to successive adjacent joints. In some cases, an alternative force triangle solution may be convenient.
4. Although not essential, it is good practice to check the calculations. Select a joint not used in the calculations and verify that $\Sigma F_V = 0$ and $\Sigma F_H = 0$.
5. After all member forces have been determined, draw a force summary diagram.

5–5 THE METHOD OF SECTIONS

The method of sections is another technique for determining the forces acting in the various members of a truss. This method has some advantages when the analysis of only a few members is desired, even when those members are situated far from the supports of the truss. Unlike the method of joints, the analysis need not proceed from joint to joint, analyzing the entire truss. With the proper approach, and a suitable truss layout, a rapid analysis of selected members may be accomplished.

In the method of sections, the truss is divided into two parts by a cutting plane. One of the two parts is isolated as a free body. The procedure involves cutting through a number of members in which the unknown forces are acting. Not more than three members with unknown forces may be cut along any cutting plane, except where the lines of action of all but one of the members intersect at a common point.

Since the truss as a whole is in external equilibrium, any isolated portion of it must likewise be in equilibrium. Therefore, the portion of the truss isolated as a free body must be in equilibrium under the action of the externally applied loads and the internal forces of the members intersected by the cutting plane. The forces (known and unknown) acting on the free body constitute a coplanar nonconcurrent force system. Therefore, the three equations of equilibrium ($\Sigma F_V = 0$, $\Sigma F_H = 0$, and $\Sigma M = 0$) are applicable.

In using this method, the truss may be cut at any location and the force in any member may be determined independently of all others. In general, when selecting a cutting plane, the plane should cut the least number of members. One of these members must be the truss member, the internal force of which is the unknown value being sought.

Depending on the forces to be found, more than one cutting plane may be required. Moreover, in some problems, it may be necessary to use a combined analysis approach utilizing both the method of joints and the method of sections. The method of sections, utilizing a single cutting plane, is illustrated in Example 5–2.

☐ **EXAMPLE 5–2**   Determine the forces acting in members BD, CD, and CE of the Howe truss shown in Fig. 5–9(a). Use the method of sections. The truss is supported by a pin at A and a roller at J.

Solution   The reactions at A and J will be computed first. Figure 5–9(b) is a free-body diagram of the entire truss which will be used for the calculation of the reactions. Forces upward and to the right and counterclockwise moments are considered positive. For the vertical reaction at A,

$$\Sigma M_J = -R_{A_V}(48) + 2400(12) + 4000(24) + 1600(36) = 0$$

from which

$$R_{A_V} = +3800 \text{ lb}$$

For the reaction at J,

$$\Sigma M_A = +R_{J_V}(48) - 1600(12) - 4000(24) - 2400(36) = 0$$

from which

$$R_{J_V} = +4200 \text{ lb}$$

Checking the calculations using a summation of vertical forces,

$$\Sigma F_V = +3800 + 4200 - 2400 - 4000 - 1600 = 0 \qquad \textbf{OK}$$

There are no horizontal forces or horizontal components of diagonal forces; therefore, $R_{A_H} = 0$.

Now, with the truss reactions known, the internal forces in members BD, CD, and CE can be calculated. This can be accomplished by passing a cutting plane $a–a$ through these three members, as shown in Fig. 5–9(b), and isolating the left portion of the truss as a free body.

FIGURE 5–9 Truss for Example 5–2.

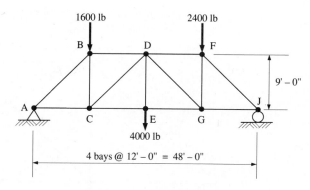

(a) Simply supported Howe truss

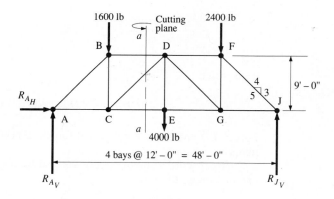

(b) Free-body diagram

The free-body diagram for the left portion of the truss is shown in Fig. 5–10. Note that the applicable external forces, as well as the internal forces of the cut members, are shown applied on the free body. The vertical and horizontal components of *CD* are shown as dashed arrows. The internal forces of the cut members are unknown in magnitude and sense. Since all of the members are two-force members, however, the lines of action of the forces are known to be collinear with the members themselves and are established by the geometry of the truss. Assume member *CE* to be in tension and members *BD* and *CD* to be in compression. Note that an arrow pointing away from the free body (such as *CE*) means that member *CE* is assumed to pull on the body and therefore to be in tension.

Any of the three equations of equilibrium can be applied to the free-body diagram, since the force system is a coplanar nonconcurrent force system. The relationships between the vertical and horizontal components of *CD*, using the slope triangle, can be written

$$\frac{CD}{5} = \frac{CD_H}{4} = \frac{CD_V}{3}$$

FIGURE 5–10 Free-body diagram.

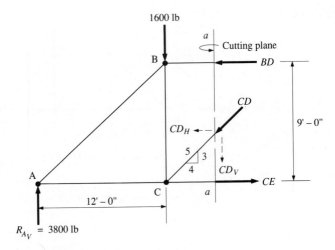

from which

$$CD_H = \frac{4}{5} CD \quad \text{and} \quad CD_V = \frac{3}{5} CD$$

Forces *BD* and *CE* are horizontal forces; therefore, they have no vertical components.

Next write a summation of vertical forces for the free body, from which you can determine the force *CD*:

$$\sum F_V = +3800 - 1600 - CD_V = 0$$
$$= +3800 - 1600 - \left(\frac{3}{5} CD\right) = 0$$

from which

$$CD = +3667 \text{ lb (compression)}$$

Since the result is a positive value, the sense for force *CD* is as assumed. (Member CD is in compression.)

The force in member BD can be calculated by summing moments of all the known and unknown forces about point C. This in effect will eliminate forces *CD*, *CE*, and the 1600 lb load from the computation, since their lines of action pass through point C.

$$\sum M_C = -3800(12) + BD(9.0) = 0$$

from which

$$BD = +5067 \text{ lb (compression)}$$

The result is again a positive value, indicating that the sense for force *BD* is as assumed. (Member BD is in compression.)

The force in member CE can be calculated by summing the horizontal forces. Assuming forces acting to the right are positive,

$$\Sigma F_H = +CE - BD - CD_H = 0$$

$$= +CE - 5067 - \frac{4}{5}(3667) = 0$$

from which

$$CE = +8000 \text{ lb (tension)}$$

The result is positive; therefore, the sense for force CE is as assumed. (Member CE is in tension.)

A check on the calculations can be accomplished by summing moments about some point not used thus far in the calculations. Point B is such a point. Moving the components of CD to point C (principle of transmissibility),

$$\Sigma M_B = -3800(12) + CE(9.0) - CD_V(9.0) = 0$$

$$= -3800(12) + 8000(9.0) - \frac{4}{5}(3667)(9.0) = 0 \qquad \text{(very close) } \textbf{OK}$$

Summary of Procedure—Method of Sections

1. Calculate the reactions for the truss using a free-body diagram for the entire truss.
2. Isolate a portion of the truss by passing a cutting plane through the truss, cutting no more than three members in which the forces are unknown (unless all but one of the lines of action of the cut members intersect at a point).
3. Apply the three equations of equilibrium to the isolated portion of the truss and solve for the unknown forces.
4. Although not essential, it is good practice to check the calculations by verifying $\Sigma F_H = 0$, $\Sigma F_V = 0$, and $\Sigma M = 0$. Use a moment center not previously used.

5-6 ANALYSIS OF FRAMES

Up to this point in the chapter, we have discussed trusses composed of two-force members. When forces are applied at more than two points along the length of a member, and when the forces are not collinear with the axis of the member, the member will be subject to bending. We will now turn our attention to members of this type, sometimes called *multiple-force members*. Straight two-force and multiple-force members are shown in Fig. 5–11.

First consider the straight two-force member acted upon by the forces shown in Fig. 5–11(a). If this member is to be in equilibrium, the resultant of the force system at B must be equal to, opposite to, and collinear with the resultant of the force system at A; hence, the two resultants R_A and R_B act collinearly with the longitudinal axis of the member. The force effect of a straight two-force member in contact with any other body must, therefore, act in the direction established by the axis of the two-force member.

FIGURE 5–11 Types of members.

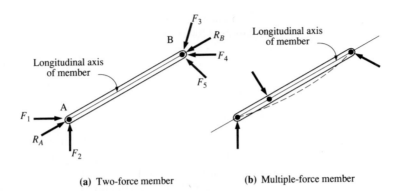

(a) Two-force member   (b) Multiple-force member

For the multiple-force member of Fig. 5–11(b), the resultant force at each end does not act along the axis of the member. A member that supports a transverse load, or for which the weight is not negligible, is always a multiple-force member.

Since the resultant forces at the ends of multiple-force members are not directed along the axis of the member but are usually unknown in direction and sense, it is customary to work with the rectangular components of each resultant force.

Structures composed partially or totally of pin-connected multiple-force members are called *frames*. As with trusses, the pins are assumed to be frictionless. The frames we will consider are coplanar; that is, all of the members lie in a common plane.

Our analysis of frames (with multiple-force members) differs significantly from the analysis of trusses. We make no attempt to determine the internal forces in the members of these frames; rather, we limit our analysis to a determination of the forces that act *on* the member. The forces resulting at the pins are called *pin reactions*. The pin reactions are normally determined as rectangular component forces. They may be combined into a resultant force using Eq. (2–3).

The usual procedure to determine the pin reactions is as follows:

1. Compute the support reactions, or their components, considering the frame as a whole. If all of the support reactions (or reaction components) cannot be determined, determine as many as possible and then proceed with the next steps. The unknown reactions will be determined at a later time.
2. Isolate each multiple-force member as a free body, showing all known forces and unknown pin reactions (or components of the pin reaction).
3. Consider each multiple-force member individually. Apply the three equations of equilibrium. Compute the unknown pin reactions or components of the pin reaction. A maximum of three unknown forces (or components) can be found using any one free body. Any unknown forces remaining must be identified and carried forward.
4. Proceed to an adjacent member, carrying forward the newly determined

forces, carefully accounting for the appropriate senses. Then repeat step 3.

5. Although not essential, it is good practice to check the calculations. Select a member not utilized in the analysis (or a member supporting many of the forces that have been determined) for the check. Verify that $\Sigma F_V = 0$, $\Sigma F_H = 0$, and $\Sigma M = 0$ for that member.

Step 4 requires some additional explanation. Once a force has been determined in magnitude and sense and is carried forward to other members, the magnitude will not change, but the sense will. If we consider the hung weight shown in Fig. 4–1, the force F acts upward on the weight (Fig. 4–1(c)). If this force were carried forward to the next member, the rope (Fig. 4–1(b)), it would have to be shown acting downward. The rope pulls upward on the weight. The weight pulls downward on the rope. This illustrates the fact that forces occur in pairs. No single force can exist as an isolated force. Newton's third law of motion, to every action there is an equal and opposite reaction, describes this condition.

Since the procedure we have described considers each member individually, it is generally called the *method of members*.

☐ **EXAMPLE 5–3** A crane consisting of a vertical post, a horizontal boom, and an inclined brace supports a vertical load of 10,000 lb as shown in Fig. 5–12(a). The crane is supported by a pin connection at A. The support at B permits only a horizontal reaction (smooth vertical surface). The members of the crane are pin connected at points C, D, and E. The members have weights as follows: post, 1400 lb; boom, 1500 lb; and brace, 900 lb. The member weights may be considered to be acting at their respective midpoints. Calculate all of the forces acting on each of the three members.

Solution First, consider the entire frame as a free body, as shown in Fig. 5–12(b). Assume the external reactions at A and B to act as shown. Apply the three equations of equilibrium, considering forces upward and to the right and counterclockwise moments positive. First,

$$\Sigma M_A = +B_H(14.0) - 1500(7.0) - 900(5.0) - 10,000(14.0) = 0$$

from which

$$B_H = +11,070 \text{ lb} \leftarrow$$

Next,

$$\Sigma F_H = +A_H - B_H = +A_H - 11,070 = 0$$

from which

$$A_H = +11,070 \text{ lb} \rightarrow$$

Finally,

$$\Sigma F_V = +A_V - 1500 - 1400 - 900 - 10,000 = 0$$

from which

$$A_V = +13,800 \text{ lb} \uparrow$$

FIGURE 5–12   Framework for
Example 5–3.

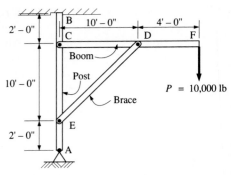

(a) Crane framework

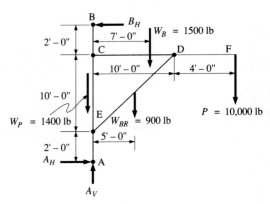

(b) Free-body diagram
of entire frame

The results are all positive values; therefore, the senses of the reactions are as assumed. Arrows indicate the sense in each case.

Next consider member CF as a free body, as shown in Fig. 5–13. Note that member CF is a multiple-force member with pin connections at points C and D. The effects of the brace and the post on the boom are unknown; therefore, vertical and horizontal component forces are indicated at C and D. Solve for these component forces. The senses of the components are assumed in Fig. 5–13. If the senses are as assumed, positive signs will result. Apply the three equations of equilibrium. First,

$$\Sigma M_C = +D_V(10.0) - 10,000(14.0) - 1500(7.0) = 0$$

from which

$$D_V = +15,050 \text{ lb } \uparrow$$

Next,

$$\Sigma F_V = -C_V + D_V - 1500 - 10,000 = 0$$

from which

$$C_V = 15,050 - 1500 - 10,000 = +3550 \text{ lb } \downarrow$$

FIGURE 5–13   Free-body diagram of boom.

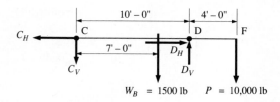

Finally,

$$\Sigma F_H = +D_H - C_H = 0$$

from which

$$D_H = C_H$$

These two components cannot be determined from this free body (there were four unknown forces and only three equations of equilibrium). Therefore, proceed to the free body for member DE. Note that the results for the preceding calculations are all positive, indicating that the senses for the components are as assumed.

The free-body diagram for member DE is shown in Fig. 5–14. Member DE is a multiple-force member with pin connections at points D and E. Note the senses of D_V and D_H which have been carried forward from the free body of member CDF. For example, D_V acted upward on member CDF; therefore, it must, of necessity, act downward on member DE. The senses of the components acting at point E are assumed as shown.

FIGURE 5–14   Free-body diagram of brace.

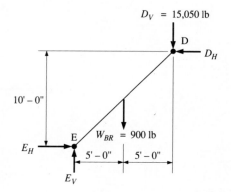

Apply the three equations of equilibrium. First,

$$\Sigma M_E = +D_H(10.0) - 15,050(10.0) - 900(5.0) = 0$$

from which

$$D_H = +15,500 \text{ lb} \leftarrow$$

Next,

$$\Sigma F_H = +E_H - D_H = 0$$

from which

$$E_H = +15,500 \text{ lb} \rightarrow$$

Finally,

$$\Sigma F_V = +E_V - 900 - 15,050 = 0$$

from which

$$E_V = +15,950 \text{ lb} \uparrow$$

Furthermore, since $C_H = D_H$, then $C_H = 15,500$ lb, which acts to the left on member CDF.

Member BCEA may now be used as a final check of the calculations. The free-body diagram with all of the forces that have been determined is shown in Fig. 5–15. You can verify that $\Sigma F_V = 0$, $\Sigma F_H = 0$, and $\Sigma M = 0$.

FIGURE 5–15   Free-body diagram of post.

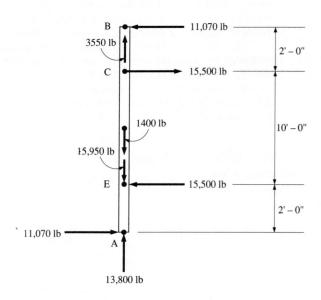

TABLE 5–1

Pin	Vertical Component (lb)	Horizontal Component (lb)	Pin Reaction (lb)
A	13,800	11,070	17,690
C	3,550	15,500	15,900
D	15,050	15,500	21,600
E	15,950	15,500	22,200

FIGURE 5–16   Analysis results.

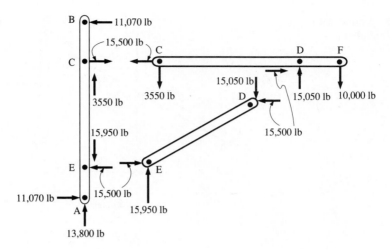

Table 5–1 lists the vertical and horizontal forces that were calculated in this example. The pin reactions (refer to Eq. (2–3)) are also indicated. The results are also shown in Fig. 5–16.

□ **EXAMPLE 5–4**   The frame shown in Fig. 5–17(a) is pin connected at A, B, and C. Calculate all pin reactions. Neglect the weights of the members.

Solution   Using the whole-frame free-body diagram of Fig. 5–17(b), calculate reaction components A_V and C_V. Since there are four unknowns, it will not be possible to find all of the reaction components with this free body. Forces upward and to the right and counterclockwise moments are considered positive.

Summing moments about point A,

$$\Sigma M_A = -10(6) - 4(2.5) + C_V(12) = 0$$

from which

$$C_V = +5.83 \text{ kips } \uparrow$$

Summing moments about point C,

$$\Sigma M_C = +10(6) - 4(2.5) - A_V(12) = 0$$

from which

$$A_V = +4.17 \text{ kips } \uparrow$$

A summation of horizontal forces yields

$$\Sigma F_H = +4 - A_H - C_H = 0 \qquad \text{(Eq. 1)}$$

A free-body diagram of member AB is shown in Fig. 5–18(a). Note that the vertical component A_V of 4.17 kips upward and the 10 kip downward load are known

FIGURE 5–17 Frame with multiple-force members.

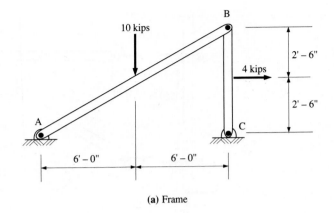

(a) Frame

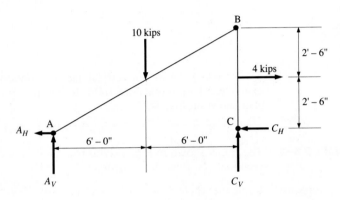

(b) Free-body diagram

forces. You can then find all three unknown forces. Summing vertical forces,

$$\Sigma F_V = +4.17 - 10 + B_V = 0$$

from which

$$B_V = +5.83 \text{ kips}$$

Summing moments around point A,

$$\Sigma M_A = -10(6) + 5.83(12) - B_H(5) = 0$$

from which

$$B_H = +2.00 \text{ kips}$$

Summing horizontal forces,

$$\Sigma F_H = -A_H + 2.00 = 0$$

from which

$$A_H = +2.00 \text{ kips}$$

From Eq. 1,

$$4 - A_H - C_H = 0$$
$$C_H = 4 - A_H = 4 - 2.00 = +2.00 \text{ kips}$$

Member BC (Fig. 5–18(b)) will be used as a check. By inspection, it is seen that $\Sigma F_V = 0$, $\Sigma F_H = 0$, and $\Sigma M = 0$ all are verified. The results are summarized in Table 5–2.

FIGURE 5–18 Free-body diagrams.

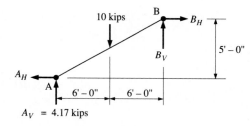

(a) Member AB

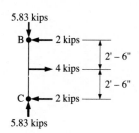

(b) Member BC

TABLE 5–2

Pin	Vertical Component (kips)	Horizontal Component (kips)	Pin Reaction (kips)
A	4.17	2.00	4.63
B	5.83	2.00	6.16
C	5.83	2.00	6.16

**5–7
SI SYSTEM
EXAMPLES**

□ **EXAMPLE 5–5**

The frame shown in Fig. 5–19(a) is pin connected at A, B, D, and E. Find the forces in members AD, DB, and ED. Check the calculations by an equilibrium check for member ABC.

Solution   The free-body diagram for the entire frame is shown in Fig. 5–19(b). Note that the pin supports at A and E have been replaced with vertical and horizontal reactions. Senses have been assumed. Forces acting upward or to the right and counterclockwise moments are considered positive. First, solve for E_H by summing moments about A:

$$\sum M_A = +E_H(7 \text{ m}) - 800 \text{ kN}(12 \text{ m}) = 0$$

$$E_H = \frac{+800 \text{ kN}(12 \text{ m})}{7 \text{ m}} = +1371 \text{ kN} \rightarrow$$

FIGURE 5–19   Framework for Example 5–5.

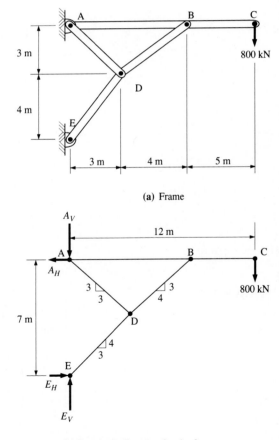

(a) Frame

(b) Free-body diagram of entire frame

The positive sign indicates that the sense is as assumed. (Member ED is in compression.)

Since member ED is a two-force member, the reaction at E must be directed along the longitudinal axis of ED. Therefore,

$$\frac{E_V}{4} = \frac{E_H}{3}$$

from which

$$E_V = \frac{4E_H}{3} = \frac{4(1371 \text{ kN})}{3} = 1828 \text{ kN} \uparrow$$

The force in member ED is then

$$ED = \sqrt{E_V^2 + E_H^2} = \sqrt{1828^2 + 1371^2} = 2285 \text{ kN}$$

Calculating A_V and A_H,

$$\Sigma F_V = 0 = 1828 \text{ kN} - A_V - 800 \text{ kN} = 0$$
$$A_V = +1028 \text{ kN} \downarrow$$
$$\Sigma F_H = 0 = -A_H + 1371 \text{ kN} = 0$$
$$A_H = +1371 \text{ kN} \leftarrow$$

Next isolate the pin at joint D (Fig. 5–20) and solve for AD and BD. Assume that both are in compression as shown. Writing the relationships between the vertical and horizontal components,

$$\frac{AD}{4.24} = \frac{AD_H}{3} = \frac{AD_V}{3}$$
$$\frac{BD}{5} = \frac{BD_H}{4} = \frac{BD_V}{3}$$

from which

$$AD_V = AD_H = \frac{3(AD)}{4.24} = 0.7076 AD$$

$$BD_V = \frac{3(BD)}{5} = 0.60 BD$$

$$BD_H = \frac{4(BD)}{5} = 0.80 BD$$

Summing vertical and horizontal forces at D yields

$$\Sigma F_H = +1371 \text{ kN} + AD_H - BD_H = 0$$
$$= +1371 \text{ kN} + 0.7076 AD - 0.80 BD = 0 \qquad \text{(Eq. 1)}$$
$$\Sigma F_V = +1828 \text{ kN} - AD_V - BD_V = 0$$
$$= +1828 \text{ kN} - 0.7076 AD - 0.60 BD = 0 \qquad \text{(Eq. 2)}$$

Equations 1 and 2 are simultaneous equations that may be solved to yield

$$AD = +646 \text{ kN}$$
$$BD = +2285 \text{ kN}$$

FIGURE 5–20   Free-body diagram of joint D.

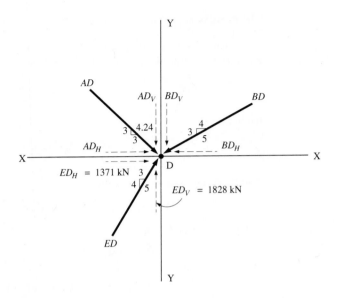

The components of *AD* and *BD* may be summarized as in Table 5–3 (where all forces and components are in units of kN).

TABLE 5–3

Member	Compressive Force (kN)	Vertical Components (kN)	Horizontal Components (kN)
ED	2285	1828	1371
AD	646	457	457
BD	2285	1371	1828

Finally, check equilibrium of member ABC (see Fig. 5–21):

$$\Sigma F_H = -1371 \text{ kN} - 457 \text{ kN} + 1828 \text{ kN} = 0 \quad \textbf{OK}$$
$$\Sigma F_V = -1028 \text{ kN} + 457 \text{ kN} + 1371 \text{ kN} - 800 \text{ kN} = 0 \quad \textbf{OK}$$
$$\Sigma M_A = +1371 \text{ kN}(7 \text{ m}) - 800 \text{ kN}(12 \text{ m}) = -3 \text{ kN·m} \quad \textbf{Say OK}$$

The final moment check reflects some slight round-off error.

FIGURE 5–21   Free-body diagram of member ABC.

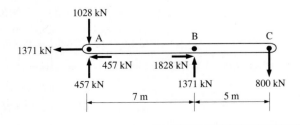

SUMMARY—BY SECTION NUMBER	**5–1** In trusses all members are two-force members. In frames all members, or some of them, are multiple-force members.

5–2 and **5–3** A truss is a structural framework consisting of straight individual members connected so as to form a triangle or series of triangles. Truss members are connected at joints (or panel points) with pins assumed to be frictionless. Loads are assumed to be applied at the joints. In the context of our presentation, all members are straight and are assumed to lie in the same plane.

5–4 The method of joints is used to determine the internal force in truss members. Each joint is isolated sequentially, beginning with one in which only two member forces are unknown, then proceeding to adjacent joints. No more than two unknown member forces can be determined at any one joint. Unknown forces are found by the method of components, applying the two equations of force equilibrium, or by solution of the force triangle.

5–5 The method of sections is also used to determine the internal force in a truss member. A portion of the truss is isolated as a free body by passing a cutting plane through the truss, cutting no more than three members in which the forces are unknown. Unknown forces are found by applying the three equations of equilibrium to the isolated portion of the truss.

5–6 A frame is a structure composed partially, or totally, of pin-connected multiple-force members. The analysis involves the determination of the forces acting at the pins joining the members. The procedure is to isolate each member as a free body, showing all known and unknown forces (or components of forces), and then to apply the three equations of equilibrium.

PROBLEMS

Section 5–4   The Method of Joints

1–5. Calculate the forces in all members of the trusses shown in Figs. 5–22 through 5–26 using the method of joints. Note the typical designation of pin and roller supports in Fig. 5–22.

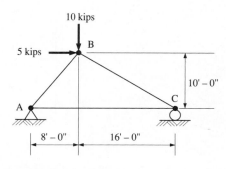

FIGURE 5–23   Problem 2.

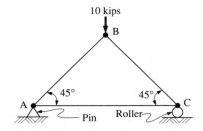

FIGURE 5–22   Problem 1.

FIGURE 5–24   Problem 3.

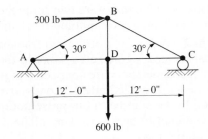

FIGURE 5–25   Problem 4.

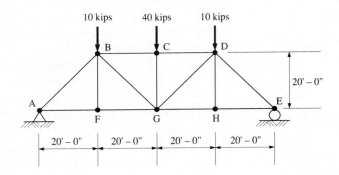

FIGURE 5–26   Problem 5.

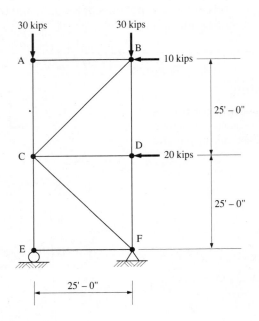

Section 5–5   The Method of Sections

For Problems 6–9, use the method of sections.

6. Determine the forces in members CD, DH, and HI for the truss shown in Fig. 5–27.

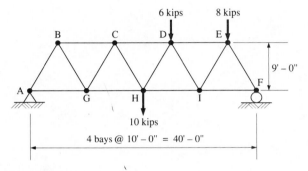

FIGURE 5–27   Problem 6.

7. Determine the forces in members BC, BE, and FE for the truss shown in Fig. 5–28.

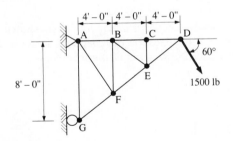

FIGURE 5–28   Problem 7.

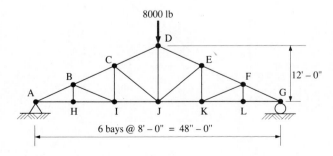

FIGURE 5–29   Problem 8.

8. For the Howe roof truss in Fig. 5–29, determine the forces in members BC, CI, and IJ.

9. Calculate the forces in members BC, BG, and FG for the cantilever truss in Fig. 5–30.

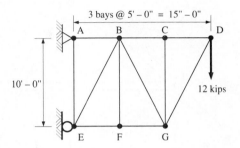

FIGURE 5–30   Problem 9.

Section 5–6   Analysis of Frames

10. A pin-connected A-frame supports a load as shown in Fig. 5–31. Compute the pin reactions at all of the pins. Neglect the weight of the members.

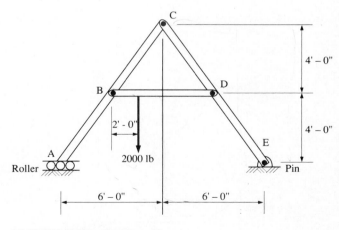

FIGURE 5–31   Problem 10.

11. Calculate the pin reactions at each of the pins in the frame shown in Fig. 5–32.

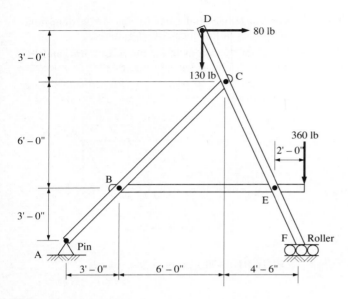

FIGURE 5–32   Problem 11.

12. A simple frame is pin connected at points A, B, and C and is subjected to loads as shown in Fig. 5–33. Compute the pin reactions at A, B, and C. Neglect the weights of the members.

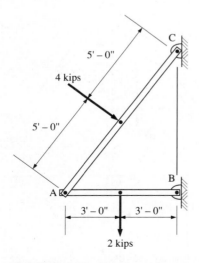

FIGURE 5–33   Problem 12.

13. A bracket is pin connected at points A, B, and D and is subjected to loads as shown in Fig. 5–34. Calculate the pin reactions. Neglect the weights of the members.

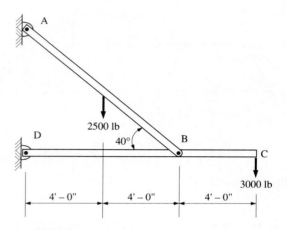

FIGURE 5–34   Problem 13.

14. A pin-connected frame is loaded as shown in Fig. 5–35. Calculate the pin reactions at A, B, and C. Neglect the weights of the members.

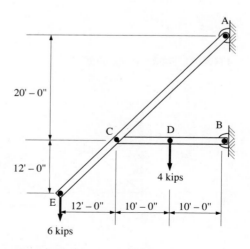

FIGURE 5–35   Problem 14.

15. The cylinder in Fig. 5–36 weighs 200 lb. Determine the force in member AB and the pin reaction at C. All surfaces are smooth. A, B, and C are frictionless pins.

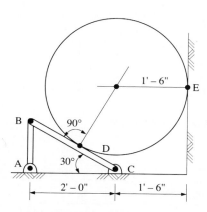

FIGURE 5–36 Problem 15.

SI System Problems

16. Rework the truss of Problem 9 solving for all of the member forces using the method of joints. Change the depth of the truss to be 10 m, the width of the bays to 5 m each, and the load to 100 kN.

17. Using the method of joints, determine all of the member forces in the vertical truss in Fig. 5–37.

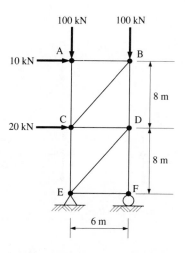

FIGURE 5–37 Problem 17.

The support at E is a pin. The support at F is a roller.

18. Using the method of sections, determine the forces in CE and ED in the vertical truss of Problem 17.

19. Using the method of sections, determine the forces in members BD and BE of the truss in Fig. 5–38.

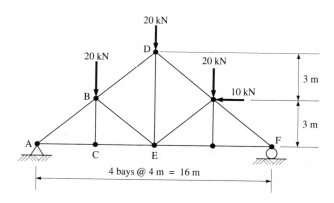

FIGURE 5–38 Problem 19.

20. The A-frame in Fig. 5–39 is pin connected at A, B, C, and D. The surface at E is level and frictionless. Calculate the reaction at E and the reaction components at the pins.

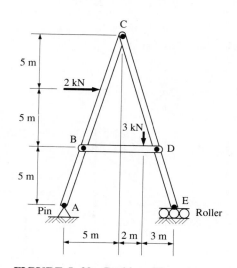

FIGURE 5–39 Problem 20.

Supplemental Problems

21–28. Calculate the forces in all members of the truss indicated in Figs. 5–40 through 5–47 using the method of joints.

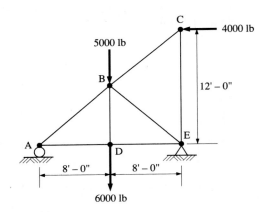

FIGURE 5–40   Problem 21.

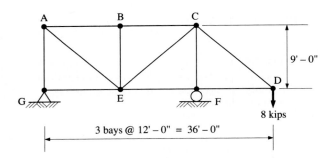

FIGURE 5–41   Problem 22.

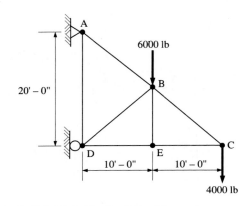

FIGURE 5–42   Problem 23.

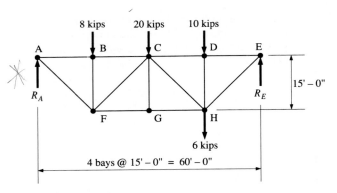

FIGURE 5–43   Problem 24.

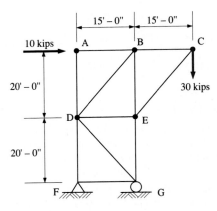

FIGURE 5–44   Problem 25.

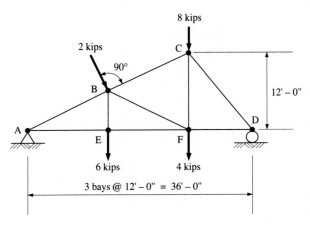

FIGURE 5–45   Problem 26.

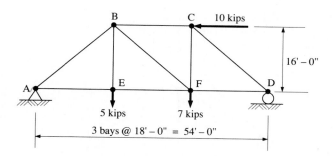

FIGURE 5–46 Problem 27.

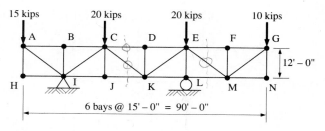

FIGURE 5–48 Problem 29.

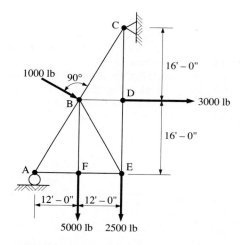

FIGURE 5–47 Problem 28.

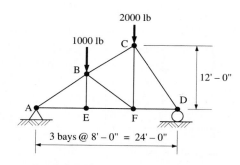

FIGURE 5–49 Problem 30.

For Problems 29–35, calculate the forces in the indicated members of the trusses shown in Figs. 5–48 through 5–54. Use the method of sections.

29. Members CD, CK, and EM.
30. Members BF, BC, and EF.
31. Members AF, BG, and FG.
32. Members CD, BH, and HG.
33. Members BE, CE, and CF.
34. Members EH, CE, CB, and AB.
35. Members BC, CD, and AD.

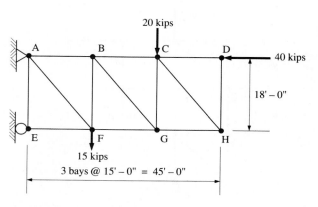

FIGURE 5–50 Problem 31.

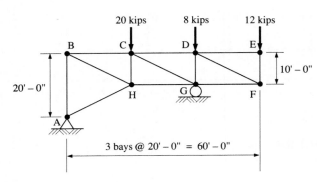

FIGURE 5–51   Problem 32.

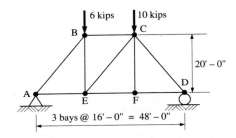

FIGURE 5–52   Problem 33.

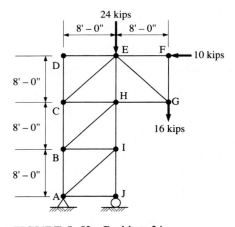

FIGURE 5–53   Problem 34.

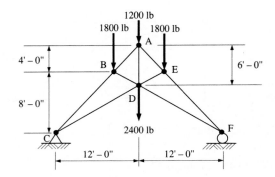

FIGURE 5–54   Problem 35.

36. A pin-connected crane framework is loaded and supported as shown in Fig. 5–55. The member weights are: post, 700 lb; boom, 800 lb; and brace, 600 lb. These weights may be considered to be acting at the midpoint of the respective members. Calculate the pin reactions at pins A, B, C, D, and E.

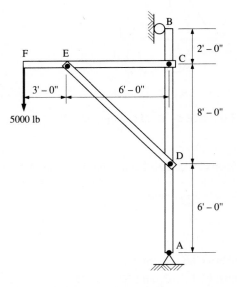

FIGURE 5–55   Problem 36.

37. Calculate the pin reactions at each of the pins in the frame shown in Fig. 5–56.

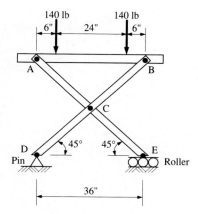

FIGURE 5–56 Problem 37.

38. Calculate the pin reactions at pins A, B, and D in Fig. 5–57. Neglect the weights of the members.

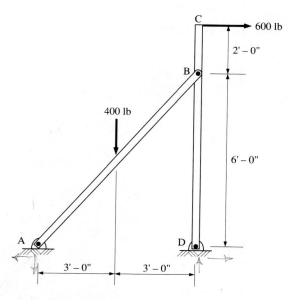

FIGURE 5–57 Problem 38.

39. The wall bracket in Fig. 5–58 is pin-connected at points A, B, and C. Calculate the pin reactions at these points. Neglect the weights of the members.

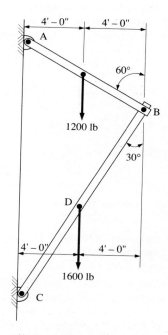

FIGURE 5–58 Problem 39.

40. The tongs shown in Fig. 5–59 are used to grip an object. For an input force of 12 lb on each handle, determine the forces exerted on the object and the forces exerted on the pin at A.

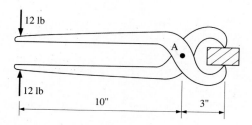

FIGURE 5–59 Problem 40.

41. A toggle joint is a mechanism by which a comparatively small force P produces or balances a force F
which increases rapidly as the angle θ approaches
180°. For a given angle of θ, the toggle joint becomes equivalent to a simple truss. In Fig. 5–60,
force $P = 10$ lb, $a = 2$ ft, $b = 5$ ft, and $c = 1$ ft. Find
force F. Neglect friction and the weights of the
members.

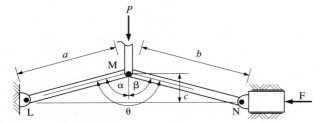

FIGURE 5–60   Problem 41.

42. In the toggle joint of Problem 41, assume that a and
b are equal and $\theta = 160°$. What force P is required
to balance a force of 100 lb at F?

43. A beverage can crusher, shown in Fig. 5–61, can
aid in the reduction of volume of these recyclable
items. For a constant force of 30 lb applied at right
angles to the handle, determine the force applied to
the beverage can for values of θ of (a) 90° and (b)
135°. Neglect friction. Link AB is 9 in. long.

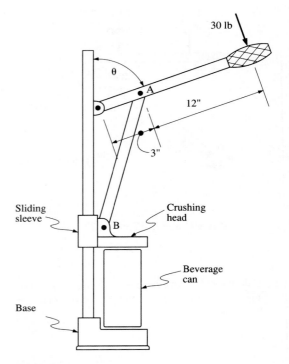

FIGURE 5–61   Problem 43.

6 Friction

6–1
INTRODUCTION

Over the centuries, many renowned scientists, among them Leonardo da Vinci (1452–1519) and Charles Augustin de Coulomb (1736–1806), have studied friction. Friction is one of the most familiar concepts in the field of engineering mechanics, yet it remains one of the least understood. Friction is a very complicated phenomenon to which considerable research is still being devoted within the scientific community.

Friction is defined as a retarding force that resists the relative movement of two bodies in contact with each other. Its action is always opposite in direction to the actual or impending motion between the two bodies. In addition, its action is always parallel or tangent to the surfaces in contact. This retarding or resisting force is commonly called *friction force* or *frictional resistance*. Its magnitude is primarily dependent on the condition of the contact surface between the two bodies. Every surface, as shown in Fig. 6–1, no matter how polished or carefully finished, will reveal projections and depressions when microscopically examined. It is the interlocking of these irregularities on the two surfaces that resists the sliding of one surface with respect to the other.

Research has indicated that other factors, such as temperature, molecular adhesion, electrostatic attraction, lubrication, and relative velocities, in addition to surface irregularities, all contribute to frictional resistance in some measure. Thus, the friction phenomenon depends on, and is a function of, a combination of complex factors. In this chapter, we will discuss friction only between dry, unlubricated surfaces. Dry friction is sometimes called *Coulomb friction*; it was Coulomb who conducted friction research in the eighteenth century and who developed some of its so-called empirical laws.

Dry friction may be categorized as either static or kinetic (dynamic). When the surfaces in contact are at rest (not moving) with respect to each other, the resistance to motion is called *static friction*. When the surfaces in contact are moving with respect to each other, a resistance to motion still exists; this resistance is called *kinetic friction* or *dynamic friction*.

Another distinct type of friction (beyond the scope of this text) is fluid friction. This is friction that develops between layers of a fluid moving at different velocities. It is generally termed *viscosity*.

FIGURE 6–1 Contact surfaces magnified.

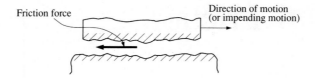

6–2
FRICTION THEORY

In spite of the complexity of the friction phenomenon, its study can be approached conceptually in a very simple manner. As mentioned previously, our discussion applies only to dry, unlubricated contact surfaces.

Consider a block of weight W supported on a horizontal surface. There is no motion, and the block is said to be "at rest." As the block rests on the supporting surface, the only forces acting are (a) the weight W of the block acting vertically downward and (b) a force perpendicular to the contact surface, which is the reaction of the supporting surface, equal and opposite to W. This normal force is designated N. There is no frictional resistance at this time, since frictional resistance is essentially a reaction. It will develop only if there is a force producing motion, or tending to produce motion.

Next, assume that a horizontal force P is applied to the block as shown in Fig. 6–2. As P increases from zero, the frictional resistance F that is developed also increases. This continues up to the point of motion impending (motion on the verge of occurring) and may be observed graphically in Fig. 6–3. Note that up to the point of impending motion, the frictional

FIGURE 6–2 Free-body diagram.

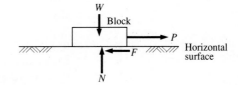

FIGURE 6–3 Frictional resistance.

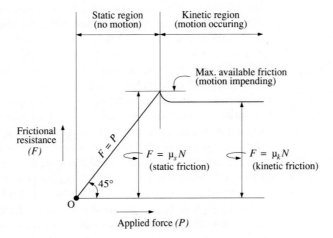

resistance F that is developed is numerically equal to the horizontally applied force P. The frictional resistance that is developed prior to motion is static friction.

After motion occurs, the frictional resistance F decreases rapidly to a value which then remains relatively constant, for all practical purposes. The frictional resistance that develops while the block is in motion is kinetic or dynamic friction. As may be observed in Fig. 6–3, kinetic friction is always less than the maximum available friction that occurs when motion is impending.

While the block remains at rest, the frictional resistance must equal the resultant force tending to cause motion. This is valid up to the instant at which the frictional resistance can no longer balance the resultant of the applied forces and motion occurs. Once motion takes place, the frictional resistance drops to a value below that which was in effect when the motion started. Thus, frictional resistance increases only up to a certain point and no further; that is, a limiting value of friction is developed, beyond which it cannot increase. The block will not move until the applied horizontal force P is greater than the limiting value of F. In other words, in the "at rest" state of equilibrium, $W = N$ and $F = P$. But when P is greater than the available frictional resistance, the block will move, implying that the maximum value of F has been reached.

The ratio of the maximum frictional resistance F to the normal force N is called the *coefficient of static friction* and is designated μ_s (μ is the Greek lowercase letter mu). This is expressed mathematically as

$$\mu_s = \frac{F}{N} \tag{6–1}$$

where μ_s = the coefficient of static friction (unitless)
$\quad\quad F$ = the maximum frictional resistance (lb, kips) (N)
$\quad\quad N$ = the normal force perpendicular to the contact surface (lb, kips) (N)

Equation (6–1) can be rewritten as

$$F = \mu_s N \tag{6–2}$$

where F is the available maximum frictional resistance (lb, kips) (N) and μ_s and N are as previously defined.

Research indicates that the coefficient of static friction (μ_s) depends primarily on the nature of the contacting surfaces and is independent of the normal force. In other words, in a particular application, as N increases, F will increase proportionally, the factor of proportionality being μ_s. Note that in Eqs. (6–1) and (6–2), the surface contact area does not enter into the problem. This is true for a wide range of values and conditions. However, in some applications, such as in brake and tire design, the contact area will affect the coefficient of friction. In these cases, the small contact area will not permit the heat developed during operation to be adequately dissipated.

Chapter 6 Friction

The increase in temperature will result in a lower coefficient of friction and frictional resistance.

Kinetic frictional resistance may also be determined by Eq. (6–2), using the coefficient of kinetic friction μ_k in place of the coefficient for static friction μ_s. It is usually assumed that kinetic friction is constant in value. However, as shown in Fig. 6–3, when a body in motion is brought to rest, the kinetic friction gradually increases at very low speeds up to the maximum value (static friction) as the body finally comes to rest.

Table 6–1 indicates a range of approximate values of coefficients of static and kinetic friction for combinations of some common materials. These values may vary appreciably due to contact surface conditions or extremes of temperature or velocity.

TABLE 6–1 Typical values of coefficients of friction for dry surfaces.

Materials	Static μ_s	Kinetic μ_k
Steel on steel	0.4–0.7	0.3–0.5
Steel on brass	0.3–0.6	0.2–0.4
Aluminum on steel	0.4–0.6	0.3–0.5
Wood on steel	0.3–0.5	0.2–0.4
Teflon on steel	0.03–0.05	0.03–0.05
Wood on wood	0.3–0.6	0.2–0.5

6–3
ANGLE OF FRICTION

We have discussed the general nature of friction problems with the illustration of a block resting on a supporting surface. It is convenient, at times, to represent the reaction of the supporting surface on the block as a single force, R, instead of two forces, F and N. It is apparent in Fig. 6–4 that F and N are really components of the total reaction R. The angle between R and N depends on the value of the frictional resistance F. If F is zero, the angle will be zero. As F increases, so does the angle. When the maximum frictional resistance is acting, the value of the angle is termed the *angle of static*

FIGURE 6–4 Angle of friction.

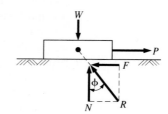

friction and is termed ϕ_s (Greek lowercase phi). The maximum value of ϕ_s develops *only* when motion is impending.

Recalling that $\mu_s = F/N$, the angle of friction (in Fig. 6–4) can be defined by

$$\tan \phi_s = \frac{F}{N} = \mu_s \tag{6-3}$$

Therefore, the coefficient of friction can be obtained by determining the angle of friction. The angle of friction is easily obtained as follows: Place a block (of weight W) on a plane inclined to the horizontal at an angle θ, as shown in Fig. 6–5. Gradually increase the angle of inclination θ from zero to some maximum value at which the block will be on the verge of sliding down the incline. At this point, the angle of inclination θ will be equal to the angle of friction ϕ_s. The angle of inclination θ when motion impends is called the *angle of repose*. Note that in order for equilibrium to exist, it is necessary that the weight W and the reaction R be collinear.

FIGURE 6–5   Block on incline.

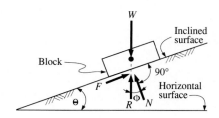

6–4 FRICTION APPLICATIONS

In this section, we will use various introductory examples to provide a methodology that may be applied to friction problems. The general approach we will use for "motion impending" problems is as follows:

1. Draw the free-body diagram(s). Show all applicable forces. Decide which way motion is impending and show frictional forces opposing the impending motion.
2. When forces are concurrent, apply the equations of force equilibrium and Eq. (6–2):
$$\Sigma F_y = 0 \qquad \Sigma F_x = 0 \qquad F = \mu_s N$$
3. If forces are *not* concurrent, in addition to the three preceding equations, apply the equation of moment equilibrium:
$$\Sigma M = 0$$

Note that if motion is not impending, the preceding steps are still valid, except that the frictional force F will be less than $\mu_s N$. F will be equal and opposite to the resultant of the forces tending to cause motion.

□ **EXAMPLE 6–1**   A 400 lb block is supported by a horizontal floor. The coefficient of static friction between the block and the floor is assumed to be 0.40. Calculate the force P required to cause motion to impend. The force is applied to the block (a) horizontally, (b) downward at an angle of 30° with the horizontal.

Solution   (a) The block is assumed to be in static equilibrium and the free-body diagram is shown in Fig. 6–6. The forces acting on the block are its own weight W of 400 lb acting vertically downward, unknown force P acting horizontally, and the reaction of the floor which consists of the normal component N acting vertically upward and the frictional resistance component F acting to the left to oppose sliding of the block. Note that since the forces W, P, and the reaction of the floor R must be concurrent, the normal component N of the floor reaction will not be collinear with the W force. Actually, the exact position of N is immaterial, since the solution is affected only by the horizontal and vertical force equilibrium conditions. This applies to subsequent problems as well, in which the body under consideration is a block or similar object, the dimensions of which are not specified.

FIGURE 6–6   Free-body diagram.

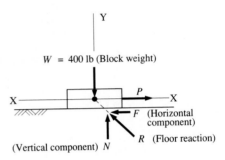

There are three unknowns in this problem: N, F, and P. They can be determined using the two equations of force equilibrium in addition to Eq. (6–2). Using the free-body diagram of Fig. 6–6 which shows all the forces and reference axes, first compute N by summing vertical forces:

$$\Sigma F_y = +N - 400 = 0$$

from which

$$N = 400 \text{ lb}$$

The available maximum frictional force is calculated from Eq. (6–2):

$$F = \mu_s N = 0.40(400) = 160 \text{ lb}$$

Lastly, compute the horizontal force P that will cause motion to impend by summing horizontal forces. A positive sign (+) will be used for the direction of impending motion.

$$\Sigma F_x = +P - F = 0$$

from which

$$P = 160 \text{ lb}$$

FIGURE 6–7 Free-body diagram.

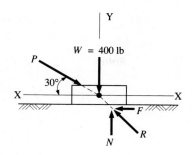

(b) This part is similar to part (a), except that the force P is applied downward at an angle. The free-body diagram is shown in Fig. 6–7 and the block is assumed to be in static equilibrium.

As in part (a), the three unknown forces N, F, and P can be calculated using the two equations of force equilibrium and Eq. (6–2). Note, however, that P now has a vertical component. First sum vertical forces (upward-acting forces are positive):

$$\Sigma F_y = +N - 400 - P \sin 30° = 0$$
$$= +N - 400 - 0.5P = 0$$

from which

$$N = 0.5P + 400$$

Similarly, summing horizontal forces,

$$\Sigma F_x = +P \cos 30° - F = 0$$

from which

$$F = 0.866P$$

Next compute P using Eq. (6–2), substituting for F and N:

$$F = \mu_s N$$
$$0.866P = 0.40(0.5P + 400)$$
$$0.866P = 0.20P + 160$$

from which

$$P = 240 \text{ lb}$$

□ **EXAMPLE 6–2**   A body weighing 300 lb is supported on an inclined plane that forms an angle of 30° with the horizontal. As shown in Fig. 6–8, a force P parallel to and acting up the plane is applied to the body. The coefficient of static friction is 0.40. (a) Find the value of P to cause motion to impend up the plane. (b) Find the value of P to cause motion to impend down the plane. (c) If $P = 100$ lb, determine the magnitude and direction of the resisting friction force.

Chapter 6 Friction

FIGURE 6–8 Body on inclined plane.

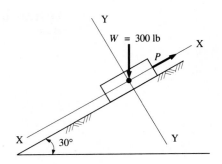

Solution (a) The body is assumed to be in static equilibrium and the free-body diagram is shown in Fig. 6–9(a). Note that since the body has motion impending up the plane, the maximum friction force F will act down the plane to resist motion. Reference axes are selected so that the X axis is parallel to the inclined plane and forces are assumed positive in the direction of impending motion.

The three unknown forces N, F, and P can be calculated using Eq. (6–2) along with the two equations of equilibrium for concurrent forces. Using the free-body diagram of Fig. 6–9(a) and summing forces perpendicular to the plane,

$$\Sigma F_y = +N - W \cos 30° = 0$$
$$= +N - 300(0.866) = 0$$

from which

$$N = 260 \text{ lb}$$

The available maximum friction force is calculated using Eq. (6–2):

$$F = \mu_s N = 0.40(260) = 104 \text{ lb}$$

Lastly, the force P is calculated by summing forces parallel to the plane:

$$\Sigma F_x = +P - F - W \sin 30° = 0$$

from which

$$P = 104 + 300(0.5) = 254 \text{ lb}$$

(b) In this part, the body rests on the plane in static equilibrium, but motion is impending down the plane. The free-body diagram is shown in Fig. 6–9(b). Note that the maximum friction force F is acting up the plane to oppose the impending motion. The magnitudes of N and F are calculated in exactly the same way as in part (a) and have the same values as in part (a). You can then compute the force P that will prevent motion down the plane by summing forces parallel to the plane. (Note that the positive direction is taken in the direction of impending motion.)

$$\Sigma F_x = -P - F + W \sin 30° = 0$$
$$= -P - 104 + 300(0.5) = 0$$

from which

$$P = 46 \text{ lb}$$

FIGURE 6–9   Free-body diagrams.

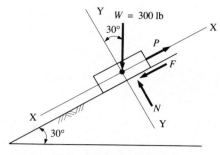

(a) Impending motion—upward

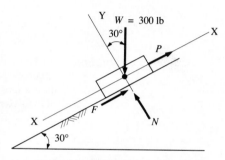

(b) Impending motion—downward

(c) Based on the results of parts (a) and (b) the 300 lb body will remain at rest for all values of P between 46 lb and 254 lb.

When P is equal to 100 lb acting up the plane, the body will tend to slide down the plane, since 100 lb < 300 sin 30° lb. Therefore, the friction force F will act up the plane, opposing the tendency of the block to move. The friction force will not be the maximum available friction, as will be shown by summing the forces in the X direction. Using the free-body diagram of Fig. 6–9(b),

$$\Sigma F_x = -P - F + 300 \sin 30° = 0$$
$$= -100 - F + 300(0.5) = 0$$

from which

$$F = 50 \text{ lb}$$

This represents the frictional resistance necessary for the body to be in static equilibrium when subjected to a force P of 100 lb acting upward along the inclined plane.

Note that if $P > 300 \sin 30°$, the body would tend to slide up the plane, and the friction force F would then act down the plane.

☐ **EXAMPLE 6–3**   A body weighing 200 lb is supported on an inclined plane forming an angle of 40° with the horizontal. A horizontal force P is applied to the body as shown in Fig. 6–10. If the coefficient of static friction between the body and the plane is 0.25, (a) calculate

FIGURE 6–10   Body on inclined plane.

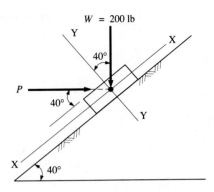

the value of P required to prevent the body from sliding down the plane and (b) calculate the value of P required to cause motion to impend up the plane.

Solution   (a) The value of P required to keep the body from sliding down the plane will be the smallest value required to maintain an "at rest" condition. Using the free-body diagram of Fig. 6–11(a), first calculate N in terms of P by summing forces perpendicular to the plane:

$$\Sigma F_y = +N - W \cos 40° - P \sin 40° = 0$$
$$= +N - 200(0.766) - P(0.643) = 0$$

from which

$$N = 153.2 + 0.643P$$

Next compute F in terms of P by summing forces parallel to the plane. Recall that forces are positive (+) when acting in the direction of impending motion which, in this case, is down the plane. This means that the friction force F is acting up the plane, opposing the motion.

$$\Sigma F_x = -F - P \cos 40° + W \sin 40° = 0$$
$$= -F - 0.766P + 200(0.643) = 0$$

from which

$$F = 128.6 - 0.766P$$

Lastly, compute the horizontal force P using Eq. (6–2) and substituting for F and N:

$$F = \mu_s N$$
$$128.6 - 0.766P = 0.25(153.2 + 0.643P)$$
$$128.6 - 0.766P = 38.3 + 0.161P$$
$$0.927P = 90.3$$

from which

$$P = 97.4 \text{ lb}$$

(b) Using the free-body diagram of Fig. 6–11(b), calculate the force P required to cause motion to impend up the plane. N is calculated exactly as in part (a). F is

FIGURE 6-11 Free-body diagrams.

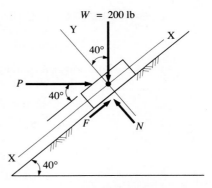

(a) Motion impending—downward

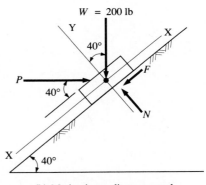

(b) Motion impending—upward

calculating by summing forces parallel to the plane. Note that since impending motion is up the plane, the friction force will act down the plane, opposing the motion.

$$\Sigma F_x = -F + P \cos 40° - W \sin 40° = 0$$
$$= -F + P(0.766) - 200(0.643) = 0$$

from which

$$F = -128.6 + 0.766P$$

Lastly, compute the horizontal force P using Eq. (6-2) and substituting for F and N:

$$F = \mu_s N$$
$$-128.6 + 0.766P = 0.25(153.2 + 0.643P)$$
$$-128.6 + 0.766P = 38.3 + 0.161P$$
$$0.605P = 166.9$$

from which

$$P = 276 \text{ lb}$$

☐ **EXAMPLE 6–4**   A 250 lb block rests on a horizontal surface and is acted on by an inclined force *P* as shown in Fig. 6–12. Determine the magnitude of the force *P* that will cause the block to move. The coefficient of friction between the two contact surfaces is 0.25.

Solution   Assuming that the force is gradually applied, motion may occur in two ways: (a) The block may tend to slide to the left or (b) it may tend to tip about point O.

(a) Initially compute the force *P* required to cause sliding (this neglects tipping). Figure 6–13(a) shows the free-body diagram for the sliding consideration.

FIGURE 6–12   Block on horizontal plane.

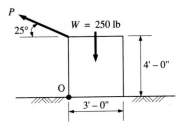

FIGURE 6–13   Free-body diagrams.

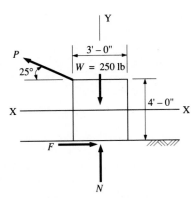

(a) Free-body diagram—sliding

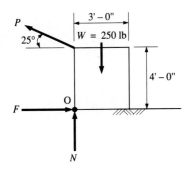

(b) Free-body diagram—tipping

The three unknown forces N, F, and P can be determined using Eq. (6–2) and the two equations of force equilibrium. First solve for N in terms of P by summing forces in the vertical direction:

$$\Sigma F_y = +N - W + P \sin 25° = 0$$
$$= +N - 250 + 0.423P = 0$$

from which

$$N = 250 - 0.423P$$

Next, compute F in terms of P by summing forces in the horizontal direction. Note that impending motion is to the left; therefore, the friction force F acts to the right, opposing motion.

$$\Sigma F_x = -F + P \cos 25° = 0$$

from which

$$F = 0.906P$$

Finally, compute the force P using Eq. (6–2) and substituting for F and N:

$$F = \mu_s N$$
$$0.906P = 0.25(250 - 0.423P)$$
$$0.906P = 62.5 - 0.106P$$
$$1.012P = 62.5$$

from which

$$P = 61.8 \text{ lb}$$

(b) Now compute the force P that will cause the block to tip about point O (this neglects sliding). Figure 6–13(b) shows the free-body diagram for the tipping consideration.

Since the block is assumed to tip about point O, the last contact point prior to motion will be point O, and the reaction of the supporting surface will pass through point O. Therefore, the normal force N and the friction force F (components of the supporting surface reaction) will pass through point O, as shown in Fig. 6–13(b).

There are three unknown forces in this problem. However, if a summation of moments is taken with respect to point O, the effects of forces F and N will be eliminated and the unknown force P can be calculated.

Using the free-body diagram of Fig. 6–13(b) and summing moments about point O, calculate the value of the force P that will cause the block to tip:

$$\Sigma M_O = +P \cos 25°(4) - 250\left(\frac{3}{2}\right) = 0$$

from which

$$P = 103.5 \text{ lb}$$

Since a force of 61.8 lb will cause the block to slide, it is evident that sliding will occur first as P increases from zero to a maximum value.

☐ **EXAMPLE 6–5**   The ladder shown in Fig. 6–14(a) is supported by a horizontal floor and a vertical wall. It is 16 ft long and weighs 48 lb. The weight may be considered to be concentrated at its midlength. The ladder must support a person weighing 200 lb at point D. The coefficient of static friction between the ladder and the wall is 0.25; between the ladder and the floor it is 0.40. The base of the ladder has been moved to the left and the ladder is on the verge of slipping. Calculate (a) the horizontal and vertical reactions at points A and B and (b) the angle θ.

FIGURE 6–14 Ladder supported by wall and floor.

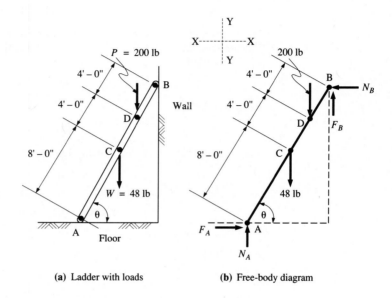

(a) Ladder with loads   **(b)** Free-body diagram

Solution   (a) The horizontal and vertical reactions at points A and B can be computed using the free-body diagram of Fig. 6–14(b) along with Eq. (6–2) and the two force equations of equilibrium. The horizontal reaction at A designated F_A and the vertical reaction at B designated F_B represent the frictional resistance at the respective points opposing any movement of the ladder.

Using Eq. (6–2), F_A and F_B can be expressed as follows:

$$F_A = \mu_s N_A = 0.40 N_A$$
$$F_B = \mu_s N_B = 0.25 N_B$$

Using the X–Y coordinate system shown and applying the two force equations of equilibrium, the reactions at A and B can be calculated. Note that motion impends to the left at point A and down at point B. As shown in Fig. 6–14(b), the direction of the friction forces F_A and F_B oppose the direction of motion. Forces will be assumed positive if acting in the direction of impending motion.

Summing forces in the X direction and substituting for F_A,

$$\Sigma F_x = +N_B - F_A = +N_B - 0.40 N_A = 0$$

from which

$$N_B = 0.40 N_A$$

Summing forces in the Y direction and substituting for F_B,

$$\Sigma F_y = +P + W - N_A - F_B = 0$$
$$= +200 + 48 - N_A - 0.25 N_B = 0$$

from which

$$N_A = 248 - 0.25 N_B$$

Substituting for N_B in the preceding expression,

$$N_A = 248 - 0.25(0.40 N_A)$$

from which

$$N_A = 225.5 \text{ lb}$$

Having determined N_A, all other reactions can now be calculated:

$$N_B = 0.40 N_A = 0.40(225.5) = 90.2 \text{ lb}$$
$$F_A = 0.40 N_A = 90.2 \text{ lb}$$
$$F_B = 0.25 N_B = 0.25(90.2) = 22.6 \text{ lb}$$

(b) The angle θ at which motion is impending to the left can be computed using the third equation of equilibrium ($\Sigma M = 0$) with respect to point A. Use a sign convention of positive for counterclockwise moments and negative for clockwise moments.

$$\Sigma M_A = -W(8 \cos \theta) - P(12 \cos \theta) + F_B(16 \cos \theta) + N_B(16 \sin \theta) = 0$$
$$= -48(8)(\cos \theta) - 200(12)(\cos \theta) + 22.6(16)(\cos \theta) + 90.2(16)(\sin \theta) = 0$$
$$= 384 \cos \theta - 2400 \cos \theta + 361.6 \cos \theta + 1443.2 \sin \theta = 0$$

$$1443.2 \sin \theta = 1654.4 \cos \theta$$

$$\frac{\sin \theta}{\cos \theta} = \tan \theta = \frac{1654.4}{1443.2} = 1.146$$

from which

$$\theta = 48.9°$$

☐ **EXAMPLE 6-6** A sliding block system is shown in Fig. 6-15. Compute the horizontal force P necessary to cause motion of the 100 lb block to impend to the left. The coefficient of friction for the contact surfaces is 0.25; the pulley is assumed to be frictionless.

FIGURE 6-15 Blocks on two planes.

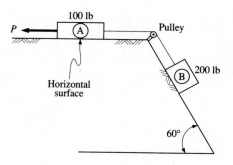

Solution   Free-body diagrams of both blocks are shown in Fig. 6–16. The free-body diagram of block A (Fig. 6–16(a)) shows the unknown force P, the weight W, the tension T in the rope, the reaction N normal to the supporting surface, and the frictional resistance F which acts to the right (since the block tends to move to the left). Similarly, the forces acting on block B are shown in the free-body diagram of Fig. 6–16(b). Since block B tends to move up the plane, the frictional resistance F acts down the plane. The other forces shown are the tension T in the rope, the weight W acting vertically, and the reaction N, normal to the supporting surface. Reference axes X and Y are selected as shown.

FIGURE 6–16   Free-body dia-grams.

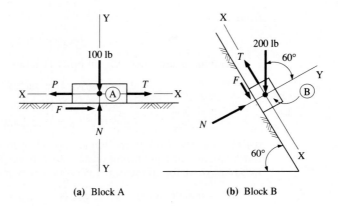

(a) Block A   (b) Block B

Considering block B and applying Eq. (6–2) along with the two force equations of equilibrium, the forces N, F, and T can be calculated:

$$\Sigma F_y = +N - W \cos 60° = +N - 200(0.5) = 0$$

from which

$$N = 100 \text{ lb}$$

From Eq. (6–2),

$$F = \mu_s N = 0.25(100) = 25 \text{ lb}$$

Next, compute T by summing forces parallel to the plane. Assume forces positive in the direction of impending motion, which is up the plane.

$$\Sigma F_x = +T - F - W \sin 60° = 0$$
$$= +T - 25 - 200(0.866) = 0$$

from which

$$T = 198.2 \text{ lb}$$

The free-body diagram of block A (Fig. 6–16(a)) may now be considered. Equation (6–2) and the two equations of force equilibrium will provide the solutions for forces N, F, and P.

$$\Sigma F_y = +N - 100 = 0$$

from which

$$N = 100 \text{ lb}$$

Next, compute the maximum friction force F acting on the horizontal surface. Equation (6–2) yields

$$F = \mu_s N = 0.25(100) = 25 \text{ lb}$$

Finally, compute P by summing forces in the X direction. Assume forces positive in the direction of impending motion, which is to the left.

$$\Sigma F_x = +P - T - F = 0$$
$$= +P - 198.2 - 25 = 0$$

from which

$$P = 223.2 \text{ lb} \leftarrow$$

**6–5
WEDGES**

A typical application of the wedge is shown in Fig. 6–17. Its use makes possible the overcoming (lifting, in this case) of a large load W by means of a relatively small applied force P.

FIGURE 6–17 The wedge.

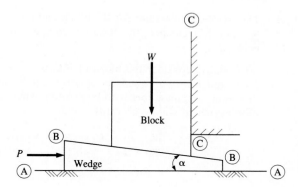

In order that P may start the wedge inward and thus move the load upward, the frictional resistance along planes A–A, B–B, and C–C must be overcome. To determine the necessary value of P, the block and the wedge must be considered separately as bodies in equilibrium with motion impending. The forces can then be determined using a method similar to that used previously in this chapter.

Free-body diagrams for the block and wedge are shown in Fig. 6–18. Note that the N_2 and F_2 forces are common to both free bodies.

The following example demonstrates an algebraic approach for this type of problem.

FIGURE 6–18   Free-body diagrams.

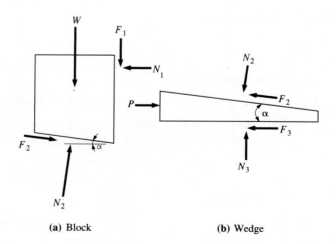

(a) Block   (b) Wedge

□ **EXAMPLE 6–7**   The block shown in Fig. 6–19 supports a load W of 700 lb and is to be raised by forcing a wedge under it. The coefficient of friction on each of the three contact surfaces is 0.25. Calculate the force P required to start the wedge in motion under the block. Neglect the weight of the block and the wedge.

Solution   The free-body diagrams for the block and the wedge are shown in Fig. 6–20(a) and (b). For convenience, the X and Y components of the sloping forces are shown in each case.

(a) Initially, consider the block of Fig. 6–20(a), and calculate F_2 and N_2 using Eq. (6–2) and the two equations of force equilibrium. Summing the forces in the Y direction, assuming forces to be positive when acting in the direction of impending motion (which will be upward), yields

$$\Sigma F_y = -W - F_1 - F_2 \sin 9° + N_2 \cos 9° = 0$$
$$= -700 - F_1 - F_2(0.156) + N_2(0.988) = 0 \qquad \text{(Eq. 1)}$$

FIGURE 6–19   Block-and-wedge arrangement.

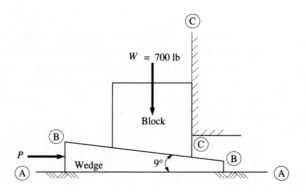

FIGURE 6–20 Free-body diagrams.

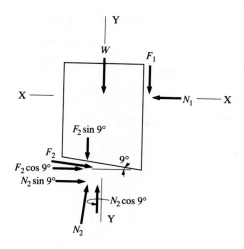

(a) Block

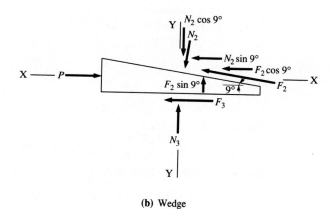

(b) Wedge

Summing forces in the X direction,

$$\Sigma F_x = -N_1 + F_2 \cos 9° + N_2 \sin 9° = 0$$
$$= -N_1 + F_2(0.988) + N_2(0.156) = 0 \qquad \text{(Eq. 2)}$$

From Eq. (6–2), obtain F_1 and F_2:

$$F = \mu_s N$$
$$F_1 = \mu_s N_1 = 0.25 N_1 \qquad \text{(Eq. 3)}$$
$$F_2 = \mu_s N_2 = 0.25 N_2 \qquad \text{(Eq. 4)}$$

Substituting Eqs. 3 and 4 into Eq. 1,

$$-700 - 0.25 N_1 - 0.25 N_2(0.156) + N_2(0.988) = 0$$
$$-700 - 0.25 N_1 - 0.0390 N_2 + 0.988 N_2 = 0$$
$$-700 - 0.25 N_1 + 0.949 N_2 = 0 \qquad \text{(Eq. 5)}$$

Substituting Eqs. 3 and 4 into Eq. 2,

$$-N_1 + 0.25N_2(0.988) + N_2(0.156) = 0$$
$$-N_1 + 0.247N_2 + 0.156N_2 = 0$$
$$-N_1 + 0.403N_2 = 0$$
$$N_1 = 0.403N_2 \qquad \text{(Eq. 6)}$$

Substituting Eq. 6 into Eq. 5 and solving for N_2,

$$-700 - 0.25(0.403N_2) + 0.949N_2 = 0$$
$$-700 - 0.1008N_2 + 0.949N_2 = 0$$
$$-700 + 0.848N_2 = 0$$
$$N_2 = 825 \text{ lb}$$

Finally, Eq. 4 yields

$$F_2 = \mu_s N_2 = 0.25(825) = 206 \text{ lb}$$

(b) Having computed F_2 and N_2 now consider the free body of the wedge, as shown in Fig. 6–20(b). Again apply Eq. (6–2) and the two equations of force equilibrium. Summing forces in the Y direction yields

$$\Sigma F_y = +N_3 - N_2 \cos 9° + F_2 \sin 9° = 0$$
$$= +N_3 - 825(0.988) + 206(0.156) = 0$$

from which

$$N_3 = 783 \text{ lb}$$

Equation (6–2) yields

$$F_3 = \mu_s N_3 = 0.25(783) = 196 \text{ lb}$$

Summing forces in the X direction,

$$\Sigma F_x = +P - F_3 - N_2 \sin 9° - F_2 \cos 9° = 0$$
$$= +P - 196 - 825(0.156) - 206(0.988) = 0$$

from which

$$P = 528 \text{ lb}$$

**6–6
BELT FRICTION**

When a flexible belt, rope, or band is wrapped around a circular pulley or a cylindrically shaped drum and then subjected to a tensile force, a frictional resistance is developed between the contact surfaces. This friction can be used either to transmit power from one shaft to another (for example, in belt-driven machinery) or to retard motion (for example, in band brakes or the tying up of a ship at a pier).

Figure 6–21(a) shows a flexible belt wrapped around a fixed drum with an arc length of contact subtending an angle β (Greek lowercase beta). This angle is commonly called the *angle of wrap*. The coefficient of friction between the belt and the drum is designated μ.

FIGURE 6–21 Flexible belt on fixed drum.

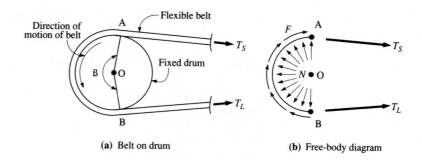

(a) Belt on drum   (b) Free-body diagram

A free-body diagram of the segment of belt in contact with the drum (segment AB) is shown in Fig. 6–21(b). The drum is stationary (fixed against rotation) and the belt is on the verge of slipping (motion is impending). As shown, the forces acting on this piece of belt are the belt tensions T_L and T_S (with the subscripts L and S indicating large and small belt tensions, respectively); the distributed normal forces N (which are really radial, normal-acting to a tangent line at any given point); and tangential friction forces F, which oppose the direction of the impending motion. Since this free-body must be in equilibrium, the sum of the moments of the previously mentioned forces about the center of the drum O must equal zero. The moment of T_L must equal the moment of T_S plus the moment of the friction forces F when motion is impending. Note that the friction forces are acting to retard motion of the belt around the fixed drum.

In order to establish a relationship between the belt tensions and the friction between the belt and the drum, the previously mentioned forces are resolved into X and Y components and equations of equilibrium applied. The derivation of the resulting mathematical expression is beyond the scope of this text. The expression itself, however, is not complex:

$$T_L = T_S e^{\mu\beta} \qquad \textbf{(6–4)}$$

where T_L = the larger belt tension (lb) (N)
  T_S = the smaller belt tension (lb) (N)
  μ = the coefficient of friction between the belt and the drum. (Since motion impends, this would be the coefficient of static friction μ_s)
  β = the angle of contact between the belt and the drum (radians). (Recall that one radian = $180°/\pi$ = $57.3°$)
  e = the base of natural logarithms (2.718). The term $e^{\mu\beta}$ represents an exponential power of the number 2.718

Equation (6–4) may also be expressed in alternate forms:

$$\ln \frac{T_L}{T_S} = \mu\beta \qquad \textbf{(6–5)}$$

or

$$\ln T_L - \ln T_S = \mu\beta \qquad\qquad (6\text{--}6)$$

where ln indicates the natural logarithm (base 2.718) and the other terms are as previously defined.

Belts are used extensively in transmitting power from one shaft to another. In Fig. 6–22 we see a belt that passes around two pulleys, which are in turn, mounted on shafts. Pulley B is the driving pulley; therefore, the lower belt has more tension in it than the upper belt. If friction did not exist between the belt and the pulleys, the driving pulley could not drive the belt, nor could the driven pulley be turned by the belt. Note that both sides of the belt are in tension. The tension T_L on the tight side is greater than the tension T_S on the loose or slack side, thus resulting in a net driving force on the pulleys equal to

$$T_{\text{net}} = T_L - T_S$$

FIGURE 6–22 Open belt drive.

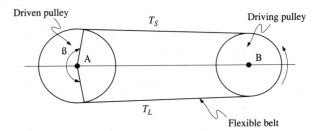

For pulley-and-shaft applications, the term *torque* is commonly used. Torque is a twisting action applied in a plane perpendicular to the longitudinal axis of a shaft. The torque that a belt can transmit can be computed by a moment summation about the center of the pulley of the two forces T_L and T_S:

$$\text{Torque} = T_L r - T_S r = (T_L - T_S)r = T_{net}r$$

where r represents the radius of the pulley.

Note that a free-body diagram of pulley A in Fig. 6–22 would be identical to the free-body diagram of Fig. 6–21(b). Therefore, it can be concluded that the same type of analysis with the same resulting equations as that given for the fixed drum may be applied.

Equations (6–4), (6–5), and (6–6) are not applicable to all of the many types of belts available. Our discussion to this point included flat belts or ropes; however, the most widely used type of belt drive is the V-belt drive. This type makes contact on two sloping sides of a grooved pulley, as shown in Fig. 6–23. The belt acts like a wedge and the normal forces on the contact surfaces are increased, thereby increasing the tractive force developed by the belt.

FIGURE 6–23 V-belt on pulley.

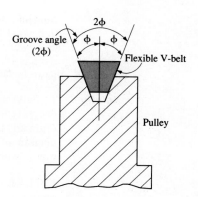

The relationship between the belt tensions and the friction forces is established using an analysis similar to that used for the flat belt and rope. The only additional parameter included in the analysis is the groove angle (2ϕ). The resulting equation, applicable to only the flexible V-belt where motion is impending, is

$$T_L = T_S e^{\mu\beta/\sin\phi} \tag{6–7}$$

where ϕ = one-half the groove angle (see Fig. 6–23) and the other terms are as previously defined.

Equation (6–7) may also be expressed in alternate forms:

$$\ln\frac{T_L}{T_S} = \frac{\mu\beta}{\sin\phi} \tag{6–8}$$

or

$$\ln T_L - \ln T_S = \frac{\mu\beta}{\sin\phi} \tag{6–9}$$

☐ **EXAMPLE 6–8** A 300 lb weight is suspended from a rope that passes over a rough cylindrical stationary drum, as shown in Fig. 6–24. The coefficient of static friction between the

FIGURE 6–24 Rope-and-drum arrangement.

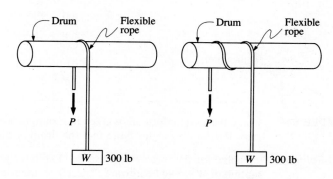

(a) Angle of wrap = 180°   **(b)** Angle of wrap = 540°

rope and the drum is 0.30. Assuming that motion is impending, calculate the force P that must be applied to the free end of the rope to keep the weight from slipping downward. (a) The rope is in contact with half of the cylindrical surface (angle of wrap is 180°). (b) The rope is wrapped around the cylindrical drum $1\frac{1}{2}$ times (angle of wrap is 540°).

Solution   To determine the value of P to keep W just from falling, consider that with a slight decrease in P the weight W will fall. Therefore, the frictional resistance between the drum and the rope will act in a counterclockwise direction, as shown in the free-body diagram of Fig. 6–25. This will assist P in holding the weight aloft. It is evident, then, that P will be numerically less than W.

FIGURE 6–25   Free-body diagram.

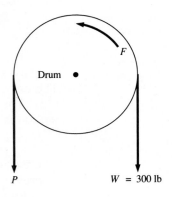

(a) From Eq. (6–4),

$$T_L = T_s e^{\mu\beta}$$
$$300 = P e^{0.30(\pi)}$$
$$300 = 2.566P$$

from which

$$P = 116.9 \text{ lb}$$

(b) From Eq. (6–4),

$$T_L = T_s e^{\mu\beta}$$
$$300 = P e^{0.30(3\pi)}$$
$$300 = 16.9P$$

from which

$$P = 17.8 \text{ lb}$$

☐ **EXAMPLE 6–9**   Using the basic information given in Example 6–8, calculate the force P required to raise the 300 lb weight. Note that the drum is stationary.

Solution   To solve for the force P to raise the 300 lb weight, it is evident that the impending motion of W would be upward. Therefore, the frictional resistance between the drum and the rope will act in a clockwise direction to oppose the motion. This frictional resistance would have to be overcome. It is logical, then, that P should be numerically larger than W.

(a) From Eq. (6–4),

$$T_L = T_S e^{\mu\beta}$$
$$P = 300 e^{0.30(\pi)} = 770 \text{ lb}$$

(b) From Eq. (6–4),

$$T_L = T_S e^{\mu\beta}$$
$$P = 300 e^{0.30(3\pi)} = 5070 \text{ lb}$$

☐ **EXAMPLE 6–10**   A flat belt makes contact with a 24 in. diameter pulley through half its circumference. If the coefficient of static friction is 0.25 and a torque of 4000 in.-lb is to be transmitted, calculate the belt tensions.

Solution   Since there are two unknowns, T_L and T_S, two equations must be developed to establish a relationship between the unknowns. From Eq. (6–4), substituting π radians for β,

$$T_L = T_S e^{\mu\beta} = T_S e^{0.25(\pi)} = 2.19 T_S$$

The torque to be transmitted is 4000 in.-lb and can be expressed as

$$\text{Torque} = (T_L - T_S)(r)$$
$$4000 = (T_L - T_S)(12)$$

from which

$$T_L - T_S = 333.3 \text{ lb}$$

Substituting the expression for T_L into the preceding equation,

$$2.19 T_S - T_S = 333.3 \text{ lb}$$

from which

$$T_S = 280.1 \text{ lb}$$

And, solving for T_L,

$$T_L = 2.19 T_S = 2.19(280.1) = 613 \text{ lb}$$

☐ **EXAMPLE 6–11**   Using the basic information given in Example 6–10, calculate the belt tensions if the belt used is a V-belt with a groove angle (2ϕ) of 60°.

Solution   From Eq. (6–7),

$$T_L = T_S e^{\mu\beta/\sin\phi} = T_S e^{0.25(\pi)/\sin 30°} = 4.81 T_S$$

As in Example 6–10,

$$\text{Torque} = (T_L - T_S)(r)$$
$$4000 = (T_L - T_S)(12)$$

from which

$$T_L - T_S = 333.3 \text{ lb}$$

And, substituting for T_L,

$$4.81T_S - T_S = 333.3 \text{ lb}$$

from which

$$T_S = 87.5 \text{ lb}$$

Finally, solving for T_L,

$$T_L = 4.81T_S = 4.81(87.5) = 421 \text{ lb}$$

6–7
SQUARE-THREADED
SCREWS

Screws are generally used for fastening purposes. However, in many types of machines and equipment, screws called *power screws* are used to translate rotary motion into uniform longitudinal motion, thereby playing an important part in transmitting force or power from one part of the machine to another. Square-threaded screws, one type of a broad category of power screws, are generally used where large loads (forces) are to be transmitted. The square thread is the most efficient of the power screw threads; however, it is relatively costly to manufacture. Common applications of square-threaded screws include jacks, valves, machine tools, and presses.

A square-threaded screw may be regarded as an inclined plane wrapped around a cylinder. The usual proportions for a square-threaded screw, as well as the concept of the inclined plane, are shown in Fig. 6–26. A block moving up (or down) the plane is then equivalent to turning the screw through a set of fixed threads (a nut) or turning a nut on a fixed screw.

FIGURE 6–26 Square-threaded screw.

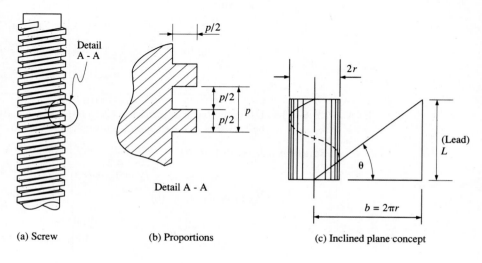

(a) Screw   (b) Proportions   (c) Inclined plane concept

A useful mechanical device that is illustrative of the square-threaded screw is the ordinary jackscrew, shown in Fig. 6–27. The jackscrew generally has a square-threaded screw that turns in a threaded hole of a stationary frame or base. As a force Q is applied to the handle (lever) of the jack at a distance a from the point of rotation, the screw turns in its fixed base and moves upward, lifting the axial load W. The thread in the fixed base represents the inclined plane and the applied load W, supported by the screw, is transmitted to this base.

FIGURE 6–27 Jackscrew.

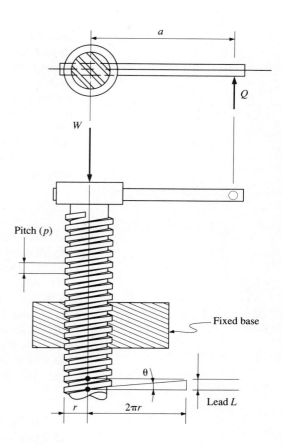

Figure 6–28 shows a block on an inclined plane. The block represents the thread of the screw that may slide up or down the threads of the fixed base, represented by the inclined plane. The horizontal force P represents the minimum force which, if applied at the mean radius r of the screw, would slide the block up the plane thus raising the load W.

The slope of the inclined plane θ depends on the mean radius r and the lead L of the screw. As shown in Fig. 6–26(c) and Fig. 6–27, by theoretically unwinding one turn of the screw thread onto a flat surface, a triangle is

FIGURE 6–28   Force acting up the plane to raise load *W*.

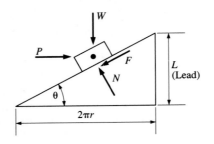

created with the base equal to the circumference of the mean radius of the thread ($2\pi r$) and the rise equal to the pitch (or *lead*) of the thread *L*. The mean radius is equal to one-half the sum of the outer radius and the root radius of the thread. The lead of a screw is the distance that a nut will advance along the screw in one revolution. In the usual case of a single-threaded screw, the lead is the same as the pitch *p*, which is the distance between similar points on adjacent threads. All square-threaded screws considered in this text are single threaded.

Therefore, the slope of the inclined plane, also called the *lead angle θ*, may be computed from

$$\theta = \tan^{-1}\frac{L}{2\pi r} = \tan^{-1}\frac{p}{2\pi r} \tag{6–10}$$

As discussed in Section 6–2, *F* represents the frictional resistance on the inclined plane ($F = \mu_s N$) and *N* represents a normal force perpendicular to the inclined plane.

After computing the angle θ, the horizontal force *P*, as shown in Fig. 6–28, can be calculated using the equations of equilibrium and friction theory as discussed in Section 6–4. Once force *P* is computed, the required applied force *Q* at the end of the jack handle can be calculated based on the principle that the moment effect of *Q*, with respect to the longitudinal axis of the screw, must be equal to the moment effect of *P* with respect to the same axis. This is expressed as

$$Qa = Pr \tag{6–11}$$

If the lead angle θ is very steep, the frictional resistance may not be sufficient to overcome the tendency for the screw thread (or block) to slide under the action of the load. This would result in the load being lowered without any applied force *Q*. However, θ is usually small and the frictional resistance is large enough to prevent this unaided movement. Such a jackscrew is called *self-locking* and this is a desirable characteristic for jacks and similar devices. In order for a screw to be self-locking, the coefficient of static friction must be greater than the tangent of the lead angle ($\mu_s > \tan\theta$).

To determine the force *P* required to lower the applied load *W* (which is equivalent to moving the block down the inclined plane), the free-body diagram as shown in Fig. 6–29 is applicable.

FIGURE 6–29 Force acting down the plane to lower the load W.

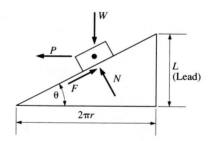

□ **EXAMPLE 6–12** The mean diameter of the jackscrew shown in Fig. 6–27 is 3.50 in. and the pitch of the square-threaded screw is 0.50 in. The coefficient of static friction is 0.15. (a) Calculate the force Q required at the end of a 24 in. jack handle ($a = 24$ in.) to raise a weight of 4000 lb. (b) Calculate the force Q required (if any) at the end of the jack handle to lower a weight of 4000 lb.

Solution (a) The applicable free-body diagram is shown in Fig. 6–30.

1. Using Eq. (6–10), compute the lead angle:

$$\theta = \tan^{-1}\left(\frac{0.50}{11.00}\right) = 2.60°$$

Therefore,

$$\sin \theta = 0.0454$$
$$\cos \theta = 0.999$$

2. Compute the horizontal force P. Taking a summation of vertical forces,

$$\Sigma F_y = -W + N\cos\theta - F\sin\theta = 0$$
$$= -4000 + N(0.999) - 0.15(N)(0.0454) = 0$$

from which

$$N = 4031 \text{ lb}$$

Taking a summation of horizontal forces,

$$\Sigma F_x = +P - N\sin\theta - F\cos\theta = 0$$
$$= +P - 4031(0.0454) - 0.15(4031)(0.999) = 0$$

from which

$$P = 787 \text{ lb}$$

FIGURE 6–30 Free-body diagram.

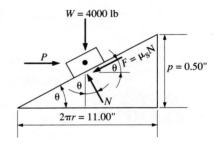

FIGURE 6–31   Free-body diagram.

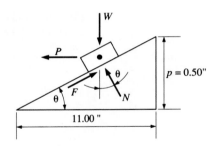

3. Using Eq. (6–11), compute force Q on the jack handle:

$$Qa = Pr$$

$$Q = \frac{787(1.75)}{24} = 57.4 \text{ lb}$$

(b) The applicable free-body diagram is shown in Fig. 6–31.

1. Check whether the jackscrew is self-locking.

$$\mu_s = 0.15 > \tan \theta = 0.0454$$

Therefore, it is self-locking.

2. The lead angle θ was calculated in part (a) to be 2.60°. Compute the horizontal force P. Taking a summation of vertical forces,

$$\Sigma F_y = -W + N \cos \theta + F \sin \theta = 0$$
$$= -4000 + N(0.999) + 0.15(N)(0.0454) = 0$$

from which

$$N = 3977 \text{ lb}$$

Taking a summation of horizontal forces,

$$\Sigma F_x = -P - N \sin \theta + F \cos \theta = 0$$
$$= -P - 3977(0.0454) + 0.15(3977)(0.999) = 0$$

from which

$$P = 415 \text{ lb}$$

3. Compute force Q on the jack handle:

$$Qa = Pr$$

$$Q = \frac{415(1.75)}{24} = 30.3 \text{ lb}$$

6–8
SI SYSTEM
EXAMPLES

☐ **EXAMPLE 6–13**

A block having a mass of 200 kg rests on a plane inclined at 25° with the horizontal, as shown in Fig. 6–32(a). The coefficient of friction is 0.2. Calculate the force P at which motion is impending down the plane.

FIGURE 6-32 Block on inclined plane.

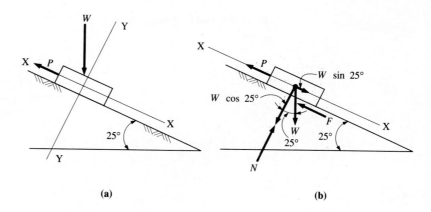

(a)   (b)

Solution   Since motion is impending down the plane, the frictional force acts up the plane, as shown. First calculate the weight of the block (force due to gravity):

$$W = mg = 200 \text{ kg}(9.81 \text{ m/s}^2) = 1962 \frac{\text{kg·m}}{\text{s}^2}$$

$$= 1.96 \text{ kN}$$

The available maximum friction force is calculated using Eq. (6-2):

$$F = \mu_s N = \mu_s W \cos 25°$$
$$= 0.2(1.96 \text{ kN})(0.906) = 0.355 \text{ kN}$$

With reference to Fig. 6-32(b), the force P can be calculated by summing forces parallel to the inclined plane (X direction). Recall that forces are assumed positive in the direction of impending motion (which is down the plane).

$$\Sigma F_x = -P - F + W \sin 25° = 0$$

from which

$$P = W \sin 25° - F$$
$$= (1.96 \text{ kN})(0.423) - 0.355 \text{ kN} = 0.474 \text{ kN}$$

☐ **EXAMPLE 6-14**   A machine having a mass of 1000 kg is to be raised by driving two 4° wedges beneath it, as shown in Fig. 6-33. The coefficient of friction on all wedge surfaces is 0.20. Determine the required force P.

FIGURE 6-33 Wedge application.

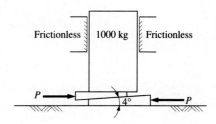

FIGURE 6–34 Free-body diagram.

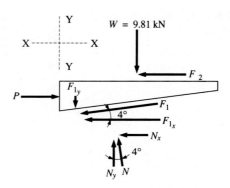

Solution   The free-body diagram for one of the wedges is shown in Fig. 6–34. (The other wedge is identical and could be analyzed in a similar way.) The weight of the machine is calculated from

$$W = mg = 100 \text{ kg}(9.81 \text{ m/s}^2) = 9810 \frac{\text{kg·m}}{\text{s}^2}$$

$$= 9.81 \text{ kN}$$

There are four unknown forces shown on the free body. Frictional force F_1 may be determined from $\mu_s N$. Frictional force F_2 may be obtained immediately from $\mu_s W$. Equilibrium of forces in the X and Y directions will yield equations from which the other two unknown forces can be determined. Note that the frictional forces are shown opposing the impending motion of the wedge.

$$F_1 = \mu_s N = 0.2N \qquad \text{(Eq. 1)}$$
$$F_2 = \mu_s W = 0.2(9.81 \text{ kN}) = 1.96 \text{ kN}$$

Summing forces in the X direction,

$$\Sigma F_x = +P - F_{1_x} - N_x - F_2 = 0$$
$$= +P - F_1 \cos 4° - N \sin 4° - 1.96 \text{ kN} = 0$$
$$= +P - 0.998 F_1 - 0.070N - 1.96 \text{ kN} = 0 \qquad \text{(Eq. 2)}$$

Summing forces in the Y direction,

$$\Sigma F_y = -9.81 \text{ kN} - F_{1_y} + N_y = 0$$
$$= -9.81 \text{ kN} - F_1 \sin 4° + N \cos 4° = 0$$
$$= -9.81 \text{ kN} - 0.070 F_1 + 0.998N = 0 \qquad \text{(Eq. 3)}$$

Substituting F_1 from Eq. 1 into Eq. 3 yields

$$-9.81 \text{ kN} - 0.070(0.2N) + 0.998N = 0$$

$$-9.81 \text{ kN} = +0.014N - 0.998N = -0.984N$$

from which

$$N = 9.97 \text{ kN}$$

From Eq. 1,

$$F_1 = \mu_s N = 0.2N = 0.2(9.97 \text{ kN}) = 1.99 \text{ kN}$$

From Eq. 2, solving for P,

$$\begin{aligned} P &= +0.998F_1 + 0.070N + 1.96 \text{ kN} \\ &= +0.998(1.99 \text{ kN}) + 0.070(9.97 \text{ kN}) + 1.96 \text{ kN} \\ &= +4.64 \text{ kN} \end{aligned}$$

□ **EXAMPLE 6–15**   A mass of 100 kg is suspended by a rope. The line is wrapped $1\frac{1}{4}$ times around a fixed shaft. The coefficient of friction between the rope and the shaft is 0.42. Calculate the required force P on the free end of the rope to keep the mass suspended.

Solution   This situation is similar to that illustrated in Fig. 6–24, except that the angle of wrap will be 450°, or 2.5π radians. Designating the tension in the rope that supports the mass as W,

$$W = mg = 100 \text{ kg}(9.81 \text{ m/s}^2) = 981 \, \frac{\text{kg·m}}{\text{s}^2}$$

$$= 981 \text{ N}$$

Since P is less than W, from Eq. (6–4), substituting and solving for P,

$$T_L = T_S e^{\mu\beta}$$
$$W = P e^{\mu\beta}$$
$$P = \frac{W}{e^{\mu\beta}}$$

from which

$$P = \frac{981 \text{ N}}{e^{0.42(2.5\pi)}} = 36.2 \text{ N}$$

SUMMARY—BY SECTION NUMBER

6–1   Friction is defined as a retarding force that resists the relative movement of two bodies in contact. It acts to oppose motion and always acts parallel to the surfaces in contact.

6–2   The coefficient of static friction is the ratio of the maximum available frictional resistance F to the normal force N between the contacting surfaces:

$$\mu_s = \frac{F}{N} \qquad\qquad \textbf{(6–1)}$$

The maximum available frictional force is developed when motion is impending. It is dependent on the type of materials and the nature of the contact surfaces. It is independent of the contact area and proportional to the normal force N:

$$F = \mu_s N \qquad\qquad \textbf{(6–2)}$$

After motion begins, $F = \mu_k N$, where μ_k is the coefficient of kinetic friction.

6–3 The reaction of the supporting surface may be expressed as a single force R which is the resultant of F and N. The angle ϕ_s between R and N is called the angle of static friction and is a maximum when motion is impending and F is a maximum:

$$\tan \phi_s = \frac{F}{N} = \mu_s \qquad (6\text{–}3)$$

6–4 The general procedure for the solution of motion-impending friction problems in this text involves drawing the free-body diagram and deciding which way motion is impending. This is followed by the application of Eq. (6–2) and the equations of force equilibrium for concurrent force systems, or the equations of force equilibrium and moment equilibrium for nonconcurrent force systems.

6–5 Wedges are devices used to overcome (lift or move) large loads by the means of relatively small applied forces. Wedge problems are solved using free bodies, friction considerations, and the equations of equilibrium.

6–6 The difference in belt tensions T_L and T_S caused by wrapping a belt around a cylindrical surface depends on the angle of contact (wrap) β and the coefficient of friction μ between the contact surfaces. When slip is impending, the following equations apply: for flat belts,

$$T_L = T_S e^{\mu\beta} \qquad (6\text{–}4)$$

or

$$\ln \frac{T_L}{T_S} = \mu\beta \qquad (6\text{–}5)$$

or

$$\ln T_L - \ln T_S = \mu\beta \qquad (6\text{–}6)$$

and for V-belts,

$$T_L = T_S e^{\mu\beta/\sin\phi} \qquad (6\text{–}7)$$

or

$$\ln \frac{T_L}{T_S} = \frac{\mu\beta}{\sin \phi} \qquad (6\text{–}8)$$

or

$$\ln T_L - \ln T_S = \frac{\mu\beta}{\sin \phi} \qquad (6\text{–}9)$$

6–7 Square-threaded screws are an application of the inclined plane and can be used to move heavy loads or transmit power. Single-threaded screws only are considered in this text. Performance of the screw is a function of the coefficient of friction between the parts, the lead, and mean radius of the thread. The lead angle is calculated from

$$\theta = \tan^{-1}\frac{L}{2\pi r} = \tan^{-1}\frac{p}{2\pi r} \qquad \text{(6–10)}$$

The equations of equilibrium and frictional considerations can be used to determine the force P required to move a screw within a threaded block. The force Q required at the end of a handle to turn the screw can be calculated from

$$Qa = Pr \qquad \text{(6–11)}$$

PROBLEMS

Section 6–4   Friction Applications

1. A 150 lb block rests on a horizontal floor. The coefficient of friction between the block and the floor is 0.30. A pull of 40 lb, acting upward at an angle of 30° to the horizontal, is applied to the block. Determine whether or not the block will slide.

2. A 200 lb block rests on a horizontal surface. The coefficient of static friction between the block and the supporting surface is 0.50. Calculate the force P required to cause motion to impend if the force applied to the block is (a) horizontal and (b) upward at an angle of 20° with the horizontal.

3. A body weighing 100 lb rests on an inclined plane as shown in Fig. 6–35. The coefficient of static friction between the body and the plane is 0.40. Compute the force P that will cause impending motion (a) up the inclined plane and (b) down the inclined plane.

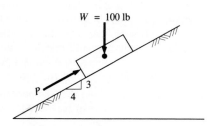

FIGURE 6–35   Problem 3.

4. For Problem 3, compute the friction force F when the force P acting up the plane is equal to (a) 40 lb, (b) 60 lb, and (c) 70 lb.

5. Compute the horizontal force P required to cause motion to impend up the plane for the 100 lb block in Fig. 6–36. The coefficient of static friction between the block and the plane is 0.25.

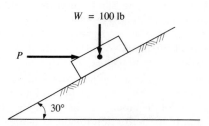

FIGURE 6–36   Problem 5.

6. Compute the horizontal force P required to prevent the block from sliding down the plane for the 175 lb block in Fig. 6–37. Assume the coefficient of static friction to be 0.65.

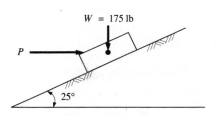

FIGURE 6–37   Problem 6.

7. Calculate the magnitude of the force P, acting as shown in Fig. 6–38, that will cause the 200 lb crate to move, either sliding to the right or tipping. Assume the coefficient of static friction to be 0.30.

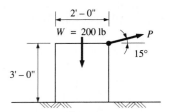

FIGURE 6–38   Problem 7.

Section 6–5   Wedges

8. For the block-and-wedge system shown in Fig. 6–17, calculate the force P required to initiate upward motion of the block. The block supports a load of 700 lb, the slope angle of the wedge is 9°, and the coefficient of static friction on the two surfaces of the wedge is 0.25. The vertical surface (C–C) is frictionless. Compare the result with Example 6–7.

9. Calculate the value of the horizontal force P required to start the V-shaped wedge in motion to the right, raising the 200 lb block, as shown in Fig. 6–39. The coefficient of static friction for the contact surfaces of the wedge is 0.364. The vertical surface is frictionless.

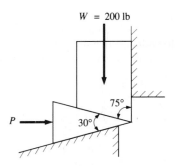

FIGURE 6–39   Problem 9.

10. Two blocks, each weighing 250 lb and resting on a horizontal surface, are to be pushed apart using a 30° wedge, as shown in Fig. 6–40. The coefficient of static

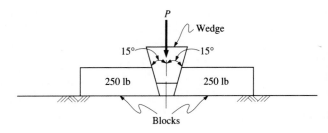

FIGURE 6–40   Problem 10.

friction for all contact surfaces is 0.25. Calculate the vertical force P required to start the wedge and the blocks in motion.

11. Calculate the force P required to move the wedges and raise the 1000 lb block in Fig. 6–41. The coefficient of static friction for all contact surfaces is 0.18.

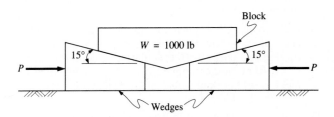

FIGURE 6–41   Problem 11.

Section 6–6   Belt Friction

12. A heavy machine is lowered into a pit by means of a rope wrapped around an 8 in. diameter stationary pole placed horizontally across the top of the pit. The coefficient of static friction for the rope on the pole is 0.35. The rope makes $1\frac{1}{2}$ turns around the pole. Calculate the maximum weight that can be sustained if a person exerts a force of 50 lb on the end of the rope.

13. Calculate the maximum weight that the person in Problem 12 can sustain if $2\frac{1}{2}$ turns of rope are taken around the pole.

14. A flat belt passes halfway around a 6 ft diameter pulley. The maximum permissible tension in the belt is 400 lb. The coefficient of static friction for the belt on the pulley is 0.3. Calculate the maximum torque the belt can transmit to the pulley under these conditions.

15. A belt-and-pulley arrangement has a maximum belt tension of 150 lb on the tight side and 75 lb on the loose side. If the coefficient of static friction between the flat belt and the pulley is 0.30, calculate the minimum angle of contact required between the pulley and the belt.

16. Rework Problem 15 for a V-belt with a 40° groove angle, rather than a flat belt.

17. A weight of 700 lb is prevented from falling by a rope wrapped around a horizontal stationary circular pole. A force of 60 lb is exerted on the free end of the rope. The coefficient of static friction between the pole and the rope is 0.25. How many turns of rope around the pole are necessary to keep the weight aloft?

18. A belt is wrapped around a pulley for 180°. The loose-side belt tension is 50 lb and the coefficient of static friction is 0.30. Calculate the maximum belt tension if motion is impending. Assume (a) a flat belt and (b) a V-belt with a groove angle of 60°.

6–7 Square-Threaded Screws

19. A jackscrew has a square thread with a pitch of 0.50 in. The mean diameter of the thread is 2 in. The coefficient of static friction is 0.30. Calculate the force required at the end of a 15 in. jack handle to raise 2000 lb.

20. The mean diameter of a square-threaded jackscrew is 1.80 in. The pitch of the thread is 0.40 in. and the coefficient of static friction is 0.20. (a) Calculate the force which must be applied at the end of an 18 in. long jack handle to raise a load of 5000 lb. (b) Calculate the force required (if any) on the handle to start the load down.

21. A woodworking vise is designed for a maximum applied force of 50 lb at each end of the handle. The square-threaded screw has a pitch of 0.25 in. and a mean diameter of 0.875 in. The coefficient of static

friction is estimated to be 0.14. Determine the maximum clamping force. See Fig. 6–42.

22. A square-threaded screw is used in a press to exert a pressure of 4000 lb. The screw has a mean diameter of 3 in. and a pitch of 0.25 in. The coefficient of static friction is 0.15. Calculate the force that must be applied at the end of the press handle. See Fig. 6–43.

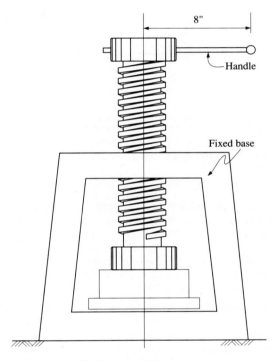

FIGURE 6–43 Problem 22.

SI System Problems

23. A tool locker having a mass of 140 kg rests on a wooden pallet. Assuming a coefficient of static friction of 0.38, determine the horizontal force that would cause sliding motion to impend.

24. A block having a mass of 1000 kg rests on a level floor. The coefficient of static friction between the block and the floor is 0.30. Calculate the force required to cause motion to impend if the force applied is (a) horizontal, (b) upward at an angle of 10° with the horizontal.

25. Work Problem 36, changing the block to one having a mass of 20 kg.

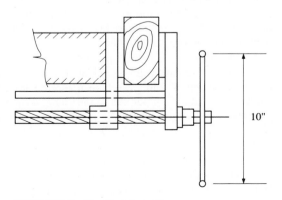

FIGURE 6–42 Problem 21.

26. The tool locker of Problem 23 is 0.8 m by 0.8 m square (in plan) and 2 m high. Determine the height above which a horizontal force would have to be applied so that the locker would tip, rather than slide.

27. A machine having a mass of 500 kg is to be raised using a wedge, as shown in Fig. 6–44. The coefficient of friction on all surfaces is 0.25. Determine the force P required to raise the machine.

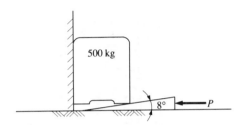

FIGURE 6–44 Problem 27.

28. A sailor wraps a hawser around a bollard (a vertical post) to secure a drifting ship to a pier. The sailor can apply a force of 400 N to the free end of the hawser. With what force can the ship pull on the hawser before another sailor will have to be called? Consider (a) one wrap, (b) two wraps, (c) three wraps. Assume a coefficient of friction of 0.35.

29. A V-belt with a groove angle of 60° drives a 0.37 m diameter pulley on a small concrete mixer. The contact angle is 210° and the coefficient of friction is 0.33. The mixer requires a torque of 50 N·m. Determine the required belt tensions.

Computer Problems

For the following computer problems, any appropriate programming language may be used. Input prompts should fully explain what is required of the user (the program should be "user friendly"). The resulting output should be well labeled and self-explanatory.

30. Write a program that will solve Problem 40. User input is to be the angle of the inclined plane with the horizontal (not to exceed 90°), the weight of the block, and the coefficient of static friction.

31. A weight W is to be held aloft by wrapping a rope around a horizontal stationary pole. The coefficient of

static friction is 0.33. A force of 20 lb will be applied to the free end of the rope. Write a program to generate a tabulation of weights that could be so supported as a function of the number of wraps around the pole, ranging from 1 wrap to 3 wraps, in $\frac{1}{4}$-wrap steps.

32. Write a program that will solve Problem 42. User input is to be width and weight of the crate. The output should state whether sliding or tipping governs the solution.

Supplemental Problems

33. A horizontal force of 18 lb is required to just start a 45 lb block in motion on a horizontal surface. Determine the coefficient of static friction between the block and the supporting surface.

34. A 90 lb block lying on a rough horizontal surface is subjected to a horizontal force P. The coefficient of static friction is 0.33. Calculate the maximum value that P can have before motion impends.

35. A one-ton weight rests on a horizontal floor. The coefficient of static friction between the weight and the floor is 0.35. Calculate the force required to cause motion to impend if the force applied to the weight is (a) horizontal and (b) downward at an angle of 15° with the horizontal.

36. A 50 lb block rests on a rough inclined plane. If the coefficient of friction between the block and the plane is 0.50, calculate the inclination of the plane with the horizontal when motion impends.

37. If in Problem 36 the plane has an inclination with the horizontal of 20°, calculate (a) the frictional resistance when the block is at rest, (b) the force parallel to the plane necessary for motion to impend up the plane, and (c) the force parallel to the plane necessary to prevent motion down the plane.

38. A 325 lb block rests on a plane inclined 25° with the horizontal. Calculate the coefficient of static friction between the block and the plane if motion impends when a force of 225 lb, parallel to the plane, is applied. The impending motion is up the plane.

39. A 47 lb body is supported on a plane inclined 33° to the horizontal, as shown in Fig. 6–45. The coefficient of static friction between the body and the plane is 0.15. The body is subjected to a 29 lb force, acting as shown. Determine whether the body will move up the plane, down the plane, or remain at rest.

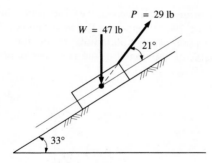

FIGURE 6–45   Problem 39.

40. Compute the horizontal force P required to prevent the 900 lb block in Fig. 6–46 from sliding down the plane. The coefficient of friction between the block and the plane is 0.20.

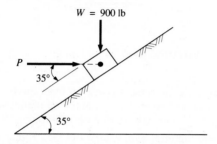

FIGURE 6–46   Problem 40.

41. Block A in Fig. 6–47 weighs 200 lb and block B weighs 400 lb. The coefficient of static friction for all contact surfaces is 0.35. Block A is anchored to the wall with a flexible cable. Calculate the force P required for motion to impend for block B.

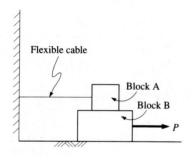

FIGURE 6–47   Problem 41.

42. Calculate the magnitude of the horizontal force P, acting as shown in Fig. 6–48, that will cause motion to impend for the 450 lb crate. (The motion may be either sliding or tipping.) Assume the coefficient of static friction to be 0.40.

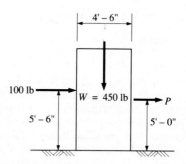

FIGURE 6–48   Problem 42.

43. In Fig. 6–49, the coefficient of static friction between the 160 lb crate and the floor is 0.35. Calculate the value of the force P and the distance h that will cause the crate to tip and slide simultaneously.

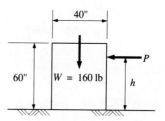

FIGURE 6–49   Problem 43.

44. A 500 lb block rests on a horizontal surface, as shown in Fig. 6–50. The coefficient of static friction is 0.25. Calculate the maximum value of the horizontal force P so that neither sliding nor tipping will occur. Assume that P is gradually applied.

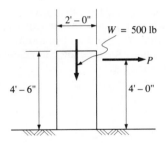

FIGURE 6–50   Problem 44.

45. The ladder in Fig. 6–51 is supported by a horizontal floor and a vertical wall. It is 12 ft long, weighs 30 lb (assumed to be concentrated at its midlength), and supports a person weighing 175 lb at point D. The vertical wall is smooth and offers no frictional resistance. The coefficient of static friction at point A is 0.30. Calculate the minimum angle θ at which the ladder will stand without slipping to the left.

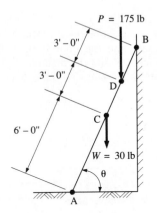

FIGURE 6–51   Problem 45.

46. A 16 ft ladder weighing 62 lb (assumed concentrated at its midlength) rests on a horizontal floor and is supported by a vertical wall, as shown in Fig. 6–52. The lower end is prevented from slipping by friction and by

a rope attached to the base of the wall. The ladder supports a 200 lb person at the top. The coefficient of static friction at the wall is 0.25; at the floor it is 0.30. Calculate the tension P in the rope necessary to prevent slipping.

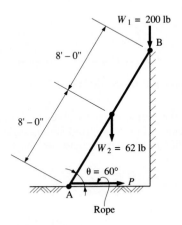

FIGURE 6–52   Problem 46.

47. Assume that the rope is removed from the base of the ladder in Problem 46. Calculate the minimum angle θ at which the ladder will stand without slipping to the left.

48. Compute the minimum weight of block B that will prevent the two-block system of Fig. 6–53 from sliding. Block A weighs 100 lb and the coefficient of static friction for the contact surfaces is 0.20. Assume the pulley to be frictionless.

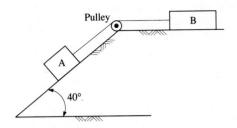

FIGURE 6–53   Problem 48.

49. Compute the minimum weight of block A in Fig. 6–54 in order for motion to be impending down the plane. Assume the pulley to be frictionless. Block B weighs 500 lb. The coefficient of static friction for all contact surfaces is 0.20.

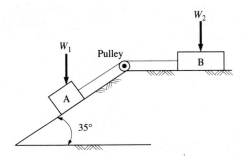

FIGURE 6–54   Problem 49.

50. The blocks shown in Fig. 6–55 are separated by a solid strut attached to the blocks with frictionless pins. The coefficient of static friction for all surfaces is 0.325. Determine the value of the horizontal force *P* that will cause motion to impend to the right. Neglect the weight of the strut.

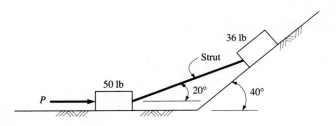

FIGURE 6–55   Problem 50.

51. For the block-and-wedge arrangement of Fig. 6–56, calculate the value of the horizontal force *P* required to

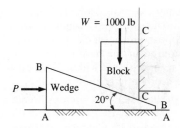

FIGURE 6–56   Problem 51.

start the wedge in motion and raise the 1000 lb block. The angle of the wedge is 20°. The coefficients of static friction at contact surfaces A–A, B–B, and C–C are 0.19, 0.25, and 0.29, respectively.

52. Rework Problem 9 assuming the coefficient of static friction to be 0.364 on all contact surfaces.

53. When a large rope is wrapped twice around a post, a pull of 100 lb at the free end will withstand a pull of 2 tons at the other end. What pull would the 100 lb withstand if the rope were wrapped only once around the post?

54. A band brake is in contact with drum C through an angle of 180° and is connected to the horizontal lever at A and B, as shown in Fig. 6–57. The drum diameter is 16 in. The coefficient of kinetic friction between the brake band and the drum is 0.40. Force *P* is 90 lb. Determine the tensions at A and B in the band brake if the drum is rotating (a) clockwise, (b) counterclockwise.

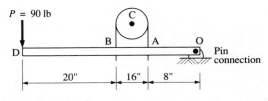

FIGURE 6–57   Problem 54.

55. A ship may exert an estimated pull of 8000 lb on its hawser (a heavy rope), which is wrapped around a mooring post on the dock. Determine the number of turns required of the hawser around the post so that the pull at the free end will not exceed 40 lb. Assume a coefficient of static friction of 0.35.

56. The mean diameter of a square-threaded jackscrew is 3.00 in., the pitch is 0.40 in., and the coefficient of static friction is 0.15. Calculate the maximum load that can be raised by a force of 75 lb applied at the end of a 14 in. jack handle.

57. The manually operated apple cider press shown in Fig. 6–58 has a 36 in. handle. The square-threaded screw has a mean diameter of 1.700 in. The screw must be turned through $2\frac{1}{2}$ turns to advance the head of the press 1.00 in. The coefficient of static friction is estimated to be 0.13. (a) Find the force on the stack if a force of 60 lb is applied at each end of the handle. (b) How much is the resulting force decreased if the coefficient of static friction goes to 0.18 due to lack of lubrication on the screw?

FIGURE 6–58   Problem 57.

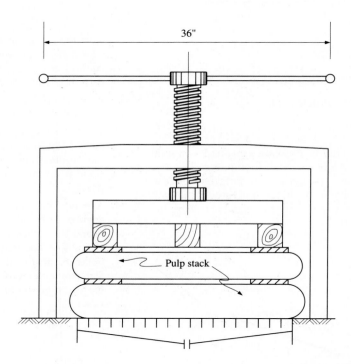

7 Centroids and Centers of Gravity

7–1 INTRODUCTION

All bodies may be considered to be composed of a multitude of small particles, each being acted upon by a gravitational force. When algebraically added, these forces exerted on the particles of a body represent the weight of the body. For all practical purposes, these forces are assumed to be parallel and to act vertically downward. Hence, the force system may be categorized as a parallel force system, with the algebraic sum (which is the resultant of the system) being called the *weight* of the body. The resultant of these individual gravity forces will always act through a definite point, regardless of how the body is oriented. This point is called the *center of gravity*.

Weight is a force, and it may be treated and represented as a vector. Therefore, it must have magnitude, direction, sense, and a point of application, all of which describe a force vector. Since the direction and sense of the force of gravity are always known, only the magnitude and the point of application must be determined. The magnitude and location of this resultant can be determined experimentally; however, for the purpose of analysis and/or design, we will limit our discussion to an analytical determination of both magnitude and location. In effect, the problem of locating the center of gravity of a body becomes one of determining the point through which the resultant weight of the body acts.

7–2 CENTER OF GRAVITY

The procedure for determining the magnitude and location of the center of gravity is the same as that for determining the magnitude and location of the resultant force of a parallel force system, as described in Chapter 3. As an example, consider the irregularly shaped flat plate of uniform thickness and homogeneous material shown in Fig. 7–1. The plate is divided into infinitesimal elements, the typical element being located a distance x from the reference Y–Y axis and a distance y from the reference X–X axis. The weight w of each element may be thought of as being concentrated at its center. The weights of the elements form a parallel force system, the resultant of which is the total weight W of the plate. The magnitude of the total weight can be written mathematically as

$$W = \Sigma w$$

The weight W (the weight of the plate), by definition, acts through the center of gravity of the plate. The coordinates of the center of gravity will be designated $\bar{x}$ and $\bar{y}$. To determine the location of W, and thus to determine

FIGURE 7–1   Center of gravity.

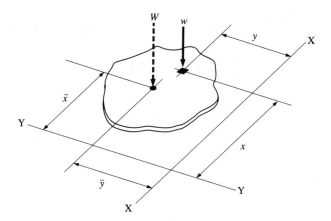

the location of the center of gravity, moments of the weights of the individual elements are taken with respect to each of the axes shown. Using Varignon's theorem that the moment of the resultant about any point or axis must equal the algebraic sum of the moments of the individual weights about the same point or axis, the following expressions can be established to locate the resultant:

$$W\bar{x} = \Sigma wx$$
$$W\bar{y} = \Sigma wy$$

Solving for the center of gravity locations,

$$\bar{x} = \frac{\Sigma wx}{W} \quad \text{or} \quad \frac{\Sigma wx}{\Sigma w} \qquad \textbf{(7–1)}$$

$$\bar{y} = \frac{\Sigma wy}{W} \quad \text{or} \quad \frac{\Sigma wy}{\Sigma w} \qquad \textbf{(7–2)}$$

It may be readily apparent, and can easily be shown, that if the plate has an axis of symmetry, then the center of gravity will lie somewhere on that axis of symmetry. If the plate has two mutually perpendicular axes of symmetry (in the case, for example, of a rectangular or a circular plate), then the center of gravity will lie at the intersection of the axes of symmetry.

Note that our discussion in this section applies strictly to bodies that have mass and, therefore, weight. Frequently, however, the center of gravity of an area is desired. (This may be thought of as the plate in Fig. 7–1 having a zero thickness.) Since an area does not have mass, it cannot have weight or, theoretically, a center of gravity. However, the point in the area that would be analogous to the center of gravity of a body having mass is commonly called the *centroid* of the area. (It is not uncommon in the applied engineering fields to use *centroid* and *center of gravity* interchangeably.)

☐ **EXAMPLE 7-1**   A 10 in. diameter steel sphere is anchored firmly to the top of a 12 in. square concrete pedestal. The pedestal is 18 in. high. The two bodies are considered to be a single unit. The resulting member, shown in Fig. 7-2, may be called a *built-up member*. Locate the center of gravity of the member.

Solution   The unit is symmetrical with respect to the Y–Y axis. Therefore, the center of gravity will lie on the Y–Y axis, which is the vertical axis of the unit. It is necessary to compute only $\bar{y}$, which in this problem represents the distance from the bottom of the concrete pedestal to the center of gravity of the member.

The unit weights of the materials can be obtained from Appendix G. Denote the concrete pedestal as w_1 and the steel sphere as w_2. The weight of each component part is calculated as the product of the volume (in cubic ft) and the unit weight (in pcf) as follows:

$$w_1 = 12(12)(18)\left(\frac{1}{1728}\right)(150) = 225 \text{ lb}$$

$$w_2 = \left(\frac{4}{3}\pi R^3\right)(490) = \frac{4}{3}\pi 5^3\left(\frac{1}{1728}\right)(490) = 148.5 \text{ lb}$$

The total weight is then

$$W = w_1 + w_2 = 225 + 148.5 = 373.5 \text{ lb}$$

Using the bottom of the concrete pedestal as the reference axis (shown as axis X–X in Fig. 7-3), $\bar{y}$ is calculated from Eq. (7-2). Note that the Z–Z axis shown in Fig. 7-2 could also serve as a reference axis. The y distances, from the reference axis to the centers of gravity of each component, are shown in Fig. 7-3.

$$\bar{y} = \frac{\Sigma wy}{W} = \frac{w_1y_1 + w_2y_2}{W}$$

$$= \frac{225(9) + 148.5(23)}{373.5} = 14.57 \text{ in.}$$

FIGURE 7-2   Two-body unit.

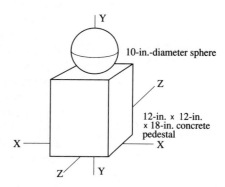

10-in.-diameter sphere

12-in. × 12-in. × 18-in. concrete pedestal

FIGURE 7–3 Location of center of gravity.

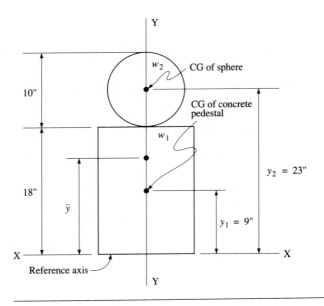

1. Sketch the member showing all dimensions. Note any axes of symmetry.
2. Divide up the member into component parts. Each part must be proportioned so that its weight can be determined and its center of gravity can be located.
3. Select a reference axis.
4. Apply Eq. (7–1) and/or Eq. (7–2) to determine $\bar{x}$ and/or $\bar{y}$. (A similar equation may be written for $\bar{z}$ if necessary.)

7–3 CENTROIDS AND CENTROIDAL AXES

If we assume that the irregularly shaped flat plate shown in Fig. 7–1 is homogeneous and has a uniform thickness, the weight of the plate would be directly proportional to the area. Thus, we can use areas instead of weights (forces) in the equations of Section 7–2 to determine the location of the centroid of the area. This is equivalent to allowing the thickness of the plate to approach zero and finding its center of gravity.

The procedure for finding the centroid of an area is exactly the same as the procedure described in Section 7–2 for finding the center of gravity, except for the following substitutions: a replaces w; A replaces W. The term a represents an infinitesimal component of area and A represents Σa (or, the total area).

Applying Varignon's theorem, the moment of the total area about any axis will be equal to the algebraic sum of the moments of the component areas about the same axis. Note that the moment of an area is analogous to the moment of a force, except that the moment of a force has physical meaning, whereas the moment of an area is a mathematical concept. Moment of area has units of length cubed (in.^{3}, mm^3). Referring to Fig. 7–1, and

TABLE 7–1 Areas and centroids of areas.

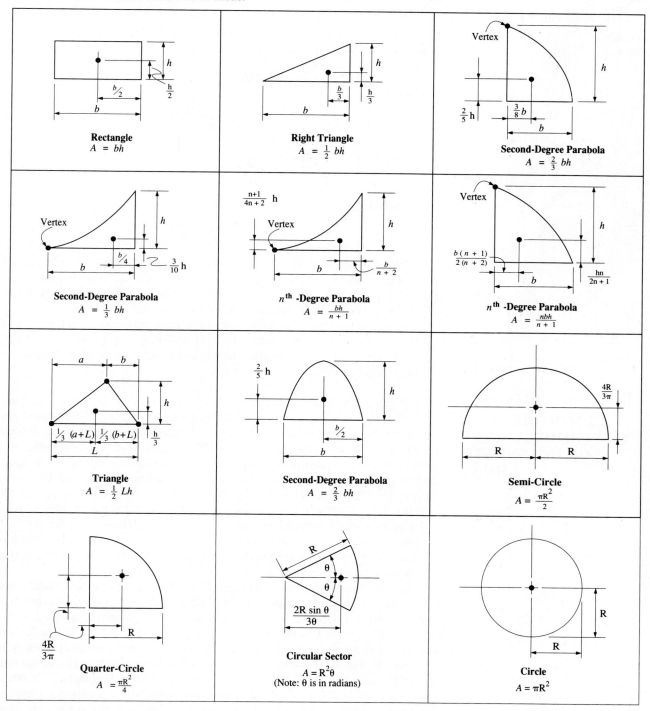

Rectangle
$A = bh$

Right Triangle
$A = \frac{1}{2} bh$

Second-Degree Parabola
$A = \frac{2}{3} bh$

Second-Degree Parabola
$A = \frac{1}{3} bh$

n^{th} **-Degree Parabola**
$A = \frac{bh}{n+1}$

n^{th} **-Degree Parabola**
$A = \frac{nbh}{n+1}$

Triangle
$A = \frac{1}{2} Lh$

Second-Degree Parabola
$A = \frac{2}{3} bh$

Semi-Circle
$A = \frac{\pi R^2}{2}$

Quarter-Circle
$A = \frac{\pi R^2}{4}$

Circular Sector
$A = R^2\theta$
(Note: θ is in radians)

Circle
$A = \pi R^2$

making the substitutions just mentioned, equations for $\bar{x}$ and $\bar{y}$ may be written as follows: With respect to the Y-Y axis,

$$A\bar{x} = \Sigma ax$$

from which

$$\bar{x} = \frac{\Sigma ax}{A} \quad \text{or} \quad \frac{\Sigma ax}{\Sigma a} \qquad (7\text{--}3)$$

With respect to the X–X axis,

$$A\bar{y} = \Sigma ay$$

from which

$$\bar{y} = \frac{\Sigma ay}{A} \quad \text{or} \quad \frac{\Sigma ay}{\Sigma a} \qquad (7\text{--}4)$$

The terms $\bar{x}$ and $\bar{y}$ represent coordinates of the centroid of an area. An axis that passes through the centroid is generally termed a *centroidal axis*. A centroidal axis, then, of great significance in statics and strength of materials, is an axis on which the centroid of the area lies.

Areas and centroids for some frequently encountered geometric shapes have been determined mathematically and are shown in Table 7–1. Examples for locating the centroids of a triangle and a semicircle using integration (calculus) are provided in Appendix K.

7–4 CENTROIDS AND CENTROIDAL AXES OF COMPOSITE AREAS

A composite, or built-up, area may be described as one made up of a number of simple geometric areas or standardized shapes. Many of the members used in structures and machines are combinations of various shapes so connected (usually by welding) as to act as a single unit. To determine the location of the centroid of a composite area, the area is generally divided into two or more component areas, each with a known centroid location. The centroid location for the composite area can then be determined by applying Eqs. (7–3) and (7–4). The centroidal axes are two axes, at right angles to each other, which intersect at the centroid. These axes are utilized in many applications. The centroidal axes of importance normally coincide with axes of symmetry or are parallel/perpendicular to major elements of the composite area.

In determining the location ($\bar{x}$ and $\bar{y}$) of the centroid or centroidal axes of a composite area, reference axes must be established. It is customary that a conventional reference X–Y coordinate axes system be utilized. One technique is to establish the axes in such a manner that the entire composite area lies in the upper right quadrant (the first quadrant) of the coordinate axes system. The lowest edge of the area should lie on the X axis and the left edge of the area should lie on the Y axis (see Fig. 7–5 in Example 7–2). This technique avoids any need for a sign convention, since x distances and y distances measured from the reference axes are upward and to the right and therefore considered positive values.

After establishing the reference axes, the composite area should be divided into simple geometric areas, such as rectangles, triangles, or standard shapes. Moments of these component areas are then taken with respect to each reference axis.

If a hole is cut out of the area, it must be treated as a *negative* area. The negative sign will then remove the effect of that area (which is absent) in the summation of Σax or Σay and Σa. Similarly, if an area is added to the composite area, a positive sign in the summations will include the effect of that area.

☐ **EXAMPLE 7–2**   Determine the location of the centroid of the area shown in Fig. 7–4.

FIGURE 7–4   Composite area.

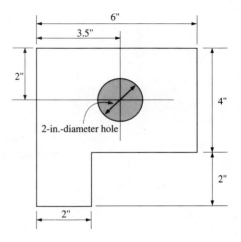

Solution   First establish a reference X–Y coordinate axes system as shown in Fig. 7–5. The entire composite area is placed in the first quadrant and then divided into three component areas as shown. The areas are then determined:

$$a_1 \text{ (rectangle)} = 6(4) = 24 \text{ in.}^2$$
$$a_2 \text{ (rectangle)} = 2(2) = 4 \text{ in.}^2$$
$$a_3 \text{ (circle)} = 0.7854(2)^2 = -3.14 \text{ in.}^2$$

Note that the area of the circle carries a negative sign since it represents a cutout from the composite area and, in effect, reduces the total area. The total area is calculated:

$$A = a_1 + a_2 + a_3$$
$$= 24 + 4 + (-3.14) = 24.86 \text{ in.}^2$$

The component areas are all simple geometric shapes with the location of the centroid of each area known. These locations are shown in Fig. 7–5 with respect to the reference axes.

FIGURE 7–5   Location of centroidal axes.

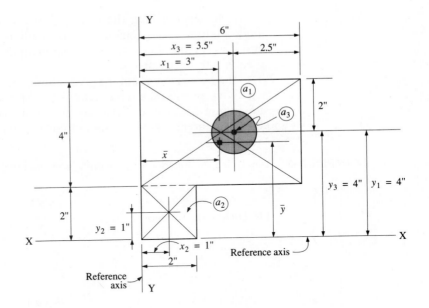

Next, Eq. (7–3) is used to determine $\bar{x}$. It may be helpful to visualize this process as one of pivoting the component areas about reference axis Y–Y.

$$\bar{x} = \frac{\sum ax}{A} = \frac{a_1 x_1 + a_2 x_2 + a_3 x_3}{A}$$

$$= \frac{24(3) + 4(1) + (-3.14)(3.5)}{24.86} = 2.62 \text{ in.}$$

Equation (7–4) yields

$$\bar{y} = \frac{\sum ay}{A} = \frac{a_1 y_1 + a_2 y_2 + a_3 y_3}{A}$$

$$= \frac{24(4) + 4(1) + (-3.14)(4)}{24.86} = 3.52 \text{ in.}$$

Table 7–2 shows a tabular format that can be used.

TABLE 7–2   Tabular format for Example 7–2.

Component Area	a (in.²)	x (in.)	ax (in.³)	y (in.)	ay (in.³)
a_1	24.0	3.0	72.0	4.0	96.0
a_2	4.	1.0	4.0	1.0	4.0
a_3	−3.14	3.5	−10.99	4.0	−12.56
$\sum$	24.86		65.01		87.44

Therefore, substituting values from the table,

$$\bar{x} = \frac{\Sigma ax}{\Sigma a} = \frac{65.01}{24.86} = 2.62 \text{ in.}$$

and

$$\bar{y} = \frac{\Sigma ay}{\Sigma a} = \frac{87.44}{24.86} = 3.52 \text{ in.}$$

Thus, the centroid of the area has been located. Note that as shown in Fig. 7–5 the centroid actually falls within the cut-out circle. Other common cross sections in which this would occur are such standard shapes as pipes, tubes, angles, and channels.

If we were to superimpose an X–Y coordinate axes system on the area so that the origin coincided with the centroid of the area, we would establish the X–Y centroidal axes. (Recall that the centroidal axes intersect at the centroid of a cross section.) These axes would normally be parallel to their respective X–X and Y–Y reference axes.

☐ **EXAMPLE 7–3**   A built-up steel member, shown in Fig. 7–6, is fabricated from two C15 × 40 American Standard channels, a 16 in. by 1 in. top plate, and a 14 in. by ½ in. bottom plate. All of the components are welded together securely so as to act as a single unit. Locate the X–X centroidal axis.

Solution   In this problem, the member is symmetrical with respect to the vertical Y–Y axis shown. Recall that an axis of symmetry is always a centroidal axis. Only $\bar{y}$, locating the X–X centroidal axis, must be determined. The lower edge of the member (the bottom of the bottom plate), is chosen as the reference X–X axis.

FIGURE 7–6   Built-up steel member.

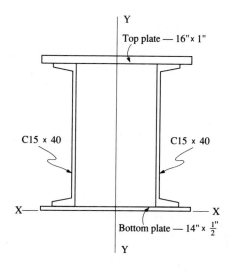

FIGURE 7–7   Location of centroidal axis.

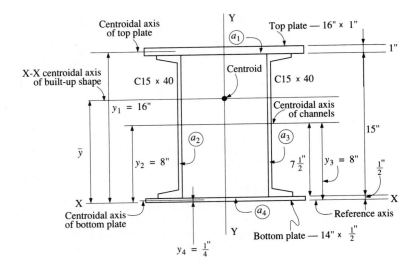

The component areas are shown in Fig. 7–7. Areas are determined as follows:

$$a_1 = 16(1) = 16 \text{ in.}^2$$
$$\left.\begin{array}{l} a_2 = 11.8 \text{ in.}^2 \\ a_3 = 11.8 \text{ in.}^2 \end{array}\right\} \text{ (see Appendix C)}$$
$$a_4 = 14(0.5) = 7 \text{ in.}^2$$
$$A = \Sigma a = 46.6 \text{ in.}^2$$

Note that the components include two rectangular plates and two standard channels. All necessary properties and dimensions of the channels may be found in Appendix C. The locations of the centroids of the components (the y dimensions) are calculated as follows and are shown in Fig. 7–7:

$$y_1 = 0.5 + 15 + 0.5 = 16 \text{ in.}$$
$$y_2 = y_3 = 7.5 + 0.5 = 8 \text{ in.}$$
$$y_4 = 0.25 \text{ in.}$$

Applying Eq. (7–4),

$$\bar{y} = \frac{\Sigma ay}{A} = \frac{a_1 y_1 + a_2 y_2 + a_3 y_3 + a_4 y_4}{A}$$
$$= \frac{16(16) + 11.8(8) + 11.8(8) + 7(0.25)}{46.6} = 9.58 \text{ in.}$$

Therefore, the X–X centroidal axis lies 9.58 in. above the reference axis.

Table 7–3 shows a tabular format can be used as well.

TABLE 7–3 Tabular format for Example 7–3.

Component Area	a (in.^{2})	y (in.)	ay (in.^{3})	Notes
a_1	16.0	16.0	256.0	Top plate
a_2	11.8	8.0	94.4	Channel
a_3	11.8	8.0	94.4	Channel
a_4	7.0	0.25	1.75	Bottom plate
Σ	46.6		446.6	

Therefore, substituting values from Table 7–3,

$$\bar{y} = \frac{\Sigma ay}{\Sigma a} = \frac{446.6}{46.6} = 9.58 \text{ in.}$$

Computer spreadsheeting is a tool that can be readily adapted to these calculations, following the concept of the tabular format. The power of the spreadsheet for this type of application is in the ability to see very quickly the effects of changes made to the original cross section. There are many packages available that would serve nicely, such as Lotus 1-2-3.[1] The tabular approach will be further extended for the calculations of Chapter 8.

□ **EXAMPLE 7–4**   Determine the location of the X–X and Y–Y centroidal axes for the area shown in Fig. 7–8.

Solution   First establish, a reference X–Y coordinate axes system and place the entire area in the first quadrant, as shown in Fig. 7–9. In dividing the area into simple geometric component areas, assume a rectangular area (a_1) 18 in. by 24 in. in size, and then remove a cut-out triangular area (a_2) and semicircular area (a_3). Areas a_2 and a_3 will be negative areas. There are several other equally acceptable ways in which the area could be divided up.

The area calculations are as follows:

$$a_1 = 18(24) = 432 \text{ in.}^2$$
$$a_2 = 0.5(8)(24) = -96 \text{ in.}^2$$
$$a_3 = 0.5(0.7854)(6)^2 = -14.1 \text{ in.}^2$$
$$A = \Sigma a = 321.9 \text{ in.}^2$$

[1]   Lotus 1-2-3 is a product of The Lotus Development Corporation, 55 Cambridge Parkway, Cambridge, MA 02142.

FIGURE 7–8   Composite area.

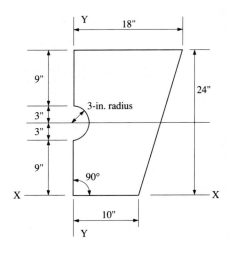

FIGURE 7–9   Location of centroidal axes.

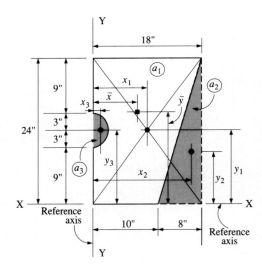

Next compute $\bar{x}$ with respect to the reference Y–Y axis where the x dimensions for each component area are determined as follows (refer to Table 7–1 for properties of the triangle and semicircle):

$$x_1 = 9 \text{ in.}$$

$$x_2 = 10 + \left(\frac{2}{3}\right)(8) = 15.33 \text{ in.}$$

$$x_3 = \frac{4R}{3\pi} = \frac{4(3)}{3\pi} = 1.27 \text{ in.}$$

Equation (7–3) yields

$$\bar{x} = \frac{\sum ax}{A} = \frac{a_1 x_1 + a_2 x_2 + a_3 x_3}{A}$$

$$= \frac{432(9) + (-96)(15.33) + (-14.1)(1.27)}{321.9} = 7.45 \text{ in.}$$

Next compute $\bar{y}$ with respect to the reference X–X axis where the y dimensions for each component area, shown in Fig. 7–9, are determined as follows:

$$y_1 = y_3 = 12 \text{ in.}$$

$$y_2 = \left(\frac{1}{3}\right)(24) = 8 \text{ in.}$$

Equation (7–4) yields

$$\bar{y} = \frac{\sum ay}{A} = \frac{a_1 y_1 + a_2 y_2 + a_3 y_3}{A}$$

$$= \frac{432(12) + (-96)(8) + (-14.1)(12)}{321.9} = 13.2 \text{ in.}$$

A tabular format (not shown) could also be used.

Summary of Procedure— Centroids of Areas

1. Sketch the area showing all known dimensions.
2. Note any axis of symmetry. Establish a reference X–Y coordinate axes system.
3. Divide the area up into component areas. Each area must be proportioned so that its area and the location of its centroid can be determined.
4. Apply Eq. (7–3) and/or Eq. (7–4).

**7–5
SI SYSTEM
EXAMPLES**

□ **EXAMPLE 7–5**

The performance of an aircraft is affected by the positioning of the cargo load it carries. The location of the center of gravity is critical. Assume that the load shown in Fig. 7–10(a) is composed of five crates (numbered 1 through 5) with masses, in kilograms, as follows: 500, 350, 200, 800, and 700, respectively. Assume that the center of gravity of each individual crate is at the center of that crate. Locate the center of gravity of the load with reference to point A.

Solution

The weight of each crate could be determined; however, this is unnecessary since weights will be directly proportional to the known masses. The given masses will be used. The resultant mass R will be

$$R = \sum m = 500 + 350 + 200 + 800 + 700 = 2550 \text{ kg}$$

R is shown in Fig. 7–10(b). The location of the resultant (which passes through the center of gravity) is calculated from Varignon's theorem,

$$R\bar{x} = \sum mx$$

FIGURE 7–10   Center of gravity.

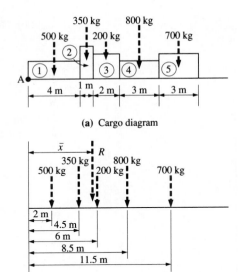

(a) Cargo diagram

(b) Load diagram

where $\bar{x}$ is the distance from point A to the resultant (or to the center of gravity):

$$\bar{x} = \frac{\Sigma\, mx}{R} = \frac{500(2) + 350(4.5) + 200(6) + 800(8.5) + 700(11.5)}{2550}$$

$$= 7.30 \text{ m}$$

□ **EXAMPLE 7–6**   The built-up structural steel member in Fig. 7–11 is fabricated from one C230 × 0.219 American Standard channel and one L127 × 127 × 22.2 angle. The two shapes are welded together securely so as to act as a single unit. Locate the centroid.

Solution   Appendices C and D furnish the properties (dimensions and areas) necessary for this solution. The required dimensions are shown in Fig. 7–11. The area for the angle is 5.15×10^{-3} m² and the area for the channel is 2.85×10^{-3} m². The approximate position of the centroid is indicated with the $\bar{x}$ and $\bar{y}$ dimensions. Using Eq. (7–3), calculate $\bar{x}$:

$$\bar{x} = \frac{\Sigma\, ax}{\Sigma\, a} = \frac{(2.85 \times 10^{-3} \text{ m}^2)(48.2 \text{ mm}) + (5.15 \times 10^{-3} \text{ m}^2)(103.0 \text{ mm})}{(2.85 \times 10^{-3} \text{ m}^2) + (5.15 \times 10^{-3} \text{ m}^2)}$$

$$= 83.5 \text{ mm}$$

And using Eq. (7–4), calculate $\bar{y}$:

$$\bar{y} = \frac{\Sigma\, ay}{\Sigma\, a} = \frac{(2.85 \times 10^{-3} \text{ m}^2)(114.3 \text{ mm}) + (5.15 \times 10^{-3} \text{ m}^2)(87.1 \text{ mm})}{(2.85 \times 10^{-3} \text{ m}^2) + (5.15 \times 10^{-3} \text{ m}^2)}$$

$$= 96.8 \text{ mm}$$

A tabular format (not shown) could also be used.

FIGURE 7–11 Built-up member.

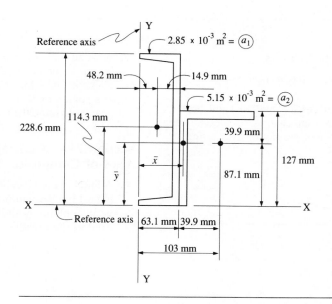

SUMMARY—BY SECTION NUMBER

7–1 The center of gravity of a body is the point through which the line of action of its total weight passes.

7–2 The center of gravity of a body is located in the same way the resultant force of a parallel force system is located. The location may be determined using

$$\overline{x} = \frac{\Sigma wx}{W} \quad \text{or} \quad \frac{\Sigma wx}{\Sigma w} \tag{7–1}$$

$$\overline{y} = \frac{\Sigma wy}{W} \quad \text{or} \quad \frac{\Sigma wy}{\Sigma w} \tag{7–2}$$

The term *centroid* is used when referring to the center of gravity of an area, which may be thought of as a plate having zero thickness.

7–3 The location of a centroid may be determined using

$$\overline{x} = \frac{\Sigma ax}{A} \quad \text{or} \quad \frac{\Sigma ax}{\Sigma a} \tag{7–3}$$

$$\overline{y} = \frac{\Sigma ay}{A} \quad \text{or} \quad \frac{\Sigma ay}{\Sigma a} \tag{7–4}$$

7–4 A composite area is made up of a number of simple geometric areas or standardized shapes. The location of the centroid of the composite area may be determined by using Eqs. (7–3) and (7–4).

PROBLEMS

Section 7–2   Center of Gravity

For the following problems, refer to Appendix G for information on unit weights.

1. A cylindrical cast-iron casting has an axial hole extending partway through the casting, as shown in Fig. 7–12. Locate the center of gravity of the casting.

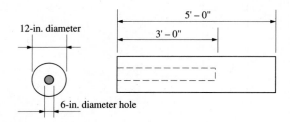

FIGURE 7–12   Problem 1.

2. Locate the center of gravity of the cast-iron casting of Problem 1 assuming the hole is filled with a magnesium alloy.

3. A 6 in. diameter steel sphere is rigidly attached to the end of a 24 in. long 1½ in. diameter aluminum rod. Locate the center of gravity of the composite member.

4. A solid steel shaft is fabricated as shown in Fig. 7–13. Locate the center of gravity with respect to the left end.

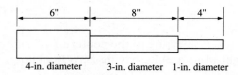

FIGURE 7–13   Problem 4.

5. A wood mallet has a cylindrical head 6 in. long and 3 in. in diameter. If the cylindrical handle is 1 in. in diameter, how long must it be for the mallet to balance at a point 6 in. from where the handle enters the mallet head? (Assume the unit weight of the wood to be 40 lb/ft³.)

Section 7–4   Centroids and Centroidal Axes of Composite Areas

6. A lintel in the form of an inverted tee is made by welding a 12 in. by ½ in. horizontal plate to a 10 in. by ½ in. vertical plate, as shown in Fig. 7–14. Locate the centroid of the area.

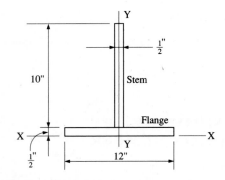

FIGURE 7–14   Problem 6.

7. Locate the X–X and Y–Y centroidal axes for the areas shown in Fig. 7–15.

8. A built-up steel member is composed of a W18 × 50 wide flange section with a 12-in. by ½-in. plate welded to its top flange, as shown in Fig. 7–16. Locate the X–X centroidal axis.

9. Locate the X–X and Y–Y centroidal axes for the areas shown in Fig. 7–17.

FIGURE 7–15 Problem 7.

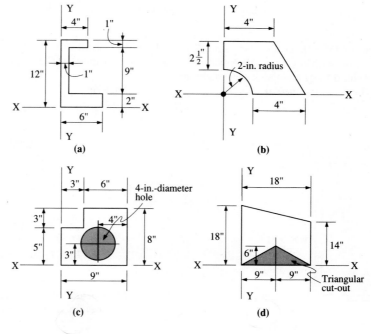

(a)

(b)

(c)

(d)

FIGURE 7–16 Problem 8.

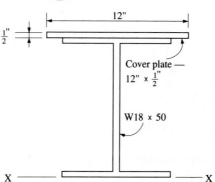

FIGURE 7–17 Problem 9.

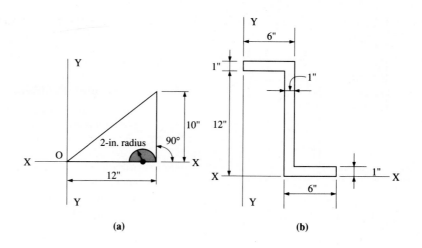

(a)

(b)

10. A built-up steel member is composed of a W21 × 62 wide flange section with a C12 × 25 American Standard channel welded to its top flange, as shown in Fig. 7–18. Locate the X–X centroidal axis.

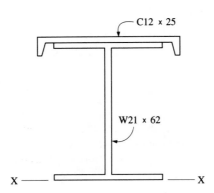

FIGURE 7–18   Problem 10.

SI System Problems

11. Find the center of gravity for a three-axle truck. The front axle supports a mass of 3.5 Mg, the middle axle supports a mass of 12.0 Mg, and the rear axle supports a mass of 14.0 Mg. The middle axle is 4.5 m from the front axle and 10 m from the rear axle. Locate the center of gravity with respect to the front axle.

12. Locate the X–X centroidal axis for the cross section shown in Fig. 7–19.

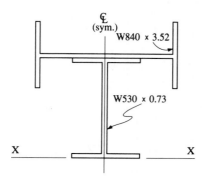

FIGURE 7–19   Problem 12.

13. Locate the X–X centroidal axis for the cross section shown in Fig. 7–20.

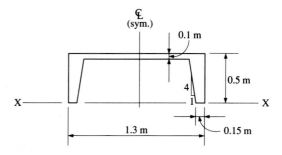

FIGURE 7–20   Problem 13.

14. Locate the X–X and Y–Y centroidal axes for the cross section shown in Fig. 7–21.

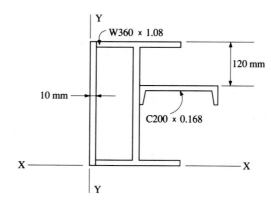

FIGURE 7–21   Problem 14.

Computer Problems

For the following computer problems, any appropriate programming language may be used. Input prompts should fully explain what is required of the user (the program should be "user friendly"). The resulting output should be well labeled and self-explanatory. For spreadsheet problems, any appropriate software may be used.

15. Write a program that will calculate the location of the centroid for an inverted tee shape similar to that of

Problem 6. User input is to be the width and thickness of each of the plates (stem and flange).

16. Write a program that will calculate the location of the centroid of a semicircle (measured from the diameter line). This may be accomplished by dividing the area into slices of uniform width, either parallel or perpendicular to the diameter, and applying Eq. (7–4). User input is to be the radius and width of the slice. Check the result with that obtained from Table 7–1. Note how the accuracy varies with the choice of the slice width.

17. Utilize a spreadsheet program to solve for the centroid location of built-up areas such as shown in Problems 8 and 24. The user should be able to input the appropriate geometric properties for any shapes of the types shown and the program should then determine the centroid location and display the result. Vary a single shape or dimension and observe how the location of the centroid is affected.

Supplemental Problems

For the following problems, where unit weights are not stated refer to Appendix G.

18. A 2 in. diameter hole, 5 in. long, is drilled into the center of the top face of a steel cube that is 6 in. on each side. The hole is filled with lead, which has a unit weight of 710 pcf. Locate the center of gravity of the cube with respect to the bottom.

19. Locate the center of gravity of the cube in Problem 18 assuming the hole is left empty.

20. The head of a maul is made of wrought iron and is 3 in. in diameter and 5 in. long. It has a wood handle 1 in. in diameter and 3 ft long. The unit weight of the wood is 80 pcf. Calculate the distance from the end of the handle to the center of gravity.

21. Locate the X–X and Y–Y centroidal axes for the areas shown in Fig. 7–22.

FIGURE 7–22 Problem 21.

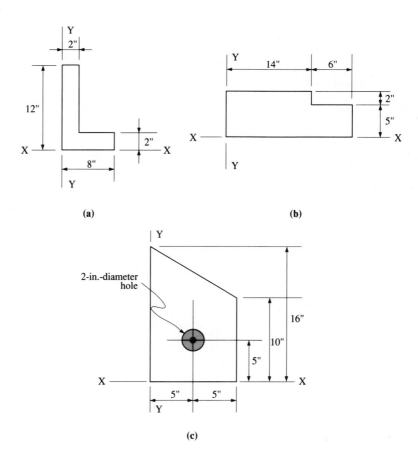

(a)

(b)

(c)

22. For the area shown in Fig. 7–23, the X–X centroidal axis is required to be located as shown. Calculate the required dimension *h*.

23. Locate the X–X and Y–Y centroidal axes for the areas shown in Fig. 7–24.

24. Locate the X–X and Y–Y centroidal axes for the built-up structural steel areas shown in Fig. 7–25.

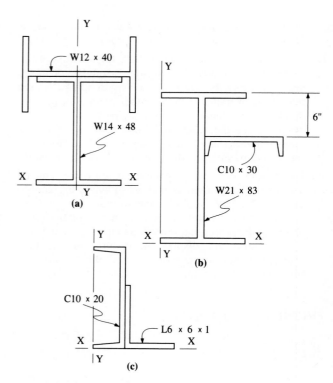

(a)

(b)

(c)

FIGURE 7–25   Problem 24.

FIGURE 7–23   Problem 22.

FIGURE 7–24   Problem 23.

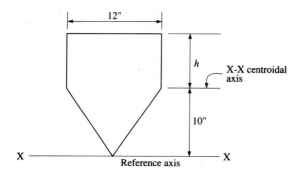

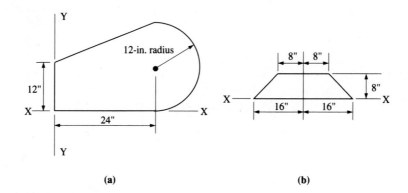

(a)   (b)

8 Area Moments of Inertia

8–1
INTRODUCTION AND DEFINITIONS

In Chapters 9 through 21 of this text, we consider aspects of the strength of machine and structural members and parts. In the development of the equations dealing with the bending of beams, the buckling of columns, and the twisting of circular shafts, we make use of several quantities that are dependent on the size and shape of the cross sections of the members. These quantities are used so widely that they have been given special names and are tabulated in numerous sources as common properties. Foremost among them is the property called the *moment of inertia*, which is denoted I.

The moment of inertia of an area represents one of the more abstract concepts in engineering mechanics. It is not a readily observable property of the area; rather, it is a purely mathematical quantity, albeit a very important one.

The moment of inertia of an area may be defined by considering a plane area designated A as shown in Fig. 8–1. Let X–X and Y–Y be any set of rectangular axes in the same plane as the area. Area A is divided into small areas (represented by a), and each small area is located with respect to the axes. The coordinates of a are distances x and y. A moment of inertia must always be computed with respect to a specific axis. Therefore, in Fig. 8–1, we may have a moment of inertia with respect to axis X–X, denoted I_x, or with respect to axis Y–Y, denoted I_y. The moment of inertia is defined as the sum of all the small areas, each multiplied by the square of its distance (moment arm) from the axis being considered.

Thus, as shown in Fig. 8–1, the moment of inertia about the X–X axis is the summation of the products of each area a and the square of its moment arm y. This gives

$$I_x = \sum ay^2 \tag{8–1}$$

Similarly, the moment of inertia about the Y–Y axis is given by

$$I_y = \sum ax^2 \tag{8–2}$$

These mathematical expressions are sometimes referred to as the *second moment* of the area, since each small area, when multiplied by its moment arm, gives the moment of the area (or first moment of the area). When multiplied a second time by the moment arm, the result is the second moment of the area (or moment of inertia). These moments of inertia are also referred to as *rectangular moments of inertia*.

FIGURE 8–1 Moment of inertia of an area.

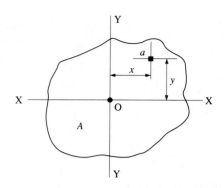

The expression "moment of inertia of an area" is actually a misnomer, since plane areas have no thickness and, therefore, no mass or inertia. It is a traditional expression, however, and we will use it throughout this text.

Since the moment of inertia is an area multiplied by the square of a distance, the resulting units will be length to the fourth power. The moment of inertia is generally expressed in in.^{4} in the U.S. Customary System. In the SI system, the recommended units are mm^4 or m^4. Note that the moment of inertia is always a positive quantity.

The magnitude of the moment of inertia is a measure of the ability of a cross-sectional area to resist bending or buckling. If we consider two beams of the same material but of different cross sections, the beam having the cross-sectional area with the greater moment of inertia would have the greater resistance to bending. However, the beam with the greater moment of inertia does not necessarily have the greater cross-sectional area. It is the distribution of the area relative to the reference axis that will determine the magnitude of the moment of inertia.

Generally, the moment of inertia desired is with respect to axes that pass through the centroid of a section. We are usually interested in either the maximum or the minimum moment of inertia, which, for symmetrical sections, will be with respect to the centroidal axis (or axes), coinciding with the axis (or axes) of symmetry. A comprehensive treatment of moment of inertia about inclined axes and of maximum/minimum moments of inertia for shapes with no axis of symmetry is beyond the scope of this text.

8–2
MOMENT OF
INERTIA

Using the calculus form of Eqs. (8–1) and (8–2) and assuming a total area divided into infinitesimal component areas, exact theoretical formulas have been mathematically derived for determining the moment of inertia of simple geometric shapes. Derivations of the moment-of-inertia formulas for rectangular and triangular areas are given in Appendix L. Moment-of-inertia formulas for the most commonly used geometric areas with respect to the designated axes are given in Table 8–1.

TABLE 8–1 Properties of areas.

Shape	Area (A)	Moment of Inertia (I)	Radius of Gyration (r)	Polar Moment of Inertia (J)
 Rectangle	$A = bh$	$I_{x_o} = \dfrac{bh^3}{12}$ $I_{y_o} = \dfrac{hb^3}{12}$ $I_x = \dfrac{bh^3}{3}$	$r_{x_o} = \sqrt{\dfrac{h}{12}}$ $r_{y_o} = \sqrt{\dfrac{b}{12}}$ $r_x = \sqrt{\dfrac{h}{3}}$	$J_{CG} = \dfrac{bh}{12}(h^2 + b^2)$
 Triangle	$A = \dfrac{bh}{2}$	$I_{x_o} = \dfrac{bh^3}{36}$ $I_x = \dfrac{bh^3}{12}$	$r_{x_o} = \sqrt{\dfrac{h}{18}}$ $r_x = \sqrt{\dfrac{h}{6}}$	
 Circle	$A = \dfrac{\pi d^2}{4}$ $= 0.7854 d^2$	$I_{x_o} = I_{y_o} = \dfrac{\pi d^4}{64}$	$r_{x_o} = r_{y_o} = \dfrac{d}{4}$	$J_{CG} = \dfrac{\pi d^4}{32}$

TABLE 8–1 (*Continued*)

Shape	Area (A)	Moment of Inertia (I)	Radius of Gyration (r)	Polar Moment of Inertia (J)
Semicircle	$A = \dfrac{\pi R^2}{2}$ $= 1.571R^2$	$I_{x_o} = 0.1098R^4$ $I_{y_o} = I_x = \dfrac{\pi R^4}{8}$ $= 0.3927R^4$	$r_{x_o} = 0.264R$ $r_{y_o} = r_x = \dfrac{R}{2}$	$J_{CG} = I_{x_o} + I_{y_o}$ $= 0.5025R^4$ $J_o = \dfrac{\pi R^4}{4}$
Hollow Circle	$A = \dfrac{\pi(d^2 - d_1^2)}{4}$ $= 0.7854(d^2 - d_1^2)$	$I_{x_o} = \dfrac{\pi(d^4 - d_1^4)}{64}$ $I_{y_o} = I_{x_o}$	$r_{x_o} = \dfrac{\sqrt{d^2 + d_1^2}}{4}$ $r_{y_o} = r_{x_o}$	$J_{CG} = \dfrac{\pi(d^4 - d_1^4)}{32}$
Hollow Rectangle	$A = bd - b_1 d_1$	$I_{x_o} = \dfrac{bd^3 - b_1 d_1^3}{12}$ $I_{y_o} = \dfrac{db^3 - d_1 b_1^3}{12}$	$r_{x_o} = \sqrt{\dfrac{bd^3 - b_1 d_1^3}{12A}}$ $r_{y_o} = \sqrt{\dfrac{db^3 - d_1 b_1^3}{12A}}$	$J_{CG} = I_{x_o} + I_{y_o}$
Quarter-Circle	$A = \dfrac{\pi R^2}{4}$	$I_{x_o} = I_{y_o} = 0.0549R^4$ $I_x = \dfrac{\pi R^4}{16}$	$r_{x_o} = r_{y_o} = 0.2644R$ $r_x = 0.5R$	$J_{CG} = 0.1098R^4$

An approximate determination of the moment of inertia of an area can be obtained by dividing the total area into finite component areas. The moment of inertia of each component area can then be calculated using Σay^2 or Σax^2. The moment of inertia of the total area is then equal to the sum of the moments of inertia of the component areas. This will result in an approximate moment of inertia, with the degree of accuracy a function of the size selected for the component areas. The smaller the size of the component areas, the greater the accuracy.

This approximate method is used chiefly with irregular areas for which exact derived formulas are either not applicable or excessively cumbersome to use. Example 8–1 illustrates the technique of the approximate method and compares the result with the theoretically exact method.

□ **EXAMPLE 8–1**   Calculate the moment of inertia with respect to the X–X centroidal axis for the area shown in Fig. 8–2. (a) Use the exact formula. (b) Use the approximate method and divide the area into four similar horizontal strips parallel to the X–X axis. (c) Use the approximate method, but use eight similar horizontal strips. For parts (b) and (c), compare the result with part (a) and calculate the percent error.

FIGURE 8–2   Rectangular area.

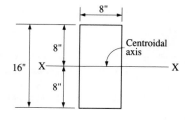

Solution   (a) Using the exact formula from Table 8–1,

$$I_x = \frac{bh^3}{12} = \frac{8(16)^3}{12} = 2731 \text{ in.}^4$$

Note in this application that b is the dimension of the side of the rectangle parallel to the axis about which the moment of inertia is being calculated, and that h is the dimension of the side perpendicular to that axis. The terms will always be defined in this manner throughout this text. If we were to calculate I_y for this rectangle, b would take the value of 16 in. and h would take the value of 8 in.

(b) Divide the area into four equal horizontal strips as shown in Fig. 8–3. Each strip has an area of 32 in.^{2} The perpendicular distance from the centroid of each strip (component areas a_1 and a_2) to the X–X centroidal axis is shown in Fig. 8–3; these distances are designated y_1 and y_2.

$$y_1 = 6 \text{ in.} \quad \text{and} \quad y_2 = 2 \text{ in.}$$

Note that due to symmetry with respect to axis X–X, the moment of inertia of the upper half of the area equals that of the lower half. Therefore, it is necessary to

FIGURE 8–3   Approximate moment of inertia.

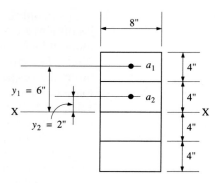

compute only the moment of inertia for either the upper or lower half and then multiply by 2 to obtain the moment of inertia for the total area. Using Eq. (8–1),

$$I_x = \sum ay^2$$
$$= 2(a_1 y_1^2 + a_2 y_2^2)$$
$$= 2[32(6)^2 + 32(2)^2] = 2560 \text{ in.}^4$$

Comparing with the exact moment of inertia, the percent error is

$$\frac{2560 - 2731}{2731} \times 100 = -6.3 \text{ percent}$$

(c) Divide the area into eight equal horizontal strips as shown in Fig. 8–4. Each strip has an area of 16 in.^{2}. The perpendicular distance from the centroid of each strip to the X–X centroidal axis is shown in Fig. 8–4. The y dimensions are noted. Equation (8–1) yields

$$I_x = \sum ay^2$$
$$= 2(a_1 y_1^2 + a_2 y_2^2 + a_3 y_3^2 + a_4 y_4^2)$$
$$= 2[16(1)^2 + 16(3)^2 + 16(5)^2 + 16(7)^2] = 2688 \text{ in.}^4$$

Comparing with the exact moment of inertia, the percent error is

$$\frac{2688 - 2731}{2731} \times 100 = -1.6 \text{ percent}$$

FIGURE 8–4   Approximate moment of inertia.

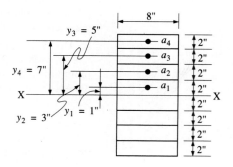

As the component area decreases in size, the moment of inertia approaches the exact theoretical value obtained by using the exact formula.

Note that in the following example, the fact that the member is concrete plays no role. The result would be the same if the material were steel or wood. Moment of inertia is a geometric property of a shape; it is not dependent on material.

☐ **EXAMPLE 8–2**   Compute the moment of inertia with respect to the X–X centroidal axis of the hollow-core precast concrete member shown in Fig. 8–5.

FIGURE 8–5   Hollow-core precast concrete member.

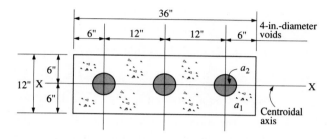

Solution   Let us define a_1 and I_1 as the area and moment of inertia of a solid rectangular area 12 in. by 36 in. in cross section and a_2 and I_2 as the area and moment of inertia of one of the voids. Note that the centroids of the circular voids lie on the centroidal X–X axis of the member, and that the centroidal X–X axis of area a_1 and the member coincide. Express the moment of inertia of the member as

$$I_x = I_{x_1} - 3I_{x_2}$$

where all moments of inertia are with respect to the X–X axis. Also note that since the circular areas represent voids, effectively reducing the rectangular area, the moments of inertia of the circles must be subtracted from the moment of inertia of the rectangle.

Computing the moments of inertia using formulas of Table 8–1,

$$I_{x_1} = \frac{bh^3}{12} = \frac{36(12)^3}{12} = 5184 \text{ in.}^4$$

$$I_{x_2} = \frac{\pi d^4}{64} = \frac{\pi (4)^4}{64} = 12.57 \text{ in.}^4$$

Therefore, the moment of inertia of the total unit is

$$I_x = I_{x_1} - 3I_{x_2}$$
$$= 5184 - 3(12.57) = 5146 \text{ in.}^4$$

8–3
THE TRANSFER
FORMULA

It is frequently necessary to determine the moment of inertia of an area about a noncentroidal axis, but one that is parallel to a centroidal axis. This is accomplished using an expression called the *transfer formula*. With reference to Fig. 8–6, the moment of inertia of an area about any axis (X′–X′ in

FIGURE 8–6 Moment of inertia of an area with respect to a non-centroidal axis.

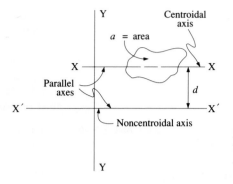

this case) that is parallel to a centroidal axis may be determined by

$$I = I_o + ad^2 \qquad \textbf{(8–3)}$$

where, as illustrated in Fig. 8–6, the terms may be defined as follows:

I = the moment of inertia of an area with respect to any axis (in.⁴) (mm⁴)
I_o = the moment of inertia of an area with respect to its own centroidal axis (in.⁴) (mm⁴)
a = the area under consideration (in.²) (mm²)
d = the perpendicular distance between parallel axes, referred to as the *transfer distance* (in.) (mm)

Note that the transfer may be made only between parallel axes. Because the axes involved are parallel, Eq. (8–3) is also called the *parallel axis theorem*.

8–4 MOMENT OF INERTIA OF COMPOSITE AREAS

As discussed previously in Section 7–4, a composite area is one made up of a number of simple geometric component areas or standardized shapes. Each of the component areas may have a centroidal axis different from that for the total composite area. If an area is composed of n component areas, where the areas are designated a_1, a_2, etc., the transfer formula (Eq. [8–3]) is applied to each component area. The moment of inertia of the entire area is then the sum of the moments of inertia of all the component areas. Mathematically,

$$I = (I_{o_1} + a_1d_1^2) + (I_{o_2} + a_2d_2^2) + \cdots + (I_{o_n} + a_nd_n^2)$$
$$I = \Sigma(I_o + ad^2) \qquad \textbf{(8–4)}$$

In order to determine the moment of inertia of the composite area with respect to its centroidal axes, the following sequence of steps is recommended:

1. Divide the composite area into common simple geometric component areas or shapes and designate them a_1, a_2, etc.
2. Determine the location of the centroidal axes for the composite area (as described in Section 7–4).

3. Determine the transfer distances from the centroidal axis of the composite area to the centroidal axis of each of the component areas and designate them d_1, d_2, etc. Note that the axes must be parallel.
4. Compute the moment of inertia of each component area with respect to its own centroidal axis and designate these moments of inertia I_{o_1}, I_{o_2}, etc. Use Table 8–1 for formulas and/or standardized tables for the moments of inertia.
5. Compute the moment of inertia of the composite area about its centroidal axis using Eq. (8–4).

☐ **EXAMPLE 8–3**   Compute the moments of inertia with respect to the X–X and Y–Y centroidal axes for the composite area shown in Fig. 8–7.

Solution   The vertical Y–Y axis is a centroidal axis because it is an axis of symmetry. In order to determine the location of the X–X centroidal axis, a reference axis is selected at the bottom of the composite area which has been divided into three rectangular component areas, as shown in Fig. 8–8. Table 8–2 shows a tabular format that can be used to organize the information.

FIGURE 8–7   Composite area.

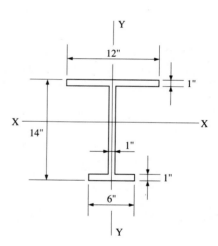

FIGURE 8–8   Location of centroidal axis.

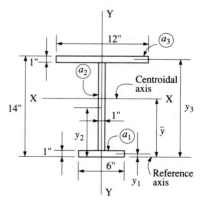

TABLE 8–2   Tabular format for
Example 8–3.

Component Area	a (in.^{2})	y (in.)	ay (in.^{3})
a_1	6	0.5	3.0
a_2	12	7.0	84.0
a_3	12	13.5	162.0
Σ	30		249

Therefore, from Table 8–2,

$$\bar{y} = \frac{\Sigma ay}{\Sigma a} = \frac{249}{30} = 8.30 \text{ in.}$$

Now compute the moment of inertia of the composite area with respect to the X–X centroidal axis. With reference to Fig. 8–9, the transfer distances are

$$d_1 = 8.3 - 0.5 = 7.8 \text{ in.}$$
$$d_2 = 8.3 - 7 = 1.3 \text{ in.}$$
$$d_3 = 14 - 8.3 - 0.5 = 5.2 \text{ in.}$$

The moment of inertia of each component area (a rectangle) with respect to its own centroid is determined with reference to Table 8–1:

$$I = \frac{bh^3}{12}$$

from which

$$I_{o_1} = \frac{6(1)^3}{12} = 0.5 \text{ in.}^4$$

$$I_{o_2} = \frac{1(12)^3}{12} = 144 \text{ in.}^4$$

$$I_{o_3} = \frac{12(1)^3}{12} = 1.0 \text{ in.}^4$$

FIGURE 8–9   Determination of transfer distances.

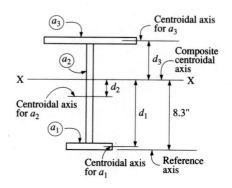

Finally, calculating the moment of inertia of the composite area about the centroidal X–X axis, using Eq. (8–4),

$$I_x = \Sigma(I_o + ad^2)$$
$$= [0.5 + 6(7.8)^2] + [144 + 12(1.3)^2] + [1.0 + 12(5.2)^2]$$
$$= 855.3 \text{ in.}^4$$

Table 8–3 shows how the entire solution can be accomplished through the use of a tabular format.

TABLE 8–3 Tabular format for Example 8–3.

Component Area	a (in.^{2})	y (in.)	ay (in.^{3})	d (in.)	ad^2 (in.^{4})	I_o (in.^{4})
a_1	6	0.5	3.0	7.8	365.0	0.5
a_2	12	7	84.0	1.3	20.3	144.0
a_3	12	13.5	162.0	5.2	324.5	1.0
Σ	30		249		709.8	145.5

Therefore from Table 8–3,

$$\bar{y} = \frac{\Sigma ay}{\Sigma a} = \frac{249}{30} = 8.30 \text{ in.}$$

and

$$I_x = \Sigma(I_o + ad^2)$$
$$= 145.5 + 709.8 = 855.3 \text{ in.}^4$$

The moment of inertia with respect to the Y–Y centroidal axis is somewhat easier to calculate since the centroidal axis for each component area coincides with the composite Y–Y centroidal axis. Therefore, the ad^2 term for each component area is zero. The transfer formula shows, then, that the moment of inertia of the composite area is the sum of the moments of inertia of the component areas about their own centroidal axes which are coincident with and parallel to the composite Y–Y centroidal axes. The moment of inertia about the Y–Y centroidal axis is

$$I_y = \Sigma I_o = \Sigma \frac{bh^3}{12}$$
$$= \frac{1(6)^3}{12} + \frac{12(1)^3}{12} + \frac{1(12)^3}{12}$$
$$= 163 \text{ in.}^4$$

Note that for the determination of I_y a tabular format offers no real advantage; hence, it is not shown.

You may again recognize, as was discussed briefly in Chapter 7, that computer spreadsheeting has application for this type of problem.

☐ **EXAMPLE 8–4**   Calculate the moment of inertia with respect to the X–X centroidal axis for the built-up (composite) structural steel member shown in Fig. 8–10.

Solution   This member is symmetrical with respect to both the X–X and Y–Y axes. Therefore, no centroidal axis computations are necessary. From Appendix A, the cross-sectional area of the W18 × 71 is 20.8 in.^{2} and the moment of inertia with respect to the X–X centroidal axis is 1170 in.^{4} The latter is the I_o term for the wide-flange shape.

The member has additional area and additional moment of inertia due to the cover plates welded to the top and bottom flanges of the wide-flange shape. The area of each cover plate is 15 in.^{2}. The vertical dimensions shown in Fig. 8–10 (parallel to the web) can be readily verified.

For the application of Eq. (8–4), the built-up area may be considered to be composed of three component areas: one standard structural steel shape and two rectangular steel plates. The summation is simplified because the plates are identical in their effect, and the transfer distance d for the standard shape is zero. For each plate, the moment of inertia about its own centroidal axis (parallel to the X–X axis) is

$$I_o = \frac{bh^3}{12} = \frac{15(1)^3}{12} = 1.25 \text{ in.}^4$$

Equation (8–4) then yields

$$\begin{aligned} I_x &= \Sigma (I_o + ad^2) \\ &= 1170 + 2[1.25 + 15(9.735)^2] = 4016 \text{ in.}^4 \end{aligned}$$

Notice how little the centroidal moment of inertia I_o for each cover plate contributes to the overall moment of inertia. Since I_o for a cover plate with respect to the axis parallel to its long side is so small, it is common practice to neglect it.

FIGURE 8–10   Built-up steel member.

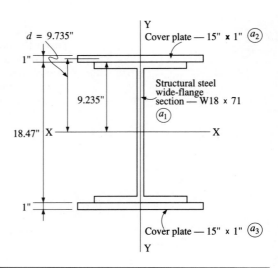

□ **EXAMPLE 8–5**   A rectangular area with a semicircular portion removed is shown in Fig. 8–11. Calculate the moment of inertia with respect to (a) the X–X centroidal axis, (b) the Y–Y centroidal axis, and (c) the X′–X′ reference axis.

Solution   The centroid of the composite area must first be located. Consider two component areas, a_1 being a rectangular area 16 in. by 24 in. and a_2 being the semicircular cut-out area having a radius of 6 in. The latter is considered a negative area. The area calculations are

$$a_1 = 16(24) = 384 \text{ in.}^2$$

$$a_2 = \frac{1}{2}(\pi R^2) = \frac{\pi(6)^2}{2} = -56.5 \text{ in.}^2$$

$$A = \Sigma a = 327.5 \text{ in.}^2$$

Computing the distance from the reference axis (axis X′–X′) to the centroid of each component area,

$$y_1 = 8 \text{ in.}$$

$$y_2 = 16 - \frac{4R}{3\pi} = 16 - \frac{4(6)}{3\pi} = 13.45 \text{ in.}$$

And applying Eq. (7–4),

$$\bar{y} = \frac{\Sigma ay}{A} = \frac{a_1y_1 + a_2y_2}{A}$$

$$= \frac{384(8) + (-56.5)(13.45)}{327.5} = 7.06 \text{ in.}$$

(a) The moment of inertia of the composite area will be determined by calculating the moment of inertia of the full rectangle (16 in. by 24 in.) and then subtracting the moment of inertia of the semicircular area (both are with respect to the X–X centroidal axis of the composite area).

FIGURE 8–11 Composite area.

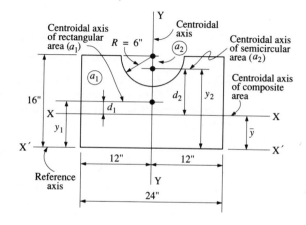

The moments of inertia of a_1 and a_2 about their respective centroidal axes parallel to axis X–X (refer to Table 8–1) are

$$I_{o_1} = \frac{bh^3}{12} = \frac{24(16)^3}{12} = 8192 \text{ in.}^4$$

$$I_{o_2} = 0.1098 R^4 = 0.1098(6)^4 = -142.3 \text{ in.}^4$$

The transfer distances (d dimensions) are

$$d_1 = y_1 - \bar{y} = 8 - 7.06 = 0.94 \text{ in.}$$

$$d_2 = y_2 - \bar{y} = 13.45 - 7.06 = 6.39 \text{ in.}$$

From Eq. (8–4), the moment of inertia for the composite area with respect to its X–X centroidal axis is

$$I_x = \Sigma(I_o + ad^2)$$
$$= [8192 + 384(0.94)^2] + [-142.3 + (-56.5)(6.39)^2]$$
$$= 6082 \text{ in.}^4$$

Table 8–4 shows a tabular format of the preceding computations.

TABLE 8–4   Tabular format for Example 8–5.

Component Area	a (in.²)	y (in.)	ay (in.³)	d (in.)	ad² (in.⁴)	I_o (in.⁴)
a_1	384	8	3072	0.94	339	8192
a_2	−56.5	13.45	−760	6.39	−2307	−142.3
Σ	327.5		2312		−1968	8050

Therefore from Table 8–4,

$$\bar{y} = \frac{\Sigma ay}{\Sigma a} = \frac{2312}{327.5} = 7.06 \text{ in.}$$

and

$$I_x = \Sigma(I_o + ad^2)$$
$$= 8050 + (-1968) = 6082 \text{ in.}^4$$

(b) The Y–Y centroidal axis coincides with the vertical axis of symmetry and is shown in Fig. 8–11. The ad^2 term for each component area will be zero. Calculating the centroidal moments of inertia of the component areas about the Y–Y centroidal axis for the rectangle yields

$$I_{o_1} = \frac{bh^3}{12} = \frac{16(24)^2}{12} = 18,430 \text{ in.}^4$$

And for the semicircle, with reference to Table 8–1,

$$I_{o_2} = 0.393 R^4 = (0.393)(6)^4 = 509 \text{ in.}^4$$

FIGURE 8–12 Moment of inertia about the base.

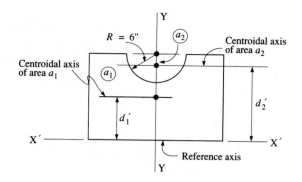

The total moment of inertia is then

$$I_y = 18{,}430 - 509 = 17{,}920 \text{ in.}^4$$

(c) The moment of inertia of the composite area with respect to the reference axis X'–X' is calculated using Eq. (8–4). In Fig. 8–12, d' is the distance to the particular component area centroid from the X'–X' axis:

$$d'_1 = 8 \text{ in.} \quad \text{and} \quad d'_2 = 13.45 \text{ in.}$$

Equation (8–4) then yields

$$\begin{aligned} I_{x'} &= \Sigma(I_o + ad^2) \\ &= [8192 + 384(8)^2] + [-142.3 + (-56.5)(13.45)^2] \\ &= 22{,}400 \text{ in.}^4 \end{aligned}$$

Table 8–5 shows a tabular format for part (c).

TABLE 8–5 Tabular format for Example 8–5, part (c).

Component Area	a (in.^{2})	d (in.)	ad^2 (in.^{4})	I_o (in.^{4})
a_1	384	8	24,576	8192
a_2	−56.5	13.45	−10,221	−142.3
Σ	327.5		14,355	8050

Therefore from Table 8–5,

$$\begin{aligned} I_{x'} &= \Sigma I_o + ad^2 \\ &= 8050 + 14{,}355 = 22{,}400 \text{ in.}^4 \end{aligned}$$

**8–5
RADIUS OF
GYRATION**

The *radius of gyration* of an area is difficult to describe in a physical sense. It is often described as the distance from a reference axis at which the entire area may be assumed to be located without changing its moment of inertia.

This description is of questionable significance. A more practical interpretation of the radius of gyration of an area with respect to a given axis is that it is a convenient concept devised to replace a mathematical relationship between the moment of inertia and the area as encountered in the analysis and design of columns. It is usually denoted by the symbol r and expressed as

$$r = \sqrt{\frac{I}{A}} \qquad\qquad (8-5)$$

where r = the radius of gyration with respect to a given axis (in.) (mm)
$\quad\quad I$ = the moment of inertia with respect to the same given axis (in.^{4}) (mm^4)
$\quad\quad A$ = the cross-sectional area (in.^{2}) (mm^2)

Note that the radius of gyration is usually expressed in units of inches in the U.S. Customary System and in millimeters in the SI system.

The radius of gyration is a function of the moment of inertia. The magnitude of the moment of inertia may be different with respect to different axes of a member. Therefore, the magnitude of the radius of gyration may also be different with respect to different axes of a member. Formulas for the radius of gyration of commonly encountered simple geometric areas are given in Table 8–1. Chapter 18, which deals with columns, includes some practical applications of the radius of gyration.

☐ **EXAMPLE 8–6**   Calculate the radius of gyration with respect to the X–X centroidal axis of the area shown in Fig. 8–13.

FIGURE 8–13   Composite area.

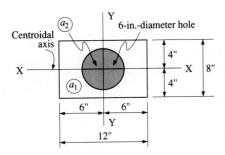

Solution   Consider the composite area to be composed of a rectangular area and a (negative) circular area. After determining the area of the composite figure and calculating its moment of inertia with respect to the X–X centroidal axis, then calculate the radius of gyration with respect to the X–X centroidal axis. The area calculations are as follows:

$$a_1 = 12(8) = 96 \text{ in.}^2$$
$$a_2 = 0.7854d^2 = 0.7854(6)^2 = -28.3 \text{ in.}^2$$
$$A = \Sigma a = 67.7 \text{ in.}^2$$

The moment of inertia for each component area about its own centroidal axis (which coincides with the X–X centroidal axis for the composite area) is calculated from

$$I_{o_1} = \frac{bh^3}{12} = \frac{12(8)^3}{12} = 512 \text{ in.}^4$$

$$I_{o_2} = \frac{\pi d^4}{64} = \frac{\pi (6)^4}{64} = -63.6 \text{ in.}^4$$

The moment of inertia for the composite area is then

$$I_x = \Sigma I_o = 512 + (-63.6) = 448 \text{ in.}^4$$

Therefore, Eq. (8–5) yields

$$r_x = \sqrt{\frac{I_x}{A}} = \sqrt{\frac{448}{67.7}} = 2.57 \text{ in.}$$

8–6
POLAR MOMENT OF INERTIA

We have previously discussed the calculation of rectangular moments of inertia, which are second moments of an area taken about axes that lie in the plane of the area. The moment of inertia of an area calculated with respect to an axis perpendicular to the plane of the area is called the *polar moment of inertia*.

In Fig. 8–14, the Z–Z axis represents an axis perpendicular to the plane of the given area. Thus, the moment of inertia about the Z–Z axis is the summation of the product of each area a and the square of its moment arm r. The polar moment of inertia is normally denoted as J. Therefore,

$$J = \Sigma a r^2 \qquad \textbf{(8–6)}$$

Since for any right triangle,

$$r^2 = x^2 + y^2$$

and, substituting in Eq. (8–6),

$$J = \Sigma a(x^2 + y^2)$$
$$= \Sigma a x^2 + \Sigma a y^2$$

FIGURE 8–14   Polar moment of inertia.

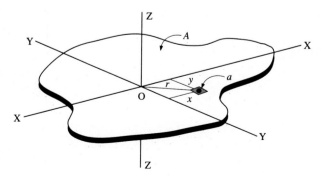

With reference to Eqs. (8–1) and (8–2), this may be written as

$$J = I_x + I_y \qquad (8–7)$$

Therefore, we see that the polar moment of inertia of a given area with respect to an axis perpendicular to its plane is equal to the sum of the moments of inertia about any two mutually perpendicular axes in its plane that intersect the polar axis. Formulas for the polar moment of inertia of solid and hollow circular areas are given in Table 8–1. The polar moment of inertia of circular members is a property required for the solution of problems involving shafts subjected to torsional loading. For further discussion on the applications of the polar moment of inertia, refer to Chapter 12.

□ **EXAMPLE 8–7**   Calculate the polar moment of inertia for a hollow circular shaft with an outside diameter of 4 in. and an inside diameter of 3 in.

Solution   From Table 8–1, the expression for the polar moment of inertia taken about the center of gravity is

$$J_{CG} = \frac{\pi}{32} (d^4 - d_1^4)$$

Substituting the given data,

$$J_{CG} = \frac{\pi}{32} (4^4 - 3^4) = 17.2 \text{ in.}^4$$

□ **EXAMPLE 8–8**   For the T-shaped area shown in Fig. 8–15, calculate the following: (a) the centroidal (rectangular) moments of inertia, (b) the radii of gyration with respect to the centroidal axes, and (c) the polar moment of inertia with respect to an axis perpendicular to the plane of the area through its centroid.

Solution   The X–X centroidal axis of the composite area has been located as shown. The reader may wish to verify that $\bar{y} = 8$ in. The vertical axis of symmetry coincides with the Y–Y centroidal axis.

FIGURE 8–15   Composite area.

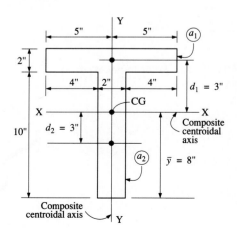

(a) First calculate I_x. The moments of inertia of a_1 and a_2 about their own centroidal axes, which are parallel to the X–X centroidal axis of the composite area, are

$$I_{o_1} = \frac{bh^3}{12} = \frac{10(2)^3}{12} = 6.67 \text{ in.}^4$$

$$I_{o_2} = \frac{bh^3}{12} = \frac{2(10)^3}{12} = 166.7 \text{ in.}^4$$

The transfer distances (d dimensions) are shown in Figure 8–15. Equation (8–4) yields

$$\begin{aligned} I_x &= \Sigma(I_o + ad^2) \\ &= [6.67 + 20(3)^2] + [166.7 + 20(3)^2] \\ &= 533.4 \text{ in.}^4 \end{aligned}$$

For the moment of inertia about the Y–Y axis, Eq. (8–4) is again applied. The ad^2 terms are zero.

$$I_y = \Sigma(I_o + ad^2) = \Sigma I_o = \Sigma \frac{bh^3}{12}$$

$$= \frac{2(10)^3}{12} + \frac{10(2)^3}{12} = 173.3 \text{ in.}^4$$

(b) The total area of the T-shaped member is

$$A = a_1 + a_2 = 20 + 20 = 40 \text{ in.}^2$$

The radii of gyration with respect to the centroidal axes are then calculated from Eq. (8–5):

$$r_x = \sqrt{\frac{I_x}{A}} = \sqrt{\frac{533.4}{40}} = 3.65 \text{ in.}$$

$$r_y = \sqrt{\frac{I_y}{A}} = \sqrt{\frac{173.4}{40}} = 2.08 \text{ in.}$$

(c) The polar moment of inertia with respect to the Z–Z axis through point CG is calculated from Eq. (8–7):

$$\begin{aligned} J_{CG} &= I_x + I_y \\ &= 533.4 + 173.3 = 707 \text{ in.}^4 \end{aligned}$$

8–7
SI SYSTEM
EXAMPLES

For these examples and the SI problems, the calculation of moment of inertia involves raising the unit lengths to the fourth power. Values can be shown either in mm^4 or m^4. Use of the centimeter (cm) unit should be avoided. In this book, we have chosen to express moment of inertia in units of m^4 when using SI units.

☐ **EXAMPLE 8–9**   The cross section of the timber beam shown in Fig. 8–16 is made up of planks 50 mm thick. Determine the moment of inertia with respect to the X–X centroidal axis.

FIGURE 8–16   Timber box
beam.

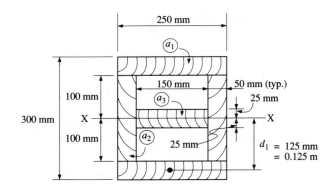

Solution   The X–X centroidal axis is shown. It is an axis of symmetry. Component areas have been designated a_1, a_2 (two of each), and a_3. Area calculations yield

$$a_1 = (250)(50) = 12\ 500\ \text{mm}^2 = 0.0125\ \text{m}^2$$
$$a_2 = (200)(50) = 10\ 000\ \text{mm}^2 = 0.0100\ \text{m}^2$$
$$a_3 = (150)(50) = 7500\ \text{mm}^2 = 0.0075\ \text{m}^2$$

Only one transfer distance is required ($d_1 = 125$ mm or 0.125 m). The centroids of a_2 and a_3 lie on the centroidal X–X axis (the transfer distances are zero).

The moment of inertia of each component area with respect to its own centroidal axis, parallel to the X–X centroidal axis, may be calculated using the appropriate formula from Table 8–1:

$$I_{o_1} = \frac{bh^3}{12} = \frac{(250)(50)^3}{12} = 2.60 \times 10^6\ \text{mm}^4 = 2.60 \times 10^{-6}\ \text{m}^4$$

$$I_{o_2} = \frac{bh^3}{12} = \frac{(50)(200)^3}{12} = 33.3 \times 10^6\ \text{mm}^4 = 33.3 \times 10^{-6}\ \text{m}^4$$

$$I_{o_3} = \frac{bh^3}{12} = \frac{(150)(50)^3}{12} = 1.56 \times 10^6\ \text{mm}^4 = 1.56 \times 10^{-6}\ \text{m}^4$$

Calculating the moment of inertia of the composite area about the centroidal X–X axis using Eq. (8–4) and substituting all length units in meters (m), we have

$$\begin{aligned}
I_x &= \Sigma(I_o + ad^2) \\
&= 2[(2.60 \times 10^{-6}) + (0.0125)(0.125)^2] + 2(33.33 \times 10^{-6}) + (1.56 \times 10^{-6}) \\
&= 464 \times 10^{-6}\ \text{m}^4
\end{aligned}$$

☐ **EXAMPLE 8–10**   Two C300 × 0.302 structural steel channel shapes are welded together at their flange tips as shown in Fig. 8–17. Determine (a) the least radius of gyration and (b) the polar moment of inertia.

Solution   The necessary properties (from Appendix C) of a single C300 × 0.302 channel are given in Fig. 8–17. For convenience, dimensions are shown in the diagram of the cross section in units of meters (m). To calculate the least radius of gyration, it first will be necessary to determine the least moment of inertia; therefore, both I_x and I_y

FIGURE 8–17 Built-up structural steel cross section.

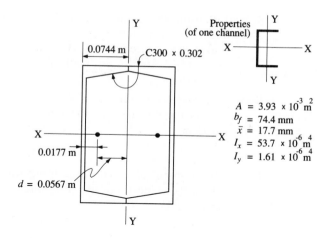

will be found. Both are required for the determination of the polar moment of inertia.

(a) The transfer distance d for the determination of I_y is calculated from $(b_f - \bar{x})$ and is shown in Fig. 8–17. From Eq. (8–4),

$$I_y = \Sigma(I_o + ad^2)$$
$$= 2(1.61 \times 10^{-6} \text{ m}^4) + 2(3.93 \times 10^{-3} \text{ m}^2)(0.0567 \text{ m})^2$$
$$= 28.5 \times 10^{-6} \text{ m}^4$$

and

$$I_x = \Sigma(I_o + ad^2)$$
$$= 2(53.7 \times 10^{-6} \text{ m}^4) = 107.4 \times 10^{-6} \text{ m}^4$$

Therefore, the Y–Y axis is the weak axis and the radius of gyration r will be determined for this axis:

$$r_y = \sqrt{\frac{I_y}{A}} = \sqrt{\frac{28.5 \times 10^{-6} \text{ m}^4}{2(3.93 \times 10^{-3} \text{ m}^2)}} = 0.0602 \text{ m} = 60.2 \text{ mm}$$

(b) The polar moment of inertia is calculated from Eq. (8–7), where the I_x and I_y values are for the built-up cross section:

$$J = I_x + I_y$$
$$= 107.4 \times 10^{-6} \text{ m}^4 + 28.5 \times 10^{-6} \text{ m}^4$$
$$= 135.9 \times 10^{-6} \text{ m}^4$$

SUMMARY—BY SECTION NUMBER

8–1 The moment of inertia of an area (also called the *second moment of an area*) is a mathematical quantity that affects the load-carrying capacity of beams and columns. Units of moment of inertia are in.^{4}, mm^4, or m^4. With respect to an X–Y rectangular axes system, moment of inertia is expressed as

$$I_x = \Sigma ay^2 \tag{8–1}$$
$$I_y = \Sigma ax^2 \tag{8–2}$$

8–2 Equations (8–1) and (8–2) yield an approximate moment of inertia, the accuracy of which is a function of the size of the small component areas, represented by a's. An exact moment of inertia may be obtained using the formulas in Table 8–1 or by using integral calculus (see Appendix L).

8–3 The moment of inertia of an area with respect to a noncentroidal axis that is parallel to a centroidal axis is determined using the transfer formula

$$I = I_o + ad^2 \qquad\qquad \textbf{(8–3)}$$

8–4 The moment of inertia of a composite area is found by summing the moments of inertia of the component parts as determined by the transfer formula:

$$I = \Sigma(I_o + ad^2) \qquad\qquad \textbf{(8–4)}$$

8–5 The radius of gyration (r) of an area is a mathematical quantity expressed as

$$r = \sqrt{\frac{I}{A}} \qquad\qquad \textbf{(8–5)}$$

Its most significant application is in the design and analysis of compression members (see Chapter 18).

8–6 The polar moment of inertia (J) of an area with respect to a polar axis perpendicular to its plane is defined as the sum of the moments of inertia about any two mutually perpendicular axes in its plane that intersect the polar axis:

$$J = I_x + I_y \qquad\qquad \textbf{(8–7)}$$

PROBLEMS

Section 8–2 Moment of Inertia

1. Calculate the moment of inertia with respect to the X–X centroidal axes in Fig. 8–18.

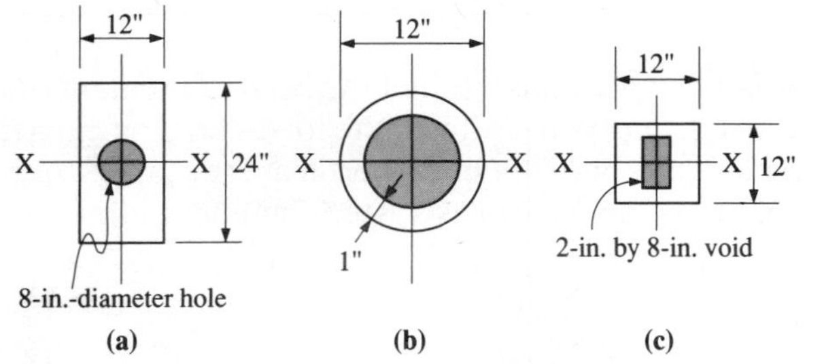

2. Calculate the moment of inertia of the triangular area in Fig. 8–19 with respect to the X–X centroidal axis and with respect to the base of the triangle.

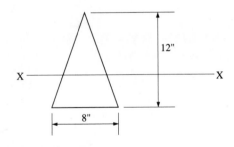

12"
24"
8-in.-diameter hole
(a)

12"
1"
(b)

12"
12"
2-in. by 8-in. void
(c)

FIGURE 8–18 Problem 1.

12"
X ——— X
8"

FIGURE 8–19 Problem 2.

3. A structural steel wide-flange section is reinforced with two steel plates attached to the web of the member as shown in Fig. 8–20. Calculate the moment of inertia of the built-up member with respect to the X–X centroidal axis.

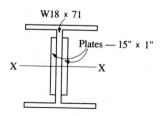

FIGURE 8–20 Problem 3.

4. Compute the moment of inertia with respect to the X–X centroidal axis for the area shown in Fig. 8–21.

5. A rectangle has a base of 6 in. and a height of 12 in. Using the approximate method, determine the moment of inertia with respect to the base. (a) Divide

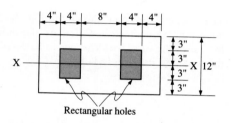

FIGURE 8–21 Problem 4.

the area into four equal horizontal strips. (b) Divide the area into six equal horizontal strips.

6. For the area of Problem 5, calculate the exact moment of inertia and then calculate the percent error relative to the results obtained using the approximate method.

Section 8–4 Moment of Inertia of Composite Areas

7. Calculate the moments of inertia with respect to both centroidal axes for the areas shown in Fig. 8–22.

FIGURE 8–22 Problem 7.

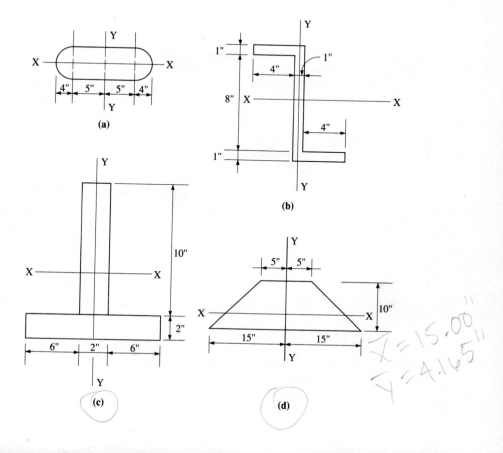

8. The rectangular area shown in Fig. 8–23 has a square hole cut from it. Calculate the moments of inertia of the area with respect to its X–X centroidal axis and its base (X′–X′).

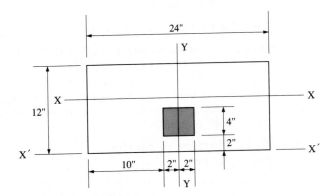

FIGURE 8–23   Problem 8.

9. For the built-up structural steel member in Fig. 8–24, calculate the moments of inertia with respect to its X–X and Y–Y centroidal axes.

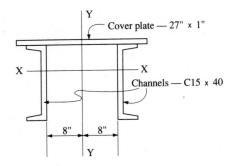

FIGURE 8–24   Problem 9.

10. Calculate the moment of inertia with respect to the X–X centroidal axis of the built-up timber member in Fig. 8–25. Use the rough dimensions shown.

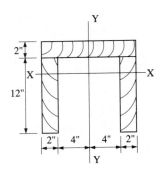

FIGURE 8–25   Problem 10.

11. For the two channels in Fig. 8–26, calculate the spacing *s* required for the moment of inertia with respect to the X–X centroidal axis to be equal to the moment of inertia with respect to the Y–Y centroidal axis.

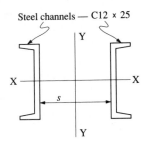

FIGURE 8–26   Problem 11.

Section 8–5   Radius of Gyration

12. Compute the radii of gyration about both centroidal axes for the following structural steel shapes and compare your results with tabulated values: (a) W10 × 54, (b) C10 × 15.3.

13. Two C10 × 15.3 channels are welded together at their flange tips to form a boxlike section. Compute the radii of gyration about the X–X and Y–Y centroidal axes for the built-up member. Compare your results with the radius of gyration values for a single C10 × 15.3.

14. Compute the radii of gyration with respect to the X–X and Y–Y centroidal axes for the built-up steel member of Problem 9.

15. Compute the radii of gyration with respect to the X–X and Y–Y centroidal axes for the aluminum extruded shape shown in Fig. 8–27.

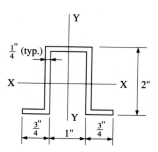

FIGURE 8–27   Problem 15.

16. Compute the radii of gyration with respect to the X–X and Y–Y centroidal axes for the built-up steel member of Problem 3.

17. Compute the radii of gyration with respect to the X–X and Y–Y centroidal axes for the built-up timber member of Problem 10.

Section 8–6   Polar Moment of Inertia

18. Calculate the polar moment of inertia for a circular solid steel shaft 3 in. in diameter.

19. Calculate the polar moment of inertia for a circular hollow steel shaft with an outside diameter of 3 in. and an inside diameter of $2\frac{1}{2}$ in.

20. For the areas (a) and (b) of Problem 7, calculate the polar moment of inertia with respect to an axis perpendicular to the plane of the area through its centroid.

SI System Problems

21. Check the tabulated moment of inertia for a 300 × 610 nominal-size timber cross section.

22. Calculate the moments of inertia about both centroidal axes for the built-up cross section of Fig. 7–19.

23. Calculate the least radius of gyration for the areas shown in Fig. 8–28. Use nominal timber sizes.

24. Calculate the polar moment of inertia for the following shapes: (a) a standard-weight steel pipe with a 150-mm nominal diameter, (b) a W530 × 1.07

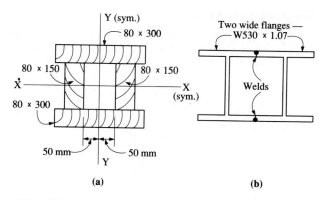

FIGURE 8–28   Problem 23.

wide-flange steel shape, and (c) a 200 × 200 timber cross section.

25. Determine the polar moment of inertia for the areas of Problem 23.

Computer Problems

For the following computer problems, any appropriate programming language may be used. Input prompts should fully explain what is required of the user (the program should be "user friendly"). The resulting output should be well labeled and self-explanatory. For spreadsheet problems, any appropriate software may be used.

26. Write a program that will calculate rectangular moments of inertia, radii of gyration, and polar moment of inertia for a rectangular area. Both the X–X (horizontal) and Y–Y (vertical) axes should be considered. User input is to be width and height of the rectangular area.

27. Write a program that will calculate the moments of inertia about the X–X and Y–Y centroidal axes for the semicircle of Problem 16 in Chapter 7. The calculation is to be based on Eq. (8–4) and a summation of the effects of component areas. User input is to be as stated in Problem 16 of Chapter 7.

28. Utilize a spreadsheet program to solve one or more of the following: Example 8–4, Problem 30, 33, or 35. The user should be able to input the appropriate geometric properties for any shapes of the types shown and the program should then determine the appropriate quantities and display the result(s). Vary a single shape or dimension and observe the effect on the properties of the built-up cross-section.

Supplemental Problems

29. Calculate the moments of inertia of the area shown in Fig. 8–29 with respect to both centroidal axes.

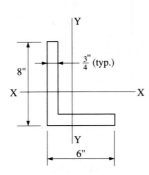

FIGURE 8–29   Problem 29.

30. Calculate the moments of inertia of the built-up steel member in Fig. 8–30 with respect to both centroidal axes.

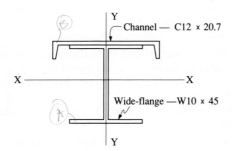

FIGURE 8–30   Problem 30.

31–33. For the cross-sectional areas shown in Figs. 8–31 through 8–33, calculate the moment of inertia with respect to the horizontal X–X centroidal axis.

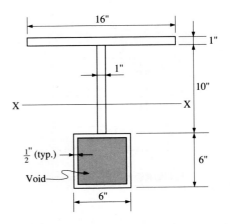

FIGURE 8–31   Problem 31.

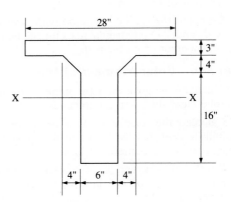

FIGURE 8–32   Problem 32.

$\overline{X} = 6''$ $I_x = 341.97$

$\overline{Y} = 6.51''$ $I_y = 182.4$

W10X45: A = 13.3
d = 10.10
tw = .350
bf = 8.02
$I_x = 248.0$
$I_y = 53.4$

C12X20.7: A = 6.09
d = 12.00
tw = 0.282
$\overline{x} = .698$
$I_x = 129$
$I_y = 3.88$

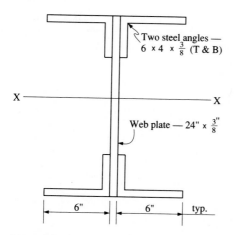

FIGURE 8–33 Problem 33.

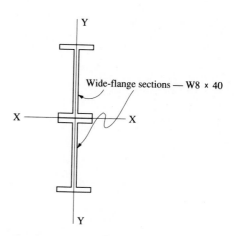

FIGURE 8–34 Problem 35.

34. Compute the radii of gyration with respect to the X–X and Y–Y centroidal axes for the areas indicated in Problem 30.

35. A structural steel built-up section is fabricated as shown in Fig. 8–34. Calculate (a) the moments of inertia and (b) the radii of gyration with respect to the X–X and Y–Y centroidal axes.

36. Calculate the polar moment of inertia about the centroid of an area that is hexagonal in shape and 6 in. on each side.

9 Stresses and Strains

As previously described, statics is a study of forces and force systems acting on rigid bodies at rest. *Strength of materials* may be described as a study of the relationships between external forces acting on elastic bodies and the internal stresses and strains generated by these forces. Based on the principles of strength of materials, we can establish the internal conditions that exist in an elastic body when it is subjected to various loading conditions.

In our study of statics, we neglected any dimensional changes (bodies were assumed rigid). For our study of strength of materials, however, the bodies will no longer be assumed rigid. Deformation and dimensional changes will be important considerations. We will consider machine and structural elements that have application in various fields of engineering technology with respect to both analysis (investigation) and design (selection) of these elements. Our approach will be rational and analytical, and it will be based on the principles of strength of materials.

9-2
TENSILE AND COMPRESSIVE STRESSES

Figure 9–1(a) represents an initially straight metal bar BC of constant cross section throughout its length. A bar of constant cross section such as this is termed a *prismatic bar*. The bar is acted on (loaded) at its ends by two equal and opposite axial forces (loads) P. An axial force, as shown in Fig. 9–1(a), is one that coincides with the longitudinal axis of the bar and acts through the centroid of the cross section. These forces, called *tensile forces*, tend to stretch or elongate the bar. The bar is said to be in tension.

Similarly, Fig. 9–2(a) represents a straight prismatic bar acted on by two equal and opposite forces P that are directed toward each other. These forces, called *compressive forces*, tend to shorten or compress the bar. The bar is said to be in compression.

Under the action of the two forces (whether tensile or compressive), internal resisting forces are developed within the bar and may be determined by imagining that a transverse plane is passed through bar BC (that is, perpendicular to its longitudinal axis), cutting it into two parts at point A. We will consider the segment of the bar on the left of point A as a free body, as shown in Figs. 9–1(b) and 9–2(b). In order for the segment to be in equilibrium, a force P_1, equal and opposite to force P, must exist. The force P_1, which shows as an external force acting on the segment, is in reality an internal force in the original bar. It is considered to be an internal resisting force because it resists the action of external force P. In addition, the inter-

FIGURE 9–1   Prismatic metal
bar in tension.

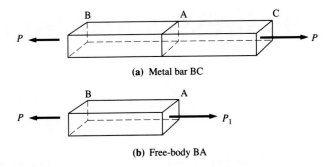

FIGURE 9–2   Prismatic metal
bar in compression.

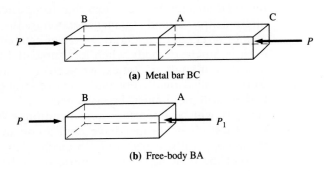

nal resisting force is assumed to be uniformly distributed over the cross
section of the bar. This, in effect, makes P_1 an axial force and also the
resultant of the system of uniformly distributed forces that compose the
internal resistance of the section.

To permit comparisons with standard and acceptable values (i.e., varying material breaking strengths), the total internal resistance P_1 acting on
cross section A is converted to a unit basis and is expressed in terms of force
per unit area or intensity of force. This is termed *unit stress*. Throughout the
remainder of this book, the word *stress* will be used to signify unit stress.
The stress represents the internal resistance developed by a unit area of the
cross section of the member. In the U.S. Customary System of units, stress
generally is expressed in pounds per square inch (psi) or kips per square inch
(ksi). In the SI system, stress is preferably expressed in pascals (Pa) or
megapascals (MPa). Stress may be computed from the expression

$$s = \frac{P}{A}$$ **(9–1)**

where s = the average computed stress (psi, ksi) (Pa, MPa)
   P = the external applied load or force (lb, kips) (N)
   A = the cross-sectional area over which stress develops (in.^{2}) (m^2,
       mm^2)

The preceding equation for stress is commonly termed the *direct stress formula*.

The tensile forces of Fig. 9–1 produce internal tensile stresses and the compressive forces of Fig. 9–2 produce internal compressive stresses. The maximum values of these stresses act on a plane that is perpendicular (or normal) to the line of action of the applied forces (see Fig. 9–5[a] and [b]); hence they are sometimes called *normal stresses*. They are also referred to as *direct stresses*. It is usually necessary to limit stresses so as not to exceed the capability of materials to support them without failure. *Allowable* (or *permissible*) *stress*, $s_{(all)}$, is simply that level of stress judged to be acceptable. It is an upper limit that should not be exceeded. (For a more comprehensive discussion of allowable stress, refer to Section 10–6).

The direct stress formula may be rewritten in several ways for use in various applications. For analysis problems in which the capacity of a member is to be determined,

$$P = s_{(all)}A \qquad (9–2)$$

where P = the axial load capacity (maximum allowable axial load)
$s_{(all)}$ = the permissible or allowable axial stress
A = the cross-sectional area of the axially loaded member

For design-type problems in which it is necessary to support a given load without exceeding an allowable stress,

$$A = \frac{P}{s_{(all)}} \qquad (9–3)$$

where A = the required cross-sectional area of the axially loaded member being designed
P = the external applied axial load or force
$s_{(all)}$ = the permissible or allowable axial stress

It should be noted that Eqs. (9–1), (9–2), and (9–3) may be used directly with tension members. However, in the analysis and design of compression members, other than short, stocky ones, we will see that the length of a member plays an important role. For instance, a typical wooden yardstick, loaded in compression, will buckle before it fractures. Therefore, for the present, when we deal with compression members, we will limit our discussion to short compression members, unaffected by length considerations. Compression members are further discussed in Chapter 18.

In the design of axially loaded members, commercially available structural shapes of various materials are commonly used. Sizes and properties for some timber and structural steel shapes are provided in Appendices A through E. Both U.S. Customary System and SI system information is given. The U.S. Customary System portion of Appendix A consists of a table from the *Manual of Steel Construction*, published by the American

Institute of Steel Construction (AISC).[1] It lists sizes and properties of wide-flange (W) shapes, the most commonly used hot-rolled structural steel shapes. A typical designation is W8 × 31, indicating the basic shape (W), the approximate depth (8 in.), and the weight (31 lb/ft). Appendix E lists sizes and properties of solid rectangular timber members. The sizes by which the sections are designated are *nominal sizes* (that is, sizes "in name only"), such as 2 in. by 4 in. or 75 mm by 200 mm. The *dressed sizes* are the actual dimensions of the finished product after surfacing (planing). Most structural timbers are surfaced on all four sides, a treatment designated as S4S (surfaced four sides).

So far, we have discussed tensile and compressive stresses, which imply an internal condition. Another type of stress is a *bearing stress*, which we will designate s_p. A bearing stress is basically a compressive stress exerted on an external surface of a body. It may be considered a contact pressure between separate bodies, rather than a stress. We usually speak of pressure with respect to gases or fluids, such as air pressure in an automobile tire or water pressure on a submerged submarine, but the terms *bearing pressure* and *bearing stress* are used almost interchangeably when they refer to situations in which a body bears on soil. When the footing of a concrete foundation bears on supporting soil, the bearing stress (pressure) is obtained by dividing the applied load by the contact area between the footing and the soil. The previous expressions for stress are also applicable for bearing stress (pressure) problems.

☐ **EXAMPLE 9–1**   (a) Compute the tensile stress developed in a steel bar 2 in. by 2 in. in cross section if it is subjected to an axial tensile load of 95 kips (see Fig. 9–1[a]). (b) Determine the tensile stress, s_t, if the member is a structural steel W8 × 31. (The load is still 95 kips.)

Solution   (a) Using the direct stress formula,

$$s_t = \frac{P}{A} = \frac{95}{2^2} = 23.75 \text{ ksi}$$

(b) From Appendix A, the cross-sectional area of a W8 × 31 is 9.13 in.[2] Therefore,

$$s_t = \frac{P}{A} = \frac{95}{9.13} = 10.41 \text{ ksi}$$

☐ **EXAMPLE 9–2**   Steel rod suspenders are to support steam pipes in a power plant. The rods are $\frac{1}{2}$ in. in diameter and have an allowable axial tensile stress of 24,000 psi. Calculate the allowable axial tensile load in the rods.

---

[1]   American Institute of Steel Construction, Inc., *Manual of Steel Construction—Allowable Stress Design*, 9th ed. (Chicago: AISC, 1989).

Solution   The cross-sectional area of the $\frac{1}{2}$ in. diameter rod is 0.196 in.² Therefore, the allowable axial tensile load is

$$P = s_{t_{(all)}}A = 24,000(0.196) = 4700 \text{ lb}$$

☐ **EXAMPLE 9–3**   Compute the required size of a short square dressed (S4S) timber post subjected to a compressive load of 24,000 lb if the allowable axial compressive stress is 800 psi.

Solution   This is a design-type problem, since the size of a member is to be found. The required cross-sectional area is

$$A = \frac{P}{s_{c_{(all)}}} = \frac{24,000}{800} = 30.0 \text{ in.}^2$$

From Appendix E, a nominal 6 in. by 6 in. square timber member has a cross-sectional area of 30.25 in.² This is based on dressed dimensions of $5\frac{1}{2}$ in. by $5\frac{1}{2}$ in. Therefore, select a 6 in. by 6 in. timber post.

☐ **EXAMPLE 9–4**   An S4S timber column (nominal 6 in. by 6 in.) is subjected to a compressive load of 22,000 lb, as shown in Fig. 9–3. The column is supported by a 2 ft square concrete footing, which is, in turn, supported by the soil. Compute (a) the bearing stress on the contact surface between the column and the footing and (b) the bearing stress at the base of the footing. Neglect the weight of the footing and the column.

FIGURE 9–3   Column on footing.

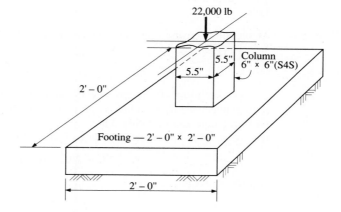

Solution   (a) The dressed dimensions of the column are $5\frac{1}{2}$ in. by $5\frac{1}{2}$ in. and the dressed area is 30.25 in.² The bearing stress on the column-footing contact surface is

$$s_p = \frac{P}{A} = \frac{22,000}{30.25} = 727 \text{ psi}$$

(b) The bearing stress at the base of the footing is

$$s_p = \frac{P}{A} = \frac{22,000}{24^2} = 38.2 \text{ psi}$$

Bearing pressure (stress) on soil is generally expressed in pounds per square foot (psf) rather than psi. Therefore,

$$s_p = 38.2(12)(12) = 5500 \text{ psf}$$

When calculating stress, as in the preceding examples, the area in question must be determined carefully. If there is a decrease in the cross-sectional area, such as occurs at section B–B in the bar shown in Fig. 9–4, there will naturally be a higher stress at that section. We will neglect, for now, the stress concentrations produced in this situation. Stress concentrations will be covered later, in Section 11–4.

FIGURE 9–4　Flat bar with holes.

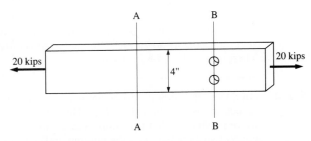

□ EXAMPLE 9–5　　A flat steel bar, $\frac{1}{2}$ in. thick and 4 in. wide, is subjected to a 20 kip tensile load. Two $\frac{3}{4}$ in. diameter holes are located as shown in Fig. 9–4. Determine the average tensile stress at sections A–A and B–B. Neglect stress concentrations adjacent to each hole.

Solution　　At section A–A the cross-sectional area is 2.0 in.². The stress is calculated from

$$s_t = \frac{P}{A} = \frac{20}{2.0} = 10.0 \text{ ksi}$$

At section B–B the cross-sectional area is reduced by the two holes. The area reduction caused by each hole is a rectangle, with one dimension equal to the diameter of the hole and the other dimension equal to the thickness of the bar. Therefore, the cross-sectional area at section B–B is calculated as

$$A = 2.0 - 2(0.75)(0.50) = 1.25 \text{ in.}^2$$

and the stress is

$$s_t = \frac{P}{A} = \frac{20}{1.25} = 16.0 \text{ ksi}$$

If a vertical bar rests on a horizontal base and supports no load except its own weight, the stress at any cross section may be calculated by dividing the weight of the part of the bar *above* the cross-section by the cross-sectional area. If a bar is suspended from its upper end and supports no load except its own weight, the method of solution is similar, but the weight of the bar *below* the cross section must be considered.

□ EXAMPLE 9–6　　A 100 ft long steel bar is suspended vertically from its upper end. The cross-sectional area of the bar is 4.0 in.² (2 in. by 2 in.). The unit weight of the steel is 490 pounds per cubic foot (pcf). Compute (a) the maximum tensile stress due to the bar's own weight and (b) the maximum load P (the allowable load) that can be safely supported at the lower end of the bar if the allowable tensile stress is 24,000 psi.

Solution Calculate the weight per foot for this bar as follows:

$$\text{Weight per foot} = (\text{volume})(\text{unit weight})$$

$$= \frac{2(2)(12)}{1728}\,(490) = 13.61 \text{ lb/ft}$$

Note that the volume calculated is for a length of one foot and must be determined in cubic feet, hence the use of the conversion factor of 1728 in.^{3}/ft^3.

(a) The maximum tensile stress in the bar due to its own weight will occur at its upper end.

$$\text{Bar weight} = 100(13.61) = 1361 \text{ lb}$$

The induced tensile stress is

$$s_t = \frac{P}{A} = \frac{1361}{4.0} = 340 \text{ psi}$$

(b) Since the allowable tensile stress is 24,000 psi and the induced stress due to the bar's own weight is 340 psi, the difference between the two represents a stress that can be induced by P. Hence, the allowable load P the bar can safely support at its lower end is calculated as follows:

$$P = s_t A$$
$$= (24,000 - 340)(4.0)$$
$$= 94,640 \text{ lb}$$

9–3
SHEAR STRESSES

In Section 9–2 we discussed how tensile and compressive stresses are developed in a direction perpendicular (or normal) to the surfaces on which they act. They are sometimes called *normal stresses*. Another type of stress, called *shear stress*, is developed in a direction parallel to the surface on which it acts. It is also called *tangential stress*. Normal stresses and shear stresses are illustrated in Fig. 9–5.

Another example of shear stress is illustrated in Fig. 9–6(a). When equal and opposite forces P are applied to two flat plates bonded together by an adhesive, the contact surface, shown shaded, is subjected to a shearing action. In the absence of the adhesive, the two surfaces would tend to slide past one another. The shear force is assumed to be uniformly distributed across the contact area. The result is the development of a shear stress, the magnitude of which is computed from the expression

$$s_s = \frac{P}{A} \tag{9–4}$$

where s_s = the average computed shear stress (psi, ksi) (Pa, MPa)
  P = the external applied shear force (lb, kips) (N)
  A = the area over which shear stress develops (in.^{2}) (m^2, mm^2)

FIGURE 9–5 Types of stress.

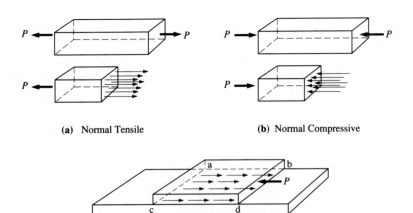

(a) Normal Tensile   (b) Normal Compressive

(c) Shear

FIGURE 9–6 Shear stress examples.

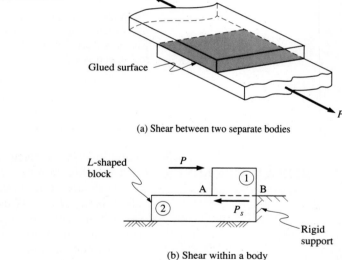

(a) Shear between two separate bodies

(b) Shear within a body

The uniformly distributed shear stress, as assumed here, is less likely to occur than is uniformly distributed tensile or compressive stress. This being the case, the shear stress computed from $s_s = P/A$ should be interpreted as an *average* value. In this particular application, along with other applications in this chapter, any nonuniform stress distribution is neglected.

Shear stress can also develop within a body when various layers of the material tend to slide with respect to each other. This action is illustrated in Fig. 9–6(b). A force P is applied as shown. A resisting force P_s acts in plane AB to prevent a sliding action between component parts 1 and 2. This

FIGURE 9–7 The punching process.

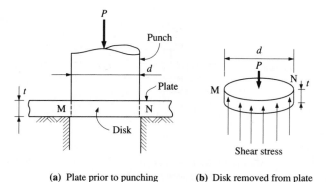

(a) Plate prior to punching   **(b)** Disk removed from plate

resisting force constitutes an internal shear force. Assuming that the resistance is uniformly distributed, the resistance per unit area, or shear stress, can be computed using Eq. (9–4).

Since the applied force P and the resisting force P_s are equal and parallel, all horizontal planes located between them have the same tendency to slide with respect to one another and each plane will develop the same intensity of shear stress.

Another shear stress illustration is shown in Fig. 9–7 where a force P is applied to a punch in order to punch a hole through a metal plate. Since there is no support directly under the punch, and since the summation of the vertical forces must equal zero, shear stresses are developed over a resisting area equal to the circumference of the punch multiplied by the plate thickness. The material removed by the punch will have a disklike shape and a thickness equal to the plate thickness. The metal around the circumference of the disk must fail in shear as the disk is separated from the rest of the plate. The magnitude of the shear stress achieved in the material when it fails in shear (when the hole is punched) is called the *ultimate shear strength*. If d is the diameter of the punch (or of the disk) and t is the thickness of the plate, then the sheared area is

$$A = \pi d t$$

and the average shear stress is

$$s_s = \frac{P}{A} = \frac{P}{\pi d t}$$

☐ **EXAMPLE 9–7**   The lap joint in Fig. 9–8 consists of two steel plates connected by two $\frac{3}{4}$ in. diameter bolts. For a tensile load P of 18,000 lb, compute the average shear stress in the bolts.

Solution   Assume that the load is resisted equally by each bolt and that the shear stress developed is uniformly distributed across the cross section of each bolt. Since there is only one plane of shear per bolt, the resisting shear (circular) area of each $\frac{3}{4}$ in.

FIGURE 9–8   Bolted joint.

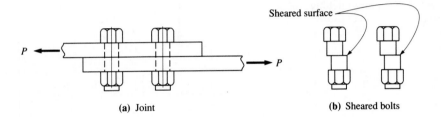

(a) Joint   (b) Sheared bolts

diameter bolt is 0.44 in.^{2}. Each bolt resists 9000 lb (one-half of the total load). The average shear stress is, therefore,

$$s_s = \frac{P}{A} = \frac{9000}{0.44} = 20,500 \text{ psi}$$

☐ **EXAMPLE 9–8**   Three pieces of wood are glued together to form an assembly (shown in Fig. 9–9) which can be used to test the shear strength of a glued joint. A load P of 10,000 lb is applied. Compute the average shear stress in each joint.

FIGURE 9–9   Glued wood joints.

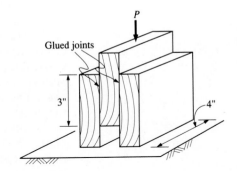

Solution   The resisting shear area in each joint is 12 in.^{2}. Assume that the load P is equally resisted by each joint and that the shear stress developed is uniformly distributed throughout the joint. Therefore, each joint resists 5000 lb (one-half the total load). The average shear stress is calculated as

$$s_s = \frac{P}{A} = \frac{5000}{12} = 417 \text{ psi}$$

☐ **EXAMPLE 9–9**   A punching operation is being planned in which a $\frac{3}{4}$ in. diameter hole is to be punched in an aluminum plate $\frac{1}{4}$ in. thick (refer to Fig. 9–7). For this aluminum alloy, the ultimate shear strength is 27,000 psi. Compute the force P that must be applied to the punch. Assume that the shear stress is uniformly distributed.

Solution   The resisting shear area is the circumference of the punch multiplied by the thickness of the plate:

$$A = \pi dt = \pi(0.75)(0.25) = 0.589 \text{ in.}^2$$

Next, rewrite Eq. (9–4) and by substituting the ultimate shear stress for s_s, solve for P, the required applied force that will induce an ultimate shear stress of 27,000 psi:

$$P = As_{s(\text{ult})} = 0.589(27,000) = 15,900 \text{ lb}$$

The previous group of problems is representative of analysis-type problems. Design-type problems are those in which the size and/or shape of a member must be determined. The designed member must support expected loads without exceeding an allowable stress. In this case, Eq. (9–4) must be rewritten to provide the required shear area. This is the same procedure that was presented initially in Section 9–2 (see Eqs. (9–1) and (9–3)). For shear considerations, the definitions of two of the three terms are slightly different:

$$A = \frac{P}{s_{s(\text{all})}} \tag{9–5}$$

where A = the required area that will limit the shear stress to the allowable shear stress
  P = the applied (or expected) load or force
  $s_{s(\text{all})}$ = the allowable shear stress

The allowable shear stress depends on the application and on material properties, which will be discussed in Chapter 10.

□ **EXAMPLE 9–10**   The rod shown in Fig. 9–10 is to support a load P of 20,000 lb. The material of which the rod is to be made will be a steel designated by the American Iron and Steel

FIGURE 9–10   Clevis connection.

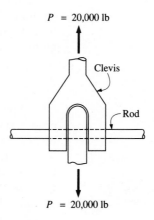

P = 20,000 lb

Clevis

Rod

P = 20,000 lb

Institute as AISI 1020. The steel has an allowable shear stress of 7500 psi. Determine the required diameter and select a diameter to use, assuming available bar diameters vary by $\frac{1}{8}$ in.

Solution   Since there are two planes of shear, each shear plane will resist 20,000/2 or 10,000 lb. The required cross-sectional area per plane is

$$A = \frac{P}{s_{s(\text{all})}} = \frac{10,000}{7,500} = 1.33 \text{ in.}^2$$

Since $A = \pi d^2/4$, or $0.7854 d^2$, the required diameter is calculated from

$$d = \sqrt{\frac{A}{0.7854}} = \sqrt{\frac{1.33}{0.7854}} = 1.30 \text{ in.}$$

Use a $1\frac{3}{8}$ in. diameter rod.

**9–4
TENSILE AND
COMPRESSIVE
STRAIN AND
DEFORMATION**

The terms *deformation* and *strain* both represent dimensional change. A body subjected to either a tensile force or a compressive force will undergo a change in length. It will be elongated (lengthened) by the tensile force and compressed (shortened) by the compressive force.

With some materials (rubber, for example), small loads produce relatively large deformations. Other engineering materials respond in a similar way, although the amount of the deformation may be extremely small. Even such very rigid materials as steel, when subjected to load, will undergo a small deformation. The total deformation, or change in length, of a member is generally designated δ (Greek lowercase delta).

To permit comparisons with standard values, the total deformation is converted to a unit basis and is expressed in terms of deformation per unit length. It is generally termed *unit strain*. For the remainder of this book, the word *strain* will be used to signify unit strain.

In the determination of tensile or compressive strain, the assumption is made that each unit of length of a member will elongate or shorten the same amount. Strain, represented in this text by ε (Greek lowercase epsilon), can be computed by dividing the total deformation by the original length of the member. Mathematically, this is written

$$\epsilon = \frac{\text{total deformation}}{\text{original length}} = \frac{\delta}{L} \qquad \textbf{(9–6)}$$

Since the strain is a ratio of two lengths, it is dimensionless and considered a pure number. However, it is common practice to express strain in terms of inches per inch in the U.S. Customary System, since both numerator and denominator are expressed in inches. In the SI system, strain may be expressed in terms of millimeters per millimeter or meters per meter. The units in the numerator and denominator must be consistent. In a perfectly straight bar (member) of homogeneous material, of constant cross section and under constant load, ε represents the theoretical strain in the bar. With any variation in these parameters, the quantity δ/L represents only the average strain along the length L.

If a member is suspended from its upper end and supports only its own weight, the strain varies uniformly from zero at the lower end to a maximum value at the upper end. The average strain is given by the expression δ/L, where δ and L are as previously defined. But the *maximum* strain occurs at the upper end, where the force in the member is the greatest. Since the strain variation is linear, the strain at the top of the member will be twice the average, or $2\delta/L$.

☐ **EXAMPLE 9–11**   Compute the total elongation (deformation) for a steel wire 60 ft in length if the strain is 0.00067 in./in.

Solution

$$L = 60(12) = 720 \text{ in.}$$
$$\epsilon = 0.00067 \text{ in./in.}$$

$$\epsilon = \frac{\delta}{L}$$

Rewriting,

$$\delta = \epsilon L = 0.00067(720) = 0.48 \text{ in.}$$

**9–5
SHEAR STRAIN**

When an axial tensile load is applied to a body, it will cause a longitudinal tensile deformation—an elongation. Similarly, an axial compressive load will cause a longitudinal compressive deformation—a shortening. When a shear force is applied to a body, it will cause a shear deformation in the same direction as the applied force. Rather than an elongation or shortening, this deformation will be an angular distortion.

A motor mount is shown in Fig. 9–11(a). The motor mount is composed of a block of elastic material with attachments to allow for connection to the base of the motor and the supporting structure. A force P is applied at the top of the block. This action subjects the block (of height L) to a pair of

FIGURE 9–11 Shear strain.

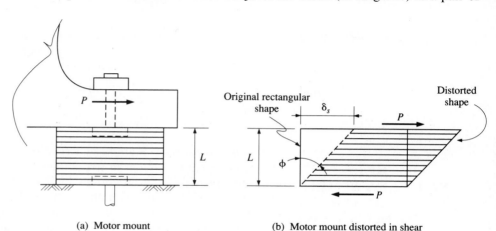

(a) Motor mount

(b) Motor mount distorted in shear

shear forces as shown in Fig. 9–11(b). If we imagine that the block is composed of many thin layers, and that each layer will slide slightly with respect to its neighbor, we can visualize how the angular distortion will develop. As may be observed, the total shear deformation in the length L is δ_s and the shear strain, similar to tensile and compressive strains, is the total shear deformation divided by the length L:

$$\epsilon_s = \frac{\delta_s}{L} \qquad\qquad (9\text{–}7)$$

Note also that in Fig. 9–11 an angular relationship exists in which

$$\tan \phi = \frac{\delta_s}{L} = \epsilon_s$$

For small angles, which is generally the case, the tangent of the angle is approximately equal to the angle expressed in radians. A radian is defined as the central angle subtended by an arc length equal to the radius of the circle (see Fig. 9–12). From this definition, we see that the angle, in radians, is the arc length divided by the radius of the circle. For extremely short arc lengths, the arc approaches a straight line and the line becomes almost perpendicular to the radius. Therefore, the tangent of the angle (opposite side/adjacent side) is approximately equal to the angle in radians (arc length/radius). Therefore, the angle ϕ in radians is very nearly equal to the shear strain.

FIGURE 9–12   Radian definition.

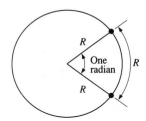

☐ EXAMPLE 9–12

In Fig. 9–5(c), assume that load P is applied at the top of the block and that it displaces the top a horizontal distance of 0.0024 in. with respect to plane *abcd*. Assume the height of the top block to be 1.4 in. Compute the shear strain.

Solution

$$\epsilon_s = \frac{\delta_s}{L} = \frac{0.0024}{1.4} = 0.0017 \text{ in./in.}$$

**9–6
THE RELATION
BETWEEN STRESS
AND STRAIN
(HOOKE'S LAW)**

For most engineering materials, a relationship exists between stress and strain. For each increment in stress there is a closely proportional increase in strain, provided that a certain limit of stress is not exceeded. If the induced stress exceeds this limiting value, the corresponding strain will no longer be proportional to the stress. This limiting value is called the *proportional limit* and may be concisely defined as that value of stress up to which strain is proportional to stress.

The proportional relationship between stress and strain was originally stated by Robert Hooke in 1678 and became known as *Hooke's law*. In effect, for material that obeys Hooke's law, applied loads P_A and P_B will produce stresses s_A and s_B and strains ϵ_A and ϵ_B, and the ratio of the two values will be a constant such that

$$\frac{s_A}{\epsilon_A} = \frac{s_B}{\epsilon_B} = \text{constant}$$

This constant is now known as the *modulus of elasticity* or *Young's modulus* (after Thomas Young, who is credited with having defined it in 1807). The modulus of elasticity for members in tension or compression is generally represented by the symbol E and is expressed by the equation

$$E = \frac{\text{stress}}{\text{strain}} = \frac{s}{\epsilon} \qquad \textbf{(9–8)}$$

Since strain is a pure number, it is evident that E has the same units as does stress (s), which, in the U.S. Customary System, is usually pounds per square inch (psi) and, in the SI system, pascals (Pa) or megapascals (MPa). For most of the common engineering materials, the modulus of elasticity in compression is equal to that found in tension for all practical purposes. In the case of steel and other ductile metals, tension tests are more easily performed than compression tests. Therefore, it is generally the tensile modulus of elasticity that is determined and used.

The standard tension test used to determine the modulus of elasticity will be discussed in Chapter 10, along with the stress limitations within which modulus of elasticity is valid. We will briefly discuss modulus of elasticity in this section because of the importance of the constant ratio between stress and strain in the design process for many engineering materials.

The modulus of elasticity of steel (in tension and compression) is commonly assumed to be 29,000,000 psi or 30,000,000 psi in the U.S. Customary System. In the SI system, these values convert to 200 000 MPa or 207 000 MPa. The particular values depend on the type of steel. For the purposes of this book, values of the modulus of elasticity for steel will be taken as 30,000,000 psi (207 000 MPa). (Note that when values are shown in this form, the first number will be the value in the U.S. Customary System and the parenthetical value will be its approximate equivalent in the SI system.)

For other materials, the values of the modulus of elasticity may be as low as 1,000,000 psi (7000 MPa) or less. Average values of modulus of elasticity for some common engineering materials are given in Appendix G (see also Appendices Notes).

Physically, the modulus of elasticity is a measure of the stiffness of a material in its response to an applied load and represents a definite property of that material. Material stiffness may be defined as the property that enables a material to withstand high stress without great strain. In Chapter 8 we discussed geometric properties of cross-sectional areas and introduced the concept of stiffness as a direct result of those geometric properties. We now see that the total stiffness of a member can be attributed to a combination of both material properties and geometric properties.

As with axially loaded bodies in tension and compression, the shear stress is proportional to the shear strain, as long as the proportional limit in shear has not been exceeded. This constant of proportionality is known as the *modulus of elasticity in shear* or the *modulus of rigidity*. It is denoted G and expressed as

$$G = \frac{\text{shear stress}}{\text{shear strain}} = \frac{s_s}{\epsilon_s} \tag{9-9}$$

Average values of the modulus of rigidity for some common engineering materials are given in Appendix G. Note that these values and the tabulated values of modulus of elasticity are significantly different. The modulus of rigidity and its relationship to other material properties is further discussed in Chapter 11.

☐ **EXAMPLE 9–13**   A 12 in. long bar with a cross-sectional area of 1.0 in.^{2} is subjected to an axial tensile load of 1000 lb. Compute the stress, strain, and the total elongation if the bar material is (a) steel, with $E_{ST} = 30,000,000$ psi; (b) aluminum, with $E_{AL} = 10,000,000$ psi; and (c) wood, with $E_W = 1,500,000$ psi. The proportional limit for each material is as follows: steel = 34,000 psi, aluminum = 35,000 psi, and wood = 6000 psi.

Solution   1. The tensile stress for *all* materials is

$$s_t = \frac{P}{A} = \frac{1000}{1.0} = 1000 \text{ psi}$$

This is less than the proportional limit for all materials; therefore, Hooke's law applies.

2. Compute the strains for the three materials.
(a) Steel:

$$\epsilon_{ST} = \frac{s_t}{E_{ST}} = \frac{1000}{30,000,000} = 0.0000333 \text{ in./in.}$$

(b) Aluminum:

$$\epsilon_{AL} = \frac{s_t}{E_{AL}} = \frac{1000}{10,000,000} = 0.0001 \text{ in./in.}$$

(c) Wood:

$$\epsilon_W = \frac{s_t}{E_W} = \frac{1000}{1,500,000} = 0.000667 \text{ in./in.}$$

3. Compute the total elongation for each material.
(a) Steel:

$$\delta_{ST} = \epsilon_{ST}L = 0.0000333(12) = 0.000400 \text{ in.}$$

(b) Aluminum:

$$\delta_{AL} = \epsilon_{AL}L = 0.0001(12) = 0.0012 \text{ in.}$$

(c) Wood:

$$\delta_W = \epsilon_W L = 0.000667(12) = 0.008 \text{ in.}$$

The strain and total deformation in the aluminum bar are approximately three times larger than those of the steel bar. The strain in the wood is approximately twenty times larger than that of the steel. This indicates that the stiffness of the steel is significantly greater than that of either the aluminum or wood.

We have established expressions for stress, strain, and modulus of elasticity E. These may now be combined and a convenient expression developed to determine directly the total deformation δ for a homogeneous axially loaded prismatic member. We begin with the definition of modulus of elasticity and substitute for stress and strain:

$$E = \frac{s}{\epsilon} = \frac{P/A}{\delta/L} = \frac{PL}{A\delta}$$

Solving for δ,

$$\delta = \frac{PL}{AE} \qquad \text{(9–10)}$$

where δ = the total axial deformation (in.) (m, mm)
P = the total applied external axial load (lb, kips) (N)
L = the length of the member (in.) (m, mm)
A = the cross-sectional area of the member (in.²) (m², mm²)
E = the modulus of elasticity (psi, ksi) (Pa, MPa)

This expression is valid only when the stress in the member does not exceed the proportional limit.

☐ **EXAMPLE 9–14**   A tensile member in a machine is subjected to an axial load of 5000 lb. It has a length of 30 in. and is made from a steel tube having an O.D. of $\frac{3}{4}$ in. and an I.D. of $\frac{1}{2}$ in. Compute the tensile stress in the tube and the total axial deformation. Assume $E = 30,000,000$ psi and a proportional limit of 34,000 psi.

Solution   1. The cross-sectional area of the hollow tube is calculated from

$$A = \frac{\pi}{4}(d^2 - d_1^2) = 0.7854(0.75^2 - 0.50^2) = 0.245 \text{ in.}^2$$

2. Calculate the axial deformation (elongation):

$$\delta = \frac{PL}{AE} = \frac{5000(30)}{0.254(30,000,000)} = 0.0204 \text{ in.}$$

3. Calculate the tensile stress developed and verify that it is not in excess of the proportional limit:

$$s_t = \frac{P}{A} = \frac{5000}{0.245} = 20,400 \text{ psi} < 34,000 \text{ psi} \qquad \textbf{OK}$$

☐ **EXAMPLE 9–15**   A weight of 1500 lb is to be suspended by a steel wire 24 ft in length. The tensile stress in the wire must not exceed 20,000 psi and the total elongation (deformation) must not exceed 0.18 in. Compute the required diameter of the wire. Neglect the weight of the wire. Use $E = 30,000,000$ psi and a proportional limit of 34,000 psi.

Solution   This is a design-type problem. The length of the wire is 24(12) = 288 in. The given maximum tensile stress is actually an allowable tensile stress $(s_{t(\text{all})})$. The required area is calculated based on the allowable tensile stress:

$$\text{Required } A = \frac{P}{s_{t(\text{all})}} = \frac{1500}{20,000} = 0.075 \text{ in.}^2$$

Next, the required area is calculated based on total deformation allowed, using

$$\delta = \frac{PL}{AE}$$

which is rewritten for the required area

$$\text{Required } A = \frac{PL}{\delta E} = \frac{1500(288)}{0.18(30,000,000)} = 0.080 \text{ in.}^2$$

The larger of the two areas (0.080 in.^{2}) must be furnished because it will satisfy both conditions. Therefore, the required diameter is calculated from

$$\text{Required } A = \frac{\pi d^2}{4} = 0.080 \text{ in.}^2$$

$$\text{Required } d = \sqrt{\frac{0.080(4)}{\pi}} = 0.32 \text{ in.}$$

Finally, confirm that the tensile stress developed is less than the proportional limit:

$$s_t = \frac{P}{A} = \frac{1500}{0.80} = 18,750 \text{ psi} < 34,000 \text{ psi} \qquad \textbf{OK}$$

☐ **EXAMPLE 9–16**   A copper wire 150 ft long and $\frac{1}{8}$ in. in diameter is suspended vertically from its upper end. The unit weight of the copper is 550 pcf. Compute (a) the total elongation of the wire due to its own weight, (b) the average strain due to its own weight, (c) the total elongation of the wire δ (which includes the elongation due to its own weight) if a weight of 70 lb is attached to its lower end, and (d) the maximum load P that this wire can safely support at its lower end if the allowable tensile stress is 12,000 psi.

Solution   For a wire suspended vertically from its upper end, the total elongation produced by the weight of the wire is equal to that produced by a load of half its weight applied at the end (this is the same as the average load being applied throughout the length of the wire).

(a) The weight of the wire is equal to the total volume times the unit weight of the copper:

$$\text{Total weight} = \frac{\pi d^2}{4}(L)(550)$$

$$= \frac{\pi (0.125)^2}{4(144)}(150)(550) = 7.03 \text{ lb}$$

The area of $\frac{1}{8}$ in. diameter wire is 0.0123 in.^{2}. The total elongation is calculated using E from Appendix G:

$$\delta = \frac{(P/2)L}{AE} = \frac{(7.03/2)(150)(12)}{0.0123(15,000,000)} = 0.0343 \text{ in.}$$

(Note that the length is converted to inches.)

(b) The average strain is

$$\epsilon = \frac{\delta}{L} = \frac{0.0343}{150(12)} = 0.000019 \text{ in./in.}$$

(c) The additional elongation due to a 70 lb weight is

$$\delta = \frac{PL}{AE} = \frac{70(150)(12)}{0.0123(15,000,000)} = 0.683 \text{ in.}$$

from which

$$\text{Total } \delta = 0.683 + 0.0343 = 0.717 \text{ in.}$$

(d) To find the maximum load P that can be safely supported (see Example 9–6), calculate the maximum stress due to the weight of the wire:

$$s_t = \frac{P}{A} = \frac{7.03}{0.0123} = 572 \text{ psi}$$

Therefore, the portion of the allowable stress that remains to resist the load P is

$$12,000 \text{ psi} - 572 \text{ psi} = 11,430 \text{ psi}$$

and the maximum (or allowable) load is

$$P = As_t = 0.0123(11,430) = 140.6 \text{ lb}$$

□ **EXAMPLE 9–17**   A steel bar having a cross-sectional area of 1.0 in.^{2} is suspended vertically and subjected to loads as shown in Fig. 9–13. The total length of the bar is 7 ft. Compute the total elongation of the bar. Use $E = 30,000,000$ psi. Neglect the weight of the bar.

FIGURE 9–13 Suspended steel bar.

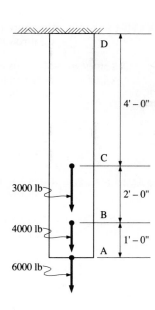

Solution   The elongation of AB is

$$\delta = \frac{PL}{AE} = \frac{6000(1)(12)}{1.0(30,000,000)} = 0.0024 \text{ in.}$$

The elongation of BC is

$$\delta = \frac{PL}{AE} = \frac{10,000(2)(12)}{1.0(30,000,000)} = 0.0080 \text{ in.}$$

The elongation of CD is

$$\delta = \frac{PL}{AE} = \frac{13,000(4)(12)}{1.0(30,000,000)} = 0.0208 \text{ in.}$$

Thus,

$$\text{Total } \delta = 0.0024 + 0.0080 + 0.0208 = 0.0312 \text{ in.}$$

9–7
SI SYSTEM
EXAMPLES

□ **EXAMPLE 9–18**

Compute the tensile stress developed in a steel bar 50 mm by 50 mm in cross section if it is subjected to an axial tensile load of 400 kilonewtons (kN). Refer to Fig. 9–1(a).

Solution   Using the direct stress formula (Eq. (9–1)),

$$P = 400 \text{ kN} = 400 \times 10^3 \text{ N}$$
$$A = (50 \text{ mm})(50 \text{ mm}) = 2500 \text{ mm}^2 = 2.5 \times 10^{-3} \text{ m}^2$$
$$s_t = \frac{P}{A} = \frac{400 \times 10^3 \text{ N}}{2.5 \times 10^{-3} \text{ m}^2} = 160 \times 10^6 \text{ N/m}^2 = 160 \text{ MPa}$$

□ **EXAMPLE 9–19**   Compute the allowable axial tensile load that may be applied to a 19 mm diameter steel rod if the allowable axial tensile stress is 150 megapascals (MPa).

Solution   The cross-sectional area of the rod is 284 mm². The allowable axial tensile load is obtained from Eq. (9–2). Since the units of the resulting force should be newtons (N), make the substitutions in terms of newtons and meters (m):

$$P = s_{t(\text{all})} A = (150 \times 10^6 \text{ N/m}^2)(284 \times 10^{-6} \text{ m}^2)$$
$$= 42\,600 \text{ N} = 42.6 \text{ kN}$$

□ **EXAMPLE 9–20**   Calculate the required diameter of an aluminum rod subjected to an axial tensile load of 50 kN, assuming an allowable tensile stress of 125 MPa.

Solution   The required cross-sectional area is obtained from Eq. (9–3). Substituting in terms of newtons and meters,

$$\text{Required } A = \frac{P}{s_{t(\text{all})}} = \frac{50 \times 10^3 \text{ N}}{125 \times 10^6 \text{ N/m}^2} = 0.400 \times 10^{-3} \text{ m}^2$$

The required diameter may be calculated using $A = \pi d^2/4$ or $A = 0.7854d^2$:

$$\text{Required } d = \sqrt{\frac{A}{0.7854}} = \sqrt{\frac{0.400 \times 10^{-3}}{0.7854}} = 0.0226 \text{ m}$$

In order to present the answer within the desired numerical range (0.1 to 1000), convert the result to millimeters:

$$\text{Required } d = 0.0226 \text{ m} = 22.6 \text{ mm}$$

☐ **EXAMPLE 9–21**   A 25 mm diameter bolt extends vertically through a timber beam, as shown in Fig. 9–14. The maximum tensile stress developed in the bolt is 83 MPa. Determine the minimum required diameter (d) for a circular steel plate which is to be placed under the head of the bolt. The bearing stress on the timber is not to exceed 3.4 MPa.

FIGURE 9–14   Bolt through timber beam.

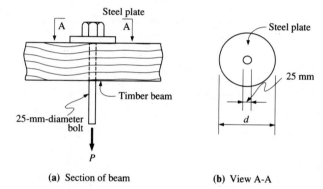

(a) Section of beam   (b) View A-A

Solution   The cross-sectional area of the bolt is 491 mm². The applied tensile load P that induces the tensile stress of 83 MPa in the bolt is obtained from Eq. (9–2). Substituting in terms of newtons and meters,

$$P = s_t A = (83 \times 10^6 \text{ N/m}^2)(491 \times 10^{-6} \text{ m}^2) = 40\ 753 \text{ N}$$

The required bearing area between the steel plate and the timber beam is then computed from Eq. (9–3):

$$\text{Required } A = \frac{P}{s_{p(\text{all})}} = \frac{40\ 753 \text{ N}}{3.4 \times 10^6 \text{ N/m}^2} = 11\ 986 \times 10^{-6} \text{ m}^2$$

$$= 11\ 986 \text{ mm}^2$$

Assuming the plate has a 25 mm diameter hole for the bolt, the net contact bearing area between the plate and the timber beam is

$$A = 0.7854 d^2 - 0.7854(25)^2$$

Equating this to the required area and solving for d,

$$11\ 986 = 0.7854 d^2 - 0.7854(25)^2$$

from which

$$\text{Required } d = 126 \text{ mm}$$

□ **EXAMPLE 9–22**   A tractor drawbar is connected to an implement as shown in Fig. 9–15. The tractor pulls with a force of 50 kN that must be transmitted by the shear bolt. (a) Calculate the shear stress in the bolt if the bolt diameter is 19 mm. (b) Calculate the percent increase in shear stress if the bolt diameter is reduced to 16 mm.

Solution   In this connection there are two planes of shear resisting the load. Each shear plane resists 25 kN, one-half of the 50 kN total load.

(a) The cross-sectional area of the 19 mm bolt, per shear plane, is 284 mm². The average shear stress is

$$s_s = \frac{P}{A} = \frac{25 \times 10^3 \text{ N}}{284 \times 10^{-6} \text{ m}^2} = 0.0880 \times 10^9 \text{ N/m}^2 = 88.0 \text{ MPa}$$

(b) The cross-sectional area of the 16 mm bolt, per shear plane, is 201 mm². The average shear stress is

$$s_s = \frac{P}{A} = \frac{25 \times 10^3 \text{ N}}{201 \times 10^{-6} \text{ m}^2} = 124.4 \times 10^9 \text{ N/m}^2 = 124.4 \text{ MPa}$$

The percent increase in shear stress is

$$\frac{124.4 - 88.0}{88.0} (100) = 41 \text{ percent}$$

FIGURE 9–15   Implement hitch.

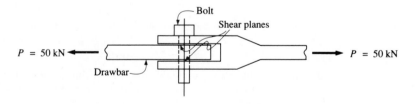

$P = 50$ kN $P = 50$ kN

□ **EXAMPLE 9–23**   A boiler plate steel has an ultimate shear strength of 290 MPa. Compute the force required to punch a 25 mm diameter hole if the steel plate thickness is 13 mm. Assume the shear stress is uniformly distributed (refer to Fig. 9–7).

Solution   The resisting shear area is the circumference of the punch multiplied by the thickness of the plate:

$$A = \pi dt = \pi(25)(13) = 1021 \text{ mm}^2$$

Solve for P, the required applied force that will induce an ultimate shear stress of 290 MPa, by rewriting Eq. (9–4) and substituting the ultimate shear stress for s_s:

$$P = As_{s(\text{ult})} = (1021 \times 10^{-6} \text{ m}^2)(290 \times 10^6 \text{ N/m}^2)$$
$$= 296\,090 \text{ N} = 296 \text{ kN}$$

EXAMPLE 9–24   Calculate the distance H required for the lower joint of a timber truss so as not to exceed an allowable shear stress parallel to the grain of 825 kPa along plane a–b. Use the nominal dimensions shown in Fig. 9–16.

Solution   First calculate the horizontal component of P which is parallel to the shear plane:

$$P_H = P \cos 30° = (25 \text{ kN})(0.866) = 21.7 \text{ kN}$$

From Eq. (9–5), then calculate the shear area required along plane a–b:

$$\text{Required } A = \frac{P}{s_{s(\text{all})}} = \frac{21.7 \times 10^3 \text{ N}}{825 \times 10^3 \text{ N/m}^2}$$
$$= 0.0263 \text{ m}^2 = 26\ 300 \text{ mm}^2$$

Since the shear area along plane a–b is the product of H and the thickness of the timber (100 mm),

$$\text{Required } H = \frac{A}{100 \text{ mm}} = \frac{26\ 300 \text{ mm}^2}{100 \text{ mm}} = 263 \text{ mm}$$

FIGURE 9–16   Timber truss joint.

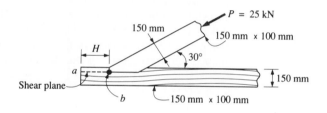

EXAMPLE 9–25   A weight of 6700 N is to be suspended by a steel wire 7.5 m in length. The tensile stress in the wire must not exceed 138 MPa and the total elongation (deformation) must not exceed 4.5 mm. Compute the required diameter of the wire. Neglect the weight of the wire. Use $E = 207\ 000$ MPa and a proportional limit of 234 MPa.

Solution   The required cross-sectional area is calculated based on the allowable tensile stress of 138 MPa. Making the substitutions in terms of newtons and meters,

$$\text{Required } A = \frac{P}{s_{t(\text{all})}} = \frac{6700 \text{ N}}{138 \times 10^6 \text{ N/m}^2} = 48.6 \times 10^{-6} \text{ m}^2 = 48.6 \text{ mm}^2$$

The required cross-sectional area is calculated based on allowable total deformation using Eq. (9–10) rewritten for the required area:

$$\text{Required } A = \frac{PL}{\delta E} = \frac{(6700 \text{ N})(7.5 \text{ m})}{(4.5 \times 10^{-3})(207\ 000 \times 10^6 \text{ N/m}^2)}$$
$$= 0.0539 \times 10^{-3} \text{ m}^2 = 53.9 \text{ mm}^2$$

The larger of the two areas (53.9 mm²) must be provided. Therefore, the required diameter is calculated from

$$d = \sqrt{\frac{A}{0.7854}} = \sqrt{\frac{53.9}{0.7854}} = 8.28 \text{ mm}$$

Checking to ensure that the tensile stress developed is less than the proportional limit,

$$s_t = \frac{P}{A} = \frac{6700 \text{ N}}{53.9 \times 10^{-6} \text{ m}^2}$$

$$= 124 \times 10^6 \text{ Pa} = 124 \text{ MPa} < 234 \text{ MPa} \qquad \textbf{OK}$$

SUMMARY—BY SECTION NUMBER

9–2   *Stress* (as used herein) represents an internal resistance developed by a unit area of a member. An axial force is collinear with the longitudinal axis of a member, acts through its centroid, and develops a uniform axial stress distribution. An axial tensile force tends to elongate a member and develops uniform internal tensile stress over the cross-section of the member. Similarly, an axial compressive force tends to shorten the member and develops uniform internal compressive stress. A bearing stress is a compressive stress exerted on external surfaces of separate bodies in contact with each other, and which transmits force. All of these stresses are called *normal stresses* (also *direct stresses*), since the resisting area is normal to the direction of the force. They may be computed using the direct stress formula:

$$s = \frac{P}{A} \qquad \qquad \textbf{(9–1)}$$

9–3   A shear force develops a shear stress over an area in a direction parallel to the surface on which it acts. An average shear stress may be computed from

$$s_s = \frac{P}{A} \qquad \qquad \textbf{(9–4)}$$

Ultimate shear strength is the magnitude of the shear stress that will cause a material to fail in shear.

9–4   *Deformation* (δ), as used herein, represents a total dimensional change of a stressed member. *Strain* (ϵ), as used herein, represents a unit strain, or a dimensional change per unit length of a stressed member, and may be computed from

$$\epsilon = \frac{\text{total deformation}}{\text{original length}} = \frac{\delta}{L} \qquad \qquad \textbf{(9–6)}$$

9–5   Axial tensile or compressive forces applied to a body cause longitudinal tensile or compressive deformations. A shear force causes a shear deformation (δ_s), which is an angular distortion in the same direction as the applied force. Shear strain (ϵ_s) may be computed from

$$\epsilon_s = \frac{\delta_s}{L} \qquad \qquad \textbf{(9–7)}$$

9–6 Hooke's law refers to the relationship in which stress is proportional to strain. This applies to most engineering materials and is valid provided that the stress does not exceed the proportional limit of the material. This constant ratio of stress to strain is called the *modulus of elasticity* (*E*) and is defined for members in tension or compression, as

$$E = \frac{\text{stress}}{\text{strain}} = \frac{s}{\epsilon} \qquad (9\text{–}8)$$

The modulus of elasticity is a measure of a material's resistance to deformation when loaded.

The shear modulus of elasticity (*G*) is defined as

$$G = \frac{\text{shear stress}}{\text{shear strain}} = \frac{s_s}{\epsilon_s} \qquad (9\text{–}9)$$

Equations (9–1), (9–6), and (9–8) may be combined to yield an expression for total deformation:

$$\delta = \frac{PL}{AE} \qquad (9\text{–}10)$$

PROBLEMS

Section 9–2   Tensile and Compressive Stresses

1. Write the direct stress formula in its three forms and explain the meaning of each one.

2. The following members are subjected to axial tensile loads of 16,000 lb. Determine the stress induced. (a) Steel bar 2 in. by 2 in., (b) W4 × 13, (c) six-inch-diameter round wood post (use nominal size).

3. Determine the stresses in the two segments of the bar in Fig. 9–17. Segment A has a diameter of $1\frac{1}{4}$ in. and segment B has a cross section 2 in. by 2 in.

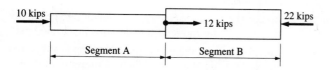

FIGURE 9–17   Problem 3.

4. A bin weighing 8 tons is supported by three steel tension rods. Each rod is $\frac{5}{8}$ in. in diameter. Assume that the rods support equal loads. Find the stress in the rods.

5. Diameters of small commercially available steel rods vary by sixteenths of an inch. Select the required commercial size of the rod required to support a tensile load of 35,000 lb if the tensile stress cannot exceed 20,000 psi.

Section 9–3   Shear Stresses

6. In the bolted connection of Fig. 9–8, assume that the bolts are $\frac{7}{8}$ in. in diameter and that the allowable shear stress for the bolts is 14,500 psi. Based on bolt shear only, compute the safe allowable load *P* that may be applied.

7. Compute the force required to punch a 1 in. diameter hole through a $\frac{1}{2}$ in. thick boiler plate. The ultimate shear strength for the material is 42,000 psi.

8. The $\frac{3}{4}$ in. diameter bolt in Fig. 9–18 is subjected to a 6000 lb tensile load. The bolt passes through a $\frac{3}{4}$ in. thick plate. Compute (a) the tensile stress in the bolt and (b) the shear stress in the head of the bolt.

9. A $\frac{3}{4}$ in. diameter punch is used to punch a hole through a steel plate $\frac{1}{2}$ in. thick. The force necessary to drive the punch through the plate is 60,000 lb. Compute the shear stress developed in the plate.

FIGURE 9–18   Problem 8.

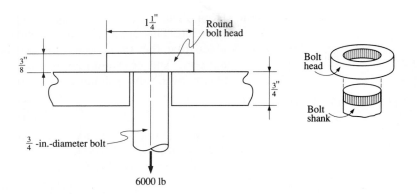

6000 lb

10. A control arm is keyed to a one-inch diameter shaft as shown in Fig. 9–19. A link transmits a 40 lb force to the end of the arm as shown. The key is $\frac{1}{4}$ in. × $\frac{1}{4}$ in. and has a length of $\frac{7}{8}$ in. Find the shear stress in the key.

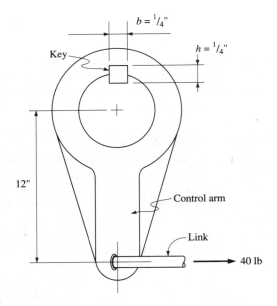

FIGURE 9–19   Problem 10.

11. Calculate the required width *b* for the key of Problem 10 if the link transmits a 60 lb force and the length of the key is $1\frac{1}{2}$ in. The allowable shear stress for the key is 4000 psi.

Section 9–4   Tensile and Compressive Strain and Deformation

12. Why is strain unitless (or dimensionless)?

13. (a) Given δ = 1.2 in. and L = 100 ft, calculate ϵ;
    (b) Given δ = 0.5 ft and ϵ = 0.00030, calculate L;
    (c) Given L = 1000 in. and ϵ = 0.00042, calculate δ.

14. A short compression member 2 in. by 2 in. in cross section and 8 in. long is subjected to a 40,000 lb axial compressive load. As a result, the length of the member is shortened to 7.85 in. Compute the strain in the member.

15. A 100 ft long rod is suspended from one end. The diameter of the rod is 1 in. and the total elongation is 0.80 in. Calculate the maximum strain.

Section 9–6   The Relation Between Stress and Strain (Hooke's Law)

For the following problems, assume the modulus of elasticity for steel to be 30,000,000 psi. Refer to Appendix F or G for other materials.

16. Compute the total elongation of a steel bar, originally 25 in. long, if the induced tensile stress is 15,000 psi.

17. A steel rod 1 in. in diameter and 20 ft long is subjected to an axial tensile load of 15,000 lb. Compute (a) stress, (b) strain, and (c) total elongation.

18. An aluminum rod 1 in. in diameter and 10 ft long is subjected to an axial tensile load of 15,000 lb. Compute (a) stress, (b) strain, and (c) total elongation.

19. A steel rod 10 ft long is made up of two 5 ft lengths, one $\frac{3}{4}$ in. in diameter and the other $\frac{1}{2}$ in. in diameter. How much will the bar elongate when subjected to an axial tensile load of 5000 lb?

20. A wrought-iron bar elongates 0.142 in. when subjected to an axial tensile load of 45,000 lb. The bar is 16 ft long and has a cross-sectional area of 2.25 in.^{2}. Compute the modulus of elasticity E. Use a proportional limit of 24,000 psi.

SI System Problems

For the following problems, assume the modulus of elasticity of steel to be 207 000 MPa. Refer to Appendix G for other materials. Use a proportional limit of 234 MPa for all steel members.

21. Calculate the stress developed in the following members which are each subjected to an axial tensile load of 70 kN: (a) steel bar, 25 mm by 50 mm; (b) 150 mm diameter wood post (use the given diameter); (c) 25 mm diameter steel tie rod.

22. A hopper weighing 75 kN is supported by three 19 mm diameter steel rods. Calculate the tensile stress developed in the rods, assuming that the rods support equal loads.

23. A steel bar has a rectangular cross section 25 mm by 100 mm and an allowable tensile stress of 140 MPa. Calculate the allowable axial tensile load that may be applied.

24. Compute the required nominal size of a short square timber post subjected to an axial compressive load of 250 kN. The allowable compressive stress is 8 MPa.

25. Calculate the required diameter of steel tie rods necessary to support a balcony. Each rod, suspended from a roof truss, supports a floor area of 7.0 m^2. The combined balcony floor load is 10.5 kPa, including dead load and live load. The allowable tensile stress in the steel tie rods is 140 MPa.

26. A 25 mm diameter aluminum rod, 3 m long, is subjected to an axial tensile load of 67 kN. Assume a proportional limit of 207 MPa. Compute (a) stress, (b) strain, and (c) total elongation.

27. A 5 mm diameter steel wire, 18 m in length, is used in the manufacture of prestressed concrete beams. Calculate the tensile load P required to stretch the wire 45 mm, and calculate the stress developed in the wire.

28. A 19 mm diameter steel rod, 30 m in length, is hanging vertically. Compute the total elongation of the rod due to its own weight and a suspended 36 kN load applied at its free end. Use a unit weight of steel of 77 kN/m^3.

29. Calculate the total elongation of a steel bar 635 mm in length if there is an induced tensile stress of 103 MPa in the bar.

30. Calculate the tensile force acting on a round steel rod 25 mm in diameter if there is an induced strain in the rod of 0.0007.

31. The bolted lap joint shown in Fig. 9–8 consists of two steel plates connected by two 25 mm diameter bolts. Assume that the load is equally resisted by the two bolts. Calculate the average shear stress in the bolts when the joint is subjected to a tensile load of 120 kN.

32. Calculate the force a punch press must exert if it is to punch out a large washer from a strip of 5 mm thick sheet steel with an ultimate shear strength of 400 MPa. The washer dimensions are as follows: hole diameter = 27 mm, outside diameter = 64 mm.

Computer Problems

For the following computer problems, any appropriate programming language may be used. Input prompts should fully explain what is required of the user (the program should be "user friendly"). The resulting output should be well labeled and self-explanatory.

33. Write a program that will calculate the allowable axial tensile load that may be applied on a round rod. User input is to be the rod diameter and the allowable axial tensile stress.

34. Write a program that will compute stress, strain, and total deformation for an aluminum bar. Use E = 10,000,000 psi. User input is to be cross-sectional area, length, and axial tensile load. An error message is to be displayed if the stress exceeds the proportional limit (35,000 psi).

35. The Viking Bin Company manufactures suspended bins weighing from 4 tons to 20 tons (in 2 ton increments). Each bin is suspended by three steel rods. Two grades of steel are available. For one, the allowable tensile stress is 22,000 psi; for the other, 29,000 psi. Write a program that will generate a table of required rod diameters (to the next $\frac{1}{8}$ in.) as a function of bin weight and allowable tensile stress.

36. Write a computer program that will calculate the allowable load for a single-shear bolted lap connection similar to that shown in Fig. 9–8. The computed allowable load is to be based on bolt shear only. User input is to be number of bolts, bolt diameter, and allowable shear stress for the bolts.

Supplemental Problems

For the following problems, assume the modulus of elasticity E for steel to be 30,000,000 psi and a proportional limit of 34,000 psi. Refer to Appendix F or G for other materials.

37. The column in Fig. 9–20 is supported by a base plate, pedestal, and footing as shown. The load applied on the column is 60 kips. Neglecting their weights, find the bearing stress under each of the four elements.

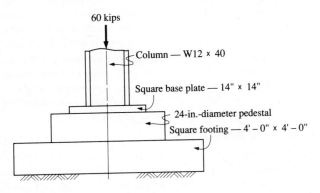

FIGURE 9–20   Problem 37.

38. If the soil pressure under the footing of Problem 37 were limited to 2.5 ksf, what would be the new required dimensions for the square footing? (Use 2 in. increments.)

39. A steel wire is suspended vertically from its upper end. The wire is 400 ft long and has a diameter of $\frac{3}{16}$ in. The unit weight of steel is 490 pcf. Compute (a) the maximum tensile stress due to the weight of the wire and (b) the maximum load P that could be supported at the lower end of the wire. Allowable tensile stress is 24,000 psi.

40. A W12 × 40 shape is subjected to a tensile load of 190 kips, as shown in Fig. 9–21. Find the stress at a transverse section where a total of six holes have been drilled. The holes are $\frac{3}{4}$ in. in diameter. There are two holes in each flange and two in the web. (Neglect stress concentration.)

41. The inclined member in Fig. 9–22 is braced with a glued block as shown. The ultimate shear stress in the glued joint is 1050 psi. Determine the dimension x at which the glued joint will fail.

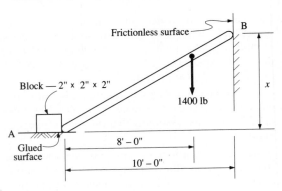

FIGURE 9–22   Problem 41.

42. A short timber post of Douglas fir is subjected to an axial compressive load of 40,000 lb. The nominal dimensions of the post area are 8 in. by 8 in. Its length is 8 ft–6 in. Using *dressed* dimensions, compute (a) stress, (b) strain, and (c) total deformation. Assume a proportional limit of 6000 psi.

FIGURE 9–21   Problem 40.

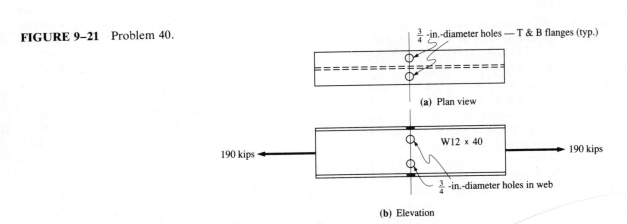

43. A 100 ft surveyor's steel tape with a cross-sectional area of 0.005 in.^{2} is subjected to a 15 lb pull when in use. Compute (a) the total elongation of the tape and (b) the tensile stress developed in the tape.

44. An 18 in. long steel rod is subjected to a tensile load of 12,000 lb. The rod has a diameter of $1\frac{1}{4}$ in. Compute (a) the tensile stress, (b) the strain, and (c) the total elongation.

45. Compute the magnitude of the tensile load that would produce a strain of 0.0007 in./in. in a 1 in. diameter steel rod.

46. A structural steel rod $1\frac{1}{2}$ in. in diameter and 20 ft long supports a balcony and is subjected to an axial tensile load of 30,000 lb. Compute (a) the total elongation and (b) the diameter of rod required if the total elongation must not exceed 0.10 in.

47. A rectangular structural steel eyebar $\frac{3}{4}$ in. thick and 24 ft long between centers of end pins is subjected to an axial tensile load of 50,000 lb. If this bar must not elongate more than 0.20 in., compute (a) the required width of bar and (b) the tensile stress developed at maximum elongation, assuming the required width is provided.

48. Compute the total elongation (in inches) of a straight steel bar that is 40 ft long and hanging vertically. The bar does not support any load other than its own weight. Use a unit weight of steel of 490 pcf.

49. A $\frac{3}{4}$ in. diameter steel rod, 100 ft long, is hanging vertically. Compute the total elongation of the rod due to its own weight and a suspended 8000 lb load applied at its free end. Use a unit weight of 490 pcf.

50. For the truss in Fig. 9–23, compute the total deformation of member CD due to the applied load of 100 kips.

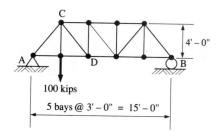

FIGURE 9–23 Problem 50.

Member CD is made of steel and has a cross-sectional area of 2.5 in.^{2}. Indicate whether the member is in tension or compression.

51. A steel bar with a cross-section of $\frac{1}{2}$ in. by $\frac{1}{2}$ in. is subjected to the axial loads shown in Fig. 9–24. Compute the total elongation of the bar.

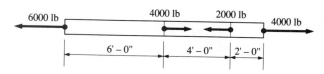

FIGURE 9–24 Problem 51.

52. Rework Problem 51, changing the second load from the left to 14,000 lb and the third load from the left to 12,000 lb.

53. A hook is suspended by two steel wires as shown in Fig. 9–25. The wires are $\frac{1}{4}$ in. in diameter. A load of one ton is hung on the hook. Compute the distance that the hook will drop due to the stretch in the wires.

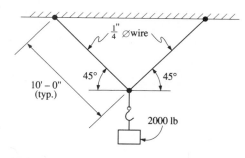

FIGURE 9–25 Problem 53.

54. The steel piston rod to the master cylinder has a diameter of $\frac{5}{16}$ in. Determine the stress in the piston rod when a force of 29 lb is applied to the brake pedal as shown in Fig. 9–26.

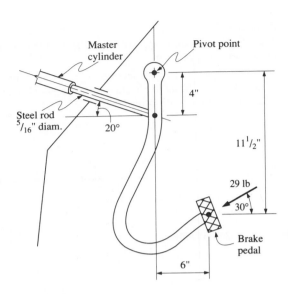

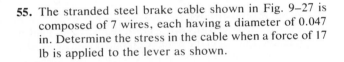

FIGURE 9–26   Problem 54.

55. The stranded steel brake cable shown in Fig. 9–27 is composed of 7 wires, each having a diameter of 0.047 in. Determine the stress in the cable when a force of 17 lb is applied to the lever as shown.

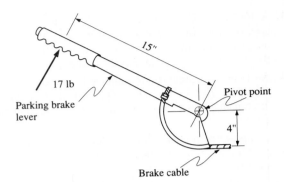

FIGURE 9–27   Problem 55.

56. The piston of a steam engine is 16 in. in diameter and its stroke is 24 in. The maximum steam pressure is 250 psi. Calculate the required diameter of the piston rod if the allowable stress is 10,000 psi.

57. A locking mechanism utilizes a slotted slide plate and stud as shown in Fig. 9–28. The slot is $\frac{9}{32}$ in. wide. The stud is tightened until the force in the $\frac{1}{4}$ in. diameter shank is 200 lb tension. (a) Find the shear stress in the head of the stud. (b) Find the bearing stress between the stud and the slotted slide plate.

FIGURE 9–28   Problem 57.

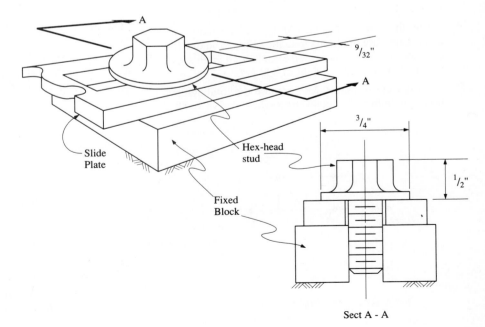

10 Properties of Materials

10–1
THE TENSION TEST

In the area of strength of materials, problems are primarily of two types: analysis and design. Problems of analysis could involve finding the greatest load that may be applied to a given body without exceeding specified limiting values of stress and strain. They could also involve determining load-induced stresses and strains for comparison with limiting values. Problems of design involve determining the required size and shape of a member to support given loads without exceeding specified limiting stress and/or strain.

Various properties of materials must be considered at this point, since both analysis and design are essentially problems of materials. In particular, the mechanical properties will be important to our study, since they are the properties that affect the limiting values of stress and strain. Appendix G may be used as a reference source for the mechanical properties of the materials we will discuss in this text.

Probably the simplest and most informative test for establishing the mechanical properties of a large number of engineering materials is the tension test. Metals and plastics are normally subjected to a tension test, whereas a material such as concrete is tested in compression due to its low tensile strength and brittleness. The tension test is categorized as a static-loaded destructive-type test and has been standardized by the American Society for Testing and Materials (ASTM). It involves the axial tensile loading of a material specimen, one type of which is shown in Fig. 10–1. The specimen is prepared for the test by placing two small punch marks (gage points) on it, which are a known distance apart (the gage length). Frequent simultaneous readings of the increasing applied load P and total change in gage length δ are made during the test. These are recorded in a tabular format. The specimen is loaded to failure. Additional information is obtained by taking the specimen from the machine, carefully fitting the ends together, and measuring the length between the gage points. The diameter of the specimen is also measured at the cross section at which failure occurs. Stress and strain values are computed and then plotted to form a stress-strain curve, which is discussed in Section 10–2.

The tension test is conducted in a laboratory using any one of several types of testing machines. Loads are read from dials or digital displays. Elongations are determined using measuring devices such as the extensometer shown in Fig. 10–2. If elongations are large, hand-held dividers and scales may be used. Some testing machines have the capability of automatically reading and recording the data and will also produce plotted results.

FIGURE 10–1 Standard tension test specimen.

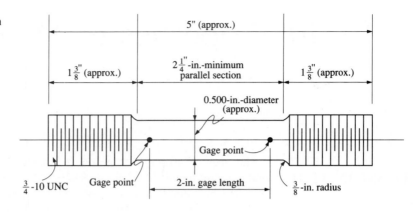

5" (approx.)

$1\frac{3}{8}"$ (approx.)

$2\frac{1}{4}"$-in.-minimum parallel section

$1\frac{3}{8}"$ (approx.)

0.500-in.-diameter (approx.)

Gage point

$\frac{3}{4}$-10 UNC

Gage point

2-in. gage length

$\frac{3}{8}$-in. radius

FIGURE 10–2 Extensometer on typical 0.505 tensile specimen (Courtesy of Tinius Olsen Testing Machine Co., Inc., Willow Grove, PA).

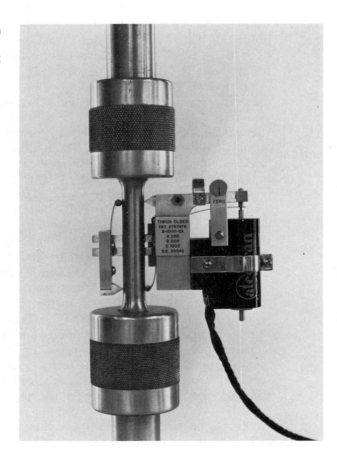

10–2
THE STRESS-STRAIN DIAGRAM

After the tension test has been conducted, the data are converted to stress (based on original area) and strain (based on original length). These values are then plotted to form a stress-strain curve, or diagram, with the ordinate corresponding to stress and the abscissa corresponding to strain.

A smooth average curve is then drawn (rather than one that passes through every plotted point). The deviations of the smooth curve from the plotted points are mostly due to errors of instrumentation and observation.

A typical stress-strain diagram for a low-carbon, mild ductile structural steel is shown in Fig. 10–3. Ductile materials are defined as materials that can undergo considerable plastic deformation under tensile load before actual rupture. (Brittleness is the opposite of ductility.) Figure 10–3(a) shows the full stress-strain curve from the beginning of the test through failure at point F. Figure 10–3(b) shows the left-most portion of the curve from a strain of zero to a strain of about 0.02. In Fig. 10–3(b), from the origin 0 to point A, the curve is a straight line, indicating that the steel conforms to Hooke's law (stress is proportional to strain). In that portion of the curve, the steel is in the elastic range (its behavior is said to be *elastic*). The slope of the straight-line portion of the curve is called the *modulus of elasticity*, E, which was defined in Chapter 9 as E = stress/strain. The stress corresponding to the upper end of the straight-line portion of the curve (point A) is the *proportional limit* of the steel. The proportional limit may be defined as the greatest stress a material is capable of developing without deviation from straight-line proportionality between stress and strain.

FIGURE 10–3 Stress-strain diagram for low-carbon, ductile steel.

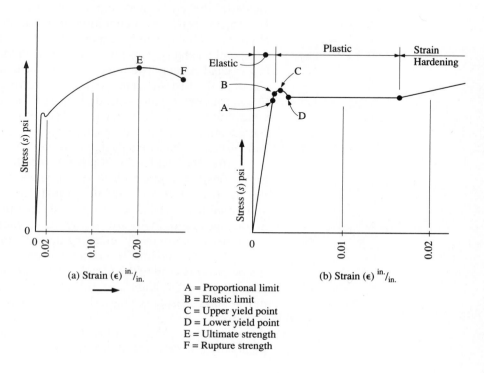

(a) Strain (ε) $^{in.}/_{in.}$

(b) Strain (ε) $^{in.}/_{in.}$

A = Proportional limit
B = Elastic limit
C = Upper yield point
D = Lower yield point
E = Ultimate strength
F = Rupture strength

For all practical purposes, the proportional limit is also generally accepted as the limit of elasticity, and is termed the *elastic limit*. The elastic limit (point B) may be defined as the greatest stress a material is capable of developing without a permanent elongation remaining upon complete unloading of the specimen. Strains associated with stresses up to the elastic limit (therefore, within the elastic range) are small and reversible.

Determining the elastic limit is a tedious, trial-and-error procedure; therefore, it is not done very often. For most materials, the elastic limit and the proportional limit are almost identical in numerical value, and the terms are often used synonymously. Where a difference does exist, the elastic limit is generally higher than the proportional limit. Unless otherwise stated, the term *elastic limit*, as used in this text, will mean the proportional limit.

Once the steel has been stressed past the elastic limit, it passes into the *plastic range* (its behavior is said to be *plastic*). This implies that strain is no longer reversible; that is, should the steel be unloaded, it will not return to its original length. Rather, it will retain a permanent elongation (sometimes called a *permanent set*). For stresses above the elastic limit, the strain increases at a faster rate until, at point C on the curve shown in Fig. 10–3(b), the strain continues with little or no increase in applied load or stress. The point at which this occurs is called the *yield point* and the stress at this point is called the *yield stress*. The yield stress may be defined as the stress at which there occurs a marked increase in strain without an increase in stress.

After the yield point has been reached (the steel is said to have yielded), the curve may dip downward to point D, also shown in Fig. 10–3(b), representing a condition in which the steel transmits less load as it elongates. This phenomenon occurs only with a low-carbon steel; it is not typical of all steels. This abrupt decrease in load (stress) results in upper and lower yield points. The upper yield point is influenced considerably by the shape of the test specimen and by the testing machine, itself, and is sometimes completely suppressed; hence, it is relatively unstable. The lower yield point, point D, is much less sensitive and is considered to be more representative of the true characteristics of the steel—characteristics that can serve as a basis for design criteria.

After the yield point is passed, the steel stretches still more at essentially constant stress up to a strain of about 0.015 (1.5%) at which time it begins to recover some of its strength and becomes capable of resisting additional load. The steel has passed from the plastic range into the strain-hardening range. With stress and strain increasing at different rates, the curve rises continuously, climbing until the maximum ordinate is reached at point E in Fig. 10–3(a), where the tangent to the curve is horizontal. This is the point representing the *ultimate strength* or *tensile strength*. The ultimate strength may be defined as the maximum stress a material is capable of developing. Tensile strength is similarly defined but it, of course, refers only to tension.

At point E, shown in Fig. 10–3(a), the steel continues to stretch accompanied by a decreasing ability to transmit load. Also, after passing point E, the steel visibly begins to decrease in diameter and to increase in length over a localized segment of the specimen. The localized decrease in the diameter is called *necking* and is depicted in Fig. 10–4. Necking progresses rapidly until the steel suddenly ruptures in the plane of the reduced cross section. Point F on the curve in Fig. 10–3(a) represents the breaking point or *rupture strength* of the steel. The rupture strength may be defined as the stress at which the specimen actually breaks.

FIGURE 10–4 Necking of ductile steel specimen.

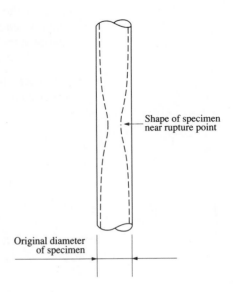

Shape of specimen near rupture point

Original diameter of specimen

As previously mentioned, the stresses plotted on the stress-strain diagram are calculated on the basis of the original cross-sectional area of the specimen, despite the fact that the area is generally reduced appreciably by the time the ultimate and rupture points are reached. In addition, the calculated unit strain is based on the original length of the specimen (which traditionally, using the U.S. Customary System, is either 2 in. or 8 in.).

Variations occur in the shape of stress-strain diagrams for different steels. However, the straight-line (elastic) portion of the curve, which represents the modulus of elasticity E, is approximately the same for all structural steels, whether low-carbon steels or high-strength steels. Some high-carbon steels and certain alloy steels exhibit no yield point when tested in tension and will rupture after much less elongation than the more ductile steels.

Materials other than steel exhibit different shapes of the stress-strain diagram. Many ductile metals, such as brass, copper and aluminum, have gradually curving stress-strain diagrams beyond the proportional limit, rather than a definite yield point. Brittle materials, such as concrete and cast iron, are incapable of much plastic deformation. They do not exhibit a necking process and will rupture without warning. A tensile stress-strain curve for a brittle material ends before it becomes horizontal; consequently, the ultimate strength and the rupture strength are the same.

For materials without a well-defined yield point, such as higher strength steels and nonferrous metals, a property (stress value) analogous to the yield point is established by the offset method. It is common to define this stress as that which would cause a permanent set of 0.2%. This is shown in Fig. 10–5, in which a line is drawn parallel to the straight-line portion of the stress-strain curve, but offset from it by 0.2% (0.002 in./in.). The stress read at the intersection of this line with the stress-strain curve is called the *yield strength*. It is also called the *yield point*, or, simply, the *yield stress*—it is not a true yield point.

FIGURE 10–5   Offset method of determining yield strength.

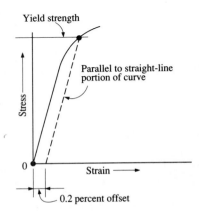

The stress-strain diagram, in the context of our discussion, represents properties at normal temperatures. When materials are to be used at temperatures much above or below the normal range, it is necessary to establish stress-strain curves for the anticipated temperature range.

☐ **EXAMPLE 10–1**   During a stress-strain test of a steel specimen, the strain at a stress of 5000 psi was calculated to be 0.000167 in./in. (point A, Fig. 10–3) and at a stress of 20,000 psi to be 0.000667 in./in. (point B). If the proportional limit is 30,000 psi, calculate the modulus of elasticity E using the slope between the two points. Also, compute the stress corresponding to a strain of 0.0002 in./in.

Solution
$$s_A = 5000 \text{ psi} \qquad \epsilon_A = 0.000167 \text{ in./in.}$$
$$s_B = 20,000 \text{ psi} \qquad \epsilon_B = 0.000667 \text{ in./in.}$$

Since both points lie on the straight-line portion of the stress-strain curve, the modulus of elasticity will be equal to the slope of the line between the points (or between either of the two points and zero).

$$E = \frac{\text{change in stress}}{\text{change in strain}} = \frac{20,000 - 5000}{0.000667 - 0.000167} = 30,000,000 \text{ psi}$$

Calculating the stress corresponding to a strain of 0.0002 in./in., since

$$E = \frac{s}{\epsilon}$$

then

$$s = \epsilon E = 0.0002(30,000,000) = 6000 \text{ psi}$$
$$6000 \text{ psi} < 30,000 \text{ psi} \quad \textbf{OK}$$

**10–3
MECHANICAL
PROPERTIES OF
MATERIALS**

As previously stated, the numerical stress values that may be obtained from a tension test are the proportional limit, elastic limit, yield stress, ultimate stress, and rupture stress. In addition, the modulus of elasticity, percent elongation, and percent reduction in cross-sectional area are also obtained. These values define those mechanical properties or qualities of a material that are significant in the applications of strength of materials.

In addition to the mechanical properties defined by numerical stress values, there are some other mechanical properties that describe how a material responds to load and deformation (see Appendix G). These properties may be defined and summarized as follows:

1. *Stiffness* is the property that enables a material to withstand high stress without great strain. It is a resistance to any sort of deformation. Stiffness of a material is a function of the modulus of elasticity E. A material having a high value of E such as steel, for which $E = 30,000,000$ psi (207 000 MPa), will deform less under load (thereby exhibiting a greater stiffness) than a material with a lower value of E such as wood, where E may be equal to $1,000,000$ psi (7000 MPa) or less.
2. *Strength* is the property determined by the greatest stress that the material can withstand prior to failure. It may be defined by the proportional limit, yield point, or ultimate strength. No single value is adequate to define strength, since the behavior under load differs with the kind of stress and the nature of the loading.
3. *Elasticity* is that property of a material enabling it to regain its original dimensions after removal of a deforming load. There is no known material that is completely elastic in all ranges of stress. However, most engineering materials are elastic, or very nearly elastic, over large ranges of stress. Steel is an elastic material only up to the elastic limit. Hence, the determination of the elastic limit establishes the elastic range or the limit of elasticity.
4. *Ductility* is that property of a material enabling it to undergo considerable plastic deformation under tensile load before actual rupture. A ductile material is one that can be drawn into a long thin wire by a tensile force without failure. Ductility is characterized by the percent elongation of the gage length of the specimen during the tensile test and by the percent reduction in area of the cross section at the plane of fracture. These

quantities are defined by the following expressions:

$$\text{Percent elongation} = \frac{\text{increase in gage length}}{\text{original gage length}} (100) \qquad \textbf{(10–1)}$$

$$\text{Percent reduction in area} = \frac{\text{orig. area} - \text{final area}}{\text{original area}} (100) \qquad \textbf{(10–2)}$$

A high percent elongation indicates a highly ductile material. Most ASTM material specifications for metals have a required minimum percent elongation for either of the two standard specimen gage lengths (2 in. and 8 in.). A lower limit value commonly used to determine ductility is about 5% elongation. That is, a metal is considered to be ductile if its percent elongation is greater than about 5%.

5. *Brittleness* implies the absence of any plastic deformation prior to failure. A brittle material is neither ductile nor malleable and will fail suddenly without warning. A brittle material exhibits no yield point or necking-down process and has a rupture strength approximately equal to its ultimate strength. Brittle materials, such as cast iron, concrete and stone, are comparatively weak in tension and are generally not subjected to a tension test. They are usually tested in compression.

6. *Malleability* is that property of a material enabling it to undergo considerable plastic deformation under compressive load before actual rupture. Most materials that are very ductile are also quite malleable. A hammering or rolling operation would require a malleable material due to the extensive compressive deformation that accompanies the process.

7. *Toughness* is that property of a material enabling it to endure high-impact loads or shock loads. When a body is subjected to an impact load, some of the energy of the blow is transmitted and absorbed by the body. In absorbing the energy, work is done on the body. This work is the product of the strain and the average stress up to the rupture point. Hence, the measure of toughness is equal to the area under the stress-strain curve from the origin through the rupture point, as shown in Fig. 10–6. A body

FIGURE 10–6   Resilience and toughness.

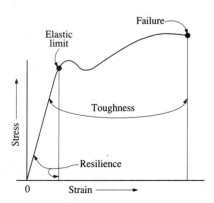

that can be both highly stressed and greatly deformed without rupture is capable of withstanding a heavy blow and is said to be tough.

8. *Resilience* is that property of a material enabling it to endure high-impact loads without inducing a stress in excess of the elastic limit. It implies that the energy absorbed during the blow is stored and recovered when the body is unloaded. The measure of resilience is furnished by the area under the elastic portion of the stress-strain curve from the origin through the elastic limit, as shown in Fig. 10–6.

Many materials have been developed to satisfy the requirements of innumerable engineering applications. The variations in mechanical properties result, in part, from differing chemical compositions as well as from differing manufacturing processes. Various classification systems enable clear distinctions to be made between grades and/or types of a given material.

The various types of steels are classified by two systems, either ASTM (American Society for Testing and Materials) or AISI (American Iron and Steel Institute). In the ASTM system, each structural steel has a number designation referring to the standard that defines the required minimum properties (both mechanical and chemical). The designation has the prefix letter A followed by one, two, or three numerals (e.g., ASTM A6, ASTM A441). Structural steel specified and manufactured according to the designated standard is expected to meet both the mechanical and the chemical property requirements.

The second way of designating steel is by means of the AISI number. The AISI uses a four-digit code to define each steel. The last two digits indicate the average percentage of carbon in the steel. For example, if the last two digits are 35, the steel has a carbon content of about 0.35%. The first two digits denote the major alloying elements in the steel other than carbon. For example, AISA 1020 is a plain carbon steel with approximately 0.20% carbon; AISI 2340 is a nickel steel with approximately 3% nickel and 0.40% carbon. The first digit (2) signifies that the steel is a nickel steel.

Other materials are classified according to similar coding systems, such as that developed for aluminum alloys by the Aluminum Association. The aluminum coding system was developed to identify the major alloying element and to indicate the temper imparted to the alloy.

□ **EXAMPLE 10–2** A $\frac{1}{2}$ in. diameter aluminum specimen is subjected to a tension test. After rupture, the two pieces are fitted back together. The distance between gage points, originally 2 in., is measured to be 2.83 in. and the final diameter of the specimen is measured to be 0.38 in. Calculate the percent elongation and the percent reduction in area.

Solution

$$\text{Percent elongation} = \frac{\text{increase in gage length}}{\text{original gage length}}(100)$$

$$= \frac{2.83 - 2.0}{2.0}(100)$$

$$= 41.5$$

and

$$\text{Percent reduction in area} = \frac{\text{orig. area} - \text{final area}}{\text{original area}}(100)$$

$$\text{Original area} = 0.7854(0.5)^2 = 0.196 \text{ in.}^2$$

$$\text{Final area} = 0.7854(0.38)^2 = 0.113 \text{ in.}^2$$

$$\text{Percent reduction in area} = \frac{0.196 - 0.113}{0.196}(100) = 42.3$$

10–4 ENGINEERING MATERIALS— METALS

In this section, we will consider some of the more common of the metals used in machines and structures. These are listed in Fig. 10–7. The list is not all-inclusive and we will not consider any of the many exotic and specialized metals that have been developed for specific and peculiar purposes.

Metals are traditionally classed as ferrous and nonferrous. The term *ferrous*, derived from the Latin word *ferrum*, meaning iron, refers to those metals which contain a large percentage of iron. Nonferrous metals have special properties that make their use advantageous in some circumstances. The ferrous metals are, at the present, the primary metals of engineered structures.

FIGURE 10–7 Some important metals for machines and structures.

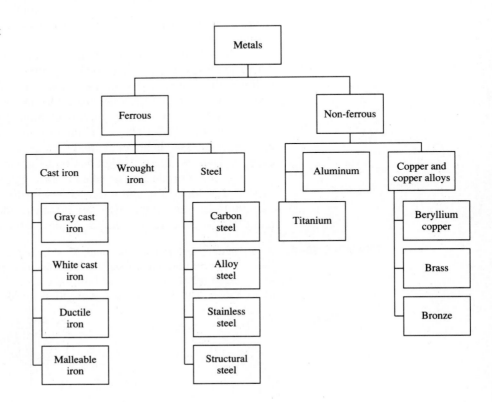

FERROUS METALS   The basic and predominant component of the ferrous metals is iron, one of the most common chemical substances found in the earth's crust. Because of its great affinity for combination with other elements, iron is never found in its pure form in nature. The iron must be extracted from iron ores, which are mineral or rock deposits containing concentrations of iron. Most of the important iron ores found in the United States are in the form of iron oxides. These ores usually contain small amounts of impurities (phosphorus and silica, among others) most of which must be removed during the production of iron.

Iron is extracted from iron ore in a blast furnace. The production process requires a combination of iron ore, fuel, and a flux of crushed limestone to remove the impurities. The fuel used is coke, a coal product. The high temperature (in excess of 3000°F) achieved in the furnace results in the iron portion of the ore melting and the limestone flux being liquified. The end result is a molten iron combined with some carbon obtained from the coke and a molten slag formed by the limestone combining with the various impurities in the ore. The molten slag floats above the molten iron and is drawn off as a waste byproduct. The molten iron is then also drawn off and cast into shallow molds called *pigs*. The cooled iron is called *pig iron*. Since pig iron is relatively brittle, additional processing and refining must be performed before it becomes a useful ferrous metal.

Most of the iron produced in the blast furnace is further processed to make steel. That portion not used to make steel is used to make cast iron and wrought iron. Cast iron, wrought iron, and steel are the three common forms of ferrous metals. All three are principally iron-carbon alloys (mixtures) containing small amounts of sulfur, phosphorus, silicon and manganese. In addition, other elements such as nickel and chromium may be added to alter the physical and mechanical properties. (An alloy is defined as a substance with metallic properties, composed of two or more elements of which at least one is a metal.)

Cast Iron   *Cast iron* is the generic name for a group of metals that are alloys of carbon and silicon with iron. Included are gray cast iron, white cast iron, ductile iron, and malleable iron. Most cast irons have at least 3% total carbon with an upper limit ranging from 3.8% to 4%. The manufacture of cast iron involves the mixing of pig iron, scrap cast iron, and a limestone flux in a small furnace fired with coke and an air blast. The properties of the resulting cast iron are established by differences in cooling rates, compositions, and subsequent heat treatment of the product. The suitability of a particular metal for an intended use is best determined by laboratory or service tests.

Gray Cast Iron

Gray cast iron (also called *gray iron*) is the most commonly used type of cast iron. It is characterized by high percentages of carbon (2% to 4%) and silicon (1% to 3%). It is cheap, easily cast, and relatively easy to machine. It is available in grades having a tensile strength ranging from 20 ksi to 60 ksi. Its ultimate compressive strength is much higher, ranging three to five times greater than its tensile strength. Gray iron is brittle, does not exhibit a yield point, and should not be subjected to dynamic-type loadings. It has excellent wear and corrosion resistance and has good vibration damping abilities. Gray iron is used in automotive engine blocks, gears, brake parts, clutch plates, machinery bases, and pipe installations.

White Cast Iron

In white cast iron (also called *white iron*), most of the carbon is combined chemically with the iron in the form of cementite rather than existing as free carbon (graphite). A fractured surface appears white. White iron's advantage over gray iron is that it is harder and more resistant to abrasion. However, it is more brittle and more difficult to machine and to cast. It is also less resistant to corrosion. White iron is used for freight car wheels, rolling-mill rolls, and plow shares.

Ductile Iron

Ductile iron (also called *nodular cast iron*) is similar to gray iron and has good wear and corrosion resistance. It is also easily cast and relatively easy to machine. However, it is far more ductile than gray iron and has a greater tensile strength and percent elongation. It thus has good toughness (shock resistance). Uses of ductile iron include crankshafts, casings, gears, hubs, and rolls.

Malleable Iron

Malleable iron has moderate to high strength with a tensile strength comparable to ductile iron and an ultimate compressive strength greater than that of the ductile iron. It also has good machinability and good wear resistance. Malleable iron is used for pipe fittings, guard rail fittings, construction machinery, and automotive and truck parts.

Wrought Iron Wrought iron is a soft, easily worked material that has a high resistance to corrosion. It possesses good ductility, malleability, and toughness and is easily machined. Wrought iron is produced from nearly pure molten iron mechanically mixed with molten slag. A forging operation then strings the slag out into fibers, and the resulting fibrous structure creates a material with different mechanical properties parallel and perpendicular to the fibers. Except for the slag (usually less than 3%), the composition is similar to that of a low-carbon steel. The carbon content is generally less than 0.1%. Wrought iron is used extensively for ornamental iron work, gratings, and steam and water pipes.

Steel Steel is an alloy consisting almost entirely of iron in combination with small quantities of various elements. A wide range of properties may be achieved by varying the composition of the steel. Carbon is the element that has the greatest effect on the properties of the steel. Up to a point, increasing the carbon content increases the hardness, strength, and abrasion resistance of steel. However, ductility, toughness, impact properties, and machinability will be decreased.

The making of steel from pig iron is essentially a refining (purifying) process. Carbon, silicon, phosphorus, and sulfur levels of the pig iron are reduced to levels permissible by the steel specification. Other alloying elements may then be added to the mix to produce a steel having the desired properties. The most commonly used processes to change the pig iron to steel are the open-hearth process, the electric furnace process, and the basic oxygen process. Of these, the basic oxygen process produces most of the steel in the United States.

Many types of steels are available. We will briefly discuss four general categories of steel: carbon steel, alloy steel, stainless steel, and structural steel.

Carbon Steel

Steel is considered to be carbon steel when no minimum content is specified or required for the recognized alloying elements such as aluminum, chromium, nickel, and numerous others. Carbon steels range from a low-carbon grade (with roughly 0.1% carbon) to a high-carbon grade (with roughly 1.0% carbon.) These steels are sometimes designated *plain carbon steel* or *machine steel* or *machinery steel*. All carbon steels contain small amounts of various elements such as manganese, phosphorus, silicon, and in many cases, other elements. Carbon steel is considered an alloy despite the fact that it is not designated an alloy steel.

It should be noted that while strength and hardness increase with increasing carbon content, the steel becomes less ductile and more brittle. The mechanical properties of carbon steel are a function of the carbon content, method of manufacture, and heat-treating processes performed on the steel. As an example, carbon steel may have an ultimate tensile strength ranging from roughly 43,000 psi to 122,000 psi.

Alloy Steel

In addition to carbon, alloy steel contains significant quantities of recognized alloying metals, the most common being aluminum, chromium, copper, manganese, molybdenum, nickel, phosphorus, silicon, titanium, and vanadium. Alloys are used to improve the hardenability of steel; to increase toughness, ductility, and tensile strength; and to improve low- and high-temperature properties.

Alloy steels are practically always heat-treated to develop specified properties. Heat treatment involves raising the temperature of the steel to some prescribed level and then cooling it rapidly by quenching. This procedure alters the crystalline structure of the steel and therefore alters its physical properties.

The method of designating carbon and alloy steels is by means of an ASTM number designation or an AISI number designation as discussed in

TABLE 10–1 AISI numbering system for steel.

Steel	AISI No.	Steel	AISI No.
Plain Carbon	10XX	Molybdenum-Nickel 1.75% NI	46XX
Plain Carbon*	11XX	Molybdenum-Nickel-Chromium	47XX
Manganese	13XX	Molybdenum-Nickel 3.5% NI	48XX
Boron	14XX	Chromium	5XXX
Nickel	2XXX	Chromium-Vanadium	6XXX
Nickel-Chromium	3XXX	Nickel-Chromium-Molybdenum	8XXX
Molybdenum-Chromium	41XX	Silicon-Manganese	92XX
Molybdenum-Chromium-Nickel	43XX	Nickel-Chromium-Molybdenum (Except 92XX)	9XXX

* With greater sulfur content for free-cutting.

Section 10–3. Table 10–1 furnishes the AISI two-digit code for identifying various types of alloy steels.

Stainless Steel

Stainless steel is the designated name for a widely used grouping of iron-chromium alloys known for their corrosion resistance (notably their nonrusting quality). This ability to resist corrosion is attributable to a surface chromium oxide film that forms in the presence of oxygen. The film is essentially insoluble, self-healing, and nonporous. A minimum chromium content of 12% is required for the formation of the film, and 18% is sufficient to resist the most severe atmospheric corrosive conditions. For other reasons, the chromium content may go as high as 30%. Other elements, such as nickel, aluminum, silicon, and molybdenum may also be present.

Some applications for the stainless steels are chemical processing and oil processing equipment, cutlery, and automotive trim.

Structural Steel

The term *structural steel* applies to hot-rolled steel of various shapes and forms and of varying alloy elements utilized to resist assorted types of loads and forces to which a structure may be subjected. The member may be a tension, compression, bending, or torsional member, or a combination of these. The structure may be a building, bridge, a transmission tower, or some other specialized type of structure. The steel shapes, as standardized by the AISC, are W shapes (wide-flange members), HP shapes (bearing pile members), S shapes (American standard beams—formerly called *I-beams*), M shapes (miscellaneous), C shapes (American standard channels), MC shapes (miscellaneous channels), and L shapes (angles). In addition, structural steel also includes plates, bars, steel pipe, and structural tubing.

Structural steel includes several types of steel. The five categories, as designated by the AISC, are carbon steel, high-strength low-alloy steel, corrosion resistant high-strength low-alloy, quenched and tempered low-alloy, and quenched and tempered alloy. The common method of specifying structural steel is based on a standardized ASTM designation described in Section 10–3.

Carbon steel is considered as the basic structural steel. The widely used carbon steel, designated A36, has a minimum yield stress of 36 ksi, except for plates in excess of 8 in. thick, which have a minimum yield stress of 32 ksi. A36 steel is ductile, weldable, and can be used in all types of structures.

High-strength low-alloy steels are designated A441 and A572. These steels have a corrosion resistance roughly twice that of A36. Their yield stresses vary from 40 ksi to 65 ksi for individual shapes, plates, and bars depending on the type of steel and on the thickness of the elements of the cross section, trending downward for the thicker elements. Both steels are suitable for welded or bolted structures.

Corrosion resistant high-strength low-alloy steels are designated A242 and A588. These steels have a corrosion resistance ranging from four to eight times that of A36 steel. Yield stresses vary from 42 ksi to 50 ksi depending on the thickness of the elements of the cross section, trending downward for the thicker elements. A242 and A588 steels can be used as exposed unpainted steel in structures and are suitable for welded or bolted structures.

Both categories of quenched and tempered steel (low-alloy and alloy) designated A852 and A514 are applicable for only high-strength plates. They are strong, tough steels with yield strengths in the 70 ksi to 100 ksi range and are primarily used in welded structures.

Eight other ASTM grades of structural steel are approved for use as per AISC Specification for Structural Steel Buildings: A53, A500, A501, A570, A606, A607, A618 and A709. These grades include pipe, tubing, sheet, strip, and structural steel for bridges.

Nonferrous Metals

Nonferrous metals and their alloys represent an important classification of engineering materials. Some have a high strength-to-weight ratio, whereas others have excellent resistance to corrosion. The mechanical properties of the primary nonferrous metals are a function of the principal element and the quantity and type of alloying element(s), as well as the method of manufacturing and the heat-treating process. A few of the more important nonferrous metals are briefly discussed next.

Aluminum

Aluminum is one of the most abundant metals found in the crust of the earth, occurring as an oxide in most clay. The basic raw material from which aluminum is produced is bauxite, an ore containing a high percentage of aluminum oxide (also termed *alumina*). The total extraction process of obtaining aluminum from the ore results in a metallic aluminum of better than 99% purity. This high-purity aluminum is soft, weak, and ductile with an ultimate tensile strength of approximately 10,000 psi. Aluminum is light in weight and has a high resistance to corrosion under most service conditions. It also has good thermal conductivity and high electrical conductivity. Aluminum of high purity is used principally for electric-conductor purposes. Even for this application, however, small percentages of certain alloying elements usually are added to improve the mechanical properties and other characteristics with only a slight reduction in the electrical conductivity.

Most aluminum is used in the alloyed state. Mechanical properties can be considerably improved by alloying it with small amounts of other metals. Aluminum alloys are generally harder and stronger than the high-purity aluminum. The alloys are widely used for structural and mechanical applications. Among its attractive properties are light weight (approximately one-third that of steel), good corrosion resistance, relative ease of machining and a pleasing silver-white appearance. It has two (perhaps significant) disadvantages: a high coefficient of thermal expansion (approximately twice that of steel) and a modulus of elasticity of 10,000,000 psi (approximately one-third that of steel). The latter property indicates a less-stiff material. Similar to steel, a multitude of aluminum alloys are available that provide a wide range of properties. Aluminum can be alloyed to provide strengths exceeding that of some carbon steels. However, these strong alloys have a lesser ductility and a lesser resistance to corrosion. Ultimate tensile strength levels up to approximately 80,000 psi may be obtained with suitable alloys resulting in a high strength-to-weight ratio.

Aluminum and its alloys are divided into two classes according to how they are formed, whether wrought or cast. Roughly 75% of the aluminum produced in the United States is fabricated into wrought products, which include sheet and plate, tube, pipe, rolled structural and other shapes, extruded shapes, rod, bar, and wire. These products have application within specific fields. Examples are the aerospace industry, architectural buildings, highway structures, tanks, pressure vessels, piping, and transportation structures.

Titanium

Titanium production in commercial quantities began in the immediate post–World War II years. Titanium and its alloys have attractive engineering properties. They are about 45% lighter than steel and 70% heavier than aluminum. They also possess a very high strength, which may reach an ultimate tensile strength of 200,000 psi depending upon the alloying element. The combination of moderate weight and high strength gives titanium alloys the highest strength-to-weight ratio of any structural metal—roughly 30% greater than aluminum or steel. This high strength-to-weight ratio is roughly maintained within a temperature range of from −400°F to +1000°F.

Other notable properties for titanium and its alloys are excellent corrosion resistance to atmospheric and sea environments as well as to a wide range of chemicals, low thermal conductivity, low coefficient of thermal expansion, high melting point (higher than iron), and high electrical resistivity. In addition, the fatigue resistance of titanium and its alloys is good. The modulus of elasticity (stiffness) is 16,000,000 psi which is 1.6 times the value used for aluminum. However, the high cost of the metal limits its usefulness and range of applications.

The majority of the present applications for titanium and its alloys are in the aerospace industry. Examples are the structures of aircraft and spacecraft, sections of jet engines and allied components, landing gears, fuselage parts, and skins. Other applications have been pressure vessels, roofing, and various architectural building items such as fascias, flashing, and gravel stops.

Copper and Copper Alloys

The term *copper*, in the United States, means copper containing less than 0.5% of impurities or alloying element. The most significant properties of copper are its high electrical conductivity (second only to that of silver), high thermal conductivity, good resistance to corrosion, and good malleability, formability, and strength.

Unalloyed copper is used widely as a structural material, for roofing and sheathing, in heat-exchange equipment, large vessels and kettles, and in various kinds of equipment used in the production of chemicals, foods, and beverages. However, the greatest single used for unalloyed copper is in the electrical field, where its electrical conductivity, corrosion resistance, and formability make it ideal for use in electrical devices and equipment of all kinds.

A few of the more important alloys of copper are beryllium copper, brass, and bronze. Each consists of copper in combination with different elements. All have different properties and all have a variety of specific applications. All of these alloys generally possess good strength and corrosion resistance.

10–5 ENGINEERING MATERIALS— NONMETALS

There is no one simple classification for nonmetallic engineering materials. Some of the more common and significant materials such as concrete, wood, and plastics are briefly introduced here. These three materials constitute a diverse grouping with respect to structure, sources, and characteristics. Wood and concrete have broad application in the civil-architectural-construction field, whereas plastics have application in virtually all fields of engineering and architecture.

Concrete

Concrete consists principally of a mixture of cement and fine and coarse aggregates (sand, gravel, crushed rock, and/or other materials) and water.

The water is added as a necessary ingredient for the hardening of the mixture. The bulk of the mixture consists of the fine and coarse aggregates. The resulting concrete strength and durability are a function of the proportions of the mix as well as other factors such as the concrete placing, finishing, and curing history.

The compressive strength of concrete is relatively high. However, it is a relatively brittle material with little tensile strength compared with its compressive strength. Steel reinforcing rods (which have high tensile and compressive strength) are used in combination with the concrete; the steel will resist the tension and the concrete will resist the compression. The result of this combination of steel and concrete is called *reinforced concrete*. In some instances, steel and concrete are positioned in members so that they both resist compression.

Structural concrete utilizes, almost exclusively, hydraulic cement, which requires water for the chemical reaction of hydration. In this process, the cement sets and bonds the fresh concrete into one mass. Portland cement, the most commonly used cement, consists chiefly of calcium and aluminum silicates. In fresh concrete, the ratio of the amount of water to the amount of cement, by weight, is termed the *water/cement ratio*. For complete hydration, a water/cement ratio of 0.35 to 0.40 is required. Higher water/cement ratios, while leading to lower strengths, are generally used to expedite mixing, handling, and placing of the concrete.

In ordinary structural concrete, the aggregates occupy approximately 70% to 75% of the hardened mass. Gradation of aggregate size to produce close-packing is desirable since this will generally result in better strength and durability. Aggregates are classified as fine or coarse. Fine aggregate is generally sand consisting of particles that will pass a No. 4 sieve (0.187 in. nominal opening). Coarse aggregate consists of particles that would be retained on a No. 4 sieve, with a maximum size governed by code requirements.

Concrete is designated according to its compressive strength, which may range from 2500 psi to 9000 psi. Much higher strengths are possible with good quality control. Under favorable conditions the strength increases with age, the specified strength usually being that which occurs 28 days after the placing of the concrete.

Wood

Wood is one of the oldest natural construction materials. It is a cellular organic material composed principally of cellulose (about 60%), which constitutes the longitudinal structural cells, and lignin (about 28%), which cements the structural cells together. The structural cells are hollow, very small in diameter, and oriented vertically in the growing tree.

Wood is divided into two classes, softwood and hardwood. The terms are misleading in that there is no direct relationship between these designations and the hardness or softness of the wood. Softwood comes from conifers (trees with needlelike or scalelike leaves) and hardwoods come from deciduous trees having broad leaves. Hardwoods shed their leaves at the end of each growing season, whereas most softwoods are evergreens. Most of the wood used in the United States for structural purposes is softwood with Douglas fir and southern pine as the most common.

Wood contains natural growth characteristics such as knots and slope of grain that may, depending on their peculiarities, adversely affect the strength properties of that member. Structural grading rules, in establishing allowable design stress values, take into account the effects of these growth characteristics on the strength of wood. Structural lumber is graded according to its intended use. Each piece is assigned a stress grade, designed to meet exacting requirements and strength values. Various lumber associations establish grade requirements and grade markings procedures for species of wood produced in their regions. A stamp, usually placed on the lumber at the mill, identifies the grade of each piece of lumber. Structural lumber is usually graded on the basis of visual inspection. Lumber that has been individually pretested by nondestructive means which supplements the visual grading is classed as *machine-graded* lumber. The allowable stress values established by grading should be adjusted for moisture service conditions and duration of applied load.

Plastics

Plastics, as used herein, refers to a group of synthetic organic materials derived by a process called *polymerization*. There are many grades and formulations for each plastic and each of these has its unique combination of characteristics and properties. Therefore, our discussion will be very general.

All plastics fall into two broad classifications: thermoplastics and thermosetting plastics. A thermoplastic material can be repeatedly softened and made to flow by heating. The thermoplastics may be formulated in such a manner as to be capable of large plastic deformation or to be flexible. Some thermoplastics such as polyvinyl chloride (PVC) and polystyrene are rigid. The thermosetting plastics have no melting or softening point, though they may be damaged by heat. All thermosetting plastics are rigid, such as phenol-formaldehyde (Bakelite). The thermosetting plastics are brittle, hard, and strong, whereas the thermoplastics are generally ductile, low in strength, and resistant to impact. The thermal expansion of most thermoplastics is approximately ten times that of steel. Most plastics have high creep characteristics that cause them to deform gradually under constant load.

Most thermoplastics will degrade due to weathering and exposure to the sun. This results in brittleness, hardness, cracking, and yellowing. Special formulations, when added to the plastic, will aid in resisting this degradation.

Some additional thermoplastics that are common are polyethylene, teflon, nylon, plexiglass, lucite, and delrin. Thermosetting plastics include epoxies, polyesters, silicones, urethanes, and urea-formaldehyde.

10–6
ALLOWABLE
STRESSES AND
ACTUAL STRESSES

The design and analysis of machines and structural elements are based on limiting values of materials with respect to stress and strain. The limiting values are, in turn, based partly on the mechanical properties of materials. The tension test and resulting stress-strain diagram is the most common test that provides information on the mechanical properties. After the various values of the stress-strain curve of a material are obtained, it is then possible to establish the magnitude of the stress that may be considered as a safe or limiting stress for a given condition or problem. This stress is generally called the *allowable stress*. The allowable stress for a material may be defined as the maximum stress considered to be safe when a member made of that material is subjected to a particular loading condition.

Appropriate values for allowable stress depend on several factors, including

1. Material properties defined by numerical stress values, such as proportional limit, yield stress, and ultimate strength
2. Ductility of the material
3. Confidence in the prediction of loads
4. Type of loading: static, cyclic, or impact
5. Confidence in the analysis and design methods
6. Possible deterioration during the design life of the structure due to such factors as corrosion or chemical attack
7. Possible danger to life and property as a result of a failure
8. Design life of the structure, whether permanent or temporary

Allowable stresses for materials used as structural components in buildings, bridges, and other civil engineering structures are established by the appropriate authoritative specification-writing agencies such as the AISC (American Institute of Steel Construction)[1] for structural steel buildings; AASHTO (American Association of State Highway and Transportation Officials)[2] for highway bridges of various materials; AREA (American Railway Engineering Association)[3] for railway bridges of various materials; ACI (American Concrete Institute)[4] for reinforced concrete structures;

---

[1]   American Institute of Steel Construction, Inc., *Specification for Structural Steel Buildings, Allowable Stress Design* (Chicago: AISC, 1989). (Also included in the AISC *Manual of Steel Construction—Allowable Stress Design.*)

[2]   American Association of State Highway and Transportation Officials, *Standard Specifications for Highway Bridges*, 13th ed. (Washington, D.C.: AASHTO, 1983).

[3]   American Railway Engineering Association, *Specification for Steel Railway Bridges*, latest edition (Washington, D.C.: AREA).

[4]   American Concrete Institute, *Building Code Requirements for Reinforced Concrete* (Detroit: ACI, 1989), 318–89.

NFPA (National Forest Products Association)[5] for timber structures; and the Aluminum Association[6] for aluminum structures. The established specifications, including allowable stresses, are generally incorporated into the building codes of states or municipalities; therefore, they become part of the legality governing the various types of construction in a particular area.

In machine design, as well as in the aircraft and spacecraft industries, many different materials with widely varying mechanical properties are used under extreme and varied loading and environmental conditions. For these situations, allowable stresses may or may not be designated by standard specifications. In the absence of a specified allowable stress, a predetermined factor of safety may be used, as determined by the engineering department of the responsible company or manufacturing organization. Factor of safety is discussed further in Section 10–7.

The term *actual stress* may be defined as the calculated stress (or the computed stress) induced in a member as a result of applied loads. The actual stress may vary depending upon the magnitude of the loads. While at times it may be only a small fraction of the allowable stress, the actual stress must never exceed the allowable stress.

☐ **EXAMPLE 10–3**   An ASTM A36 steel rod 20 ft long must be capable of supporting a tensile load of 2400 lb without elongating more than 0.25 in. or exceeding an allowable tensile stress of 22,000 psi. Calculate the required rod diameter to an eighth of an inch. The proportional limit is 34,000 psi. Refer to Appendix G for necessary mechanical properties.

Solution   First calculate the required cross-sectional area of the rod based on stress:

$$\text{Required } A = \frac{P}{s_{t(\text{all})}} = \frac{2400}{22,000} = 0.109 \text{ in.}^2$$

Equation (9–10), from Section 9–6, is then used to calculate the required cross-sectional area based on allowable elongation. Recall that Eq. (9–10) is valid only if the stress does not exceed the proportional limit. In this situation, the condition is satisfied, since the allowable stress to be used in the calculation is less than the proportional limit.

$$\delta = \frac{PL}{AE}$$

from which

$$\text{Required } A = \frac{PL}{\delta E} = \frac{2400(20)(12)}{0.25(30,000,000)} = 0.0768 \text{ in.}^2$$

---

[5]   National Forest Products Association, *National Design Specifications for Wood Construction* (Washington, D.C.: NFPA, 1982).

[6]   The Aluminum Association, *Specifications for Aluminum Structures* (Washington, D.C.: The Aluminum Association, 1986).

The largest required area is based on the allowable tensile stress. Since

$$A = \pi d^2/4 = 0.7854 d^2$$

the required diameter is calculated as

$$\text{Required } d = \sqrt{\frac{0.109}{0.7854}} = 0.373 \text{ in.}$$

Use a ⅜ in. rod.

☐ **EXAMPLE 10–4**   A tracked military vehicle, shown in Fig. 10–8, is to operate on terrain where the bearing pressure under the tracks is not to exceed 10 psi. The vehicle weighs a maximum of 30 tons and each of the two tracks is 20 in. wide. Determine the minimum required contact length L of the tracks.

FIGURE 10–8   Tracked vehicle.

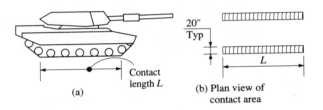

(a)

Contact length L

20" Typ

L

(b) Plan view of contact area

Solution   The maximum bearing pressure under the tracks is to be 10 psi. Consider this to be an allowable bearing pressure. The bearing length L required will be based on this maximum pressure. The contact area A for two tracks is calculated from

$$A = 2L(20) = 40L \text{ in.}^2$$

Based on a maximum bearing pressure of 10 psi,

$$\text{Required } A = \frac{P}{10} = \frac{30(2000)}{10} = 6000 \text{ in.}^2$$

Therefore,

$$\text{Minimum } L = \frac{6000}{40} = 150 \text{ in.} = 12.5 \text{ ft}$$

The tracks must have a 12.5 ft contact length.

**10–7
FACTOR OF SAFETY**   In reaching a decision as to just how safe a structural element should be, safety may be expressed in terms of factor of safety. *Factor of safety* is defined in many ways, but in a broad sense it is the ratio of a failure stress to an allowable stress. In an elastic design approach, such as that used for the structural steels and aluminum, the attainment of yield stress in a member is considered to be analogous to failure. Although the steel or aluminum will not actually fail (rupture) at yield, significant and unacceptable deformations

are on the verge of occurring. These deformations may render the structural element unusable. The factor of safety, then, is a factor of safety against yielding.

As an example, assume a structural steel with a yield stress of 36,000 psi and an allowable tensile stress of 24,000 psi. The factor of safety against yielding would then be

$$\text{F.S.} = \frac{\text{yield stress}}{\text{allowable stress}} \qquad \textbf{(10–3)}$$

$$= \frac{36,000}{24,000} = 1.5$$

Another way to consider the case would be to think of the member as having a 50 percent reserve of strength against yielding in this particular application.

Since the factor of safety and allowable stress are interrelated, the recommended factor of safety values set forth in various specifications and codes depend on the same factors as discussed for allowable stress values. Recommended factors of safety and/or allowable stresses are the result of cumulative pooled experiences and history. They are limiting values, traditionally accepted as good practice.

Factors of safety may be based on a material yield strength, as previously discussed, or on a material ultimate strength, and values may range from 1.5 to 20. For example, for such ductile metals as steel, which are subjected to static loads, a factor of safety of 1.5 based on the yield strength is often suggested. For brittle metals, such as cast iron or wood, which are subjected to shock or impact loads, a factor of safety of 20 based on the ultimate strength of the material may be suggested.

☐ **EXAMPLE 10–5**   Test results of an ASTM A36 steel specimen indicated an ultimate tensile strength of 75,000 psi and a yield stress of 36,000 psi. If the tensile allowable stress per design specifications is 22,000 psi, compute the factor of safety based on (a) yield stress and (b) tensile strength.

Solution   For part (a),

$$\text{F.S.} = \frac{\text{yield stress}}{\text{allowable stress}} = \frac{36,000}{22,000} = 1.64$$

For part (b),

$$\text{F.S.} = \frac{\text{tensile strength}}{\text{allowable stress}} = \frac{75,000}{22,000} = 3.41$$

☐ **EXAMPLE 10–6**   Design a 10 ft long rod subjected to a tensile load of 15,000 lb. Using a factor of safety of 2.5 based on the yield stress, calculate the required rod diameter if it is to be made of (a) steel with a yield stress of 50,000 psi and (b) aluminum alloy with a yield stress of 40,000 psi.

Solution   For part (a),

$$\text{Allowable stress} = \frac{\text{yield stress}}{\text{F.S.}} = \frac{50,000}{2.5} = 20,000 \text{ psi}$$

$$\text{Required } A = \frac{P}{s_{t(\text{all})}} = \frac{15,000}{20,000} = 0.75 \text{ in.}^2$$

Since $A = 0.7854 d^2$,

$$\text{Required } d = \sqrt{\frac{A}{0.7854}} = \sqrt{\frac{0.75}{0.7854}} = 0.977 \text{ in.}$$

For part (b),

$$\text{Allowable stress} = \frac{40,000}{2.5} = 16,000 \text{ psi}$$

$$\text{Required } A = \frac{P}{s_{t(\text{all})}} = \frac{15,000}{16,000} = 0.938 \text{ in.}^2$$

$$\text{Required } d = \sqrt{\frac{A}{0.7854}} = \sqrt{\frac{0.938}{0.7854}} = 1.09 \text{ in.}$$

10–8 ELASTIC-INELASTIC BEHAVIOR

In Section 9–2 we discussed the analysis and design of axially loaded tension members. Design involves determining the required cross-sectional area of the member and then selecting the actual cross section to be used, subject to all the constraints such as the shape of the member. Note that the design process as previously discussed was based on some allowable axial stress which in turn was based on some margin of safety against failure. It is evident then, that the basis for design hinges on some definition of failure.

The term *failure*, as used herein, means the condition that renders the load-resisting member unfit for resisting further increase in loads. In general, a member fails by inelastic deformation (yielding) if it is made of ductile material or by fracture (rupture or breaking) if the material is brittle. For ductile materials, the yield point stress has commonly been established as the stress at which inelastic deformation starts. From this stress, one can obtain the upper limit of the load that may be applied to a member without causing it to fail. That is, if the load is increased and the yield stress is reached, failure is said to be imminent. The rationale for this approach is that undesirable deformations will occur rendering the member unusable. When allowable stresses are utilized in the proportioning of members of a structural system, the approach may be called *allowable stress design* or *elastic design*. This implies that stresses and the accompanying strains will all lie within the elastic range.

In other words, if we consider a three-bar structure of ductile steel, as shown in Fig. 10–9(a), when one bar is stressed to its yield point, the structure cannot carry more load despite the fact that the other bars may not be stressed to their yield points. Note that in this structure, all bars must elon-

FIGURE 10–9 Elastic-inelastic behavior.

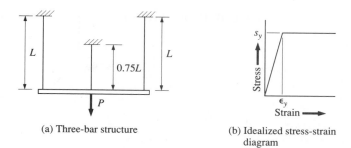

(a) Three-bar structure

(b) Idealized stress-strain diagram

gate the same amount and that the horizontal member can only translate (no rotation) in a vertical plane.

An alternate design approach holds that failure is not considered to have occurred until all bars of the structure have been stressed to the yield point. From our study of the stress-strain curve for ductile materials, we recognize that the material will not be on the verge of rupture at this point, though all bars may have elongated significantly. When at least part of the structure is strained beyond the yield strain, the resulting behavior is said to be elastic-inelastic, meaning that the material strain of one bar has passed through the elastic range and into the inelastic range.

This alternate design approach, where all members are stressed to the yield point, will determine the maximum possible load, called the *ultimate load*, that can be applied to the structure. Note that the assumption is made that the stress-strain curve is of the type shown in Fig. 10–9(b) and the material has infinite ductility. This approach is sometimes referred to as *Ultimate Strength Design* or *Limit Design*. This approach is supposedly more rational, resulting in greater economy and a more uniform factor of safety. It has been used for several decades in the analysis and design of reinforced concrete and is currently being introduced in the structural steel design field as *Load and Resistance Factor Design*.

To illustrate the difference between the two approaches, we will analyze the three-bar structure shown in Fig. 10–9(a).

□ **EXAMPLE 10–7**   Calculate the maximum load P that can be applied to the three-bar structure shown in Fig. 10–9(a). All bars are vertical. The symmetrical arrangement allows the rigid horizontal member to deflect vertically without rotation while the three bars elongate by the same amount. The cross-sectional area and the modulus of elasticity are the same for all three bars. Assume a ductile material.

Solution   Assume an idealized stress-strain relationship as shown in Fig. 10–9(b). Note that for strains less than (or equal to) the yield strain ϵ_Y the stress is proportional to strain. For strains greater than the yield strain, stress is constant and is equal to the yield stress s_Y.

Initially, compute a maximum load based on an elastic design approach in which the bars will not be stressed beyond the elastic range. In this approach, the load limit of P will be based on one bar reaching its yield-point stress along with its yield strain. The other two bars will have their stress and strain within the elastic range.

FIGURE 10–10   Free-body diagram.

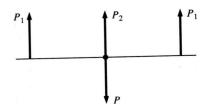

Refer to the free-body diagram of Fig. 10–10. As the load P increases from an initial value of zero, an equilibrium equation can be expressed as

$$P = 2P_1 + P_2 \qquad \text{(Eq. 1)}$$

where P_1 represents the resisting force in each of the outer bars and P_2 is the resisting force in the middle bar. Assume equal elongation in each bar; therefore,

$$\delta_1 = \delta_2$$

Substituting,

$$\frac{P_1 L}{AE} = \frac{P_2(0.75\,L)}{AE} \qquad \text{(Eq. 2)}$$

Solving for P_1,

$$P_1 = 0.75P_2 \qquad \text{(Eq. 3)}$$

Substituting P_1 from Eq. 3 into Eq. 1,

$$P = 2(0.75P_2) + P_2$$
$$P_2 = 0.40P$$

Also, from Eq. 1,

$$2P_1 = P - P_2$$
$$2P_1 = P - 0.40P$$
$$P_1 = 0.30P$$

Since the middle bar (P_2) resists most of the applied load, it will reach its yield-point stress and strain before the outer bars. This, in effect, constitutes failure despite the fact that the outer bars are stressed and strained within the elastic range. Therefore, the resisting force P_2 can be expressed as

$$P_2 = s_Y A$$

Since $P_2 = 0.40P$, an expression can be written to compute P:

$$0.40P = s_Y A$$
$$P = 2.5 s_Y A$$

This represents the maximum load that the structure can support. The stress in the middle bar has reached its yield while the stress in outer bars is still within the elastic range.

As a next step, assume that load P is further increased. The strain in the center bar will increase beyond the yield strain (at constant stress s_Y), while the two outer bars will still be straining within the elastic range. The horizontal member will be supported by the elastically acting outer bars together with a constant resisting force $(s_Y A)$ furnished by the middle bar. The value of P will increase until yielding begins in each of the outer bars, that is, when $P_1 = s_Y A$, and failure is assumed to have occurred.

Therefore, the maximum load that can be applied to the structure will exist when all three bars have reached their yield point stress. Note that this maximum load constitutes an ultimate load based on an elastic-inelastic behavior. This load is calculated from

$$P = 2P_1 + P_2 = 2s_Y A + s_Y A = 3s_Y A$$

Note that this ultimate load is 20% greater than the maximum load that was based only on an elastic behavior.

In actual design, no matter which design approach is used, elastic or ultimate strength (inelastic), appropriate factors of safety are introduced to provide adequate margins of safety.

10–9
SI SYSTEM
EXAMPLES

□ **EXAMPLE 10–8**

A 14 mm diameter steel rod is tested in tension and elongates 0.182 mm in a length of 200 mm under a load of 29 kN. Compute (a) the stress, (b) the strain, and (c) the modulus of elasticity based on this one reading. The proportional limit of this steel is 228 MPa.

Solution
(a) The cross-sectional area of the rod is calculated from

$$A = 0.7854(14)^2 = 153.9 \text{ mm}^2 = 153.9 \times 10^{-6} \text{ m}^2$$

The stress is calculated from

$$s_t = \frac{P}{A} = \frac{29 \times 10^3 \text{ N}}{153.9 \times 10^{-6} \text{ m}^2} = 0.1884 \times 10^9 \text{ N/m}^2$$

$$= 188.4 \text{ MPa} < 228 \text{ MPa} \qquad \textbf{OK}$$

(b) The strain is calculated from

$$\epsilon = \frac{\delta}{L} = \frac{0.182 \text{ mm}}{200 \text{ mm}} = 0.000\ 91 \text{ mm/mm}$$

(c) The modulus of elasticity is calculated from

$$E = \frac{s_t}{\epsilon} = \frac{188.4 \text{ MPa}}{0.000\ 91} = 207\ 000 \text{ MPa}$$

□ **EXAMPLE 10–9**

A 6 m long ASTM A36 steel rod is subjected to a tensile load of 10.7 kN. Calculate the required rod diameter if the allowable tensile stress is 150 MPa and the maximum elongation cannot exceed 6.5 mm. The proportional limit of the steel is 234 MPa.

Solution The required cross-sectional area of the rod based on stress (substituting in terms of newtons and meters) is calculated:

$$\text{Required } A = \frac{P}{s_{t(\text{all})}} = \frac{10.7 \times 10^3 \text{ N}}{150 \times 10^6 \text{ N/m}^2} = 0.0713 \times 10^{-3} \text{ m}^2 = 71.3 \text{ mm}^2$$

The required cross-sectional area based on allowable elongation is calculated from Eq. (9–10). Rewriting for required area, and substituting in terms of newtons and meters,

$$\text{Required } A = \frac{PL}{\delta E} = \frac{(10.7 \times 10^3 \text{ N})(6 \text{ m})}{(6.5 \times 10^{-3} \text{ m})(207\,000 \times 10^6 \text{ N/m}^2)}$$

$$= 0.000\,047\,7 \text{ m}^2$$

$$= 47.7 \text{ mm}^2$$

The largest required area controls (because it satisfies both requirements) and is based on the allowable tensile stress, which is less than the proportional limit. Therefore, the required diameter is calculated from

$$\text{Required } d = \sqrt{\frac{A}{0.7854}} = \sqrt{\frac{71.3}{0.7854}} = 9.53 \text{ mm}$$

☐ **EXAMPLE 10–10** Calculate the required diameter of a 3 m long steel rod subjected to an axial tensile load of 67 kN. The yield stress is 345 MPa. Use a factor of safety of 2.5 based on the yield stress.

Solution

$$\text{Allowable stress} = \frac{\text{yield stress}}{\text{F.S.}} = \frac{345 \text{ MPa}}{2.5} = 138 \text{ MPa}$$

$$\text{Required } A = \frac{P}{s_{t(\text{all})}} = \frac{67 \times 10^3 \text{ N}}{138 \times 10^6 \text{ N/m}^2}$$

$$= 0.4855 \times 10^{-3} \text{ m}^2$$

$$= 485.5 \text{ mm}^2$$

$$\text{Required } d = \sqrt{\frac{A}{0.7854}} = \sqrt{\frac{485.5}{0.7854}} = 24.9 \text{ mm}$$

SUMMARY—BY SECTION NUMBER

10–1 The tension test is a static-loaded destructive-type laboratory test used to establish the mechanical properties of many engineering materials. During the test, applied axial tensile loads and elongations are measured simultaneously.

10–2 A stress-strain diagram is a graphic representation of the results of the tension test. Some of the values obtained from the stress-strain diagram are modulus of elasticity, proportional limit, elastic limit, yield stress, ultimate stress, and rupture stress.

10–3 Values obtained from the stress-strain diagram establish those mechanical properties of a material describing how a material responds to an applied load. Significant mechanical properties are stiffness, strength, elasticity, ductility, brittleness, malleability, toughness, and resilience.

10–4 and **10–5** There are many different types of engineering materials, all of which have application in particular situations. They may be broadly classified as ferrous and nonferrous metals and as nonmetals. Iron and steel of various types comprise the ferrous metals. Aluminum is one of the principal nonferrous metals, being the most abundant metal found in the crust of the earth. Concrete, wood, and plastics are among the principal nonmetal engineering materials.

10–6 Allowable stress represents a safe (or limiting) stress for purposes of design and/or analysis. Actual stress represents the stress developed in a member as a result of applied loads. The actual stress developed must not exceed the allowable stress.

10–7 Factor of safety is the ratio of a failure stress to an allowable stress. The failure stress could be based on the yield stress or the ultimate stress of the material.

10–8 Elastic design considers that a structure has been loaded to capacity when it attains initial yielding, on the theory that inelastic deformation would terminate the utility of the structure. Ultimate strength design (inelastic design), on the other hand, recognizes that a structure may be loaded beyond initial yielding of some part of the structure provided that other parts that remain in the elastic stress range are capable of resisting the additional load.

PROBLEMS

For the following problems, unless noted otherwise refer to Appendices F and G for necessary mechanical properties of materials.

Section 10–2   The Stress-Strain Diagram

1. A $\frac{9}{16}$ in. diameter steel rod is tested in tension and elongates 0.00715 in. in a length of 8 in. under a tensile load of 6500 lb. Compute (a) the stress, (b) the strain, and (c) the modulus of elasticity, E, based on this one reading. The proportional limit of this steel is 34,000 psi.

2. A concrete cylinder 6 in. in diameter was tested in compression and observed to have shortened an amount of 0.0029 in. over a gage length of 12 in. Total load at the time of the reading was 20,000 lb. Compute the modulus of elasticity, E.

3. A mild steel, known to have a proportional limit of 34,000 psi, was subjected to a tension test. Stresses and strains were calculated at three points as follows: (stress) 10,000 psi, 25,000 psi, 40,000 psi; (associated strains) 0.000343, 0.000858, and 0.035 in./in. Calculate the modulus of elasticity, E, and comment on the data.

4. The data from the tension test of a steel specimen are given in Table 10–2. The gage length was 2.000 in. and the original diameter was 0.505 in. The final diameter was 0.397 in. Calculate stress and strain at each point and draw the stress-strain diagram. Estimate the value of the modulus of elasticity, upper and lower yield points, ultimate strength, rupture strength, and percent reduction in area. **Hint:** draw a second curve enlarging the plot of the first eight points.

TABLE 10–2

Load (kips)	Deformation (in.)	Load (kips)	Deformation (in.)
0	0.0000	10.0	0.1100
2.0	0.0006	13.0	0.2320
4.0	0.0013	13.9	0.3200
6.0	0.0019	14.0	0.3600
8.0	0.0025	12.9	0.4000
7.83	0.0043	10.8	0.4200
7.80	0.0120		
8.0	0.0361		

Section 10–6   Allowable Stresses and Actual Stresses

5. An 18 in. long AISI 1040 steel rod is subjected to a tensile load of 12,000 lb. If the allowable tensile stress is 22,000 psi and the allowable total elongation is not to exceed 0.008 in., compute the required rod diameter. The proportional limit of the steel is 40,000 psi.

6. ASTM A36 steel rods are used to support a balcony. Each rod is suspended from a roof beam and supports a balcony floor area of 75 ft². If the combined dead load and live load to be carried by the floor is 250 psf, compute the rod diameter required. Use an allowable tensile stress of 22,000 psi.

7. A tension member in a roof truss is composed of two ASTM A36 structural steel angles which together have a net cross-sectional area of 8.62 in.². (a) Compute the allowable total load if the allowable tensile stress is 22,000 psi. (b) If the total tensile load in the member is 150,000 lb, compute the actual tensile stress.

Section 10–7   Factor of Safety

8. A main cable in a large bridge is designed for an actual tensile force of 2,600,000 lb. The cable consists of 1470 parallel wires, each 0.16 in. in diameter. The wires are cold-drawn steel with an average ultimate strength of 230,000 psi. What factor of safety was used in the design of the cable?

9. A concrete canoe in storage is supported by two rope slings, as shown in Fig. 10–11. Each sling supports 48 lb. The rope has a tensile breaking strength of 252 lb. Determine the maximum value of θ if there is to be a factor of safety of 4.0 against breaking.

Section 10–8   Elastic-Inelastic Behavior

10. A load is applied to a rigid bar that is symmetrically supported by three steel rods as shown in Fig. 10–12.

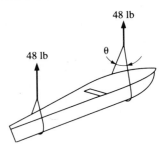

FIGURE 10–11   Problem 9.

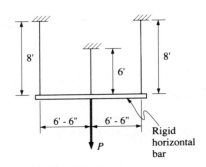

FIGURE 10–12   Problem 10.

The cross-sectional area of each of the rods is 1.2 in.². Calculate the maximum load P that may be applied (a) using an elastic approach with an allowable stress of 22,000 psi and (b) using an ultimate strength approach (inelastic) with a load factor (factor of safety) of 1.85. Assume a ductile material with a yield stress of 36,000 psi.

SI System Problems

11. A concrete cylinder 150 mm in diameter was tested in compression and found to be shortened an amount of

0.074 mm over a length of 300 mm. The load at the time of the reading was 89 kN. Compute the modulus of elasticity.

12. A 450 mm long AISI 1020 steel rod is subjected to a tensile load of 55 kN. The allowable tensile stress is 140 MPa and the allowable total elongation is not to exceed 0.2 mm. Calculate the required rod diameter. The proportional limit is 175 MPa.

13. Test results of a steel specimen indicated an ultimate tensile strength of 827 MPa and a yield stress of 350 MPa. If the tensile allowable stress per design specifications is 210 MPa, compute the factor of safety based on (a) yield stress and (b) tensile strength.

14. A 50 mm diameter steel tie rod used in a machine is subjected to an axial tensile load of 180 kN. The rod length is 1.75 m prior to load application. The proportional limit is 225 MPa. Calculate (a) the stress, (b) the strain, and (c) the total elongation.

15. Calculate the load applied to a square wrought iron bar if the total elongation was measured to be 1.50 mm. The bar is 3 m long and its cross-sectional dimensions are 50 mm × 50 mm.

16. Compute the modulus of elasticity of a copper alloy wire that stretches 14.0 mm when subjected to a load of 320 N. The wire is 4.00 meters long and has a diameter of 1.00 mm.

17. An aluminum alloy rod is subjected to a tensile load of 30 kN. If the rod is elongated 3.5 mm, calculate the original length of the rod. The diameter of the rod is 25 mm.

Computer Problems

For the following computer problems, any appropriate programming language may be used. Input prompts should fully explain what is required of the user (the program should be "user friendly"). The resulting output should be well labeled and self-explanatory.

18. Write a program that will calculate stress, strain, and modulus of elasticity for a rod of circular cross section that is loaded in tension. User input is to be rod diameter, load, original gage length, elongation, and the proportional limit for the material. Use the program to solve Problem 1.

19. Write a program that will allow a user to input the initial and final diameters and gage length for a tension test specimen along with a specified number of load-elongation combinations. The program should then calculate the stress and strain for each data set as well

as the percent reduction in area. Use the program to check the calculations of Problem 4.

Supplemental Problems

20. A $\frac{1}{2}$ in. diameter structural nickel steel specimen was subjected to a tension test. After rupture it was determined that the 2 in. standard gage length had stretched to 2.42 in. The minimum diameter at the fracture was measured to be 0.422 in. Compute the percent elongation and percent reduction in area.

21. If the allowable tensile stress for steel plates is 24,000 psi, how many plates 9 in. by $1\frac{1}{4}$ in. will be required to carry a load of 1,600,000 lb? With this number of plates, compute the tensile stress in each, assuming they are all stressed the same.

22. A pair of wire cutters is designed to operate under a maximum 35 lb force applied as shown in Fig. 10–13. Determine the required diameter of the pin (to the next higher $\frac{1}{32}$ in.) The allowable shear stress in the pin is 12,000 psi.

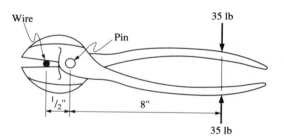

FIGURE 10–13   Problem 22.

23. Calculate the end bearing length required for a 10 in. by 16 in. timber beam (dressed) that is supported on a reinforced concrete wall as shown in Fig. 10–14. The

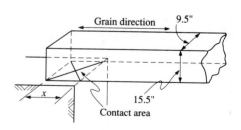

FIGURE 10–14   Problem 23.

beam reaction is 15,000 lb and the allowable compressive stress perpendicular to the grain for the timber member is 300 psi.

24. In a tension test of steel, the ultimate load is 13,100 lb and the elongation is 0.52 in. The original diameter of the specimen is 0.50 in. and the gage length is 2.00 in. Calculate (a) the ultimate tensile stress and (b) the ductility of the material in terms of percent elongation.

25. A standard steel specimen having a diameter of 0.505 in. and a 2.00 in. gage length is used in a tension test. At what load P will the extensometer read 0.002 in. deformation? Assume a proportional limit of 34,000 psi.

26. An aluminum bar 2 in. $\times \frac{1}{2}$ in. in cross section is subjected to a tensile load of 16,000 lb. At this load, the axial strain is 1650×10^{-6} in./in. Assuming a proportional limit of 21,000 psi, calculate the modulus of elasticity for the material.

27. A 10 ft long steel member is subjected to a tensile load of 200,000 lb. The steel has an ultimate tensile stress of 95,000 psi. (a) Calculate the cross-sectional area required using a factor of safety of 5.0. (b) Calculate the required cross-sectional area required assuming that the maximum elongation is to be 0.10 in.

28. The collar bearing shown in Fig. 10–15 is subjected to a compressive load P of 60,000 lb. Calculate the required diameter D_c of the collar. The diameter D_s of the shaft is 4 in. and the allowable bearing stress for the collar on its support is 300 psi. Assume a $\frac{1}{32}$ in. clearance all around between the shaft and the support.

29. A concrete slab of uniform thickness weighs 20,000 lb and is supported by two steel rods as shown in Fig. 10–16. The initial length of rod AB is 2 ft. The initial length of rod CD is 3 ft. Rod AB has an area of 1 in.² and rod CD has an area of 2 in.². Calculate (a) the

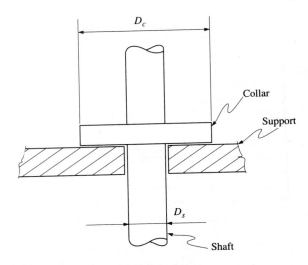

FIGURE 10–15   Problem 28.

elongation of each rod and (b) the required ratio of the areas of AB and CD so that the elongation of each bar is the same.

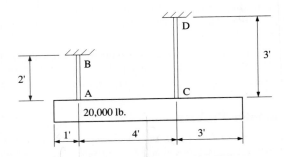

FIGURE 10–16   Problem 29.

11 Stress Considerations

11-1
POISSON'S RATIO

Tests have shown that if a body of elastic material is subjected to a tensile load, its transverse (or lateral) dimensions decrease at the same time that its axial dimensions in the direction of the load increase. This is shown pictorially in Fig. 11-1. Similarly, if a body of elastic material is subjected to a compressive load, its transverse dimensions increase at the same time that its axial dimensions in the direction of the load decrease.

At stresses below the proportional limit, the transverse strain is proportional to the axial stress; similarly, the axial (longitudinal) strain is proportional to the axial stress. Since both lateral strain and axial strain are proportional to the axial stress, their ratio must be constant (and positive) for a given material. This ratio of lateral strain to axial strain is called *Poisson's ratio* and is represented by μ (Greek lowercase mu) and expressed by the equation

$$\mu = \frac{\text{transverse strain}}{\text{axial strain}} \tag{11-1}$$

Commonly used values for Poisson's ratio are given in Appendix G.

□ **EXAMPLE 11-1**

A 10 ft long rectangular ASTM A441 steel plate 1 in. by 12 in. in cross section is used as a hanger and subjected to a tensile load of 240,000 lb. The proportional limit of the steel is 34,000 psi. Compute (a) axial stress, (b) axial strain, (c) transverse strain, (d) total axial dimensional change, and (e) total transverse (12 in.) dimensional change. Refer to Appendix G for necessary mechanical properties.

Solution

(a) Axial stress (s_t):

$$s_t = \frac{P}{A} = \frac{240,000}{12(1)} = 20,000 \text{ psi}$$

$$20,000 \text{ psi} < 34,000 \text{ psi} \quad \textbf{OK}$$

(b) Axial strain (ϵ): Since

$$E = \frac{s}{\epsilon} = \frac{\text{stress}}{\text{strain}}$$

then

$$\epsilon = \frac{s_t}{E} = \frac{20,000}{30,000,000} = 0.000667 \text{ in./in.}$$

FIGURE 11–1 Dimensional changes due to axial tensile load.

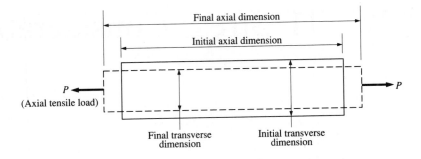

(c) Transverse strain (ϵ):

$$\text{Transverse } \epsilon = \mu(\text{axial } \epsilon)$$
$$= 0.25(0.000667) = 0.000167 \text{ in./in.}$$

(d) Total axial elongation (δ): Since

$$\epsilon = \frac{\delta}{L}$$

then

$$\delta = \epsilon L = 0.000667(10)(12) = 0.080 \text{ in.}$$

(e) Total transverse change in the 12 in. dimension:

$$\delta = \epsilon L = 0.000167(12) = 0.0020 \text{ in.}$$

The transverse strain that accompanies the axial stress does not result from a transverse stress and does not cause a transverse stress. However, if the transverse strain is prevented in some way, a transverse stress will develop.

If an elastic body is loaded in two directions as shown in Fig. 11–2, a stress s is induced in both the x and y directions. The strains in the x and y directions can be summed and expressed as follows:

$$\epsilon_x = \frac{s_x}{E} - \mu\left(\frac{s_y}{E}\right) = \frac{1}{E}(s_x - \mu s_y) \qquad \textbf{(11–2)}$$

$$\epsilon_y = \frac{s_y}{E} - \mu\left(\frac{s_x}{E}\right) = \frac{1}{E}(s_y - \mu s_x) \qquad \textbf{(11–3)}$$

where s_x/E and s_y/E are the axial effects of the loads, and $\mu(s_y/E)$ and $\mu(s_x/E)$ are the transverse effects of the loads. Note that the negative signs in the equations result from the loads both being tensile.

☐ **EXAMPLE 11–2**   A 12 in. long 1 in. by 3 in. ASTM A36 steel bar is loaded as shown in Fig. 11–3. The proportional limit is 34,000 psi. Compute (a) the strains in the x and y directions and

FIGURE 11–2 Tension member axially loaded in two directions.

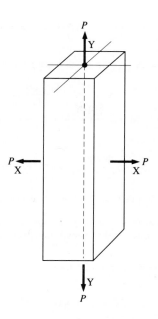

FIGURE 11–3 Load diagram.

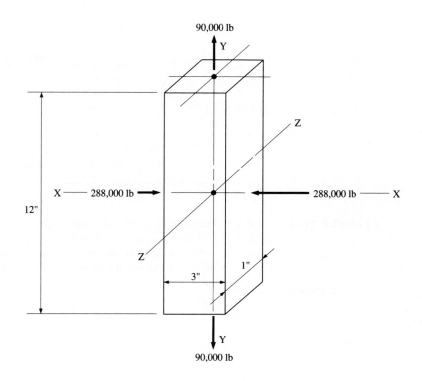

(b) the total dimensional changes in the x and y directions. Refer to Appendix G for necessary mechanical properties:

Solution   First the axial stresses (s_x and s_y) are computed:

$$s_y = \frac{P}{A} = \frac{90,000}{3(1)} = 30,000 \text{ psi}$$

30,000 psi < 34,000 psi (tension)   **OK**

and

$$s_x = \frac{P}{A} = \frac{288,000}{12(1)} = 24,000 \text{ psi}$$

24,000 psi < 34,000 psi (compression)   **OK**

(a) Calculate the strains. (Note that the positive signs in the expressions for strain are due to the fact that one load is tensile and the other load is compressive.)

$$\epsilon_x = \frac{1}{E}(s_x + \mu s_y)$$

$$= \left(\frac{1}{30,000,000}\right)(24,000 + 0.25(30,000))$$

$$= 0.00105 \text{ in./in. (decrease)}$$

and

$$\epsilon_y = \frac{1}{E}(s_y + \mu s_x)$$

$$= \left(\frac{1}{30,000,000}\right)(30,000 + 0.25(24,000))$$

$$= 0.00120 \text{ in./in. (increase)}$$

(b) Calculating the total dimensional change in the x and y directions,

$$\delta = \epsilon L$$
$$\delta_x = 0.00105(3) = 0.00315 \text{ in. (decrease)}$$
$$\delta_y = 0.00120(12) = 0.0144 \text{ in. (increase)}$$

☐ **EXAMPLE 11–3**   A $1\frac{1}{2}$ in. diameter ASTM A36 steel bar is subjected to a tension test. Under a tensile load of 58,000 lb it was observed that the original gage length of 2 in. increased in length by 0.0022 in. and the diameter decreased by 0.00042 in. If the proportional limit is 34,000 psi, compute the modulus of elasticity E and Poisson's ratio μ.

Solution   Axial stress:

$$s_t = \frac{P}{A} = \frac{58,000}{0.7854(1.5)^2} = 32,824 \text{ psi}$$

32,824 psi < 34,000 psi   **OK**

Axial strain:

$$\epsilon = \frac{\delta}{L} = \frac{0.0022}{2} = 0.0011 \text{ in./in.}$$

Modulus of elasticity:

$$E = \frac{s_t}{\epsilon} = \frac{32{,}824}{0.0011} = 29{,}800{,}000 \text{ psi}$$

Poisson's ratio:

$$\text{Transverse strain } \epsilon = \frac{\delta}{L} = \frac{0.00042}{1.5} = 0.00028 \text{ in./in.}$$

$$\mu = \frac{\text{transverse } \epsilon}{\text{axial } \epsilon} = \frac{0.00028}{0.0011} = 0.255$$

There is a relationship between the modulus of elasticity, modulus of rigidity, and Poisson's ratio.[1] In Section 9–6 we discussed the modulus of rigidity G as the ratio of shear stress to shear strain. For homogeneous elastic materials, the modulus of rigidity can be determined by means of a tensile test. However, both the longitudinal (axial) strain and the transverse strain must be measured. The modulus of rigidity can then be calculated from

$$G = \frac{E}{2(1 + \mu)} \tag{11–4}$$

where G = the modulus of rigidity (psi, ksi) (Pa, MPa)
     E = the modulus of elasticity (tensile or compressive) (psi, ksi) (Pa, MPa)
     μ = Poisson's ratio

Note that the three preceding properties are not independent of each other. Also, note that the modulus of rigidity G will always be less than E, since Poisson's ratio is always positive.

□ **EXAMPLE 11–4**   A 2 in. diameter metal specimen is subjected to an axial compressive load of 40,000 lb. Transverse and longitudinal dimensional changes are measured with the use of electronic strain gages and the strains are determined to be 0.0012 longitudinally and 0.0004 transversely. Compute (a) Poisson's ratio μ, (b) the modulus of elasticity E, and (c) the modulus of rigidity G.

Solution   (a) For Poisson's ratio,

$$\mu = \frac{\text{transverse strain}}{\text{longitudinal strain}} = \frac{0.0004}{0.0012} = 0.333$$

---

[1]   S. Timoshenko, *Strength of Materials*, Part I (New York: Van Nostrand, 1950).

(b) For the modulus of elasticity, first calculating stress,

$$s_c = \frac{P}{A} = \frac{40{,}000}{0.7854(2)^2} = 12{,}732 \text{ psi}$$

$$E = \frac{s_c}{\epsilon} = \frac{12{,}732}{0.0012} = 10{,}610{,}000 \text{ psi}$$

(c) For the modulus of rigidity, Eq. (11–4) yields

$$G = \frac{E}{2(1 + \mu)} = \frac{10{,}610{,}000}{2(1 + 0.333)} = 3{,}980{,}000 \text{ psi}$$

☐ **EXAMPLE 11–5**   Compute the modulus of rigidity for steel using the normal values of $E = 30{,}000{,}000$ psi and $\mu = 0.25$.

Solution   Using Eq. (11–4),

$$G = \frac{E}{2(1 + \mu)} = \frac{30{,}000{,}000}{2(1 + 0.25)} = 12{,}000{,}000 \text{ psi}$$

11–2
THERMAL EFFECTS

Materials commonly used in engineering exhibit dimensional changes when subjected to temperature changes. For any particular material, the amount of dimensional change per unit temperature change is constant over moderate temperature ranges. Most materials expand as their temperatures rise and contract as their temperatures fall.

For most materials, standard values for dimensional change per degree of temperature change have been established by test. Such a value is called the *linear coefficient of thermal expansion* and is denoted by the Greek letter α. It is a measure of the change in length per unit of length per degree of temperature change. It will have the same numerical value for any particular material, no matter what unit of length is used. The coefficient is frequently given units such as in./in./F° in the U.S. Customary System and mm/mm/C° (where C denotes Celsius) in the SI system. However, since *any* convenient unit of length may be used, it is not uncommon for the coefficient to be expressed in units of 1/F° or 1/C°. For typical values, see Appendix G.

If a body is free to expand or contract due to temperature variations, there will be no stress induced in the member. The magnitude of the dimensional change can be expressed by

$$\delta = \alpha L(\Delta T) \tag{11–5}$$

where δ = the total change in length (in.) (mm)
     α = the linear coefficient of thermal expansion; also, strain per degree change in temperature (1/F°) (1/C°)
     L = the original length of the member (in.) (mm)
    ΔT = the change in temperature (F°)(C°)

If a body is somehow partially or fully restrained so as to prevent a dimensional change due to temperature variations, internal stresses will develop. These are generally termed *temperature stresses* or *thermal stresses*. An expression for these stresses can be developed as follows:

1. Assume that a total dimensional change of δ is allowed to occur due to a temperature change:

$$\delta = \alpha L(\Delta T)$$

2. The member is actually restrained. Therefore, apply an axial force P to the member to restore it to its original length. This dimensional change is written as

$$\delta = \frac{PL}{AE} = s\left(\frac{L}{E}\right)$$

3. Equating the two values of δ,

$$s\left(\frac{L}{E}\right) = \alpha L(\Delta T)$$

from which

$$s = E\alpha(\Delta T) \qquad \text{(11–6)}$$

where s = the temperature stress developed in a restrained member due to a temperature variation (valid only if s does not exceed the proportional limit) (psi) (MPa)
E = the modulus of elasticity (psi) (MPa)
α = the coefficient of thermal expansion (1/F°) (1/C°)
ΔT = the change in temperature (F°) (C°)

If a member is fully restrained and then cooled, the stress induced is tension. If the member is fully restrained and then heated, the stress induced is compression.

☐ **EXAMPLE 11–6** A 100 in. long ASTM A36 steel rod with a cross-sectional area of 2.0 in.^{2} is secured between rigid supports. If there is no stress in the rod at a temperature of 70°F, compute the stress when the temperature drops to 0°F. The proportional limit of the steel is 34,000 psi. Refer to Appendix G for necessary mechanical properties.

Solution Since no yielding occurs,

$$s = E\alpha(\Delta T)$$
$$= (30,000,000 \text{ lb/in.}^2)(0.0000065 \text{ 1/F°})(70 \text{ F°})$$
$$= 13,650 \text{ psi (tension)}$$
$$13,650 \text{ psi} < 34,000 \text{ psi} \qquad \textbf{OK}$$

☐ **EXAMPLE 11–7** Compute the stress in the rod of the previous example if the supports yield and move together a distance of 0.02 in. as the temperature drops. Refer to Fig. 11–4.

FIGURE 11–4 Support for steel rod.

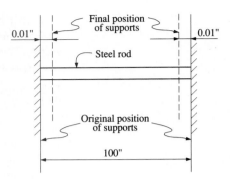

Solution Note that in Example 11–6 the stress developed is independent of the length of the member. However, when yielding of the supports occurs, the length of the member does affect the stress developed. If the rod were free to contract, the amount that it would shorten could be calculated from

$$\delta = \alpha L(\Delta T)$$
$$= (0.0000065 \; 1/\text{F}°)(100 \text{ in.})(70 \text{ F}°)$$
$$= 0.0455 \text{ in.}$$

If the supports were rigid, δ would be the restrained change in length. Since the supports yield 0.02 in., the restrained change in length is

$$\delta = 0.0455 - 0.02 = 0.0255 \text{ in.}$$

Therefore, the strain is

$$\epsilon = \frac{\delta}{L} = \frac{0.0255}{100} = 0.000255 \text{ in./in.}$$

from which

$$s = E\epsilon$$
$$= 30,000,000(0.000255) = 7650 \text{ psi (tension)}$$
$$7650 \text{ psi} < 34,000 \text{ psi} \quad \textbf{OK}$$

☐ **EXAMPLE 11–8** An AISI 1040 steel fence wire 0.148 in. in diameter is stretched between rigid end posts with a tension of 300 lb when the temperature is 90°F. The proportional limit of the wire is 40,000 psi. Calculate the temperature drop that could occur without causing a permanent set in the wire. Refer to Appendix G for necessary properties.

Solution The cross-sectional area of the wire is

$$A = 0.7854(0.148)^2 = 0.0172 \text{ in.}^2$$

The stress in the wire due to the tensile load of 300 lb is

$$s = \frac{P}{A} = \frac{300}{0.0172} = 17,440 \text{ psi}$$

The additional temperature stress to reach the proportional limit is

$$s = 40,000 - 17,440 = 22,560 \text{ psi}$$

The temperature change that would induce this stress in the wire is calculated from

$$s = E\alpha(\Delta T)$$

Substituting the numerical values,

$$22,560 = 30,000,000(0.0000065)(\Delta T)$$

from which

$$\Delta T = 115.7 \text{ F}° \text{ (decrease)}$$

Therefore, the temperature would be

$$90 - 115.7 = -25.7°\text{F}$$

**11–3
MEMBERS
COMPOSED OF
TWO OR MORE
MATERIALS**

In some situations, structural members may be composed of two (or more) different materials. One example is a building column composed of a steel shape encased in concrete. Another is a wood post strengthened with steel plates or channels, as shown in Fig. 11–5. We will assume that the materials in the reinforced post have different modulus of elasticity values and are so connected as to act as a single unit, each deforming equally under load. In this case, the stresses developed in the two materials by the applied load will be proportional to their moduli of elasticity.

FIGURE 11–5 Wood post with steel plates.

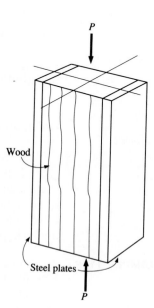

For the same deformation, the stress developed in the material with the higher modulus of elasticity (material A) will be greater than the stress developed in the material with the lower modulus of elasticity (material B). Assuming the two materials are of the same length and deform equally,

$$\delta_A = \delta_B \quad \text{and} \quad \epsilon_A = \epsilon_B$$

Since

$$\epsilon = \frac{s}{E}$$

it follows that

$$\frac{s_A}{E_A} = \frac{s_B}{E_B}$$

The preceding expression can be rearranged to yield the stress in material A (the higher stress):

$$s_A = \frac{E_A}{E_B} s_B$$

The ratio of the modulus of elasticity values is generally called the *modular ratio* and is denoted as n. Therefore,

$$s_A = n s_B \tag{11-7}$$

This concept may be carried one step further. Based on a condition of static equilibrium, the total load P must be resisted (carried) in part by each material. Therefore,

$$P = P_A + P_B$$

Substituting,

$$P = A_A s_A + A_B s_B \tag{11-8}$$

Substituting for s_A from Eq. (11-7),

$$P = A_A(n s_B) + A_B s_B$$

which may be rearranged as

$$P = s_B(n A_A + A_B) \tag{11-9}$$

The quantity $n A_A$ is sometimes called an *equivalent area*. It is a hypothetical area that may be considered a replacement for area A_A. The resulting hypothetical cross section is then composed of a single material having the lower modulus of elasticity value. This concept sometimes facilitates the analysis process.

□ **EXAMPLE 11-9**   A short post consisting of a 6 in. diameter standard weight steel pipe (see Appendix B) is filled with concrete, which has an ultimate compressive strength (s'_c) of 3000 psi. The pipe is made of ASTM A501 steel. The post is subjected to an axial compressive load of 100,000 lb. Assuming both materials deform equally, compute the stress developed in the steel and the concrete. Refer to Appendix G for necessary properties.

Solution   The pipe has a cross-sectional area of 5.58 in.^{2} and an inside diameter of 6.065 in. The cross-sectional area of the concrete is calculated from

$$A_{CON} = 0.7854(6.065)^2 = 28.89 \text{ in.}^2$$

The modular ratio is

$$n = \frac{E_{ST}}{E_{CON}} = \frac{30,000,000}{3,120,000} = 9.62$$

Using Eq. (11–9),

$$P = s_{CON}(nA_{ST} + A_{CON})$$

from which

$$s_{CON} = \frac{P}{nA_{ST} + A_{CON}} = \frac{100,000}{9.62(5.58) + 28.89} = 1211 \text{ psi}$$

From Eq. (11–7),

$$s_{ST} = ns_{CON} = 9.62(1211) = 11,650 \text{ psi}$$

□ **EXAMPLE 11–10**   A 4 by 4 (S4S) Douglas fir wood truss tension member is strengthened by the addition of two ASTM A36 steel plates, as shown in Fig. 11–6. Compute the allowable load for the composite member. In addition to mechanical properties from Appendices F and G, allowable tensile stresses are 1000 psi and 22,000 psi, respectively, for this wood and steel. Assume that the materials are of equal lengths and are so connected as to act as a single unit and deform equally.

Solution   Assume that the stress in the wood reaches its allowable stress of 1000 psi before the steel reaches its allowable stress. The stress in the steel is computed using Eq. (11–7):

$$s_{ST} = ns_W = \frac{30,000,000}{1,700,000}(1000) = 17,650 \text{ psi}$$

FIGURE 11–6   Composite member.

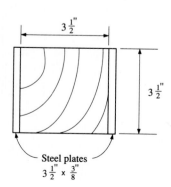

$3\frac{1}{2}"$

$3\frac{1}{2}"$

Steel plates
$3\frac{1}{2}" \times \frac{3}{8}"$

Therefore, when the stress in the wood is 1000 psi, the stress in the steel is 17,650 psi. This is acceptable because the steel stress of 17,650 psi is less than the allowable stress of 22,000 psi. The stresses may not increase beyond this point; if they did, the stress in the wood would exceed the allowable stress of the wood. This calculation shows then, that the allowable stress of the wood limits the load-carrying capacity of the post.

The area of the wood is calculated from

$$A_W = (3.5 \text{ in.})^2 = 12.25 \text{ in.}^2$$

Using Eq. (11–8) to calculate the allowable load,

$$
\begin{aligned}
P &= P_{ST} + P_W \\
&= A_{ST}s_{ST} + A_W s_W \\
&= 3.5(0.375)(2)(17,650) + 12.25(1000) \\
&= 58,600 \text{ lb}
\end{aligned}
$$

We can also consider a type of system in which an axial load is simultaneously applied to two or more members of different materials and of different lengths. The analysis methodology is similar to the case of one member composed of two or more materials.

Assuming the total deformation of the members to be the same, but the lengths of the members *not* the same, Eq. (11–7) cannot be used. We can express the equality of the total deformation of each member as follows (the two different members are denoted by subscripts A and B):

$$\left(\frac{PL}{AE}\right)_A = \left(\frac{PL}{AE}\right)_B \tag{11–10}$$

The use of this equation is similar to the use of Eq. (11–7).

□ **EXAMPLE 11–11**   The structural system of Fig. 11–7 consists of a horizontal plate suspended by three rods. A load of 50 kips is applied to the plate. The plate is perfectly level prior to the application of the load and remains level after the load has been applied.

FIGURE 11–7 Members of different materials.

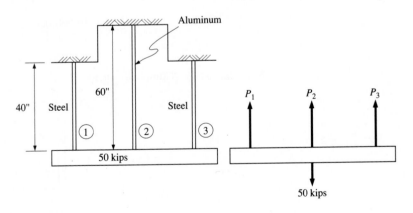

(a) Structural system   (b) Free-body diagram

The steel rods are of AISI 1020 steel. Each is 40 in. long and has a cross-sectional area of 1.0 in.^{2}. The aluminum rod is 60 in. long and has a cross-sectional area of 1.5 in.^{2}. Calculate the load that each rod will carry. Refer to Appendix G for necessary mechanical properties.

Solution   Designating the rods as 1, 2, and 3 and assuming all three rods elongate the same amount,

$$\delta_1 = \delta_2 = \delta_3$$

or

$$\left(\frac{PL}{AE}\right)_1 = \left(\frac{PL}{AE}\right)_2 = \left(\frac{PL}{AE}\right)_3$$

Substituting (where E is in units of ksi, A is in units of in.^{2}, and L is in in.),

$$\frac{P_1(40)}{(1.0)(30,000)} = \frac{P_2(60)}{(1.5)(10,000)} = \frac{P_3(40)}{(1.0)(30,000)} \qquad \text{(Eq. 1)}$$

Due to symmetry, $P_1 = P_3$. Therefore, work with and solve for P_1 and P_2. Cross-multiplying the first two parts of Eq. 1,

$$P_1(40)(1.5)(10,000) = P_2(60)(1.0)(30,000)$$

or

$$600,000P_1 - 1,800,000P_2 = 0 \qquad \text{(Eq. 2)}$$

With reference to Fig. 11–7(b), a summation of vertical forces ($\Sigma F_V = 0$) yields

$$P_1 + P_2 + P_3 = 50 \text{ kips}$$

or since $P_1 = P_3$,

$$2P_1 + P_2 = 50 \text{ kips} \qquad \text{(Eq. 3)}$$

Equations 2 and 3 are simultaneous equations which, when solved for P_1 and P_2, will yield

$$P_1 = 21.43 \text{ kips} \qquad P_2 = 7.14 \text{ kips}$$

Therefore, the steel rods each carry 21.43 kips and the aluminum rod carries 7.14 kips.

☐ **EXAMPLE 11–12**   A solid rolled brass cylinder with a cross-sectional area of 4.0 in.^{2} is inserted in a hollow steel tube with a cross-sectional area of 8.0 in.^{2}. The brass cylinder is 10.005 in. in length and the steel tube is 10.000 in. in length, as shown in Fig. 11–8. The cylinder and tube are supported on a flat rigid surface. A compressive axial load of 100 kips is applied by means of a rigid cap plate. Compute the stresses that will be developed in the two materials. See Appendix G for values of modules of elasticity E.

FIGURE 11–8 Statically indeterminate system.

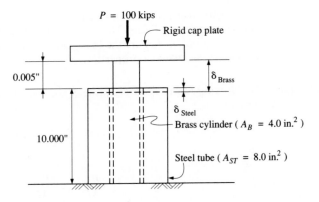

Solution   For this example, units will be in inches and kips. Assuming that both the steel and
the brass will share in carrying the load, the brass cylinder will shorten 0.005 in. more
than will the steel tube. Mathematically,

$$\delta_B = \delta_{ST} + 0.005$$

Substituting, using Eq. (9–10),

$$\left(\frac{PL}{AE}\right)_B = \left(\frac{PL}{AE}\right)_{ST} + 0.005 \qquad \text{(Eq. 1)}$$

This equation has two unknowns (P_B and P_{ST}). For a solution, a second equation
must be established containing the two unknowns. From statics, a summation of
vertical forces ($\Sigma F_V = 0$) yields

$$P_B + P_{ST} = 100$$

from which

$$P_{ST} = 100 - P_B \qquad \text{(Eq. 2)}$$

Substituting P_{ST} from Eq. 2 into Eq. 1 yields

$$\frac{P_B L_B}{A_B E_B} = \frac{(100 - P_{ST})L_{ST}}{A_{ST} E_{ST}} + 0.005$$

Substituting numerical values,

$$\frac{P_B(10.005)}{4.0(14,000)} = \frac{(100 - P_B)(10.000)}{8.0(30,000)} + 0.005$$

Multiplying both sides by 1000 for convenience and simplifying yields,

$$0.1787 P_B = 4.167 - 0.04167 P_B + 5$$

from which

$$0.2204 P_B = 9.167$$
$$P_B = 41.6 \text{ kips}$$

This represents the load that is carried by the brass. The load carried by the steel can be obtained from Eq. 2:

$$P_{ST} = 100 - P_B = 100 - 41.6 = 58.4 \text{ kips}$$

The stresses developed in the brass and the steel, respectively, are

$$s_B = \frac{P_B}{A_B} = \frac{41.6}{4.0} = 10.4 \text{ ksi}$$

$$s_{ST} = \frac{P_{ST}}{A_{ST}} = \frac{58.4}{8.0} = 7.3 \text{ ksi}$$

11–4
STRESS
CONCENTRATION

As previously shown, when an axially loaded prismatic member is subjected to a tensile load, a tensile stress (P/A) will develop. This stress is assumed to be uniformly distributed over the cross-sectional area perpendicular to the direction of the load. This uniform stress distribution will occur on all planes except those in the vicinity of any load application points. Should abrupt changes (sometimes called *stress-raisers*) exist in the cross-section of the member, however, large irregularities in the uniform stress distribution will develop. Examples of stress-raisers in flat axially loaded members and the resulting stress distributions are shown in Fig. 11–9.

Figure 11–9(a) indicates the presence of a circular hole centrally located in the member. Figure 11–9(b) indicates the presence of symmetrically placed semicircular side notches in the member. Figure 11–9(c) shows a member composed of two segments of different lateral dimensions joined with fillets. It is generally accepted that the tensile stress distribution as shown at the reduced cross section will return to a uniform stress distribution a short distance away.

In each case, as may be observed, a localized stress concentration develops immediately adjacent to the stress-raiser. This stress concentration represents a maximum tensile stress, the magnitude of which may be considerably in excess of an average tensile stress acting on the net cross-sectional area in the plane of the reduced cross section.

The calculation of this maximum tensile stress has been simplified with an experimentally determined *stress concentration factor* denoted by the letter k. The value of this factor depends on the geometric proportions of the member as well as on the type and size of the stress-raiser. Information on stress concentration factors is available in the technical literature in the form of tables and curves.

Figure 11–10 shows curves indicating stress concentration factors for flat axially loaded members with three types of change in cross section. With a stress concentration factor from the curves, the maximum tensile stress immediately adjacent to the stress-raiser can be calculated from

$$s_{t(max)} = k\left(\frac{P}{A_{net}}\right) \tag{11–11}$$

FIGURE 11–9 Tensile stress distributions.

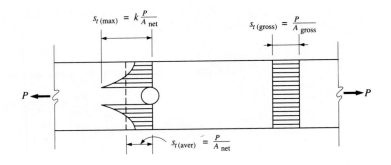

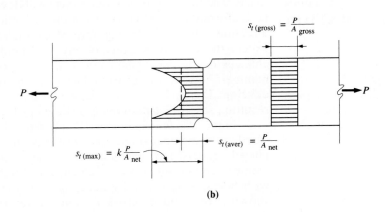

(a)

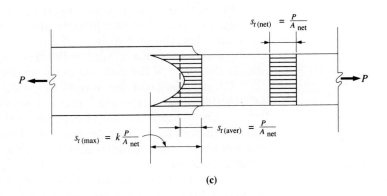

(b)

(c)

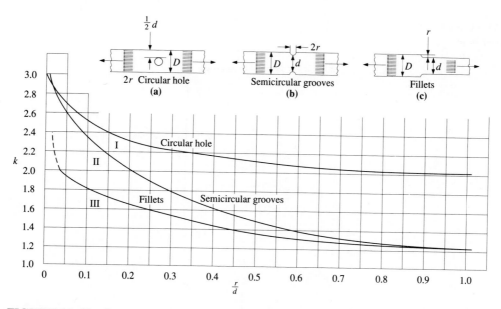

FIGURE 11–10 Stress concentration factors for flat bars (Source: M. M. Frocht. 1935. *Factors of Stress Concentration Photoelastically Determined*. ASME Journal of Applied Mechanics 2:A67–A68).

where $s_{t(max)}$ = the maximum tensile stress immediately adjacent to the discontinuity (psi, ksi) (Pa, MPa)

k = the stress concentration factor

P = the applied axial tensile load (lb, kips) (N)

A_{net} = the net cross-sectional area in the plane of the reduced cross section (in.²) (m², mm²)

This high stress concentration is not necessarily dangerous for ductile metals due to the fact that plastic yielding and subsequent stress redistribution will occur. However, in the case of brittle materials, stress concentrations are much more serious. Cracks may occur in the material in areas of high localized stress due to the inability of the brittle material to deform plastically. Ductile materials, when subjected to repetitive-type loads, fare only slightly better. Should stress concentrations be unavoidable in a member, a reduction in allowable stresses should be considered.

☐ **EXAMPLE 11–13**   A $\frac{3}{4}$ in. diameter hole is drilled on the centerline of a flat steel bar as shown in Fig. 11–11. The bar is subjected to a tensile load of 4000 lb. Calculate the average stress in the plane of the reduced cross section and the maximum tensile stress immediately adjacent to the hole.

Solution   The radius of the hole is 0.375 in. Therefore,

$$\frac{r}{d} = \frac{0.375}{2.5 - 0.75} = 0.214$$

FIGURE 11–11 Long bar with centrally located hole.

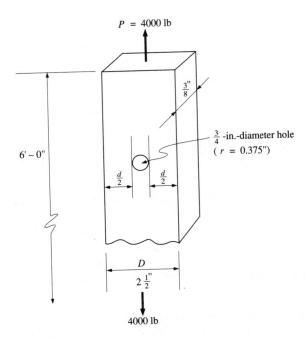

From Fig. 11–10, the value of k is approximately 2.3. The net area in the plane of the reduced cross section is then calculated:

$$A_{net} = 1.75(0.375) = 0.656 \text{ in.}^2$$

The average stress is then calculated as

$$s_t = \frac{P}{A_{net}} = \frac{4000}{0.656} = 6098 \text{ psi}$$

and the maximum tensile stress is

$$s_{t(max)} = k\left(\frac{P}{A_{net}}\right) = 2.3(6098) = 14,000 \text{ psi}$$

□ **EXAMPLE 11–14**   A 36 in. long flat bar 3 in. wide by $\frac{1}{4}$ in. thick has a circular hole 1.0 in. in diameter, centrally located. With an allowable tensile stress of 24,000 psi, calculate the axial tensile load that may be applied to the bar.

Solution

$$\frac{r}{d} = \frac{0.5}{3.0 - 1.0} = 0.25$$

From Fig. 11–10, the value of k is approximately 2.3.

$$s_{t(max)} = k\left(\frac{P}{A_{net}}\right)$$

from which

$$P = \frac{s_{t(max)} A_{net}}{k} = \frac{24{,}000(2)(0.25)}{2.3} = 5200 \text{ lb}$$

11–5 STRESSES ON INCLINED PLANES

When a prismatic member is subjected to a uniaxial tensile or compressive force, maximum tensile and compressive stresses are developed on a plane perpendicular (normal) to the longitudinal axis of the member. This was shown in Fig. 9–5(a) and (b). Additionally, tensile and compressive stresses of lesser intensity, along with shear stresses, will also be developed on planes inclined to the cross section. If a member is composed of a material that does not exhibit the same strength in all directions, this consideration may be important.

In Fig. 11–12(a), a prismatic member is subjected to an axial tensile force P. The member is cut into two parts by plane CD in such a manner that the normal to the plane makes an angle θ with the longitudinal axis of the member. The bottom part is shown as a free body in Fig. 11–12(b). At section C–D the force P is resolved into two components, one parallel to plane C–D and the other perpendicular (normal) to plane C–D. The components have values of $P \sin \theta$ and $P \cos \theta$, respectively.

If the cross-sectional area of the member is denoted as A, then the area of the inclined plane C–D is equal to $A/\cos \theta$. The parallel component of the applied force acting on the inclined plane causes a shear stress of

$$s'_s = \frac{P \sin \theta}{(A/\cos \theta)} = \frac{P}{A} \sin \theta \cos \theta = \frac{P}{2A} \sin 2\theta \qquad \textbf{(11–12)}$$

FIGURE 11–12 Forces and stresses on an inclined plane.

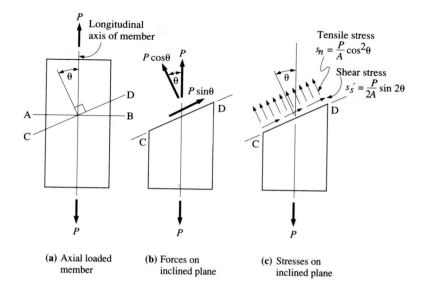

(a) Axial loaded member

(b) Forces on inclined plane

(c) Stresses on inclined plane

This expression becomes a maximum when $\sin 2\theta$ is a maximum, which occurs when $\sin 2\theta$ is 1. This, in turn, occurs when $2\theta = 90°$ and $\theta = 45°$. When $\sin 2\theta = 1$, the previous expression becomes

$$s'_{s(\text{max})} = \frac{P}{2A} \qquad\qquad (11\text{--}13)$$

and represents a maximum shear stress developed on a plane inclined at 45° to the cross section of the member.

The perpendicular component of the applied force acting on the inclined plane C–D causes a tensile stress of

$$s_n = \frac{P \cos \theta}{(A/\cos \theta)} = \frac{P}{A} \cos^2\theta \qquad\qquad (11\text{--}14)$$

This expression becomes a maximum when $\cos^2\theta$ is a maximum. The maximum value of $\cos^2\theta$ is 1 and occurs when $\theta = 0°$. Note in Eq. (11–13) that the maximum shear stress on the 45° inclined plane equals one-half the maximum normal tensile stress on the 0° plane.

Hence, we can conclude that at any point in a body subjected to load, there are different stresses or combinations of stresses developed on planes of different inclinations. If the applied force P were compressive, the same expressions would be obtained.

□ **EXAMPLE 11–15**   A square steel bar, 1 in. by 1 in. in cross section, is subjected to a tensile load of 20,000 lb. (a) Compute the shear stress and the tensile stress on inclined planes whose normals make angles of 30°, 45°, and 60° with the longitudinal axis of the bar. (b) Compute the maximum tensile stress and indicate on which plane this would be developed.

Solution   (a) Normal line at 30°:

$$s'_s = \frac{P}{2A} \sin 2\theta = \frac{20,000}{2(1)} \sin 60° = 8660 \text{ psi}$$

$$s_n = \frac{P}{A} \cos^2\theta = \frac{20,000}{1} \cos^2 30° = 15,000 \text{ psi}$$

Normal line at 45°:

$$s'_s = \frac{20,000}{2(1)} \sin 90° = 10,000 \text{ psi}$$

$$s_n = \frac{20,000}{1} \cos^2 45° = 10,000 \text{ psi}$$

Normal line at 60°:

$$s'_s = \frac{20,000}{2(1)} \sin 120° = 8660 \text{ psi}$$

$$s_n = \frac{20,000}{1} \cos^2 60° = 5000 \text{ psi}$$

(b) The maximum tensile stress occurs when $\theta = 0°$, since $\cos 0° = 1$. The expression can then be written as

$$s_n = \frac{P}{A} \cos^2\theta = \frac{P}{A} (1) = \frac{20,000}{1} = 20,000 \text{ psi}$$

☐ **EXAMPLE 11-16** A square steel bar, 2 in. by 2 in. in cross section, is subjected to an axial tensile load. The maximum shear stress caused by this load is 10,000 psi. Compute the magnitude of the applied load.

Solution The shear stress on an inclined plane is given by

$$s'_s = \frac{P}{2A} \sin 2\theta$$

and is a maximum when $\theta = 45°$. Solving for P,

$$P = \frac{s'_s(2A)}{\sin 2\theta} = \frac{10,000(2)(4)}{\sin 90°} = 80,000 \text{ lb}$$

☐ **EXAMPLE 11-17** A steel specimen having a diameter of 0.505 in. is subjected to a 10,000 lb axial tensile load. (a) Compute the maximum shear stress. (b) Compute the tensile stress acting on the same plane at which the shear stress is a maximum.

Solution (a) The maximum shear stress will occur on a plane whose normal is inclined at 45° to the longitudinal axis of the member.

$$s'_{s(max)} = \frac{P}{2A} \sin 2\theta = \frac{10,000}{2(0.7854)(0.505)^2} \sin 90° = 25,000 \text{ psi}$$

(b) The tensile stress on the plane whose normal is inclined at 45° to the longitudinal axis is calculated from

$$s_n = \frac{P}{A} \cos^2\theta = \frac{10,000}{0.7854(0.505)^2} \cos^2 45° = 25,000 \text{ psi}$$

11-6
SHEAR STRESSES
ON MUTUALLY
PERPENDICULAR
PLANES

In this section we will show that at any point in a stressed member where shear stresses exist on a plane, there must simultaneously exist shear stresses of equal intensity on a perpendicular plane.

Let us consider a member subjected to a shear force as shown in Fig. 11–13(a). The shear force results in a shear stress s_{s1} acting on the right-hand face of the member. An infinitesimal element ABCD at the face of the member is then removed as a free body in equilibrium (see Fig. 11–13(b)). Element ABCD is assumed to have a thickness of unity (1). If there is a shear stress of s_{s1} acting on the right-hand face of the element, the shear force on this face is $s_{s1}(h)(1)$. Furthermore, there must be an equal and oppositely directed shear force on the left face, since the summation of vertical forces must equal zero.

FIGURE 11–13 Shear stresses on perpendicular planes.

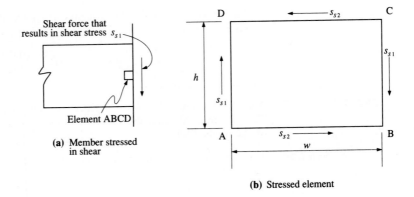

(a) Member stressed in shear

(b) Stressed element

The two vertical forces described constitute a couple. To prevent rotation of the element there must exist another couple made up of shear forces $s_{s2}(w)(1)$ acting on the top and bottom faces of element ABCD. These two couples must be numerically equal and acting in opposite directions, as shown in Fig. 11–13(b).

Taking moments of the forces with respect to point A and equating the two couples,

$$s_{s1}(h)(1)(w) = s_{s2}(w)(1)(h)$$

from which

$$s_{s1} = s_{s2}$$

Therefore, we see that a shear stress on a plane cannot exist alone but induces an equal shear stress on a perpendicular plane.

In this discussion, no stresses other than shear stresses were considered. This condition is identified as a state of *pure shear*. Pure shear can exist in an element of a member which is being sheared (such as in a punching operation shown in Fig. 9–7), or in circular shafts and in some beams. These latter two will be discussed in later chapters.

11–7
TENSION AND COMPRESSION CAUSED BY SHEAR

Figure 11–14(a) shows a stressed element subjected to pure shear. In the previous section, it was proved that shear stresses on mutually perpendicular planes are equal. The shear stresses s_s are shown on each of the four faces of the element. However, the arrows shown in Fig. 11–14(a) represent shear forces, which are obtained by multiplying the shear stress by the area on which the shear stress acts, in each case.

A section R-R is cut through the element from corner to corner, and the upper left half is shown as a free body in Figure 11–14(b). The angle θ is defined by the dimensions w and h. Angle θ is also the angle between the

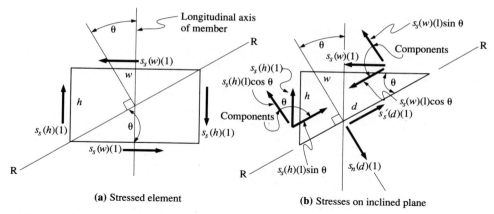

(a) Stressed element **(b)** Stresses on inclined plane

FIGURE 11-14 Stress at a point.

perpendicular to the diagonal plane and the longitudinal axis of the member. If d is the length of the diagonal, the forces acting on the diagonal surface are the shear force $s'_s(d)(1)$ and the tensile force $s_n(d)(1)$. In these expressions, s'_s is the shear stress acting on the diagonal and s_n is the tensile stress acting normal to the diagonal.

Since the free body showing the cut element must be in equilibrium the summation of forces perpendicular to the diagonal surface must equal zero. (Recall that the cutting of small elements into smaller free bodies is justified on the basis of equilibrium. If a body is in equilibrium, every portion of that body, no matter how small, must also be in equilibrium.) Taking the algebraic summation of forces perpendicular to the diagonal surface, we have

$$s_n(d)(1) = s_s(h)(1)\cos\theta + s_s(w)(1)\sin\theta$$

Dividing through by $(d)(1)$,

$$s_n = s_s\left(\frac{h}{d}\right)\cos\theta + s_s\left(\frac{w}{d}\right)\sin\theta$$

Since $\sin\theta = h/d$ and $\cos\theta = w/d$,

$$s_n = s_s\sin\theta\cos\theta + s_s\cos\theta\sin\theta$$
$$= 2s_s\sin\theta\cos\theta$$
$$s_n = s_s\sin 2\theta \tag{11-15}$$

This expression becomes a maximum when $\sin 2\theta$ is a maximum. The maximum value of $\sin 2\theta$ is 1. This occurs when $2\theta = 90°$ and $\theta = 45°$.

If it is desired to compute the shear stress on the diagonal surface, an algebraic summation of forces may be taken parallel to that surface. The signs of the forces in this summation are obtained by observation from Fig. 11–14(b):

$$s'_s(d)(1) = s_s(w)(1) \cos \theta - s_s(h)(1) \sin \theta$$

Dividing through by $(d)(1)$,

$$s'_s = s_s \left(\frac{w}{d}\right) \cos \theta - s_s \left(\frac{h}{d}\right) \sin \theta$$

Since $\sin \theta = h/d$ and $\cos \theta = w/d$,

$$\begin{aligned} s'_s &= s_s \cos^2 \theta - s_s \sin^2 \theta \\ &= s_s(\cos^2 \theta - \sin^2 \theta) \\ s'_s &= s_s \cos 2\theta \end{aligned} \qquad \textbf{(11–16)}$$

Note that when $\theta = 45°$, the shear stress on the diagonal plane is zero.

A similar situation occurs on the other diagonal plane of the element subjected to pure shear. However, the stress normal to this diagonal plane will be compressive rather than tensile. An equilibrium analysis will show that Eq. (11–15) will yield the magnitude of this compressive stress. It will also reach a maximum value when $\theta = 45°$.

We see, then, that for a stressed element subjected to pure shear, tensile and compressive stresses are developed on diagonal planes as a result of shear stresses. The maximum tensile and compressive stresses will develop on planes at angles of 45° with the applied shear stress and will be of an intensity equal to the shear stress.

The tensile stress that develops on the diagonal surface is generally designated as *diagonal tension*. This stress is of great significance in the design of reinforced concrete members, since the capacity of concrete to resist tension is very limited.

☐ **EXAMPLE 11–18**   A shear stress (pure shear) of 8000 psi exists in a member. Compute the tensile normal stress developed on diagonal planes whose normals are oriented at angles of 15°, 30°, and 45° with the longitudinal axis as shown in Fig. 11–15.

Solution $$s_n = s_s \sin 2\theta$$

For $\theta = 15°$,

$$s_n = 8000 \sin 30° = 4000 \text{ psi}$$

For $\theta = 30°$,

$$s_n = 8000 \sin 60° = 6930 \text{ psi}$$

For $\theta = 45°$,

$$s_n = 8000 \sin 90° = 8000 \text{ psi}$$

FIGURE 11–15 Stressed element.

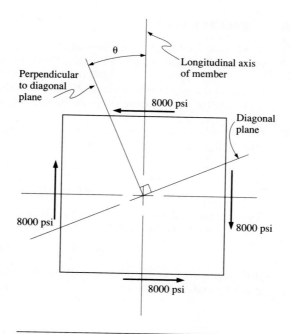

□ **EXAMPLE 11–19**   An element taken from a wood block (Fig. 11–16(a)) is subjected to shear stresses on horizontal and vertical planes as shown in Fig. 11–16(b). The wood grain is at an angle of 20° with the axis of the member. Compute the shear stress and compressive normal stress developed on plane A–A which is parallel to the grain.

Solution   The compressive normal stress on plane A–A is

$$s_n = s_s \sin 2\theta = 200 \sin 140° = 128.6 \text{ psi}$$

The shear stress on plane A–A is

$$s'_s = s_s \cos 2\theta = 200 \cos 140° = 153.2 \text{ psi}$$

FIGURE 11–16 Wood block subjected to pure shear.

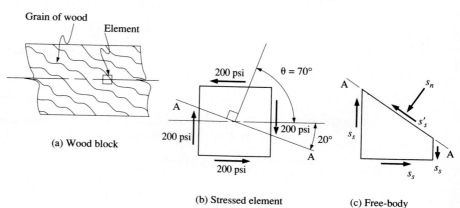

(a) Wood block

(b) Stressed element

(c) Free-body

11–8

SI SYSTEM EXAMPLES

□ **EXAMPLE 11–20**

Steel crane rails are laid with their adjacent ends 3.2 mm apart when the temperature is 15°C. The length of each rail is 18 m. Refer to Appendix G for mechanical properties. (a) Calculate the temperature at which the rails will touch end to end. (b) Calculate the gap between adjacent ends when the temperature drops to −10°C. (c) Calculate the compressive stress in the rails when the temperature reaches 45°C.

Solution

(a) The rails will touch end to end when the 3.2 mm gap is closed by a temperature rise. Using Eq. (11–5) and solving for ΔT,

$$\Delta T = \frac{\delta}{\alpha L} = \frac{3.2 \times 10^{-3} \text{ m}}{\left(0.000\ 011\ 7 \ \frac{1}{\text{C}°}\right)(18 \text{ m})} = 15.2 \text{ C}°$$

This represents the temperature change (increase) that will make the rail ends touch. The actual temperature will be

$$15°\text{C} + 15.2 \text{ C}° = 30.2°\text{C}$$

(b) The gap between the adjacent ends when the temperature drops to −10°C, which is a temperature change of −25 C°, is also calculated from Eq. (11–5):

$$\delta = \alpha L(\Delta T)$$
$$= \left(0.000\ 011\ 7 \ \frac{1}{\text{C}°}\right)(18 \times 10^3 \text{ mm})(25 \text{ C}°)$$
$$= 5.27 \text{ mm}$$

The gap would then become

$$3.2 \text{ mm} + 5.27 \text{ mm} = 8.47 \text{ mm}$$

(c) A compressive stress in the rails will develop when the temperature rises above 30.2°C [from part (a)]. Therefore, if the temperature rose to 45°C, the temperature change would be

$$\Delta T = 45°\text{C} - 30.2°\text{C} = +14.8 \text{ C}°$$

And the stress, calculated from Eq. (11–6), is

$$s_c = E\alpha(\Delta T)$$
$$= (207\ 000 \text{ MPa})\left(0.000\ 011\ 7 \ \frac{1}{\text{C}°}\right)(14.8 \text{ C}°)$$
$$= 35.8 \text{ MPa}$$

□ **EXAMPLE 11–21**

Compute the modulus of rigidity for an aluminum alloy that has a modulus of elasticity E of 70 000 MPa and a Poisson's ratio μ of 0.33.

Solution

Using Eq. (11–4),

$$G = \frac{E}{2(1 + \mu)} = \frac{70\ 000 \text{ MPa}}{2(1 + 0.33)} = 26\ 300 \text{ MPa}$$

☐ **EXAMPLE 11–22**   A 50 mm diameter steel rod, 254 mm in length, is placed inside a brass tube having an inside diameter of 50 mm and an outside diameter of 75 mm, as shown in Fig. 11–17. The member is subjected to an axial tensile load of 300 kN. The two materials are so connected to the rigid plates as to act as a unit, and both will elongate the same amount. Calculate (a) the stress developed in the steel and in the brass, (b) the magnitude of load supported by each material, and (c) the elongation of the system.

Solution   The cross-sectional area of the steel rod is calculated from

$$A_{ST} = 0.7854(50)^2 = 1963.5 \text{ mm}^2$$

And for the brass tube,

$$A_{BR} = 0.7854(d_o^2 - d_i^2)$$
$$= 0.7854(75^2 - 50^2) = 2454.4 \text{ mm}^2$$

The modular ratio is

$$n = \frac{E_{ST}}{E_{BR}} = \frac{200\ 000}{97\ 000} = 2.06$$

(a) From Eq. (11–9),

$$P = s_{BR}(nA_{ST} + A_{BR})$$

from which

$$s_{BR} = \frac{P}{nA_{ST} + A_{BR}} = \frac{300 \times 10^3 \text{ N}}{[2.06(1963.5) + 2454.4] \times 10^{-6} \text{ m}^2}$$
$$= 0.0462 \times 10^9 \text{ N/m}^2$$
$$= 46.2 \text{ MPa (tension)}$$

And from Eq. (11–7),

$$s_{ST} = ns_{BR} = 2.06(46.2) = 95.2 \text{ MPa (tension)}$$

(b) The load supported by each material can be calculated from

$$P_{BR} = s_{BR}A_{BR} = (46.2 \times 10^6 \text{ N/m}^2)(2454.4 \times 10^{-6} \text{ m}^2)$$
$$= 113\ 400 \text{ N}$$
$$= 113.4 \text{ kN}$$
$$P_{ST} = s_{ST}A_{ST} = (95.2 \times 10^6 \text{ N/m}^2)(1963.5 \times 10^{-6} \text{ m}^2)$$
$$= 186\ 900 \text{ N}$$
$$= 186.9 \text{ kN}$$

FIGURE 11–17 Load diagram.

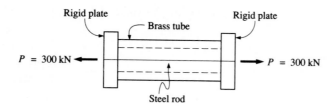

(c) Elongation of the system can be obtained using the previous results for either the brass or the steel. Using the steel values, the strain is calculated from

$$\epsilon = \frac{s_{ST}}{E_{ST}} = \frac{95.2 \text{ MPa}}{200\ 000 \text{ MPa}} = 0.000\ 476$$

The total deformation (elongation) is then calculated from

$$\delta = \epsilon L = (0.000\ 476)(254 \text{ mm}) = 0.121 \text{ mm}$$

☐ **EXAMPLE 11–23**   A flat steel bar 100 mm wide and 10 mm thick is reduced in width to 75 mm. There are circular fillets of 12.5 mm radius on each side, as shown in Fig. 11–18. The bar is subjected to an axial tensile load of 55 kN. Calculate (a) the average tensile stress in the wide portion of the bar some distance from the change in section, (b) the average tensile stress in the narrow part of the bar, and (c) the maximum tensile stress adjacent to the fillet.

FIGURE 11–18   Stress concentration example.

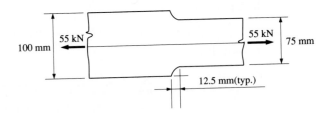

Solution   (a) The average tensile stress in the wide portion of the bar is

$$s_t = \frac{P}{A} = \frac{55 \times 10^3 \text{ N}}{[100(10)] \times 10^{-6} \text{ m}^2} = 0.055 \times 10^9 \text{ N/m}^2 = 55.0 \text{ MPa}$$

(b) The average tensile stress in the narrow portion of the bar is

$$s_t = \frac{P}{A} = \frac{55 \times 10^3 \text{ N}}{[75(10)] \times 10^{-6} \text{ m}^2} = 0.0733 \times 10^9 \text{ N/m}^2 = 73.3 \text{ MPa}$$

(c) The maximum tensile stress at the fillet is calculated using Eq. (11–11). First obtain the value of k:

$$\frac{r}{d} = \frac{12.5}{75} = 0.167$$

From Fig. 11–10, the value of k is approximately 1.7. Then, from Eq. (11–11),

$$s_{t(max)} = k\left(\frac{P}{A}\right) = 1.7\left(\frac{55 \times 10^3 \text{ N}}{[75(10)] \times 10^{-6} \text{ m}^2}\right)$$
$$= 0.125 \times 10^9 \text{ N/m}^2$$
$$= 125 \text{ MPa}$$

EXAMPLE 11–24

A short 150 mm diameter compression member is made of a material with ultimate compressive and ultimate shear strengths of 82.7 MPa and 34.5 MPa, respectively. Calculate the maximum axial load that may be placed on this member before failure occurs.

Solution

Recognizing that the maximum compressive stress occurs on a section normal to the longitudinal axis of the member ($\theta = 0°$), we write Eq. (11–14) as

$$s_{n(\text{max})} = \frac{P}{A} \cos^2 \theta = \frac{P}{A} \quad (1)$$

Substituting the ultimate compressive strength of 82.7 MPa (in terms of newtons and meters) for the maximum normal stress and solving for P,

$$P = As_{n(\text{max})} = 0.7854(150^2 \times 10^{-6} \text{ m}^2)(82.7 \times 10^6 \text{ N/m}^2)$$
$$= 1\ 461\ 000 \text{ N}$$
$$= 1461 \text{ kN}$$

The maximum shear stress occurs on a plane inclined at 45° to the cross section of the member. Using Eq. (11–13), substitute the ultimate shear strength of 34.5 MPa for the maximum shear stress and solve for P:

$$P = 2As'_{s(\text{max})} = 2(0.7854)(150^2 \times 10^{-6} \text{ m}^2)(34.5 \times 10^6 \text{ N/m}^2)$$
$$= 1\ 219\ 000 \text{ N}$$
$$= 1219 \text{ kN}$$

Therefore, the maximum load that may be placed on the member is 1219 kN, the lesser of the two values.

EXAMPLE 11–25

An element removed from a thin-walled cylindrical member loaded in torsion (to be discussed in Chapter 12) is subjected to pure shear stresses as shown in Fig. 11–19. Calculate the magnitudes of the normal and shear stresses developed on diagonal planes defined by angles θ of 30° and 45°.

Solution

Use Eqs. (11–15) and (11–16).
For $\theta = 30°$,

$$s_n = s_s \sin 2\theta = (70 \text{ MPa}) \sin 60° = 60.6 \text{ MPa}$$
$$s'_s = s_s \cos 2\theta = (70 \text{ MPa}) \cos 60° = 35.0 \text{ MPa}$$

FIGURE 11–19 Stressed-element design.

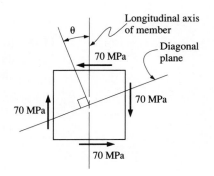

For $\theta = 45°$,

$$s_n = s_s \sin 2\theta = (70 \text{ MPa}) \sin 90° = 70.0 \text{ MPa}$$
$$s'_s = s_s \cos 2\theta = (70 \text{ MPa}) \cos 90° = 0$$

SUMMARY—BY SECTION NUMBER

11–1 The ratio of lateral strain to axial strain for an unrestrained member is called *Poisson's ratio* and is expressed as

$$\mu = \frac{\text{transverse strain}}{\text{axial strain}} \qquad \textbf{(11–1)}$$

11–2 Materials commonly used in engineering tend to change their dimensions (either expanding or contracting) when subjected to temperature changes. The linear coefficient of thermal expansion (α) is a measure of the change in length per unit of length per degree temperature change. For a body free to expand or contract, the magnitude of the dimensional change is

$$\delta = \alpha L (\Delta T) \qquad \textbf{(11–5)}$$

If a body is restrained against dimensional change due to temperature change, an internal stress (temperature stress) will develop. This is calculated from

$$s = E\alpha(\Delta T) \qquad \textbf{(11–6)}$$

11–3 In members made of two (or more) materials so connected that each deforms equally under load, the stresses developed in the two materials are proportional to their moduli of elasticity. The total load carried can be expressed as

$$P = A_A(ns_B) + A_B s_B = s_B(nA_A + A_B) \qquad \textbf{(11–9)}$$

where $n = E_A/E_B$, E_A is the larger of the two moduli of elasticity, and s_B is the lower of the two stresses.

11–4 Localized stress concentrations occur at abrupt changes in the cross section of axially loaded prismatic members in tension. The maximum stress at this location is obtained from

$$s_{t(\max)} = k \left(\frac{P}{A_{\text{net}}} \right) \qquad \textbf{(11–11)}$$

11–5 In axially loaded members, shear stresses and normal stresses are developed on planes inclined to the cross section of the member. The shear stress (developed parallel to the inclined plane) is

$$s'_s = \frac{P}{2A} \sin 2\theta \qquad \textbf{(11–12)}$$

The normal stress (developed normal to the inclined plane) is

$$s_n = \frac{P}{A} \cos^2 \theta \qquad \textbf{(11–14)}$$

11–6 At any point in a stressed member where a shear stress exists on a plane, there must also exist a shear stress of equal intensity on a perpendicular plane.

11–7 At any point in a stressed member that is subjected to a pure shear condition, tensile and compressive stresses are developed on diagonal planes as a result of the shear stresses. The stress normal to the diagonal surface is

$$s_n = s_s \sin 2\theta \qquad \qquad (11–15)$$

The stress parallel to the diagonal surface is

$$s'_s = s_s \cos 2\theta \qquad \qquad (11–16)$$

PROBLEMS

For the following problems, unless noted otherwise, refer to Appendices F and/or G for necessary mechanical properties.

Section 11–1 Poisson's Ratio

1. A 2 in. diameter AISI 1020 steel rod is 10 ft long. Under an applied load, the rod elongates by 0.48 in. (a) Compute the axial (longitudinal) strain in the bar. (b) If the transverse deformation of the rod is 0.0024 in., compute Poisson's ratio for the material.

2. A rectangular ASTM A36 steel bar 2 in. by 6 in. in cross section is subjected to an axial tensile load of 300,000 lb. The proportional limit of the steel is 34,000 psi. Compute the change in the transverse 6 in. dimension.

3. Modulus of elasticity, modulus of rigidity, and Poisson's ratio are interrelated. (a) Calculate G for $E = 30,000,000$ psi and $\mu = 0.28$. (b) Calculate μ for $E = 16,000,000$ psi and $G = 6,000,000$ psi.

Section 11–2 Thermal Effects

4. A surveyor's steel tape is exactly 100 ft long between end markings at 70°F. What error is made in measuring a distance of 1000 ft when the temperature of the tape is 32°F?

5. An aluminum wire is stretched between two rigid supports. If the tensile stress in the wire is 6000 psi at 68°F, what is it at 32°F and 90°F? Assume that the proportional limit is 35,000 psi.

6. A concrete roadway pavement is placed in 50 ft long sections. An expansion joint between the ends of the sections is established as $\frac{1}{2}$ in. at 70°F. Calculate the width of the joint at temperatures of 30°F and 110°F.

7. An ASTM A36 structural steel beam 10 ft long is placed between two rigid supports when the temperature is 60°F. Compute the stress developed in the beam if the temperature rises to 120°F. Assume that the proportional limit is 34,000 psi.

Section 11–3 Members Composed of Two or More Materials

8. A 4 in. by 8 in. short wood post is reinforced on all four sides by ASTM A36 steel plates. Two plates are 4 in. by $\frac{1}{8}$ in. thick and two plates are 8 in. by $\frac{1}{4}$ in. thick. Using nominal dimensions, calculate the maximum axial compressive load that the member can safely carry. The wood is southern pine. The allowable stress is 20,000 psi for the steel.

9. The cables of a power line are copper-coated steel wire. The overall diameter of the wire is $\frac{3}{4}$ in. The steel core has a diameter of $\frac{5}{8}$ in. If the maximum tension in a wire is 10,000 lb, what are the stresses in the steel and the copper?

10. A $5\frac{1}{2}$ in. by $11\frac{1}{2}$ in. Douglas fir column and a Hem-fir column of the same size are bolted together to form a short composite column, as shown in Fig. 11–20. What portion of a total load of 70,000 lb will each material carry?

11. For the short column in Fig. 11–21, assuming that lateral buckling is prevented, (a) calculate the magnitude of the axial load P that will cause the total length of the member to decrease by 0.01 in. and (b) calculate the compressive stress in the steel.

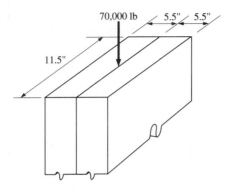

FIGURE 11–20 Problem 10.

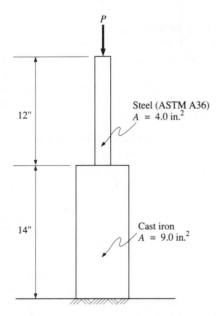

FIGURE 11–21 Problem 11.

Section 11–4   Stress Concentration

12. A 1.0 in. diameter hole is drilled on the centerline of a long, flat steel bar which is $\frac{1}{2}$ in. thick and 4 in. wide. The bar is subjected to a tensile load of 30,000 lb. Calculate the average stress in the plane of the reduced cross section and the maximum tensile stress immediately adjacent to the hole.

13. A long, flat steel bar 4 in. wide and $\frac{3}{8}$ in. thick is reduced in width to 3 in. There are circular fillets of $\frac{1}{2}$ in. radius on each side. If the bar is subjected to an axial tensile load of 12,000 lb, calculate (a) the average tensile stress in the wide portion of the bar some distance from the change in section, (b) the average tensile stress in the narrow part of the bar, and (c) the maximum tensile stress adjacent to the circular fillet.

14. A long, flat steel bar 5 in. wide and $\frac{3}{8}$ in. thick has a circular hole 2 in. in diameter, centrally located. The allowable tensile stress for the steel is 22,000 psi. Calculate the axial tensile load that may be applied to the bar.

Section 11–5   Stresses on Inclined Planes

15. An aluminum specimen of circular cross section, 0.500 in. in diameter, ruptured under a tensile load of 12,000 lb. The plane of failure was found to be at 48° with a plane perpendicular to the longitudinal axis of the specimen. (a) Compute the shear stress on the failure plane. (b) Compute the maximum tensile stress. (c) Compute the tensile stress on the failure plane.

16. A prismatic bar, 2 in. by 3 in. in cross section, is subjected to an axial tensile load of 110,000 lb. (a) Compute the maximum shear stress developed in the bar. (b) Compute the shear stress and tensile stress on a plane whose normal is inclined at 70° to the line of action of the axial load.

17. A short, square steel bar, 1 in. by 1 in. in cross section, is subjected to an axial tensile load of 8000 lb. Compute the shear stress and tensile stress acting on a plane whose normal is inclined at 60° to the longitudinal axis of the member.

Section 11–7   Tension and Compression Caused by Shear

18. An element in a member is subjected to a pure shear of 10,000 psi. (a) Sketch the element and show the shear forces. (b) Determine the tensile normal stress on a plane at 35° with the horizontal. Show this plane on the sketch.

19. For the element of Problem 18, (a) locate the plane on which the shear stress is 6000 psi and (b) verify your answer by showing that a summation of forces parallel to the plane does equal zero.

SI System Problems

20. A 50 mm diameter ASTM A36 steel rod is subjected to an axial tensile load of 270 kN. The proportional limit

FIGURE 11–22 Problem 21.

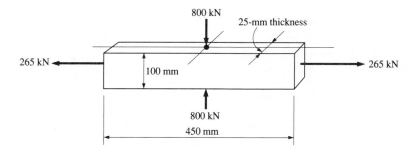

of this steel is 234 MPa. (a) Compute the axial (longitudinal) strain and the transverse strain. (b) If the rod is one meter in length, compute the total elongation.

21. Compute all the dimensional changes for the steel bar in Fig. 11–22 when subjected to the loads shown. The proportional limit of the steel is 230 MPa.

22. A 3 m long steel member is set snugly between two walls and then heated so that the temperature rise is 27 C°. If each wall yields 0.38 mm, what is the compressive stress developed in the member?

23. A 1 m long copper bar is placed between two rigid, unyielding walls as shown in Fig. 11–23, with only one end attached. Calculate the stress developed in the bar due to a temperature rise of 60 C°.

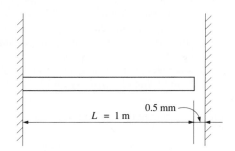

FIGURE 11–23 Problem 23.

24. A horizontal steel member is anchored at each end. Just prior to installation it was heated. After anchoring, it was allowed to cool to 20°C, at which time a stress of 70 MPa developed. If the member is 3 m in length, to what temperature was it heated in order to have developed this stress upon cooling?

25. Calculate Poisson's ratio for a cast iron that has a modulus of elasticity E of 83 000 MPa and a modulus of rigidity G of 33 000 MPa.

26. A structural steel bar 100 mm in width and 12.7 mm in thickness has a copper bar 100 mm wide and 3.2 mm thick securely attached on each side. The assembly acts as a single unit. Calculate the stress and strain in each material caused by an axial tensile load of 225 kN.

27. Three rods support a weight as shown in Fig. 11–24. The supports are rigid and the load is applied uniformly through a rigid block. If the nuts on the three threaded rods were in the same horizontal plane before and after the applied loading, calculate the load carried by each rod.

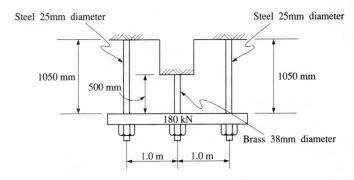

FIGURE 11–24 Problem 27.

28. A solid brass cylinder with a cross-sectional area of 3226 mm² is placed inside a steel tube having a cross-sectional area of 5162 mm². Both the cylinder and the tube are 216 mm long. They stand on end on a flat rigid surface and they support a rigid block. A load of 580 kN is applied to the block. Calculate the stress in each material.

29. A 19 mm diameter hole is drilled on the centerline of a long, flat steel bar. The bar is 10 mm by 75 mm in cross section and is subjected to a tensile load of 18 kN.

Calculate the average tensile stress in the plane of the reduced cross section and the maximum tensile stress immediately adjacent to the hole.

30. A long, flat steel bar 125 mm wide and 10 mm thick has a circular hole of 38 mm diameter centrally located. The allowable tensile stress for the steel is 150 MPa. Calculate the axial tensile load that may be applied to the bar.

31. A 25 mm diameter rod is subjected to an axial tensile load of 80 kN. Compute (a) the normal and shear stresses developed on an inclined plane at an angle of 30° with the cross section of the rod and (b) the maximum normal and shear stresses developed in the rod.

32. An axially loaded 50 mm by 75 mm steel bar has a shear stress of 138 MPa developed on a plane at an angle of 50° with the cross section of the bar. Calculate (a) the applied axial load on the bar and (b) the normal stress developed on this plane.

33. A short timber compression block of rectangular cross section, 150 mm by 200 mm, has an allowable compressive stress of 6 MPa and an allowable shear stress of 1.6 MPa. Calculate the allowable load that may be applied to the member.

34. A wood block, subjected to a tensile load, fails on a plane whose perpendicular is at an angle θ of 30° with the longitudinal axis as shown in Fig. 11–12(a). The shear stress on the failure plane is 35 MPa. Compute the diagonal tension (normal stress) on this plane.

Computer Problems

For the following computer problems, any appropriate programming language may be used. Input prompts should fully explain what is required of the user (the program should be "user friendly"). The resulting output should be well labeled and self-explanatory.

35. Write a program that will compute the allowable load for a short composite member composed of two materials. User input is to be area, modulus of elasticity, and allowable stress for each material. The output should include the allowable load, the final stresses in each material, and an indication as to which of the two materials controls (has reached its allowable stress).

36. Write a computer program that will solve for the shear stress and tensile stress developed in the plate of Problem 67. Have the program generate a table of these stresses for angles ranging from 0° to 75° in increments of 5°.

37. A metal bar is subjected to a temperature change. Write a program that will calculate (a) the total change in the length of the bar and (b) the stress developed in the bar if the ends were rigidly fixed and the bar were short enough to prevent compression buckling. The user should be prompted to input the original length of the bar, the temperature change, and the material of which the bar is made (steel, aluminum, or copper).

38. A surveyor's steel tape is exactly 100.000 ft long at 68°F. Write a program that will generate a table of errors (to 0.001 ft) that will occur in measurements from 70 ft to 100 ft (in 5 ft increments) and in the temperature range of 40°F to 80°F (in increments of 5 F°).

Supplemental Problems

39. A 4 ft long square ASTM A36 steel bar, 2 in. by 2 in., is subjected to an axial tensile load of 64,000 lb parallel to its length. Calculate the dimensional changes in the bar's lateral and longitudinal dimensions.

40. A concrete test cylinder is 6 in. in diameter and 12 in. in length. During an axial compression test, the diameter increased by 0.0005 in. and the length decreased by 0.011 in. Compute the value of Poisson's ratio and the magnitude of the compressive load. The concrete has an ultimate compressive strength of 3000 psi.

41. A 14 in. long steel rod, $1\frac{1}{2}$ in. in diameter, was subjected to an axial tensile load of 60,000 lb in a universal testing machine. It was observed that a gage length of 2 in. near the midpoint of the rod increased in length by 0.0023 in. and the rod diameter decreased by 0.00043 in. Assume the steel proportional limit to be 34,000 psi. Calculate the modulus of elasticity and Poisson's ratio for the steel.

42. Determine the change in the diameter of an ASTM A36 steel rod subjected to an axial compressive load that results in a compressive stress of 30,000 psi. The original diameter is 3 in. and the proportional limit is 34,000 psi.

43. The steel bar shown in Fig. 11–3 (see Example 11–2) is subjected to a compressive load of 648,000 lb in the z direction in addition to the forces shown. Calculate the new dimensions of the bar.

44. Compute the change in the thickness of the ASTM A36 steel bar when subjected to the loads shown in Fig. 11–25. Assume a proportional limit of 34,000 psi.

45. The steel rails of a railroad track are laid in the winter at a temperature of 15°F with gaps of 0.01 ft between

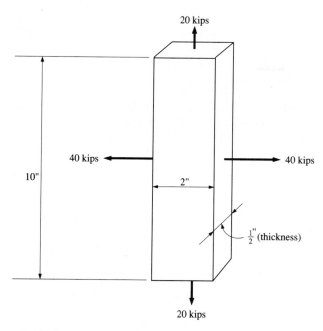

FIGURE 11–25 Problem 44.

the rod if the temperature rises 30 F° and if it falls 50 F°. Assume that the walls do not move.

50. A copper wire is held taut between two supports 20 ft apart. How much may the temperature drop before a stress of 15,000 psi is reached (a) if the supports are immovable and (b) if one of the supports yields 0.05 in. while the temperature is dropping.

51. The rod in Fig. 11–26 is firmly attached to rigid supports. If there is initially no stress in the rod, compute the stress in each material if the temperature drops 100 F° and there is no movement of the supports.

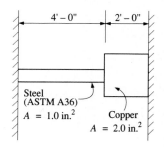

FIGURE 11–26 Problem 51.

the ends of the rails. Each rail is 33 ft long. At what temperature will the rails touch end to end? What stress will result in the rails if the temperature rises to 110°F?

46. A thin ring of hard drawn copper is heated to 250°F. It is then placed over a 6 in. diameter steel cylinder (snug fit) which is at 70°F. Then the entire assembly cools to 70°F. Assume that the diameter of the cylinder does not change. Find the stress in the copper ring.

47. The distance between two fixed points on a missile range in the desert was measured at a temperature of 112°F with a steel tape that was 100.00 ft long at 68°F. The distance was measured as 2208.56 ft. What is the distance corrected for temperature? What distance would have been recorded if the temperature at the time of measurement had been 20°F?

48. A surveyor's steel tape has a cross-sectional area of 0.016 in.^{2} and is exactly 100.00 ft long at 70°F under a tensile pull of 15.0 lb. Compute the pull required to make it 100.00 ft long at 25°F.

49. A 1 in. diameter ASTM A36 steel tie rod, 30 ft long, is used to tie together two walls of a building. The rod is tensioned to 10,000 lb. Compute the tensile stress in

52. Three vertical steel wires are loaded as shown in Fig. 11–27. At a temperature of 68°F, the bottom ends of the wires were at the same elevation with no load on the bar. A downward load of 5000 lb was then centrally applied on the bar and the temperature was increased by 100 F°. What load is carried by each wire assuming the load bar remains horizontal?

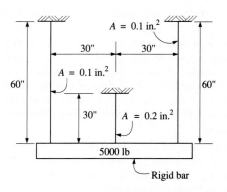

FIGURE 11–27 Problem 52.

53. Assume for Problem 52 that the same load is applied on the bar. To what temperature must the wires be heated before the load is entirely carried by the middle wire?

54. A California redwood timber member having a 16 in. square cross section is reinforced with an ASTM A441 structural steel angle (L6 × 6 × ½) at each corner, as shown in Fig. 11–28. The materials are so attached as to act as a single unit. Compute the total allowable axial compressive load that may be applied through a rigid cap plate that covers the entire cross-sectional area. Use dressed dimensions and assume the materials have the same length. The allowable compressive stress for the steel is 20,000 psi.

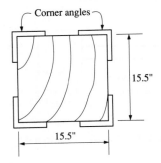

FIGURE 11–28   Problem 54.

55. An ASTM A501 steel pipe has an outside diameter of 16 in. and an inside diameter of 15 in. A 3 in. diameter solid cast-iron cylinder is placed in the pipe and is concentric with the pipe. The space between the two is filled with concrete (ultimate compressive strength = 3000 psi). Compute the stress in each material when the assembly is subjected to an axial load of 600,000 lb.

56. A short 14 in. square concrete pier is reinforced with four longitudinal #7 bars (⅞ in. diameter). The ultimate compressive strength of the concrete is 4000 psi and the modulus of elasticity for the steel bars is 30,000,000 psi. The pier supports a load of 100 kips. Compute the portion of the load carried by the concrete and the portion carried by the steel. Compute the stress in each material.

57. A 12 in. long ASTM A441 steel rod, 2 in. in diameter, is placed within a 12 in. long brass tube with an inside diameter of 2 in. and an outside diameter of 3 in. The tube and rod are so connected as to act as a single unit. An axial tensile load of 65,000 lb is applied through a

rigid cap plate. Compute (a) the tensile stress in each material, (b) the loads carried by each material, and (c) the total elongation of the unit.

58. An aluminum rod with an area of 1.5 in.^{2} and an AISI 1020 steel rod with an area of 1.0 in.^{2} support a rigid bar as shown in Fig. 11–29. The bar is horizontal prior to the application of a 24,000 lb load. Find the distance a and the stresses in the two rods if the bar remains horizontal after the load has been applied.

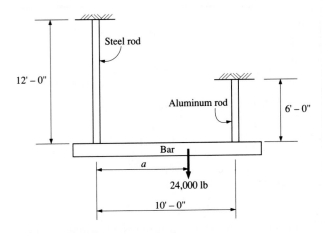

FIGURE 11–29   Problem 58.

59. Three ¼ in. diameter wires are symmetrically spaced and 20 ft long, as shown in Fig. 11–30. The outer wires are steel and the middle wire is bronze. The wires support a rigid horizontal steel box weighing 2000 lb. Calculate (a) the stresses in the wires and (b) the total elongation of the wires.

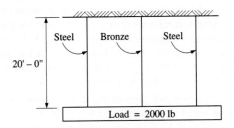

FIGURE 11–30   Problem 59.

60. Rework Problem 59 with outer wires of aluminum, a middle wire of steel, and a 1500 lb box.

61. Three rods support a weight as shown in Fig. 11–31. The supports are rigid and the load is applied centrally through a rigid block. What load W will develop a stress of 12,000 psi in the AISI 1020 steel rod?

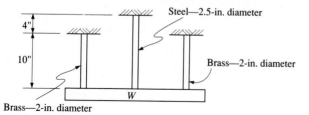

FIGURE 11–31 Problem 61.

62. Referring to Problem 61, compute the stress in the middle rod if the load W is 90,000 lb.

63. A flat steel bar 4 in. wide and $\frac{1}{2}$ in. thick must be reduced to a $2\frac{1}{2}$ in. width as shown. The allowable tensile stress is 15,000 psi. If the applied axial tensile load is 12,000 lb, compute the minimum size fillet radius r that may be used. (See Fig. 11–32.)

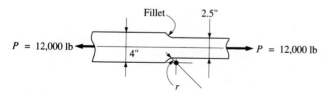

FIGURE 11–32 Problem 63.

64. A flat bar is $\frac{3}{8}$ in. thick and has a centrally located drilled hole of diameter D. The bar is subjected to a 4000 lb tensile load. Compute the maximum tensile stress adjacent to the hole (a) if the bar is 2 in. wide and $D = 0.25$ in., (b) if the bar is 2.5 in. wide and $D = 0.50$ in., and (c) if the bar is 3 in. wide and $D = 1.00$ in.

65. A short 6 in. diameter compression member is made of a material with an ultimate shear strength of 5000 psi and a compressive strength of 12,000 psi. Compute the axial compressive load that may be applied before failure occurs.

66. A rectangular block of wood, 2 in. by 2 in. in cross section, has an ultimate shear strength of 1200 psi. The block is subjected to an increasing tensile load until failure by shear occurs along the grain, which is at an angle of 10° with the longitudinal axis of the member. Compute the magnitude of the load P at which failure occurs. Use nominal dimensions. (See Fig. 11–33.)

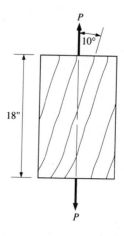

FIGURE 11–33 Problem 66.

67. The rectangular plate in Fig. 11–34 is subjected to a uniaxial tensile stress of 2000 psi. Compute the shear stress and the tensile stress developed on a plane forming an angle of 30° with the longitudinal axis of the member. (**Hint:** Assume a cross-sectional area of unity.)

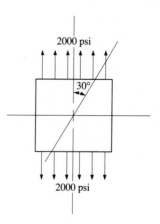

FIGURE 11–34 Problem 67.

68. A shear stress (pure shear) of 5000 psi exists on an element. (a) Determine the maximum tensile and compressive stresses caused in the element due to this shear. (b) Sketch the element showing the planes on which the maximum tensile and compressive stresses act.

69. A shaft in a speed-reduction mechanism is loaded in torque. An element at the surface of the shaft is stressed to 9200 psi (pure shear). (a) Determine the maximum tensile and compressive stresses caused in the element due to this shear. (b) Sketch the element showing the planes on which the maximum tensile and compressive stresses act.

12 Torsion in Circular Sections

12–1
INTRODUCTION

In previous chapters, our discussion centered on the analysis and design of members subjected to axial (concentric) loads or loads that caused direct shear stresses. In this chapter, we will turn our attention to members subjected to a twisting action caused by a couple or a twisting moment. The twisting action, applied in a plane perpendicular to the longitudinal axis of the member, is commonly called *torque*, which is the terminology we will use in this chapter. An example of this type of action may be observed in Fig. 12–1, where the jaws of the bench vise are tightened by applying forces to the handle. A torque is applied to the threaded screw of the vise, turning it, which causes the jaws to tighten. An applied torque such as this is called an *external torque*.

12–2
MEMBERS IN
TORSION

First, we will consider members in static equilibrium when subjected to a twisting action caused either by a pair of externally applied equal and oppositely directed couples acting in parallel planes or by a single external couple applied to a member that has one end fixed against rotation. The fixed end, in effect, furnishes an internal resisting torque. The portion of the member between the two externally applied couples, or between the single externally applied couple and the fixed end, is subjected to a torque and is said to be in torsion, or under torsional load. This would occur in the screw of the bench vise when the jaws were fully tightened and the forces still applied to the handle. Next, we will consider constant-speed rotating shaft-and-pulley systems that are in "dynamic" or "steady-state" equilibrium.

The case of two equal externally applied couples acting in opposite directions on parallel planes perpendicular to the longitudinal axis of the member is shown in Fig. 12–2(a). The magnitude of the torque in the bar is simply equal to that of one of the two couples (*Fd*). In Fig. 12–2(b) the bar has been cut and the right-hand portion shown as a free body. It is apparent that for equilibrium there must be an internal resisting torque equal to the external torque.

In Fig. 12–2(c) the bar is rigidly fixed against rotation at one end, and only one external couple is applied. Equilibrium exists in this case due to an equal and opposite internal resisting torque at the fixed end. The magnitudes of the external and internal torques are, again, *Fd*. In fact, the free-body diagram of Fig. 12–2(b) would apply for either of the two members shown.

FIGURE 12–1 Bench vise.

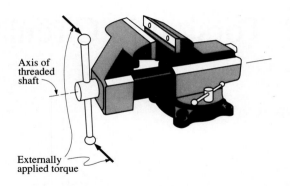

FIGURE 12–2 Members in torsion.

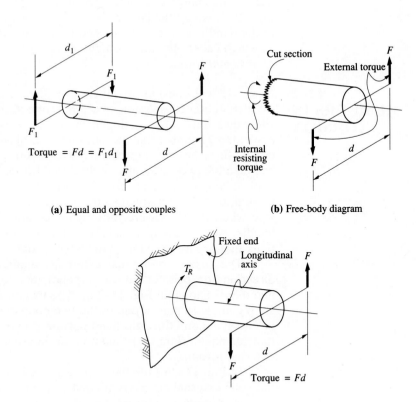

(a) Equal and opposite couples

(b) Free-body diagram

(c) Rigidly fixed bar

The couple, or torque, is generally expressed in units of in.-lb in the U.S. Customary System and N·m in the SI system. In the field of machine design, members subjected to torques or couples are generally shafts used for the transmission of power. Shafts are usually circular in cross section and may be either solid or hollow. The torque to which a shaft is subjected is generally applied through the use of pulleys or gears.

A shaft will commonly have several pulleys mounted on it. One of the pulleys (sometimes called the *driver pulley*) provides the torque to drive the shaft. The shaft then transmits torque to the other pulleys (sometimes called *power take-off pulleys*) which, in turn, provide the necessary torque to drive machines or equipment. In such applications, the torque along the shaft length varies, depending on location and the magnitudes of the torques associated with the various types of pulleys. The torque at any cross section may be determined with the use of a free-body diagram. The internal torque must be equal to the algebraic sum of external torques on either side of the cross section in question.

One objective of this chapter is to determine the relationship between the torque and the resulting stresses and strains in shafts. The shafts will be assumed weightless, thereby making the effect of any bending negligible.

☐ **EXAMPLE 12–1**   Calculate the internal torque at sections R–R and S–S for the shaft shown in Fig. 12–3. The shaft is acted upon by the four torques indicated. Assume negligible bearing friction.

FIGURE 12–3   Shaft and pulleys.

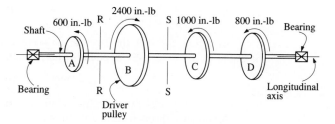

Solution   Note that pulley B is the driver pulley. The other pulleys are power take-off pulleys. The torque of 2400 in.-lb at B is balanced by the three torques of 600 in.-lb, 1000 in.-lb, and 800 in.-lb at A, C, and D, respectively, which are opposite in direction. Therefore, the entire system may be thought of as being in steady-state equilibrium; that is, neither gaining nor losing speed. If the system as a whole is in equilibrium, every segment of the system must also be in equilibrium.

To determine the torque at section R–R, cut section R–R perpendicular to the longitudinal axis of the shaft anywhere between pulleys A and B and consider the left portion a free body, as shown in Fig. 12–4. For equilibrium, the summation of the torques must equal zero ($\Sigma T = 0$). This is the same as saying that the externally applied torque must be balanced by an internal resisting torque that is numerically equal but opposite in direction.

FIGURE 12–4   Free body.

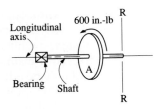

Since the externally applied torque for this free body is 600 in.-lb counterclockwise, when observing from the left end of the shaft, the internal resisting torque between pulleys A and B must also be 600 in.-lb (but clockwise). A sign convention could be defined to describe the actual direction of the internal resisting torque, but for our purposes the use of the terms *clockwise* and *counterclockwise* will be adequately descriptive.

Using the same approach to calculate the torque at section S–S, between pulleys B and C, cut section S–S and consider the left portion of the shaft a free body. This is shown in Fig. 12–5.

FIGURE 12–5   Free body.

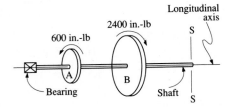

Applying $\Sigma T = 0$, the internal resisting torque between pulleys B and C must be equal to the externally applied torque, which is clockwise. Therefore,

$$T_{int} = T_{ext} = 2400 - 600 = 1800 \text{ in.-lb}$$

which is a counterclockwise torque when observing from the left end of the shaft.

A similar approach results in an internal resisting torque of 800 in.-lb between pulleys C and D and zero torque between pulley D and the frictionless end bearing.

12–3
TORSIONAL
SHEAR STRESS

Let us consider a torsionally loaded member of circular cross section fixed against rotation at one end and subjected to a torque at the other end, as shown in Fig. 12–2(c). Since couples cause neither bending nor direct tension or compression, this condition of loading develops pure shear stresses on each cross-sectional plane that lies between the couple and the fixed end.

If the torsionally loaded member is assumed to be made up of a series of thin plates clamped together, each thin plate tends to slip by, or shear, across the contact surface of the adjacent plate. However, since the member is in equilibrium (and does not fracture), it is evident that some internal resistance is developed which, in effect, prevents slippage. This internal resistance (per unit area) is termed the *torsional shear stress*. The resultant, or total, of these resisting stresses on any cross-sectional plane constitutes an internal resisting torque.

Since all materials have limited shear strength, it is necessary for design purposes to develop a mathematical relationship between the torsional shear stress, the torque, and the physical properties of the member. Prior to this development, we will evaluate a torsionally loaded circular member to establish a cross-sectional shear stress distribution based on stress-strain relationships.

Figure 12–6 shows a segment of a circular shaft that lies between two parallel planes A and B perpendicular to the longitudinal axis of the shaft. CD represents a straight line on the surface of the shaft, parallel to the longitudinal axis and extending from plane A to plane B. Since plane A is fixed against rotation, if the shaft is subjected to a torque applied at plane B, plane B will rotate slightly. The radius OD will then assume the position OD′ and line CD will become CD′, part of a helix. Hence, the shear distortion of line CD is equal to DD′ and the shear strain is (DD′)/L.

FIGURE 12–6 Circular shaft.

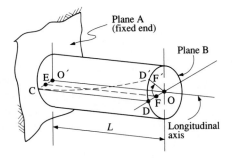

Experiments have found that a plane cross section of a circular shaft will remain a plane after the shaft has twisted and also that a straight-line radius such as OD will remain a straight line as the shaft is twisted. Both of the foregoing conditions will hold, provided that the maximum stress developed in the shaft does not exceed the proportional limit of the material. Therefore, if we consider a line EF parallel to line CD, but halfway between the longitudinal axis and the outer surface, the distortion of line EF, equal to FF′, will be one-half that of line CD. The shear strain, (FF′)/L, will also be one-half of the shear strain at the outer surface.

Assuming that the developed stresses are below the proportional limit and that the shaft material conforms to Hooke's law (stress is proportional to strain), the shear stress at the location of line EF (radial distance OF from the longitudinal axis) will be equal to one-half of the shear stress developed at the outer surface, line CD. Therefore, it may be concluded that the shear stress developed in a torsionally loaded circular shaft is proportional to the distance from the longitudinal axis of the member. Further, the stress is equal to zero at the longitudinal axis and varies linearly to a maximum at the outer surface of the shaft.

Figure 12–7 shows an enlarged view of the cross section of the shaft of Fig. 12–6. The cross section is taken somewhere between plane A and plane B. Point O represents the centroidal longitudinal axis of the shaft. The variation of the shear stress on the cross section, which develops from an externally applied torque, is drawn using radius OD as a reference line. The radial distance from point O to the outer surface is denoted c. An infinitesimal area located a radial distance r from point O is denoted a. The shaft has a diameter d.

FIGURE 12–7   Cross-sectional stress distribution.

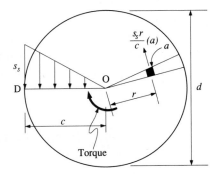

As a result of a torsional loading, a shear stress s_s is developed on the cross section of the shaft at its outer surface (c distance from O). The shear stress developed at a radial distance r from point O can be calculated by proportion from

$$\frac{s_s(r)}{c}$$

Recalling that a force is equal to the product of stress and area, the resisting shear force developed on area a can be calculated from

$$\frac{s_s(r)}{c}(a)$$

Considering only area a, an internal resisting torque is developed with respect to point O by the resisting shear force acting on this area. This resisting torque can be written as

$$\frac{s_s(r^2)}{c}(a)$$

Finally, the total internal resisting torque about point O, from all of the forces acting on the infinitesimal areas, can be calculated from

$$\frac{s_s}{c}\sum ar^2$$

The mathematical quantity of $\sum ar^2$ represents the moment of inertia of the circular shaft with respect to the centroidal longitudinal axis, which is perpendicular to the plane of the area. Recall from Chapter 8 that this is called the polar moment of inertia and is represented by the symbol J.

Therefore, the expression for the total internal resisting torque with respect to point O can be written as

$$\text{Resisting torque} = \frac{s_s J}{c}$$

This total resisting torque must be equal and opposite in direction to the externally applied torque T. Therefore,

$$T = \frac{s_s J}{c} \qquad\qquad\qquad \textbf{(12–1)}$$

Solving for the shear stress,

$$s_s = \frac{Tc}{J} \qquad\qquad\qquad \textbf{(12–2)}$$

where s_s = the computed shear stress (psi, ksi) (Pa, MPa)
$\quad T$ = the externally applied torque (in.-lb, in.-kips) (N·m)
$\quad c$ = the radial distance from the centroidal longitudinal axis to the outer surface (in.) (mm, m)
$\quad J$ = the polar moment of inertia (in.⁴) (mm⁴, m⁴) (see Table 8–1)

We could also rewrite Eq. (12–1) to find the maximum resisting torque or allowable torque for a member. To use this formula, the allowable shear stress must be known:

$$T_R = \frac{s_{s(\text{all})} J}{c} \qquad\qquad\qquad \textbf{(12–3)}$$

where T_R = the allowable torque (in.-lb, in.-kips) (N·m)
$\quad s_{s(\text{all})}$ = allowable shear stress (psi, ksi) (Pa, MPa)

Note that T_R (the allowable torque) is the maximum torque that can exist without exceeding the allowable shear stress. It is an upper limit of the torque that should be applied to the member. It depends on the geometry and material of the member, itself, not on the actual torque applied. Equations (12–1), (12–2), and (12–3) may be used for both solid and hollow circular shafts. The appropriate value for J can be obtained from Table 8–1.

For purposes of design of solid circular shafts, Eq. (12–3) can be rewritten so as to enable direct calculation of the required diameter of a circular shaft to resist a given applied torque. We must provide a shaft diameter d such that T_R will be equal to or greater than the applied torque. Naturally, the allowable shear stress must be known.

Rewriting Eq. (12–3) for the required centroidal polar moment of inertia J and substituting the applied torque T for T_R,

$$\text{Required } J = \frac{Tc}{s_{s(\text{all})}}$$

For a solid circular shaft, $c = d/2$. Also, from Table 8–1,

$$J = \frac{\pi d^4}{32}$$

Substituting the two quantities,

$$\frac{\pi d^4}{32} = \frac{T(d/2)}{s_{s(\text{all})}}$$

Solving for the required d results in

$$\text{Required } d = \sqrt[3]{\frac{16 T}{\pi s_{s(\text{all})}}} \qquad\qquad \textbf{(12–4)}$$

where all the terms have been previously defined. Note that this design expression is valid only for solid circular shafts. For hollow shafts, J depends on both the inner and the outer diameters (see Table 8–1). Therefore, there are two unknown diameters to be determined. One solution to this problem is to establish a ratio between the two diameters, reducing the problem to one unknown diameter, which then can be determined. This approach will be demonstrated in Example 12–5.

Design problems normally include the determination of a required size of the member. For circular sections this would be a diameter (or diameters, in the case of a hollow section). The natural conclusion is to select, or specify, the actual diameters to use. Shafts and bars are available in various sizes and can, in fact, be machined to almost any size within reason. For our purposes, assuming machinery shafting, we will adopt the practice of selecting diameters based on the commonly available increments given in Table 12–1.

TABLE 12–1

Diameter Range (in.)	Increment (in.)
$\frac{1}{2}$ to $2\frac{1}{2}$	$\frac{1}{16}$
$2\frac{5}{8}$ to 4	$\frac{1}{8}$
$4\frac{1}{4}$ to 6	$\frac{1}{4}$

□ **EXAMPLE 12–2**   Calculate the allowable torque that can be applied to a circular shaft if the allowable shear stress is 12,000 psi. (a) Assume the shaft is solid and has a 6 in. diameter. (b) Assume the shaft is hollow and has an outside diameter of 6 in. and an inside diameter of 5 in.

Solution   (a) For the solid shaft, the centroidal polar moment of inertia is calculated from

$$J = \frac{\pi d^4}{32} = \frac{\pi (6)^4}{32} = 127.2 \text{ in.}^4$$

Calculating the allowable torque, using Eq. (12–3),

$$T_R = \frac{s_{s(\text{all})} J}{c} = \frac{12,000(127.2)}{3} = 509,000 \text{ in.-lb}$$

(b) For the hollow shaft, the centroidal polar moment of inertia is calculated with an outer diameter d of 6 in. and an inner diameter d_1 of 5 in.:

$$J = \frac{\pi(d^4 - d_1^4)}{32} = \frac{\pi(6^4 - 5^4)}{32} = 65.9 \text{ in.}^4$$

Then, from Eq. (12–3), the allowable torque is

$$T_R = \frac{s_{s(\text{all})}J}{c} = \frac{12,000(65.9)}{3} = 264,000 \text{ in.-lb}$$

□ **EXAMPLE 12–3**   Calculate the maximum shear stress developed in a circular steel shaft when subjected to a torque T of 95,000 in.-lb. Assume the shaft is (a) solid, with a diameter of 4 in. (b) hollow, with an outside diameter of 4 in. and an inside diameter of 2 in.

Solution   (a) For the solid shaft, the centroidal polar moment of inertia is calculated from

$$J = \frac{\pi d^4}{32} = \frac{\pi(4)^4}{32} = 25.1 \text{ in.}^4$$

The maximum shear stress is calculated using Eq. (12–2):

$$s_s = \frac{Tc}{J} = \frac{95,000(2)}{25.1} = 7570 \text{ psi}$$

(b) For the hollow shaft, the centroidal polar moment of inertia is calculated from

$$J = \frac{\pi(d^4 - d_1^4)}{32} = \frac{\pi(4^4 - 2^4)}{32} = 23.6 \text{ in.}^4$$

The maximum shear stress is calculated using Eq. (12–2):

$$s_s = \frac{Tc}{J} = \frac{95,000(2)}{23.6} = 8050 \text{ psi}$$

The shear stress varies linearly from 0 at the longitudinal center of the shaft to 8050 psi at the outer surface. The shear stress at the inner surface can be calculated by proportion, where

$$s_s = 8050 \left(\frac{1}{2}\right) = 4025 \text{ psi}$$

□ **EXAMPLE 12–4**   Calculate the required diameter of a solid circular steel shaft which must resist a torque T of 200,000 in.-lb. The allowable shear stress in the shaft is 12,000 psi. Select a shaft diameter to use.

Solution   Using Eq. (12–4),

$$\text{Required } d = \sqrt[3]{\frac{16T}{\pi s_{s(\text{all})}}} = \sqrt[3]{\frac{16(200,000)}{\pi(12,000)}} = 4.39 \text{ in.}$$

Therefore, use a $4\frac{1}{2}$ in. diameter shaft.

☐ **EXAMPLE 12–5**   A circular AISI 1040 hot-rolled steel shaft transmits a torque of 300,000 in.-lb under varying load conditions. Using a factor of safety of 3.0, calculate the required size of the hollow shaft if the inside diameter d_1 is to be three-quarters times the outside diameter d. Select the diameters to use.

Solution   First determine the allowable shear stress. Use a shear yield strength of one-half the tensile yield strength:

$$0.5(42,000) = 21,000 \text{ psi}$$

The allowable shear stress is then calculated from

$$s_{s(\text{all})} = \frac{\text{shear yield strength}}{\text{F.S.}} = \frac{21,000}{3.0} = 7000 \text{ psi}$$

Next, calculate the centroidal polar moment of inertia in terms of the outside diameter d:

$$J = \frac{\pi(d^4 - d_1^4)}{32} = \frac{\pi(d^4 - (0.75\,d)^4)}{32} = 0.0671\,d^4$$

The required outside diameter is then calculated using Eq. (12–3) and equating the applied torque to the allowable torque:

$$T_R = \frac{s_{s(\text{all})}J}{c}$$

$$300,000 = \frac{7000(0.0671\,d^4)}{(d/2)}$$

Therefore,

$$\text{Required } d^3 = \frac{300,000}{7000(0.0671)(2)}$$

from which the required outside diameter is obtained:

$$d = 6.84 \text{ in.}$$

The required inside diameter is then calculated as

$$d_1 = 0.75(6.84) = 5.13 \text{ in.}$$

Therefore, use a hollow steel shaft with an outside diameter of 7 in. and an inside diameter of $5\frac{1}{4}$ in.

☐ **EXAMPLE 12–6**   In Fig. 12–8, pulleys B, C, and D are attached to the solid shaft supported on bearings at A and E. The shaft is driven at a uniform speed by pulley C. The shaft, in turn, drives pulleys B and D. The diameters of pulleys B, C, and D are 10 in., 12 in., and 14 in., respectively. Belt tensions are shown. The diameter of the shaft is $1\frac{1}{2}$ in. (a) Calculate the belt tension F_3. (b) Calculate the torque in the shaft between pulleys C and D. (c) Calculate the maximum shear stress developed from the torque of part (b).

FIGURE 12–8   Shaft and pulley system.

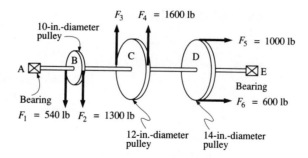

Solution   Note that all of the belt tensions are known except F_3. An imbalance between the belt tensions on either side of a pulley may be used to determine how much torque is transmitted between the pulley and the shaft. In each case, the torque equals the difference between the belt tensions times the radius of the pulley.

(a) Calculate belt tension F_3. Since the entire system is in equilibrium, the torque at C must be balanced by the torque at B and D. The torque from pulley B is

$$(F_2 - F_1)(5) = (1300 - 540)(5) = 3800 \text{ in.-lb (clockwise)}$$

The torque from pulley D is

$$(F_5 - F_6)(7) = (1000 - 600)(7) = 2800 \text{ in.-lb (clockwise)}$$

Pulley C drives the shaft. It must deliver a torque opposite in direction to the sum of the torques from pulleys B and D:

$$2800 + 3800 = 6600 \text{ in.-lb}$$

The torque from pulley C is

$$(F_4 - F_3)(6) = (1600 - F_3)(6)$$

Equating,

$$(1600 - F_3)(6) = 6600 \text{ in.-lb}$$

from which

$$F_3 = 500 \text{ lb (belt tension)}$$

(b) Calculate the torque between pulleys C and D. A section S–S is cut between the two pulleys and the left portion taken as a free body, as shown in Fig. 12–9.

Applying $\Sigma T = 0$, the external applied torque must equal the internal resisting torque:

$$6600 - 3800 = T_{\text{int}} = 2800 \text{ in.-lb}$$

(c) Calculate the maximum shear stress developed in the shaft between pulleys C and D. The shaft diameter is $1\frac{1}{2}$ in.; therefore, $c = 0.75$ in. Calculating the centroidal polar

FIGURE 12–9   Free body.

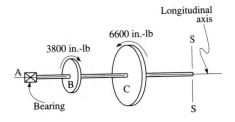

moment of inertia and the shear stress,

$$J = \frac{\pi d^4}{32} = \frac{\pi (1.5)^4}{32} = 0.497 \text{ in.}^4$$

$$s_s = \frac{Tc}{J} = \frac{2800(0.75)}{0.497} = 4255 \text{ psi}$$

12–4
ANGLE OF TWIST

If a circular shaft of length L is subjected to a torque T throughout its length, one end of the shaft will twist about its longitudinal axis relative to the other end. In Fig. 12–10, part of a circular shaft is shown. AB represents a straight line on the surface of the untwisted shaft parallel to the longitudinal axis of the shaft. AB′ represents a curve (part of a helix) that line AB assumes after the torque is applied. As a result of the applied torque, radius OB, shown on the end of the shaft, rotates and assumes a position OB′. The angle BOB′ is called the *angle of twist* and is generally expressed in radians (see Section 9–5) and designated θ.

FIGURE 12–10   Angle of twist of a shaft.

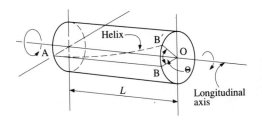

The deformation of a line on the surface of a shaft subjected to a torque is a shear deformation. Hence, the total shear deformation (δ_s) of line AB in the length L is BB′. The shear strain may then be expressed as

$$\epsilon_s = \frac{\delta_s}{L} = \frac{BB'}{L}$$

Since the shaft is circular in cross section, the magnitude of BB′ is equal to $c\theta$ where c is the radius of the shaft and θ is the angle of twist expressed in radians. Therefore,

$$\epsilon_s = \frac{BB'}{L} = \frac{c\theta}{L}$$

Assuming that Hooke's law is applicable, and using the shear stress-strain relationships, the modulus of rigidity G (see Chapter 9) is expressed as

$$G = \frac{s_s}{\epsilon_s}$$

Substituting for ϵ_s, the expression becomes

$$G = \frac{s_s}{(c\theta/L)} = \frac{s_s L}{c\theta}$$

Solving for θ,

$$\theta = \frac{s_s L}{Gc} \qquad\qquad \textbf{(12–5)}$$

which gives the angle of twist in terms of the maximum shear stress occurring at the outer surface.

Since

$$s_s = \frac{Tc}{J}$$

the angle of twist may also be expressed in terms of the torque T. Substituting for s_s in Eq. (12–5),

$$\theta = \frac{TcL}{JGc} = \frac{TL}{JG} \qquad\qquad \textbf{(12–6)}$$

where θ = the angle of twist (radians, where one radian = 57.3°)
   T = the torque (in.-lb, in.-kips) (N·m)
   L = the length of the shaft subjected to the torque (in.) (mm, m)
   J = the polar moment of inertia (in.^{4}) (mm^4, m^4)
   G = the modulus of rigidity (or modulus of elasticity in shear) (psi, ksi) (MPa, Pa)

Equations (12–5) and (12–6) are applicable to both solid and hollow circular shafts.

In some design cases, the size of the shaft necessary to transmit a given torque may be governed by the allowable angle of twist, rather than by the allowable shear stress. A shaft may be strong enough to function properly but be entirely too flexible.

☐ **EXAMPLE 12–7**   A $1\frac{1}{2}$ in. diameter solid steel shaft, 6 ft long, is subjected to a torque of 5000 in.-lb. The steel is AISI 1020 hot-rolled. Calculate (a) the maximum shear stress and (b) the total angle of twist.

Solution   The centroidal polar moment of inertia J is calculated from

$$J = \frac{\pi d^4}{32} = \frac{\pi (1.5)^4}{32} = 0.497 \text{ in.}^4$$

(a) Calculating the maximum shear stress (from Eq. (12–2)),

$$s_s = \frac{Tc}{J} = \frac{5000(0.75)}{0.497} = 7550 \text{ psi}$$

(b) Using Eq. (12–6) and G from Appendix G, the angle of twist is calculated from

$$\theta = \frac{TL}{JG} = \frac{5000(6)(12)}{0.497(11,500,000)} = 0.0630 \text{ radians}$$

Since one radian equals 57.3°,

$$\theta = 0.0630(57.3) = 3.61°$$

☐ **EXAMPLE 12–8**   A solid steel shaft is to resist a torque of 300,000 in.-lb. The angle of twist is not to exceed 1° in 5 ft and the maximum shear stress is not to exceed 12,000 psi. Calculate the required shaft diameter and select a diameter to use. Assume $G = 12,000,000$ psi.

Solution   To calculate the required diameter based on the allowable angle of twist, rewrite Eq. (12–6) and solve for J:

$$J = \frac{TL}{\theta G}$$

For a solid circular shaft,

$$J = \frac{\pi d^4}{32}$$

Substituting this into the preceding expression and solving for the required d will yield

$$\text{Required } d^4 = \frac{32TL}{\pi \theta G}$$

The maximum angle of twist is to be 1°. This must be converted to radians:

$$\theta = \frac{1°}{57.3} = 0.01745 \text{ radians}$$

Therefore,

$$\text{Required } d = \sqrt[4]{\frac{32(300,000)(5)(12)}{12,000,000(\pi)(0.01745)}} = 5.44 \text{ in.}$$

Next, use Eq. (12–4) to calculate the required diameter based on the allowable shear stress:

$$\text{Required } d = \sqrt[3]{\frac{16T}{\pi s_{s(\text{all})}}} = \sqrt[3]{\frac{16(300,000)}{\pi (12,000)}} = 5.03 \text{ in.}$$

It is obvious, then, that the limitation on the angle of twist controls the design. Use a $5\frac{1}{2}$ in. diameter shaft.

☐ **EXAMPLE 12–9**   A 0.80 in. diameter rod of an aluminum alloy was tested in a torsion testing machine. When the applied torque was 1135 in.-lb, the angle of twist in a length of 8 in. was 3.21°. Calculate the modulus of rigidity G.

Solution   The centroidal polar moment of inertia is calculated from

$$J = \frac{\pi d^4}{32} = \frac{\pi (0.80)^4}{32} = 0.0402 \text{ in.}^4$$

Using Eq. (12–6) and solving for G,

$$G = \frac{TL}{J\theta} = \frac{1135(8)}{0.0402(3.21/57.3)} = 4{,}030{,}000 \text{ psi}$$

**12–5
TRANSMISSION OF
POWER BY A SHAFT**

Rotating shafts are commonly used for transmitting power. If an applied torque turns a shaft, work is done by the torque. Recall from physics that work is defined as the energy developed by a force acting through a distance against a resistance. When the distance is linear, work can be expressed as

$$\text{Work} = \text{force} \times \text{distance}$$

This definition must be changed somewhat with regard to a rotating shaft, in which case an applied torque turns the shaft through a circular distance. Here, work can be expressed as

$$\text{Work} = \text{torque} \times (\text{angular distance}) = T\theta$$

To verify this expression, consider the bar shown in Fig. 12–11. The bar is pivoted at O and is acted on by two equal and opposite forces separated by a distance d. The two forces constitute a couple (or a torque) T having a magnitude Fd. If the bar moves through an angle θ (in radians), then the distance through which each force moves will be $\theta(d/2)$. The work done by the two forces will be

$$\text{Work} = 2F\theta \left(\frac{d}{2}\right) = F(d)(\theta) = T\theta$$

FIGURE 12–11   Work done by a couple.

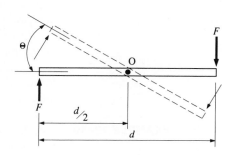

where work = work (in in.-lb or in.-kips)

T = the torque (in.-lb or in.-kips)

θ = the angle through which the rotating body turns (radians)

If a shaft is being rotated at a constant speed against a resistance, the work done in one revolution will be $2\pi T$, since θ equals 2π radians per revolution (360°).

If T is in units of in.-lb and if n_r is defined as the number of revolutions per minute (rpm), then the work done in one minute is

$$\text{Work per minute} = 2\pi T n_r \text{ (in.-lb per minute)}$$

Power is defined as work done per unit time:

$$\text{Power} = \frac{\text{work}}{\text{time}}$$

The common unit of power in the U.S. Customary System is the horsepower (hp), the value of which is 33,000 ft-lb per minute, or 396,000 in.-lb per minute. In the SI system, the unit of power is the watt (W) and is discussed later in Section 12–6.

Since 396,000 in.-lb of work per minute constitutes one horsepower, the number of horsepower may be determined from

$$\text{HP} = \frac{\text{work per minute}}{396,000} = \frac{2\pi T n_r}{396,000}$$

Therefore,

$$\text{HP} = \frac{T n_r}{63,025} \qquad \text{(12–7)}$$

where HP = the horsepower developed

T = the torque (in.-lb)

n_r = the number of revolutions per minute

With this relationship, the horsepower delivered by a shaft can be computed knowing the speed of rotation and the torque applied to the shaft. The torque which a shaft delivers may be determined by using the speed of rotation and the horsepower of the driving motor or engine. As is usual with equations of this kind in which special constants have been introduced, one must be very careful to substitute data with prescribed units.

In the three examples that follow, friction considerations have been neglected. This is convenient for the purpose of discussion, but is unrealistic. In all applications, some power will be lost in overcoming frictional forces. The treatment of frictional considerations in these applications belongs to the realm of machine design and is beyond the scope of this text.

□ **EXAMPLE 12–10**   Calculate the maximum horsepower that can be transmitted by a $2\frac{1}{2}$ in. diameter solid steel shaft operating at 300 rpm. The allowable shear stress in the shaft is 9000 psi.

Solution   The centroidal polar moment of inertia is calculated from

$$J = \frac{\pi d^4}{32} = \frac{\pi (2.5)^4}{32} = 3.835 \text{ in.}^4$$

Using Eq. (12–3), the allowable torque can be calculated. This is the maximum torque that could be transmitted without exceeding the allowable shear stress.

$$T_R = \frac{s_{s(all)} J}{c} = \frac{9000(3.835)}{1.25} = 27,610 \text{ in.-lb}$$

Using Eq. (12–7), the maximum horsepower is

$$\frac{T n_r}{63,025} = \frac{27,610(300)}{63,025} = 131.4 \text{ hp}$$

□ **EXAMPLE 12–11**   A hollow steel shaft with an outside diameter of $3\frac{1}{4}$ in. and an inside diameter of 3 in. transmits 280 hp at 900 rpm. Calculate the maximum torsional shear stress developed in the shaft.

Solution   The centroidal polar moment of inertia is calculated from

$$J = \frac{\pi (d^4 - d_1^4)}{32} = \frac{\pi (3.25^4 - 3.0^4)}{32} = 3.00 \text{ in.}^4$$

Using Eq. (12–7), the torque developed can then be calculated:

$$\text{HP} = \frac{T n_r}{63,025}$$

from which

$$T = \frac{63,025(\text{HP})}{n_r} = \frac{63,025(280)}{900} = 19,610 \text{ in.-lb}$$

Using Eq. (12–2), the torsional shear stress is

$$s_s = \frac{Tc}{J} = \frac{19,610(3.25/2)}{3.00} = 10,620 \text{ psi}$$

□ **EXAMPLE 12–12**   A solid steel shaft is to transmit 120 hp. The allowable shear stress is 8000 psi. (a) Select the diameter if the shaft speed is 3000 rpm. (b) Select the diameter if the shaft speed is 300 rpm. (c) Calculate the angle of twist of each shaft in a length of 10 ft. Use $G = 12,000,000$ psi.

Solution   (a) From Eq. (12–7), calculate the torque developed:

$$T = \frac{63,025(\text{HP})}{n_r} = \frac{63,025(120)}{3000} = 2521 \text{ in.-lb}$$

Using Eq. (12–4), calculate the required diameter for a solid shaft:

$$\text{Required } d = \sqrt[3]{\frac{16 T}{\pi s_{s(all)}}} = \sqrt[3]{\frac{16(2521)}{\pi (8000)}} = 1.17 \text{ in.}$$

Therefore, use a $1\frac{3}{16}$ in. diameter shaft ($d = 1.1875$ in.)

(b) If the shaft rotates at 300 rpm, the torque developed is

$$T = \frac{63{,}025(\text{HP})}{n_r} = \frac{63{,}025(120)}{300} = 25{,}210 \text{ in.-lb}$$

Using Eq. (12–4), obtain the required diameter:

$$\text{Required } d = \sqrt[3]{\frac{16\,T}{\pi\,s_{s(\text{all})}}} = \sqrt[3]{\frac{16(25{,}210)}{\pi\,(8000)}} = 2.52 \text{ in.}$$

Therefore, use a $2\frac{5}{8}$ in. diameter shaft ($d = 2.625$ in).

(c) The angle of twist in a 10 ft length for each of the two selected shafts is calculated as follows:

For the $1\frac{3}{16}$ in. diameter shaft,

$$J = \frac{\pi d^4}{32} = \frac{\pi(1.1875^4)}{32} = 0.1952 \text{ in.}^4$$

From Eq. (12–6),

$$\theta = \frac{TL}{JG} = \frac{2521(10)(12)}{0.1952(12{,}000{,}000)} = 0.1291 \text{ radians}$$

Converting to degrees,

$$\theta = 0.1291(57.3) = 7.40°$$

For the $2\frac{5}{8}$ in. diameter shaft,

$$J = \frac{\pi d^4}{32} = \frac{\pi(2.625)^4}{32} = 4.66 \text{ in.}^4$$

$$\theta = \frac{TL}{JG} = \frac{25{,}210(10)(12)}{4.66(12{,}000{,}000)} = 0.054 \text{ radians}$$

Converting to degrees,

$$\theta = 0.054(57.3) = 3.09°$$

Note the difference in the required shaft diameters depending on the speed at which the shaft rotates. The amount of power transmitted is the same. The higher the speed, the smaller the required shaft diameter. This is of importance in machine design.

12–6
SI SYSTEM
EXAMPLES

☐ **EXAMPLE 12–13**

Calculate the allowable torque that may be applied to a hollow circular shaft with an outside diameter of 150 mm and an inside diameter of 120 mm. The allowable shear stress is 75 MPa.

Solution The centroidal polar moment of inertia for the shaft is calculated from

$$J = \frac{\pi(d^4 - d_1^4)}{32} = \frac{\pi(150^4 - 120^4)}{32} = 29\,343\,457 \text{ mm}^4$$

$$= 29.3 \times 10^{-6} \text{ m}^4$$

Using Eq. (12–3), we obtain the allowable torque:

$$T_R = \frac{s_{s(\text{all})}J}{c} = \frac{(75 \times 10^6 \text{ N/m}^2)(29.3 \times 10^{-6} \text{ m}^4)}{75 \times 10^{-3} \text{ m}} = 29.3 \times 10^3 \text{ N} \cdot \text{m}$$

$$= 29.3 \text{ kN} \cdot \text{m}$$

☐ **EXAMPLE 12–14** A 2 m long solid steel shaft, 38 mm in diameter, is subjected to a torque of 565 N · m. The modulus of rigidity for this material is 77 000 MPa. Calculate the maximum shear stress and the total angle of twist.

Solution The centroidal polar moment of inertia is calculated from

$$J = \frac{\pi d^4}{32} = \frac{\pi(38 \text{ mm})^4}{32} = 204.7 \times 10^3 \text{ mm}^4$$

$$= 204.7 \times 10^{-9} \text{ m}^4$$

Calculating the maximum shear stress from Eq. (12–2),

$$s_s = \frac{Tc}{J} = \frac{(565 \text{ N} \cdot \text{m})(19 \times 10^{-3} \text{ m})}{204.7 \times 10^{-9} \text{ m}^4} = 52.4 \times 10^6 \text{ N/m}^2$$

$$= 52.4 \text{ MPa}$$

Using Eq. (12–6), we obtain the angle of twist:

$$\theta = \frac{TL}{JG} = \frac{(565 \text{ N} \cdot \text{m})(2 \text{ m})}{(204.7 \times 10^{-9} \text{ m}^4)(77\,000 \times 10^6 \text{ N/m}^2)}$$

$$= 0.000\,071\,7 \times 10^3 \text{ radians}$$
$$= 0.0717 \text{ radians}$$

Converting to degrees,

$$\theta = 0.0717(57.3) = 4.11°$$

The recommended unit for power in the SI system is the watt (W). One watt is defined as one newton meter per second:

$$1 \text{ W} = 1 \frac{\text{N} \cdot \text{m}}{\text{s}}$$

In Section 12–4, power is defined as work per unit time and is expressed further as work per minute, where

$$\text{Power} = \text{work per minute} = 2\pi T n_r$$

with n_r expressed in units of revolutions per minute (rpm) and T expressed in units of in.-lb. This relationship is still valid in the SI system. However, since power is expressed in watts and defined as N · m/s, the term n_r must be divided by 60 (to convert it to revolutions per second). In addition, T must be expressed in units of N·m. Equation (12–7) may then be written as

$$\text{Power} = W = \frac{2\pi T n_r}{60} \qquad\qquad \textbf{(12–7[SI])}$$

where W = power in watts (or kilowatts)
$\quad\ T$ = torque (N·m)
$\quad\ n_r$ = revolutions per minute (r/min)

Equation (12–7[SI]) is one of those special equations in which one must carefully substitute terms having the prescribed units.

   Note that the symbol for revolutions per minute is "r/min" in the SI system, whereas in the U.S. Customary System it is "rpm." Also note that the symbol for revolutions per second in the SI system is "r/s."

   Speed of rotation may also be expressed in terms of radians per second (rad/s). For use in this book, we prefer the more familiar r/min (or r/s).

□ EXAMPLE 12–15   A hollow shaft with an inside diameter of 25 mm and an outside diameter of 35 mm transmits power of 20 kilowatts. The speed of rotation of the shaft is 500 r/min. Determine the torsional shear stress developed in the shaft.

Solution   From Table 8–1, the centroidal polar moment of inertia for the shaft is calculated from

$$J = \frac{\pi}{32}(d^4 - d_1^4) = \frac{\pi}{32}(35^4 - 25^4) = 109\,000 \text{ mm}^4$$

$$= 109 \times 10^{-9} \text{ m}^4$$

Solving Eq. (12–7[SI]) for torque and substituting in terms of prescribed units yields

$$T = \frac{60\,W}{2\pi\,n_r} = \frac{60(20 \times 10^3)}{2\pi(500)} = 382 \text{ N·m}$$

From Eq. (12–2), the shear stress is calculated from

$$s_s = \frac{Tc}{J} = \frac{382 \text{ N·m }(35/2 \times 10^{-3} \text{ m})}{109 \times 10^{-9} \text{ m}^4}$$

$$= 61.3 \times 10^6 \text{ N/m}^2$$

$$= 61.3 \text{ MPa}$$

□ EXAMPLE 12–16   A solid AISI 1020 steel shaft is required to transmit power of 50 kW. The speed of the shaft will be 6 r/s. The allowable shear stress is 67 MPa and the allowable angle of twist (per meter of shaft length) is not to exceed 0.065 radians. Determine the required diameter of the shaft.

Solution   Determine the required diameter based on shear stress and on twist of the shaft. The final required diameter will be the larger of the two. First the magnitude of the torque to be transmitted is calculated. From Eq. (12–7[SI]) rewriting for torque, and converting r/s to r/min,

$$T = \frac{60\ W}{2\pi n_r} = \frac{60(50 \times 10^3)}{2\pi(6 \times 60)} = 1.326 \times 10^3\ \text{N} \cdot \text{m}$$

Based on allowable shear stress, using Eq. (12–4),

$$\text{Required } d = \sqrt[3]{\frac{16T}{\pi s_{s(\text{all})}}} = \sqrt[3]{\frac{16(1.326 \times 10^3\ \text{N} \cdot \text{m})}{\pi(67 \times 10^6\ \text{N/m}^2)}}$$
$$= 0.465 \times 10^{-1}\ \text{m}$$
$$= 46.5\ \text{mm}$$

Based on the allowable angle of twist, from Eq. (12–6),

$$\theta = \frac{TL}{JG} = \frac{TL}{(\pi d^4/32)G}$$

from which

$$\text{Required } d = \sqrt[4]{\frac{32TL}{\pi \theta G}} = \sqrt[4]{\frac{32(1.326 \times 10^3\ \text{N} \cdot \text{m})(1\ \text{m})}{\pi(0.065)(77\ 000 \times 10^6\ \text{N/m}^2)}}$$
$$= 0.4053 \times 10^{-1}\ \text{m}$$
$$= 40.5\ \text{mm}$$

Therefore, the required diameter is 46.5 mm.

SUMMARY—BY SECTION NUMBER

12–1 and **12–2** Torque is a twisting action (moment) applied in a plane perpendicular to the longitudinal axis of a member. A member subjected to a torque is said to be in *torsion*. Examples of such members are shafts used for the transmission of power.

12–3 Torsional shear stresses are developed on cross-sectional planes of a shaft as a result of an applied torque. In a shaft of circular cross section, the shear stress varies linearly from zero at the centroidal longitudinal axis to a maximum at the outer surface of the shaft and can be obtained from

$$s_s = \frac{Tc}{J} \tag{12–2}$$

The maximum resisting (or allowable) torque can be computed from

$$T_R = \frac{s_{s(\text{all})}J}{c} \tag{12–3}$$

Solid shafts of circular cross section can be designed using

$$\text{Required } d = \sqrt[3]{\frac{16T}{\pi s_{s(\text{all})}}} \tag{12–4}$$

12–4 The design of a solid or hollow shaft of circular cross section may be governed by an allowable angle of twist rather than by an allowable shear stress. The angle of twist is computed from

$$\theta = \frac{TL}{JG} \tag{12-6}$$

and is indicative of the flexibility of a shaft. The torsional rigidity or stiffness of a shaft is the product JG.

12-5 Rotating shafts of circular cross section are commonly used for transmitting power. In the U.S. Customary System, the common unit of power is the *horsepower*. The horsepower developed by a shaft rotating at n_r rpm as a result of an applied torque is computed from

$$\mathrm{HP} = \frac{Tn_r}{63,025} \tag{12-7}$$

PROBLEMS

For the following problems, unless noted otherwise use a modulus of rigidity G of 12,000,000 psi for steel and 4,000,000 psi for aluminum.

Section 12-2   Members in Torsion

1. Determine the internal resisting torque in the shaft in Fig. 12-12 at A, B, and C. Show the free-body diagrams.

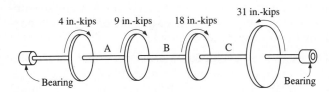

FIGURE 12-12   Problem 1.

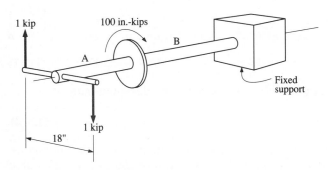

FIGURE 12-13   Problem 2.

2. Determine the internal resisting torque in the shaft in Fig. 12-13 at A and B. Show the free-body diagrams. Assume the shaft is fixed against rotation at the fixed support.

Section 12-3   Torsional Shear Stress

3. Calculate the maximum shear stress developed in a $3\frac{1}{2}$ in. diameter circular solid steel shaft subjected to an applied torque of 5000 ft-lb.

4. Calculate the allowable torque that may be applied to a 5 in. diameter solid circular steel shaft if the allowable shear stress is 10,000 psi.

5. A hollow circular steel shaft has a 4 in. outside diameter and a 3 in. inside diameter. Calculate the allowable torque that can be transmitted if the allowable shear stress is 9000 psi. When the allowable torque is applied, calculate the shear stress at the inner surface of the shaft.

6. Design a solid circular steel shaft to transmit an applied torque of 25,000 ft-lb. The allowable shear stress is 10,000 psi.

7. Calculate the shear stresses at the outer and inner surfaces of a hollow circular steel shaft subjected to a torque of 350,000 in.-lb. The outside diameter of the shaft is 6 in. and the inside diameter is 3 in.

8. A hollow shaft is produced by boring a 6 in. diameter concentric core in a 9 in. diameter solid circular shaft. Compute the percentage of the torsional strength lost.

9. Pulleys C and D are attached to shaft AB as shown in Fig. 12-14. The shaft is supported on bearings at each end. The shaft rotates at a uniform speed. Pulley D is

the driver and pulley C is the power take-off. The diameter of the shaft is $2\frac{1}{2}$ in. Calculate (a) the belt tension P_3 and (b) the maximum shear stress in the shaft.

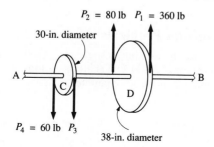

$P_2 = 80\text{ lb}$ $P_1 = 360\text{ lb}$

30-in. diameter

A ──────── C ──── D ──────── B

$P_4 = 60\text{ lb}$ P_3

38-in. diameter

FIGURE 12–14 Problem 9.

Section 12–4   Angle of Twist

10. Calculate the angle of twist θ in a 3 in. diameter solid steel shaft 4 ft long. The shaft is stressed to its allowable shear stress of 9000 psi.

11. Calculate the angle of twist θ in a 4 in. diameter solid steel shaft 20 ft long. The shaft is subjected to a torque of 40,000 in.-lb.

12. A 1 in. diameter solid steel shaft, 60 in. long, was tested in a large torsion machine. An angle of twist of 5° was measured when the shaft was subjected to a torque of 1700 in.-lb. Calculate the modulus of rigidity G.

13. A hollow aluminum tube has an outside diameter of 1 in. and an inside diameter of 0.50 in. The tube is subjected to a maximum shear stress of 7000 psi. Calculate the angle of twist per foot of length.

14. A solid steel shaft is to resist a torque of 9000 in.-lb. The angle of twist is not to exceed 0.12° per foot of length, and the allowable shear stress is 8000 psi. Calculate the required diameter of the shaft and select an appropriate diameter to use.

15. A hollow steel shaft has a 2 in. outside diameter and a 1.6 in. inside diameter. The shaft is 8 ft long. Compute the angle of twist when the maximum shear stress is 7500 psi.

16. If the shaft of Problem 15 were solid, with the same outside diameter and with the same maximum shear stress, what would be the angle of twist? Compute the percentage increase in torque that the solid shaft could

transmit. Compute the percentage increase in weight (use a unit weight of steel of 490 pcf).

Section 12–5   Transmission of Power by a Shaft

17. An automobile engine develops 90 hp at 3500 rpm. What torque is developed?

18. Calculate the speed (rpm) at which a 1 in. diameter solid steel shaft must operate in order to transmit 10 hp without exceeding an allowable shear stress of 8000 psi.

19. Select the diameter of a solid circular steel shaft for a 10 hp motor operating at 1800 rpm. The allowable shear stress in the shaft is 7000 psi.

20. Select the diameter for a hollow steel shaft which is to transmit 36 hp at 1200 rpm. The allowable shear stress is 9000 psi and the inside diameter of the shaft is to be three-fourths of the outside diameter.

21. A 6 ft long solid steel shaft with a diameter of 4 in. transmits 250 hp at a speed of 250 rpm. Determine whether the following two requirements are satisfied: (a) the maximum shear stress is not to exceed 10,000 psi and (b) the angle of twist is not to exceed 1°.

22. The outside and inside diameters of a hollow steel shaft are 6 in. and 4 in., respectively. Determine the maximum horsepower that can be transmitted if the shaft rotates at 1200 rpm and has an allowable shear stress of 9000 psi.

23. Calculate the maximum shear stress developed in a $1\frac{1}{2}$ in. diameter solid steel shaft that transmits 20 hp at a speed of 400 rpm.

SI System Problems

For the following SI problems, unless noted otherwise use a modulus of rigidity G of 83 000 MPa for steel and 28 000 MPa for aluminum.

24. Calculate the allowable torque for a hollow steel shaft. The inside diameter is 40 mm and the outside diameter is 85 mm. The allowable shear stress is 68 MPa.

25. The 65 mm diameter solid shaft in Fig. 12–15 is subjected to torques of 600 N·m and 1400 N·m at points B and C respectively. Determine the maximum shear stress in the shaft.

26. Rework Problem 25, changing the diameter of segment II of the shaft to 55 mm.

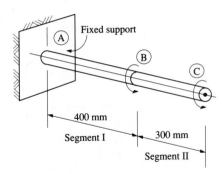

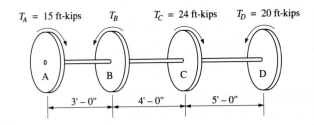

FIGURE 12–15   Problem 25.

27. A 25 mm diameter solid shaft with an allowable shear stress of 60 MPa rotates at a speed of 15 r/s. Determine the maximum power that can be transmitted by this shaft.

28. Calculate the maximum power that may be transmitted by a 60 mm diameter solid steel shaft operating at 300 r/min. The allowable shear stress in the shaft is 65 MPa.

29. A 32 mm diameter solid shaft transmits 100 kW of power at a speed of 28 r/s. Determine the maximum shear stress in the shaft.

30. A solid steel shaft is to transmit power of 58 kW at a speed of 60 r/s. The allowable shear stress is 90 MPa. Determine the minimum permissible diameter for this shaft.

Computer Problems

For the following computer problems, any appropriate programming language may be used. Input prompts should fully explain what is required of the user (the program should be "user friendly"). The resulting output should be well labeled and self-explanatory.

31. Write a program that will calculate the allowable torque that may be applied to either a solid or a hollow circular shaft. User input is to be the allowable shear stress, the type of shaft, and the diameter (or diameters).

32. Write a program that will generate a table of required diameters (to $\frac{1}{8}$ in.) of solid circular shafts to resist torques ranging from 50 to 200 in.-kips (in increments of 25 in.-kips). Allowable shear stresses range from 10,000 psi to 12,000 psi (in increments of 500 psi). Note that there is no user input. Check your result with Example 12–4.

33. Write a program that will generate a table of power (hp) transmitted by a shaft as a function of shaft speed and torque. The shaft speed is to vary from 100 to 500 rpm (in increments of 50 rpm) and the torque is to vary from 1000 to 3000 in.-kips (in increments of 500 in.-kips).

Supplemental Problems

34. Compute the maximum shear stress in the hollow steel shaft in Fig. 12–16. The shaft has a 4 in. outside diameter and a 2 in. inside diameter.

FIGURE 12–16   Problem 34.

35. Design a hollow steel shaft to transmit a torque of 13,000 in.-lb. The allowable shear stress in the shaft is 9000 psi. The outside diameter is to be twice the inside diameter.

36. A 32 in. long solid steel circular shaft, 3 in. in diameter, is twisted through an angle of 0.012 radians. Calculate the maximum shear stress developed in the shaft.

37. A solid aluminum shaft, 6 ft in length, is to transmit a torque of 350 ft-lb. The allowable shear stress is 5000 psi and the angle of twist must not exceed 4° in a 6 ft length. Select the required diameter.

38. Determine the allowable torque a hollow steel shaft can transmit if its outside diameter is 4 in. and its inside diameter is 3 in. The allowable shear stress is 10,000 psi. Compute the angle of twist per foot of length.

39. A 20 in. long steel wire, 0.18 in. in diameter, is twisted through an angle of 18°–30′ by a torque of 20 in.-lb. Determine its modulus of rigidity G.

40. Compute the maximum shear stress in the circular steel shaft in Fig. 12–17 if the shaft is subjected to the torques indicated. The shaft is solid and 2 in. in diameter for 21 in. of its length and is hollow with an outside diameter of 2 in. and an inside diameter of 1 in. for 21 in. of its length, as shown.

FIGURE 12–17   Problem 40.

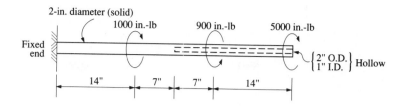

41. Select the outside diameter for a hollow steel shaft subjected to a torque of 1,000,000 in.-lb. The maximum shear stress developed in the shaft is 8000 psi. The inside diameter is $\frac{2}{3}$ of the outside diameter. Calculate the angle of twist, in degrees, for a 14 ft length of this shaft.

42. Select the diameter for a solid steel shaft which is to transmit 60 hp at 850 rpm without exceeding an allowable shear stress of 7000 psi.

43. Compute (a) the maximum shear stress developed in a 6 in. diameter solid steel shaft that transmits 600 hp at a speed of 80 rpm and (b) the shear stress if the speed were increased to 300 rpm.

44. What horsepower can a solid steel shaft 6 in. in diameter transmit at 160 rpm if the allowable shear stress is 5000 psi?

45. A solid steel shaft is to transmit 100 hp at a speed of 1000 rpm. The allowable shear stress is 9000 psi. What is the required diameter of the shaft?

46. A small ski lift has a main cable driving wheel 11 ft in diameter. The cable speed is to be 500 ft per minute. The cable tension on one side of the driving wheel is 25,000 lb and on the other side is 28,800 lb, as shown in Fig. 12–18. Calculate the horsepower required to turn the main cable driving wheel.

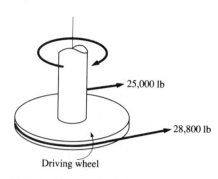

Driving wheel

FIGURE 12–18   Problem 46.

47. Two shafts—one a hollow steel shaft with an outside diameter of 4 in. and an inside diameter of $1\frac{1}{2}$ in., the other a solid steel shaft with a diameter of 4 in.—are to transmit 120 hp each. Compare the shear stresses in the two shafts if both operate at 150 rpm.

48. A $1\frac{1}{2}$ in. diameter solid steel shaft is 40 ft in length. If the torque necessary to operate a piece of equipment is 1500 in.-lb, what horsepower must be delivered to the shaft to maintain a shaft speed of 1000 rpm? Calculate the maximum shear stress developed in the shaft.

13 Shear and Bending Moment in Beams

**13-1
TYPES OF BEAMS
AND SUPPORTS**

Beams are among the most common structural members. They carry loads applied at right angles to the longitudinal axis of the member, which causes the member to bend. A plank placed across a trench supporting a person passing over it is an example of a beam subjected to a load applied at a right angle.

Beams are generally oriented in a horizontal or near horizontal position, although there are exceptions. They may also exist in a vertical or sloping orientation and be subjected to loads that will produce bending.

In this chapter, we will concern ourselves only with straight horizontal beams subjected to loads that will cause bending. Beams are sometimes called by other names, indicative of some specialized function. They may be called girders, stringers, floor beams, joists, lintels, spandrels, purlins, or girts. Many machine members with a specialized function, such as rotating shafts, are also subjected to bending.

A beam's type, as well as its behavior when subjected to load, is a function of the type and number of its supports. In Chapter 4, we discussed supports for beams and trusses. Recall that a roller support provides a reaction perpendicular to the contact surface. (For the beams considered here, a horizontal supporting surface for the roller is assumed; therefore, the reaction is in a vertical direction.) A pin support provides two reactions. (Again, for the beams considered here, one reaction is vertical and one is horizontal.) The pin and roller supports are sometimes called *simple supports*. These two supports are shown in Fig. 13–1, which summarizes the real supports, the idealized supports, and the associated reactions. A fixed support is also indicated in Fig. 13–1. This type of support may be constructed in a variety of ways. One easily visualized example would be a beam sufficiently anchored into a solid mass of concrete so as not to allow movement, as shown in Fig. 13–1(c). Note that the fixed support provides a vertical reaction, a horizontal reaction, and a moment reaction. The fixed support will theoretically allow no movement whatsoever: no horizontal, no vertical, and no rotational movement. The roller support, on the other hand, is assumed to permit horizontal and rotational movement, but no vertical movement. The pin support is assumed to permit rotational movement, but no horizontal or vertical movement.

The various types of beams in common use are shown in Fig. 13–2, along with the deflected shape of the loaded beam. It should be noted that it

FIGURE 13–1 Beam supports and reactions.

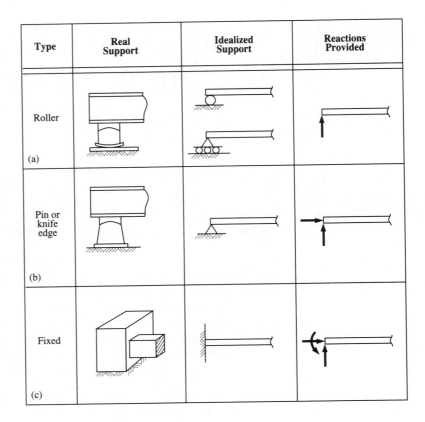

Type	Real Support	Idealized Support	Reactions Provided
Roller (a)			
Pin or knife edge (b)			
Fixed (c)			

is convenient to represent a beam with a single line. We will follow this convention for representing beams in load diagrams (which are also free-body diagrams). A diagram in which the beam depth is shown will be used in connection with our discussion of internal stresses in beams.

Beams are categorized according to type and/or number of supports. A beam supported at the ends by simple supports and used to carry any system of loads between the supports is called a *simple beam*. A *fixed beam* (or totally restrained beam) is supported by two fixed supports that do not permit any end rotation or translation as the beam is loaded. A beam with a fixed support at one end, with no other support along its length, is called a *cantilever beam*. Again, the fixed support does not permit any end rotation or translation as the beam deflects under load.

Any beam with one or two simple supports that are not located at the ends of the beam is called an *overhanging beam*. A beam that is fixed at one end, similar to a cantilever beam, with a simple support at the other end, is called a *propped cantilever*. A beam supported on three or more supports is called a *continuous beam*.

The simple, cantilever, and overhanging beams are categorized as *statically determinate* because their reactions can be determined using the three

FIGURE 13-2 Types of beams.

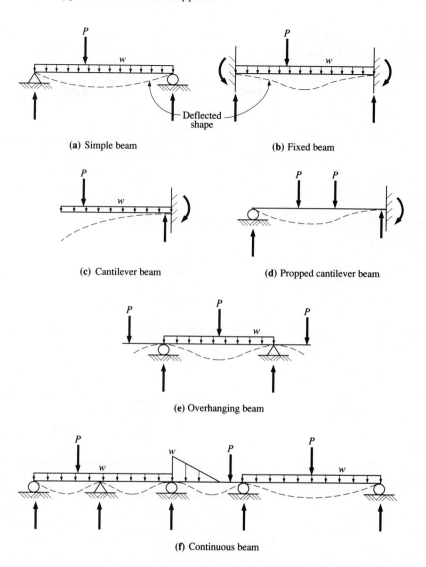

(a) Simple beam

(b) Fixed beam

(c) Cantilever beam

(d) Propped cantilever beam

(e) Overhanging beam

(f) Continuous beam

basic laws of equilibrium discussed in Chapter 4: $\Sigma F_y = 0$, $\Sigma F_x = 0$, and $\Sigma M = 0$. Since the beams considered here will be horizontal beams, for convenience we will alter the notation slightly. Instead of subscripts y and x for the force summation equations, we will use subscripts V and H to indicate vertical and horizontal: $\Sigma F_V = 0$ and $\Sigma F_H = 0$.

The fixed beam, the propped cantilever beam, and the continuous beam are categorized as *statically indeterminate* because their reactions cannot be determined by the three laws of equilibrium alone. Additional relationships based on the deflection of the beam must be introduced. This is discussed in Chapter 21. This chapter deals only with statically determinate beams.

13–2
TYPES OF LOADS
ON BEAMS

Loads on beams are classified as *concentrated* and *distributed*. In addition, distributed loads may be categorized as *uniformly* or *nonuniformly* distributed.

    Concentrated loads have been dealt with previously in Chapters 2 and 3. Recall that they are assumed to act at a definite point. Although such loads are actually distributed over a small area (or a short length) of the beam, the distance is usually small in comparison with the length of the beam, so that the load is considered to be concentrated at a single point. A single arrow is used to indicate the location, direction, and sense of the concentrated load, as shown in Fig. 13–3. The unit for the concentrated load is generally pounds (lb) or kips in the U.S. Customary System and newtons (N) in SI.

FIGURE 13–3   Types of loads.

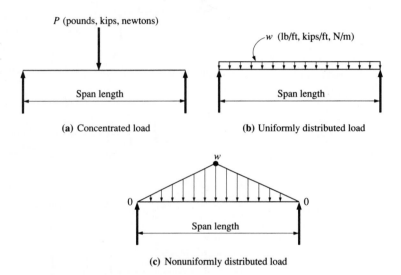

(a) Concentrated load   (b) Uniformly distributed load

(c) Nonuniformly distributed load

    Distributed loads have been discussed in Chapter 3. Recall that a distributed load is one that is spread out over a length of the beam. If the distributed load is of equal magnitude for each unit of length, the load is a uniformly distributed load. It may exist over the entire length of the beam or over a portion of it. The load diagram usually used to indicate a uniformly distributed load is a rectangular block. The block can be shaded or can show a system of arrows to indicate load direction as shown in Fig. 13–3(b). This diagram defines both the intensity of the distributed load and the length of the beam over which it exists. The length (whether full span or a portion of the span) is indicated with a dimension, and the load intensity is indicated with a notation. For the intensity of the uniformly distributed load, w is commonly used. The units for the intensity are pounds per linear foot (lb/ft) or kips per linear foot (kips/ft) in the U.S. Customary System and newtons per meter (N/m) in SI. Assuming that a beam has a uniform cross section throughout its length, the weight of the beam would be a uniformly distributed load.

The distributed load may have a varying intensity. In this case it is referred to as a nonuniformly distributed load. Generally, a nonuniformly distributed load increases or decreases at a definite and known rate along the length of the beam. The diagram for this type of distributed load results in a triangular or trapezoidal block, as shown in Fig. 13–3(c). Note that, in this case, the maximum distributed load intensity is w at midspan and decreases to zero at the supports.

In structural design applications (bridges, buildings, etc.), loads are usually categorized as *dead loads* or *live loads*. Dead loads are static loads that produce vertical forces due to gravity and include the weight of the structural framework and all materials permanently attached to it and supported by it. Reasonable estimates of the weight of the structure (or its component parts) can usually be made based on some preliminary computations. Live loads may be defined as all loads that are not dead loads. Examples of live loads are vehicles, people, stored materials, snow, ice, wind, fluid pressure, earth pressure, earthquake, impact, and explosive blast. Some of these loads may be vertical; some may be lateral. The applicable load is a function of the type of structure and its intended use as well as its geographic location.

Loads can occur in combinations and the probability of such combinations must be considered. Live loads on standard-type structures are generally specified by applicable building codes.

13–3
BEAM REACTIONS

In order to determine internal stresses at various points along the length of a beam, it is generally necessary to first compute the external reactions for the beam. The determination of reactions was discussed in Chapter 4. At this point, we will briefly review the procedure. The determination of external reactions is accomplished for statically determinate beams by using the three laws of equilibrium: $\Sigma F_H = 0$, $\Sigma F_V = 0$, and $\Sigma M = 0$. The algebraic sum of all the external applied loads and reactions must equal zero. Also, the algebraic sum of moments about any point due to the externally applied loads and reactions must equal zero.

Recall that for the purpose of determining reactions, distributed loads may be replaced by an equivalent concentrated resultant load. As we pointed out in Section 3–5, this equivalent concentrated resultant load may be represented as a dashed arrow to distinguish it from the given concentrated loads and is assumed to act through the centroid of the distributed load. This replacement is shown graphically in Fig. 13–4 where a 1 kip/ft uniformly distributed load extends full length on a 40 ft long beam. The uniformly distributed load is replaced by its equivalent concentrated resultant load, denoted W and shown as a dashed arrow at the center of the uniformly distributed load. Therefore,

$$W = 1 \text{ kip/ft} \times 40 \text{ ft} = 40 \text{ kips}$$

It is important to note that this replacement is only for purposes of finding the external reactions. The internal effect of the uniformly distributed

FIGURE 13–4 Determination of beam reactions.

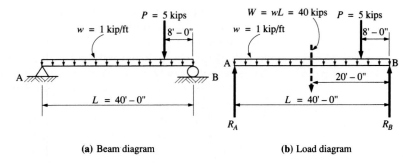

(a) Beam diagram   (b) Load diagram

load of 40 kips is very different from that of the concentrated resultant load of 40 kips.

The reactions of a simply supported beam will always be vertically upward if the applied loads are vertically downward. This does not necessarily hold for overhanging beams, in which case one of the reactions may be downward, even if all the loads are downward.

☐ **EXAMPLE 13–1**   Compute the reactions for the simply supported beam AB in Fig. 13–4(a).

Solution   Figure 13–4(a), called a *beam diagram*, defines the type of beam; the types of supports; the span; and the types, magnitudes, and locations of the loads.

The *load diagram* is shown in Fig. 13–4(b). Note that the supports at points A and B are replaced with the expected reactions from those supports. The pin at A could provide a horizontal reaction, but since there is no horizontally applied load, this reaction would be zero and can be neglected. In reality, the load diagram is a free-body diagram of the beam.

Assuming counterclockwise rotation to be positive, the reaction at point B can be computed by summing the moments about point A:

$$\Sigma M_A = +R_B(40) - (5)(32) - (1)(40)(20) = 0$$

from which

$$R_B = +24 \text{ kips} \uparrow$$

Recall that the positive sign indicates that the correct direction (sense) was assumed for the reaction at point B. No inference as to clockwise or counterclockwise moment should be made. Had this result turned out to be negative, it would mean only that the incorrect direction had been assumed and that the reaction was actually downward.

The reaction at point A can be computed by taking moments about B:

$$\Sigma M_B = -R_A(40) + (5)(8) + (1)(40)(20) = 0$$

from which

$$R_A = +21 \text{ kips} \uparrow$$

An algebraic summation of all the vertical forces serves as a check. In order that the beam be in equilibrium, it is required that this sum be equal to zero. Assuming upward to be positive,

$$+24 + 21 - 5 - (1)(40) = 0 \quad \textbf{OK}$$

An alternate solution would be to compute the reaction at point A by taking moments about point B and then to compute the reaction at B using $\Sigma F_V = 0$. However, if an error is made in computing the first reaction, the second computation also will be incorrect. At this point, it is recommended that each reaction be computed by means of the moment equations and then the calculations checked using a summation of vertical forces.

☐ **EXAMPLE 13–2**   Compute the reactions for the overhanging beam in Fig. 13–5.

FIGURE 13–5   Overhanging beam.

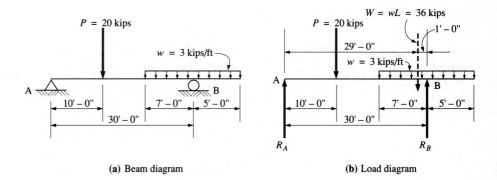

(a) Beam diagram   (b) Load diagram

Solution   Calculating the resultant for the distributed load,

$$W = 3(12) = 36 \text{ kips}$$

The reaction at point B can be computed by taking moments about point A. Assuming counterclockwise positive,

$$\Sigma M_A = +R_B(30) - (20)(10) - 36(29) = 0$$

from which

$$R_B = +41.5 \text{ kips} \uparrow$$

The reaction at point A can be computed by taking moments about point B:

$$\Sigma M_B = -R_A(30) + (20)(20) + (36)(1) = 0$$

from which

$$R_A = +14.5 \text{ kips} \uparrow$$

Using $\Sigma F_V = 0$ to check, assuming upward to be positive,

$$+41.5 + 14.5 - 20 - (3)(12) = 0 \quad \textbf{OK}$$

☐ **EXAMPLE 13–3**   Compute the reactions for the overhanging beam in Fig. 13–6.

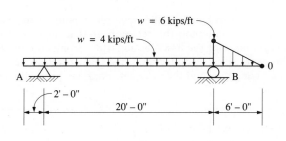

(a) Beam diagram

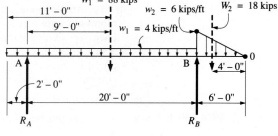

(b) Load diagram

FIGURE 13–6   Overhanging beam.

Solution   Calculating the distributed load resultants,

$$W_1 = w_1(22) = 4(22) = 88 \text{ kips}$$

$$W_2 = \frac{1}{2} w_2(6.0) = \frac{1}{2}(6.0)(6.0) = 18 \text{ kips}$$

Note that the total nonuniformly distributed load is equal to the area of the triangle and is assumed to act at the centroid of the triangle, which is 2 ft to the right of point B.

The reaction at point B can be computed using $\Sigma M_A = 0$:

$$+R_B(20) - 88(9) - 18(22) = 0$$

from which

$$R_B = +59.4 \text{ kips} \uparrow$$

The reaction at point A can be computed using $\Sigma M_B = 0$:

$$-R_A(20) + (88)(11) - (18)(2) = 0$$

from which

$$R_A = +46.6 \text{ kips} \uparrow$$

Using $\Sigma F_V = 0$ to check, assuming upward to be positive,

$$+59.4 + 46.6 - 88 - 18 = 0 \quad \textbf{OK}$$

13–4
SHEAR FORCE AND
BENDING MOMENT

We will now investigate the internal effects of the externally applied loads and forces. When a beam is subjected to applied loads and the resulting reactions, bending of the beam will occur and internal stresses will develop. These stresses, termed *shear stresses* and *bending stresses*, will be evaluated in Chapter 14 to establish their effect in the design and analysis of the bending members.

Our immediate objective concerning bending members is to evaluate the effects of the loading. These effects take the form of shear force and bending moment (usually called simply "shear" and "moment") which are developed internally by the externally applied loads and reactions. The magnitude of the shear and moment will directly affect the magnitude of the shear stress and the bending stress.

The method used to determine the shears and moments involves equilibrium considerations. External reactions must first be calculated. These will put the beam in external equilibrium. Recall that if the entire beam is in equilibrium, every segment must be in equilibrium. If we isolate a small segment of the beam for analysis purposes, equilibrium considerations will enable us to determine the shears and moments that must exist on the cut faces of the segment. These are the internal shears and moments in the beam. Depending on the loading and the conditions of support, shear and/or moment may vary throughout the length of the beam. We will develop methods with which we can obtain a complete picture of shear and moment variations along the length of the beam.

□ **EXAMPLE 13–4**   Compute the shear force and bending moment (a) at 7 ft and (b) at 17 ft from the left end of the beam having the load diagram shown in Fig. 13–7. Neglect the weight of the beam.

FIGURE 13–7   Load diagram.

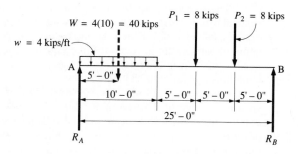

Solution   We first compute the beam reactions. Assuming counterclockwise to be positive,

$$\Sigma M_A = R_B(25) - (40)(5) - (8)(15) - (8)(20) = 0$$

from which

$$R_B = +19.2 \text{ kips } \uparrow$$

and

$$\Sigma M_B = -R_A(25) + (40)(20) + (8)(10) + (8)(5) = 0$$

from which

$$R_A = +36.8 \text{ kips } \uparrow$$

We check by summing vertical forces ($\Sigma F_V = 0$):

$$+36.8 + 19.2 - 40 - 8 - 8 = 0 \qquad \textbf{OK}$$

(a) *At 7 ft from the left end:*
To compute the shear force and bending moment at a location 7 ft from the left end, we cut the beam at that location (designated point *x* in Fig. 13–8) and consider the part on the left of the cutting plane as a free body, applying all external loads acting on the free-body portion of the beam. If we compare Fig. 13–8 with Fig. 13–7, we see that the resultant load of 40 kips from Fig. 13–7 is *not* used when analyzing the 7 ft long segment of Fig. 13–8. It was used to determine the beam reactions only.

FIGURE 13–8   Free-body diagram for shear and moment determination.

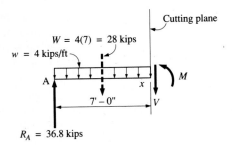

Actually, the free body may be taken from either end of the beam. However, to be consistent with sign conventions (to be discussed later) it is recommended that the free body always be taken between the left end of the beam and the cutting plane.

The free body must be in equilibrium. In other words, the algebraic sum of the vertical forces must equal zero and the algebraic sum of the moments about any point must equal zero. Since there are no external horizontal forces or horizontal components of forces, the horizontal force consideration does not enter the problem.

Considering the free body of Fig. 13–8, if we sum the external vertical forces (36.8 kips upward and 28 kips downward) we see that there is an unbalanced upward force (36.8 kips − 28 kips, or 8.8 kips). Therefore, there must exist internally, at section *x*, a vertical force of 8.8 kips that will put the beam segment in vertical equilibrium. This is the *internal shear*, which we can now define as *the algebraic sum of all the external forces on one side of a cutting plane*. We could achieve the same numerical result using the beam segment to the right of the section. However, the sense of the resulting 8.8 kips shear would be opposite to that determined using the segment to the left of the section. To eliminate this ambiguity, we will define a sign convention commonly used for shear. With reference to Fig. 13–9,

FIGURE 13–9   Shear sign convention.

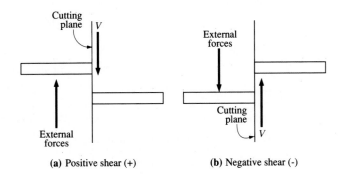

(a) Positive shear (+)   **(b)** Negative shear (-)

Shear is considered positive (+) if the segment of the beam to the left of a cutting plane tends to move upward with respect to the segment to the right. Negative shear is indicative of the left segment moving downward with respect to the right segment.

In taking a summation of vertical external forces on a freebody that has been taken from the left end of the beam, the correct sign for the shear (in accordance with the preceding convention) will always result if upward acting forces are taken to be positive and downward acting forces negative. This is the procedure we will use whenever calculating shear.

Computing the shear due to the external loads (using upward acting forces positive),

$$V = +36.8 - 4(7) = +8.8 \text{ kips}$$

The shear at this location (7 ft from the left end) is a positive 8.8 kips, indicating that the part of the beam to the left of the cutting plane tends to move upward with respect to the right part of the beam. It should be noted that the V force shown in Fig. 13–8 represents the *internal resisting shear*, which must be equal and opposite to the computed shear found from the external loads. Shear of other magnitudes will occur at other locations along the length of the beam, as will be shown in part (b) of the solution.

In order for the free body to be in equilibrium, it is necessary that all laws of equilibrium be satisfied. With reference to Fig. 13–8, this implies that the sum of the moments of the external forces about any point must equal zero. As with the shear, we may define the *moment at any location along the beam as the algebraic sum of the moments of all the external forces acting on one side of the cutting plane.* Therefore, the bending moment about point x (at the cutting plane) can be determined considering all of the external forces acting on the free body.

With reference to Fig. 13–10, a sign convention for bending moment is defined as follows:

The bending moment is positive when the bottom fibers are in tension and the top fibers are in compression. The bending moment is negative when the top fibers are in tension and the bottom fibers are in compression.

FIGURE 13–10 Moment sign convention.

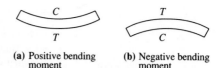

(a) Positive bending moment   **(b)** Negative bending moment

When determining the bending moment due to the external loads using segments as described in this example, the correct sign for the moment will always result if the moments due to the external upward acting forces are taken to be positive and moments due to the external downward acting forces are taken to be negative. This is the procedure we will use when calculating bending moments using cut sections along a beam. Note that there is no consideration here as to whether the moment is clockwise or counterclockwise as was the case when determining reactions. The same sign for moment will result whether a beam segment to the left or to the right of the cutting plane is considered.

Computing the bending moment about point x due to the external loads (assuming upward acting forces producing positive moments and downward acting forces producing negative moments),

$$M = +36.8(7) - 28(3.5) = +159.6 \text{ ft-kips}$$

Therefore, the bending moment at a point 7 ft from the left end of the beam is a positive 159.6 ft-kips. The positive sign indicates that the bottom fibers are in tension and the top fibers are in compression.

It should be noted that the bending moment M shown in Fig. 13–8 represents the *internal resisting moment,* which must be equal and opposite to the computed moment found from the external loads. Bending moments of other magnitudes will occur at other locations along the length of the beam as will be shown later in this problem.

(b) *At 17 ft from the left end:*
To compute the shear and bending moment at a location 17 ft from the left end, we cut the original beam again (but this time at the 17 ft location) and take out the left part as a free body, as shown in Fig. 13–11. (This same procedure can be followed to find shear and moment at *any* desired point along the length of the beam.)

FIGURE 13–11   Free-body diagram.

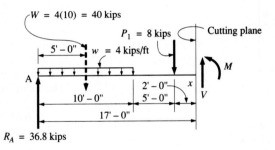

Computing the shear due to the external loads (assuming upward acting forces positive),

$$V = +36.8 - 40 - 8 = -11.2 \text{ kips}$$

The shear at this location due to the external loads is -11.2 kips. The negative sign implies that the part of the beam to the left of the cutting plane tends to move downward with respect to the right part. Note that the internal resisting shear shown on the cutting plane in Fig. 13–11 must be 11.2 kips acting upward.

Computing the bending moment due to the external loads (assuming upward acting forces producing positive moments and downward acting forces producing negative moments),

$$M = +36.8(17) - 40(12) - 8(2) = +129.6 \text{ ft-kips}$$

Therefore, the bending moment at a point 17 ft from the left end of the beam is a positive 129.6 ft-kips. The positive sign indicates that the bottom fibers are in tension and the top fibers are in compression.

In the preceding example, the calculated bending moment due to the external loads represents the tendency of the left segment of the beam to rotate about x. Since the free body must be in equilibrium, and internal resisting moment must exist at point x which, in effect, resists the tendency to rotate. This internal resisting moment M, as shown in Fig. 13–11, must be equal and opposite to the moment computed from the external loads. It is actually a couple (acting on the cut face) developed by the bending action of the beam.

As shown in Fig. 13–12, a loaded simply supported beam will deflect between the end supports. As a result, the bottom fibers of the beam are lengthened and placed in tension, whereas the top fibers are shortened and placed in compression. Somewhere between the top and bottom fibers, a surface or plane must exist where there is no tension or compression and which remains at its original unloaded length. This plane is called the *neutral plane* of the beam.

FIGURE 13–12   Shear and bending moment in beams.

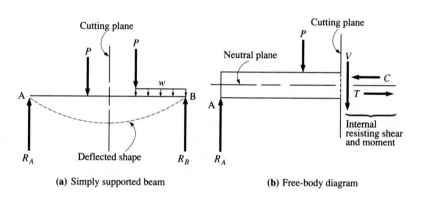

(a) Simply supported beam   **(b)** Free-body diagram

Below the neutral plane, the beam is in tension. On the free body (Fig. 13–12(b)), the total tension acting on the cutting plane may be denoted by T. Above the neutral plane, the beam is in compression and the total compression acting on the cutting plane may be denoted by C. Since the free body shown must be in equilibrium, the algebraic summation of all the horizontal forces must equal zero ($\Sigma F_H = 0$). Therefore, C must be equal to T. Equal and opposite forces that are parallel constitute a *couple*. This particular couple is equal to (identical with) the internal resisting moment, and it is referred to as the *internal couple*.

☐ **EXAMPLE 13–5**   Compute the shear and bending moment (a) at 5 ft, (b) at 10 ft, and (c) at 20 ft from the left end of the beam having the load diagram shown in Fig. 13–13.

Solution   The location and magnitude of the resultant of the uniformly distributed load is shown on the load diagram. Computing the beam reactions,

$$\Sigma M_A = R_B(20) - 10(5) - 50(12.5) - 5(25) = 0$$

FIGURE 13–13 Load diagram.

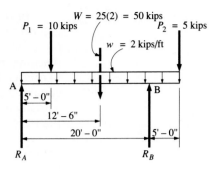

from which

$$R_B = +40.0 \text{ kips } \uparrow$$

and

$$\Sigma M_B = -R_A(20) + 10(15) + 50(7.5) - 5(5) = 0$$

from which

$$R_A = +25 \text{ kips } \uparrow$$

Using $\Sigma F_V = 0$ to check,

$$+40 + 25 - 10 - 50 - 5 = 0 \qquad \textbf{OK}$$

(a) *At 5 ft from the left end:*
As may be observed, a concentrated load exists at this point. Therefore, the shear at this point will change abruptly and will have one value at an infinitesimal distance to the left of the load and another value at an infinitesimal distance to the right of the load. Two free bodies are shown in Fig. 13–14 for purposes of computing the two shear values. Either free body may be used to compute the bending moment. In each case, note that the internal resisting shear V and the internal resisting moment M are shown acting on the cutting plane.

FIGURE 13–14 Free-body diagrams (5'–0").

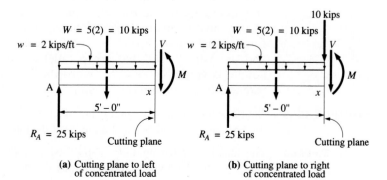

(a) Cutting plane to left
of concentrated load

(b) Cutting plane to right
of concentrated load

Calculating the shear force due to the external loads from Fig. 13–14(a),

$$V = +25 - 2(5) = +15 \text{ kips}$$

Calculating the shear force due to the external loads from Fig. 13–14(b),

$$V = +25 - 2(5) - 10 = +5 \text{ kips}$$

Calculating the bending moment due to the external loads at the cutting plane (point x),

$$M = +25(5) - 2(5)(2.5) = +100 \text{ ft-kips}$$

(b) *At 10 ft from the left end:*
The free body is shown in Fig. 13–15.

FIGURE 13–15   Free-body diagram (10'–0").

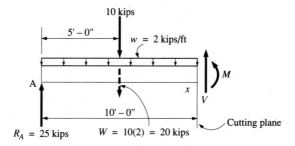

The shear and bending moment are calculated from

$$V = +25 - 2(10) - 10 = -5 \text{ kips}$$
$$M = +25(10) - 10(5) - 2(10)(5) = +100 \text{ ft-kips}$$

(c) *At 20 ft from the left end:*
A sudden change in the shear will exist because of the support. We calculate the shear just to the left (Fig. 13–16(a)) and just to the right (Fig. 13–16(b)) of the support. From Fig. 13–16(a),

$$V = +25 - 10 - 2(20) = -25 \text{ kips}$$

From Fig. 13–16(b),

$$V = +25 - 10 - 2(20) + 40 = +15 \text{ kips}$$

The moment can be calculated from either diagram:

$$M = +25(20) - 10(15) - 2(20)(10) = -50 \text{ ft-kips}$$

The moment at the 20 ft point is negative, indicating that the top fibers are in tension and the bottom fibers are in compression. This may be verified by observing Fig. 13–13; it is apparent that the cantilevered end at the right side of the beam will deflect downward.

FIGURE 13–16   Free-body
diagrams (20'–0").

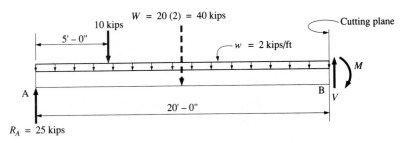

(a) Cutting plane to left of support

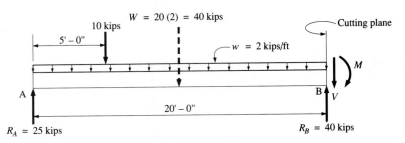

(b) Cutting plane to right of support

13–5
SHEAR DIAGRAMS

In the previous section, shear and bending moments were computed at arbitrary locations along the span length of a beam. In the design and analysis of beams, however, it is more important to compute the maximum values of the shear and bending moment. In addition, it is also important to determine the variation of the shear and bending moment along the length of a beam. This may be accomplished using graphical representations known as *shear diagrams* and *moment diagrams*.

In this section, we will consider only shear diagrams. The shear diagram is usually drawn directly below a sketch of the load diagram. (The beam diagram and the load diagram may be combined by superimposing the supports and the reactions.) A shear baseline, which represents zero shear, is drawn parallel to the beam. The abscissa along the baseline represents the locations of successive cross sections of the beam. The ordinate of the diagram represents the value of the shear at that particular cross section.

The drawing and construction of the shear diagram is best illustrated through a variety of examples. Generally, shear diagrams are not drawn to scale. Approximate proportions are satisfactory.

□ **EXAMPLE 13–6**   Draw the shear diagram for the simply supported beam having the load diagram shown in Fig. 13–17. The reactions have been computed and are indicated. Neglect the weight of the beam.

Solution   We will initially draw shear diagrams by computing the shear at various points along the beam. As demonstrated in Section 13–4, the shear at any point along the length of

FIGURE 13–17 Load diagram.

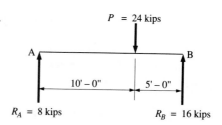

the beam may be computed utilizing free-body diagrams and performing an algebraic summation of vertical forces (ΣF_V). this may be done at one-foot intervals if so desired, and the computed values plotted as ordinates either above or below the zero shear line. These points, when connected, will result in a graphic representation portraying the shear variation along the length of the beam. This method, however, is a tedious procedure. Instead, if the shear is computed at specific locations, sufficient shear values may be obtained from which to draw a shear diagram. The shear should be computed at the following locations: (a) at the beginning and end of all distributed-type loads and (b) at an infinitesimal distance to the right and/or left of each concentrated load and reaction.

Working from the left end of the beam, we will first compute the shear at an infinitesimal distance to the right of the left reaction. Drawing a free-body diagram of infinitesimal length and summing vertical forces as shown in Fig. 13–18, the shear is computed to be

$$V = +8 \text{ kips}$$

This positive shear value is plotted above the zero shear line (refer to Fig. 13–21) directly under the left reaction.

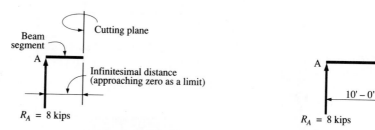

FIGURE 13–18 Free body.   **FIGURE 13–19** Free body.

Now we will compute the shear an infinitesimal distance to the left of the concentrated load. Using a 10 ft long free-body diagram as shown in Fig. 13–19, the shear is computed to be

$$V = +8 \text{ kips}$$

This value is plotted directly under the concentrated load and, since it is positive shear, above the zero shear line. As may be observed, the shear at the left reaction and the shear just to the left of the concentrated load are both equal to +8 kips. This is due to the fact that there is no change of load between the two points. Therefore,

FIGURE 13–20   Free body.

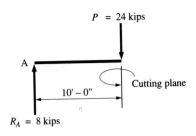

for the length of beam from the left reaction to the concentrated load, there is a positive shear equal to 8 kips. This is shown on the shear diagram (see Fig. 13–21) as a horizontal straight line between the two points.

We will now compute the shear an infinitesimal distance to the right of the concentrated load. Drawing a free-body diagram as shown in Fig. 13–20 and summing the vertical forces, the shear is computed to be

$$V = +8 - 24 = -16 \text{ kips}$$

This value is plotted directly under the concentrated load, and since it is negative, below the zero shear line (Fig. 13–21). A positive shear of 8 kips exists just to the left of the concentrated load, and a negative shear of 16 kips exists just to the right. The shear diagram shows a sudden change in shear at this location. The shear is said to "go through zero."

FIGURE 13–21   Load and shear diagrams.

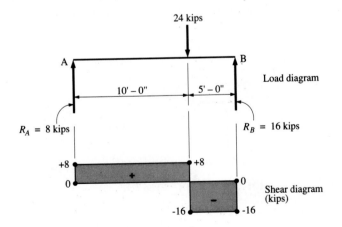

As may be observed, there is no load change between the concentrated load and the right reaction. Therefore, the shear will not change and will remain at negative 16 kips over that length of the beam. This is shown on the shear diagram as a horizontal straight line between the two points. The completed shear diagram is shown in Fig. 13–21 portraying the variation of shear along the length of the beam. Note that the positive and negative shear areas are equal.

☐ **EXAMPLE 13–7**   Draw the shear diagram for the simply supported beam in Fig. 13–22. The reactions have been calculated and are shown in the load diagram. Neglect the weight of the beam.

Solution   Working from the left end, as in the previous example, the shear just to the right of the left reaction is equal to +50 kips. The magnitude is equal to the left reaction. The sign (+) is in accordance with our sign convention for shear from Section 13–4.

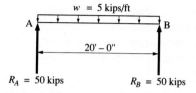

FIGURE 13–22   Load diagram.

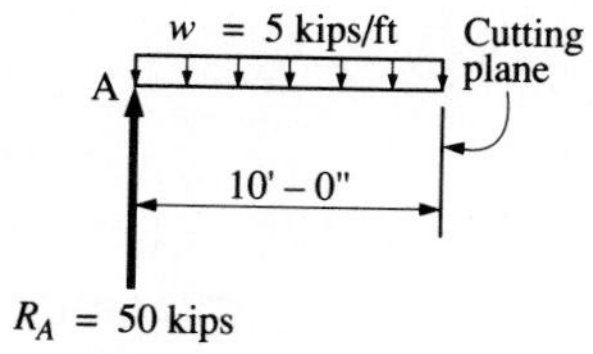

FIGURE 13–23   Free body.

If free-body diagrams were drawn at every foot along the length of the beam, we would note a change in the shear values of 5 kips per linear foot (kips/ft). A 10 ft long free body is shown in Fig. 13–23. The shear at the 10 ft point can be calculated as

$$V = +50 - 50 = 0$$

The rate of change of shear is 5 kips/ft (equal to the uniformly distributed load). If an 11 ft long free-body diagram is drawn and a summation of vertical forces performed, the shear will be found to be −5 kips. The negative shears will then continue to increase (become more negative) until, at the right reaction, the shear will be −50 kips. Note that the right reaction is 50 kips.

Plotting the positive shear values above the zero shear line and the negative shear values below the zero shear line results in the graphic representation of Fig. 13–24. This diagram shows the shear variation along the length of the beam. Notice that the shape of the shear diagram is that of a sloping straight line. Also note that the positive and negative shear areas are equal.

FIGURE 13–24   Load and shear diagrams.

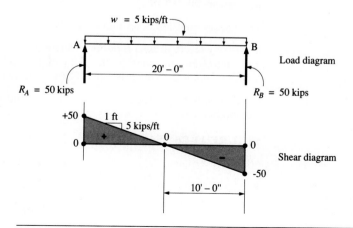

Based on the two previous examples, the following general rules applicable to subsequent construction of shear diagrams can now be stated:

1. For any part of a beam where there are no loads, the shape of the shear diagram will be that of a horizontal straight line.
2. The shear diagram at the point of application of a concentrated load will be a vertical line; that is, there will be a sudden change in the shear.
3. For any part of a beam where there is a uniformly distributed load, the shape of the shear diagram will be a straight sloping line with a slope equal to the intensity of the load.
4. For a simply supported beam subjected to vertical loads, the positive area of the shear diagram will be equal to the negative area.

☐ **EXAMPLE 13–8**   Draw the shear diagram for the beam in Fig. 13–25. The reactions have been computed and are shown. Neglect the weight of the beam. (This is the beam of Example 13–4.)

FIGURE 13–25   Load diagram.

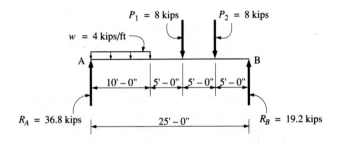

Solution   Working from the left reaction, we will compute the shear at the following points where changes will occur in the shear diagram:

(a) Just to the right of the left reaction
(b) At the end of the uniformly distributed load
(c) Just to the left *and* right of each concentrated load
(d) Just to the left of the right reaction

The shears are then determined as follows:

(a) If we were to draw a free-body diagram and sum the vertical forces, the shear at an infinitesimal distance to the right of the left reaction would be found to be a positive 36.8 kips, which is equal to the reaction.

(b) To compute the shear at the end of the uniformly distributed load, a 10 ft long free-body diagram is drawn, as shown in Fig. 13–26. A summation of vertical forces

FIGURE 13–26   Free body.

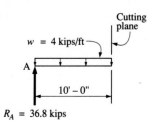

will result in

$$V = +36.8 - 4(10) = -3.2 \text{ kips}$$

(c) To compute the shear just to the left and right of the concentrated loads, free-body diagrams can be drawn and vertical forces summed as usual. The results of these calculations will yield

to the left of P_1: $V = -3.2$ kips

to the right of P_1: $V = -11.2$ kips

to the left of P_2: $V = -11.2$ kips

to the right of P_2: $V = -19.2$ kips

(d) Similarly, the shear can be calculated an infinitesimal distance to the left of the right reaction. This results in a shear of -19.2 kips, which is numerically equal to the right reaction. The completed shear diagram, obtained by connecting the shears at the various points, is shown in Fig. 13–27 along with the load diagram.

FIGURE 13–27 Load and shear diagrams.

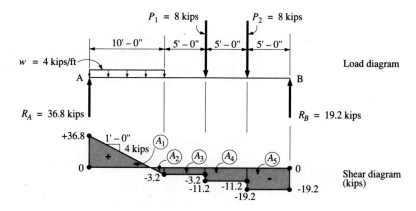

A check on the shear diagram and calculations is performed by comparing the positive shear area with the negative shear area. They should be equal. The areas are calculated by breaking up the shear diagram into simple geometric shapes. We will first determine the location of the zero shear point. Since the slope of the shear diagram in the uniformly distributed load portion of the beam is equal to 4 kips/ft, the distance to zero shear from the left reaction must be

$$\frac{36.8 \text{ kips}}{4 \text{ kips/ft}} = 9.2 \text{ ft}$$

Therefore, for positive shear area,

$$A_1 = \left(\frac{1}{2}\right)(36.8)(9.2) = 169.3 \text{ ft-kips}$$

and for negative shear areas,

$$A_2 = \left(\frac{1}{2}\right)(3.2)(0.8) = 1.3 \text{ ft-kips}$$

$$A_3 = 3.2(5.0) = 16.0 \text{ ft-kips}$$
$$A_4 = 11.2(5.0) = 56.0 \text{ ft-kips}$$
$$A_5 = 19.2(5.0) = 96.0 \text{ ft-kips}$$

The total negative shear area is 169.3 ft-kips, which is equal to the positive shear area. Therefore, the shear diagram and calculations check.

☐ **EXAMPLE 13–9**   Draw the shear diagram for the cantilever beam in Fig. 13–28. Neglect the weight of the beam.

FIGURE 13–28   Beam diagram.

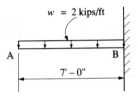

Solution   Reactions for a cantilever beam include both vertical and moment reactions. Although only the vertical reaction will be required for the construction of the shear diagram, we will calculate both reactions. Figure 13–29 shows the load diagram for the given beam with the two reactions to be found. From a summation of vertical forces,

$$R_B = wL = 14 \text{ kips}$$

From a summation of moments about the right support,

$$M_B = -14(3.5) = -49 \text{ ft-kips}$$

The negative sign indicates tension in the top. Free-body diagrams may next be drawn from the free left end to any location along the beam length and a summation of vertical forces performed, as previously done. Recall that the shear should be computed at the beginning and end of all distributed loads. The beginning of the uniformly distributed load is at the free end, and since no load (or beam) exists to the left of this point, or at this point, the shear at the free end is zero. The shear at the right end of the load can be computed by using the load diagram (which is the free-body diagram of the whole beam), as shown in Fig. 13–29, and summing vertical forces:

$$V = -2(7) = -14 \text{ kips}$$

FIGURE 13–29   Load diagram.

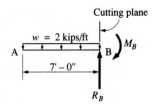

Note that this is equal in magnitude to the vertical reaction at the support. The negative sign for shear indicates that the left side of the section moves down with respect to the right.

After computing the two shear values, the shear diagram may be drawn as shown in Fig. 13–30. The slope is equal to the magnitude of the uniformly distributed load (2 kips/ft). Note that the shear for a cantilever beam will always be negative assuming that the loads are vertically downward and that the beam is drawn with the free end on the left. Note, too, that the maximum shear in a cantilever beam will always occur at the fixed end.

FIGURE 13–30   Load and shear diagrams.

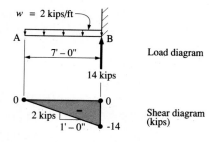

By now it may be evident that the drawing of free bodies may actually be neglected in the process of constructing shear diagrams. Once reactions have been determined, the shear diagram may be drawn by successive summations of forces (and reactions, as shown on a load diagram) beginning at the left side. If upward acting forces are considered positive and downward acting forces negative, the correct signs for shear will result. An electronic calculator is particularly useful for this procedure. The method is illustrated in the following example.

☐ **EXAMPLE 13–10**   Draw the shear diagram for the beam having the load diagram shown in Fig. 13–31.

FIGURE 13–31   Load and shear diagrams.

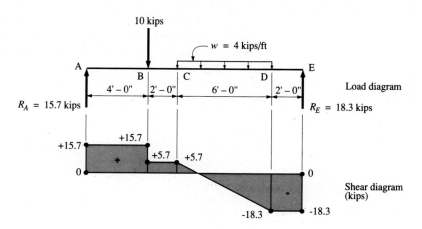

Solution   For convenience, the load diagram has been labeled with points A through E. Since the reactions are given, the first step would be to draw the horizontal baseline for the shear diagram.

Starting at the left, point A, the shear diagram goes up to a value of +15.7 kips due to the reaction. Proceeding to the right, there is no change in the shear until the load at point B is encountered. At that point, the sum of the vertical forces decreases (the load is downward) by 10 kips, and the shear diagram drops to +5.7 kips. Again proceeding to the right, there is no change until the beginning of the uniform load at C is encountered. The total change in the sum of the vertical forces between C and D is −24 kips, resulting in shear of −18.3 kips at D. The shear diagram between C and D must be a sloping straight line. Again proceeding to the right, there is no change in shear between D and E, resulting in a shear of −18.3 kips at E. Note that the right reaction is 18.3 kips. Such details as locating the point of zero shear and checking negative and positive shear areas may then be accomplished in the usual way.

The procedure for construction of the shear diagram may be summarized as follows:

1. Sketch the load diagram.
2. Compute the reactions and note these on the load diagram.
3. Sketch lines vertically downward from selected points along the beam to use as a guide for the shear diagram. (For convenience and clarity, the shear diagram should be drawn directly below the load diagram.)
4. Draw a horizontal baseline representing zero shear. The baseline is the same length as the beam.
5. Starting at the left end, compute the shears at selected points using free-body diagrams and sketch the shear diagram. *Or*, sketch the shear diagram directly using a continuous algebraic summation of vertical forces and reactions from the load diagram.
6. Locate points of zero shear.

13–6
MOMENT DIAGRAMS

Just as a shear diagram shows the amount of shear at any point along the length of a beam, a moment diagram shows the amount of bending moment at any location. The moment diagram is drawn directly under the shear diagram. A moment baseline representing zero bending moment is drawn parallel to the shear base line. The abscissa along the baseline represents the locations of successive cross sections of the beam. The ordinate of the diagram represents the value of the bending moment at that particular cross section.

As with shear diagrams, moment diagrams need not be drawn to scale. Moment diagrams are most conveniently drawn using moments that have been determined at certain points where moment changes occur. The points to consider are as follows:

1. At each concentrated load and reaction
2. At points of zero shear and where the shear diagram goes through zero
3. At the beginning and end of all distributed loads

The following important general rules concerning moment diagrams should be noted:

1. The bending moment at the ends of a simply supported beam will always be equal to zero.
2. With loads acting vertically downward, the bending moment at the free end of a cantilever beam will always be equal to zero and the maximum moment will always occur at the fixed end. The shear will also be maximum at the fixed end.
3. The bending moment will always be positive for a simply supported beam and negative for a cantilever beam, assuming that all loads act vertically downward.
4. Except for cantilever beams, the maximum bending moment will always occur at a point of zero shear or where the shear diagram goes through zero.

The construction of the moment diagram is best illustrated through a variety of examples.

□ **EXAMPLE 13–11**   Draw the bending moment diagram for the beam in Fig. 13–32. (This is the beam of Example 13–6.)

Solution   The load diagram and shear diagram are given. The procedure for computing the bending moment at any point along the length of a beam is discussed in Section 13–4. Each value of moment is computed by considering the loads acting on the free body

FIGURE 13–32 Load, shear, and moment diagrams.

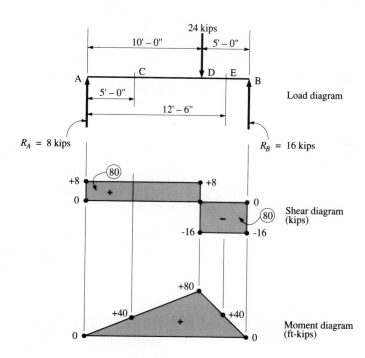

taken from the left end of the beam. The computed values are then plotted as ordinates from the zero moment line (positive values above and negative values below).

Since the bending moment at the simple supports equals zero, it is necessary to compute the bending moment at the concentrated load only. (This is also the point where the shear goes through zero, as shown in Fig. 13–32.) However, in order to illustrate the shape of the moment diagram graphically, we will also calculate the moments at sections 5 ft and 12.5 ft from the left end. Referring to the load diagram of Fig. 13–32, the three points have been labeled C, D, and E:

$$\text{At point C:} \quad M = 8(5) = +40 \text{ ft-kips}$$
$$\text{At point D:} \quad M = 8(10) = +80 \text{ ft-kips}$$
$$\text{At point E:} \quad M = 8(12.5) - 24(2.5) = +40 \text{ ft-kips}$$

These three values are plotted on the moment diagram of Fig. 13–32, resulting in two straight sloping lines.

With reference to Fig. 13–32, several interesting relationships between shear and moment diagrams may be observed:

1. The maximum moment occurs at the point where the shear goes through zero.
2. The change in magnitude of the moment between any two points is equal to the area of the shear diagram between those two points. This means that the moment at any point can be determined by the summation of the shear area up to that point.
3. The slope of a line tangent to the moment diagram at any point is equal in sign and magnitude to the shear at that point. (This will be further explained in Example 13–12.)

☐ **EXAMPLE 13–12**   Draw the bending moment diagram for the beam in Fig. 13–33. (This is the beam that was analyzed in Example 13–7.)

FIGURE 13–33 Load, shear, and moment diagrams.

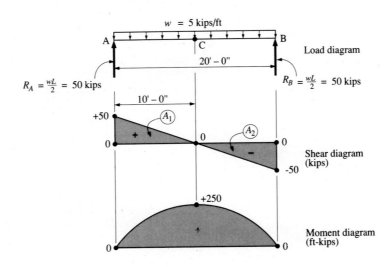

FIGURE 13–34 Signs of slopes.

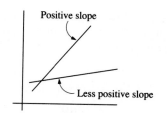
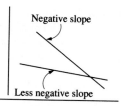

Solution The load and shear diagrams are given. The moments at the supports are zero. The only other moment needed is that at C, midspan, where the shear is zero. We will determine the moment at C by noting that the change in moment between A and C is equal to the area of shear diagram between A and C. This triangular area, designated A_1 on the shear diagram, is calculated from

$$A_1 = \frac{1}{2}(10 \text{ ft})(+50 \text{ k}) = +250 \text{ ft-kips}$$

Since the moment at A is zero, the moment at C is +250 ft-kips. Note that since A_2 is equal to A_1 in magnitude, but opposite in sign, the summation of shear areas from the left to point B is zero, which is the moment at point B.

Sketching in the actual curve of the moment diagram, we note that proceeding to the right from point A to point C the shear is positive and decreasing. This means that the slope of a line tangent to the moment diagram is positive, but decreasingly positive as point C is approached. Also, from point C to point B, the shear is negative and is increasingly negative as point B is approached. This means that the slope of a line tangent to the moment diagram is increasingly negative proceeding from C to B. (Signs of slopes are reviewed in Fig. 13–34.) At midspan, the shear is zero. Therefore, the slope of a line tangent to the moment diagram at midspan is zero (the tangent to the curve is horizontal). We see then that the uniform load results in a moment diagram that is concave downward. This shape could also be verified by computing moments at arbitrary intermediate points.

□ **EXAMPLE 13–13** Draw the bending moment diagram for the beam in Fig. 13–35. (This is the beam of Example 13–9.)

FIGURE 13–35 Load, shear, and moment diagrams.

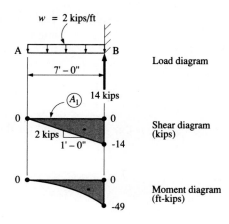

Solution   The load and shear diagrams are given. Since the moment at the free end is zero and the maximum moment will always occur at the fixed end, the only calculation necessary is that for the moment at the fixed end. Calculating the total area of the shear diagram A_1,

$$A_1 = \frac{1}{2} (7 \text{ ft})(-14 \text{ k}) = -49 \text{ ft-kips}$$

This is the change in moment from A to B and, therefore, the moment at B. Note that the shear, starting at point A, is zero and then becomes increasingly negative. The curve of the moment diagram is sketched in accordingly using relative slopes from Fig. 13–34 as a guide.

☐ **EXAMPLE 13–14**   Draw the moment diagram for the beam in Fig. 13–36. (This is the beam of Example 13–8.)

FIGURE 13–36 Load, shear, and moment diagrams.

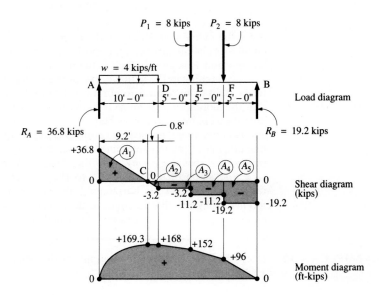

Solution   The load and shear diagrams are given. For convenience, intermediate points of interest between the endpoints A and B have been labeled C through F. Calculating the various areas of the shear diagram in units of ft-kips, we obtain for the triangular areas,

$$A_1 = +169.3$$
$$A_2 = -1.3$$

and for the rectangular areas,

$$A_3 = -16.0$$
$$A_4 = -56.0$$
$$A_5 = -96.0$$

The moment values can now be calculated using the shear diagram areas. Working from the left end of the beam where the moment is equal to zero, the moments at all of the points of interest can be computed by adding algebraically the successive areas of the shear diagram. The steps are as follows:

1. The moment at A is zero.
2. The moment at C, the point of zero shear, is equal to the shear area between A and C:

$$M = +169.3 \text{ ft-kips}$$

3. The moment at D, the end of the distributed load, is equal to the moment at C plus the shear area between C and D:

$$M = 169.3 - 1.3 = +168.0 \text{ ft-kips}$$

4. The moment at E, under the load P_1, is equal to the moment at D plus the shear area between D and E:

$$M = 168.0 - 16.0 = +152.0 \text{ ft-kips}$$

5. The moment at F, under the load P_2, is equal to the moment at E plus the shear area between E and F:

$$M = 152.0 - 56.0 = +96.0 \text{ ft-kips}$$

6. The moment at the right support should be zero and can be checked by adding the shear area between F and B to the moment at F:

$$M = 96.0 - 96.0 = 0 \quad \textbf{OK}$$

With all necessary moment values computed, the shape of the moment diagram between the points may be determined by inspection of the shear diagram. The moment diagram will be parabolic from the left end to the end of the distributed load and will be composed of sloping straight lines otherwise. Note that the moment is always positive. This is typical for simply supported beams subjected to vertical downward loads.

In summary, the procedure for the construction of the moment diagram, once the shear diagram has been completely defined, may be stated as follows:

1. Extend guidelines used for the shear diagram vertically downward. (For convenience and clarity, the moment diagram should be drawn directly under the shear diagram.)
2. Draw a horizontal baseline representing zero moment. The baseline has the same length as the beam and shear diagrams.
3. Starting from the left end, use free bodies to compute the moments due to the external loads and reactions at all concentrated loads and reactions, at the beginning and end of distributed loads, at points of zero shear, and where the shear goes through zero. *Or*, calculate the shear areas between the preceding points and compute the moments by adding algebraically the successive areas starting at the left end.
4. Plot the moment values. (While the moment diagram need not be to scale, using approximately correct proportions is recommended for clarity.)

5. Sketch the correct shape of the moment diagram between the plotted points by referring to the shear diagram.

13–7
SECTIONS OF
MAXIMUM MOMENT

In the case of moment, as in the case of shear, the largest numerical value, regardless of sign, is called the maximum. The positive sign merely indicates that the beam, or beam segment, is concave upward, while the negative sign indicates that it is concave downward.

As previously stated, except for the cantilever beam, the maximum bending moment occurs at some location on the beam where the shear is equal to zero or the shear diagram goes through zero. The location of maximum bending moment is sometimes called the *critical section*. This term is based on the fact that the bending moment will develop a bending stress, which is generally the critical or controlling stress with respect to beam design and analysis.

In the case of a beam subjected to distributed load and concentrated loads, the shear diagram may or may not go through zero under a concentrated load. The exact location of zero shear must be calculated. In overhanging beams subject to various types of loads, the shear may equal zero and/or go through zero more than once. Hence, there may exist two or more locations of a possible maximum moment. When this occurs, the moment must be computed at each location, with the largest numerical value representing the maximum moment. In the design process, however, it is sometimes necessary that both a maximum positive and a maximum negative moment be utilized. This has particular relevance for reinforced concrete design.

In simply supported beams subjected to vertical downward loads, only one critical location will occur resulting in a maximum positive moment. In cantilever beams, both maximum moment and maximum shear occur at the support or fixed end.

□ **EXAMPLE 13–15**

Draw the shear and moment diagrams for the overhanging beam in Fig. 13–37. The reactions are given.

Solution

(a) We draw the shear diagram by summing the vertical forces beginning at point A and proceeding to the right. The shear at A is zero. Therefore, the shear diagram begins at zero (see Fig. 13–38). Between A and B, the shear goes from zero to a value of −5 kips as a result of the uniformly distributed load between these two points. At

FIGURE 13–37   Load diagram.

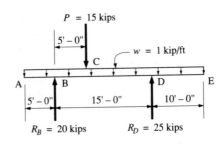

FIGURE 13–38 Load, shear, and moment diagrams.

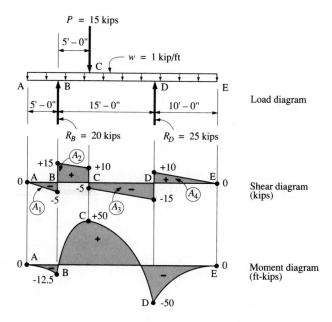

Load diagram

Shear diagram (kips)

Moment diagram (ft-kips)

point B, the upward acting reaction of 20 kips produces an abrupt change in the shear, causing it to go from a value of −5 kips to a value of +15 kips.

This procedure continues across the beam and the shears are computed by successive algebraic addition of each load (or reaction) to the shear already calculated to the left of the load. The shears are plotted as they are computed. Note that the shear must become zero at point E and that the maximum shear value is 15 kips.

(b) In order to draw the moment diagram, we next calculate the shear areas (in units of ft-kips). The negative areas are

$$A_1 = \frac{1}{2}(5)(-5) = -12.5$$

$$A_3 = \frac{1}{2}(-5 - 15)(10) = -100.0$$

And for the positive areas,

$$A_2 = \frac{1}{2}(15 + 10)(5) = +62.5$$

$$A_4 = \frac{1}{2}(10)(10) = +50$$

A check on the calculations can be made by ensuring that negative and positive shear areas are equal:

$$50 + 62.5 = 100 + 12.5$$
$$112.5 = 112.5 \quad \textbf{OK}$$

This indicates that the moment at the right end of the beam will be zero, as it should be in this case.

The moments can now be calculated at those points recommended in Section 13–6. Utilizing the shear diagram areas, the moments are found by adding algebraically the successive areas, starting at the left end:

$$M_A = 0$$

$$M_B = M_A + A_1 = 0 - 12.5 = -12.5 \text{ ft-kips}$$

$$M_C = M_B + A_2 = -12.5 + 62.5 = +50.0 \text{ ft-kips}$$

$$M_D = M_C + A_3 = +50.0 - 100.0 = -50.0 \text{ ft-kips}$$

$$M_E = M_D + A_4 = -50 + 50 = 0$$

The moment diagram is drawn in Fig. 13–38. The curvature between all the points may be observed as concave downward based on inspection of the shear diagram:

A to B—negative shear, increasing in negative value

B to C—positive shear, decreasing

C to D—negative shear, increasing in negative value

D to E—positive shear, decreasing

Note that at point C, a maximum positive moment of 50 ft-kips occurs and at point D a maximum negative moment of −50 ft-kips occurs. Note also that the moment diagram crosses the zero moment baseline at two locations. These points are called *points of inflection* and represent points of zero moment. They can be located by setting up an expression for the bending moment and equating it to zero.

**13–8
MOVING LOADS**

The structural design of bridge members, as well as that of crane girders in industrial buildings, is based on moving live loads. Thus far, all of our discussion has assumed fixed loads.

If a system of concentrated loads with fixed distances between the loads moves across a simple beam, the moment under any given load will increase from zero to a maximum value and then decrease to zero again as the load moves across the beam and approaches the support. If we neglect the weight of the beam, the maximum moment will always occur directly under a load since the shear goes through zero at a load point. It is evident that the magnitude of both the shear and bending moment will change as the load system moves across the beam. The position of the load system that will develop a maximum bending moment will always be different from the position that will develop a maximum shear.

Taking an arbitrary moving load system as shown in Fig. 13–39, the location of load P_2 when the moment under P_2 is a maximum has been established mathematically. It can be shown that when the moment under P_2 is a maximum, the centerline of the span length or beam bisects the distance between P_2 and the resultant of the load system. This is true for any of the

FIGURE 13–39 Moving load system.

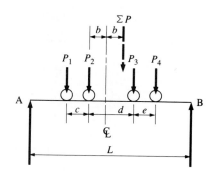

loads. Therefore, we state general rule 1 as follows:

1. The moment under any one load of a given system of moving loads is maximum when the centerline of the beam length bisects the distance between that one load and the resultant of the load system.

Each load of the load system may be examined for its maximum moment in accordance with general rule 1. The largest of these moments is the *absolute maximum*, which must be used in the design of the beam. Additional general rules are as follows:

2. It is usually necessary to examine the maximum moment only under the two loads nearest the resultant of the load system to obtain the absolute maximum moment. If the largest load on the span is nearest the resultant, it is necessary to examine under this load only.
3. With a load system of two equal moving loads having a distance of $0.586L$ or greater between them, the maximum moment is obtained by placing one of the two loads at the centerline of the span and finding the moment under the load at that point.
4. In most cases, to compute the maximum shear due to the moving load system, the maximum load is placed over one support with as many of the other loads as possible on the beam span. The maximum shear will be the maximum reaction. In some cases, the absolute maximum shear must be obtained by trial, in which each load is successively placed over the support and the reaction calculated.

The preceding are termed "general" rules. This implies that exceptions may exist under certain conditions.

□ **EXAMPLE 13–16**   A two-axle truck weighing a total of 60 kips crosses a bridge that spans 36 ft. The front axle carries 16 kips and is 9 ft from the rear axle. Assuming that this loading is distributed equally to each of two parallel bridge girders, compute the absolute maximum moment and shear. Neglect the weight of the girder.

Solution   Using loads to one girder (one-half the total load) as shown in Fig. 13–40, we compute the magnitude and location of the resultant load. In this parallel force

FIGURE 13–40   Resultant of a force system.

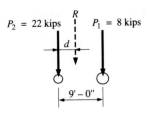

system, the magnitude of the resultant is

$$22 \text{ kips} + 8 \text{ kips} = 30 \text{ kips}$$

The location of the resultant is determined by summing moments with respect to P_2,

$$8(9.0) = 30d$$

$$d = 2.40 \text{ ft}$$

To obtain the absolute maximum moment, we place the centerline of the span midway between the resultant force R and the nearest load P_2, as shown in Fig. 13–41. The absolute maximum moment will be directly under P_2.

FIGURE 13–41   Load diagram.

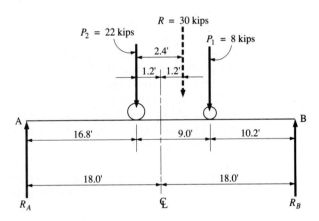

Computing R_A,

$$\Sigma M_B = -R_A(36) + 8(10.2) + 22(19.2) = 0$$

from which

$$R_A = 14 \text{ kips}$$

It is not necessary to compute R_B since only the left beam segment up to P_2 will be taken out as a free body. (Refer to Fig. 13–42.)

Calculating the moment at P_2,

$$M_{P_2} = +14(16.8) = +235 \text{ ft-kips}$$

FIGURE 13–42   Free-body
diagram.

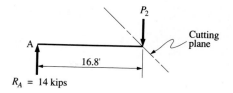

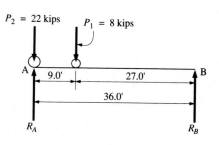

FIGURE 13–43   Load diagram.

The absolute maximum shear can be obtained by moving the load system as shown in Fig. 13–43. The load of 22 kips should be considered to be at an infinitesimal distance to the right of the support at A.

Computing R_A,

$$\Sigma M_B = -R_A(36) + 22(36) + 8(27) = 0$$

from which

$$R_A = 28 \text{ kips}$$

This represents the absolute maximum shear.

☐ **EXAMPLE 13–17**   The load system shown in Fig. 13–44 is designated as an HS 20–44 truck load and is commonly used in the design of highway bridges. Compute the absolute maximum shear and moment that would occur in a simple span bridge subjected to this load system. The bridge span is 40 ft. Neglect the weight of the bridge, itself.

Solution   The magnitude of the resultant load is

$$32 \text{ kips} + 32 \text{ kips} + 8 \text{ kips} = 72 \text{ kips}$$

FIGURE 13–44   Standard truck
HS 20–44 load.

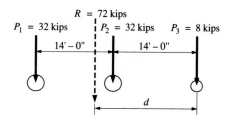

FIGURE 13–45 Load diagram.

R
2.33'
P1 = 32 kips
P3 = 8 kips
8.33'
9.34
14.0'
3.67'
P2 = 32 kips
A
B
20.0'
20.0'
RA
RB
℄

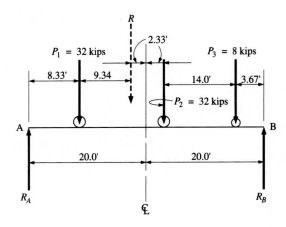

The location of the resultant is determined by summing moments with respect to P_3:

$$32(28) + 32(14) = 72d$$
$$d = 18.67 \text{ ft}$$

The centerline of the span is placed midway between the resultant force R and the nearest load P_2. (Refer to Fig. 13–45.) The absolute maximum moment will be directly under P_2.

Computing R_A,

$$\sum M_B = -R_A(40) + 8(3.67) + 32(17.67) + 32(31.67) = 0$$

from which

$$R_A = 40.2 \text{ kips}$$

Calculating the moment under P_2 using as a free body the beam segment to the left of P_2,

$$M_{P_2} = 40.2(22.33) - 32(14) = +450 \text{ ft-kips}$$

The absolute maximum shear may be obtained by moving the load system as shown in Fig. 13–46.

Again, computing R_A,

$$\sum M_B = -R_A(40) + 32(40) + 32(26) + 8(12) = 0$$

FIGURE 13–46 Load diagram.

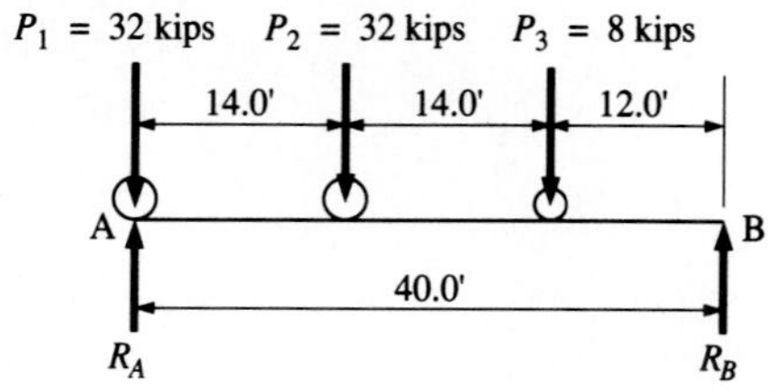

P1 = 32 kips
P2 = 32 kips
P3 = 8 kips
14.0'
14.0'
12.0'
A
B
40.0'
RA
RB

from which

$$R_A = 55.2 \text{ kips}$$

This represents the absolute maximum shear.

13–9
SI SYSTEM
EXAMPLES

□ **EXAMPLE 13–18**

Compute the reactions for the overhanging beam in Fig. 13–47(a).

Solution The load diagram is shown in Fig. 13–47(b). The supports at A and B are replaced with the reactions that we expect from those supports. The pinned support at A could provide a horizontal reaction, but since there is no horizontally applied load, this reaction is zero and may be neglected.

FIGURE 13–47 Overhanging beam.

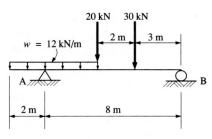

(a) Beam diagram

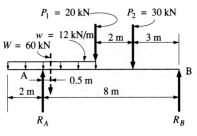

(b) Load diagram

Calculating the equivalent concentrated resultant load for the distributed load,

$$W = 12 \text{ kN/m}(5 \text{ m}) = 60 \text{ kN}$$

The reaction at B can be computed by summing moments about point A. Assuming counterclockwise moments positive,

$$\Sigma M_A = +R_B(8) - 30(5) - 20(3) - 60(0.5) = 0$$

from which

$$R_B = +30 \text{ kN} \uparrow$$

The reaction at A can be computed by summing moments about point B. Assuming counterclockwise moments positive,

$$\Sigma M_B = -R_A(8) + 30(3) + 20(5) + 60(7.5) = 0$$

from which

$$R_A = +80 \text{ kN } \uparrow$$

Recall that a positive sign for the computed reaction indicates that the correct sense was assumed. Had the result turned out to be negative, it would have meant only that the incorrect sense had been assumed.

Checking the calculations by summing the vertical forces and assuming upward positive,

$$\Sigma F_V = +80 + 30 - 60 - 20 - 30 = 0 \quad \textbf{OK}$$

☐ **EXAMPLE 13–19**   The load diagram for a beam is shown in Fig. 13–48. Compute the shear and bending moment (a) at 5 m from the left end of the beam and (b) at 10 m from the left end of the beam. Neglect the weight of the beam.

FIGURE 13–48   Load diagram.

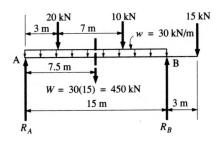

Solution   The location and magnitude of the resultant of the uniformly distributed load is shown on the load diagram. The beam reactions are computed from

$$\Sigma M_A = +R_B(15) - 15(18) - 10(10) - 20(3) - 450(7.5) = 0$$

from which

$$R_B = +253.7 \text{ kN } \uparrow$$

and

$$\Sigma M_B = -R_A(15) - 15(3) + 10(5) + 20(12) + 450(7.5) = 0$$

from which

$$R_A = +241.3 \text{ kN } \uparrow$$

Checking the calculations by summing vertical forces,

$$\Sigma F_V = +253.7 + 241.3 - 15 - 10 - 20 - 450 = 0 \quad \textbf{OK}$$

FIGURE 13-49 Free-body diagram for shear and moment determination.

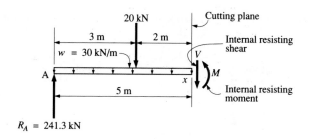

(a) *At 5 m from the left end:*
We cut the beam at 5 m from the left end (designated point x in Fig. 13-49) and consider the part on the left of the cutting plane as a free body. The shear force is obtained by summing the vertical external forces acting on the free body of Fig. 13-49:

$$V = +241.3 - 20 - 30(5) = +71.3 \text{ kN}$$

Recall that the positive sign indicates that the part of the beam to the left of the cutting plane tends to move upward with respect to the right.

The bending moment due to the external loads (at the cutting plane—point x) is obtained by using upward acting forces as producing positive moments and summing moments about point x:

$$M = +241.3(5) - 20(2) - 30(5)(2.5) = +791.5 \text{ kN} \cdot \text{m}$$

The positive sign indicates a positive moment (the bottom fibers are in tension and the top fibers are in compression).

(b) *At 10 m from the left end:*
As may be observed, a concentrated load exists at this point; therefore, the shear will change abruptly. Two free bodies are shown in Fig. 13-50 for purposes of computing the two shear values. Either free body may be used to compute the bending moment. In each case, note that the internal resisting shear V and the internal resisting moment M are shown acting on the cutting plane.

We will calculate the shear force due to the external loads for each case shown. With reference to Fig. 13-50(a),

$$V = +241.3 - 20 - 30(10) = -78.7 \text{ kN}$$

With reference to Fig. 13-50(b),

$$V = +241.3 - 20 - 30(10) - 10 = +88.7 \text{ kN}$$

Calculating the bending moment due to the external loads (at the cutting plane—point x),

$$M = +241.3(10) - 20(7) - 30(10)(5) = +773 \text{ kN} \cdot \text{m}$$

FIGURE 13–50   Ten-meter free-body diagrams.

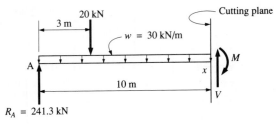

(a) Cutting plane to left of concentrated load

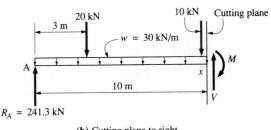

(b) Cutting plane to right of concentrated load

SUMMARY—BY SECTION NUMBER

13–1   Beams are categorized according to type and/or number of supports. Types of beams in common use are the simple, cantilever, and overhanging beams, all of which are statically determinate, since their reactions can be computed using the three laws of equilibrium.

13–2   Loads on beams are either concentrated or distributed. Distributed loads may be uniform or nonuniform. Concentrated loads act at a single point; distributed loads are spread out over a length of the beam.

13–3   External beam reactions may be computed by using a free-body diagram of the beam (called a *load diagram*) and applying the three laws of equilibrium. The algebraic sum of all externally applied loads and reactions must equal zero.

13–4   The effect of the externally applied loads and reactions on a beam is to develop internal shear force and bending moment. The shear force is the algebraic sum of the external vertical forces acting on one side of a cutting plane. The bending moment is the algebraic sum of the moments of all the external forces acting on one side of a cutting plane.

13–5   A shear diagram is a graphical representation showing how the shear varies along the length of a beam. It is drawn directly below the load diagram. The shape of the shear diagram is a function of the load diagram.

13–6   A moment diagram is a graphical representation showing how the moment varies along the length of a beam. It is drawn directly below the shear diagram. The shape of the moment diagram is a function of the shear diagram.

13–7 The critical section of a beam occurs where the moment is a maximum. In beams other than the cantilever, the maximum moment will occur at a point of zero shear or where the shear goes through zero.

13–8 Moving load systems create varying patterns of moment and shear within beams. The absolute maximum moment and the absolute maximum shear are created when the load system is at specific positions on the beam.

PROBLEMS

Section 13–3   Beam Reactions

1–5. For Problems 1 through 5, calculate the reactions at points A and B for the beams shown in Figs.

13–51 through 13–55. Draw the complete load diagram in each case.

FIGURE 13–51   Problem 1.

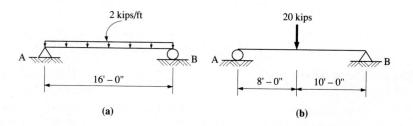

(a)   (b)

FIGURE 13–52   Problem 2.

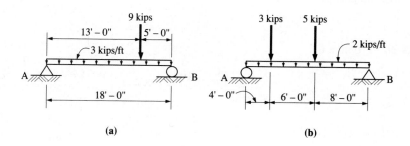

(a)   (b)

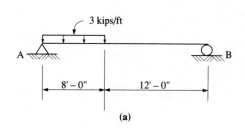

(a)

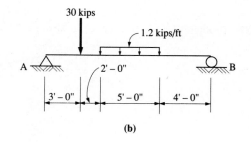

(b)

FIGURE 13–53   Problem 3.

FIGURE 13–54　　Problem 4.

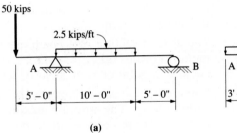

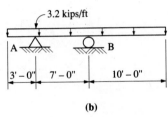

(a)　　　　　　　　　　　　　　　　　　(b)

FIGURE 13–55　　Problem 5.

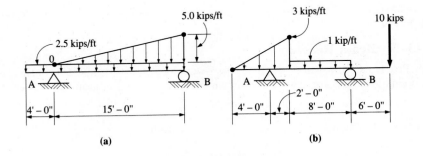

(a)　　　　　　　　　　　　　　　　　　(b)

Section 13–4　Shear Force and Bending Moment

6. Calculate the shear and moment at 3 ft and at 8 ft from the left for the beams in Fig. 13–56. Show free-body diagrams.

7. Calculate the shear and moment at midspan for the beams in Fig. 13–57. Show free-body diagrams.

8. Calculate the shear and moment at 5 ft and at 15 ft from the left for the beams in Fig. 13–58. Show free-body diagrams.

FIGURE 13–56　　Problem 6.

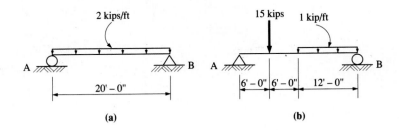

(a)　　　　　　　　　　　　　　　　　　(b)

FIGURE 13–57　　Problem 7.

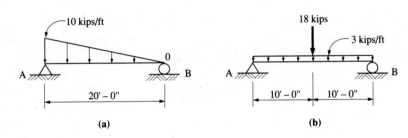

(a)　　　　　　　　　　　　　　　　　　(b)

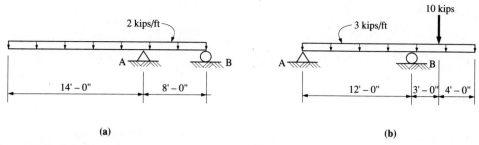

(a)

(b)

FIGURE 13–58   Problem 8.

Section 13–5   Shear Diagrams

9–12. For Problems 9 through 12, draw complete shear diagrams for the beams shown in Figs. 13–59 through 13–62.

Section 13–6   Moment Diagrams

13–15. For Problems 13 through 15, draw complete shear and moment diagrams for the beams shown in Figs. 13–63 through 13–65.

FIGURE 13–59   Problem 9.

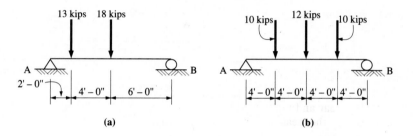

(a)

(b)

FIGURE 13–60   Problem 10.

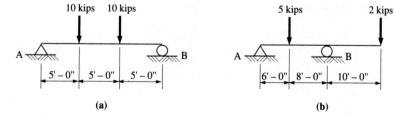

(a)

(b)

FIGURE 13–61   Problem 11.

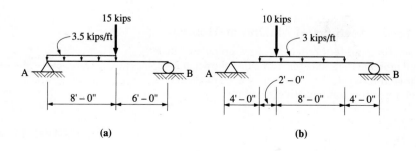

(a)

(b)

FIGURE 13–62   Problem 12.

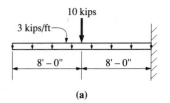

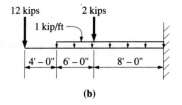

(a)

(b)

FIGURE 13–63   Problem 13.

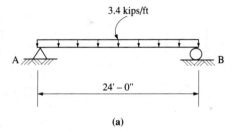

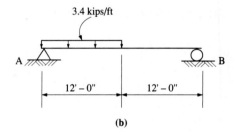

(a)

(b)

FIGURE 13–64   Problem 14.

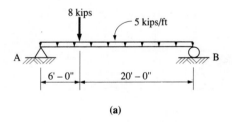

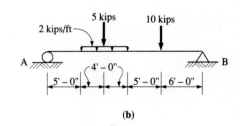

(a)

(b)

FIGURE 13–65   Problem 15.

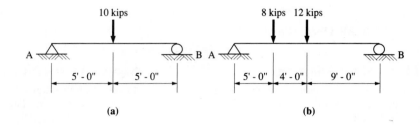

(a)

(b)

Section 13–7   Sections of Maximum Moment

16–18. For Problems 16 through 18, draw complete shear and moment diagrams for the beams shown in Figs. 13–66 through 13–68. State the values of the maximum positive and negative moments.

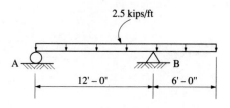

FIGURE 13–66   Problem 16.

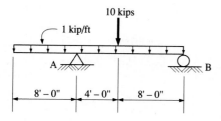

FIGURE 13–67 Problem 17.

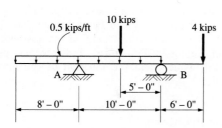

FIGURE 13–68 Problem 18.

Section 13–8 Moving Loads

19. A moving load system is composed of two concentrated loads, each 20 kips, separated by a distance of 10 ft. The loads are to cross a 30 ft simple span. Calculate the absolute maximum shear and bending moment.

20. A moving load system is composed of two concentrated loads separated by 16 ft. One load is 26 kips

and the other is 12 kips. The loads are to cross a 40 ft simple span. Calculate the absolute maximum shear and bending moment.

21. An HS 15–44 standard truck load is composed of three concentrated loads, as shown in Fig. 13–69. Calculate the absolute maximum shear and moment produced in a simple bridge span having a length of 50 ft.

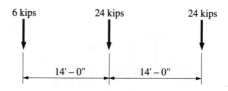

FIGURE 13–69 Problem 21.

SI System Problems

22. and **23.** Calculate the reactions at points A and B for the beams in Figs. 13–70 and 13–71. Draw the complete load diagram (free-body diagram) in each case.

24. Calculate the shear and bending moment at points 2 m and 3.5 m from the left end of the beam in Fig. 13–72. Use free-body diagrams.

25. Refer to beam (a) in Fig. 13–70 and beam (b) in Fig. 13–71. Calculate the shear and bending moment at points 10 m and 16 m from the left end of (a) the beam in Fig. 13–70(a) and (b) the beam in Fig. 13–71(b).

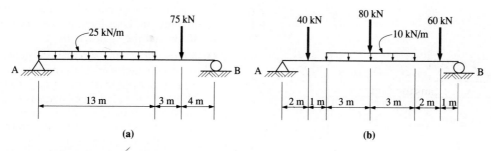

(a) (b)

FIGURE 13–70 Problem 22.

FIGURE 13–71   Problem 23.

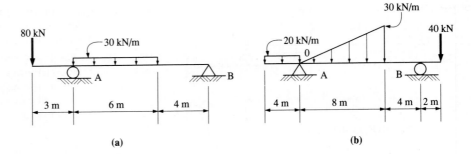

(a)   (b)

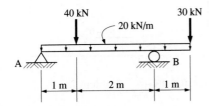

FIGURE 13–72   Problem 24.

For Problems 26 through 29, draw complete load, shear, and moment diagrams for the beams indicated. Ordinates should be labeled at all maximum and minimum points and where loads occur or change.

26. For beam (b) of Problem 22.

27. For beam (a) of Problem 23.

28. For the beam of Problem 24.

29. For the beam in Fig. 13–73.

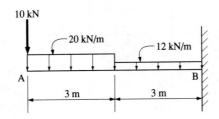

FIGURE 13–73   Problem 29.

30. A two-axle roller with axles 5 m apart passes over a 15 m simply supported beam bridge. The load is 200 kN on each axle. Compute the absolute maximum moment and shear. Indicate the position of the wheels and the location of the maximum values.

Computer Problems

For the following computer problems, any appropriate programming language may be used. Input prompts should fully explain what is required of the user (the program should be "user friendly"). The resulting output should be well labeled and self-explanatory.

31. Write a computer program that will calculate the shear and moment at any point along the length of a simply supported beam subjected to a uniformly distributed load. User input is to be beam span length, intensity of distributed load, and the location (with respect to the left support) at which the shear and moment are to be calculated.

32. Write a program that will calculate the shear and moment at the tenth points (the increment along the span is to be one-tenth of the overall length) for a simply supported beam subjected to a full-span uniformly distributed load and a concentrated load at midspan. User input is to be the magnitude of the loads and the span length.

33. Viking Consultants wishes to generate a table of maximum moments for a range of simply supported beams subjected to uniformly distributed loads. The table is to have beam spans ranging from 10 ft to 50 ft (10-ft increments) on the horizontal axis and loads of 0.5 kips/ft to 5 kips/ft on the vertical axis. Write the program that will generate this table. (*Note:* no user input for this program.)

Supplemental Problems

34. Calculate the reactions for the simple beams in Fig. 13–74.

35. Calculate the reactions for the overhanging beams in Fig. 13–75.

36. For the beam in Fig. 13–76, calculate the shear and bending moment at points 6 ft and 16 ft from the left end using free-body diagrams.

37. Calculate the shear and bending moment at points 4 ft and 10 ft from the left end for beams (a) and (b) of Problem 35 using free-body diagrams.

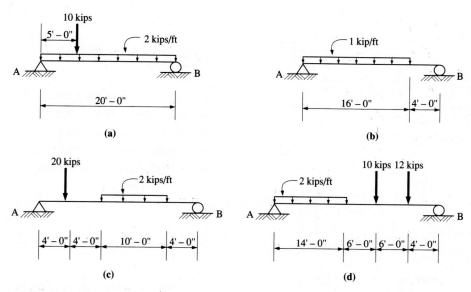

FIGURE 13–74   Problem 34.

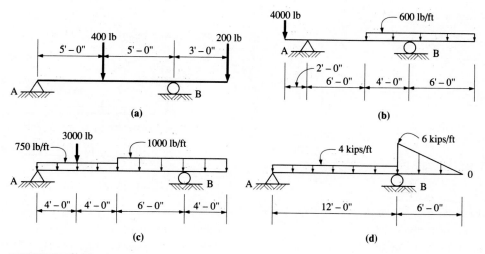

FIGURE 13–75   Problem 35.

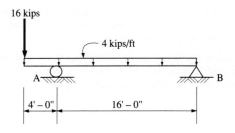

FIGURE 13–76   Problem 36.

For Problems 38 through 50, draw the load, shear, and moment diagrams showing ordinates at change-of-load points and indicating maximum shear and moment. Neglect the weight of the beam.

38. For the beam in Fig. 13–77.

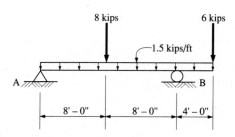

FIGURE 13–77   Problem 38.

39. For the beam in Fig. 13–78.

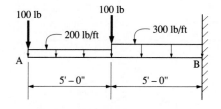

FIGURE 13–78   Problem 39.

40. For the beam in Fig. 13–79.
41. For beam (b) of Problem 34.
42. For beam (c) of Problem 34.

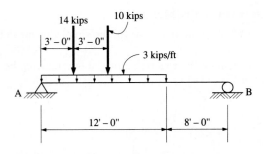

FIGURE 13–79   Problem 40.

43. For beam (d) of Problem 34.
44. For beam (b) of Problem 35.
45. For beam (c) of Problem 35.
46. For the beam of Problem 36.
47. For the beam in Fig. 13–80.

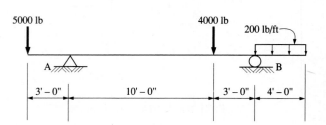

FIGURE 13–80   Problem 47.

48. For the beam in Fig. 13–81.

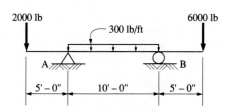

FIGURE 13–81   Problem 48.

49. For the beam in Fig. 13–82.
50. For the beam in Fig. 13–83.

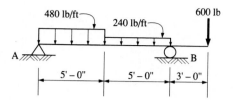

FIGURE 13–82   Problem 49.

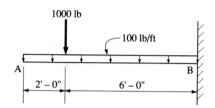

FIGURE 13–83   Problem 50.

51. A two-axle roller with axles 12 ft apart passes over a 30 ft simply supported beam bridge. The load is 20 tons on each axle. Compute the absolute maximum moment and shear. Indicate the position of the wheels and the location of the maximum values.

52. A moving load, as shown in Fig. 13–84, with wheels at fixed distances apart crosses a 40 ft simply supported beam bridge. Compute the absolute maximum moment and shear.

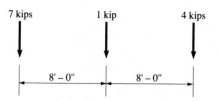

FIGURE 13–84   Problem 52.

53. The sloping roof joist shown in Fig. 13–85 supports the loads shown. Its left support provides vertical and horizontal reactions. Its right support provides a reaction perpendicular to the joist. Determine the reactions. Neglect the weight of the joist.

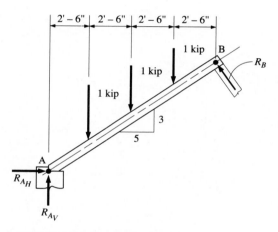

FIGURE 13–85   Problem 53.

14 Stresses in Beams

**14–1
TENSILE AND
COMPRESSIVE
STRESSES DUE
TO BENDING**

In Chapter 13 we saw that on every vertical section of a loaded horizontal beam a shear force and/or bending moment will occur, the magnitudes of which can be determined by calculation or from shear and moment diagrams. Therefore, at every vertical section an internal resisting shear and/or moment must be developed in order for a free body of any segment of the beam to be in equilibrium.

These internal resistances are functions of the shape and area of the cross section of the beam and can be expressed as internal shear stress and bending stress. The stresses may be thought of as representing the effect of the adjacent portion of the beam on the section under consideration.

For the design and analysis of a beam, it is necessary to calculate the induced stresses that occur at specific locations in order to compare these values with some allowable stress for the material used. Since the bending moment is generally the basis for beam design, our discussion at first will deal exclusively with the bending stresses developed.

Consider the straight horizontal beam of rectangular cross section shown in Fig. 14–1, which is subjected to equal vertical loads P. The beam is simply supported and will bend (or deform) as shown by the dashed line. The shear and moment diagrams for the beam are also shown. Assume the beam to be symmetrical with respect to the X–X and Y–Y axes as shown. The loads are applied in the plane of the Y–Y axis.

The intersection of the two axes represents the centroid of the cross section. Therefore, axis X–X may be termed a centroidal axis. In addition, assume the beam to be homogeneous, of a material that obeys Hooke's law, and with a modulus of elasticity of equal value in both tension and compression.

We will consider a segment of the beam between the two equal loads where no shear and, therefore, no shear stresses exist. The straight unloaded condition and the bent loaded condition are shown in Fig. 14–2. Initially, in the unloaded beam segment, line segments AB, JF, and CD are of equal lengths.

As the beam is loaded and bends, segment AB shortens to A′B′ and segment CD lengthens to C′D′. The top of the beam is in compression and the bottom is in tension. Points J and F remain the same distance apart in both the loaded or unloaded condition, indicating no shortening or elongating and, therefore, zero compression and tension. The plane in the member

FIGURE 14–1 Load, shear, and moment diagrams.

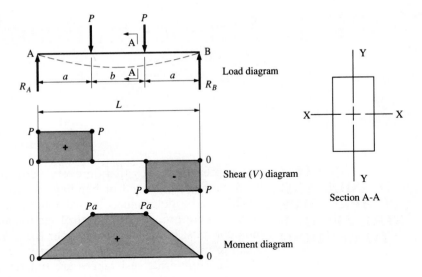

Load diagram

Shear (V) diagram

Moment diagram

Section A-A

FIGURE 14–2 Beam subject to pure bending.

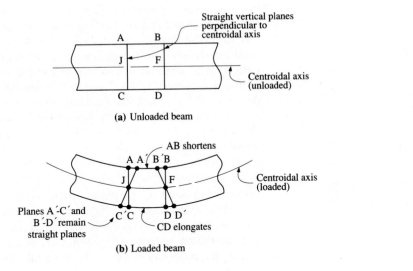

on which the zero tension or compression occurs is called the *neutral plane*, and the intersection of the neutral plane with a cross-sectional plane is called the *neutral axis*. In a homogeneous member, the neutral plane passes through the centroid of any cross section and becomes a centroidal axis of the cross section. Hence, planes A–C and B–D, which were originally straight and vertical in the unloaded beam, have rotated about their intersection with the neutral plane and have become planes A′–C′ and B′–D′ in the loaded beam. Note that the planes remain straight, even if not vertical. Many experiments and much research have shown that a plane section before bending remains a plane section after bending.

FIGURE 14–3   Stress and strain distribution.

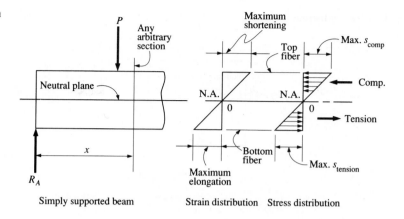

Simply supported beam   Strain distribution   Stress distribution

The dimensional changes between the original unloaded plane and the rotated plane may be observed in Fig. 14–3. The zero change at the neutral axis and the maximum change at the outer fibers in combination with the straight planes create a triangular strain distribution, which, in turn, indicates that the change in length of any fiber is proportional to the distance of the fiber from the neutral axis.

Assuming that the stress in any fiber does not exceed the proportional limit of the material, it follows from Hooke's law that the stress in any fiber at a given section is proportional to the distance from the neutral axis to that fiber. Therefore, the stress distribution, like the strain distribution, is triangular in shape. The stress varies from zero at the neutral axis to a maximum compressive stress at the top outer fiber and a maximum tensile stress at the bottom outer fiber.

14–2
THE FLEXURE
FORMULA

For the purpose of analysis and design of beams, it is necessary to work with the relationship between bending stresses, bending moment, and the geometric properties of a cross section. Whether we are dealing with analysis, in which, perhaps, a stress is to be determined, or with a design, in which an allowable stress is used as a factor in the selection of a member, the same basic relationship, called the *flexure formula*, will apply. The derivation and application of the flexure formula requires that we know the location of the neutral (or centroidal) axis. In a symmetrical member, this is easily obtained by inspection. For unsymmetrical members, the location of the centroidal axis must be calculated in the manner described in Chapter 7.

Figure 14–4(a) shows the side view of a small part of a simply supported beam with a typical stress distribution at some arbitrary location. Figure 14–4(b) shows the cross section of the beam. The beam is symmetrical with respect to the X–X and Y–Y axes. The section shown is rectangular, but the following discussion is valid for a cross section of any shape

FIGURE 14–4 Flexure formula derivation.

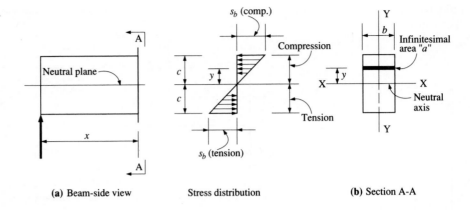

(a) Beam-side view   Stress distribution   (b) Section A-A

having a Y–Y axis of symmetry where the loading is applied in the plane of the Y–Y axis. The line XX is the neutral axis of the cross section. The width of the beam is denoted b. The distance from the neutral axis to an infinitesimal area a is denoted y. The distance from the neutral axis to the outer fiber of the cross section is denoted c.

The bending stress (either tension or compression) that develops at the outer fiber (a distance c from the neutral axis) will be referred to, for now, as $s_{b(max)}$. This maximum bending stress at the outer fiber is the bending stress that is usually of greatest importance. However, we can also calculate, by proportion, the bending stress that develops at a distance y from the neutral axis. For now, we will refer to this lesser stress as s_b. It can be expressed as

$$s_b = \frac{s_{b(max)}(y)}{c}$$

Recalling that force is equal to stress times area, we write the force developed on infinitesimal area a as

$$\frac{s_{b(max)}(y)}{c}(a)$$

The moment of the preceding force with respect to axis X–X can be calculated as

$$\frac{s_{b(max)}(y^2)}{c}(a)$$

Finally, the total moment, with respect to axis X–X, of all the internal forces acting on all the infinitesimal areas can be written

$$\frac{s_{b(max)}}{c}\sum y^2(a)$$

As shown in Chapter 8, the mathematical quantity $(\sum y^2 a)$ is the moment of inertia of a cross section about its X–X axis and is represented by the symbol I. Therefore, the expression for the total moment may be written as

$$\frac{s_{b(\text{max})}}{c} I$$

Since this total internal moment holds in equilibrium the moment due to the external loads M, it is sometimes called an *internal resisting moment*, and can be expressed as

$$M = \frac{s_{b(\text{max})}I}{c} \qquad \textbf{(14–1)}$$

where M = the bending moment due to external loads, or the internal resisting moment (in.-lb, ft-kips) (N·m)

$s_{b(\text{max})}$ = the bending stress developed at the outer fiber (psi, ksi) (Pa)

I = the moment of inertia about the neutral axis (in.^{4}) (m^4)

c = the distance from the neutral axis to the outer fiber (in.) (m)

Rewriting this expression and solving for the stress,

$$s_{b(\text{max})} = \frac{Mc}{I} \qquad \textbf{(14–2)}$$

where all terms are as previously defined.

Since stresses are proportional to distance from the neutral axis, we can also write the expression for bending stress developed at any distance y from the neutral axis:

$$s_b = \frac{My}{I} \qquad \textbf{(14–3)}$$

where s_b in this case will be less than the maximum bending stress that occurs at the outer fiber. Note that substitution of c for y in Eq. (14–3) results in Eq. (14–2). Since, for all practical purposes, it is the maximum bending stress that is of importance, we will omit the "(max)" from $s_{b(\text{max})}$, with the understanding that it is the maximum bending stress we are working with (unless otherwise noted).

We can also rewrite Eq. (14–1) to find the maximum resisting moment, or allowable moment, for a cross section. To use this expression, the allowable bending stress must be known:

$$M_R = \frac{s_{b(\text{all})}I}{c} \qquad \textbf{(14–4)}$$

where M_R = the allowable moment (in.-lb, ft-kips) (N·m)

$s_{b(\text{all})}$ = the allowable bending stress (psi, ksi) (Pa)

and I and c are as previously defined.

In these various forms of the flexure formula, note that the moment of inertia I and the distance c are both functions of the size and shape of the beam cross section. They are both geometric properties of the cross section and do not depend on the material or span length of the beam or on the type of loading on the beam. The quantity I/c, therefore, is also a geometric property. I/c is called the *section modulus* and is generally represented by the symbol S. The section modulus has units of in.^{3} and can be calculated using the moment of inertia from Table 8–1 divided by the distance c.

The flexure formula, then, can be rewritten and used in the following forms depending on whether the problem is one of analysis or design. For analysis problems,

$$s_b = \frac{M}{S} \qquad \text{(14–5)}$$

or

$$M_R = s_{b(\text{all})} S \qquad \text{(14–6)}$$

For design problems, the most convenient and the most used form is

$$\text{Required } S = \frac{M}{s_{b(\text{all})}} \qquad \text{(14–7)}$$

The values of the section modulus for standard rolled structural shapes and other standard shapes, both metallic and nonmetallic, can be found in various publications of the American Institute of Steel Construction, the National Forest Products Association, and other such organizations. Some of this material is provided in the appendices of this book.

As a physical analogy, the section modulus can be considered a measure of the comparative strength of beams. All other things being equal, if the section modulus of the cross section of beam A is twice as great as that of beam B of the same material, beam A will have double the bending strength.

Note that if the cross section of a beam is not symmetrical with respect to its X–X axis (the axis of bending), it will have two section modulus values of different magnitudes since the c values for the top and bottom fibers will be different. Also, the centroidal axis will not be located at midheight of the cross section.

Since beams are such common members in structures, and since the use of the flexure formula is basic to beam analysis and design, the limitations of the formula should be recognized. Its use is valid only under certain conditions, which may be briefly itemized as follows:

1. The beam must be straight before loading.
2. The beam must be homogeneous, obey Hooke's law, and have equal moduli of elasticity in both tension and compression.
3. The loads and reactions must lie in a vertical plane of symmetry perpendicular to the longitudinal axis of the beam.
4. The beam must be of uniform cross section.
5. The maximum bending stress must not exceed the proportional limit.
6. The beam must have adequate lateral buckling resistance.

7. All component parts of the beam must have adequate localized buckling resistance.
8. The beam must be relatively long in proportion to its depth.
9. The cross section must not be disproportionately wide.
10. The dimensional changes must not be appreciably affected by shear strains that are also present.

Despite the many limiting factors, the flexure formula may be applied quite satisfactorily in the design and analysis of most beams normally encountered.

14–3
COMPUTATION OF
BENDING STRESSES

In the analysis of beams, one type of problem involves calculating maximum bending stress. This will occur at the outer fibers at the section of the beam where the bending moment is a maximum. In computing the maximum bending stress, the location and the magnitude of the maximum bending moment must be calculated first. The value of the section modulus (or the moment of inertia and the c distance) must then be calculated or obtained from standard tables of properties of sections. By substituting these values in Eq. (14–5) (or Eq. [14–2]), we obtain the maximum bending stress.

The actual use of the flexure formula is straightforward, although the units must be carefully considered. In order that the values of the bending stress be expressed in psi or ksi, the bending moment must be used in units of inch-pounds or inch-kips. This is accomplished by converting the length units of moment to inches.

The following examples illustrate the use of the flexure formula. They are not to be interpreted as complete beam-analysis problems. Various types of more comprehensive problems will follow in Section 14–7.

☐ **EXAMPLE 14–1**

A nominal 8 in. by 12 in. timber member, shown in Fig. 14–5, is used as a beam that is subjected to a vertical loading. For maximum bending strength, the beam is oriented so that the 12 in. dimension is vertical. Calculate the section modulus of the beam with respect to the axis of bending (the X–X axis). Use dressed dimensions.

FIGURE 14–5   Beam cross section.

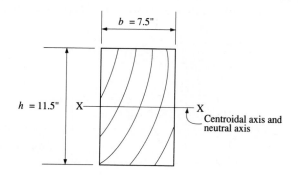

Solution   The dressed dimensions for the timber member are $7\frac{1}{2}$ in. by $11\frac{1}{2}$ in. For solid rectangular shapes, an expression for section modulus can be derived as follows:

$$S = \frac{I}{c} = \frac{(bh^3/12)}{h/2} = \frac{bh^2}{6}$$

This is a convenient expression to use in dealing with solid homogeneous rectangular shapes, such as timber members.

Substituting numerical values,

$$S_x = \frac{bh^2}{6} = \frac{7.5(11.5^2)}{6} = 165.3 \text{ in.}^3$$

This value may be verified in Appendix E.

☐ **EXAMPLE 14–2**   If the timber member of Example 14–1 were used as a simply supported beam on a 16 ft span length carrying a uniformly distributed load of 400 lb/ft, calculate the maximum induced bending stress. Neglect the weight of the beam.

Solution   The maximum stress can be calculated using Eq. (14–5). First calculate the maximum bending moment. From Appendix H,

$$M = \frac{wL^2}{8} = \frac{400(16)^2}{8} = 12,800 \text{ ft-lb}$$

Using the section modulus for the cross section from the previous example, substitute into Eq. (14–5) to find the maximum bending stress:

$$s_b = \frac{M}{S} = \frac{12,800(12)}{165.3} = 929.2 \text{ psi}$$

☐ **EXAMPLE 14–3**   Assume that, due to a construction error, the beam of the previous two examples was placed incorrectly in the field, and the small dimension ($7\frac{1}{2}$ in. dressed) was oriented vertically. For the same span length and loading, calculate the new maximum bending stress that would be developed.

Solution   The maximum bending moment remains the same at 12,800 ft-lb. The orientation of the cross section is shown in Fig. 14–6. Note that the horizontal axis is designated as

FIGURE 14–6   Beam cross section.

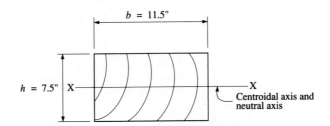

axis X–X. Calculating the section modulus,

$$S_x = \frac{bh^2}{6} = \frac{11.5(7.5)^2}{6} = 107.8 \text{ in.}^3$$

Calculating the maximum bending stress,

$$s_b = \frac{M}{S} = \frac{12,800(12)}{107.8} = 1425 \text{ psi}$$

The results of Examples 14–2 and 14–3 indicate that the orientation of a beam cross section is an important strength consideration. For maximum bending resistance, the bending axis (perpendicular to the applied loads) should be the axis about which the section modulus is the largest value.

☐ **EXAMPLE 14–4**   A W30 × 99 hot-rolled structural steel wide-flange shape is used as a simply supported beam on a span length of 32 ft. The allowable bending stress for this beam is 24.0 ksi. The beam supports a superimposed uniformly distributed load of 4.0 kips/ft in addition to its own weight. Calculate the maximum bending stress. Is the beam satisfactory?

Solution   Compute the maximum bending stress for comparison with the allowable bending stress. In order for the beam to be satisfactory for moment, the computed bending stress must be less than the allowable bending stress.

First calculate the maximum bending moment. The total load is the sum of the superimposed load (4.0 kips/ft) and the weight of the beam (0.099 kips/ft). From Appendix H,

$$M = \frac{wL^2}{8} = \frac{4.099(32)^2}{8} = 524.7 \text{ ft-kips}$$

The section modulus can be obtained from Appendix A, which is a partial tabulation of the AISC tables of dimensions and properties of structural shapes. Assume that the shape is oriented so that bending is about the strong axis. Hence, $S = 269$ in.^{3}.

Calculating the maximum bending stress,

$$s_b = \frac{M}{S} = \frac{524.7(12)}{269} = 23.4 \text{ ksi}$$

Since 23.4 ksi < 24.0 ksi, the beam is satisfactory.

☐ **EXAMPLE 14–5**   A steel beam is fabricated in the shape of a "T" by welding a 12 in. by $\frac{1}{2}$ in. steel plate to a 10 in. by $\frac{1}{2}$ in. steel plate, as shown in Fig. 14–7. A beam of this type is often used in masonry building construction as a lintel, which is a bending member that spans over wall openings such as doors and windows.

Calculate the maximum bending stresses in tension and compression if the beam supports a wall load of 700 lb/ft (which includes its own weight) and is simply supported with a span length of 16 ft.

FIGURE 14–7   Lintel beam.

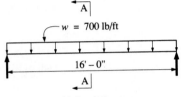

(a) Load diagram

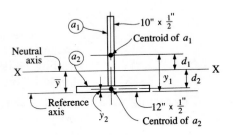

(b) Section A-A

Solution   This is a composite member made up of two simple geometric areas, a_1 and a_2. Since this member is not symmetrical with respect to the X–X bending axis, the neutral axis must first be located. In order to determine the distance $\bar{y}$, a reference axis is first established at the lower side of the cross section as shown in Fig. 14–7(b). Then, utilizing the principal of moments, the location of the neutral axis will be found using Eq. (7–4):

$$\bar{y} = \frac{\sum ay}{A}$$

where A is the total area, or the sum of the component areas $\sum a$. From Fig. 14–7,

$$a_1 = 10(0.5) = 5 \text{ in.}^2$$
$$a_2 = 12(0.5) = 6 \text{ in.}^2$$
$$A = a_1 + a_2 = 11 \text{ in.}^2$$
$$y_1 = 5.0 + 0.5 = 5.5 \text{ in.}$$
$$y_2 = 0.25 \text{ in.}$$

Substituting into Eq. (7–4),

$$\bar{y} = \frac{\sum ay}{A} = \frac{a_1 y_1 + a_2 y_2}{A} = \frac{5(5.5) + 6(0.25)}{11.0} = 2.64 \text{ in.}$$

In the flexure formula, this value represents the c distance to the bottom fiber. The c distance to the top fiber may be calculated from

$$c_{(\text{top})} = 10.5 - 2.64 = 7.86 \text{ in.}$$

Next calculate the moment of inertia about the neutral (centroidal) axis X–X using Eq. (8–4):

$$I_x = \sum (I_o + ad^2) = (I_o + ad^2)_1 + (I_o + ad^2)_2$$

where

$$d_1 = 7.86 - 5.0 = 2.86 \text{ in.}$$
$$d_2 = 2.64 - 0.25 = 2.39 \text{ in.}$$

Therefore,

$$I_x = \left[\frac{0.5(10)^3}{12} + 5(2.86)^2 \right] + \left[\frac{12(0.5)^3}{12} + 6(2.39)^2 \right]$$
$$= 82.6 + 34.4$$
$$= 117.0 \text{ in.}^4$$

The maximum bending moment from Appendix H is

$$M = \frac{wL^2}{8} = \frac{700(16)^2}{8} = 22{,}400 \text{ ft-lb}$$

Next, calculate the section moduli values. Since the member is not symmetrical, the section modulus for the top is different from that for the bottom:

$$S_{(top)} = \frac{I}{c_{(top)}} = \frac{117.0}{7.86} = 14.9 \text{ in.}^3$$

$$S_{(bot)} = \frac{I}{c_{(bot)}} = \frac{117.0}{2.64} = 44.3 \text{ in.}^3$$

Now the maximum bending stress can be calculated using Eq. (14–5):

$$s_{b(bot)} = \frac{M}{S_{(bot)}} = \frac{22{,}400(12)}{44.3} = 6070 \text{ psi}$$

$$s_{b(top)} = \frac{M}{S_{(top)}} = \frac{22{,}400(12)}{14.9} = 18{,}040 \text{ psi}$$

Assuming that this beam has an allowable bending stress of 22,000 psi, note that if the stress in the bottom were to reach the maximum allowable value, the stress in the top would be greater than that allowed and, therefore, would not be acceptable. Thus, the section modulus with respect to the top controls the load-carrying capacity of the member.

☐ **EXAMPLE 14–6**   Using a steel with an allowable bending stress of 24,000 psi, calculate the maximum allowable bending moment to which a W18 × 60 hot-rolled structural steel wide-flange section may be subjected. Assume the following conditions: (a) the bending moment is applied about the X–X axis, (b) the bending moment is applied about the Y–Y axis. (See Fig. 14–8.)

Solution   The section modulus values can be obtained from Appendix A: $S_x = 108$ in.^{3} and $S_y = 13.3$ in.^{3}. Then the allowable bending moment can be calculated using Eq. (14–6):

$$M_{R_x} = s_{b(all)} S_x = 24{,}000(108) = 2{,}592{,}000 \text{ in.-lb}$$

To reduce this to a more manageable value, convert to ft-kips:

$$\frac{2{,}592{,}000 \text{ in.-lb}}{(12 \text{ in./ft})(1000 \text{ lb/kip})} = 216 \text{ ft-kips}$$

FIGURE 14–8   Beam cross section.

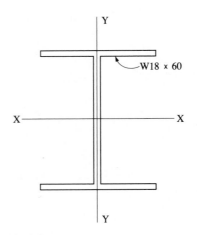

W18 × 60

Also,

$$M_{R_y} = s_{b(\text{all})} S_y = \frac{24{,}000(13.3)}{12(1000)} = 26.6 \text{ ft-kips}$$

14–4
SHEAR STRESSES

As discussed in Section 14–1, internal resisting shear stresses are developed on every vertical section of a loaded horizontal beam where the vertical shear force has a numerical value other than zero. It is the summation of these shear stresses that provides the internal resisting shear force which must be equal to the external vertical shear force in order for any free body of a segment of the beam to be in equilibrium. Chapter 13 dealt with the determination of the vertical shear force at any location along the length of a beam due to the external loads. This section deals with the shear stresses resulting from the vertical shear forces.

The distribution of the shear stress developed over the beam cross section is appreciably different from the bending stress distribution. The shear stress is zero at those points on the cross section where the bending stress is a maximum. The location of the point of maximum shear stress is almost always at the neutral axis. It should be noted that this maximum value can occur on horizontal planes other than the neutral axis for odd-shaped, impractical, and uneconomical sections that are seldom encountered.

It was shown in Section 11–6 that if at any point in a stressed member a shear stress exists on a plane, at the same time there must exist a shear stress of equal intensity on a perpendicular plane. Therefore, in a loaded horizontal beam, both vertical and horizontal shear stresses are developed at any given point.

The existence of the horizontal shear stresses is best illustrated by considering several planks of identical cross section. If the planks are stacked together to form a simply supported beam, as shown in Fig. 14–9(a),

FIGURE 14–9   Horizontal shear in beams.

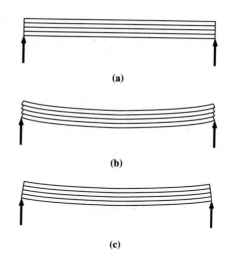

(a)

(b)

(c)

and a vertical load is applied, the planks will bend, as shown in Fig. 14–9(b). Observe that each plank is bending independently. The top fibers of each plank are shortened and the bottom fibers are lengthened. Assuming a frictionless contact surface, the planks tend to slide over each other. This is evident if one notes that the ends of the planks no longer lie in a straight line or plane. The top of one plank will slide inward relative to the bottom of the adjacent plank. If an adhesive is applied between the planks to bond them together, the loaded beam will then take the form shown in Fig. 4–9(c). The sliding that occurred between the planks is now resisted by the adhesive, and a shear stress is developed in the adhesive.

If the beam is a one-piece solid member instead of planks bonded together, the material of which the beam is made will resist the tendency for horizontal sliding of layers. Hence, a horizontal shear stress is developed on any horizontal plane of the beam cross section.

14–5
THE GENERAL
SHEAR FORMULA

Whereas the magnitude of the maximum bending stress can be obtained through the use of the flexure formula, the magnitude of the maximum shear stress can be obtained through the use of the general shear formula. This formula will yield the shear stress on any horizontal plane. As previously shown in Chapter 11, the horizontal shear stress equals the vertical shear stress at any point in a beam.

For the derivation of the general shear formula, consider the load, shear, and moment diagrams in Fig. 14–10(a) for a loaded simply supported beam. The free-body diagram for an element of this beam bounded by planes D, E, and F is shown in Fig. 14–10(b). Planes D and E are initially an infinitesimal distance x apart, and plane F lies a distance y_1 above the neutral axis. The object of the derivation is to establish an expression for the horizontal shear stress s_s on the bottom surface of the element.

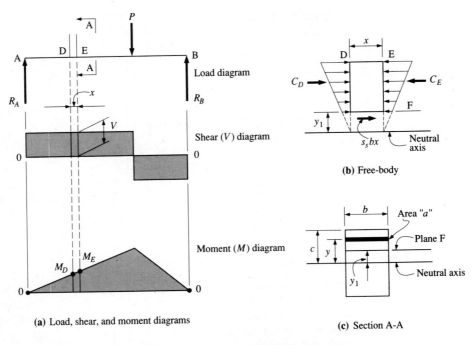

(a) Load, shear, and moment diagrams

(b) Free-body

(c) Section A-A

FIGURE 14–10   Derivation of horizontal shear formula.

Let C_D and C_E represent the resultants of the compressive bending stresses on planes D and E. Since the bending moment at E exceeds that at D, C_E is greater than C_D. Since the element must be in equilibrium and the sum of the horizontal forces must equal zero, we must conclude that a horizontal shear force is present on plane F. The horizontal shear force is the product of a horizontal shear stress and the area of the bottom surface of the element and may be expressed as $s_s bx$.

With reference to Fig. 14–10(c), consider an infinitesimal rectangular area a that lies parallel to the neutral axis on plane E of the element. The centroid of area a lies a distance y from the neutral axis. The compressive bending stress on this area, using the flexure formula, is

$$s_b = M_E \frac{y}{I}$$

and the force acting on area a is

$$s_b(a) = M_E \frac{y}{I}(a)$$

The total force acting on the total area above plane F is then equal to C_E, where

$$C_E = \Sigma \frac{M_E}{I}(ya)$$

which can be rewritten as

$$C_E = \frac{M_E}{I} \Sigma ya$$

Note in the preceding that M_E and I are fixed quantities.

The quantity Σya represents the moment of the total area of the element (above plane F) on plane E about the neutral axis. This moment is generally designated by the symbol Q and is termed the *statical moment of the area*. Therefore,

$$C_E = \frac{M_E Q}{I}$$

In the same way, the resultant force C_D acting on plane D of the element is found to be

$$C_D = \frac{M_D Q}{I}$$

The difference between the two forces, then, is

$$C_E - C_D = \frac{M_E Q}{I} - \frac{M_D Q}{I} = (M_E - M_D)\frac{Q}{I}$$

Since the element is in equilibrium, this value must equal the horizontal shear force on the bottom surface of the element (plane F). This force was noted earlier as $s_s bx$. Therefore,

$$s_s bx = (M_E - M_D)\frac{Q}{I}$$

from which

$$s_s = \frac{(M_E - M_D)Q}{Ibx}$$

As we demonstrated in Chapter 13, the change in bending moment between any two points of a beam is equal to the area of the shear diagram between the same two points. With reference to Fig. 14–10(a), the area of the shear diagram between planes D and E is equal to Vx. Substituting this for $M_E - M_D$ in the previous equation,

$$s_s = \frac{VxQ}{Ibx} = \frac{VQ}{Ib} \qquad \textbf{(14–8)}$$

where s_s = the horizontal (and vertical) computed shear stress on any given plane of a given cross section of the beam (psi, ksi) (Pa)

V = the computed vertical shear force at the given cross section (lb, kips) (N)

Q = the statical moment about the neutral axis of the cross-sectional area between the horizontal plane where the shear stress is to be calculated and the top (or bottom) of the beam (in.^{3}) (m^3)

I = the moment of inertia of the entire cross section with respect to the neutral axis (the same I used in flexure formula calculations) (in.^{4}) (m^4)

b = the width of the cross-section in the horizontal plane where the shear stress is being calculated (in.) (m)

The general shear formula can be rewritten in a form useful for the calculation of an allowable shear force (or shear capacity) for a bending member. Denoting the shear capacity as V_R, Eq. (14–8) yields

$$V_R = \frac{s_{s(all)}Ib}{Q}$$ (14–9)

where V_R = the allowable shear force (or shear capacity) at a given cross section

$s_{s(all)}$ = the allowable shear stress

and the other terms are as previously defined.

14–6
SHEAR STRESSES
IN STRUCTURAL
MEMBERS

In the analysis and design of beams in structures, flexure (bending moment) is generally the most critical factor. That is, bending stresses will reach allowable limits first. In some beams, however, due to very heavy loads and/or very short spans, shear may become more critical. The following examples illustrate the application of the general shear formula.

☐ **EXAMPLE 14–7**

The rough, solid rectangular timber beam in Fig. 14–11 is 8 in. wide and 12 in. deep. The beam is subjected to a vertical load, inducing a maximum shear V of 7000 lb. (a) Calculate the maximum shear stress at the neutral axis. (b) Calculate the shear stress at 2 in. above and below the neutral axis. (c) Calculate the shear stress at 4 in. above and below the neutral axis. (d) Plot these stresses showing the distribution of the horizontal shear stress.

Solution

Use the general shear formula, Eq. (14–8):

$$s_s = \frac{VQ}{Ib}$$

FIGURE 14–11   Beam cross section.

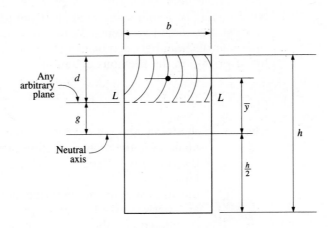

Two of the terms, V and b, are given. These two terms will remain constant in this example. Next calculate the moment of inertia with respect to the neutral axis:

$$I = \frac{bh^3}{12} = \frac{8(12)^3}{12} = 1151 \text{ in.}^4$$

The only other value that must be calculated prior to computing s_s is the statical moment of area, Q.

(a) With reference to Fig. 14–11, assume that plane L–L lies at the neutral axis ($g = 0$). Then Q is equal to the statical moment of the area above the neutral axis, with respect to the neutral axis. The area A above the neutral axis is

$$A = bd = 8(6) = 48 \text{ in.}^2$$

The distance from the centroid of area A to the neutral axis is

$$\bar{y} = \frac{h}{2} - \frac{d}{2} = 6 - 3 = 3 \text{ in.}$$

Therefore,

$$Q = A\bar{y} = 48(3) = 144 \text{ in.}^3$$

and, at the neutral axis,

$$s_s = \frac{VQ}{Ib} = \frac{7000(144)}{1152(8)} = 109.4 \text{ psi}$$

(b) With reference to Fig. 14–11, assume that plane L–L lies 2 in. above the neutral axis ($g = 2$ in.). The area above the plane is

$$A = bd = 8(4) = 32 \text{ in.}^2$$

The distance from the centroid of this area to the neutral axis is

$$\bar{y} = \frac{h}{2} - \frac{d}{2} = 6 - \frac{4}{2} = 4 \text{ in.}$$

Calculating Q,

$$Q = A\bar{y} = 32(4) = 128 \text{ in.}^3$$

and

$$s_s = \frac{VQ}{Ib} = \frac{7000(128)}{1152(8)} = 97.2 \text{ psi}$$

Placing the plane on which to calculate the horizontal shear stress at 2 in. below the neutral axis, we see in Fig. 14–12 that g remains at 2 in., d is 8 in., and that the area above plane L–L is

$$A = bd = 8(8) = 64 \text{ in.}^2$$

The distance from the centroid of the area to the neutral axis is calculated from

$$\bar{y} = \frac{d}{2} - g = \frac{8}{2} - 2 = 2 \text{ in.}$$

FIGURE 14–12 Beam cross section.

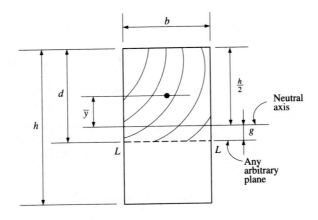

Therefore,

$$Q = A\bar{y} = 64(2) = 128 \text{ in.}^3$$

and

$$s_s = \frac{VQ}{Ib} = \frac{7000(128)}{1152(8)} = 97.2 \text{ psi}$$

As can be observed, the horizontal shear stress is the same whether 2 in. above or 2 in. below the neutral axis. This is due to symmetry with respect to the neutral axis. This situation will exist for all corresponding planes above or below the neutral axis where symmetry with respect to the neutral axis exists.

(c) Placing the plane on which to calculate the horizontal shear stress at 4 in. above (or below) the neutral axis, calculate as follows (using plane L–L above the neutral axis with $g = 4$ in., and referring to Fig. 14–11):

$$d = \frac{h}{2} - g = 6 - 4 = 2 \text{ in.}$$

$$A = bd = 8(2) = 16 \text{ in.}^2$$

$$\bar{y} = \frac{h}{2} - \frac{d}{2} = 6 - 1 = 5 \text{ in.}$$

$$Q = A\bar{y} = 16(5) = 80 \text{ in.}^3$$

$$s_s = \frac{VQ}{Ib} = \frac{7000(80)}{1152(8)} = 60.8 \text{ psi}$$

Note that at the outer fibers, Q would be zero. Therefore, the shear stress would also be zero.

(d) The diagram of the shear stress distribution is shown in Fig. 14–13. Note that for a homogeneous solid rectangular cross section the maximum horizontal shear stress occurs at the neutral axis and the minimum, which is a zero value, occurs at the outer fibers. A closer inspection of how shear stress varies with distance from the neutral axis will show that the curve is parabolic.

FIGURE 14–13 Shear stress distribution in a beam.

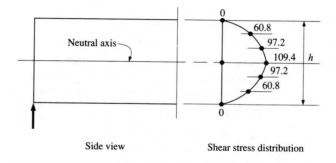

Side view   Shear stress distribution

Generally, only the maximum value of the shear stress is of any interest, since this is the value that must be compared with some allowable shear stress. It is practical, then, to develop a type of dedicated equation to apply in special cases (see Section 1–5). One such case is that of the preceding example: a solid, homogeneous rectangular cross section in which the maximum horizontal shear stress is desired. We derive the equation as follows, with reference to Fig. 14–14:

$$Q = (b)\left(\frac{h}{2}\right)\left(\frac{h}{4}\right) = \frac{bh^2}{8}$$

$$I = \frac{bh^3}{12}$$

$$s_s = \frac{VQ}{Ib} = \frac{V(bh^2/8)}{(bh^3/12)(b)} = \frac{Vbh^2}{8} \times \frac{12}{b^2h^3}$$

from which

$$s_s = \frac{12V}{8bh} = 1.5\frac{V}{A} \qquad\qquad \textbf{(14–10)}$$

FIGURE 14–14 Beam cross section.

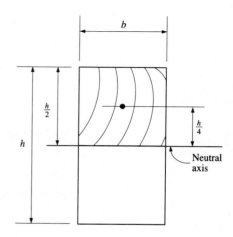

Thus, the maximum horizontal shear stress for a homogeneous solid rectangular beam is 1.5 times the average shear stress V/A, where A represents the *total* cross-sectional area.

☐ **EXAMPLE 14–8**   The bending member in Fig. 14–15 is built up of three steel plates welded together to act as a single unit. It is generally called a *plate girder*. The girder is subjected to a vertical loading, which induces a maximum shear V of 300 kips. (a) Calculate the maximum shear stress in the plane of the neutral axis. (b) Calculate the shear stress at the junction of the flange and the web (plane E–E). (c) Plot the distribution of the shear stress for the entire cross section.

FIGURE 14–15   Girder cross section.

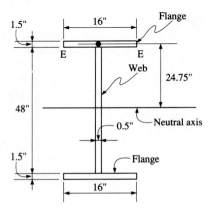

Solution   Using the general shear formula, V and I will have the same numerical value when calculating the shear stress on any horizontal plane. V and b are given. The moment of inertia with respect to the neutral axis (I_{NA}) may be computed as follows:

For the bottom or top flange (these are symmetrical with respect to the neutral axis),

$$I_{NA} = I_o + Ad^2 = \frac{bh^3}{12} + Ad^2$$

$$= \frac{16(1.5)^3}{12} + 16(1.5)(24.75)^2$$

$$= 14{,}706 \text{ in.}^4$$

For the web,

$$I_{NA} = \frac{bh^3}{12} = \frac{0.5(48)^3}{12} = 4608 \text{ in.}^4$$

Therefore, the total moment of inertia about the neutral axis is

$$\text{Total } I_{NA} = 2(14{,}706) + 4608 = 34{,}020 \text{ in.}^4$$

FIGURE 14–16 Beam cross section above the neutral axis.

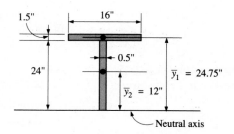

(a) Calculating the horizontal shear stress at the neutral axis with reference to Fig. 14–16, and using subscripts f and w to designate flange and web, respectively,

$$Q_f = A_f \bar{y}_1 = 16(1.5)(24.75) = 594 \text{ in.}^3$$
$$Q_w = A_w \bar{y}_2 = 24(0.5)(12) = 144 \text{ in.}^3$$
$$\text{Total } Q = 594 + 144 = 738 \text{ in.}^3$$

from which

$$s_s = \frac{VQ}{Ib} = \frac{300(738)}{34{,}020(0.5)} = 13.02 \text{ ksi}$$

(b) Calculating the horizontal shear stress at the junction of the flange and the web, with reference to Fig. 14–17,

$$\text{Total } Q = Q_f = A_f \bar{y}_1 = 16(1.5)(24.75) = 594 \text{ in.}^3$$

Note that at the junction of the web and the flange the value of b may be either 16 in. or 0.5 in., depending on whether one is considering the stress at an infinitesimal distance above or below the actual junction. The magnitude of the horizontal shear stress at this location, therefore, undergoes an abrupt change. Where b is 16 in.,

$$s_s = \frac{VQ}{Ib} = \frac{300(594)}{34{,}020(16)} = 0.327 \text{ ksi}$$

Where b is 0.5 in.,

$$s_s = \frac{VQ}{Ib} = \frac{300(594)}{34{,}020(0.50)} = 10.48 \text{ ksi}$$

FIGURE 14–17 Beam cross section above the neutral axis.

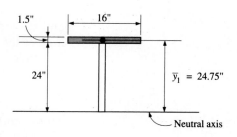

(c) The diagram of the distribution of the horizontal shear stresses is shown in Fig. 14–18.

FIGURE 14–18 Shear stress distribution in a plate girder.

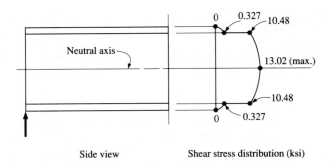

Side view   Shear stress distribution (ksi)

Even though the flange areas of the girder in Example 14–8 are quite large, the low shear stress values in the flange indicates that the flanges resist only a small portion of the total vertical shear force. It is the web that predominantly resists the shear in plate girders having the I-shape. This is also true for rolled structural steel shapes, such as wide-flange beams (W shapes), American standard I-beams (S shapes), and channels (C shapes).

It is estimated that the webs for the types of steel beams just mentioned resist 85 to 90 percent of the total vertical shear force. For this reason, design specifications (e.g., the American Institute of Steel Construction (AISC) Specification) allow the use of an "average web shear" approach for the determination of shear stress in rolled and fabricated shapes, rather than requiring the use of the general shear formula. The average web shear is calculated from

$$s_s = \frac{V}{dt_w} \qquad (14\text{–}11)$$

where s_s = the computed maximum shear stress (psi, ksi) (Pa)

$\quad\quad V$ = the vertical shear force at the section under consideration (lb, kips) (N)

$\quad\quad d$ = the *full* depth of the beam (in.) (m)

$\quad\quad t_w$ = the web thickness of the beam (in.) (m)

This method is approximate compared to the theoretically correct general shear formula and assumes that the shear is resisted by the rectangular area of the web extending the full depth of the beam. For the plate girder of Example 14–8,

$$s_s = \frac{V}{dt_w} = \frac{300}{51(0.5)} = 11.8 \text{ ksi}$$

This is almost 10 percent below the theoretically correct maximum value of 13.02 ksi. Therefore, the average web shear method results in a shear stress that is too low. This could be considered unsafe; however, allowable shear stresses (such as that of the AISC) are set intentionally low to account for the fact that the computed average shear stress will always be lower than the actual maximum shear stress.

Using the average web shear approach, we can also write an expression for the allowable shear V_R or the shear capacity of these cross sections. Rewriting Eq. (14–11),

$$V_R = s_{s(all)} dt_w \qquad (14–12)$$

This approximate approach should be used only for steel beams having symmetry with respect to the neutral axis, such as the types previously mentioned.

☐ **EXAMPLE 14–9**   A W16 × 100 (steel wide-flange beam) is subjected to a vertical shear of 80 kips. Calculate the maximum horizontal shear stress using (a) the general shear formula and (b) the average web shear approach.

Solution   The dimensions and properties of the W16 × 100 can be obtained from Appendix A. The necessary dimensions are shown in Fig. 14–19.

FIGURE 14–19   Beam cross section above the neutral axis.

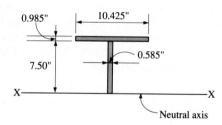

(a) Calculating Q at the neutral axis,

$$Q_f = 10.425(0.985)\left(7.50 + \frac{0.985}{2}\right) = 82.1 \text{ in.}^3$$

$$Q_w = 7.50(0.585)\left(\frac{7.50}{2}\right) = 16.50 \text{ in.}^3$$

$$\text{Total } Q = 82.1 + 16.5 = 98.6 \text{ in.}^3$$

from which

$$s_s = \frac{VQ}{Ib} = \frac{80(98.6)}{1490(0.585)} = 9.05 \text{ ksi}$$

(b) Using the average web shear approach,

$$d = 2(7.50 + 0.985) = 16.97 \text{ in.}$$

$$s_s = \frac{V}{dt_w} = \frac{80}{16.97(0.585)} = 8.06 \text{ ksi}$$

FIGURE 14–20   Circular cross section.

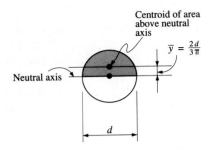

A comment is in order here concerning notation and substitution of numerical values. Note that the web thickness in structural steel members is generally denoted t_w and the flange width is denoted b_f. One of these two terms must be substituted for b in the denominator of the general shear formula, since b is defined as the width of the cross section at the level where the shear stress is being calculated.

For a member of circular cross section, such as a shaft used as a beam, the maximum shear stress also occurs at the neutral axis, despite the fact that all other horizontal planes are narrower in width (as may be observed in Fig. 14–20). To compute the maximum shear stress, the general shear formula may be used. Also, a simplified expression for the maximum shear stress for this cross section may be developed as follows. Referring to Table 8–1 for appropriate properties, and expressing all values in terms of the diameter d,

$$b = d$$

$$I_{NA} = \frac{\pi d^4}{64}$$

The area above the neutral axis is

$$\frac{\pi d^2}{8}$$

and the distance from the neutral axis to the centroid of the area above the neutral axis is

$$\bar{y} = \frac{2d}{3\pi}$$

Therefore,

$$Q = A\bar{y} = \frac{\pi d^2}{8}\left(\frac{2d}{3\pi}\right) = \frac{d^3}{12}$$

and

$$s_s = \frac{VQ}{Ib} = \frac{V(d^3/12)}{(\pi d^4/64)(d)} = \frac{16V}{12(\pi d^2/4)}$$

from which

$$s_s = \frac{4V}{3A} \qquad\qquad (14\text{–}13)$$

where A represents the total circular cross-sectional area. Thus, the maximum horizontal shear stress for a solid homogeneous beam of circular cross section is 4/3 times the average value.

14–7 BEAM ANALYSIS

The analysis problem is generally considered to be the investigation of a beam with a known cross section. Three common types of problems are

1. Computing the actual induced stresses (bending and shear) for a given beam, loading, and span length and comparing with allowable stresses
2. Computing the load-carrying capacity of a given beam based on given allowable stresses (bending and shear) and span length
3. Computing a maximum span length for a given beam, loading, and allowable stresses (bending and shear)

All of these problems are related. All make use of the flexure formula and the general shear formula. In addition, analysis-type problems may involve the connecting of component parts of a built-up member.

☐ **EXAMPLE 14–10**   A W21 × 73 steel wide-flange beam is to span 40 ft on simple supports, as shown in Fig. 14–21. The load shown is a superimposed load, meaning that it does not include the weight of the beam. The type of steel used has an allowable bending stress of 24,000 psi and an allowable shear stress of 14,500 psi. Determine whether the beam is adequate (a) by comparing the actual bending and shear stresses with the allowable bending and shear stresses and (b) by comparing the actual induced moment and shear with allowable values of moment and shear. Use the average web shear approach.

FIGURE 14–21 Load diagram.

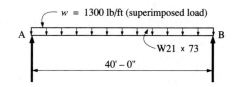

Solution   (a) The flexure formula is used to determine the actual bending stress. From Appendix A (properties of wide-flange sections), $S_x = 151$ in.^{3}. The bending moment can be determined by shear and moment diagrams or by formula. The total uniform load, including the weight of the beam, is

$$1300 + 73 = 1373 \text{ lb/ft}$$

Calculating the maximum bending moment (see Appendix H),

$$M = \frac{wl^2}{8} = \frac{1373(40)^2}{8} = 274{,}600 \text{ ft-lb}$$

Obtain the maximum computed bending stress:

$$s_b = \frac{M}{S} = \frac{274,600(12)}{151} = 21,823 \text{ psi}$$

Therefore,

$$s_b < s_{b(\text{all})} \qquad \textbf{OK}$$

The actual shear stress can be determined using the average web shear approach. From Appendix A, the following are obtained: $d = 21.24$ in. and $t_w = 0.455$ in. Calculating the maximum induced shear,

$$V = \frac{wL}{2} = \frac{1373(40)}{2} = 27,460 \text{ lb}$$

Obtain the actual shear stress:

$$s_s = \frac{V}{dt_w} = \frac{27,460}{(21.24)(0.455)} = 2841 \text{ psi}$$

Therefore,

$$s_s < s_{s(\text{all})} \qquad \textbf{OK}$$

(b) The applied or actual induced bending moment has been calculated to be 274,600 ft-lb. The allowable, or resisting, moment can be calculated from the flexure formula based on an allowable bending stress (see Eq. [14–6]):

$$M_R = s_{b(\text{all})} S_x = \frac{24,000}{12} (151) = 302,000 \text{ ft-lb}$$

Therefore,

$$274,600 \text{ ft-lb} < 302,000 \text{ ft-lb} \qquad \textbf{OK}$$

The applied or actual induced shear force has been calculated to be 27,460 lb. The allowable, or resisting, shear force can be calculated from Eq. (14–12):

$$\begin{aligned} V_R = s_{s(\text{all})} dt_w &= 14,500(21.24)(0.455) \\ &= 140,100 \text{ lb} \end{aligned}$$

Therefore,

$$27,460 \text{ lb} < 140,100 \text{ lb} \qquad \textbf{OK}$$

□ **EXAMPLE 14–11**   A factory building floor is supported by 3 in. by 16 in. (S4S) Douglas fir joists spaced 24 in. on center and on a span length of 16 ft. Assume the joists to be simply supported beams. (a) Calculate the allowable uniformly distributed load w for each joist in lb/ft. (b) Calculate the allowable uniformly distributed floor load in pounds per square foot (psf).

Solution (a) See Appendices E and F for beam properties and allowable stresses. Properties of the 3 × 16 (S4S) are

$$S_x = 96.9 \text{ in.}^3$$
$$A = 38.1 \text{ in.}^2$$
$$\text{Wt.} = 10.6 \text{ lb/ft}$$

Allowable stresses are

$$s_{b(\text{all})} = 1450 \text{ psi}$$
$$s_{s(\text{all})} = 95 \text{ psi}$$

Considering moment, calculate the allowable bending moment as

$$M_R = s_{b(\text{all})} S_x = \frac{1450}{12} (96.9) = 11{,}709 \text{ ft-lb}$$

To determine the allowable uniformly distributed load w based on allowable moment, rewrite

$$M = \frac{wL^2}{8}$$

and solve for w:

$$w = \frac{8M}{L^2} = \frac{8(11{,}709)}{16^2} = 366 \text{ lb/ft}$$

Note that in this case, lb/ft actually means "pounds per lineal foot of joist."

Considering shear, the allowable shear V_R can be calculated by rewriting Eq. (14–10) and solving for V. This yields

$$V_R = \frac{s_{s(\text{all})} A}{1.5} = \frac{95(38.1)}{1.5} = 2413 \text{ lb}$$

The maximum shear produced in a simply supported uniformly loaded beam is

$$V = \frac{wL}{2}$$

Equating these two shears and solving for w,

$$w = \frac{2V_R}{L} = \frac{2(2413)}{16} = 301.6 \text{ lb/ft}$$

The controlling value is the smaller of the two allowable loads. Therefore, the allowable total uniformly distributed load is 301.6 lb/ft. If the allowable superimposed load is desired, the weight of the joist (lb/ft) must be subtracted from the total allowable load. The allowable superimposed load is then

$$301.6 - 10.6 = 291 \text{ lb/ft}$$

(b) The allowable superimposed floor load in pounds per square foot is a function of the joist spacing. In Fig. 14–22, note that each joist supports a 24 in. (or 2 ft) width of

FIGURE 14–22   Floor cross section.

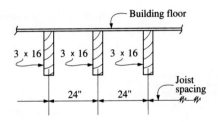

floor. It has been found that each joist can support a load of 291 lb/ft. The allowable load per foot, divided by the spacing in feet, will yield the allowable load in pounds per square foot:

$$\frac{291}{2} = 145.5 \text{ psf}$$

□ **EXAMPLE 14–12**   Calculate the maximum allowable span length for a W18 × 50 simply supported steel wide-flange beam subjected to a uniformly distributed load of 2000 lb/ft (this includes the weight of the beam). The allowable bending stress is 24,000 psi and the allowable shear stress is 14,500 psi.

Solution   From Appendix A, the necessary properties for the wide-flange shape are $d = 17.99$ in., $t_w = 0.355$ in., $S_x = 88.9$ in.^{3}. Considering moment, calculate the allowable bending moment as

$$M_R = \frac{24,000}{12}(88.9) = 177,800 \text{ ft-lb}$$

The allowable span length L is then obtained from

$$M = \frac{wL^2}{8}$$

by equating to M_R, rewriting, and solving for L:

$$L = \sqrt{\frac{8M_R}{w}} = \sqrt{\frac{8(177,800)}{2000}} = 26.7 \text{ ft}$$

Considering shear, the allowable shear, using Eq. (14–12), is

$$V_R = s_{s(\text{all})}dt_w = 14,500(17.99)(0.355) = 92,600 \text{ lb}$$

The allowable span length L is obtained from

$$V = \frac{wL}{2}$$

by equating to V_R, rewriting, and solving for L:

$$L = \frac{2V_R}{w} = \frac{2(92,600)}{2000} = 92.6 \text{ ft}$$

The controlling value is the smaller of the two span lengths, indicating that moment controls and that the maximum allowable span length is 26.7 ft.

☐ **EXAMPLE 14–13**

Three 4 in. by 6 in. timber members are bolted together so as to act as a single unit (see Fig. 14–23). The bolts are spaced 10 in. on centers. The maximum vertical shear V is equal to 750 lb. Calculate the maximum force that the cross section of each bolt must resist at each longitudinal joint.

Solution

Using the general shear formula, the horizontal shear stress will be computed in the longitudinal joints. Because of symmetry, both joints are subjected to the same stress.

$$I_{NA} = \frac{bh^3}{12} = \frac{6(12)^3}{12} = 864 \text{ in.}^4$$

$$Q = A\bar{y} = 4(6)(4) = 96 \text{ in.}^3$$

$$s_s = \frac{VQ}{Ib} = \frac{750(96)}{864(6)} = 13.89 \text{ psi}$$

In effect, the bolt is compensating for the fact that the member is not one solid unit. Therefore, the bolt may be considered to replace a horizontal area of 6 in. by 10 in. which would normally resist the calculated shear stress (see Fig. 14–23[b]). Each bolt, then, must resist a force which can be calculated from

$$P = s_s(A) = 13.89(6)(10) = 833 \text{ lb}$$

FIGURE 14–23 Bolted timber beam.

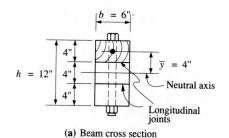

(a) Beam cross section

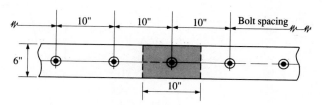

(b) Top view of beam

14–8
INELASTIC BENDING
OF BEAMS

The preceding sections of this chapter discussed ductile, homogeneous bending members. A linear bending stress distribution over the depth of the member was assumed, varying from a maximum at the outer fibers to zero at the neutral axis. This stress distribution served as a basis for the design and/ or analysis of bending members when utilizing an allowable bending stress. Since the allowable bending stress is always specified as some fraction of a proportional limit stress (or a yield stress), the induced bending stress for a properly designed member will always be in the elastic range. Hooke's law (stress is proportional to strain) was assumed to apply.

Now consider bending members so loaded that they are stressed beyond the proportional limit (and the yield point) into the plastic range. This means that permanent deformation of the material occurs and that stress is no longer proportional to strain. The applied loads considered for this condition are large enough to produce plastic deformations, but not so large that they produce failure of the member or structure. This design approach, as it related to tension members, was introduced in Section 10–8.

For purposes of discussion, a ductile, homogeneous material such as structural steel will be considered. For such a material, the stress-strain relationship may be reasonably idealized as shown in Fig. 14–24. Note that a straight-line relationship between stress and strain is assumed up to the yield point. It is also assumed that the yield point and modulus of elasticity are the same for both tension and compression. Past the yield point, the stress within the plastic range is assumed to be constant despite the fact that strain increases. Let us also assume that plane sections before bending remain plane into the plastic range, thus strains are always proportional to the distance from the neutral axis.

The beam of which the behavior will be examined is assumed to be rectangular in cross section, as shown in Fig. 14–25(a), and subjected to a gradually increasing bending moment. Further, the cross section is symmet-

FIGURE 14–24 Idealized stress-strain diagram for structural steel.

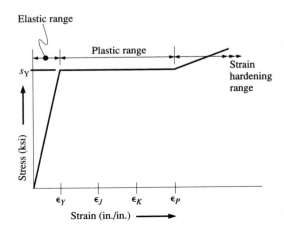

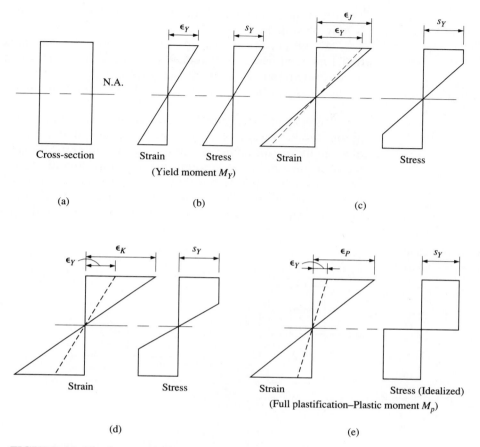

FIGURE 14–25   Stress-strain relationship.

rical about an axis which lies in the plane of the loading. As the moment is increased until the outer fiber strain reaches ϵ_Y, the relationship between stress and strain is a linear one, yielding a linear stress distribution, as shown in Fig. 14–25(b), and the outer fiber stress reaches s_Y. The moment at this point is called the *yield moment*. The moment is increased further and the outer fiber strains reach the value ϵ_J as shown in Fig. 14–25(c). The corresponding stress in the outer fibers for this strain is still s_Y. Those interior fibers that have been strained past ϵ_Y will also be stressed to s_Y. However, other interior fibers, closer to the neutral axis, that have not been strained past ϵ_Y will have a linear stress distribution decreasing to zero at the neutral axis. As the moment increases further, more and more of the cross section is stressed to the yield stress s_Y, as shown in Fig. 14–25(d). This process is called *plastification* of the cross section. Finally, almost all the fibers will be strained past ϵ_Y and the stress distribution will approach a rectangular shape.

This stress distribution is shown idealized in Fig. 14–25(e) (the cross section is fully plastified).

At full plastification of the cross section, it is assumed that no additional moment can be resisted. When this occurs, a plastic hinge is said to have formed and the moment that exists at this point is called the *plastic moment* (M_P), which in effect represents the limiting moment strength of the beam.

To evaluate M_P, consider a rectangular shape that is stressed as shown in Fig. 14–26. Assume, for discussion purposes, that this is the cross section of a simply supported beam that is subjected to vertical load only. Since the member must be in equilibrium, the algebraic sum of the internal forces C and T must be equal. Note that these are resultant forces of the bending stresses and are parallel, equal and opposite in sense, thereby forming an internal couple M_P, which must resist the applied bending moment.

FIGURE 14–26   Rectangular cross section for M_P calculation.

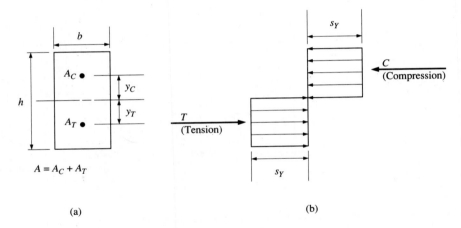

(a)   (b)

Considering the horizontal forces (see Fig. 14–26(b)),

$$C = T$$
$$s_Y A_C = s_Y A_T$$

from which

$$A_C = A_T = \frac{A}{2}$$

indicating that the neutral axis of a fully plastified cross section divides the cross section into two parts of equal area. (For cross sections having symmetry about both the X–X and Y–Y axes, the neutral axis is at the centroid of the section for both elastic and plastic loading.)

The internal resistance constitutes a couple, which can be expressed as

$$M_P = C(y_C + y_T) \qquad \text{or} \quad M_P = T(y_C + y_T)$$
$$= s_Y A_C(y_C + y_T) \qquad\qquad = s_Y A_T(y_C + y_T)$$

FIGURE 14–27 Shapes for M_P calculations.

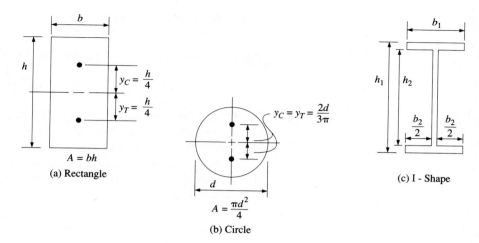

(a) Rectangle

(b) Circle

(c) I - Shape

It was previously shown that $A_C = A_T = A/2$. Therefore, by substitution,

$$M_P = s_Y \left(\frac{A}{2}\right)(y_C + y_T) \qquad (14\text{–}14)$$

Although derived for a rectangular shape, Eq. (14–14) can be used to determine M_P for various shapes. M_P can be calculated for the three common shapes shown in Fig. 14–27. For a rectangular shape (Fig. 14–27(a)),

$$M_P = s_Y \left(\frac{bh}{2}\right)\left(\frac{h}{4} + \frac{h}{4}\right) = s_Y \left(\frac{bh^2}{4}\right) \qquad (14\text{–}15)$$

For a circular shape (Fig. 14–27(b)),

$$M_P = s_Y \left(\frac{\pi d^2}{8}\right)\left(\frac{2d}{3\pi} + \frac{2d}{3\pi}\right) = s_Y \left(\frac{d^3}{6}\right) \qquad (14\text{–}16)$$

For an I-shape (Fig. 14–27(c)),

$$M_P = s_Y \left(\frac{b_1 h_1^2}{4} - \frac{b_2 h_2^2}{4}\right)$$

$$M_P = \frac{s_Y}{4}\left(b_1 h_1^2 - b_2 h_2^2\right) \qquad (14\text{–}17)$$

The total resistance to bending at the yield point is

$$M_Y = s_Y S \qquad (14\text{–}18)$$

where M_Y = yield moment
s_Y = yield stress
S = section modulus

The total resistance to bending when the plastic moment develops is

$$M_P = s_Y Z \qquad (14\text{–}19)$$

where M_P = plastic moment
        Z = plastic section modulus

Note that for a rectangular shape (Eq. 14–15),

$$M_P = s_Y\left(\frac{bh^2}{4}\right)$$

Therefore, for a rectangular shape, $bh^2/4$ can be designated as the plastic section modulus Z. Z is equal to the sum of the moments of the areas above and below the neutral axis, taken about that axis.

   Recall that the elastic section modulus S for a rectangular shape is $bh^2/6$. Therefore, the ratio of the plastic moment and the yield moment is calculated from

$$\frac{M_P}{M_Y} = \frac{6}{4} = 1.50$$

This indicates that the rectangular beam can carry 50% more moment from the time that it first yields until it reaches its full plastic moment. This ratio is called the *shape factor*.

□ **EXAMPLE 14–14**   An A36 structural steel beam is built-up using three steel plates as shown in Fig. 14–28. Calculate the yield moment M_Y, the plastic moment M_P (both with respect to the X–X axis), and the shape factor. Also calculate the uniformly distributed load that the beam can carry at yield and at full plastification for a simple span length of 40 ft.

FIGURE 14–28   Built-up structural I-shape.

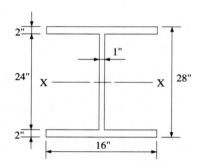

Solution   (a) *Elastic behavior*:

$$I_x = \frac{1(24)^3}{12} + \left(\frac{16(2)^3}{12}\right)2 + 16(2)(13)^2(2) = 11{,}990 \text{ in.}^4$$

$$S_x = \frac{I_x}{c} = \frac{11{,}990}{14} = 856 \text{ in.}^3$$

$$M_Y = s_Y S_x = \frac{36(856)}{12} = 2570 \text{ ft-kips}$$

From

$$M = \frac{wL^2}{8} \quad \text{or} \quad M_Y = \frac{w_Y L^2}{8}$$

solve for load at yield w_Y:

$$w_Y = \frac{8M_Y}{L^2} = \frac{8(2570)}{40^2} = 12.85 \text{ kips/ft}$$

(b) *Inelastic (plastic behavior)*:

$$Z_x = 1(12)(6)(2) + 2(16)(13)(2) = 976 \text{ in.}^3$$

$$M_P = s_Y Z_x = \frac{36(976)}{12} = 2930 \text{ ft-kips}$$

$$\text{Shape factor} = \frac{M_P}{M_Y} = \frac{2930}{2570} = 1.140$$

From

$$M = \frac{wL^2}{8}$$

solve for load at full plastification of the cross section w_P:

$$w_P = \frac{8M_P}{L^2} = \frac{8(2930)}{40^2} = 14.65 \text{ kips/ft}$$

14-9
SI SYSTEM
EXAMPLES

□ **EXAMPLE 14-15**

An extra-strong steel pipe having a nominal outside diameter of 125 mm is to be used as a simple beam with a span length of 7 m. The pipe supports a concentrated load at midspan of 6 kN. Calculate the maximum bending stress due to (a) the weight of the pipe alone, (b) the concentrated load alone.

Solution The following properties for this pipe are obtained from Appendix B:

$$\text{Wt.} = 303 \times 10^{-3} \text{ kN/m}$$

$$S = 122 \times 10^{-6} \text{ m}^3$$

For a simply supported beam, the moment due to the beam's own weight is obtained from

$$M = \frac{wL^2}{8} = \frac{(303 \times 10^{-3} \text{ kN/m})(7 \text{ m})^2}{8}$$

$$= 1.856 \text{ kN} \cdot \text{m}$$
$$= 1856 \text{ N} \cdot \text{m}$$

The maximum moment due to the applied concentrated load is obtained from

$$M = \frac{PL}{4} = \frac{(6 \text{ kN})(7 \text{ m})}{4} = 10.5 \text{ kN} \cdot \text{m}$$

$$= 10\,500 \text{ N} \cdot \text{m}$$

The flexure formula (Eq. [14-5]) is used to compute the bending stress.

(a) Due to the beam's own weight,

$$s_b = \frac{M}{S} = \frac{1856 \text{ N} \cdot \text{m}}{122 \times 10^{-6} \text{ m}^3} = 15.2 \times 10^6 \text{ N/m}^2$$

$$= 15.2 \text{ MPa}$$

(b) Due to the applied concentrated load,

$$s_b = \frac{M}{S} = \frac{10\ 500 \text{ N} \cdot \text{m}}{122 \times 10^{-6} \text{ m}^3} = 86.1 \times 10^6 \text{ N/m}^2$$

$$= 86.1 \text{ MPa}$$

We could also determine the total stress. Since the calculated stresses occur at the same point in the beam, they are additive:

$$\text{Total } s_b = 15.2 + 86.1 = 101.3 \text{ MPa}$$

□ **EXAMPLE 14–16**   An aluminum bar 10 mm in width and 100 mm in depth is placed on a simple span of 1000 mm and is loaded as shown in Fig. 14–29. (a) Compute the maximum shear stress and maximum bending stress developed in the bar. (b) Compute the horizontal shear stress on plane G–G, 25 mm above (or below) the neutral axis at the location of maximum shear. Neglect the weight of the bar.

Solution   Computing the beam reactions,

$$\Sigma M_A = R_B(1000) - 900(125) - 800(500) - 1200(875) = 0$$

from which

$$R_B = 1562.5 \text{ N}$$

and

$$\Sigma M_B = -R_A(1000) + 1200(125) + 800(500) + 900(875) = 0$$

FIGURE 14–29   Aluminum beam.

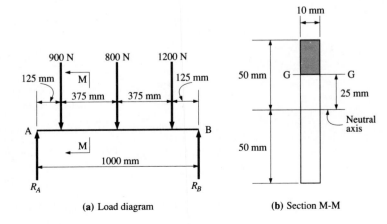

(a) Load diagram   (b) Section M-M

from which

$$R_A = 1337.5 \text{ N}$$

Checking by summing vertical forces ($\Sigma F_V = 0$),

$$+1562.6 + 1337.5 - 900 - 800 - 1200 = 0 \quad \textbf{OK}$$

Since the member is unsymmetrically loaded, shear and moment diagrams are drawn to determine maximum values for the shear and moment (see Fig. 14–30). The free bodies and calculations are not shown; however, the reader is urged to check the ordinates for each diagram. Note that the maximum shear is 1562.5 N and the maximum bending moment is 331.3×10^3 N·mm.

FIGURE 14–30 Load, shear, and moment diagrams.

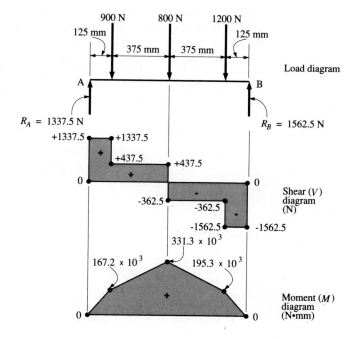

(a) The bar is a homogeneous solid rectangular member. Therefore, the maximum horizontal shear stress occurs at the neutral axis in the plane of maximum shear and can be calculated from Eq. (14–10):

$$s_s = \frac{1.5V}{A} = \frac{1.5(1562.5 \text{ N})}{(10 \text{ mm})(100 \text{ mm})} = 2.34 \text{ N/mm}^2$$

$$= 2.34 \text{ MPa}$$

Note that $1 \text{ N/mm}^2 = 1 \times 10^6 \text{ N/m}^2 = 1 \times 10^6 \text{ Pa} = 1 \text{ MPa}$.

Referring to Example 14–1, the section modulus for a bar having a rectangular cross-sectional area can be obtained from

$$S = \frac{bh^2}{6} = \frac{10(100)^2}{6} = 16.7 \times 10^3 \text{ mm}^3$$

The maximum bending stress will occur in the plane of maximum moment and can be computed from the flexure formula (Eq. [14–5]):

$$s_b = \frac{M}{S} = \frac{331.3 \times 10^3 \text{ N·mm}}{16.7 \times 10^3 \text{ mm}^3} = 19.8 \text{ N/mm}^2$$
$$= 19.8 \text{ MPa}$$

(b) The horizontal shear stress on plane G–G, 25 mm above the neutral axis, cannot be calculated using Eq. (14–10). The general shear formula (Eq. [14–8]) must be used:

$$s_s = \frac{VQ}{Ib}$$

Maximum shear V for use in the preceding formula has already been calculated (1562.5 N) and b is given as 10 mm. Calculating the moment of inertia I with respect to the neutral axis,

$$I = \frac{bh^3}{12} = \frac{10(100)^3}{12} = 833.3 \times 10^3 \text{ mm}^4$$

We must also determine the statical moment Q for use in Eq. (14–8). Q is equal to the moment of the area above plane G–G with respect to the neutral axis. With reference to Fig. 14–29(b), the area A above plane G–G is

$$A = (10 \text{ mm})(25 \text{ mm}) = 250 \text{ mm}^2$$

The distance from the centroid of area A to the neutral axis is calculated from

$$\bar{y} = 25 + \frac{25}{2} = 37.5 \text{ mm}$$

Therefore,

$$Q = A\bar{y} = (250 \text{ mm}^2)(37.5 \text{ mm})$$
$$= 9375 \text{ mm}^3$$

and substituting in Eq. (14–8),

$$s_s = \frac{VQ}{Ib} = \frac{(1562.5 \text{ N})(9375 \text{ mm}^3)}{(833.3 \times 10^3 \text{ mm}^4)(10 \text{ mm})}$$
$$= 1.76 \text{ N/mm}^2$$
$$= 1.76 \text{ MPa}$$

□ **EXAMPLE 14–17**   Calculate the allowable superimposed uniformly distributed load that may be placed on a W610 × 1.52 hot-rolled structural steel wide-flange section that is simply supported on a 12 m span length. The beam is oriented with the strong axis of its cross-section horizontal. The allowable stresses for the steel are 165 MPa in bending and 100 MPa in shear.

Solution   From Appendix A, for the W610 × 1.52,

$$S_x = 4.23 \times 10^{-3} \text{ m}^3$$
$$d = 611 \text{ mm}$$
$$t_w = 12.7 \text{ mm}$$

Calculating the allowable bending moment using Eq. (14–6),

$$
\begin{aligned}
M_{R_x} &= s_{b(\text{all})} S_x \\
&= (165 \text{ MPa})(4.23 \times 10^{-3} \text{ m}^3) \\
&= (165 \times 10^6 \text{ N/m}^2)(4.23 \times 10^{-3} \text{ m}^3) \\
&= 698 \times 10^3 \text{ N} \cdot \text{m}
\end{aligned}
$$

From Appendix H, the maximum bending moment for a simply supported beam that supports a uniformly distributed load is

$$
M = \frac{wL^2}{8}
$$

Solving for w and substituting,

$$
w = \frac{8M}{L^2} = \frac{8(698 \times 10^3 \text{ N} \cdot \text{m})}{(12 \text{ m})^2} = 38\ 800 \text{ N/m}
$$

$$
= 38.8 \text{ kN/m}
$$

Considering shear, using the average web shear approach, the allowable shear may be calculated from Eq. (14–12):

$$
\begin{aligned}
V_R = s_{s(\text{all})} dt_w &= (100 \text{ MPa})(611 \text{ mm})(12.7 \text{ mm}) \\
&= (100 \text{ N/mm}^2)(611 \text{ mm})(12.7 \text{ mm}) \\
&= 776 \times 10^3 \text{ N}
\end{aligned}
$$

From Appendix H, the maximum shear for a simply supported beam that supports a uniformly distributed load is

$$
V = \frac{wL}{2}
$$

Solving for w and substituting,

$$
w = \frac{2V}{L} = \frac{2(776 \times 10^3 \text{ N})}{12 \text{ m}} = 129.3 \times 10^3 \text{ N/m}
$$

$$
= 129.3 \text{ kN/m}
$$

$$
38.8 \text{ kN/m} < 129.3 \text{ kN/m}
$$

Therefore, moment controls, since it results in the lower value. Subtracting the beam weight, determine the allowable superimposed load to be

$$
38.8 - 1.52 = 37.3 \text{ kN/m}
$$

SUMMARY—BY SECTION NUMBER

14–1   In beams subjected to loads that produce bending, tensile stresses are developed on the convex side where the fibers have elongated; compressive stresses are developed on the concave side where the fibers have shortened. The plane on which no elongation or shortening of fibers occurs is also the plane of zero bending stress and is called the *neutral plane*. The intersection of the neutral plane with a cross-sectional plane is called a *neutral axis*. The tensile and compressive stresses, called *bending stresses*, vary linearly from zero at the neutral axis to a maximum at the outer fibers.

14–2 The maximum bending stress occurs at the outer fibers (farthest from the neutral axis) and is computed using the flexure formula:

$$s_b = \frac{Mc}{I} = \frac{M}{S} \qquad \textbf{(14–2) (14–5)}$$

where S represents the quantity I/c and is called the *section modulus*.

The maximum resisting moment (called the *allowable moment*) can be computed from

$$M_R = \frac{s_{b(\text{all})}I}{c} = s_{b(\text{all})}S \qquad \textbf{(14–4) (14–6)}$$

For design problems, the most convenient form is

$$\text{Required } S = \frac{M}{s_{b(\text{all})}} \qquad \textbf{(14–7)}$$

14–4 Horizontal and vertical shear stresses are developed in beams subjected to vertical loads. The shear stress is zero at the outer fibers and almost always is maximum at the neutral axis.

14–5 The shear stress on any horizontal plane at any cross section of a beam is calculated from the general shear stress formula:

$$s_s = \frac{VQ}{Ib} \qquad \textbf{(14–8)}$$

The maximum shear stress will occur in the plane of maximum shear. The maximum resisting shear (allowable shear) can be computed from

$$V_R = \frac{s_{s(\text{all})}Ib}{Q} \qquad \textbf{(14–9)}$$

14–6 The maximum horizontal shear stress occurs at the neutral axis for the following: For a homogeneous solid beam of rectangular cross section,

$$s_s = \frac{1.5V}{A} \qquad \textbf{(14–10)}$$

and for a homogeneous solid beam of circular cross section,

$$s_s = \frac{4V}{3A} \qquad \textbf{(14–13)}$$

The AISC allows the use of an average web shear approach to simplify shear stress calculations for hot-rolled and fabricated shapes:

$$s_s = \frac{V}{dt_w} \qquad \textbf{(14–11)}$$

Using the AISC approach, the allowable shear can be computed from

$$V_R = s_{s(\text{all})}dt_w \qquad \textbf{(14–12)}$$

14–7 In beam-analysis type problems both bending and shear stresses always must be considered.

14–8 When only the outer fibers of a beam are stressed to the yield point, the beam moment strength is

$$M_Y = s_Y S \qquad \text{(14–18)}$$

However, a considerable amount of strength remains in the interior fibers. When all the fibers are stressed to the yield point, no additional moment can be resisted and the beam moment strength becomes

$$M_P = s_Y Z \qquad \text{(14–19)}$$

where Z is the plastic section modulus and is calculated from the sum of the moments of the areas above and below the neutral axis.

PROBLEMS

Section 14–3 Computation of Bending Stresses

1. Using the dressed dimensions from Appendix E, calculate the section modulus (with respect to the X–X axis) of a solid rectangular 6 in. by 10 in. timber beam. Assume the large dimension vertical and parallel to the applied loads.

2. Calculate the section modulus (with respect to the X–X axis) for the beams in Fig. 14–31.

3. Assume that the timber member (a) of Problem 2 is used as a simply supported beam with a 12 ft span length carrying a uniformly distributed load of 500 lb/ft. The given load includes the weight of the beam. Calculate the maximum induced bending stress.

4. The structural steel built-up member (b) of Problem 2 is to be used as a simply supported beam on a span of 40 ft. It is to carry a uniformly distributed load of 3.0 kips/ft, which includes its own weight. Calculate the maximum induced bending stress.

5. Calculate the maximum bending stress in a W18 × 40 steel beam that spans 36 ft and supports two equal concentrated loads of 12 kips each. The loads are placed at the third points. Be sure to include the weight of the beam.

6. Calculate the allowable bending moment for a W36 × 300 wide-flange section. The allowable bending stress is 33 ksi. Consider (a) strong-axis bending and (b) weak-axis bending.

7. Calculate the allowable bending moment for a solid rectangular 6 in. by 16 in. timber beam if the allowable bending stress is 1000 psi. (a) Use nominal dimensions. (b) Use dressed dimensions. Assume that the large dimension is vertical and parallel to the applied loads.

FIGURE 14–31 Problem 2.

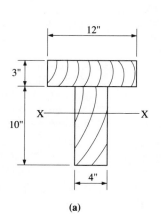

(a)

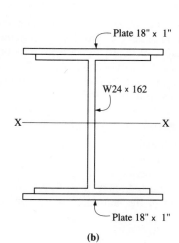

(b)

Section 14–6   Shear Stresses
in Structural Members

8. A solid rectangular simply supported timber beam 6 in. wide, 20 in. deep, and 10 ft long carries a concentrated load of 16,000 lb at midspan. (a) Compute the maximum horizontal shear stress at the neutral axis. (b) Compute the shear stress 4 in. and 8 in. above and below the neutral axis. Neglect the weight of the beam.

9. A W14 × 30 supports the loads shown in Fig. 14–32. Calculate the maximum shear stress using (a) the general shear formula and (b) the average web shear approach. Be sure to include the weight of the beam.

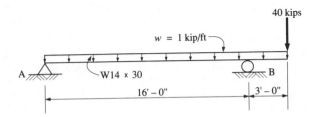

FIGURE 14–32   Problem 9.

10. If the allowable shear stress is 14,500 psi, calculate the maximum shear force V that a W18 × 50 structural steel wide-flange is capable of resisting. Use the average web shear approach.

11. A steel pin 1½ in. in diameter is subjected to a shear force of 10,000 lb in the plane of its cross section. Compute the maximum horizontal shear stress.

12. A timber power-line pole is 10 in. in diameter at its base where it is solidly embedded in concrete. The pole extends 20 ft vertically upward from its base and is subjected to a horizontal pull of 300 lb at its top. Calculate the maximum bending stress and the maximum shear stress produced in the pole.

Section 14–7   Beam Analysis

13. Calculate the allowable superimposed uniformly distributed load that may be safely carried by a Hem-fir timber beam. The beam is 12 in. by 18 in. (S4S) and is on a simply supported span of 20 ft. Be sure to consider the weight of the beam. Refer to Appendices E and F.

14. Assume the floor-joist dimensions of Example 14–11 to be changed to 3 in. by 12 in. and that they are spaced 16 in. on centers. The wood and span length remain the same. Calculate (a) the allowable superimposed uniformly distributed load for each joist in lb/ft and (b) the allowable superimposed uniformly distributed floor load in psf.

15. Calculate the allowable superimposed uniformly distributed load that may be placed on a structural steel wide-flange W14 × 34 that has a simple span length of 6.5 ft. The allowable bending stress is 24,000 psi and the allowable shear stress is 14,500 psi.

16. A 3 in. by 12 in. (S4S) scaffold timber plank is laid flatwise on supports that are 8 ft apart. Calculate the allowable uniformly distributed load that can be carried by the plank. The allowable bending stress is 1600 psi and the allowable shear stress is 100 psi. Neglect the weight of the plank.

Section 14–8   Inelastic Bending of Beams

17. Calculate the value of S and Z and the shape factor for the beam cross sections shown in Fig. 14–33.

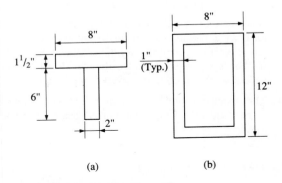

FIGURE 14–33   Problem 17.

18. For beams that have cross sections as shown in Fig. 14–33, calculate the yield moment M_Y, the plastic moment M_P and the maximum uniformly distributed load that the beams can carry on a simple span of 14 ft. Assume $s_Y = 36$ ksi.

19. Calculate the maximum load P that the beam shown in Fig. 14–34 can carry based on plastic moment M_P. Assume a yield stress s_Y of 40,000 psi. Neglect the weight of the beam.

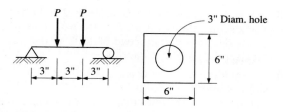

FIGURE 14–34 Problem 19.

SI System Problems

20. A round steel rod, 25 mm in diameter, is subjected to a bending moment of 160 N·m. Calculate the maximum bending stress.

21. A square steel bar, 38 mm on each side, is used as a beam and subjected to a bending moment of 460 N·m. Calculate the maximum bending stress.

22. A beam is subjected to a maximum bending moment of 2600 N·m. The cross section of beam is 100 mm by 150 mm. The 150 mm side is oriented vertically. (a) Calculate the maximum bending stress. (b) Calculate the bending stress 25 mm below the top surface.

23. Rework Problem 22 assuming that the beam is placed with the 100 mm dimension oriented vertically.

24. Calculate the maximum bending stress in a W530 × 1.07 steel beam that spans 13 m on simple supports and supports two equal concentrated loads of 54 kN each. The loads are placed at the third points. Include the weight of the beam.

25. Calculate the allowable bending moment for a W910 × 3.79 wide-flange section. The allowable bending stress is 207 MPa. Consider (a) strong axis bending and (b) weak axis bending.

26. A rectangular beam 100 mm in width and 250 mm in depth is oriented with the large dimension placed vertically. Using the general shear formula, calculate the maximum shear stress when the beam is subjected to a maximum shear force of 140 kN.

27. A W610 × 1.23 steel wide-flange section is subjected to a vertical shear force V of 525 kN. Calculate the maximum horizontal shear stress using (a) the general shear formula and (b) the average web shear approach.

28. A W250 × 0.99 steel wide-flange section supports a uniformly distributed load on a simple span of 5 m. (a) Using an allowable bending stress of 165 MPa, calculate the allowable load. (b) With the beam carrying the load from part (a), calculate the maximum horizontal shear stress using the general shear formula. Neglect the weight of the beam.

29. A rectangular hollow shape carries loads as shown in Fig. 14–35. Calculate the maximum value of P if the allowable stresses are 10 MPa for bending and 1.5 MPa for shear. Neglect the weight of the beam.

Computer Problems

For the following computer problems, any appropriate programming language may be used. Input prompts should fully explain what is required of the user (the program should be "user friendly"). The resulting output should be well labeled and self-explanatory.

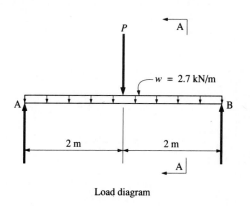

Load diagram

Section A-A

FIGURE 14–35 Problem 29.

30. Write a program that may be used to calculate the maximum bending stress and the maximum horizontal shear stress for a simply supported single-span timber beam that carries a uniformly distributed load. User input is to be uniformly distributed load intensity, span, and the nominal dimensions of the cross section. (Use nominal dimensions for the calculations and assume a unit weight for the timber of 35 pcf.)

31. Repeat Problem 30, except now assume the beam to be a hot-rolled steel wide-flange shape. User input will be span length, uniform load intensity, and the following dimensions for the shape: overall depth, thickness and width of the flange, and the web thickness. Assume the unit weight of the steel to be 490 pcf. Calculate I by approximating the cross section with three rectangles.

32. Write a program that will generate a table of maximum allowable span lengths for a group of 2 in. thick planks (ranging from 4 in. to 12 in. deep in 2 in. increments) as a function of applied uniformly distributed load. The planks are to bend about their strong axes. Both moment and shear should be considered. The load should range from 50 lb/ft to 120 lb/ft in steps of 10 lb/ft. User input is to be allowable bending stress and allowable horizontal shear stress. Assume the wood to have a unit weight of 35 pcf. Use full nominal dimensions.

Supplemental Problems

33. Calculate the section modulus with respect to the X–X axis for the sections shown in Fig. 14–36.

34. The timber box section (a) of Problem 33 is used as a simply supported beam on an 18 ft span length. The beam carries a uniformly distributed load of 500 lb/ft, which includes its own weight. Calculate the maximum induced bending stress.

35. A $\frac{1}{2}$ in. diameter steel rod projects 2 ft horizontally from a concrete wall. Calculate the maximum bending stress due to a vertical load of 12 lb at the free end. Neglect the weight of the rod.

36. A cantilever cast-iron beam is 6 ft long and has a "T" cross section, as shown in Fig. 14–37. Calculate the maximum tensile and compressive bending stresses. The applied load is 450 lb. Neglect the weight of the beam.

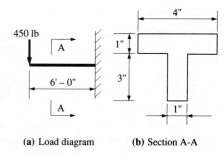

(a) Load diagram   **(b)** Section A-A

FIGURE 14–37   Problem 36.

37. A W8 × 13 steel wide-flange beam on a 20 ft span is supported and loaded as shown in Fig. 14–38. The uniformly distributed load includes the weight of the beam. Calculate the maximum bending stress.

38. A simply supported beam with a cruciform cross section is loaded as shown. Calculate the maximum horizontal shear stress. Neglect the weight of the beam. (**Hint:** Check two planes in the cross section.) (See Fig. 14–39.)

FIGURE 14–36   Problem 33.

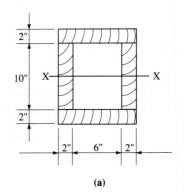

(a)

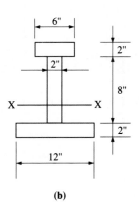

(b)

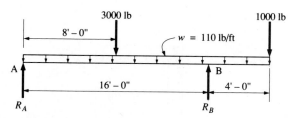

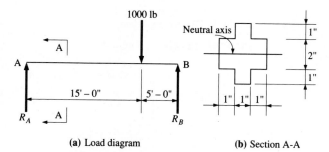

FIGURE 14–38   Problem 37.

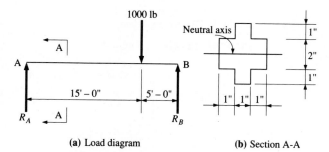

(a) Load diagram   (b) Section A-A

FIGURE 14–39   Problem 38.

39. The timber box section (a) of Problem 33 is made up of four component parts so connected as to act as a single unit. Calculate the maximum horizontal shear stress if the member is subjected to a maximum vertical shear of 4500 lb.

40. For the I-shaped timber beam in Fig. 14–40, calculate the maximum vertical shear force that will induce a maximum horizontal shear stress of 120 psi. Then calculate the shear stress in the plane of the junction between the web and the flange.

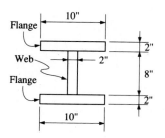

FIGURE 14–40   Problem 40.

41. A W36 × 150 steel wide-flange beam is oriented so that bending is about the weak (Y–Y) axis. The member is subjected to a maximum shear of 140 kips. Calculate the horizontal shear stress at a sufficient number of points so that a shear stress distribution diagram can be drawn. Draw the diagram.

42. A W10 × 45 steel wide-flange beam supports a uniformly distributed load on a simple span of 14 ft. (a) Using an allowable bending stress of 24 ksi, compute the allowable load (kips/ft). (b) Determine the maximum horizontal shear stress using the general shear formula.

43. A W30 × 108 steel wide-flange beam is simply supported on a span of 10 ft and is subjected to a uniformly distributed load of 42 kips/ft, which includes the weight of the beam. Calculate the maximum bending stress and the maximum horizontal shear stress using the general shear formula.

44. Four wood boards 1 in. by 6 in. in cross section are stacked as shown in Fig. 14–41 and used as an 8 ft long simply supported beam. A 400 lb concentrated load is applied at the center of the span. Calculate the following based on the full nominal (rough) dimensions: (a) the maximum bending stress (boards not glued together), (b) the maximum bending stress if the boards are glued together, and (c) the maximum shear stress in the middle glued joint. Neglect the weight of the beam.

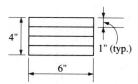

FIGURE 14–41   Problem 44.

45. A lintel consists of two 8 in. by $\frac{1}{2}$ in. steel plates welded together to form an inverted tee. Calculate the maximum bending stresses in tension and compression when the lintel carries a total uniformly distributed load of 10,000 lb on a simple span of 6 ft. In addition, calculate the maximum shear stress and plot the distribution of the shear stress for the entire cross section. Neglect the weight of the beam.

46. A 2 in. by 12 in. scaffold timber plank, placed flatwise, must carry a uniformly distributed load of 100 lb/ft without exceeding an allowable bending stress of 1600 psi and an allowable shear stress of 120 psi. Calculate the maximum allowable simple span length. Use full nominal dimensions and neglect the weight of the plank.

47. A laminated wood beam is built-up by gluing together three 2 in. by 4 in. members to form a solid beam 4 in. by 6 in. in cross section, as shown in Fig. 14–42. The allowable shear stress in the glued joints is 60 psi. The beam is a 3 ft long cantilever with a load P applied at its free end. (a) Calculate the maximum allowable load P that can be applied at the free end. (b) Calculate the maximum bending stress in the beam. Neglect the weight of the beam.

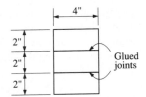

FIGURE 14–42 Problem 47.

48. For the beam in Fig. 14–43, calculate the maximum tensile and compressive bending stresses and the maximum shear stresses. Neglect the beam weight.

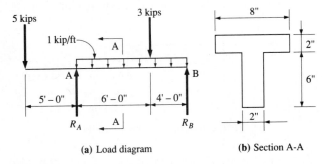

(a) Load diagram   (b) Section A-A

FIGURE 14–43 Problem 48.

49. A box beam is built up of four 1 in. by 6 in. boards as shown in Fig. 14–44. The beam is subjected to a maximum vertical shear force of 920 lb. Calculate the required spacing of wood screws if the allowable shear load on each screw is 250 lb.

50. Find the value of the loads P that can be safely supported by the 6 in. by 12 in. (S4S) timber beam in Fig. 14–45. The allowable bending stress is 1800 psi and the allowable shear stress is 140 psi. Be sure to consider the weight of the beam.

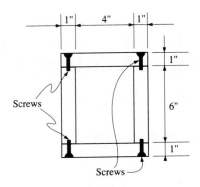

Beam cross section

FIGURE 14–44 Problem 49.

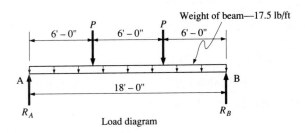

Load diagram

FIGURE 14–45 Problem 50.

51. Solve Problem 50 assuming that the timber beam is a dressed 6 in. by 18 in. member.

52. Calculate the values of S and Z and the shape factor for the cross section shown in Fig. 14–46.

53. A W18 × 50 is supported on simple supports on a 30 ft span. Calculate the yield moment M_Y and the plastic moment M_P and the maximum uniformly distributed load that the beam can carry. Assume $s_Y = 36$ ksi.

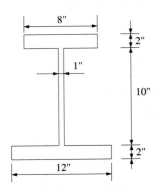

FIGURE 14–46 Problem 52.

15 Design of Beams

15–1
THE DESIGN PROCESS

The design problem involves the selection of the size and shape of a beam based on given loads and a given span length so as not to exceed established allowable stresses for a given material. The primary consideration in the design process is structural safety. Other considerations such as economy, functionality, maintainability, permanence, and the like are also of importance, but are secondary when compared to structural safety concerns.

The same principles and relationships established in Chapters 13 and 14 and applied to analysis-type problems can also be applied to the design of beams. As stated previously, a loaded beam must be capable of resisting both bending stresses and shear stresses that have been developed by the applied loads. Generally, it is the bending stress that limits the allowable load on a beam. That is, when the loads are applied so as to stress the beam to the full value of the allowable bending stress, it will generally be found that the shear stress is less than the allowable shear stress value. Therefore, in the normal process of design, it is customary to select a beam on the basis of the bending moment and then investigate or check the beam for shear. If the computed shear stress is found to be excessive, a new beam must be selected that will provide the required shear strength along with the required bending strength. Shear rarely controls a design unless loads are very heavy and close to the supports and/or spans are very short.

Our discussion of beam design involves no new principles. Rather, it will focus on the normally used techniques of design for metal and timber beams. The following general procedure is a useful step-by-step approach:

1. Establish the location, magnitude, and type of superimposed loads along with the span length and support conditions (both vertical and horizontal) of the beam. In addition, establish all design constraints such as allowable stresses and any physical limitations of the beam due to the material of which it is made and/or its dimensions.
2. Draw a load diagram and calculate the beam reactions.
3. Draw a shear diagram, if necessary.
4. Calculate the bending moments at points of zero shear or where the shear diagram goes through zero. If necessary, draw a moment diagram.

5. Using the largest numerical value of the bending moment, regardless of the sign, calculate the required section modulus using the flexure formula:

$$\text{Required } S_x = \frac{M}{s_{b(\text{all})}}$$

6. Select the most economical beam (from the appropriate table in the appendix) with a section modulus slightly larger than that required. (At this point, consider the most economical beam as the beam having the least weight (lb/ft).)

7. After the section has been selected, calculate the additional moment caused by the weight of the beam at the same point where the maximum moment due to the superimposed loads was calculated.

8. The total design moment to be considered will now be the sum of the moment due to the superimposed loads and the moment due to the beam's own weight. Calculate the new required section modulus and compare it with the furnished section modulus of the selected member. If the furnished section modulus is greater than the new required section modulus, then no change is necessary. If the furnished section modulus is less than the new required section modulus, then the member selected will be inadequate; therefore, select a new member with an appropriate section modulus and repeat steps 7 and 8.

9. Check the shear stress using either the general shear formula or a simplified or approximate shear expression. The total shear force should include the additional shear due to the beam's own weight. If the computed shear stress is less than the allowable shear stress, then the beam will be satisfactory with respect to shear. If, however, the computed shear stress exceeds the allowable shear stress, select a new member that has a greater shear capacity and repeat steps 8 and 9.

This design procedure can be simplified if an estimated beam weight is included in step 1. Various rules of thumb exist as guidelines for timber and steel beam weight estimates. Suggested values range from 3% to 10% of the total load. The use of such rules of thumb should be based on judgement developed through practical experience. In general, assuming a beam loaded up to its maximum allowable bending stress, the longer the span, the greater the proportion of the total load that is due to the weight of the beam. At this point, however, we suggest that the recommended steps in the preceding design procedure be followed.

Other considerations influencing the design process include beam deflection (to be discussed in Chapter 16) and lateral support. If a vertically loaded beam does not have adequate lateral support, its compression flange (or compression side, in the case of a rectangular cross section) will have a tendency to buckle or deflect laterally where it is free to do so. If we consider a simply supported beam subjected to vertically downward loads, the bending stresses developed are compressive above the neutral axis and tensile below the neutral axis. The tensile stresses tend to hold the beam in a straight line between the supports, while the compressive stresses tend to deflect the beam in a lateral direction. This is similar to the behavior of a slender column subjected to axial compressive loads. The opposing tendencies of the top and bottom portions of the beam result in a torsional buckling. This occurs concurrently with a vertical deflection resulting from the applied loads. Figure 15–1 shows a laterally unsupported beam that has deflected vertically and buckled laterally.

FIGURE 15–1 Beam deflection and lateral buckling.

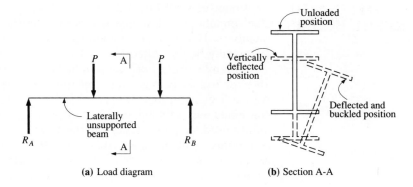

(a) Load diagram   (b) Section A-A

The lateral buckling of the compression flange can be prevented by proper beam design or by providing physical lateral support for the compression flange or the compression side of the beam. As a practical matter, most beams are constructed with adequate lateral support. However, should a beam have inadequate lateral support, a special design would be required that, in effect, reduces the load-carrying capacity of the beam. For purposes of our discussion, we will assume that all beams have adequate lateral support.

The design process described in this section is commonly termed *elastic design*. It is also called *allowable stress design*, since the design is based on a comparison of allowable stresses with the computed stresses produced by the applied loads (sometimes called *service loads*). As indicated in Chapter 10, the allowable stresses can be established through the combined use of laboratory tests and factors of safety. However, other factors must be taken into account, such as the aforementioned lateral buckling as well as other forms of localized buckling, which can indicate that the allowable stresses need to be modified.

While design methods allow for reasonable assurance of structural adequacy, in every structural member there is a finite chance for failure to occur. An unforeseen overload could lead to a beam failure. A ductile material is considered to have failed when the tensile or compressive bending stress reaches the yield point of the material. A brittle material is considered to have failed when rupture occurs and the bending stresses reach the ultimate strength of the material. Failure also occurs if the shear stress reaches the ultimate shear strength. Note that failure is not to be interpreted only as a collapse or rupture. Failure may be said to have occurred if a beam becomes unserviceable due to some structural deficiency and does not perform the function for which it was intended. For example, a beam or series of beams may not be functional if excessive vibrations result from some form of applied load.

15–2 DESIGN OF STEEL BEAMS

Since the material included in this section is of an introductory nature, the treatment of steel beam design will focus on conventional standard rolled steel members and simple geometric shapes that may be used as bending members. For a more detailed treatment of beam design, additional references may be consulted.[1,2]

Many varieties of standard hot-rolled structural steel shapes are available for use as beams. Some of the commonly available shapes shown in Appendices A through D may be used individually as beams, or they may be combined with other shapes and/or plates as built-up beams. The most commonly used shape for beams is the W shape. The maximum nominal depth commonly available is approximately 36 in., although deeper sections (up to about 40 in.) do exist. When the span length and/or load requirements become excessive for the economic use of a standard W shape, it may be strengthened by the addition of steel plates, or a section built up entirely of plates (called a *plate girder*) can be used.

The designation for the W shape was briefly discussed in Chapter 9. For American Standard channels (see Appendix C), the designation is interpreted in a similar manner. For instance, in the designation C15 × 50, the initial letter (C) indicates that the shape is a channel, 15 is the actual depth of the channel in inches, and 50 is the weight of the member in pounds per foot. Standard-weight steel pipe and extra strong steel pipe sections are included in Appendix B. Examples of the designations for these sections are Pipe 4 Std and Pipe 4 X-Strong, which designate a standard-weight and an extra-strong-weight pipe section of 4 in. nominal outside diameter. For angles (Appendix D), a typical designation is L8 × 6 × 1, which indicates an unequal leg angle with 8 in. and 6 in. legs and a thickness of 1 in. In each case, reference must be made to standard tables to obtain most properties. Appendices A through D contain only some of the shapes and sizes available. For a more complete listing, refer to manufacturers' literature and/or the AISC *Manual of Steel Construction—Allowable Stress Design*.[3]

---

[1] L. Spiegel and G. F. Limbrunner, *Applied Structural Steel Design*, 2nd ed. (Englewood Cliffs, NJ: Prentice-Hall, 1992).

[2] J. C. McCormac, *Structural Steel Design*, 4th ed. (New York: Harper & Row, 1992).

[3] American Institute of Steel Construction, Inc., *Manual of Steel Construction—Allowable Stress Design*, 9th ed. (Chicago: AISC, 1989).

The design procedure described in Section 15–1 is applicable to steel beams. However, one complicating factor is the determination of the allowable bending stress ($s_{b(\text{all})}$). A detailed discussion concerning the determination of applicable allowable bending stress values is beyond the scope of this text. For our treatment of beam design, we will assume that optimum conditions exist which will enable us to use the allowable bending stress at the upper end of the range of values as set forth in the AISC Specification.[4] This basic allowable bending stress (in both tension and compression) which can be used for rolled shapes is

$$s_{b(\text{all})} = 0.66 s_Y$$

where s_Y is the yield stress of the steel. In order for the steel shape to qualify for this allowable bending stress, however, several requirements must be satisfied including the provision of adequate lateral support for the compression flange. In addition, specified cross-sectional geometry restrictions must be satisfied to ensure that the shape will not undergo a localized buckling prematurely. Not only are these AISC design specification requirements technical and detailed, but some of the considerations require that a certain amount of engineering judgement be exercised. We will, therefore, proceed with steel beam selection on the assumption that an allowable bending stress is known or given. For further study, refer to the texts and the AISC Manual (which have been previously referenced) in which the subject is treated in detail.

The determination of an allowable shear stress is much simpler than the determination of the allowable bending stress. According to the AISC design specification, the allowable shear stress may be taken as

$$s_{s(\text{all})} = 0.40 s_Y$$

where, as before, s_Y is the yield stress of the steel. The allowable shear stress is set intentionally low to permit the use of the approximate average web shear approach (rather than the more precise general shear formula) as discussed in Chapter 14.

The AISC design specification is applicable to many different types of steel. These steels are specified according to ASTM (American Society for Testing and Materials) number. This number represents the number of the standard that defines the required minimum properties. Among these steels, the most widely used is the structural carbon steel designated ASTM A36. Its yield stress is 36 ksi, except for plates and bars in excess of 8 in. thick. Additional steels approved for use in building construction by the AISC Specification are ASTM A242 and ASTM A588, which are atmospheric corrosion-resistant, high-strength, low-alloy structural steels; ASTM A441,

---

[4]   American Institute of Steel Construction, Inc., *Specification for Structural Steel Buildings, Allowable Stress Design* (Chicago: AISC, 1989). (Also included in the AISC *Manual of Steel Construction—Allowable Stress Design.*)

which is a high-strength, low-alloy structural manganese-vanadium steel; ASTM A572, which is a high-strength, low-alloy columbium-vanadium steel; and ASTM A529, which is a structural carbon steel with a yield stress of 42 ksi. Various standardized structural products such as shapes, plates, and bars are available in some of the different types of steel.

The discussion in this section as well as Section 15–1 reflects beam design and analysis based on an elastic design method (also termed *allowable stress design*). In this method, actual applied loads (also called *service loads* or *working loads*) are used and the beam is designed or analyzed based on allowable bending and shear stresses. This method has been used for many decades with satisfactory results.

A recently developed method for structural steel design and analysis is based on an ultimate strength concept. In this method, service loads are amplified using load factors. Beams are then designed so that their practical strength at failure, which is somewhat less than the true strength at failure, is sufficient to resist the amplified loads. This new method is called *Load and Resistance Factor Design (LRFD)* and is discussed further in Section 15–4.

☐ **EXAMPLE 15–1**   Select the lightest (most economical) W shape to support a superimposed uniformly distributed load of 4 kips/ft on a simply supported span of 25 ft. Assume an allowable bending stress of 24 ksi and an allowable shear stress of 14.5 ksi.

Solution   The load diagram with the reactions is shown in Fig. 15–2. Since the beam is simply supported and carries a uniformly distributed load, maximum shear and moment can be computed using the shear and moment equations of Appendix H. The maximum moment is calculated from

$$M = \frac{wL^2}{8} = \frac{4(25)^2}{8} = 312.5 \text{ ft-kips}$$

This represents the maximum moment due to the superimposed loads and does not include the weight of the beam. The required strong-axis section modulus is then obtained from

$$\text{Required } S_x = \frac{M}{s_{b(\text{all})}} = \frac{312.5(12)}{24} = 156.3 \text{ in.}^3$$

From Appendix I (the beam-selection table), we select a W24 × 76, which has a section modulus of 176 in.³. The table facilitates the selection of W shapes used as beams on the basis of any allowable bending stress. The shapes are listed in order of decreasing section modulus (S_x) and further grouped so that the first shape in each

FIGURE 15–2   Load diagram and reactions.

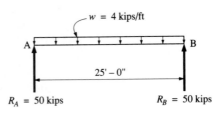

group (shown in bold print) is the lightest. Note that the W24 × 76 is the lightest W shape that furnishes at least the required section modulus. The use of the table depends on the value of the required section modulus which is calculated from the maximum bending moment and an appropriate allowable bending stress. Note that any shape having at least the required section modulus will be satisfactory for moment. If a shallower member is required, one may be chosen; however, it will be a heavier member.

The additional moment due to the beam's own weight is calculated from

$$\text{Additional } M = \frac{wL^2}{8} = \frac{0.076(25)^2}{8} = 5.94 \text{ ft-kips}$$

Therefore,

$$\text{New } M = 312.5 + 5.94 = 318.4 \text{ ft-kips}$$

from which

$$\text{New required } S_x = \frac{318.4(12)}{24} = 159.2 \text{ in.}^3$$

Since 159.2 in.^{3} < 176 in.^{3}, the W24 × 76 is satisfactory with respect to bending moment.

The maximum shear due to the superimposed load and the weight of the beam is calculated from

$$V = 50 + \frac{wL}{2} = 50 + \frac{0.076(25)}{2} = 51.0 \text{ kips}$$

Using beam dimensions from Appendix A, the shear stress is calculated as

$$s_s = \frac{V}{dt_w} = \frac{51.0}{23.92(0.440)} = 4.85 \text{ ksi}$$

$$4.85 \text{ ksi} < 14.5 \text{ ksi} \qquad \textbf{OK}$$

The W24 × 76 is satisfactory for both shear and moment.
Use a W24 × 76.

☐ **EXAMPLE 15–2** Select the lightest steel wide-flange section (W shape) for the beam shown in Fig. 15–3. Consider moment and shear. The allowable bending stress is 30 ksi and the allowable shear stress is 20 ksi.

Solution Using the superimposed loads, the reactions have been calculated and the shear and moment diagrams have been drawn. You should verify the values shown. In this case, a formula approach for shear and moment would be inappropriate because the location of the maximum moment is really unknown until the shear is seen to go through zero under the concentrated load. Based on the maximum bending moment from the moment diagram,

$$\text{Required } S_x = \frac{M}{s_{b(all)}} = \frac{832.5(12)}{30} = 333.0 \text{ in.}^3$$

From Appendix I, we select a W33 × 118 with S_x of 359 in.^{3}.

FIGURE 15–3   Steel beam design.

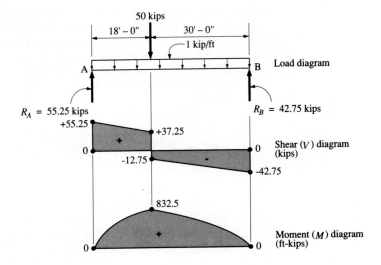

Load diagram

Considering the beam weight, we calculate the additional moment under the concentrated load (which is the previously determined point of maximum moment) due to an additional uniformly distributed load of 118 lb/ft (refer to Case 1 in Appendix H):

$$\text{Additional } M = \frac{wx}{2}(L - x) = \frac{0.118(18)}{2}(48 - 18) = 31.9 \text{ ft-kips}$$

Therefore,

$$\text{New } M = 832.5 + 31.9 = 864 \text{ ft-kips}$$

from which

$$\text{New required } S_x = \frac{864(12)}{30} = 346 \text{ in.}^3$$

$$346 \text{ in.}^3 < 359 \text{ in.}^3 \qquad \textbf{OK}$$

The W33 × 118 is satisfactory for bending moment.

The maximum shear due to the superimposed load and the weight of the beam is calculated from

$$V = 55.25 + \frac{wL}{2} = 55.25 + \frac{0.118(48)}{2} = 58.1 \text{ kips}$$

Using beam dimensions from Appendix A, the shear stress is calculated as

$$s_s = \frac{V}{dt_w} = \frac{58.1}{32.86(0.550)} = 3.21 \text{ ksi}$$

$$3.21 \text{ ksi} < 20 \text{ ksi} \qquad \textbf{OK}$$

The W33 × 118 is adequate for both shear and moment.
   Use a W33 × 118.

☐ **EXAMPLE 15–3**   A 6 in. thick reinforced concrete floor is supported on steel beams and girders with a layout for a typical building interior bay as shown in the plan view of Fig. 15–4. The floor is to be designed for a live load of 150 psf. The allowable bending stress is 24 ksi and the allowable shear stress is 14.5 ksi. Select the lightest W shapes for the beams and girders.

Solution   This is a typical structural steel beam-and-girder floor system for a commercial or industrial building. While some beams are supported by the girders and some by the columns, the span, load, and support conditions are the same. Therefore, the design will be the same for all beams.

Design of Beams (B1): As shown in Fig. 15–4, the span length of the beam is 40 ft. In a floor system of this type, it is common practice to assume the beam simply supported. The end connections of the beam determine whether it is simply supported or not. It is assumed at this point that the end connections will be so designed that the loaded beam will have a freedom of rotation at each end. This, in effect, creates a simple beam.

The loading on beam B1 includes the live load and the dead load. The dead load consists of the weight of the concrete floor and the weight of the beam itself. The live load is given as 150 psf. The weight of the reinforced concrete may be taken as 150 pcf. Therefore, the weight of a 6 in. thick slab per square foot of floor can be computed as

$$150 \left(\frac{6}{12} \right) = 75 \text{ psf}$$

The weight of the beam will be neglected for now, since the beam section is unknown. It will be considered, however, after an initial beam selection has been made.

FIGURE 15–4   Framing plan.

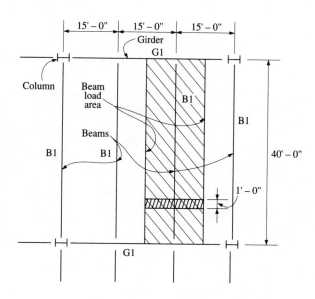

FIGURE 15–5   Beam load diagram.

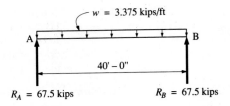

Since the beams are spaced 15 ft on centers, each beam will support a 15 ft width of floor. This load area is shown cross-hatched in Fig. 15–4. Calculating the design load per linear foot of beam (double cross-hatched area),

$$\text{Live load} = 150(15) = 2250 \text{ lb/ft}$$

$$\text{Slab dead load} = 75(15) = 1125 \text{ lb/ft}$$

from which

$$\text{Total design load} = 2250 + 1125 = 3375 \text{ lb/ft}$$
$$= 3.375 \text{ kips/ft}$$

The beam is isolated as a free body and the reactions calculated from the load diagram shown in Fig. 15–5:

$$R_A = R_B = \frac{wL}{2} = \frac{3.375(40)}{2} = 67.5 \text{ kips}$$

Since the beam is simply supported and supports a uniformly distributed load, maximum shear and moment can be computed using the shear and moment equations of Appendix H. The maximum moment is computed from

$$M = \frac{wL^2}{8} = \frac{3.375(40)^2}{8} = 675 \text{ ft-kips}$$

from which

$$\text{Required } S_x = \frac{M}{s_{b(\text{all})}} = \frac{675(12)}{24} = 338 \text{ in.}^3$$

From Appendix I, we select a W33 × 118 with S_x of 359 in.^{3}.

The additional moment due to the beam's own weight is calculated from

$$\text{Additional } M = \frac{wL^2}{8} = \frac{0.118(40)^2}{8} = 23.6 \text{ ft-kips}$$

Therefore,

$$\text{New } M = 675 + 23.6 = 699 \text{ ft-kips}$$

from which

$$\text{New required } S_x = \frac{M}{s_{b(\text{all})}} = \frac{699(12)}{24} = 350 \text{ in.}^3$$

$$350 \text{ in.}^3 < 359 \text{ in.}^3 \qquad \textbf{OK}$$

Therefore, the W33 × 118 is satisfactory for bending moment.

The maximum shear due to the superimposed load and the weight of the beam is calculated from

$$V = 67.5 + \frac{wL}{2} = 67.5 + \frac{0.118(40)}{2} = 69.9 \text{ kips}$$

Using beam dimensions from Appendix A, the shear stress is calculated as

$$s_s = \frac{V}{dt_w} = \frac{69.9}{32.86(0.550)} = 3.87 \text{ ksi}$$

$$3.87 \text{ ksi} < 14.5 \text{ ksi} \qquad \textbf{OK}$$

The W33 × 118 is adequate for both shear and moment.
   Use a W33 × 118.

Design of Girders (G1): The span length of the girder shown in Fig. 15–4 is taken as the center-to-center distance of the supporting members which, in this example, is 45 ft. The loading to G1 is symmetrical, consisting of two concentrated loads and a uniformly distributed load due to the weight of the member. The small portion of slab supported directly on the girder can be neglected, since the entire floor load was taken by the beams and is therefore transferred to the girder as a beam reaction. Since this is an interior bay, the girder will support beams framing into it from each side. The reaction from B1 is 69.9 kips. Therefore, each concentrated load supported by the girder is equal to 2(69.9) = 139.8 kips.

   The girder load diagram is shown in Fig. 15–6. Only the superimposed loads are included. Reactions are also shown. Note that the beam on the column line does not enter into the problem since it frames directly into the column, transferring its load to the column without physically touching the girder.

FIGURE 15–6 Girder load diagram.

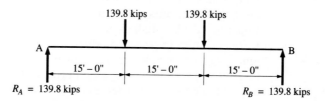

Since the beam is simply supported and symmetrically loaded, maximum shear and moment can be computed using the equations in Appendix H. The maximum moment can be calculated from

$$M = Pa = 139.8(15) = 2097 \text{ ft-kips}$$

from which

$$\text{Required } S_x = \frac{M}{s_{b(\text{all})}} = \frac{2097(12)}{24} = 1049 \text{ in.}^3$$

From Appendix I, we select a W40 × 268 with S_x of 1090 in.[3].
   The additional moment due to the girder's own weight is calculated from

$$\text{Additional } M = \frac{wL^2}{8} = \frac{0.268(45)^2}{8} = 67.8 \text{ ft-kips}$$

Therefore,

$$\text{New } M = 2097 + 67.8 = 2165 \text{ ft-kips}$$

from which

$$\text{New Required } S_x = \frac{M}{s_{b(\text{all})}} = \frac{2165(12)}{24} = 1083 \text{ in.}^3$$

$$1083 \text{ in.}^3 < 1090 \text{ in.}^3 \quad \textbf{OK}$$

Therefore, the W40 × 268 is satisfactory for bending moment.

The maximum shear due to the superimposed load and the weight of the beam is calculated from

$$V = 139.8 + \frac{wL}{2} = 139.8 + \frac{0.268(45)}{2} = 145.8 \text{ kips}$$

Using beam dimensions from Appendix A, the shear stress is calculated as

$$s_s = \frac{V}{dt_w} = \frac{145.8}{39.37(0.750)} = 4.94 \text{ ksi}$$

$$4.94 \text{ ksi} < 14.5 \text{ ksi} \quad \textbf{OK}$$

Therefore, the W40 × 268 is adequate for both shear and moment.

Use a W40 × 268.

As with structural elements, design methods for the various machine elements are founded on the theories of mechanics and strength of materials. One such element is a shaft. A shaft may be described as a machine component that transmits rotational motion or power. In the process of transmitting power, the shaft is subjected to a torsional moment, or torque. The torque results in torsional shear stress being developed in the shaft. Also, since a shaft usually carries power-transmitting components such as gears and/or belt sheaves, a loading perpendicular to the axis of the shaft may be applied. These forces (perpendicular or transverse) cause bending moments to be developed in the shaft. The combination of torsional shear stresses and bending stresses occurring simultaneously requires a consideration of the effects of the *combined* stresses.

The following example considers only the bending action on a shaft. Torsional shear stresses have been neglected. Combined stresses are discussed in Chapter 17.

□ **EXAMPLE 15–4**   Design a solid circular steel shaft that is simply supported on bearings 8 ft apart. A pulley weighing 50 lb is attached at the center of the 8 ft span. A belt pulls on the pulley with a force of 850 lb in a vertically downward direction. The allowable bending stress is 12,500 psi and the allowable shear stress is 8000 psi. Neglect the weight of the shaft.

Solution   The load diagram and reactions are shown in Fig. 15–7. Maximum shear and moment can be computed using the equations of Appendix H. The maximum moment is

FIGURE 15–7   Load diagram.

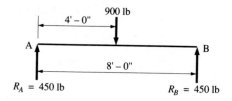

calculated from

$$M = \frac{PL}{4} = \frac{900(8)}{4} = 1800 \text{ ft-lb}$$

from which

$$\text{Required } S_x = \frac{M}{s_{b(\text{all})}} = \frac{1800(12)}{12,500} = 1.728 \text{ in.}^3$$

This is a solid circular shape. Section modulus is calculated from

$$S_x = \frac{\pi d^3}{32}$$

Solving for d,

$$\text{Required } d = \sqrt[3]{\frac{32 S_x}{\pi}} = \sqrt[3]{\frac{32(1.728)}{\pi}} = 2.60 \text{ in.}$$

Try a $2\frac{5}{8}$ in. diameter shaft.

Checking shear, the maximum shear (the reaction) from the load diagram is 450 lb. This results in a shear stress [see Eq. (14–13)] of

$$s_s = \frac{4V}{3A} = \frac{4(450)}{3[0.7854(2.625)^2]} = 111 \text{ psi}$$

Since 111 psi < 8000 psi, the shaft is satisfactory for shear, also.
Use a $2\frac{5}{8}$ in. diameter shaft.

15–3 DESIGN OF TIMBER BEAMS

Timber beam applications are generally limited to the building design and construction field. The design procedure, as itemized in Section 15–1, is applicable to timber beams as well as to steel beams.

A timber beam (like a steel beam) is designed for moment and then checked for shear. While there are several other elements that must be considered in a complete design, an important additional item for timber beam design is compression perpendicular to the grain. As shown in Fig. 15–8, a load (force) applied perpendicular to the grain, such as that which occurs under the ends of a beam, or under a column supported by a beam, induces a localized compressive or bearing stress.

As shown in Appendix F, the allowable compressive stress perpendicular to the grain is generally considerably less than the allowable compres-

FIGURE 15–8 Compression perpendicular to grain.

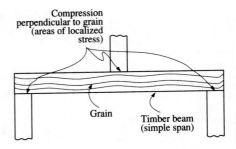

sive stress parallel to the grain. This is due to the fact that wood, being an aggregate of cells (or fibers) running primarily in one direction, is not an isotropic material. That is, the strength properties of a piece of wood are not the same in all directions. Wood is strongest when loaded to induce a stress parallel to the grain, either in tension or compression.

While the solid rectangular member shown in Fig. 15–9(a) is the most common section used when designing in wood, various shapes of built-up timber members, such as those shown in Fig. 15–9(b–d), may be used. All of these shapes (except the solid rectangular shape) are built-up of sawn timber elements connected with various types of fasteners. These are not to be confused with structural glued laminated timber members (commonly called *glulam members*). Glulam members are engineered products composed of assemblies of suitably selected and prepared wood laminations bonded together with adhesives, as shown in Fig. 15–10.

FIGURE 15–9 Timber shapes.

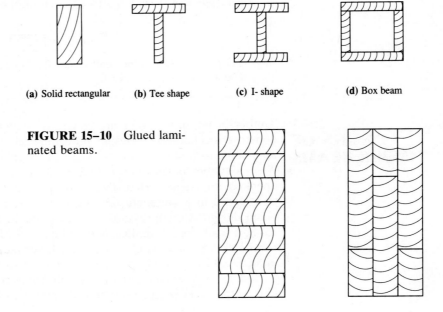

(a) Solid rectangular   **(b)** Tee shape   **(c)** I- shape   **(d)** Box beam

FIGURE 15–10 Glued laminated beams.

As is the case with steel beam design, a complicating factor in the design of timber beams is the determination of the allowable bending stress, as well as other allowable stresses. These values will vary with the multitude of different wood species, grades, and sizes, and also may vary depending on which design specification or code is being used. Our discussion is based on the *National Design Specification for Wood Construction* of the National Forest Products Association (NFPA).[5]

Appendix F contains some representative values of design stresses and modulus of elasticity values for timber. It is primarily intended as a resource to accompany the examples and problems of this text. Various modifications of the allowable stresses due to inadequate lateral support, as well as to conditions of usage, are applicable to these tabular values, but are considered to be beyond the scope of this text. Those who require more detailed information are referred to the NFPA Specification as well as other publications that treat timber design in great detail.[6,7,8]

Since the material included in this section is intended solely as an introduction, our treatment of timber beam design will be limited to solid rectangular timber members. The design of built-up timber sections is largely a matter of trial and error. This is generally accomplished using rules of thumb and approximations to create a desired cross section and then performing an analysis to check that allowable stresses will not be exceeded.

The design of timber members should be based on sizes that are readily available. These will usually be the dressed sizes, but may be the nominal sizes. As shown in Appendix E, the sections are customarily specified in terms of nominal sizes, which are the approximate sizes of rough, green, undressed pieces. The dressed size is the actual dimension of the finished product after surfacing. The dimensions and other properties of the dressed timber (S4S) are also provided in Appendix E. The S4S designation indicates that the member has been surfaced by a planing machine (for the purpose of obtaining smoothness of surface and uniformity of size) on all four sides. This table is an important reference and may be used for both analysis and design problems. It will be used for the timber design examples which follow.

Due to the many sizes of solid rectangular sections that would satisfy a design, it is sometimes difficult to select the most economical structurally

5 National Forest Products Association, *National Design Specification for Wood Construction* (Washington, D.C.: NFPA, 1986).

6 Western Wood Products Association, *Western Woods Use Book—Structural Data and Design Tables* (Portland, OR: WWPA, 1987).

7 American Institute of Timber Construction, *Timber Construction Manual* (New York: Wiley, 1966).

8 National Forest Products Association, *Wood Structural Design Data* (Washington, D.C.: NFPA, 1986).

adequate member. To facilitate the selection process, rules of thumb relative to beam depth-width ratio should be considered. Generally, a timber beam depth-width ratio should be between 2.0 and 3.0, with 1.33 as an absolute lower limit. (This does not apply to joists that are closely spaced; for example, 16 in. or 24 in. spacings. Depth-width ratios for joists may range from 4.0 to 6.0.) Another rule of thumb indicates that under normal loading conditions, the depth of a timber beam should be approximately 1 inch per foot of span where a beam is uniformly loaded, and approximately $1\frac{1}{4}$ inch per foot of span where the loads are concentrated or combined.

□ **EXAMPLE 15–5**   Design a timber beam to support a superimposed load of 1000 lb/ft on a simply supported span of 16 ft. The beam is to be a solid rectangular Douglas fir (S4S) member.

Solution   From Appendix F, the allowable bending stress is 1450 psi and the allowable shear stress is 95 psi. The load diagram is shown in Fig. 15–11. The reactions due to the superimposed loads are

$$R_A = R_B = \frac{wL}{2} = \frac{1000(16)}{2} = 8000 \text{ lb}$$

FIGURE 15–11   Load diagram and reactions.

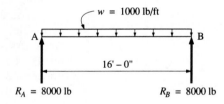

$$R_A = 8000 \text{ lb} \qquad\qquad R_B = 8000 \text{ lb}$$

From Appendix H, the maximum moment is calculated from

$$M = \frac{wL^2}{8} = \frac{1000(16)^2}{8} = 32{,}000 \text{ ft-lb}$$

Note that this does not include the weight of the beam. The required section modulus is calculated from

$$\text{Required } S_x = \frac{M}{s_{b(\text{all})}} = \frac{32{,}000(12)}{1450} = 265 \text{ in.}^3$$

From Appendix E, any member that will provide a section modulus of at least 265 in.³ will be satisfactory for moment. Among these are the 6 × 18, the 8 × 16, and the 10 × 14. Based on weight, the most economical member is the 6 × 18. Therefore, we will carry out the rest of the calculations based on the 6 × 18 as the trial section. Note the following properties for the 6 × 18:

$$S_x = 281 \text{ in.}^3$$

$$A = 96.3 \text{ in.}^2$$

$$\text{Wt.} = 26.7 \text{ lb/ft}$$

The additional moment due to the weight of the beam is calculated from

$$\text{Additional } M = \frac{wL^2}{8} = \frac{26.7(16)^2}{8} = 854 \text{ ft-lb}$$

The total design moment is then

$$\text{New } M = 32{,}000 + 854 = 32{,}854 \text{ ft-lb}$$

from which

$$\text{Required } S_x = \frac{M}{s_{b(all)}} = \frac{32{,}854(12)}{1450} = 272 \text{ in.}^3$$

$$272 \text{ in.}^3 < 281 \text{ in.}^3 \quad \textbf{OK}$$

The 6 × 18 is satisfactory with respect to bending moment.

The maximum shear is equal to the reaction and includes the effects of the superimposed load and the weight of the beam:

$$V = R_A = R_B = 8000 + \frac{wL}{2} = 8000 + \frac{26.7(16)}{2} = 8214 \text{ lb}$$

The shear stress is then calculated to be

$$s_s = \frac{1.5V}{A} = \frac{1.5(8214)}{96.3} = 127.9 \text{ psi}$$

$$127.9 \text{ psi} > 95 \text{ psi} \quad \textbf{N.G.}$$

The beam is unsatisfactory with respect to shear. Therefore, utilizing the shear stress formula, we solve for a required area A based on the allowable shear stress:

$$\text{Required } A = \frac{1.5V}{s_{s(all)}} = \frac{1.5(8214)}{95} = 129.7 \text{ in.}^2$$

and we now try an 8 × 18, which has properties of

$$S_x = 383 \text{ in.}^3$$

$$A = 131 \text{ in.}^2$$

$$\text{Wt.} = 36.4 \text{ lb/ft}$$

The additional moment due to the weight of the beam is calculated from

$$\text{Additional } M = \frac{wL^2}{8} = \frac{36.4(16)^2}{8} = 1165 \text{ ft-lb}$$

The total design moment is then

$$\text{New } M = 32{,}000 + 1165 = 33{,}165 \text{ ft-lb}$$

from which

$$\text{Required } S_x = \frac{M}{s_{b(all)}} = \frac{33{,}165(12)}{1450} = 274 \text{ in.}^3$$

$$274 \text{ in.}^3 < 383 \text{ in.}^3 \quad \textbf{OK}$$

The 8 × 18 is satisfactory with respect to bending moment.

The maximum shear is equal to the reaction and includes the effects of the superimposed load and the weight of the beam:

$$V = R_A = R_B = 8000 + \frac{wL}{2} = 8000 + \frac{36.4(16)}{2} = 8291 \text{ lb}$$

The shear stress is then calculated to be

$$s_s = \frac{1.5V}{A} = \frac{1.5(8291)}{131} = 94.9 \text{ psi}$$

$$94.9 \text{ psi} < 95 \text{ psi} \quad \textbf{OK}$$

The beam is satisfactory with respect to both shear and moment.
Use an 8 × 18 (S4S).

□ **EXAMPLE 15–6**   The most common floor-framing arrangement in timber construction consists of closely spaced joists supported by either bearing walls or beams as shown in Fig. 15–12.

FIGURE 15–12   Floor cross section.

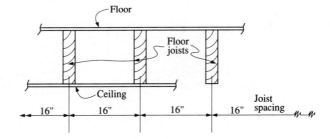

Select the Hem-fir (S4S) joist required to support the floor shown. The live load is 40 psf. There will be a superimposed dead load of 15 psf, which includes the dead weight of the floor and the ceiling, but not the weight of the joists. The joists are spaced 16 in. on center and span 16 ft. Assume the joist to be simply supported.

Solution   From Appendix F, the allowable bending stress is 1000 psi and the allowable shear stress is 75 psi. The floor loading per square foot must be converted to a loading per linear foot of joist. Since each joist supports a 16 in. wide strip of floor, the loading per foot is calculated as follows:

$$\text{Live load} = 40 \frac{\text{lb}}{\text{ft}^2} \left(\frac{16 \text{ in.}}{12 \text{ in./ft}} \right) = 53.3 \text{ lb/ft}$$

$$\text{Dead load} = 15 \left(\frac{16}{12} \right) = 20.0 \text{ lb/ft}$$

Therefore, the total load is

$$53.3 + 20.0 = 73.3 \text{ lb/ft}$$

FIGURE 15-13 Joist load diagram.

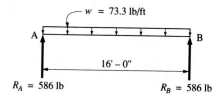

The load diagram (with the reactions due to the superimposed load) is shown in Fig. 15-13. From Appendix H, the maximum moment is calculated from

$$M = \frac{wL^2}{8} = \frac{73.3(16)^2}{8} = 2346 \text{ ft-lb}$$

from which

$$\text{Required } S_x = \frac{M}{s_{b(\text{all})}} = \frac{2346(12)}{1000} = 28.15 \text{ in.}^3$$

From Appendix E, we try a 2 × 12, which has properties of

$$S_x = 31.6 \text{ in.}^3$$
$$A = 16.9 \text{ in.}^2$$
$$\text{Wt.} = 4.68 \text{ lb/ft}$$

The additional moment due to the weight of the joist is calculated from

$$\text{Additional } M = \frac{wL^2}{8} = \frac{4.68(16)^2}{8} = 150 \text{ ft-lb}$$

The total design moment is then

$$\text{New } M = 2346 + 150 = 2496 \text{ ft-lb}$$

from which

$$\text{Required } S_x = \frac{M}{s_{b(\text{all})}} = \frac{2496(12)}{1000} = 30.0 \text{ in.}^3$$
$$30.0 \text{ in.}^3 < 31.6 \text{ in.}^3 \quad \textbf{OK}$$

The 2 × 12 joist is satisfactory with respect to bending moment.

The maximum shear is equal to the reaction and includes the effects of the superimposed load and the weight of the joist:

$$V = R_A = R_B = 586 + \frac{wL}{2} = 586 + \frac{4.68(16)}{2} = 623 \text{ lb}$$

The shear stress is then calculated to be

$$s_s = \frac{1.5V}{A} = \frac{1.5(623)}{16.9} = 55 \text{ psi}$$
$$55 \text{ psi} < 75 \text{ psi} \quad \textbf{OK}$$

The joist is satisfactory with respect to both shear and moment.

Use a 2 × 12 joist (S4S).

Assume that this joist is supported at each end by concrete masonry walls. Calculate the length of end bearing required. From Appendix F, the allowable compressive stress perpendicular to the grain is 245 psi.

In Fig. 15–14, the unknown length of end bearing is designated as x. Solve for x by equating the bearing stress on the contact surface and the allowable compressive stress. The bearing stress is determined by dividing the reaction by the area of the contact surface:

$$\frac{623 \text{ lb}}{x(1.5 \text{ in.})} = 245 \text{ lb/in.}^2$$

from which

$$\text{Required } x = \frac{623}{245(1.5)} = 1.70 \text{ in.}$$

This represents a minimum value. Most building codes require a minimum bearing length of $1\frac{1}{2}$ in. or 2 in. on wood or metal and 3 in. on masonry. Therefore, use a bearing length of 3 in.

FIGURE 15–14   End bearing.

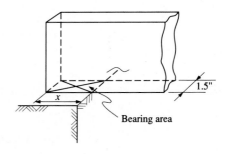

Bearing area

☐ **EXAMPLE 15–7**   Design a timber beam for the span and superimposed loading shown in Fig. 15–15. The beam is to be a solid rectangular California redwood (S4S) member. Consider moment and shear only.

Solution   From Appendix F, the allowable bending stress is 1350 psi and the allowable shear stress is 100 psi. In this example, the reactions are calculated and load, shear, and moment diagrams are as shown in Fig. 15–15. Alternative methods of computing maximum shear and moment values could be used if desired. As indicated on the diagrams, the maximum shear is 10,100 lb and the maximum moment is 49,000 ft-lb. The locations on the beam at which these maximum values occur are also apparent. These values are due to superimposed load only; they do not include the weight of the beam.

The required section modulus is calculated from

$$\text{Required } S_x = \frac{M}{s_{b(\text{all})}} = \frac{49,000(12)}{1350} = 436 \text{ in.}^3$$

FIGURE 15–15 Load, shear, and moment diagrams.

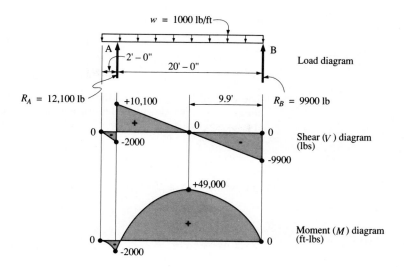

From Appendix E, the following members provide a section modulus of at least 436 in.^{3}: 8 × 20, 10 × 18, and 12 × 16. Of the three, the 8 × 20 is the lightest; therefore, it is the one we will try. We note the following properties:

$$S_x = 475 \text{ in.}^3$$

$$A = 146 \text{ in.}^2$$

$$\text{Wt.} = 40.6 \text{ lb/ft}$$

Considering the moment due to the weight of the beam, since this is an overhanging beam supporting a uniformly distributed load and since the weight of the beam is the same type of load, we will calculate the additional moment based on proportion. Therefore, no new reactions need be calculated and the diagrams need not be redrawn.

$$\text{New moment} = \left(\frac{1000 + 40.6}{1000}\right)(49,000) = 51,000 \text{ ft-lb}$$

from which

$$\text{Required } S_x = \frac{M}{s_{b(\text{all})}} = \frac{51,000(12)}{1350} = 453 \text{ in.}^3$$

$$453 \text{ in.}^3 < 475 \text{ in.}^3 \quad \textbf{OK}$$

The 8 × 20 is satisfactory with respect to bending moment.

The maximum shear due to superimposed loads is 10,100 lb. The additional shear due to the weight of the beam can be obtained by proportion:

$$\text{New total } V = \left(\frac{1000 + 40.6}{1000}\right)(10,100) = 10,510 \text{ lb}$$

from which we calculate

$$s_s = \frac{1.5V}{A} = \frac{1.5(10,510)}{146} = 108 \text{ psi}$$

$$108 \text{ psi} > 100 \text{ psi} \quad \textbf{N.G.}$$

The 8 × 20 beam is unsatisfactory with respect to shear. Therefore, utilizing the shear stress formula, we solve for a required area A based on the allowable shear stress:

$$\text{Required } A = \frac{1.5V}{s_{s(\text{all})}} = \frac{1.5(10,510)}{100} = 157.7 \text{ in.}^2$$

and we now try a 10 × 18, which has properties of

$$S_x = 485 \text{ in.}^3$$

$$A = 166 \text{ in.}^2$$

$$\text{Wt.} = 46.1 \text{ lb/ft}$$

The new total moment due to the superimposed load plus the weight of the beam is calculated from

$$\text{New moment} = \left(\frac{1000 + 46.1}{1000}\right)(49,000) = 51,300 \text{ ft-lb}$$

from which

$$\text{Required } S_x = \frac{M}{s_{b(\text{all})}} = \frac{51,300(12)}{1350} = 456 \text{ in.}^3$$

$$456 \text{ in.}^3 < 485 \text{ in.}^3 \quad \textbf{OK}$$

The 10 × 18 is satisfactory with respect to bending moment.
Checking shear,

$$\text{New total } V = \left(\frac{1000 + 46.1}{1000}\right)(10,100) = 10,570 \text{ lb}$$

from which we calculate

$$s_s = \frac{1.5V}{A} = \frac{1.5(10,570)}{166} = 95.5 \text{ psi}$$

$$95.5 \text{ psi} < 100 \text{ psi} \quad \textbf{OK}$$

The beam is satisfactory with respect to both moment and shear.
Use a 10 × 18 (S4S).

**15–4
LOAD AND
RESISTANCE FACTOR
DESIGN (LRFD) FOR
BENDING MEMBERS**

In the preceding sections, we have discussed allowable stress design where the allowable stress has represented a limit of structural usefulness that must not be exceeded. A somewhat different design approach has evolved that also establishes a limit of structural usefulness, but it is based on a limit state philosophy. The term *limit state* describes a condition at which a structure or some part of the structure ceases to perform its intended function. This approach has, at various times, been termed *ultimate strength design, limit state design,* and *load and resistance factor design.*

In this section, we will discuss load and resistance factor design (LRFD), the relatively new and modern approach to structural steel design. Our discussion is limited to bending members and is based on the inelastic behavior of bending members as discussed in Section 14–8 of this text. The LRFD method was introduced and developed in the early 1970s. In 1986 the American Institute of Steel Construction (AISC) published the first LRFD Specification[9] and recommended its use in the building design field. This method utilizes a series of factors of safety called *load factors* when applied to loads and *resistance factors* when applied to member strength or resistance. Load factors attempt to assess the possibility that prescribed loads will be exceeded while resistance factors provide for the possible understrength of the member.

The LRFD method for the design of bending members involves the following initial steps:

1. The calculation of the design load and load combinations. The design load is the product of the service load and the load factor.
2. The calculation of the required design strength expressed in terms of flexural strength M_u and shear strength V_u. Both M_u and V_u, for elementary cases, can be determined from shear and moment diagrams or by formulas.
3. The calculation of the furnished design strength expressed in terms of flexural strength (moment strength or bending strength) $\phi_b M_n$ and shear strength $\phi_v V_n$, where ϕ_b and ϕ_v represent resistance factors for bending and shear, respectively.

For any bending member, furnished flexural strength and the furnished shear strength must be greater than the required flexural strength and the required shear strength, respectively.

For flexural strength, this criterion can be expressed as

$$\phi_b M_n \geq M_u$$

where ϕ_b = resistance factor for flexure (0.90)
M_n = nominal (theoretical) flexural strength
M_u = required flexural strength

The product $\phi_b M_n$ represents the furnished or design flexural strength and is limited by

1. Lateral buckling considerations (as previously discussed in this chapter).
2. Localized buckling of one or more elements of the beam (e.g., the flange or web of the beam).

---

[9] American Institute of Steel Construction, Inc., *Manual of Steel Construction— Load and Resistance Factor Design* (Chicago: AISC, 1986).

3. The formation of a plastic hinge at a specific location. This indicates that the entire cross section of the member has yielded and the plastic moment M_P has been developed (see Section 14–8). Note that M_P represents the limit of the flexural strength M_n.

In this introduction to LRFD we will assume that the beam has adequate lateral support for its compression flange and that localized buckling of the beam flange and web will not occur. Hence, the beam will be able to attain its upper limit of flexural strength that occurs upon formation of the plastic hinge which, in turn, develops when the plastic moment M_P is reached. As shown in Chapter 14, the major axis flexural strength of doubly symmetric sections used as beams (e.g., W shapes) loaded in a plane of symmetry can be calculated from

$$M_P = s_Y Z$$

where all terms have been previously defined. Z (or Z_x) can be obtained from Table 15–1, which is a beam selection table for use in the LRFD method of

TABLE 15–1   Beam selection table (LRFD) for selected shapes.

Shape	Z_x in.^{3}	$\phi_b M_P$ ft-kips*	Shape	Z_x in.^{3}	$\phi_b M_P$ ft-kips*
W36 × 300	1260	3400	W18 × 60	123	332
W36 × 260	1080	2920	W18 × 50	101	273
W36 × 194	767	2070	W18 × 40	78.4	212
W36 × 150	581	1570			
			W16 × 45	82.3	222
W33 × 241	939	2540	W16 × 26	44.2	119
W33 × 201	772	2080			
W33 × 118	415	1120	W14 × 48	78.4	212
			W14 × 34	54.6	147
W30 × 108	346	934	W14 × 22	33.2	89.6
W30 × 99	312	842			
			W12 × 40	57.5	155
W27 × 114	343	926	W12 × 16	20.1	54.3
W27 × 94	278	751			
			W10 × 54	66.6	180
W24 × 104	289	780	W10 × 22	26.0	70.2
W24 × 76	200	540			
W24 × 62	153	413	W8 × 31	30.4	82.1
			W8 × 24	23.2	62.6
W21 × 73	172	464			
W21 × 50	110	297	W6 × 25	18.9	51.0

* For A36 steel ($s_Y = 36$ ksi).

design. This table is applicable for A36 steel ($s_Y = 36$ ksi) and it contains only a representative sample of W shapes.

LRFD is similar to elastic design (as discussed in Section 15–1) in the sense that bending member selection is also based on moment requirements and then checked for shear.

The shear strength design criterion can be expressed as

$$\phi_v V_n \geq V_u$$

where ϕ_v = shear resistance factor (0.90)
V_n = nominal (theoretical) shear strength
V_u = required shear strength

The product $\phi_v V_n$ represents the furnished or design shear strength.

As with flexural strength calculations, if we neglect buckling considerations, the nominal shear strength of a bending member may be obtained from

$$V_n = 0.60 s_Y A_w \qquad \text{AISC LRFDS Eqn. (F2–1)}$$

where A_w represents the area of the web of the shape. This represents a shear limit state resulting from yielding of the web. This is valid for all rolled W shapes made of steel with a yield stress of 65 ksi or less.

☐ **EXAMPLE 15–8**   Using LRFD, compute the allowable uniformly distributed nominal live load (lb/ft) that may be superimposed on W18 × 40 structural steel beams spaced at 8'–0" on center. The beams are simply supported on a span length of 30 ft and support a 6 in. thick reinforced concrete slab. Assume A36 steel ($s_Y = 36$ ksi) and full lateral support of the compression flange. Check shear. (Use a unit weight of reinforced concrete of 150 pcf.) Use load factors of 1.2 for dead load (*DL*) and 1.6 for live load (*LL*). The beam dimensions are such that localized buckling of the flange and web will not occur.

Solution   **1.** Dead load:
The weight of the concrete slab is calculated from

$$\frac{6}{12}(150) = 75 \text{ psf}$$

Since each beam supports a load area 8 ft wide, the weight of the slab per linear foot of beam is calculated from

$$75(8) = 600 \text{ lb/ft}$$

The beam weight is 40 lb/ft. Therefore, adding the slab weight and the beam weight results in a total dead load of 640 lb/ft.
2. Since the beam has full lateral support of its compression flange and localized buckling will not occur, the plastic moment M_P of the cross section can be developed and the design flexural strength of the section $\phi_b M_n$ is calculated as

$$\phi_b M_n = \phi_b M_P = \phi_b s_Y Z_x$$
$$= \frac{0.9(36)(78.4)}{12} = 212 \text{ ft-kips}$$

where Z_x is obtained from Table 15–1. Note that $\phi_b M_P$ can be obtained from the same table.

3. As a limit,

$$M_u = \phi_b M_n = \phi_b M_P$$

Therefore, we will compute the total factored design load w_u that will develop M_u. Rearranging

$$M_u = \frac{w_u L^2}{8}$$

and solving for w_u,

$$w_u = \frac{8M_u}{L^2} = \frac{8(212)}{30^2} = 1.884 \text{ kips/ft}$$

Solving for the nominal live load and using the given load factors,

$$w_u = (1.2)(DL) + (1.6)(LL)$$
$$1.884 = 1.2(0.640) + 1.6(LL)$$

from which

$$LL = 0.698 \text{ kips/ft}$$

4. Check shear:

$$V_u = \frac{w_u L}{2} = \frac{1.884(30)}{2} = 28.3 \text{ kips}$$

Since this is a W shape with yield stress of 36 ksi (less than 65 ksi), shear yielding governs and

$$V_n = 0.6 s_Y A_w$$
$$= 0.6(36)(17.90)(0.315) = 121.8 \text{ kips}$$
$$\phi V_n = 0.90(121.8) = 109.6 \text{ kips}$$
$$109.6 \text{ kips} > 28.3 \text{ kips} \quad \textbf{OK}$$

Therefore, the W18 × 40 is satisfactory in shear when subjected to a live load of 0.698 kips/ft.

□ **EXAMPLE 15–9**   The simply supported beam shown in Figure 15–16 supports a uniformly distributed service dead load of 0.5 kips/ft (which includes an assumed weight of beam) and a concentrated service live load of 10.0 kips. Assume A36 steel and full lateral support for the compression flange. Select the lightest shape within the limits of Table 15–1. Consider moment and shear. Use load factors of 1.2 for dead load and 1.6 for live load. Neglect localized buckling considerations.

FIGURE 15–16   Beam loading.

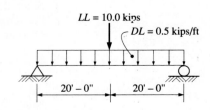

Solution   1. Loading:

$$P_u = 1.6(10.0) = 16.0 \text{ kips}$$
$$w_u = 1.2(0.5) = 0.60 \text{ kips/ft}$$

2. Maximum moment:

$$M_u = \frac{w_u L^2}{8} + \frac{P_u L}{4}$$

$$= \frac{0.60(40)^2}{8} + \frac{16.0(40)}{4} = 280 \text{ ft-kips}$$

3. Since, as a limit,

$$M_u = \phi_b M_n = \phi_b M_P = \phi Z_x s_Y$$

we can solve for Z_x:

$$\text{Required } Z_x = \frac{M_u}{\phi s_Y} = \frac{280(12)}{0.90(36)} = 103.7 \text{ in.}^3$$

From Table 15–1, select a W21 × 50 ($Z_x = 110 \text{ in.}^3$).

4. Check shear:

$$V_u = \frac{w_u L}{2} + \frac{P_u}{2} = \frac{0.60(40)}{2} + \frac{16.0}{2} = 20.0 \text{ kips}$$

For W shapes with $s_Y < 65$ ksi,

$$V_n = 0.60 s_Y A_w$$
$$= 0.60(36)(20.83)(0.380) = 171.0 \text{ kips}$$
$$\phi V_n = 0.90(171.0) = 153.9 \text{ kips}$$
$$153.9 \text{ kips} > 20.0 \text{ kips} \quad \textbf{OK}$$

Therefore, the W21 × 50 is satisfactory in shear.

15–5
SI SYSTEM EXAMPLES

□ **EXAMPLE 15–10**

A solid steel shaft, circular in cross section, is used as a machine element. It carries a load that causes a maximum bending moment of 960 N·m and a maximum shear of 8.0 kN (which includes the effect of an assumed weight of the shaft). The allowable bending stress for the steel is 75 MPa. Calculate the required minimum shaft diameter. Compute the shear stress.

Solution   Since this is a design problem, the design format of the flexure formula (Eq. 14–7) may be used:

$$\text{Required } S = \frac{M}{s_{b(\text{all})}}$$

Substituting,

$$\text{Required } S = \frac{960 \text{ N·m}}{75 \text{ MPa}} = \frac{960 \text{ N·m}}{75 \times 10^6 \text{ N/m}^2} = 12.8 \times 10^{-6} \text{ m}^3$$

The section modulus for a solid circular shape is

$$S = \frac{\pi d^3}{32}$$

Substituting and solving for d,

$$d = \sqrt[3]{\frac{32(12.8 \times 10^{-6} \text{ m}^3)}{\pi}} = 5.1 \times 10^{-2} \text{ m}$$

$$= 51 \text{ mm}$$

The area is calculated from

$$A = 0.7854d^2 = 0.7854(51^2) = 2040 \text{ mm}^2$$

The maximum horizontal shear stress (refer to Chapter 14) can be obtained from Eq. (14–13), where

$$s_s = \frac{4V}{3A} = \frac{4(8.0 \text{ kN})}{3(2040 \text{ mm}^2)} = 0.00523 \text{ kN/mm}^2$$

$$= 5.23 \text{ N/mm}^2$$
$$= 5.23 \text{ MPa}$$

☐ **EXAMPLE 15–11**   Select the lightest steel wide-flange section (W shape) for the beam in Fig. 15–17. Consider moment and shear. The allowable bending stress is 165 MPa and the allowable shear stress is 100 MPa.

Solution   Using the superimposed loads, the reactions have been calculated and the shear and moment diagrams have been drawn. You should verify the values shown.

Based on the maximum bending moment,

$$\text{Required } S_x = \frac{M}{s_{b(\text{all})}} = \frac{436 \text{ kN} \cdot \text{m}}{165 \text{ MPa}} = \frac{436 \times 10^3 \text{ N} \cdot \text{m}}{165 \times 10^6 \text{ N/m}^2} = 2.64 \times 10^{-3} \text{ m}^3$$

FIGURE 15–17 Load, shear, and moment diagrams.

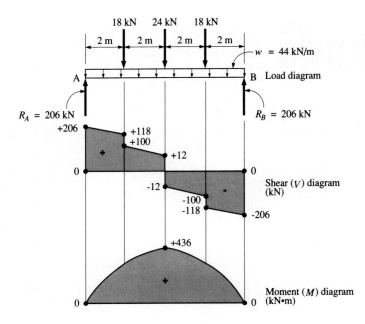

From Appendix I, we select a W610 × 1.11 with S_x of 2.88×10^{-3} m³.

Additional moment due to the beam's own weight of 1.11 kN/m is calculated from

$$M = \frac{wL^2}{8} = \frac{(1.11 \text{ kN/m})(8 \text{ m})^2}{8} = 8.88 \text{ kN} \cdot \text{m}$$

Therefore, the new total moment is

$$\text{New } M = 436 + 8.88 = 445 \text{ kN} \cdot \text{m}$$

from which

$$\text{Required } S_x = \frac{445 \text{ kN} \cdot \text{m}}{165 \text{ MPa}} = \frac{445 \times 10^3 \text{ N} \cdot \text{m}}{165 \times 10^6 \text{ N/m}^2} = 2.70 \times 10^{-3} \text{ m}^3$$

$$2.70 \times 10^{-3} \text{ m}^3 < 2.88 \times 10^{-3} \text{ m}^3 \quad \textbf{OK}$$

The W610 × 1.11 is satisfactory with respect to bending moment.

The maximum shear due to the superimposed load *and* the weight of the beam is

$$V = 206 \text{ kN} + (1.11 \text{ kN/m})(4 \text{ m}) = 210.4 \text{ kN}$$
$$= 210\ 400 \text{ N}$$

We obtain depth d and web thickness t_w for this shape from Appendix A. The shear stress can then be calculated from

$$s_s = \frac{V}{dt_w} = \frac{210\ 400 \text{ N}}{(608 \text{ mm})(11.2 \text{ mm})} = 30.9 \text{ N/mm}^2$$
$$= 30.9 \text{ MPa}$$
$$30.9 \text{ MPa} < 100 \text{ MPa} \quad \textbf{OK}$$

The W610 × 1.11 is adequate for both shear and moment.

Use a W610 × 1.11.

☐ **EXAMPLE 15–12** Design a timber beam to support the superimposed loading shown in Fig. 15–18. The 4 m span is simply supported. The beam is to be a solid rectangular southern pine (S4S) member.

FIGURE 15–18 Load diagram.

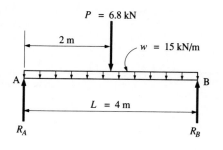

Solution   From Appendix F, the allowable bending stress is 11.0 MPa and the allowable shear stress is 0.62 MPa. The reactions are calculated from

$$R_A = R_B = \frac{wL}{2} + \frac{P}{2} = \frac{(15 \text{ kN/m})(4 \text{ m})}{2} + \frac{6.8 \text{ kN}}{2} = 33.4 \text{ kN}$$

The maximum moment can be computed using the equations in Appendix H (Cases 1 and 5):

$$M = \frac{wL^2}{8} + \frac{PL}{4}$$

$$= \frac{(15 \text{ kN/m})(4 \text{ m})^2}{8} + \frac{(6.8 \text{ kN})(4 \text{ m})}{4} = 36.8 \text{ kN} \cdot \text{m}$$

Note that this does not include the weight of the beam.

Based on the maximum bending moment,

$$\text{Required } S_x = \frac{M}{s_{b(\text{all})}} = \frac{36.8 \text{ kN} \cdot \text{m}}{11.0 \text{ MPa}} = \frac{36.8 \times 10^3 \text{ N} \cdot \text{m}}{11.0 \times 10^6 \text{ N/m}^2} = 3.35 \times 10^{-3} \text{ m}^3$$

From Appendix E, the following members furnish a section modulus of at least 3.35×10^{-3} m³: 150 × 410, 200 × 360, and 250 × 300. Of these, the 150 × 410 is the lightest; therefore, it is the one we will try. Its properties are

$$S_x = 3.63 \times 10^{-3} \text{ m}^3$$
$$A = 55.2 \times 10^{-3} \text{ m}^2$$
$$\text{Wt.} = 347 \times 10^{-3} \text{ kN/m}$$

The additional moment due to the weight of the beam is calculated from

$$M = \frac{wL^2}{8} = \frac{(347 \times 10^{-3} \text{ kN/m})(4 \text{ m})^2}{8} = 0.694 \text{ kN} \cdot \text{m}$$

Therefore, the new total moment is

$$\text{New } M = 36.8 + 0.694 = 37.5 \text{ kN} \cdot \text{m}$$

From which

$$\text{Required } S_x = \frac{37.5 \text{ kN} \cdot \text{m}}{11.0 \text{ MPa}} = \frac{37.5 \times 10^3 \text{ N} \cdot \text{m}}{11.0 \times 10^6 \text{ N/m}^2} = 3.41 \times 10^{-3} \text{ m}^3$$

$$3.41 \times 10^{-3} \text{ m}^3 < 3.63 \times 10^{-3} \text{ m}^3 \qquad \textbf{OK}$$

The 150 × 410 is satisfactory with respect to bending moment.

Making use of the previously calculated reaction, the maximum shear due to the superimposed load and the weight of the beam is

$$V = 33.4 \text{ kN} + \frac{(347 \times 10^{-3} \text{ kN/m})(4 \text{ m})}{2} = 34.1 \text{ kN}$$

The shear stress can be calculated from

$$s_s = \frac{1.5V}{A} = \frac{1.5(34.1 \times 10^3 \text{ N})}{55.2 \times 10^{-3} \text{ m}^2} = 0.927 \text{ MPa}$$

$$0.927 \text{ MPa} > 0.62 \text{ MPa} \qquad \textbf{N.G.}$$

The beam is unsatisfactory with respect to shear. Therefore, using the shear stress formula, we solve for a required area A based on an allowable shear stress of 0.62 MPa:

$$\text{Required } A = \frac{1.5V}{s_{s(\text{all})}} = \frac{1.5(34.1 \text{ kN})}{0.62 \text{ MPa}} = 82.5 \times 10^{-3} \text{ m}^2$$

From Appendix E, we now try a 200 × 460 timber beam. Its properties are

$$S_x = 6.30 \times 10^{-3} \text{ m}^3$$
$$A = 85.0 \times 10^{-3} \text{ m}^2$$
$$\text{Wt.} = 534 \times 10^{-3} \text{ kN/m}$$

Additional moment due to the weight of the beam is calculated from

$$M = \frac{wL^2}{8} = \frac{(534 \times 10^{-3} \text{ kN/m})(4 \text{ m})^2}{8} = 1.068 \text{ kN} \cdot \text{m}$$

The new total moment is calculated from

$$\text{New moment} = 36.8 + 1.068 = 37.9 \text{ kN} \cdot \text{m}$$

from which

$$\text{Required } S_x = \frac{37.9 \text{ kN} \cdot \text{m}}{11.0 \text{ MPa}} = \frac{37.9 \times 10^3 \text{ N} \cdot \text{m}}{11.0 \times 10^6 \text{ N/m}^2} = 3.45 \times 10^{-3} \text{ m}^3$$
$$3.45 \times 10^{-3} \text{ m}^3 < 6.30 \times 10^{-3} \text{ m}^3 \quad \textbf{OK}$$

The 200 × 460 is satisfactory with respect to bending moment. While the section should still be satisfactory for shear, it will be rechecked because of the increase in the shear due to the increased beam weight. The shear due to the superimposed load and the beam weight is

$$V = 33.4 \text{ kN} + \frac{(534 \times 10^{-3} \text{ kN/m})(4 \text{ m})}{2} = 34.5 \text{ kN}$$

The shear stress is calculated from

$$s_s = \frac{1.5V}{A} = \frac{1.5(34.5 \text{ kN})}{85 \times 10^{-3} \text{ m}^2} = 0.61 \text{ MPa}$$
$$0.61 \text{ MPa} < 0.62 \text{ MPa} \quad \textbf{OK}$$

The beam is satisfactory with respect to both shear and moment.
Use a 200 × 460 (S4S).

SUMMARY—BY SECTION NUMBER

15–1 Bending stress is the type of stress that usually limits the allowable load on a beam. Therefore, in the normal design process, a beam should be selected on the basis of bending moment and then checked for shear. (See the detailed step-by-step design procedure included in this section.)

15–2 Our discussion of the design of steel beams is limited to standard hot-rolled structural steel shapes and simple geometric shapes (e.g., circular shafts) used as bending members. For the design of the structural steel shapes, reference must be made to standard tables of properties included in the AISC *Manual of Steel Construction—Allowable Stress Design*. Portions of these tables appear in Appendices A through D of this text. In addition, Appendices H and I serve as references to expedite the design process.

15–3   Our discussion of the design of timber beams is limited to solid rectangular shapes, since these are, by far, the most commonly used sections in timber design. The necessary member properties and suggested design values needed for design are given in Appendices E and F. Although the design of timber members is generally based on dressed dimensions, nominal dimensions may also be used.

15–4   Our discussion of the AISC Load and Resistance Factor Design (LRFD) method is limited to W shapes and A36 steel bending members with full lateral bracing for the compression flange. Load factors used are 1.2 for dead load and 1.6 for live load. The resistance factor for flexure and shear is taken as 0.90. Localized buckling considerations are neglected. Problem solutions are based on the use of Table 15–1, which is limited in scope to selected W shapes.

PROBLEMS

In the following problems, the given loads are superimposed service loads; that is, they do not include the weights of the beams (unless noted otherwise). Consider moment and shear. For structural steel beams (allowable stress design), use an allowable bending stress of 24 ksi and an allowable shear stress of 14.5 ksi (unless noted otherwise). For structural steel beams using Load and Resistance Factor Design, neglect localized buckling considerations. For timber beams, all beams are solid, rectangular shapes and Appendices E and F are applicable.

Section 15–2   Design of Steel Beams

1. Select the lightest W shape to support a uniformly distributed load of 2.1 kips/ft on a simple span of 24 ft.

2. A simply supported beam is to support a uniformly distributed load of 10 kips/ft. Select the lightest W shape. The span length is 32 ft.

3. Rework Problem 1 given a load of 1.0 kip/ft and a span length of 30 ft.

4. Select a diameter for a solid circular steel shaft simply supported on bearings 8 ft apart. The shaft must be capable of carrying an applied load of 1200 lb acting vertically downward at any point along its length. The allowable bending stress is 12,000 psi and the allowable shear stress is 6000 psi. Neglect the weight of the shaft.

5. A simply supported beam is to span 15 ft. It will support a uniformly distributed load of 2 kips/ft over the full span and a concentrated load of 60 kips at midspan. Select the lightest W shape.

6. A simply supported beam is to span 24 ft. It will support a uniformly distributed load of 1.2 kips/ft over the full span and a concentrated load of 8 kips located 7 ft from the left support. Select the lightest W shape.

7. A rectangular aluminum bar being used as a beam is subject to a maximum bending moment of 120 ft-lb. Its height is to be three times its width. Compute the required dimensions of the bar so as not to exceed a bending stress of 8000 psi.

8. A simply supported steel beam is subjected to the loads shown in Fig. 15–19. Select the most economical

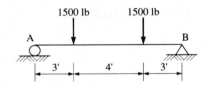

FIGURE 15–19   Problem 8.

section for the following shapes if the allowable bending stress is 20,000 psi. Neglect shear and the weight of the member. Compute the weight per foot for each shape and select the most economical (use 490 pcf for the unit weight for steel).
(a) Square bar
(b) Circular, solid section
(c) Rectangular bar with the depth three times the width
(d) Rectangular bar with the depth one-third the width
(e) Wide-flange (W shape)
(f) Standard weight pipe section

Section 15–3 Design of Timber Beams

9. Design a timber beam of Hem-fir (S4S) to support a uniformly distributed load of 600 lb/ft on a simply supported span of 15 ft.

10. Select simply supported timber beams (S4S) for the following uniformly distributed loads and spans: (a) Douglas fir, 600 lb/ft, 22 ft span; (b) Hem-fir, 500 lb/ft, 18 ft span.

11. Select a southern pine (S4S) timber beam for the conditions indicated in Fig. 15–20.

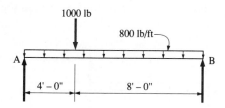

FIGURE 15–20 Problem 11.

12. Select simply supported Hem-fir (S4S) joists to support a floor live load of 40 psf on a 12 ft simple span. The superimposed dead load will be 15 psf (not including the joist weight). The joists are to be spaced 24 in. on center.

Section 15–4 Load and Resistance Factor Design (LRFD) for Bending Members

13. Calculate the furnished flexural strength $\phi_b M_n$ for the following beams and compare with the values shown in Table 15–1. The beams are all A36 steel and have adequate lateral support.
(a) W24 × 104   (b) W18 × 40   (c) W12 × 16

14. Using LRFD, calculate the uniformly distributed service live load (kips/ft) that may be superimposed on a W24 × 104 simply supported beam with a span length of 40 ft. Assume A36 steel and full lateral support of the compression flange. The beam supports a partition load (DL) of 200 lb/ft. Check shear. Use the load factors of 1.2 for dead load and 1.6 for live load.

15–17. For the beams shown (Figs. 15–21 through 15–23), select the most economical W shape using A36 steel and assuming full lateral support for the compression flange. Use LRFD. Service loads are shown (these exclude the weight of the beam). Use load factors of 1.2 for DL and 1.6 for LL.

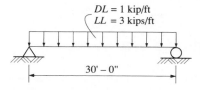

FIGURE 15–21 Problem 15.

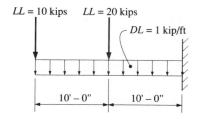

FIGURE 15–22 Problem 16.

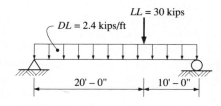

FIGURE 15–23 Problem 17.

SI System Problems

18. Select the lightest W shape to support a superimposed uniformly distributed load of 435 kN/m on a simply supported span of 5 m. Assume an allowable bending

stress of 165 MPa and an allowable shear stress of 100 MPa.

19. Select the lightest steel wide-flange section (W shape) for the beam in Fig. 15–24. Consider moment and shear. The allowable bending stress is 165 MPa and the allowable shear stress is 100 MPa.

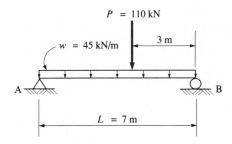

FIGURE 15–24   Problem 19.

20. Select the lightest steel wide-flange section (W shape) for the beam in Fig. 15–25. Consider moment and shear. The allowable bending stress is 165 MPa and the allowable shear stress is 100 MPa.

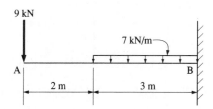

FIGURE 15–25   Problem 20.

21. Select Douglas fir (S4S) simply supported joists to carry a floor live load of 2.5 kPa and a superimposed dead load (floors and ceiling) of 1.0 kPa. The joists are to be spaced 0.4 m on center and the span length is 6 m.

22. Design simply supported timber beams (S4S) for the following uniformly distributed loads and spans: (a) eastern white pine, 9.0 kN/m, 7 m span length; (b) southern pine, 7.5 kN/m, 8 m span length.

Computer Problems

For the following computer problems, any appropriate programming language may be used. Input prompts should fully explain what is required of the user (the program should be "user friendly"). The resulting output should be well labeled and self-explanatory.

23. Write a program that will select the required depth (2 in. increments) of a 2 in. thick timber joist that will be spaced 16 in. on center and support a floor carrying a uniformly distributed load. Use nominal dimensions and properties. Consider moment and shear. Limits on the nominal depth are to be 6 in. minimum and 12 in. maximum. User input is to be allowable bending stress (psi), allowable shear stress (psi), live load (psf), dead load (psf), and span (ft). Neglect the weight of the joist.

24. Write a program that will select the lightest W shape from a limited group of W shapes. The beam is to be simply supported and will carry a uniformly distributed load. Consider moment only. You will need to create a short table of W shape properties from which the program can make the selection. (Real-world software contains more comprehensive tables of shape properties.) Choose a few shapes (perhaps six) from Appendix I and include nominal depth, weight, and section modulus in the table. User input for this program is to be uniformly distributed load intensity (kips/ft), span length (ft), and allowable bending stress (ksi).

Supplemental Problems

25. Select the lightest W shape to support a uniformly distributed load of 1600 lb/ft on a simple span of 48 ft.

26. Select the lightest W shape for the beam in Fig. 15–26.

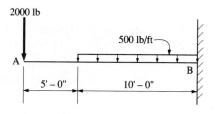

FIGURE 15–26   Problem 26.

27. Select the lightest W shape for the cantilever beam in Fig. 15–27.

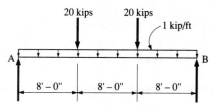

FIGURE 15–27   Problem 27.

28. Select the lightest W shape for the beams in Fig. 15–28.

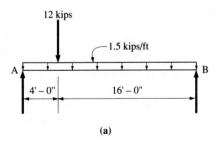

(a)

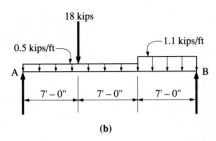

(b)

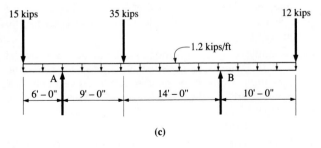

(c)

FIGURE 15–28 Problem 28.

29. The structural steel floor system of Fig. 15–29 is to support a 5 in. thick reinforced concrete slab, a live load of 100 psf, and an additional future load of 15 psf. Select the lightest W shapes for the beams and girders.

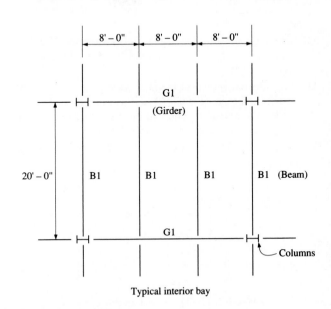

Typical interior bay

FIGURE 15–29 Problem 29.

30. The structural steel framing plan of Fig. 15–30 represents a floor system for a narrow commercial building. The floor is to support a 6 in. thick reinforced concrete slab, a live load of 150 psf, and an additional future load of 20 psf. Select the lightest W shapes for B1, B2, G1, G2, and G3.

FIGURE 15–30 Problem 30.

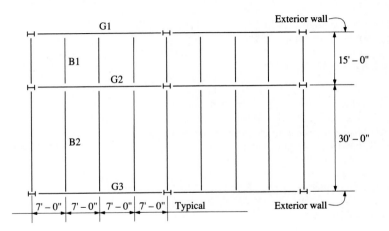

31. A cantilever beam 3 ft long is to support a concentrated load of 300 lb at its free end. Neglecting the weight of the beam, compute the required diameter of a solid circular steel bar to carry the load. The allowable bending stress is 20,000 psi and the allowable shear stress is 13,000 psi.

32. Select a southern pine (S4S) simply supported beam to support a uniformly distributed load of 400 lb/ft on a span of 14 ft.

33. A California redwood beam is to support a uniformly distributed load of 400 lb/ft and a concentrated load of 5.5 kips at midspan. The simply supported beam spans 20 ft. Select a 12 in. wide (S4S) beam.

34. Select a Douglas fir (S4S) beam for the conditions indicated in Fig. 15–31.

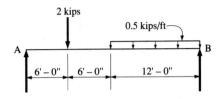

FIGURE 15–31   Problem 34.

35. Select an eastern white pine (S4S) beam for the conditions indicated in Fig. 15–32.

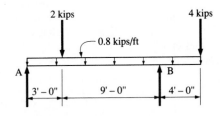

FIGURE 15–32   Problem 35.

36. A rough Douglas fir cantilever beam, 6 ft long with a circular cross section, is to support a concentrated load of 400 lb at the free end. Compute the required diameter. Neglect the weight of the beam.

37. Select southern pine (S4S) simply supported joists to carry a floor live load of 50 psf and a superimposed dead load (flooring and ceiling) of 20 psf. The joists are to be spaced 16 in. on center and the span length is to be 20 ft.

38. Rework Problem 37 using joists spaced 12 in. on center.

39. Rework Problem 37 using joists spaced 24 in. on center.

40. Select southern pine (S4S) simply supported joists to carry a floor live load of 120 psf and a superimposed dead load of 20 psf. The joist spacing will be 16 in. and the span will be 14 ft.

41. A 15 ft span simply supported Hem-fir (S4S) beam supports the contributory load of a width of 10 ft of floor. The live load is 50 psf and the superimposed dead load is 50 psf. Select a beam with a depth of not more than 18 in.

42. A series of 15 ft long Douglas fir (S4S) joists carry a floor live load of 40 psf and a superimposed dead load of 20 psf. In addition, there is a concentrated load from a partition running perpendicular to the joists 5 ft from one end. The partition load is 300 lb/ft. The joists are spaced 16 in. on center. Select the required size for the joists.

43. A cantilever beam 9 ft long is to be made from several 2 × 10 Douglas fir (S4S) planks bolted together (the long dimensions of the planks are placed vertically). A concentrated load of 1000 lb is to be carried at the free end. Determine the number of planks necessary to support the load.

44. Using LRFD, calculate the maximum service live load *P* that may be placed on the W30 × 99 beam shown in Fig. 15–33. Assume A36 steel and full lateral support

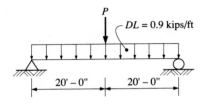

FIGURE 15–33   Problem 44.

of the compression flange. The beam also supports a uniformly distributed service dead load of 0.9 kip/ft (which excludes the weight of the beam). Check shear and use load factors of 1.2 for *DL* and 1.6 for *LL*.

45. Using LRFD, design A36 W-shape beams (shown in Fig. 15–34) to support a 6 in. reinforced concrete slab. The beams are 7 ft on center and have a span length of 50 ft. The service live load on the concrete slab is 125 psf. Assume full lateral support for the compression flange. Use load factors of 1.2 for *DL* and 1.6 for *LL*.

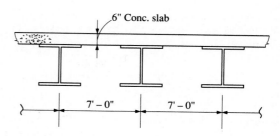

FIGURE 15–34 Problem 45.

16 Deflections of Beams

16–1
REASONS FOR CALCULATING BEAM DEFLECTIONS

When a beam is subjected to a load that creates bending, the beam will sag or deflect, as shown in Fig. 16–1. Although a beam may be satisfactory with respect to bending moment and shear, it may be unsatisfactory with respect to deflection. Large deflections are generally indicative of a lack of structural rigidity, despite adequate strength. Therefore, consideration of the deflection of beams is another part of the beam design or analysis process.

Excessive deflections are to be avoided for many reasons. With respect to buildings, excessive deflections can cause substantial cracking in ceilings, floors, and partitions, as well as in attached nonstructural elements, such as windows and doors. Other troublesome and potentially serious conditions that could result from beams that lack structural rigidity are

1. Changes in floor and wall alignment
2. Improper functioning of roof drainage systems resulting in the ponding of water on the roof and excessive roof loads
3. Undesirable vibrations from various live loads (e.g., lack of rigidity in a floor system may lead to vibrations caused by pedestrian traffic)
4. Misalignment of precision equipment housed in a building

In addition, visibly sagging beams tend to lessen one's confidence in both the strength of the structure and the skill of the designer.

The proper performance of machine parts is also affected by excessively flexible bending members. Shafts carrying gears must have adequate rigidity so that gear teeth will mesh as designed and synchronization of motion will be maintained. Various parts of metal-shaping equipment, as well as supporting frames of vehicles and machines, must also have sufficient rigidity. Without adequate rigidity, undesirable vibrations, friction, and abrasion can develop.

There are two considerations in evaluating deflection. The first is the computation of the magnitude and direction of the deflection and the second is the establishment of an allowable deflection. This chapter is primarily concerned with the computation of the magnitude of the deflection. Allowable deflections are generally established by design specifications, codes, and/or recommendations based on past acceptable practice and judgement. As an example, the AISC design specification stipulates that for beams and girders supporting plastered ceilings, the maximum live load deflection must not exceed $\frac{1}{360}$ of the span. Additional criteria that, in effect, limit the deflec-

FIGURE 16–1 Beam deflection.

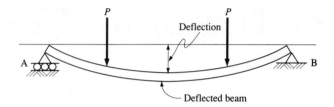

tion of a beam are also established by the AISC. Other design specifications furnished by various agencies address the deflection problem by requiring minimum depth/span ratios. For example, the AASHTO *Standard Specifications for Highway Bridges* stipulates that the ratio of depth to length of span for beams or girders preferably will not be less than $\frac{1}{25}$. This is in addition to upper limits on live load deflections.

16–2 CURVATURE AND BENDING MOMENT

As we described in Section 14–1, when a simple beam is subjected to vertical loads, tensile stresses are developed on one side of the neutral plane and compression stresses on the other side. The fibers subjected to tension are elongated and those subjected to compression are shortened. This causes the beam to curve and, therefore, to deflect from its original unloaded position. When a beam deflects, the curved position assumed by the neutral plane is generally called the *elastic curve*.

The radius of curvature at any point on the elastic curve is equal to the radius of that circle having a circumference that conforms to the shape of the elastic curve at that point. Generally, the elastic curve of a loaded beam is not circular in shape. However, an infinitesimal segment of it may be considered the arc of a circle.

With reference to Fig. 14–1, the portion of the beam between the two concentrated loads is a beam segment of zero shear and constant bending moment. As shown in Fig. 14–2(a), planes A–C and B–D located in that segment are vertical, parallel, and normal to the horizontal unloaded beam.

FIGURE 16–2   Loaded beam segment in pure bending.

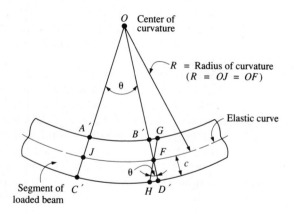

The loaded shape, which exists after the loads have been applied and bending occurs, is shown in Fig. 14–2(b).

Figure 14–2(b) is redrawn in Fig. 16–2 to show the total circle geometry. The planes $A'-C'$ and $B'-D'$ of the loaded beam (which were originally planes $A-C$ and $B-D$ of the unloaded beam) now intersect at point O and the angle between them is designated θ. R is the radius of curvature of the elastic curve and c is the distance from the neutral plane (elastic curve) to the outer fiber.

Line GFH is drawn through point F parallel to plane $A'-C'$ as shown in Figure 16–2. Line segments $C'H$ and JF are both equal to the original length of the outer fibers which have increased in length by the amount HD'. Therefore, the strain of the outer fiber is as follows:

$$\epsilon = \frac{\text{change in length}}{\text{original length}} = \frac{HD'}{C'H} = \frac{HD'}{JF}$$

By definition, the modulus of elasticity is

$$E = \frac{s}{\epsilon}$$

Therefore, the stress can be expressed as

$$s_b = E\epsilon = E\left(\frac{HD'}{JF}\right)$$

If the curvature is very slight and the distance JF is very small, $D'FH$ and JOF may be considered to be similar triangles. Therefore,

$$\frac{c}{R} = \frac{HD'}{JF}$$

and, by substitution, the stress equation becomes

$$s_b = \frac{Ec}{R} \tag{16–1}$$

where s_b = the tensile or compressive bending stress at the outer fiber (psi or ksi) (Pa)
   E = the modulus of elasticity (psi or ksi) (Pa)
   c = the distance from the neutral axis to the outer fibers (in.) (mm)
   R = the radius of curvature of the elastic curve (in.) (mm)

From the flexure formula, stress also can be written

$$s_b = \frac{Mc}{I}$$

Therefore,

$$\frac{Ec}{R} = \frac{Mc}{I}$$

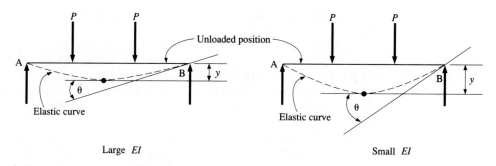

Large *EI*   Small *EI*

FIGURE 16–3 Relative beam stiffness.

and, solving for *R*,

$$R = \frac{EI}{M} \tag{16–2}$$

where all terms have been previously defined.

The equation just derived expresses the relation between the radius of curvature of the beam at any section and the bending moment at that section. Therefore, with *E* and *I* constant, *R* varies inversely with the moment *M*. If the moment *M* is zero, the radius of curvature *R* becomes infinity and the elastic curve is a straight line. As the moment *M* increases, the radius of curvature *R* decreases and the smallest value occurs where the moment is a maximum. If the bending moment is constant over part of the length of a beam, the radius of curvature is also constant and the elastic curve is the arc of a circle.

Note also that if the product *EI* is (theoretically) infinitely large, the radius of curvature will also become infinitely large and the elastic curve will be a straight line. Therefore, *EI* is an important factor in determining the shape of the elastic curve as well as in determining the displacement of points on the curve from the unloaded horizontal position.

The effect of relative magnitudes of *EI* (other parameters being equal) is shown in Fig. 16–3. Note that with a large *EI*, angle θ (defined as the angle between the tangents to two points on the elastic curve) is small, as is the vertical displacement *y* of a point on the elastic curve. With a small *EI*, both θ and *y* are relatively larger.

□ **EXAMPLE 16–1**   The steel blade for a small band saw is 0.025 in. thick and runs on pulleys that are 12 in. in diameter. Compute the maximum bending stress caused by bending the blade around the pulleys. $E = 30,000,000$ psi for the blade.

Solution   As the blade goes over the pulley, it conforms to the radius of the pulley; therefore, $R \approx 6$ in.

From Eq. (16–1) the maximum stress is written as

$$s_b = \frac{Ec}{R} = \frac{30,000,000(0.025/2)}{6} = 62,500 \text{ psi}$$

□ **EXAMPLE 16–2** A $\frac{3}{4}$ in. square steel bar is loaded as shown in Fig. 16–4. Calculate the radius of curvature of the beam segment between the supports. Use $E = 30,000,000$ psi.

FIGURE 16–4 Load, shear, and moment diagrams.

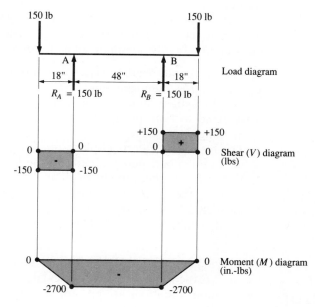

Load diagram

Shear (V) diagram (lbs)

Moment (M) diagram (in.-lbs)

Solution The shear and moment diagrams are shown. Note that constant moment exists between the supports. For the calculation of moment of inertia, see Table 8–1:

$$I = \frac{bh^3}{12} = \frac{0.75(0.75)^3}{12} = 0.0264 \text{ in.}^4$$

Calculating the radius of curvature from Eq. (16–2),

$$R = \frac{EI}{M} = \frac{30,000,000(0.0264)}{2700} = 293 \text{ in.} = 24.4 \text{ ft}$$

**16–3
METHODS OF
CALCULATING
DEFLECTIONS**

Deflections of bending members can be calculated in a number of ways. Two basic methods presented in this chapter are the *moment-area method* and the *formula method*. The formula method is clearly easier to use and should be applied whenever possible. Numerous derived formulas for beams having various support conditions and subjected to various types and combinations of loadings are given in Appendix H. These formulas have been derived for members with a constant moment of inertia using either the moment-area method or some other theoretical method.

In practice, the most commonly used method of computing deflections is the formula method. However, when support and loading conditions are such that the deflection formulas cannot be utilized, the moment-area method, which is extremely versatile, may be used. It is particularly useful for bending members with varying moments of inertia. These are frequently encountered in machine design and occasionally in building design.

Regardless of the method used, be aware that deflection calculations (like stress calculations) are based on certain assumptions. These assumptions can be briefly summarized as follows:

1. The maximum bending stress does not exceed the proportional limit.
2. The beam is homogeneous, obeys Hooke's law, and has an equal modulus of elasticity in tension and compression.
3. A plane section through the beam before bending remains a plane after bending takes place.
4. The beam has a vertical plane of symmetry and the loads and reactions act in this plane perpendicular to the longitudinal axis of the beam.
5. Deflections are relatively small and the length of the elastic curve is the same as the length of its horizontal projection.
6. Deflection due to shear is negligible. (Shear deflections are generally much smaller than the deflections due to bending moment.)

16–4
THE FORMULA
METHOD

As previously mentioned, the formula method is based on formulas derived by various theoretical means. Typical derivations for specific types of beams and loading conditions are presented in Section 16–7. In general, Appendix H provides derived formulas for calculating (among other quantities) the following:

1. The deflection of any point along the length of the beam
2. The maximum deflection
3. The location of the maximum deflection

In some cases, all three items are not included. This implies that too many complicating factors preclude the derivation of a usable formula; hence, a theoretical method should be used for deflection calculations. Generally, only the location and magnitude of the maximum deflection are of interest since it is the maximum deflection that must be compared with some allowable deflection value.

In the formulas of Appendix H, Greek uppercase delta sub x (Δ_x) represents the deflection of the beam from its straight unloaded position at any point a distance x from the left end of the beam or from a support. The designation Δ_{max} represents the maximum deflection occurring at a definite location that is a distance x from the left end of the beam or from a support.

Note that the deflection formulas in Appendix H are all shown as indicating a positive (+) deflection; that is, a downward deflection is assumed positive. This is the practice we will follow, although, from a practical viewpoint, whether the deflection is upward or downward is less important than the magnitude of the deflection. Generally, the direction of the deflection is evident and may be readily determined from the loading conditions.

Deflection is inversely proportional to E and I. The combined term EI represents the stiffness of a beam and is indicative of the resistance of the beam to deflection. The units of these two quantities are psi (or ksi) and in.^{4}, respectively, in the U.S. Customary System, and Pa (or MPa) and m^4 or mm^4 in the SI system. Other factors that enter into the problem of deflection determination are the type, magnitude, and location of the applied loads along with the span length and type of supports.

Since deflections are usually numerically small, it is desirable to state the computed deflection in inch units in the U.S. Customary System and in millimeters in the SI system. Therefore, units and conversions must be considered carefully since span lengths are usually given in feet in the U.S. Customary System and in meters in the SI system and uniformly distributed loads are in lb/ft or kips/ft in the U.S. Customary System and in N/m or kN/m in the SI system. All length units should be converted to inches or meters and, as is necessary in all formulas, other units must be consistent. The following three examples will illustrate the application of the formula method using the U.S. Customary System. SI system examples are presented in Section 16–8.

□ **EXAMPLE 16–3**   A W14 × 74 structural steel wide-flange section supports a superimposed uniformly distributed load of 2300 lb/ft on a simply supported span of 24 ft. Calculate the maximum deflection. The allowable deflection is $\frac{1}{360}$ of the span length. Determine whether or not the beam is satisfactory.

Solution   From Appendix H, Case 1, the maximum deflection for a simply supported beam subjected to a uniformly distributed load will occur at midspan ($x = L/2$) and is given by

$$\Delta_{max} = \frac{5wL^4}{384EI}$$

where w = the uniformly distributed load per unit of length (in this case, the sum of the superimposed load and the beam weight)
L = the span length (24 ft)
E = the modulus of elasticity (30,000,000 psi, from Appendix G)
I = the moment of inertia (796 in.^{4} from Appendix A)

Substituting,

$$\Delta_{max} = \frac{5((2300 + 74)\text{lb/ft})(24 \text{ ft})^4(12 \text{ in./ft})^3}{384(30,000,000 \text{ lb/in.}^2)(796 \text{ in.}^4)} = 0.74 \text{ in.}$$

The allowable deflection is calculated from

$$\Delta_{all} = \frac{span}{360} = \frac{24(12)}{360} = 0.80 \text{ in.}$$

Since 0.74 in. < 0.80 in., the beam is satisfactory with respect to deflection.

☐ **EXAMPLE 16–4**   Compute the maximum deflection for the Douglas fir beam designed in Example 15–5. The final selected member based on moment and shear was an 8 in. by 18 in. (S4S). The allowable deflection is $\frac{1}{360}$ of the span length.

Solution   From Appendix H, Case 1, the maximum deflection is given by

$$\Delta_{max} = \frac{5wL^4}{384EI}$$

From Appendix E, the moment of inertia for this timber beam is 3350 in.[4] and the weight is 36.4 lb/ft. From Appendix F, the modulus of elasticity is 1,700,000 psi. Substituting,

$$\Delta_{max} = \frac{5(1000 + 36.4)(16)^4(1728)}{384(1,700,000)(3350)} = 0.27 \text{ in.}$$

$$\Delta_{all} = \frac{span}{360} = \frac{16(12)}{360} = 0.53 \text{ in.}$$

Since 0.27 in. < 0.53 in., the beam is satisfactory with respect to deflection.

The Principle of Superposition

The applicability of the formula method may be expanded. The *principle of superposition* is used in the common situation where a beam is subjected to more than one load and/or different types of loads. The deflection at any given point is determined by computing the deflection (using the formulas) at the point under consideration due to each individual load and then adding the results.

It should be recognized that under some combined loadings the maximum deflection for each of the different loadings may not occur at the same point. As a result, it may be difficult to determine the location of the maximum deflection, or its magnitude. In such a situation, the principle of superposition may not be applicable and an alternate solution using the moment-area method may be necessary.

☐ **EXAMPLE 16–5**   A 4 in. nominal diameter standard-weight steel pipe is used as a cantilever beam projecting 12 ft from the fixed end. Compute the deflection at the free end due to the two loads shown in Fig. 16–5. Neglect the weight of the beam.

Solution   The principle of superposition will be used. The loading can be separated into two different cases as shown. The case designations (13 and 14) are from Appendix H, and are shown in Fig. 16–6.

$$\Delta_{14} = \frac{PL^3}{3EI} \quad \text{and} \quad \Delta_{13} = \frac{Pb^2}{6EI}(3L - b)$$

FIGURE 16–5 Load diagram.

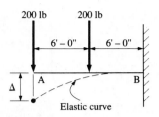

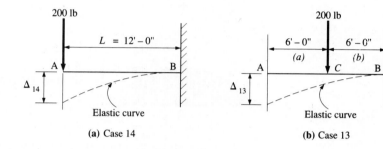

FIGURE 16–6 Principle of superposition.

(a) Case 14

(b) Case 13

We will use $E = 30,000,000$ psi (from Appendix G) and $I = 7.23$ in.^{4} (from Appendix B). Substituting,

$$\Delta_{14} = \frac{200(12)^3(1728)}{3(30,000,000)(7.23)} = 0.918 \text{ in.}$$

$$\Delta_{13} = \frac{(200 \text{ lb})(6 \text{ ft})^2(12 \text{ in./ft})^2}{6(30,000,000 \text{ lb/in.}^2)(7.23 \text{ in.}^4)} [3(12 \text{ ft}) - 6 \text{ ft}](12 \text{ in./ft}) = 0.287 \text{ in.}$$

The total deflection at the free end is calculated from

$$\Delta = \Delta_{14} + \Delta_{13} = 0.918 + 0.287 = 1.205 \text{ in.}$$

16–5 THE MOMENT-AREA METHOD

As previously mentioned, in the case of beams subjected to combined and/or unsymmetrical loadings, deflection formulas may not be available. Or, if such formulas do exist, they may be too cumbersome for quick and easy use. When this occurs, the moment-area method proves a very valuable non-calculus alternate method for computing deflections. This method is based on the relationships established in Section 16–2 and is generally presented in the form of two theorems. These theorems, as stated here, apply to beams with constant values of modulus of elasticity and moment of inertia (which is the usual situation). The application to beams with varying moments of inertia will be discussed in Section 16–7.

The simply supported beam shown in Fig. 16–7 is subjected to a uniformly distributed load. Its shear and moment diagrams are shown in Fig. 16–7(b) and (c). The deflected shape of the beam is indicated by the elastic curve which originally (before loading) was straight and horizontal.

FIGURE 16–7 Load, shear, and moment diagrams.

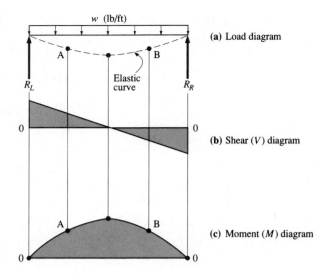

(a) Load diagram

(b) Shear (V) diagram

(c) Moment (M) diagram

We will remove an arbitrary portion of the beam between points A and B and show it in exaggerated form in Fig. 16–8. The first moment-area theorem can be stated as follows:

The angle θ between the tangents to any two points (A and B) on the elastic curve equals the area of the moment diagram that lies between those two points, divided by EI.

Expressing this as a formula,

$$\theta = \frac{A_M}{EI} \qquad\qquad \textbf{(16–3)}$$

where θ = the angle between any two tangents to the elastic curve. Its units will be *radians* (recall that one radian equals 57.3°).

FIGURE 16–8 First moment-area theorem.

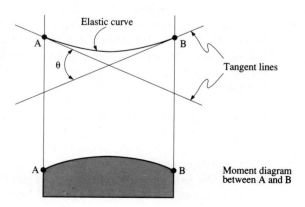

Elastic curve

Tangent lines

Moment diagram between A and B

A_M = the area of the moment diagram lying between two vertical planes that pass through points A and B. Its units in the U.S. Customary System will be ft²-lb, ft²-kips, or in.²-kips. In the SI system, its units will be kN·m² or N·m².

E = the modulus of elasticity (psi, ksi) (Pa)

I = the moment of inertia with respect to the neutral axis of the beam (in.⁴) (m⁴)

As in all other formulas, the units must be consistent. In the U.S. Customary System, the length units of moment of inertia and modulus of elasticity are always expressed in inch units. Units in the numerator should also be expressed in inch units. Either lb or kips may be used for force units. The following units are recommended for the calculation of θ (radians):

$$\theta = \frac{A_M}{EI} = \frac{(\text{in.-lb})(\text{in.})}{(\text{lb/in.}^2)(\text{in.}^4)} = \frac{\text{in.}^2\text{-lb}}{\text{in.}^2\text{-lb}}$$

In the SI system, the length units of moment of inertia and modulus of elasticity are generally expressed in meter (or millimeter) units with either kN or N used for force units. The following units are recommended for the calculation of θ (radians):

$$\theta = \frac{A_M}{EI} = \frac{(\text{N·m})(\text{m})}{(\text{N/m}^2)(\text{m}^4)} = \frac{\text{N·m}^2}{\text{N·m}^2}$$

Recall that 1 Pa = 1 N/m².

The same arbitrary portion of the beam (between points A and B) is again removed. It is shown in exaggerated form in Fig. 16–9. The second moment-area theorem can be stated as follows:

The vertical displacement y_{AB} of a point (point A) on the elastic curve from a tangent to the elastic curve at a second point (point B) equals the moment of

FIGURE 16–9   Second moment-area theorem.

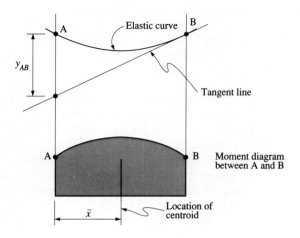

the area of the moment diagram that lies between the two points, taken about the first point (point A), divided by EI.

Expressing this as a formula,

$$y_{AB} = \frac{A_M \bar{x}}{EI} \tag{16–4}$$

where y_{AB} = the vertical displacement of point A on the elastic curve from a tangent line drawn at point B on the elastic curve

$\bar{x}$ = the distance from the centroid of the moment diagram area between any two points (A and B) to the point where the vertical displacement is desired

The other terms are as previously defined.

When using Eq. (16–4), the units must be consistent. In the U.S. Customary System, the following units are recommended for the calculation of displacement (in inches):

$$y = \frac{A_M \bar{x}}{EI} = \frac{(\text{in.-lb})(\text{in.})(\text{in.})}{(\text{lb/in.}^2)(\text{in.}^4)} = \frac{\text{in.}^3\text{-lb}}{\text{in.}^2\text{-lb}} = \text{in.}$$

In the SI system, the following units are recommended for the calculation of displacement in meters:

$$y = \frac{A_M \bar{x}}{EI} = \frac{(\text{N} \cdot \text{m})(\text{m})(\text{m})}{(\text{N/m}^2)(\text{m}^4)} = \frac{\text{N} \cdot \text{m}^3}{\text{N} \cdot \text{m}^2} = \text{m}$$

Multiplying meters (m) by 1000 will yield an answer in millimeters (mm).

The two preceding theorems, represented by Eqs. (16–3) and (16–4), are applicable between any two points on the elastic curve of any beam for any type of loading. However, it must be emphasized that only relative rotation of the tangents and relative vertical displacements are obtained directly. The displacement y calculated from Eq. (16–4) does not necessarily represent the desired deflection of the loaded beam. The distinction between y and the usually desired deflection Δ of a particular point will become apparent in the examples that follow.

In applying the moment-area method, the following sequence of steps should be followed:

1. Draw a load diagram of the beam. This does not need to be to scale, but approximate proportions should be used.
2. Directly below the load diagram, draw an approximate elastic curve exaggerating the deflections. Indicate all reference tangents to be used as well as all displacements to be calculated.
3. Below the elastic curve diagram, draw a moment diagram—either a combined moment diagram or a moment diagram by parts (which will be discussed in Section 16–6). Draw a shear diagram only if it is necessary for the drawing of the moment diagram.

The use of a horizontal tangent to the elastic curve, assuming its location is known, generally will provide the simplest solution. In cantilever beams, the tangent at the fixed end is always horizontal, thus greatly simplifying the solution of this type of problem. For simply supported beams subjected to a symmetrical loading, the maximum deflection will always occur at midspan and the tangent to the elastic curve will be horizontal at that point, thus simplifying the solution of the problem.

To aid in the application of the moment-area method, areas enclosed by curves and centroids for such areas are furnished in Table 7–1.

☐ **EXAMPLE 16–6**   A cantilever beam AB is subjected to a concentrated load of 6000 lb at its free end as shown in Fig. 16–10(a). The elastic curve of the deflected beam is shown in Fig. 16–10(b). Compute the slope of the tangent line at point A on the elastic curve. The beam is a W8 × 31 structural steel wide-flange section. Neglect the weight of the beam.

Solution   Note that the tangent line to the elastic curve at the fixed end of a cantilever beam is always horizontal. Therefore, the computed slope will actually be a change in slope between two tangent lines and can be computed directly using the first moment-area theorem, Eq. (16–3).

Using E from Appendix G and I from Appendix A,

$$\theta = \frac{A_M}{EI} = \frac{(0.5)(10)(12)(60,000)(12)}{30,000,000(110)} = 0.0131 \text{ radians}$$

FIGURE 16–10   Beam diagrams.

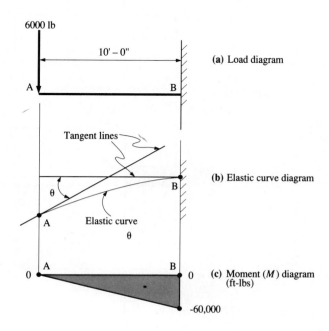

6000 lb

10' – 0"

(a) Load diagram

A   B

Tangent lines

θ

A   Elastic curve

θ

A   B

0   0   **(c)** Moment (M) diagram (ft-lbs)

B   **(b)** Elastic curve diagram

–60,000

Converting to degrees,

$$\theta = 0.0131(57.3) = 0.75°$$

□ **EXAMPLE 16–7**   Compute the angle between the tangent lines to the elastic curve at midspan and at the left end (points A and B) of the simply supported beam shown in Fig. 16–11. The beam is a W10 × 33 structural steel wide-flange section. Neglect the weight of the beam.

Solution   When a simply supported beam is symmetrically loaded, the tangent at midspan is horizontal. The computed angle between this tangent and the tangent to the elastic curve at the left end is designated θ_1 in the elastic curve diagram. The slope of the tangent line at the left end (point A) with respect to the original unloaded position of the beam is designated θ_2. This angle is generally termed the *end rotation* of the beam. Since the tangent at B is horizontal, $\theta_1 = \theta_2$.

In this problem, a shear diagram is drawn to simplify the construction of the moment diagram. Using E and I from Appendices G and A, respectively, calculate the desired angle using Eq. (16–3):

$$\theta_1 = \theta_2 = \frac{A_M}{EI} = \frac{0.5(6)(12)(60)(12) + 3(12)(60)(12)}{30,000(170)} = 0.0102 \text{ radians}$$

FIGURE 16–11 Beam diagrams.

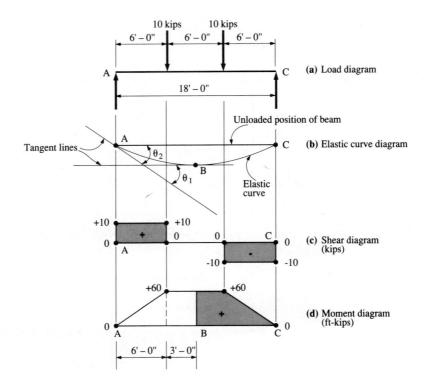

(a) Load diagram

(b) Elastic curve diagram

(c) Shear diagram (kips)

(d) Moment diagram (ft-kips)

Converting to degrees,

$$\theta_1 = \theta_2 = 0.0102(57.3) = 0.584°$$

☐ **EXAMPLE 16–8**

Compute the maximum deflection Δ of the cantilever beam shown in Fig. 16–12. The beam is a W8 × 24 structural steel wide-flange section.

Solution

The maximum deflection of a cantilever beam subjected to vertical downward loads will always occur at the free end. Since the tangent to the elastic curve at the fixed end is horizontal and parallel to the original unloaded position of the beam, the vertical displacement y_{AB} of point A on the elastic curve from the tangent to the curve at point B is equal to the deflection at the free end. The vertical displacement y_{AB} is equal to the moment, about A, of the moment diagram area between points A and B.

Since the shear diagram (not shown) is a sloping straight line, the moment curve is a second-degree parabola. The area of the moment diagram (see Table 7–1) between points A and B is calculated from

$$A_M = 0.333(12)(-36,000) = -144,000 \text{ ft}^2\text{-lb}$$

Note that the area is a negative value. The distance from the centroid of the moment area to point A (see Table 7–1) is

$$\bar{x} = \frac{3}{4}(12) = 9 \text{ ft}$$

FIGURE 16–12 Beam diagrams.

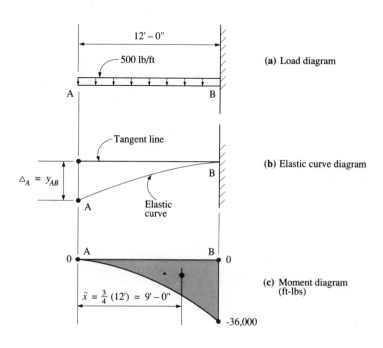

(a) Load diagram

(b) Elastic curve diagram

(c) Moment diagram (ft-lbs)

Using E and I from Appendices G and A, respectively,

$$y_{AB} = \frac{A_M \bar{x}}{EI} = \frac{(-144{,}000)(144)(9)(12)}{30{,}000{,}000(82.8)} = -0.90 \text{ in.}$$

The negative sign indicates that the vertical displacement y_{AB} from the tangent line is downward. Or, point A will lie below the tangent drawn at point B. Similarly, if we drew a tangent line at A and calculated y_{BA}, it would be negative, indicating that point B was situated below the tangent line drawn at point A. Therefore, note that the sign of the vertical displacement from one point to the tangent line drawn at another point will be the same as the sign of the moment diagram between those same two points. Further note that the distance $\bar{x}$ is always considered positive.

Completing this example problem, since we consider downward deflection of a beam from the original horizontal position to be positive,

$$\Delta_A = -y_{AB} = 0.90 \text{ in.}$$

You will discover that once a point has been located relative to the tangent line using the rationale for signs of the displacements as discussed, the sign of the final deflection can be determined by inspection of the absolute values of the calculated quantities. Therefore, a well-drawn sketch will be invaluable.

☐ **EXAMPLE 16–9** A 4 in. nominal diameter standard-weight steel pipe is used as a cantilever beam projecting 12 ft from the fixed end, as shown in Fig. 16–13. Calculate the maximum deflection at the free end. Neglect the weight of the beam.

Solution This solution is similar to that of the previous example. The vertical displacement y_{AB} of point A on the elastic curve from the tangent at point B is equal to the maximum deflection Δ_A at the free end. From the second moment-area theorem, Eq. (16–4), we see that this displacement (and, therefore, the desired deflection) is equal to the moment, about A, of the moment diagram area between points A and B. However, note that in this case the moment diagram area must be broken up into simple geometric shapes in order to determine areas and centroids.

The area of the moment diagram of Fig. 16–13(d) is broken up into two triangles and one rectangle. The areas are calculated as follows:

$$A_1 = 0.5(6)(-1200) = -3600 \text{ ft}^2\text{-lb}$$
$$A_2 = 6(-1200) = -7200 \text{ ft}^2\text{-lb}$$
$$A_3 = 0.5(6)(-2400) = -7200 \text{ ft}^2\text{-lb}$$

The distances from the centroid of each area to point A are calculated from

$$\bar{x}_1 = 0.667(6) = 4 \text{ ft}$$
$$\bar{x}_2 = 6 + 0.5(6) = 9 \text{ ft}$$
$$\bar{x}_3 = 6 + 0.667(6) = 10 \text{ ft}$$

FIGURE 16–13   Beam diagrams.

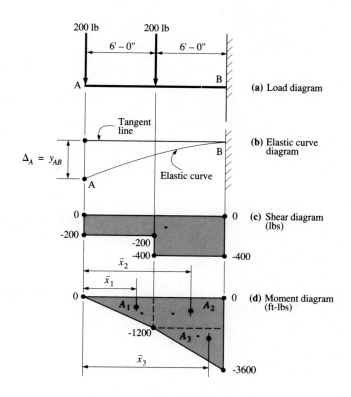

Using E and I from Appendices G and B, respectively, the vertical displacement is calculated as

$$y_{AB} = \frac{A_M \bar{x}}{EI} = \frac{(A_1 \bar{x}_1 + A_2 \bar{x}_2 + A_3 \bar{x}_3)12^3}{EI}$$

$$= \frac{-[3600(4) + 7200(9) + 7200(10)]1728}{30,000,000(7.23)}$$

$$= -1.205 \text{ in.}$$

The negative sign indicates that the vertical displacement of point A is downward from the tangent line. Since downward deflection is considered positive,

$$\Delta_A = -y_{AB} = 1.205 \text{ in.}$$

□ **EXAMPLE 16–10**   Calculate the maximum deflection at midspan for the symmetrically loaded simple beam in Fig. 16–14. The beam is a W6 × 12 structural steel wide-flange section. Neglect the weight of the beam.

Solution   The maximum deflection occurs at point B. Therefore, the tangent to the elastic curve at B is horizontal. Since the unloaded position of the beam is also horizontal, we see that $y_{AB} = \Delta_B$. Therefore, the determination of y_{AB} using the second moment-area theorem, Eq. (16–4), will also yield Δ_B.

FIGURE 16–14   Beam diagrams.

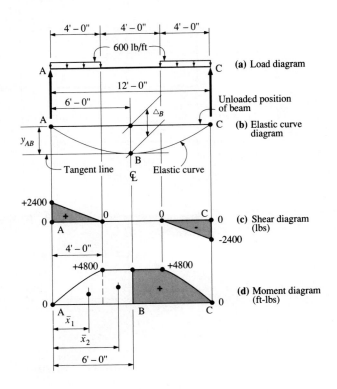

The area of the moment diagram between points A and B is broken up into two separate geometric shapes: a second-degree parabola and a rectangle, as shown in Fig. 16–14(d). These two areas are calculated as follows (refer to Table 7–1):

$$A_1 = \frac{2}{3} bh = 0.667(4)(4800) = 12,800 \text{ ft}^2\text{-lb}$$

$$A_2 = bh = 2(4800) = 9600 \text{ ft}^2\text{-lb}$$

The distances from the centroid of each area to point A are calculated from

$$\bar{x}_1 = \frac{5}{8}(4) = 2.5 \text{ ft}$$

$$\bar{x}_2 = 4 + 1 = 5 \text{ ft}$$

Using E and I from Appendices G and A, respectively, calculate the vertical displacement as

$$y_{AB} = \frac{A_M \bar{x}}{EI} = \frac{(A_1 \bar{x}_1 + A_2 \bar{x}_2)12^3}{EI}$$

$$= \frac{[12,800(2.5) + 9600(5)]1728}{30,000,000(22.1)}$$

$$= 0.21 \text{ in.}$$

Note that the y_{AB} is a positive value. This indicates that the vertical displacement of point A is upward from the tangent line at B. Therefore, B lies below A, and since we consider downward deflection positive,

$$\Delta_B = y_{AB} = 0.21 \text{ in.}$$

□ **EXAMPLE 16–11**   The beam shown in Fig. 16–15 is a 3 in. nominal diameter standard-weight steel pipe. Calculate the deflection at the free end and at midspan between supports. Neglect the weight of the beam.

FIGURE 16–15   Beam diagrams.

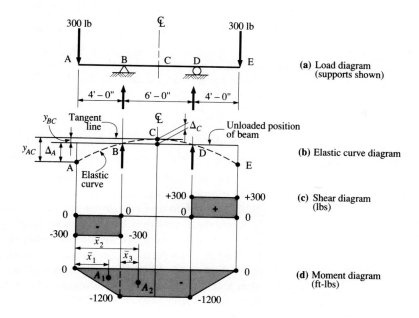

(a) Load diagram (supports shown)

(b) Elastic curve diagram

(c) Shear diagram (lbs)

(d) Moment diagram (ft-lbs)

Solution   The shear diagram is drawn to simplify the drawing of the moment diagram. The beam is symmetrically loaded and the tangent to the elastic curve at midspan is horizontal. The desired deflections are Δ_A and Δ_C. Note that these are measured from the original unloaded position of the beam. We will compute y_{AC} using the second moment-area theorem, Eq. (16–4). Values for E and I are obtained from Appendices G and B, respectively.

$$y_{AC} = \frac{A_M \bar{x}}{EI} = \frac{(A_1 \bar{x}_1 + A_2 \bar{x}_2)(12^3)}{EI}$$

$$= \frac{-[0.5(4)(1200)(0.667)(4) + 3(1200)(5.5)]1728}{30,000,000(3.02)}$$

$$= -0.500 \text{ in.}$$

Note that y_{AC} is *not* the deflection at the free end. The deflection at the free end can be calculated from

$$\Delta_A = y_{AC} - y_{BC}$$

Therefore, y_{BC} must first be calculated. Moments of the moment diagram area between points B and C are taken about point B:

$$y_{BC} = \frac{A_M \bar{x}}{EI} = \frac{(A_2 \bar{x}_3)(12^3)}{EI}$$

$$= \frac{3(-1200)(1.5)(1728)}{30,000,000(3.02)}$$

$$= -0.103 \text{ in.}$$

Therefore, the deflection of A is downward (positive). Neglecting the signs of the displacements y_{AC} and y_{BC}, calculate the deflection of point A from

$$\Delta_A = y_{AC} - y_{BC} = 0.500 - (0.103) = 0.397 \text{ in.}$$

Note that Δ_C is the upward deflection of point C and that it is equal to y_{BC}. The calculated displacement y_{BC} has a value of -0.103 in., which verifies that point B lies below the tangent line at point C. (Point B is a support point and does not move vertically.) Therefore,

$$\Delta_C = y_{BC} = -0.103 \text{ in.}$$

where the negative sign indicates an upward deflection at point C.

In this section the beams used in the examples were loaded so as to yield moment diagrams consisting of simple geometric shapes or areas that could be broken up into simple geometric shapes. When beams are subjected to combined and/or unsymmetrical loadings, however, the moment diagrams can become more difficult to analyze. In such situations, an alternate method of drawing moment diagrams is preferable. This method is presented in Section 16–6. Further applications of the moment-area method using this alternate approach to the drawing of moment diagrams are presented in Section 16–7.

16–6 MOMENT DIAGRAM BY PARTS

The areas encountered in conventional moment diagrams, particularly where unsymmetrical and/or combined loadings exist, often can be very cumbersome to work with. These areas and their centroids may be difficult to obtain without some calculus manipulations. In order to simplify the calculations of areas for the moment-area method, we may draw the moment diagrams in a somewhat different form. Essentially, we draw the various rudimentary parts that comprise the moment diagrams; hence, the descriptor "moment diagram by parts."

The moment-diagram-by-parts method offers a practical solution whereby the moment diagram results in common geometric shapes with known properties, or properties that can easily be obtained. The technique involves considering each loading separately and then drawing a bending moment diagram for that load alone, as if there were no other loads acting on the member. Actually, all loads are acting simultaneously, and the true value of the moment at any point will be the algebraic sum of the values indicated on the separate diagrams. The individual load moment diagram is based on the assumption that the beam is fixed at some location, in effect, creating a cantilever beam subject to the one load.

The moment diagram by parts can be started at either end of the beam. At each concentrated load, or reaction, a triangular area will begin. This triangular area will be positive for upward loads or reactions and negative for downward loads or reactions. At locations where a uniformly distributed loads exists, a parabolic curve will result. The method is illustrated in Example 16–12.

☐ **EXAMPLE 16–12**   Draw the bending moment diagram by parts for the beam shown in Fig. 16–16. Neglect the weight of the beam.

FIGURE 16–16 Beam diagrams.

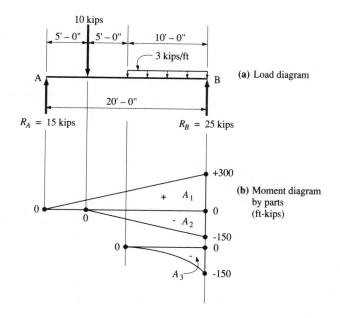

(a) Load diagram

(b) Moment diagram by parts (ft-kips)

Solution   The beam reactions are computed in the normal way and are shown on the load diagram. In this problem we will work from the left end (point A) and we will assume a fixed end condition at the other end (point B).

The moment diagram may be considered to consist of three parts: one part due to the reaction R_A, a second part due to the concentrated load, and a third due to the uniformly distributed load. Beginning at A, the moment diagram due to the reaction is designated A_1 in Fig. 16–16. As shown in Fig. 16–17, it is the same moment diagram as that for a cantilever beam that is subjected to a concentrated load at its free end. It is positive, since the load is acting upward.

The moment diagram for the applied concentrated load of 10 kips is designated A_2 in Fig. 16–16. As shown in Fig. 16–18, it is seen to be the same moment diagram as that of a cantilever beam with a 15 ft span length subjected to a concentrated load at its free end. It is negative, since the load is acting downward. (Note that in drawing

FIGURE 16–17 Theoretical
beam diagrams.

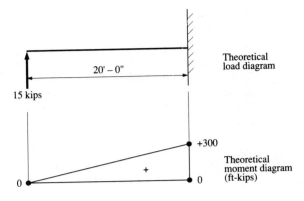

FIGURE 16–18 Theoretical
beam diagrams.

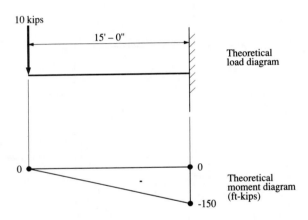

the moment diagrams of Fig. 16–16, the same horizontal reference line is used for the two diagrams.)

The moment diagram for the uniformly distributed load is designated A_3 in Fig. 16–16. As shown in Fig. 16–19, it is seen to be the same moment diagram as that of a cantilever beam of 10 ft span length subjected to a uniformly distributed load for its full length. It is negative, since the load is acting downward. (Note that this is drawn in Fig. 16–16 using another reference line. The various diagrams may be combined, superimposed, or drawn separately, whichever is preferred.)

If so desired, the moment at any point due to all the loads can be obtained by an algebraic sum of all the ordinates from all the diagrams at any given point. However, it should be noted that this method does not indicate the location or the magnitude of the maximum moment; therefore, its usage is relatively limited.

Using the same reasoning, the moment diagram by parts can be drawn working from right to left. The fixed condition may be assumed to exist at point A. The resulting diagram is shown in Fig. 16–20.

FIGURE 16–19 Theoretical beam diagrams.

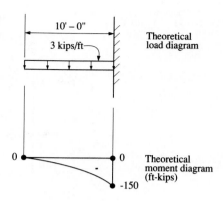

FIGURE 16–20 Beam diagrams.

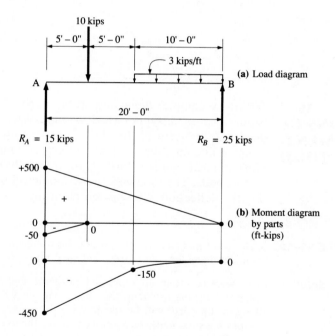

It is also possible to draw the moment diagram by parts by working both ways (left to right and right to left) to some reference location. Figure 16–21 shows the moment diagram using midspan as the location at which the beam is fixed.

You may wish to verify that each of the three moment diagrams (part [b] of Figs. 16–16, 16–20, and 16–21) yields the same result. For instance, in each case, the midspan moment determined by summing the magnitude of the applicable midspan moment from each part of the diagram is +100 ft-kips. (Note that in Fig. 16–21 the summation is made just to the left or just to the right of midspan.)

FIGURE 16–21 Beam diagrams.

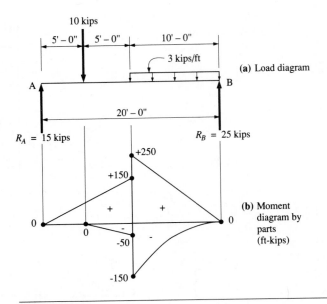

(a) Load diagram

(b) Moment diagram by parts (ft-kips)

16–7

APPLICATIONS OF THE MOMENT-AREA METHOD

The more complex deflection problems generally involve beams subjected to some form of unsymmetrical loading. When this occurs, the position of a horizontal tangent to the elastic curve is unknown. Therefore, some other tangent line must be used as a reference tangent. Generally this will be the tangent at the support farthest from an individual load or from the resultant of the loads. The following two examples illustrate the procedure of calculating the deflection at any given point for a simple beam and an overhanging beam.

☐ **EXAMPLE 16–13**

Calculate the deflection at midspan for the beam in Fig. 16–22. The beam is a 4 in. nominal diameter standard-weight steel pipe. Neglect the weight of the beam.

Solution

The beam reactions are computed in the normal manner and are shown on the load diagram. In this problem, the shear diagram will not be drawn since the moment diagram by parts will be used. With this method, the resulting moment diagram shapes are sufficiently elementary to make the use of the shear diagram unnecessary.

The bending moment diagram by parts is drawn from left to right and the reference tangent, as shown in the elastic curve diagram, is the tangent at the left support (point A). We will first calculate y_{BA}, which is the vertical displacement of point B from the tangent to the elastic curve at point A. The value of y_{BA} is equal to the moment, about B, of the moment diagram area between points A and B.

Using E from Appendix G and I from Appendix B,

$$y_{BA} = \frac{A_M \bar{x}}{EI} = \frac{[(\frac{1}{2})(10)(11{,}000)(\frac{10}{3}) - (\frac{1}{2})(5)(9000)(\frac{5}{3}) - (\frac{1}{2})(2.5)(2000)(\frac{2.5}{3})]1728}{30{,}000{,}000(7.23)}$$

$$= +1.145 \text{ in.}$$

FIGURE 16–22   Beam diagrams.

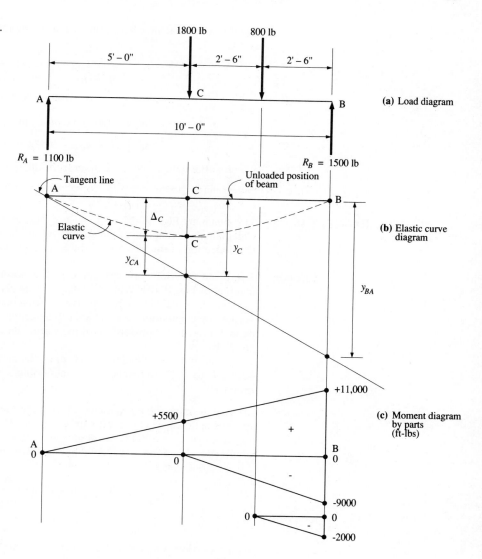

(a) Load diagram

(b) Elastic curve diagram

(c) Moment diagram by parts (ft-lbs)

The positive sign indicates that point B lies above the tangent line. By proportion, y_C can be calculated from

$$y_C = \frac{5}{10} y_{BA} = \frac{+1.145}{2} = +0.572 \text{ in.}$$

Note that the quantity y with a single subscript denotes the distance from that point on the *undeflected* beam to a line tangent to the elastic curve.

We will now calculate y_{CA}, which is the vertical displacement of point C on the elastic curve from the tangent to the elastic curve at point A. The value of y_{CA} is equal to the moment, about point C, of the moment diagram area between points A and C.

$$y_{CA} = \frac{A_M \bar{x}}{EI} = \frac{+(\frac{1}{2})(5)(5500)(\frac{5}{3})1728}{30,000,000(7.23)}$$

$$= +0.183 \text{ in.}$$

As shown on the elastic curve diagram,

$$\Delta_C = y_C - y_{CA} = 0.572 - 0.183 = 0.389 \text{ in.}$$

which is the deflection at midspan from the unloaded position of the beam.

□ **EXAMPLE 16–14**   The beam in Fig. 16–23 is a W12 × 50 structural steel wide-flange section. (a) Calculate the deflection at midspan between the two supports A and B. (b) Calculate the deflection at the free end of the beam (point C). Neglect the weight of the beam.

Solution   The beam reactions are computed in the normal way and are shown in the load diagram of Fig. 16–23. While the shape of the elastic curve for a simple beam is quite evident, the elastic curve for the overhanging beam will have to be assumed and then verified by calculation. It is difficult to determine by inspection whether the free-end deflection will be upward or downward. The assumed shape is shown in Fig. 16–23(b).

For this type of beam, we will draw the moment diagram by parts working from both left to right and right to left, assuming a fixed condition at the right support (point B).

(a) To determine the midspan deflection between points A and B, a reference tangent to the elastic curve at the left support (point A) is drawn. The vertical displacement y_{BA} is equal to the moment, about point B, of the moment diagram area between points A and B. From the second moment-area theorem, Eq. (16–4), and using E and I from Appendices G and A, respectively,

$$y_{BA} = \frac{A_M \bar{x}}{EI} = \frac{[(\frac{1}{2})(24)(276)(\frac{24}{3}) - (\frac{1}{3})(24)(288)(\frac{24}{4})]1728}{30,000(394)}$$

$$= +1.853 \text{ in.}$$

By proportion, the vertical distance from the tangent line at A to point D on the undeflected beam can be calculated from

$$y_D = \frac{12}{24}(y_{BA}) = \frac{12}{24}(1.853) = +0.926 \text{ in.}$$

We will now calculate y_{DA}, which is the vertical displacement (at midspan) of point D on the elastic curve from the tangent to the elastic curve at A. The value of y_{DA} is equal to the moment, about point D, of the moment diagram area between points A and D. From the second moment-area principle, Eq. (16–4),

$$y_{DA} = \frac{A_M \bar{x}}{EI} = \frac{[(\frac{1}{2})(12)(138)(\frac{12}{3}) - (\frac{1}{3})(12)(72)(\frac{12}{4})]1728}{30,000(394)}$$

$$= +0.358 \text{ in.}$$

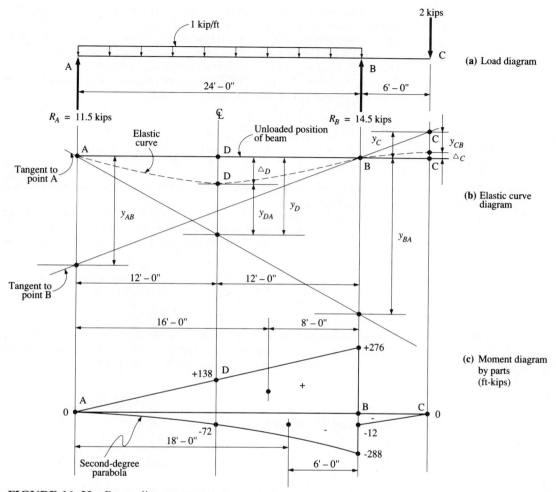

FIGURE 16–23 Beam diagrams.

As shown in the elastic curve diagram in Fig. 16–23(b), the deflection at mid-span (point D) between the two support points is calculated from

$$\Delta_D = y_D - y_{DA} = 0.926 - 0.358 = 0.568 \text{ in.}$$

(b) To obtain the deflection at the free end of the beam (point C), the same tangent line at point A could be used. However, for purposes of example, we will use a tangent to the elastic curve at point B. We first calculate y_{AB}, the vertical displacement of point A from this tangent line. The value of y_{AB} is equal to the moment, about point A, of the moment diagram area between points A and B.

$$y_{AB} = \frac{A_M \bar{x}}{EI} = \frac{[(\frac{1}{2})(24)(276)(\frac{2}{3})(24) - (\frac{1}{3})(24)(288)(\frac{3}{4})(24)]1728}{30,000(394)}$$

$$= +1.684 \text{ in.}$$

Note that y_{AB} is a positive (+) value. If a negative value had resulted, the assumed slope of the tangent line at point B would have been incorrect. A negative value would indicate that point A on the elastic curve was situated below the tangent line. This would imply a clockwise rotation of the line tangent to the beam at point B. However, since y_{AB} is a positive value, indicating that point A lies above the tangent line, the direction of rotation of the assumed tangent line at point B is correct.

By proportion, y_C can be calculated from

$$\frac{y_C}{6} = \frac{y_{AB}}{24}$$

$$y_C = \frac{6}{24}(1.684) = 0.421 \text{ in.}$$

Note that point C on the undeflected beam lies below the tangent line drawn at point B on the elastic curve.

We will now calculate y_{CB}, which is the vertical displacement of point C on the elastic curve from the tangent to the elastic curve at point B. The value of y_{CB} is equal to the moment, about point C, of the moment diagram area between points B and C. From the second moment-area theorem, Eq. (16–4),

$$y_{CB} = \frac{A_M \bar{x}}{EI} = \frac{-(\frac{1}{2})(6)(12)(\frac{2}{3})(6)(1728)}{30,000(394)}$$

$$= -0.021 \text{ in.}$$

The negative sign indicates that point C on the elastic curve lies below the tangent line drawn at point B. Therefore, by inspection of Fig. 16–23(b), and neglecting the sign of y_{CB} (using its absolute value), the free-end deflection at point C is obtained from

$$y_C - y_{CB} = 0.421 - (0.021) = 0.400 \text{ in.}$$

Recognizing that the deflection at point C is upward,

$$\Delta_C = -0.400 \text{ in.}$$

The previous two examples illustrated the procedure for computing the deflection at any given point. The maximum deflection, however, was not obtained in either case. It is the maximum deflection that is generally the value of interest. The following example will illustrate the procedure for calculating both the location and the magnitude of the maximum deflection.

□ **EXAMPLE 16–15**   A 10 ft long aluminum bar is used as a beam and loaded as shown in Fig. 16–24. The bar is 2 in. by 2 in. in cross section. Calculate the maximum deflection of the beam. Neglect the weight of the beam.

Solution   The beam reactions are computed in the usual way and are shown in the load diagram. The moment diagram by parts will be drawn from right to left. Note that it is the right support that is farthest from the resultant of the loads. The reference tangent to the elastic curve will be the tangent at the right support (point B).

The modulus of elasticity for the aluminum bar is obtained from Appendix G.

FIGURE 16–24 Beam diagrams.

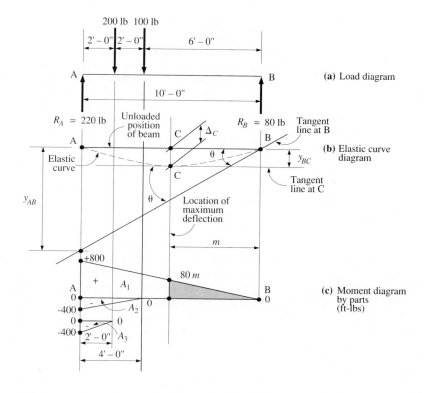

Calculating moment of inertia (from Table 8–1),

$$I = \frac{bh^3}{12} = \frac{2(2)^3}{12} = 1.333 \text{ in.}^4$$

We next compute y_{AB}, which is equal to the moment, about point A, of the moment diagram area between points A and B. Using the second moment-area theorem, Eq. (16–4),

$$y_{AB} = \frac{A_M \bar{x}}{EI}$$

$$= \frac{[(\frac{1}{2})(10)(800)(\frac{1}{3})(10) - (\frac{1}{2})(4)(400)(\frac{1}{3})(4) - (\frac{1}{2})(2)(400)(\frac{1}{3})(2)]1728}{10,000,000(1.333)}$$

$$= 1.556 \text{ in.}$$

We next compute the angle between the tangent at point B and the original unloaded position of the beam. Recalling that for small angles the tangent of an angle approximately equals the angle, itself, measured in radians,

$$\theta = \tan \theta = \frac{y_{AB}}{L} = \frac{1.556}{10(12)} = 0.0130 \text{ radians}$$

The tangent to the elastic curve at the point of maximum deflection is horizontal. Therefore, it also makes an angle of θ with the line that is tangent to the elastic curve at point B. Using the first moment-area theorem, Eq. (16–3), the angle θ between the two tangents is equal to the area of the moment diagram between the same two points (B and C), divided by EI. Therefore, recognizing that the slope of the A_1 portion of the moment diagram is 80 ft-lb/ft, and arbitrarily locating the point of maximum deflection a distance m (where $m \leq 6$ ft) from point B,

$$\theta = \frac{A_M}{EI} = \frac{(\frac{1}{2})(m)(80m)(144)}{EI}$$

Solving for m,

$$40m^2 = \frac{\theta EI}{144} = \frac{0.013(10,000,000)(1.333)}{144}$$

from which

$$m = 5.49 \text{ ft}$$

Had m been greater than 6 ft (but not greater than 8 ft), the solution would have had to be rewritten to include the effect of A_2 of the moment diagram.

Knowing the location of the maximum deflection, we can now compute y_{BC}, which is equal to the moment, about point B, of the moment diagram area between points B and C. Using the second moment-area theorem, Eq. (16–4),

$$y_{BC} = \frac{A_M \bar{x}}{EI} = \frac{(\frac{1}{2})(5.49)(80)(5.49)(\frac{2}{3})(5.49)(1728)}{10,000,000(1.333)}$$

$$= 0.572 \text{ in.}$$

Since the tangent at point C is horizontal and the unloaded position of the beam is horizontal, $\Delta_C = y_{BC}$. Therefore, the maximum deflection for the beam is 0.572 in.

The previous discussion and examples were applicable to beams with constant values for modulus of elasticity, E, and moment of inertia, I. When a beam has a varying moment of inertia, the moment-area method may still be used with only slight modifications to the procedure we have followed thus far. The two moment-area theorems are still valid. However, instead of calculating areas of moment diagrams or moments of moment diagram areas and then dividing by EI in the final step, each moment area must now be divided by the applicable value of I. This is usually accomplished by drawing an additional diagram, called an M/I diagram, in which the varying moment of inertia is introduced prior to computing any moment diagram areas. Division by E (normally constant) is then accomplished in the final calculation. The following example illustrates the procedure of calculating the deflection for a beam of varying moment of inertia.

☐ **EXAMPLE 16–16**   Calculate the maximum deflection for the structural steel cantilever beam shown in Fig. 16–25. The beam is stiffened with steel plates in such a manner that for a length of 4 ft at the fixed end, the moment of inertia is 400 in.^{4}. For the remaining 8 ft, the section has no plates and the moment of inertia is 200 in.^{4}. Use a modulus of elasticity value of 30,000,000 psi. Neglect the weight of the beam.

FIGURE 16–25   Beam diagrams.

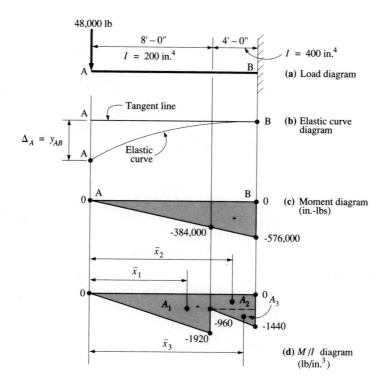

(a) Load diagram

(b) Elastic curve diagram

(c) Moment diagram (in.-lbs)

(d) M/I diagram (lb/in.^{3})

Solution   The solution is similar to that of Example 16–9. The vertical displacement y_{AB} of point A on the elastic curve from the tangent at point B is numerically equal to the maximum deflection Δ_A at the free end. The vertical displacement y_{AB}, from the second moment-area theorem, Eq. (16–4), equals the moment, about A, of the M/I diagram between points A and B, divided by E.

The M/I diagram ordinates are expressed in units of lb/in.^{3}. Note that the M/I diagram must be broken up into simple geometric shapes for the determination of areas as follows:

$$A_1 = \frac{1}{2}(8)(12)(-1920) = -92{,}160 \text{ lb/in.}^2$$

$$A_2 = 4(12)(-960) = -46{,}080 \text{ lb/in.}^2$$

$$A_3 = \frac{1}{2}(4)(12)(-480) = -11{,}520 \text{ lb/in.}^2$$

The distances from the centroid of each area to point A can be calculated next:

$$\bar{x}_1 = \frac{2}{3}(8)(12) = 64.0 \text{ in.}$$

$$\bar{x}_2 = (8)(12) + (2)(12) = 120.0 \text{ in.}$$

$$\bar{x}_3 = 8(12) + \frac{2}{3}(4)(12) = 128.0 \text{ in.}$$

Using Eq. (16–4),

$$y_{AB} = \frac{A_{(M/I)}\bar{x}}{E} = \frac{A_1\bar{x}_1 + A_2\bar{x}_2 + A_3\bar{x}_3}{E}$$

$$= \frac{-[(92,160)(64.0) + (46,080)(120.0) + (11,520)(128.0)]}{30,000,000}$$

$$= -0.430 \text{ in.}$$

The negative sign indicates that point A lies below the tangent line. Since downward deflection is assumed positive,

$$\Delta_A = 0.430 \text{ in.}$$

The previous examples in this section illustrated some of the applications of the moment-area method. In addition, the method may be used to derive formulas, such as those used in Section 16–4. The following examples illustrate the procedure for the derivation of such formulas.

☐ **EXAMPLE 16–17**   Derive expressions for the maximum deflection of a cantilever beam subjected to (a) a concentrated load at its free end and (b) a uniformly distributed load. In each case, neglect the weight of the beam.

Solution   Figures 16–26 and 16–27 show the load diagrams, elastic curve diagrams, and the moment diagrams for the beams under consideration. The maximum deflection in each case is the vertical displacement y_{AB} of the free end (point A) from the tangent to the elastic curve at B. (Refer to Table 7–1 for expressions for areas and centroid locations.)

(a)

$$y_{AB} = \frac{A_M\bar{x}}{EI} = \frac{-(\frac{1}{2})(PL^2)(\frac{2}{3}L)}{EI} = \frac{-PL^3}{3EI}$$

Since downward deflection is considered positive,

$$\Delta_A = \frac{PL^3}{3EI}$$

FIGURE 16–26 Beam diagrams.

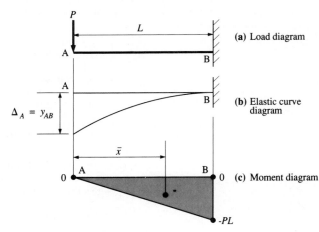

(a) Load diagram

(b) Elastic curve diagram

(c) Moment diagram

FIGURE 16–27 Beam diagrams.

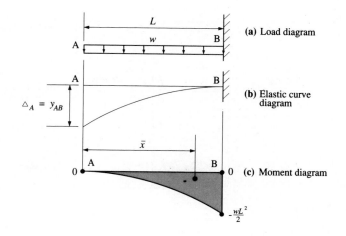

(a) Load diagram

(b) Elastic curve diagram

(c) Moment diagram

(b)

$$y_{AB} = \frac{A_M \bar{x}}{EI} = \frac{-(\frac{1}{3})(L)(\frac{wL^2}{2})(\frac{3}{4}L)}{EI} = \frac{-wL^4}{8EI}$$

Since downward deflection is considered as positive,

$$\Delta_A = \frac{wL^4}{8EI}$$

□ **EXAMPLE 16–18**   Derive expressions for the maximum deflection of a simply supported beam subjected to (a) a concentrated load at midspan and (b) a uniformly distributed load over the entire span length. Neglect the weight of the beam in each case.

Solution   Figures 16–28 and 16–29 show the load diagrams, the elastic curve diagrams, and the moment diagrams. Using the horizontal tangent at midspan, the vertical displace-

FIGURE 16–28 Beam diagrams.

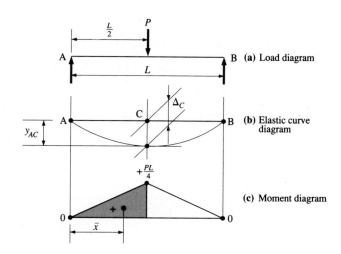

(a) Load diagram

(b) Elastic curve diagram

(c) Moment diagram

FIGURE 16–29 Beam diagrams.

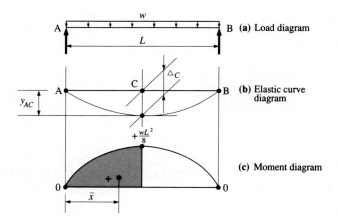

ment y_{AC} at the support equals the maximum deflection at midspan.

(a)
$$\Delta_C = y_{AC} = \frac{A_M \bar{x}}{EI} = \frac{(\frac{1}{2})(\frac{L}{2})(\frac{PL}{4})(\frac{L}{3})}{EI} = \frac{PL^3}{48EI}$$

(b)
$$\Delta_C = y_{AC} = \frac{(\frac{2}{3})(\frac{L}{2})(\frac{wL^2}{8})(\frac{5}{16}L)}{EI} = \frac{5wL^4}{384EI}$$

16–8
SI SYSTEM
EXAMPLES

☐ **EXAMPLE 16–19**

A solid, round simply supported steel shaft, having a diameter of 50 mm, supports a
concentrated load as shown in Fig. 16–30. Compute the deflection under the load and
at point C. Neglect the weight of the shaft. Use the formula method.

Solution   From Appendix H, Case 6, the deflection at the point of load (point D) is given by

$$\Delta = \frac{Pa^2b^2}{3EIL}$$

From Table 8–1, we obtain an expression for moment of inertia:

$$I = \frac{\pi d^4}{64} = \frac{\pi (0.05 \text{ m})^4}{64} = 307 \times 10^{-9} \text{ m}^4$$

FIGURE 16–30   Load diagram.

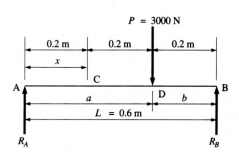

Substituting,

$$\Delta_D = \frac{(3000 \text{ N})(0.4 \text{ m})^2(0.2 \text{ m})^2}{(3)(207 \times 10^9 \text{ Pa})(307 \times 10^{-9} \text{ m}^4)(0.6 \text{ m})}$$

$$= 0.000 \ 168 \text{ m}$$

$$= 0.168 \text{ mm}$$

From Appendix H, the deflection at point C is given by

$$\Delta_C = \frac{Pbx}{6EIL} (L^2 - b^2 - x^2)$$

$$= \frac{(3000 \text{ N})(0.2 \text{ m})(0.2 \text{ m})[(0.6 \text{ m})^2 - (0.2 \text{ m})^2 - (0.2 \text{ m})^2]}{(6)(207 \times 10^9 \text{ Pa})(307 \times 10^{-9} \text{ m}^4)(0.6 \text{ m})}$$

$$= 0.000 \ 147 \text{ m}$$

$$= 0.147 \text{ mm}$$

□ **EXAMPLE 16–20** A 100 mm nominal diameter standard-weight steel pipe is used as a simple beam with a span length of 4 m. The member is loaded as shown in Fig. 16–31. Calculate the deflection at midspan. Neglect the weight of the pipe.

FIGURE 16–31 Beam diagrams.

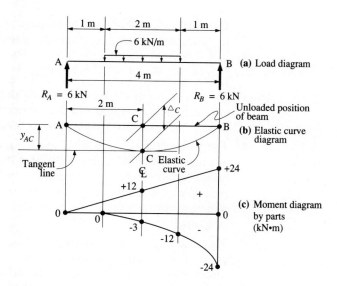

Solution The beam reactions are computed in the usual manner and are shown on the load diagram. The shear diagram is not drawn since the moment diagram is drawn by parts.

The bending moment diagram by parts is drawn from left to right. The maximum deflection occurs at midspan (point C), since the loading is symmetrical. Therefore, the tangent to the elastic curve at C is horizontal and $y_{AC} = \Delta_C$.

We will calculate y_{AC}, the vertical displacement of point A from the tangent to the elastic curve at point C. The value of y_{AC} is equal to the moment, about point A, of the moment diagram between points A and C.

Using the modulus of elasticity E from Appendix G and the moment of inertia I from Appendix B,

$$y_{AC} = \frac{A_M \bar{x}}{EI} = \frac{(\frac{1}{2})(2 \text{ m})(12\ 000 \text{ N} \cdot \text{m})(\frac{2}{3})(2 \text{ m}) - (\frac{1}{3})(1 \text{ m})(3000 \text{ N} \cdot \text{m})(1.75 \text{ m})}{(207 \times 10^9 \text{ Pa})(3.01 \times 10^{-6} \text{ m}^4)}$$

$$= \frac{(16\ 000 \text{ N} \cdot \text{m}^3) - (1750 \text{ N} \cdot \text{m}^3)}{623\ 000(\text{N/m}^2)(\text{m}^4)}$$

$$= 22.9 \times 10^{-3} \text{ m}$$

$$= 229 \text{ mm}$$

Note that y_{AC} is a positive value indicating that the vertical displacement of point A is upward from the tangent line at C. Therefore, C lies below A and since we assume downward deflection as positive,

$$\Delta_C = y_{AC} = 229 \text{ mm}$$

SUMMARY—BY SECTION NUMBER

16–2 In a beam subject to pure bending, the relationship between stress in the outer fibers and the radius of curvature of the elastic curve is

$$s_b = \frac{Ec}{R} \qquad \textbf{(16–1)}$$

The relationship between the radius of curvature of the beam at any section and the bending moment at that section is

$$R = \frac{EI}{M} \qquad \textbf{(16–2)}$$

16–3 Two basic methods for calculating deflections of bending members are the formula method and the moment-area method. The most commonly used method is the formula method.

16–4 The formula method is based on the usage of standard formulas that are readily available. Appendix H provides numerous deflection formulas for different types of beams and loading conditions. Using the principle of superposition, the formula method may also be used to compute the deflection at a point due to many and/or different types of loads.

16–5 Since deflection formulas may not be available for specific cases of combined and/or unsymmetrical loadings, the moment-area method offers a practical, noncalculus-based alternate solution. It involves the use of the bending moment diagram. The two basic moment-area theorems are

$$\theta = \frac{A_M}{EI} \qquad \textbf{(16–3)}$$

$$y_{AB} = \frac{A_M \bar{x}}{EI} \qquad \textbf{(16–4)}$$

where θ represents an angular rotation and y_{AB} represents a vertical displacement. This method may be used for beams with variable or constant EI values.

16–6 Drawing a moment diagram by parts is an alternate method of drawing a moment diagram. The technique involves creating a hypothetical cantilever beam, loading it with each load separately, and drawing a moment diagram for that load alone. The individual moment diagrams are then superimposed to represent the sum. This technique facilitates the subsequent moment-area calculations.

PROBLEMS

For the following problems:

1. *For steel members unless otherwise noted, use a modulus of elasticity of 30,000,000 psi. For other materials, refer to the appropriate appendix table.*
2. *Unless otherwise noted, the given loads are superimposed loads; neglect the beam weight; neglect shear and moment considerations.*
3. *Unless otherwise noted, use any appropriate method to calculate deflection.*

Section 16–2   Curvature and Bending Moment

1. A $\frac{1}{4}$ in. diameter aluminum rod is bent into a circular ring having a mean diameter of 125 in. Compute the maximum bending stress in the rod.

2. Calculate the maximum bending stress produced in a $\frac{1}{32}$ in. diameter steel wire when it passes around a 20 in. diameter pulley.

3. An aluminum wire has a diameter of d in. Determine the minimum diameter D of the coil in which the wire can be wound without exceeding a bending stress of 30,000 psi.

4. A 3 in. wide by $\frac{1}{2}$ in. thick board is bent to a radius of curvature of 62 in. by a bending moment of 700 in.-lb. Calculate the modulus of elasticity.

5. A Douglas fir beam is 6 in. wide and 14 in. deep. Compute the radius of curvature of the beam if it is subjected to a constant bending moment of 200,000 in.-lb.

Section 16–4   The Formula Method

For Problems 6–11, use the formula method.

6. Compute the maximum deflection of a 10 in. by 14 in. simply supported solid rectangular California redwood (S4S) beam. The span length is 15 ft and the beam is subjected to a uniformly distributed load of 1000 lb/ft. The allowable deflection is $\frac{1}{300}$ of the span length. Is the beam satisfactory?

7. Compute the maximum deflection for the beam of Problem 6 if the loading consists of one concentrated load of 5000 lb at midspan instead of the uniformly distributed load. The allowable deflection is $\frac{1}{300}$ of the span length. Is the beam satisfactory?

8. A W16 × 45 structural steel beam is simply supported on a span length of 24 ft. It is subjected to two concentrated loads of 12 kips each applied at the third points (a = 8 ft). Compute the maximum deflection.

9. A 4 in. diameter, 12 ft long solid steel shaft having a circular cross section is used as a simply supported beam. A concentrated load P is to be applied at midspan. Calculate the maximum allowable load P. Consider moment and deflection. Neglect shear. The allowable bending stress is 22,000 psi and the allowable deflection is 0.50 in.

10. Assume that a company's design criterion specifies that the deflection at the center of a simply supported solid (circular cross section) steel shaft due to its own weight must not exceed 0.010 in. per foot of span. (a) Calculate the maximum permissible span length for a 3 in. diameter shaft. (b) Calculate the bending stress caused by the weight of the shaft.

11. A steel wide-flange section is used as a cantilever beam having a span of 12 ft. The beam supports a distributed load that varies uniformly from 5 kips/ft at the support to zero at the free end. Determine the required moment of inertia of the beam if the deflection is not to exceed $\frac{1}{240}$ times the span.

Section 16–5   The Moment-Area Method

For Problems 12–22, use the moment-area method.

12. A W10 × 22 structural steel wide-flange beam on a simple span of 16 ft supports a concentrated load of 10,000 lb at midspan. Calculate the end rotation of the beam.

13. A W24 × 84 structural steel wide-flange beam on a simple span of 30 ft supports a uniformly distributed load of 3.0 kips/ft. Calculate the end rotation of the beam.

14. A steel bar 1 in. thick and $1\frac{1}{2}$ in. wide is subjected to the loads shown in Fig. 16–32. Calculate the angle between the tangent lines to the elastic curve at points B and C.

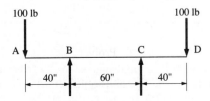

FIGURE 16–32   Problem 14.

15. For the beam of Problem 14, calculate the slope of the tangent line to the elastic curve at point A.

16. Using the moment-area method, check the deflection obtained in Problem 6.

17. Using the moment-area method, check the deflection obtained in Problem 7.

18. An 8 in. by 12 in. Douglas fir (S4S) is used as a 10 ft long cantilever beam. Compute the concentrated load at the free end that will cause a $\frac{3}{4}$ in. maximum deflection.

19. Compute the load required at the midspan of the beam of Problem 18 to cause a deflection of $\frac{3}{4}$ in. at the free end.

20. Using the moment-area method, check the deflection obtained in Problem 8.

21. A W12 × 30 structural steel beam is simply supported on a span of 12 ft. It is subjected to a uniformly distributed load of 675 lb/ft and a concentrated load of 14,000 lb at midspan. The allowable deflection is $\frac{1}{360}$ of the span. Is the beam satisfactory?

22. Compute the maximum deflection for the beam of Problem 13.

Section 16–6   Moment Diagram by Parts

23. Draw the moment diagram by parts for the beam in Fig. 16–11. (Draw the diagram from left to right.) Compare the value of the maximum moment.

24. Draw the moment diagram by parts for the beam in Fig. 16–4. (Draw the diagram from left to right.) Compare the value of the maximum moment.

25. Draw the moment diagram by parts for the beam in Fig. 16–14. (Draw the diagram from left to right.) Compare the value of the maximum moment.

26. For the beam in Fig. 16–33, draw the conventional moment diagram and left-to-right moment diagram by parts. Compare the maximum moments.

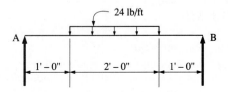

FIGURE 16–33   Problem 26.

Section 16–7   Applications of the Moment-Area Method

For Problems 27–39, use the moment-area method.

27. A steel bar is 3 in. wide and 1 in. thick. It is used as a 10 ft long simply supported beam. Bending is about the weak axis. The beam supports two concentrated loads of 120 lb each spaced 4 ft apart and placed 3 ft from each end. Calculate the maximum deflection and the deflection at one of the loads.

28. Verify the deflection at point C that was calculated in part (b) of Example 16–14 by using the tangent to the elastic curve at point A.

29. A wood test beam 1.72 in. wide by 1.75 in. deep spans 28 in. between simple supports. The beam is deflected 0.074 in. under a 200 lb load applied at midspan. Compute the modulus of elasticity E.

30. A W21 × 83 structural steel wide-flange section is used as a 40 ft long simply supported beam. It is subjected to a uniformly distributed load for its full length. The load induces a maximum bending stress of 22,000 psi. Compute the magnitude of the superimposed load (lb/ft) and determine the maximum deflection. Include the effect of the beam weight for this problem.

31. A W8 × 58 structural steel wide-flange section is loaded as shown in Fig. 16–34. Calculate the maximum deflection between the supports and the deflection of the free end.

32. An aluminum beam with a moment of inertia of 200 in.[4] is used as a cantilever beam and loaded as shown in Fig. 16–35. Calculate the maximum deflection.

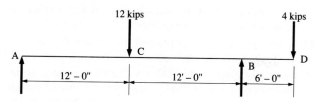

FIGURE 16–34 Problem 31.

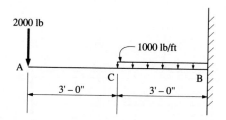

FIGURE 16–35 Problem 32.

33. For the W27 × 114 structural steel wide-flange beam shown in Fig. 16–36, calculate the maximum deflection.

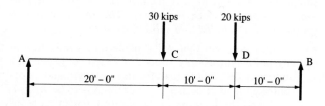

FIGURE 16–36 Problem 33.

34. Calculate the minimum required diameter for a solid circular steel cantilever beam 10 ft long. The maximum deflection due to its own weight is not to exceed 0.1 in.

35. For the steel beam in Fig. 16–37, calculate the slope at the free end and the maximum deflection. Note the varying moment of inertia.

36. Calculate the maximum deflection for the simply supported steel beam in Fig. 16–38. Neglect the weight of the beam.

37. A sign post is composed of standard-weight steel pipe. The top portion is 4 in. nominal diameter and the bottom portion is 5 in. nominal diameter, as shown in Fig.

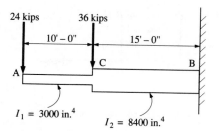

FIGURE 16–37 Problem 35.

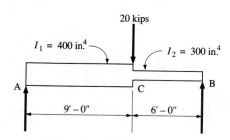

FIGURE 16–38 Problem 36.

16–39. The weight of the sign is 100 lb and its center of gravity is located as shown. Assume simple supports between sign and post. Calculate the horizontal deflection at the top of the post due to the weight of the sign.

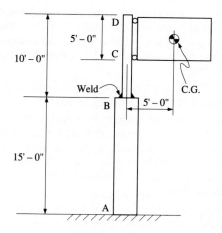

FIGURE 16–39 Problem 37.

38. Derive an expression for the maximum deflection of the cantilever beam in Fig. 16–40.

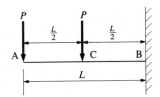

FIGURE 16–40   Problem 38.

39. Derive an expression for the maximum deflection of the simply supported beam in Fig. 16–41.

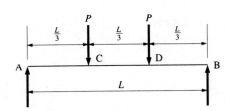

FIGURE 16–41   Problem 39.

SI System Problems

For Problems 40–45, for steel members, use a modulus of elasticity of 207 000 MPa.

40. A solid, round simply supported steel shaft has a diameter of 38 mm and a span length of 800 mm. The shaft supports a concentrated load of 3 kN at midspan. Calculate the maximum deflection of the shaft using the formula method.

41. Rework Problem 40 changing the diameter of the shaft to 25 mm and the span length to 500 mm.

42. A simply supported W410 × 0.83 structural steel wide-flange beam spans a length of 9 m. It is subjected to a uniformly distributed load of 4 kN/m and two concentrated loads of 40 kN each applied at the third points of the span. Compute the maximum deflection using the formula method.

43. A solid, round simply supported steel shaft is used as a beam with a span length of 700 mm. The shaft supports two concentrated loads of 3 kN each applied at the third points of the span. Calculate the required shaft diameter if its deflection must not exceed 0.20 mm. Using the computed diameter, compute the maximum bending stress and shear stress and compare with al-lowable stresses of 165 MPa in bending and 100 MPa in shear. Use the formula method.

44. Check the deflection obtained in Problem 42 using the moment-area method.

45. A simply supported W530 × 1.21 structural steel wide-flange section is used as a beam and supports loads as shown in Fig. 16–42. Compute the deflection at the center of the span and at the point of the concentrated load. Use the moment-area method.

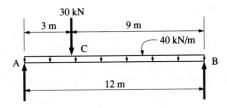

FIGURE 16–42   Problem 45.

Computer Problems

For the following computer problems, any appropriate programming language may be used. Input prompts should fully explain what is required of the user (the program should be "user friendly"). The resulting output should be well labeled and self-explanatory.

46. Write a program that will calculate the deflection at any point on a simply supported beam subjected to a uniformly distributed load for its full length and a concentrated load at midspan. User input is to be L, E, I, w, P, and the location of the desired deflection referenced to the left support.

47. Rework Problem 46, but have the program produce a list of deflections at the tenth points of the span. Input is to be the same, except that no location of the desired deflection will be given.

48. Write a program that will calculate the slope of the elastic curve at the tenth points of the span for a cantilever beam subjected to a concentrated load at its free end. User input is to be E, I, L, and P.

49. Rework Problem 48, but add a uniformly distributed load w over the full length of the beam.

Supplemental Problems

50. If the elastic limit of a steel wire is 60,000 psi, compute the diameter of the smallest circle into which a No. 14

(0.080 in. diameter) wire may be coiled without undergoing a permanent set.

51. Calculate the bending moment required to produce a radius of curvature of 1250 ft for a $1\frac{1}{2}$ in. square steel bar.

52. A 6 ft long cantilever beam is subjected to a concentrated load of 5 kips acting at its free end. Calculate the maximum deflection at the free end if the beam is (a) a solid rectangular Douglas fir (S4S), 8 in. by 12 in. beam; (b) a W10 × 22 structural steel wide-flange shape; (c) an 8 in. nominal diameter extra-strong steel pipe. Use the formula method.

53. A W16 × 36 structural steel wide-flange section is simply supported on a span length of 20 ft. It is subjected to a uniformly distributed load of 1.0 kip/ft and a concentrated load of 10 kips at midspan. (a) Compute the maximum deflection using the formula method. (b) Compute the maximum bending stress.

54. Select the lightest structural steel W shape for a beam that supports a uniformly distributed load of 400 lb/ft on a simple span of 22 ft. Use an allowable bending stress of 24 ksi and an allowable shear stress of 14.5 ksi. The maximum allowable deflection is $\frac{1}{360}$ of the span length. Consider moment, shear, and deflection. Use the formula method for deflection.

55. A W10 × 22 structural steel wide-flange shape is simply supported on a span of 20 ft. A superimposed uniformly distributed load of 1000 lb/ft is applied to the beam. As the beam deflects under load it comes into contact with a third support at midspan that is $\frac{1}{2}$ in. below the level of the two outer supports. Calculate the reaction on the center support. Use the formula method.

56. Using the moment-area method, check the deflections obtained in Problem 52.

57. A 1 in. diameter steel bar is 25 ft long and balanced at the middle in a horizontal position. Calculate the distance that the ends deflect below the middle of the bar due to the weight of the bar.

58. An 8 in. by 12 in. deep California redwood timber beam (S4S) is used as a 20 ft long simply supported beam. Compute the concentrated load at midspan that will cause a bending stress of 1350 psi. Compute the deflection at midspan and at the quarter points.

59. A solid steel shaft, 3 in. in diameter and 20 ft long, is used as a simply supported beam subjected to a 500 lb load at midspan. Calculate the maximum deflection. Neglect the weight of the member.

60. For the beam in Fig. 16–43, draw the conventional moment diagram and the moment diagram by parts (right to left and left to right).

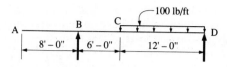

FIGURE 16–43 Problem 60.

61. Rework Problem 60 with concentrated loads of 5 kips added at points A and C. Additionally, draw the moment diagram by parts right to left and left to right to point C.

62. A solid steel shaft, 3 in. in diameter and 20 ft long, is used as a simply supported beam. A concentrated load of 500 lb is located 6 ft from one end. Calculate the maximum deflection, the deflection at midspan, and the deflection under the load.

63. A W27 × 114 structural steel wide-flange section is loaded as shown in Fig. 16–44. Calculate the slope at the free end, the deflection at the free end, and the deflection at the other end of the load.

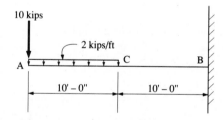

FIGURE 16–44 Problem 63.

64. A 6 in. by 10 in. Hem-fir timber beam (S4S) is loaded as shown in Fig. 16–45. Calculate the deflection under the concentrated load and at midspan between supports.

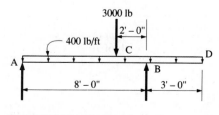

FIGURE 16–45 Problem 64.

65. Calculate the maximum permissible span length for a 3 in. diameter solid steel shaft used as a cantilever beam. The maximum deflection due to the weight of the shaft may not exceed 0.2 in.

66. A W10 × 33 structural steel wide-flange section, 10 ft long, is used as a cantilever beam supporting a uniformly distributed load of 800 lb/ft. (a) Calculate the deflection at the free end. (b) What superimposed uniformly distributed load can be placed on the beam if the allowable deflection is 0.25 in.?

67. A W10 × 68 structural steel wide-flange section supports loads as shown in Fig. 16–46. What are the maximum allowable loads P that the beam can support if the upward deflection under the 10 kip load is limited to 0.05 in.?

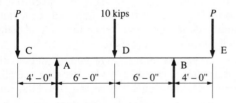

FIGURE 16–46   Problem 67.

68. Determine the deflection at point C and midway between the supports for the W36 × 194 structural steel beam in Fig. 16–47.

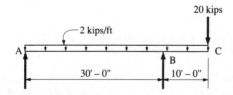

FIGURE 16–47   Problem 68.

69. Select the lightest W24 structural steel shape to support the loads for the beam of Problem 68. Consider

moment and deflection. Maximum downward deflection is limited to 0.60 in. Allowable bending stress is 27 ksi.

70. Calculate the deflection midway between the reactions and at the free end for the steel beam in Fig. 16–48.

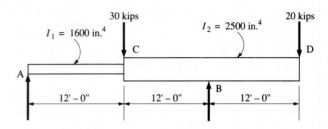

FIGURE 16–48   Problem 70.

71. Derive an expression for the maximum deflection of the beam in Fig. 16–49.

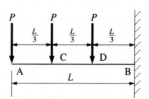

FIGURE 16–49   Problem 71.

72. Derive an expression for the maximum deflection of the beam in Fig. 16–50.

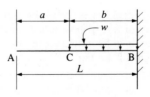

FIGURE 16–50   Problem 72.

17 Combined Stresses

<table>
<tr><td>17–1
INTRODUCTION</td><td>Thus far in our study of strength of materials, we have considered individual fundamental stresses (normal stresses and shear stresses) acting alone, without taking into account any interaction. When stresses acting at a point are collinear and of the same type, they can be added algebraically. For instance, tensile and compressive normal stresses due to bending and axial load can be added algebraically when they are collinear (acting along the same line). When stresses are not collinear, or when they are of different types (e.g., normal stresses and shear stresses), they must be added vectorially.

While some machine and structural parts can be adequately designed by considering the stresses individually, there are situations in which combinations of stresses may produce critical conditions that should be investigated. This is particularly the case when the stresses at actual failure of a member are being studied.</td></tr>
</table>

**17–2
COMBINED AXIAL
AND BENDING
STRESSES**

Many structural and machine members are subjected to both transverse loads and direct axial loads simultaneously, thereby developing a combination of bending stresses and direct axial stresses. One such example, as shown in Fig. 17–1(a), is a simply supported beam subjected to a transverse uniform loading w in combination with a direct axial load P. The transverse loading will produce a bending moment and a subsequent triangular bending stress distribution over the cross section of the beam, as shown in Fig. 17–1(c). While the cross section could be arbitrarily taken at any point, we will consider the cross section at midspan (where moment will be maximum), and we will designate this plane as plane X–X. The bending stress will vary from a maximum at the outer fibers to zero at the neutral axis. Note that since the beam in Fig. 17–1(a) is simply supported and subjected to downward-acting transverse loads, the bending stress will be tensile below the neutral axis and compressive above the neutral axis. Recall that the magnitude of the maximum bending stress can be obtained using the flexure formula,

$$s_b = \frac{Mc}{I}$$

See Chapter 14 for definitions of all terms along with the limitations of usage.

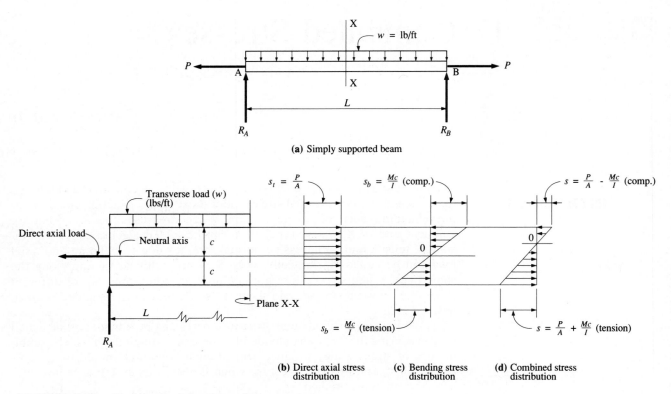

FIGURE 17–1   Combined axial and bending stresses.

Since the beam is simultaneously subjected to a direct axial load parallel to the longitudinal axis of the member (this load could be either tensile or compressive), a direct axial stress will be developed as described in Chapter 9. Recall that the direct axial stress distribution is assumed to be uniform over the cross section of the member, as shown in Fig. 17–1(b), and that the magnitude of the direct axial stress can be obtained from the expression

$$s_t = \frac{P}{A}$$

All terms were defined in Chapter 9 along with the limitations of usage.

Note that the bending results in tensile and compressive stresses that are normal to the plane of the cross section (plane X–X). Similarly, the direct axial tension results in tensile stresses normal to this plane. Since the same kind of stresses (normal) are developed, a resultant combined stress at any location can be determined by a simple algebraic sum of the individual direct axial and bending stresses produced at the same locations, as shown in Fig. 17–1(d). This process of algebraic addition of the same kind of stresses to obtain the resultant or combined stress is called *superposition*. The principle of superposition is based on the premise that the resultant stress due to several forces in any system is the algebraic sum of the effects caused by the individual forces. Superposition of stresses is permissible if the maximum stress remains within the elastic limit and if deformations are very small. Since the two types of stresses acting on any arbitrary plane are collinear, they can be added algebraically as follows:

$$s = \pm\frac{P}{A} \pm \frac{Mc}{I} \tag{17–1}$$

It is customary to use a sign convention for the stresses. The plus sign is commonly used to designate tensile stress and the minus sign is used to designate compressive stress. Note that if the axial stress is tensile, it will add to the tensile bending stress; if it is compressive it will add to the compressive bending stress. In either case, the total combined stress at any point will always equal the algebraic sum of the axial and bending stresses at that point. Therefore, for the beam of Fig. 17–1, the resultant combined stress is

$$s = +\frac{P}{A} \pm \frac{Mc}{I}$$

Note that this approach is slightly inexact, particularly for slender members or members with relatively low EI values where comparatively large deflections may be produced by the transverse loads. With reference to Fig. 17–2, the load P multiplied by the deflection Δ will induce an additional bending moment (sometimes called the ''$P\Delta$'' *moment*). For the axial compressive load P shown, this additional moment would be added to the moment caused by the transverse loads P_1 and P_2. If P were tensile, the $P\Delta$ moment would be subtracted from the transverse load moment. Unless members are long and slender, the $P\Delta$ moment is usually neglected since the deflections are very small.

FIGURE 17–2 Axial compression and bending.

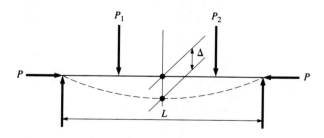

In this chapter, only members of relatively short span lengths (hence, small deflections) will be considered. Therefore, the bending stresses caused by the axial loads may be neglected without appreciable error.

☐ **EXAMPLE 17–1**   A W14 × 61 structural steel wide-flange section is used as a simply supported beam with a span length of 10 ft. It is subjected to a uniformly distributed transverse load of

6 kips/ft, including its own weight, and an axial tensile force of 100 kips. Compute the maximum combined tensile and compressive stresses.

Solution   Appendix A contains the necessary properties of the W14 × 61:

$$A = 17.9 \text{ in.}^2$$
$$I_x = 640 \text{ in.}^4$$
$$S_x = 92.2 \text{ in.}^3$$

To illustrated the method of superposition, this problem is solved by dividing it into two parts. The loaded beam is shown in Fig. 17–3(a). It is shown subjected to only the axial load in Fig. 17–3(b), while in Fig. 17–3(c) it is shown subjected to only the transverse load.

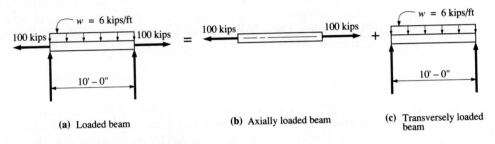

(a) Loaded beam   (b) Axially loaded beam   (c) Transversely loaded beam

FIGURE 17–3   Method of superposition.

For the axial load, the normal direct stress throughout the length of the beam is

$$s_t = \frac{P}{A} = \frac{100}{17.9} = +5.59 \text{ ksi (tension)}$$

and is shown in Fig. 17–4(a).

The normal stress due to the transverse load depends on the magnitude of the maximum bending moment, which, in this case, occurs at midspan. The maximum

FIGURE 17–4   Stress distributions at midspan.

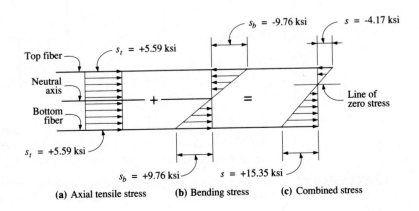

(a) Axial tensile stress   (b) Bending stress   (c) Combined stress

bending moment (see Appendix H) is

$$M = \frac{wL^2}{8} = \frac{6(10)^2}{8} = 75 \text{ ft-kips}$$

From the flexure formula, the maximum stresses at the outer fibers caused by this moment are

$$s_b = \frac{Mc}{I} = \frac{M}{S} = \frac{75(12)}{92.2} = \pm 9.76 \text{ ksi}$$

These stresses will be normal to the cross section of the beam at midspan and decrease linearly toward the neutral axis, as is shown in Fig. 17–4(b).

To obtain the resultant or combined stress, the bending stress must be added algebraically to the direct axial tensile stress. Expressing this mathematically,

$$s = +\frac{P}{A} \pm \frac{Mc}{I}$$

from which, the bottom-fiber stress is

$$s = +5.59 + 9.76 = +15.35 \text{ ksi (tension)}$$

and the top-fiber stress is

$$s = +5.59 - 9.76 = -4.17 \text{ ksi (compression)}$$

Thus, as can be seen from Fig. 17–4(c), the resultant normal stress is 15.35 ksi tension at the midspan bottom fiber and 4.17 ksi compression at the midspan top fiber. Since the bending moment varies along the length of the beam, the bending stresses and, therefore, the combined stresses, will be different at different sections. Note that in Fig. 17–4(c), the line of zero stress, which lies on the neutral axis if only flexural stresses exist, moves upward.

□ **EXAMPLE 17–2**   An 8 in. nominal diameter standard-weight steel pipe extends vertically out of a solid concrete anchorage, as shown in Fig. 17–5. The pipe is subjected to horizontal and vertical loading as shown. The vertical load includes the weight of the pipe. Compute

FIGURE 17–5 Load diagram.

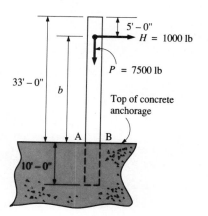

5' – 0"
H = 1000 lb

P = 7500 lb

33' – 0"

b

Top of concrete anchorage

A   B

10' – 0"

the maximum combined tensile and compressive stresses at the top of the concrete anchorage.

Solution Appendix B contains the necessary properties for the 8 in. diameter standard-weight steel pipe:

$$A = 8.40 \text{ in.}^2$$
$$S = 16.8 \text{ in.}^3$$
$$I = 72.5 \text{ in.}^4$$

The section to be investigated is in the plane of the top of the concrete anchorage (plane A–B). The direct axial load P is 7500 lb. The direct stress on plane A–B caused by the axial load is

$$s_c = \frac{P}{A} = \frac{7500}{8.40} = -893 \text{ psi (compression)}$$

This is shown in Fig 17–6(a). Note that this is a projected view of the compressive stress. The actual compressive stress block would be ring-shaped, with a hole in the middle.

The bending stress in plane A–B caused by the horizontal load H depends on the magnitude of the bending moment. The maximum bending moment (see Appendix H) is

$$M = Hb = 1000(28) = 28,000 \text{ ft-lb}$$

FIGURE 17–6 Stress distributions in plane A–B.

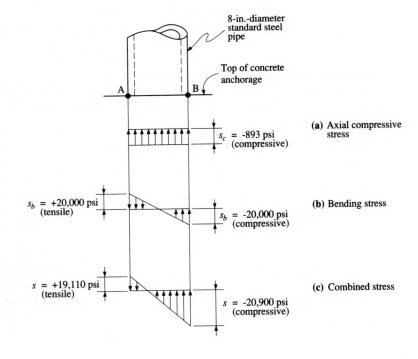

8-in.-diameter standard steel pipe

Top of concrete anchorage

A   B

$s_c = -893$ psi (compressive)

(a) Axial compressive stress

$s_b = +20,000$ psi (tensile)

$s_b = -20,000$ psi (compressive)

(b) Bending stress

$s = +19,110$ psi (tensile)

$s = -20,900$ psi (compressive)

(c) Combined stress

Using the flexure formula, the maximum stresses at the outer fibers caused by this moment are

$$s_b = \frac{Mc}{I} = \frac{M}{S} = \frac{28,000(12)}{16.8} = \pm 20,000 \text{ psi}$$

This is shown in Fig. 17–6(b).

The resultant or combined stress is the algebraic sum of the two stresses and is shown in Fig. 17–6(c):

$$s = -\frac{P}{A} \pm \frac{Mc}{I}$$

Therefore, the combined fiber stress at point A is

$$s = -893 + 20,000 = +19,110 \text{ psi (tension)}$$

and the combined fiber stress at point B is:

$$s = -893 - 20,000 = -20,900 \text{ psi (compression)}$$

17–3 ECCENTRICALLY LOADED MEMBERS

A *short* compression member may be defined as a member that will fail by crushing or yielding, as opposed to buckling, when subjected to an increasing axial compressive load. When such a member is subjected to a vertical direct axial load (a load coincident with a longitudinal axis through the centroid of the cross section), the compressive stress developed is assumed to be uniformly distributed over the cross section of the member. If the vertical load is not axial but is parallel to a longitudinal axis through the centroid, as shown in Fig. 17–7(a), the stress developed on any cross section will not be uniformly distributed. The stress at any point will then be a combined stress similar to the combined stresses developed in a member subjected to both transverse and direct axial loads, as discussed in Section 17–2.

When the vertical load is not coincident with the longitudinal axis through the centroid, it is said to be *eccentric*. The eccentricity *e* of the load is the distance measured along one of the centroidal axes from the centroid O to the line of action of the force *P*. Note that in this section the discussion is limited to the case where the load *P* lies on one of the centroidal axes and is eccentric to the other. Also note that since the member is short, its lateral deflection is assumed so small that it can be neglected in comparison with the initial eccentricity *e*. Therefore, the method of superposition may be used.

The analysis of an eccentrically loaded member involves what is commonly termed the *load-moment relationship*. In Fig. 17–7(a) we see a short compression member subjected to a vertical load *P* applied with an eccentricity *e* on one of the two centroidal axes. In Fig. 17–7(b) two equal and opposite forces *P* are applied at the centroid of the cross section, each equal in magnitude to the eccentric load *P*. The addition of these two forces does not change the problem, since the algebraic sum of the two collinear and equal additional forces is equal to zero.

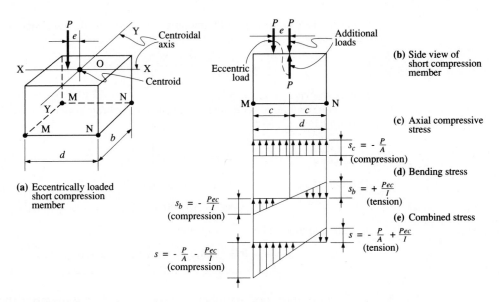

FIGURE 17–7   Combined stresses caused by eccentric load.

In effect, we have replaced the eccentric load system with a downward axial compressive force P acting at the centroid and a couple having a magnitude of Pe. As shown in Fig. 17–7(c) and (d), the compressive force P, at the centroid, produces direct axial compressive stresses of

$$s_c = -\frac{P}{A}$$

and the couple Pe, acting about the Y–Y axis, produces bending stresses of

$$s_b = \pm\frac{Mc}{I} = \pm\frac{Pec}{I}$$

The total combined stress can be expressed as follows, using the same sign convention as that used in Section 17–2:

$$s = -\frac{P}{A} \pm \frac{Pec}{I} \qquad\qquad \textbf{(17–2)}$$

The diagram of the combined stress is shown in Fig. 17–7(e). It is assumed that the maximum bending stress is larger than the direct axial stress; therefore, there is a zone of tensile stress on the right and compressive stress on the left. If the maximum bending stress is smaller in magnitude than the direct axial stress, then there will be compressive stress over the entire cross section of the member and no tensile stress will be developed.

☐ **EXAMPLE 17–3**   A full-size rectangular timber member is used as a short compression post similar to that shown in Fig. 17–7(a). The cross section of the post has dimensions of $b = 10$ in. and $d = 16$ in. The post is subjected to an eccentric load P of 40,000 lb applied 5 in.

from the Y–Y axis. Refer to Fig. 17–7(a) and calculate the combined stresses in the outer fibers at edges MM and NN.

Solution   The axial compressive stress due to the load P is

$$s_c = -\frac{P}{A} = -\frac{40,000}{10(16)} = -250 \text{ psi (compression)}$$

The moment caused by the eccentric load is

$$M = Pe = 40,000(5) = 200,000 \text{ in.-lb}$$

The bending stress developed by the moment is

$$s_b = \frac{Mc}{I} = \frac{Pec}{I}$$

where $c = d/2 = 8$ in. and I, with respect to the Y–Y bending axis, is calculated from

$$I_y = \frac{bd^3}{12} = \frac{10(16)^3}{12} = 3413 \text{ in.}^4$$

Therefore,

$$s_b = \frac{200,000(8)}{3413} = \pm 468.8 \text{ psi}$$

The combined stress is the algebraic sum of the two stresses and is shown in Fig. 17–7(e):

$$s = -\frac{P}{A} \pm \frac{Pec}{I}$$

The combined fiber stress at edge MM is

$$s = -250 - 468.8 = -719 \text{ psi (compression)}$$

The combined fiber stress at edge NN is

$$s = -250 + 468.8 = +219 \text{ psi (tension)}$$

☐ **EXAMPLE 17–4**   A press frame is shown in Fig. 17–8. When the press operates, a maximum force P of 30,000 lb is exerted between the upper and lower jaws, subjecting the arm of the frame to combined bending and axial stress. Calculate the maximum tensile and compressive stresses at points o and i on plane A–A, where these points represent the outside and inside faces of the arm of the frame, respectively.

Solution   The eccentrically applied load P results in a force reaction P and a moment reaction Pe at plane A–A, as shown on the free-body diagram of the frame (Fig. 17–9).

First calculate the properties of the cross section of the frame (refer to Fig. 17–8(b)). The cross-sectional area is

$$A = a_1 + a_2 = 27 + 36 = 63 \text{ in.}^2$$

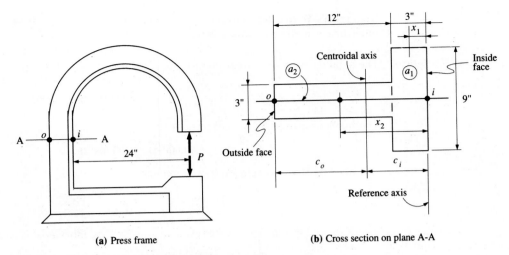

(a) Press frame

(b) Cross section on plane A-A

FIGURE 17–8 Eccentrically loaded press frame.

FIGURE 17–9 Press frame free body.

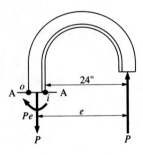

The centroidal axis is located from the designated reference axis. Note that x is the distance from the centroid of each area a to the reference axis.

$$c_i = \frac{\sum ax}{\sum a} = \frac{a_1 x_1 + a_2 x_2}{a_1 + a_2} = \frac{27(1.5) + 36(9)}{27 + 36} = 5.79 \text{ in.}$$

from which

$$c_o = 15 - 5.79 = 9.21 \text{ in.}$$

The moment of inertia is calculated with respect to the centroidal axis:

$$
\begin{aligned}
I &= \sum (I_o + ad^2) \\
&= \left[\frac{9(3)^3}{12} + 9(3)(5.79 - 1.5)^2\right] + \left[\frac{3(12)^3}{12} + 3(12)(9.21 - 6)^2\right] \\
&= 1320 \text{ in.}^4
\end{aligned}
$$

The stresses developed on plane A–A can be determined using Eq. (17–2). Note that the direct tensile force reaction P develops a tensile stress equal to P/A that is uniformly distributed over the cross section. The P/A term has a positive sign because it is in tension. The bending moment Pe will develop a tensile stress on the inside face and a compressive stress on the outer face.

$$s = +\frac{P}{A} \pm \frac{Pec}{I}$$

From the free-body diagram, the eccentricity $e = c_i + 24 = 29.79$ in. Therefore, the stress at the inside face, point i, is

$$s_i = +\frac{30,000}{63} + \frac{30,000(29.79)(5.79)}{1320} = +4396 \text{ psi (tension)}$$

and at the outside face, point o, the stress is

$$s_o = +\frac{30,000}{63} - \frac{30,000(29.79)(9.21)}{1320} = -5,760 \text{ psi (compression)}$$

17-4 MAXIMUM ECCENTRICITY FOR ZERO TENSILE STRESS

When a short compression member is subjected to an eccentric load as discussed in Section 17-3, the maximum combined stresses, either tensile or compressive, will develop at the outer edges of the cross section. In Fig. 17-10, these stresses occur at edges AB and CD, since the bending occurs with respect to the Y–Y plane.

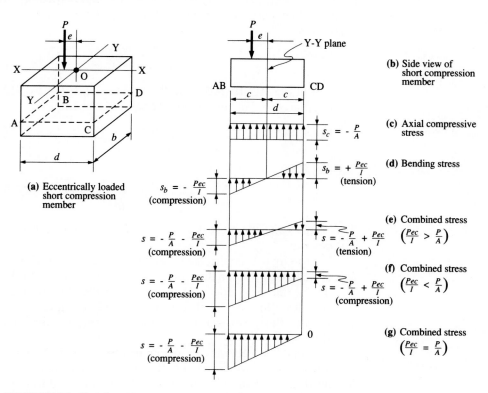

(a) Eccentrically loaded short compression member

(b) Side view of short compression member

(c) Axial compressive stress $s_c = -\dfrac{P}{A}$

(d) Bending stress $s_b = +\dfrac{Pec}{I}$ (tension)

$s_b = -\dfrac{Pec}{I}$ (compression)

(e) Combined stress $s = -\dfrac{P}{A} + \dfrac{Pec}{I}$ (tension) $\left(\dfrac{Pec}{I} > \dfrac{P}{A}\right)$

$s = -\dfrac{P}{A} - \dfrac{Pec}{I}$ (compression)

(f) Combined stress $s = -\dfrac{P}{A} + \dfrac{Pec}{I}$ (compression) $\left(\dfrac{Pec}{I} < \dfrac{P}{A}\right)$

$s = -\dfrac{P}{A} - \dfrac{Pec}{I}$ (compression)

(g) Combined stress $\left(\dfrac{Pec}{I} = \dfrac{P}{A}\right)$

$s = -\dfrac{P}{A} - \dfrac{Pec}{I}$ (compression)

FIGURE 17-10   Combined stresses caused by eccentric load.

As discussed in Section 17–2, the maximum combined stress will be

$$s = -\frac{P}{A} \pm \frac{Pec}{I}$$

Note that the stress at edge AB will always be compressive, since the eccentric load is closer to AB than it is to CD. The stress at CD could be compressive, tensile, or zero, depending on the eccentricity of the load. If the bending stress is greater than the direct axial stress, outer edge CD will develop a tensile stress, as shown in Fig. 17–10(e). On the other hand, if the bending stress is less than the direct axial stress, outer edge CD will develop a compressive stress, as shown in Fig. 17–10(f). If the bending stress is equal to the direct stress, then the stress at outer edge CD will be zero, as shown in Fig. 17–10(g).

Since several materials (such as concrete) have insignificant tensile strength, it is important to know the largest load eccentricity for which a tensile stress will not develop for given conditions. While this discussion will focus on the zero tensile stress situation, the same theory applies to the case of zero compressive stress, albeit less important and less practical.

The limiting stress distribution for zero tensile stress is shown in Fig. 17–10(g), where zero stress occurs along edge CD. This will occur when the load eccentricity reaches and does not exceed a specific value. This maximum load eccentricity for the case of zero stress can be calculated, since at zero stress

$$\frac{Pec}{I} = \frac{P}{A}$$

If, in Fig. 17–10, AB $= b$, AC $= d$, $c = d/2$, $I = bd^3/12$, and the cross-sectional area $= bd$, the preceding equation becomes

$$\frac{Pe(d/2)}{(bd^3/12)} = \frac{P}{bd}$$

Solving for e,

$$e = \frac{d}{6} \tag{17–3}$$

This represents the largest load eccentricity for which a tensile stress will not develop.

If the load eccentricity e becomes greater than $d/6$, then the bending stress becomes greater than the direct axial stress and a tensile stress will develop at the outer edge CD, as shown in Fig. 17–10(e). If the load eccentricity is less than or equal to $d/6$, then only compressive stresses will develop over the cross section of the member, as shown in Fig. 17–10(f) and (g).

Should the eccentricity occur in the other direction [on the opposite side of the Y–Y axis of Fig. 17–10(a)], then the largest eccentricity for zero tensile stress will still be $d/6$, but it will be measured to the right of the Y–Y axis. Therefore, if the eccentric load is applied within the middle one-third ($d/6 + d/6$) of the length d, tensile stresses will not be developed.

The same principles would hold if the eccentric load were applied along the Y–Y axis instead of the X–X axis, as shown in Fig. 17–10(a). The maximum eccentricity for the limiting zero tensile stress would be $b/6$.

**17–5
ECCENTRIC
LOAD NOT ON
CENTROIDAL AXIS**

In this section we will consider the case of an eccentric load that does not lie on either centroidal axis. This results in a situation sometimes termed *double eccentricity*, which is illustrated in Fig. 17–11. The combined stress at any point is the algebraic sum of

1. The direct axial stress due to the load P at the centroid O
2. The bending stress due to the moment Pe_1 with respect to the bending axis Y–Y
3. The bending stress due to the moment Pe_2 with respect to the bending axis X–X

FIGURE 17–11 Short compression member with double eccentricity.

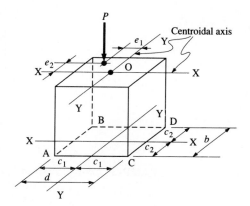

For points around the outside of the member, this can be expressed in equation form:

$$s = -\frac{P}{A} \pm \frac{Pe_1c_1}{I_y} \pm \frac{Pe_2c_2}{I_x} \qquad \textbf{(17–4)}$$

where c_1 and c_2 represent the distance from the centroidal axes to the outer fibers of the member where the combined stress is being computed. While the preceding is written for any point around the outside of the member, where the maximum and minimum stresses will occur, the stress at *any* point within the member can also be found. We simply substitute y_1 and y_2 for the respective c dimensions, where the y dimensions locate the point with respect to the appropriate centroidal axes [refer to Eq. (14–3)].

The limits of the position of the eccentric load P so that no tensile stress will occur at any location is similar to that established in Section 17–4.

The largest load eccentricity along the Y–Y axis is $b/6$ and along the X–X axis, $d/6$. However, since the eccentric load does not lie on either axis, it must fall within an area that is formed by connecting those designated limiting points in order that no tensile stress develop. As shown in Fig. 17–12, if the eccentric load on a rectangular short compression member falls within the diamond-shaped area, no tensile stress will develop anywhere over the cross section of the member. This area is commonly called the *kern* of the cross section, and its shape will depend on the shape of the cross section.

FIGURE 17–12   Kern area of rectangular short compression member.

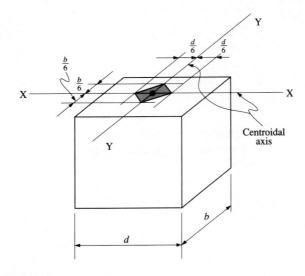

□ **EXAMPLE 17–5**   A short compression member is subjected to a compressive load of 100,000 lb. There is a double eccentricity, as shown in Fig. 17–13. The member is 12 in. by 14 in. in cross section. Calculate the combined stress at each of the four corners A, B, C, and D. Locate the line of zero stress.

Solution   The use of Eq. (17–4) will require the calculation of the moments of inertia with respect to the X–X and Y–Y axes:

$$I_x = \frac{db^3}{12} = \frac{14(12)^3}{12} = 2016 \text{ in.}^4$$

$$I_y = \frac{bd^3}{12} = \frac{12(14)^3}{12} = 2744 \text{ in.}^4$$

The various terms of the equation are evaluated as follows:

$$\frac{P}{A} = \frac{100,000}{12(14)} = 595 \text{ psi}$$

$$\frac{Pe_1c_1}{I_y} = \frac{100,000(3)(7)}{2744} = 765 \text{ psi}$$

$$\frac{Pe_2c_2}{I_x} = \frac{100,000(2.5)(6)}{2016} = 744 \text{ psi}$$

FIGURE 17–13 Rectangular short compression member with double eccentricity.

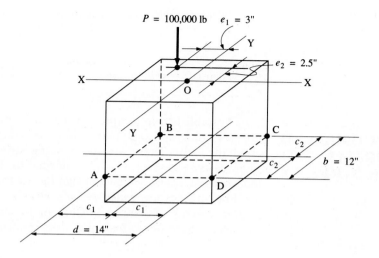

The stresses at the four points can now be calculated using Eq. (17–4):

$$s = -\frac{P}{A} \pm \frac{Pe_1c_1}{I_y} \pm \frac{Pe_2c_2}{I_x}$$

At point A,

$$s = -595 - 765 + 744 = -616 \text{ psi (compression)}$$

At point B,

$$s = -595 - 765 - 744 = -2104 \text{ psi (compression)}$$

At point C,

$$s = -595 + 765 - 744 = -574 \text{ psi (compression)}$$

At point D,

$$s = -595 + 765 + 744 = +914 \text{ psi (tension)}$$

The line of zero stress is located as shown in Fig. 17–14.

FIGURE 17–14 Line of zero stress.

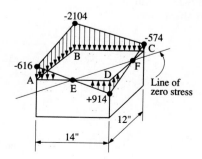

Calculating DE from similar triangles,

$$\frac{DE}{914} = \frac{14 - DE}{616}$$

from which DE = 8.36 in. Similarly, DF = 7.37 in.

**17–6
COMBINED
NORMAL AND
SHEAR STRESSES**

It was established in Chapter 14 that in homogeneous elastic beams, where stresses are proportional to strains, two kinds of stresses are developed simultaneously. These are the bending stresses, consisting of tensile and compressive stresses, which are normal to the surfaces on which they act, and shear stresses, which are parallel to the surfaces on which they act. Bending stresses are called *normal stresses*; shear stresses are called *tangential stresses*. As shown in Chapter 14, they can be calculated using the following expressions: From the flexure formula,

$$s_b = \frac{Mc}{I}$$

and from the general shear formula,

$$s_s = \frac{VQ}{Ib}$$

Consider an infinitesimal element of a simply supported beam, as shown in Fig. 17–15. The element is taken at some point where the shear and the bending moment are not equal to zero and at some location other than the extreme fibers or the neutral axis; thus, the element is subjected to both bending and shear stresses. These stresses are shown acting on the faces of the element in Fig. 17–15(b). The normal stress is represented by s_x and the shear stress is represented by s_{xy}. The combination of these stresses will result in maximum and minimum normal and shear stresses being developed on planes that are inclined with respect to the axis of the beam.

For the development of the various relationships that exist in this situation, we will consider a more general case of an element from a member that is acted on by forces from many directions. This general case is repre-

FIGURE 17–15 Combined normal and shear stress.

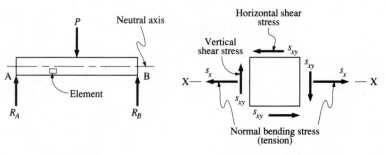

(a) Loaded beam   (b) Magnified stress element

sented by the stressed element shown in Fig. 17–16. On this stressed element, s_x represents the normal stress acting on a plane perpendicular to the X–X axis and s_y represents the normal stress acting on a plane perpendicular to the Y–Y axis. These two stresses may be either tension or compression (tension is shown), and they may result either from the bending moment or direct load. The shear stress at the same point acting on all the faces of the element is represented by s_{xy}. Recall that in Section 11–6 the shear stresses on mutually perpendicular planes were shown to be equal. The shear stress may be the result of an externally applied torsional moment (Chapter 12) or beam shear (Chapter 14).

FIGURE 17–16 Stressed element.

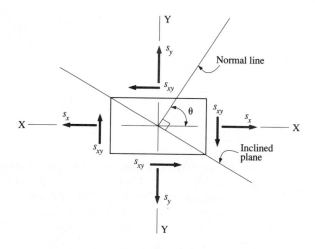

By combining the stresses shown in Fig. 17–16, the normal and shear stresses on any inclined plane can be determined. We will define the direction of a plane by specifying the angle between the X–X axis and a line that is *normal* to the plane. This convention is the same as that used in Chapter 11. If we cut the rectangular element along the inclined plane indicated in Fig. 17–16, we obtain the triangular element shown in Fig. 17–17 as a free body.

Note in Fig. 17–17 that since the top portion of the rectangular element has been removed, it must be replaced by the forces that it exerted upon the inclined plane of the lower portion. These consist of both normal and shear forces acting along the inclined plane. These unknown forces (and the resulting stresses) are the ones we will evaluate. Note that in Fig. 17–16 stresses are shown acting on the various faces of the element, rather than forces. Each of these stresses is assumed to be uniformly distributed over the area on which it acts. Prior to applying equations of equilibrium to the free body, these stresses must first be converted to forces (recall that force equals stress multiplied by area). Since the element is shown as being two-dimensional, the thickness of the element perpendicular to the plane of the paper is taken as unity (a unit thickness).

FIGURE 17–17   Free-body diagram.

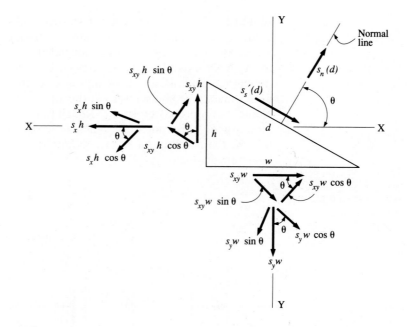

All stresses are converted to forces in the free body of Fig. 17–17, so that equations of equilibrium can now be applied. In this figure, s_n represents the normal stress (either tension or compression) on the inclined plane and s'_s represents the tangential stress (shear stress) on the inclined plane. Taking an algebraic summation of forces normal to the inclined plane with the element faces assigned length dimensions of h, w, and d as shown,

$$s_n(d)(1) = s_x(h)(1)\cos\theta + s_y(w)(1)\sin\theta - s_{xy}(h)(1)\sin\theta - s_{xy}(w)(1)\cos\theta$$

Dividing through by $(d)(1)$,

$$s_n = s_x\left(\frac{h}{d}\right)\cos\theta + s_y\left(\frac{w}{d}\right)\sin\theta - s_{xy}\left(\frac{h}{d}\right)\sin\theta - s_{xy}\left(\frac{w}{d}\right)\cos\theta$$

Since $\sin\theta = w/d$ and $\cos\theta = h/d$,

$$s_n = s_x\cos^2\theta + s_y\sin^2\theta - 2s_{xy}\sin\theta\cos\theta \qquad \textbf{(17–5)}$$

This represents the normal stress on any plane, which has a normal inclined at an angle θ with the X–X axis in terms of stresses on the planes perpendicular to the X–X and Y–Y axes.

Next, taking an algebraic summation of forces parallel to the inclined plane,

$$s'_s(d)(1) = s_{xy}(h)(1)\cos\theta + s_x(h)(1)\sin\theta - s_y(w)(1)\cos\theta - s_{xy}(w)(1)\sin\theta$$

Dividing through by $(d)(1)$,

$$s'_s = s_{xy}\left(\frac{h}{d}\right)\cos\theta + s_x\left(\frac{h}{d}\right)\sin\theta - s_y\left(\frac{w}{d}\right)\cos\theta - s_{xy}\left(\frac{w}{d}\right)\sin\theta$$

Again, since $\sin \theta = w/d$ and $\cos \theta = h/d$,

$$s'_s = (s_x - s_y)\sin \theta \cos \theta + s_{xy}(\cos^2\theta - \sin^2\theta) \qquad \textbf{(17–6)}$$

This represents the shear stress on any plane, which has a normal inclined at an angle θ with the X–X axis in terms of the stresses on planes perpendicular to the X–X and Y–Y axes.

If the inclined plane is rotated by varying the angle θ, it will reach a position for which the normal stress acting on it is a maximum or minimum. The position of the plane for maximum normal stress is perpendicular to the position of the plane for a minimum normal stress. On these planes there are no shear stresses. The planes on which the normal stresses become maximum or minimum are called the *principal planes*, and the corresponding normal stresses acting on these planes are called the *principal stresses*.

To determine the value of the angle θ_P that would locate the principal planes, we set the value of s'_s (given by Eq. [17–6]) equal to zero and solve for the angle in terms of the stresses on the planes perpendicular to the X–X and Y–Y axes:

$$0 = (s_x - s_y)\sin \theta_P \cos \theta_P + s_{xy}(\cos^2\theta_P - \sin^2\theta_P)$$

With the use of some trigonometric identities, the solution for θ_P can be shown to be

$$\tan 2\theta_P = -\frac{2s_{xy}}{s_x - s_y} \qquad \textbf{(17–7)}$$

Since $\tan 2\theta = \tan(2\theta + 180°)$, we see that the use of Eq. (17–7) will result in two possible values of θ_P, differing by 90°. On one plane thus described, the *maximum* normal stress will exist, while on the other plane the *minimum* normal stress will exist.

Having solved for $\tan 2\theta_P$, the two values of θ_P can be determined. By substituting one of the values of θ_P (from Eq. [17–7]) into Eq. (17–5), it can be determined which of the two principal stresses is acting on that plane. The other principal stress acts on the other plane. We will define the principal stresses as s_1 and s_2, where s_1 is the algebraically larger stress. Equation (17–5) may be modified to obtain the principal stresses directly. It will be shown later (in Section 17–8) that the principal stresses can be obtained from

$$s_{1,2} = \left(\frac{s_x + s_y}{2}\right) \pm \sqrt{\frac{(s_x - s_y)^2}{4} + s_{xy}^2} \qquad \textbf{(17–8)}$$

By substituting the value of θ_P as determined from Eq. (17–7) into Eq. (17–6) to obtain s'_s, it may be shown that the shear stresses on the principal planes are equal to zero.

The orientation (θ_s) of the planes of maximum shear stress can be developed mathematically from Eq. (17–6):

$$\tan 2\theta_s = \frac{s_x - s_y}{2s_{xy}} \qquad \textbf{(17–9)}$$

Substitution of one of the values of θ_s from Eq. (17–9) into Eq. (17–6) (for s'_s) will yield the maximum shear stress, which can be expressed by the direct equation

$$s'_{s(max)} = \pm \sqrt{\frac{(s_x - s_y)^2}{4} + s_{xy}^2} \qquad \textbf{(17–10)}$$

This represents the shear stress developed on the inclined plane, which has a normal inclined at an angle of θ_s with the original X–X axis of the stressed element, as defined by Eq. (17–9). Both positive and negative values of shear stress result from the calculation. They could be considered algebraic maximum and minimum values; however, they are equal in absolute magnitude.

Note that the maximum shear stress expression is the same as the radical that occurs in Eq. (17–8). Hence the principal stresses can be calculated from

$$s_{1,2} = \frac{s_x + s_y}{2} \pm s'_{s(max)}$$

Note that if θ_s, which defines the position of the maximum shear stress [Eq. (17–9)], is substituted into Eq. (17–5), it will be found that the planes of maximum shear stress are not always free of normal stresses. This is unlike the situation on the principal planes, where no shear stresses exist.

The expression for the normal stress acting on the plane of maximum shear stress is

$$s_n = \frac{s_x + s_y}{2} \qquad \textbf{(17–11)}$$

Equations (17–5) through (17–10) are valid for the directions of stresses shown in Fig. 17–16. If the stresses are opposite in direction, the signs for the various terms should be changed. For normal stresses, tensile stresses are considered positive; compressive stresses are considered negative. For shear stresses, the positive sense is indicated as shown in Fig. 17–16. Also, as shown in Fig. 17–16, θ is positive when measured counterclockwise from the horizontal and negative when measured clockwise from the horizontal.

☐ **EXAMPLE 17–6**   An element of a machine member subjected to biaxial loading is stressed as shown in Fig. 17–18. (a) Compute the normal and shear stress intensities on a plane that has a normal rotated 60° counterclockwise from the X–X axis. (b) Compute the principal stresses and the orientation of the principal planes. (c) Compute the maximum shear stress and locate the plane on which it acts.

Solution   (a) To compute the normal and shear stress on any plane, Eqs. (17–5) and (17–6) can be used. In accordance with our sign convention for tensile and compressive stresses, and with reference to Fig. 17–16 for the shear stress signs, $s_x = -10,000$ psi,

FIGURE 17–18 Original stressed element.

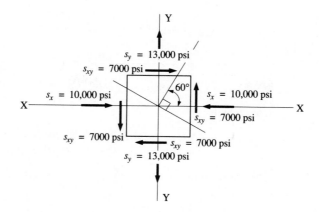

$s_y = +13,000$ psi, $s_{xy} = -7000$ psi, and $\theta = +60°$. From Eq. (17–5), the normal stress is

$$s_n = s_x\cos^2\theta + s_y\sin^2\theta - 2s_{xy}\sin\theta\cos\theta$$
$$= -10,000(0.25) + 13,000(0.75) - 2(-7000)(0.8660)(0.500)$$
$$= +13,310 \text{ psi (tension)}$$

From Eq. (17–6), the shear stress is

$$s'_s = (s_x - s_y)\sin\theta\cos\theta + s_{xy}(\cos^2\theta - \sin^2\theta)$$
$$= (-10,000 - 13,000)(0.4333) + (-7000)(-0.500)$$
$$= -6470 \text{ psi}$$

Recall that Eqs. (17–5) and (17–6) were based on shear as shown in Fig. 17–17. Therefore, the minus sign for the shear stress indicates that the direction of the computed stress on the inclined plane is that shown in Fig. 17–19.

(b) The magnitudes of the principal stresses can be determined using Eq. (17–8). The principal stresses are

FIGURE 17–19 Stresses on inclined plane.

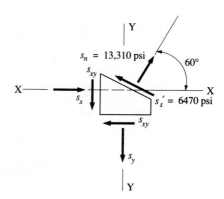

$$s_{1,2} = \frac{s_x + s_y}{2} \pm \sqrt{\frac{(s_x - s_y)^2}{4} + s_{xy}^2}$$

$$= \frac{-10{,}000 + 13{,}000}{2} \pm \sqrt{\frac{(-10{,}000 - 13{,}000)^2}{4} + (-7000)^2}$$

$$= +1500 \pm 13{,}460$$

from which

$$s_1 = +14{,}960 \text{ psi (tension)}$$

$$s_2 = -11{,}960 \text{ psi (compression)}$$

The planes on which the principal stresses act can be found using Eq. (17–7):

$$\tan 2\theta_P = -\frac{2s_{xy}}{s_x - s_y} = -\frac{2(-7000)}{-10{,}000 - 13{,}000} = -0.609$$

For convenience in understanding how θ_P is determined, Fig. 17–20 shows a plot of the tangent function. If $\tan 2\theta_P$ is negative, then $2\theta_P$ must lie in the second and fourth quadrants. Also observe in Fig. 17–20 that

$$\tan 2\theta = \tan(180° + 2\theta)$$

Further, relating angles in the first and second quadrants,

$$\tan 2\theta = -\tan(180° - 2\theta)$$

Relating angles in the first and fourth quadrants,

$$\tan 2\theta = -\tan(360° - 2\theta)$$

For $\tan 2\theta_P = -0.609$, if $2\theta_P$ were in the first quadrant (if the tangent value were positive) it would have the value of 31.34°. However, in the second quadrant,

$$2\theta_P = 180° - 31.34° = 148.66°$$

and in the fourth quadrant,

$$2\theta_P = 360° - 31.34° = 328.66°$$

FIGURE 17–20   Tangent function (0° to 360°).

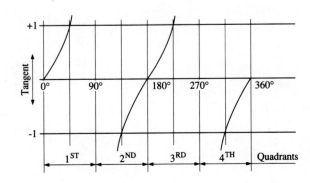

Therefore, the principal planes are defined by two planes located by θ_P measured counterclockwise from the X–X axis:

$$\theta_P = \frac{148.66}{2} = 74.33°$$

$$\theta_P = \frac{328.66}{2} = 164.33°$$

To determine on which plane the maximum principal stress acts, we substitute $\theta_P = 74.33°$ into Eq. (17–5) along with the original given stresses as shown in Fig. 17–18:

$$s_n = s_x\cos^2\theta_P + s_y\sin^2\theta_P - 2s_{xy}\sin\theta_P\cos\theta_P$$
$$= -10,000(0.0730) + 13,000(0.9270) - 2(-7000)(0.2601)$$
$$= +14,960 \text{ psi}$$

Thus the maximum principal stress of +14,960 psi occurs on the principal plane that has a normal oriented at 74.33° counterclockwise from the X–X axis, and the minimum principal stress of −11,960 psi occurs on the principal plane that has a normal oriented at 164.33° counterclockwise from the X–X axis. These results are commonly shown by rotating the element so that the X–Y axes of the original stressed element align with the principal planes and noting the principal stresses, as shown in Fig. 17–21. The shear stresses on the principal planes are zero.

FIGURE 17–21 Principal stress element.

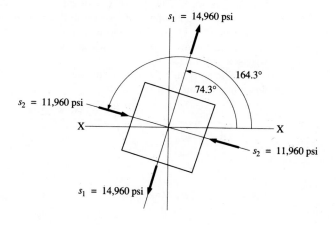

(c) The maximum shear stress can be determined from Eq. (17–10):

$$s'_s = \pm\sqrt{\frac{(s_x - s_y)^2}{4} + s_{xy}^2}$$

$$= \pm\sqrt{\frac{(-10,000 - 13,000)^2}{4} + (-7000)^2}$$

$$= \pm 13,460 \text{ psi}$$

The angles of inclination of the normals to the planes on which the maximum shear stresses occur can be obtained using Eq. (17–9):

$$\tan 2\theta_s = \frac{s_x - s_y}{2s_{xy}} = \frac{-10,000 - 13,000}{2(-7000)} = +1.643$$

The angle $2\theta_s$ is in the first and third quadrants and has values of 58.67° and 238.67°. Therefore,

$$\theta_s = \frac{58.67}{2} = 29.34°$$

and

$$\theta_s = \frac{238.67}{2} = 119.34°$$

Note that the maximum combined shear stress occurs on planes inclined at 45° to the principal planes. This is always the case.

To determine the direction of the shear stress, substitute $\theta_s = 29.34°$ into Eq. (17–6):

$$\begin{aligned}
s'_s &= (s_x - s_y)\sin \theta_s \cos \theta_s + s_{xy}(\cos^2\theta_s - \sin^2\theta_s) \\
&= (-10,000 - 13,000)(0.4900)(0.8717) + (-7000)(0.8717^2 - 0.4900^2) \\
&= -13,460 \text{ psi}
\end{aligned}$$

With reference to Fig. 17–16, the negative sign indicates that the shear stress on the plane, the normal of which is oriented at 29.34° with the original X–X axis, is as shown in Fig. 17–22. The element is rotated so that the X–Y axes of the original stressed element are parallel with the planes of maximum shear stress and the shear stresses are noted.

The normal stresses acting on the planes of maximum shear stress can be determined from Eq. (17–11):

$$s_n = \frac{s_x + s_y}{2} = \frac{(-10,000 + 13,000)}{2} = +1500 \text{ psi (tension)}$$

FIGURE 17–22 Maximum shear stress element.

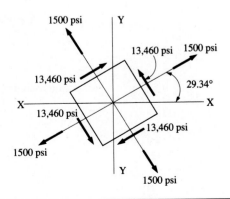

□ **EXAMPLE 17–7**   A simply supported short-span built-up steel beam is shown in Fig. 17–23. At point A, the juncture of the web with the bottom flange, (a) calculate the principal stresses, (b) locate the principal planes, and (c) calculate the maximum shear stress. Neglect the weight of the beam and the effects of stress concentrations. The loading is assumed to be a point loading.

FIGURE 17–23   Built-up steel beam.

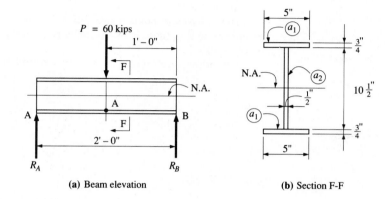

(a) Beam elevation

(b) Section F-F

Solution   (a) Prior to determining the principal stresses, the normal and shear stresses (s_x, s_y, and s_{xy}) must be determined. The bending moment at midspan is

$$M = \frac{PL}{4} = \frac{60(2)}{4} = 30 \text{ ft-kips}$$

The shear at midspan (actually an infinitesimal distance to the left of midspan) is

$$V = \frac{P}{2} = \frac{60}{2} = 30 \text{ kips}$$

Calculating the bending stress at midspan at the outer fibers, the moment of inertia about the strong axis of the cross section, neglecting I_o for the flanges, is

$$I = \Sigma I_{o2} + \Sigma a_1 d^2$$
$$= \frac{0.5(10.5)^3}{12} + 2(5)(0.75)(5.625)^2$$
$$= 285.5 \text{ in.}^4$$

from which, the maximum bending stress (at the outer fibers) is

$$s_b = \pm \frac{Mc}{I} = \pm \frac{30(12)(6)}{285.5} = \pm 7.57 \text{ ksi}$$

The bending stress at point A will be denoted s_x:

$$s_x = \frac{5.25}{6} (7.57) = +6.62 \text{ ksi (tension)}$$

Calculating the shear stress at point A (just within the web), the first statical moment of area Q is determined:

$$Q = 5(0.75)(5.625) = 21.1 \text{ in.}^3$$

from which

$$s_{xy} = \frac{VQ}{Ib} = \frac{30(21.1)}{285.5(0.5)} = 4.43 \text{ ksi}$$

The stresses at point A are shown in Fig. 17–24(a) acting on an infinitesimal element. Note that there is no normal stress on the plane perpendicular to the Y–Y axis ($s_y = 0$).

The principal stresses can now be determined. From Eq. (17–8),

$$s_{1,2} = \frac{s_x + s_y}{2} \pm \sqrt{\frac{(s_x - s_y)^2}{4} + s_{xy}^2}$$

$$= \frac{6.62}{2} \pm \sqrt{\frac{(+6.62)^2}{4} + 4.43^2}$$

$$= 3.31 \pm 5.53$$

from which

$$s_1 = +8.84 \text{ ksi (tension)}$$

$$s_2 = -2.22 \text{ ksi (compression)}$$

Note that the maximum principal stress s_1 is larger than the maximum bending stress at the outer fibers.

(b) The principal planes can be located using Eq. (17–7):

$$\tan 2\theta_P = -\frac{2s_{xy}}{s_x - s_y} = -\frac{2(4.43)}{6.62} = -1.338$$

Therefore, $2\theta_P$ lies in the second and fourth quadrant (refer to Fig. 17–20). If it were in the first quadrant, $2\theta_P$ would be 53.23°. We see, then, that in the second quadrant

$$2\theta_P = 180° - 52.23° = 126.77°$$

and in the fourth quadrant

$$2\theta_P = 360° - 53.23° = 306.77°$$

Therefore, the principal planes are defined by two planes which have normals located by θ_P measured counterclockwise from the X–X axis where

$$\theta_P = \frac{126.77}{2} = 63.4°$$

and

$$\theta_P = \frac{306.77}{2} = 153.4°$$

To determine on which plane the maximum principal stress acts, substitute 63.4° for θ_P in Eq. (17–5) along with the original computed normal and shear stress,

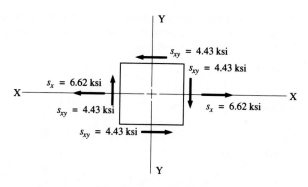

(a) Original stressed element

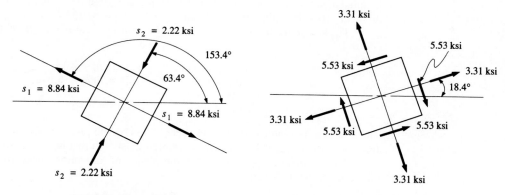

(b) Principal stress element

(c) Maximum shear stress element

FIGURE 17–24 Stress elements.

as shown in Fig. 17–24(a):

$$s_n = s_x\cos^2\theta_P + s_y\sin^2\theta_P - 2s_{xy}\sin\theta_P\cos\theta_P$$
$$= 6.62(0.2005) - 2(4.43)(0.4004)$$
$$= -2.22 \text{ ksi}$$

This is the minimum principal stress. Thus, the maximum principal stress of +8.84 ksi occurs on the principal plane which has a normal oriented 153.4° counterclockwise from the original X–X axis of the element. The principal stress element is shown in Fig. 17–24(b).

(c) The maximum shear stress can be determined from Eq. (17–10):

$$s'_s = \pm\sqrt{\frac{(s_x - s_y)^2}{4} + s_{xy}^2}$$
$$= \pm\sqrt{\frac{(-6.62)^2}{4} + 4.43^2}$$
$$= \pm5.53 \text{ ksi}$$

The angle of inclination of the normal to the plane on which the maximum shear stress occurs, measured counterclockwise from the X–X axis of the element, can be determined from Eq. (17–9):

$$\tan 2\theta_s = \frac{s_x - s_y}{2s_{xy}} = \frac{6.62}{2(4.43)} = +0.7472$$

Therefore,

$$2\theta_s = 36.77° \quad \text{and} \quad 216.77°$$

and

$$\theta_s = 18.4° \quad \text{and} \quad 108.4°$$

As expected, the principal planes and the planes of maximum shear stress are inclined at 45° to each other. You may wish to verify that the principal planes are free of shear stress by substituting θ values of 63.4° (and/or 153.4°) into Eq. (17–6).

To determine the direction of the shear stress, substitute 18.4° for θ in Eq. (17–6):

$$
\begin{aligned}
s'_s &= (s_x - s_y)\sin \theta_s \cos \theta_s + s_{xy}(\cos^2\theta_s - \sin^2\theta_s) \\
&= 6.62(0.3156)(0.9489) + 4.43(0.9004 - 0.0996) \\
&= +1.983 + 3.548 \\
&= +5.53 \text{ ksi}
\end{aligned}
$$

The positive sign indicates that the shear stress on the plane, which has a normal at 18.4° with the original X–X axis, is oriented as shown in Fig. 17–24(c).

The normal stress s_n acting on the planes of maximum shear stress is determined from Eq. (17–11):

$$s_n = \frac{s_x + s_y}{2} = \frac{+6.62}{2} = +3.31 \text{ ksi (tension)}$$

17–7
MOHR'S CIRCLE

An ingenious method for the solution of the state of stress at a point in a stressed body was presented by Otto Mohr, a German engineer, in 1882. The method, called *Mohr's circle*, is graphical (or can be partially graphical) and can be used to evaluate the variation of shear and normal stresses on all inclined planes at a point in a body.

In this section, we will discuss a simple uniaxial application to determine the normal and shear stresses developed on an inclined plane of an axially loaded prismatic member. More complex applications are discussed in Section 17–8.

With reference to Fig. 17–25(a), the member being considered is a bar of cross-sectional area A, loaded in tension with an axial load P. Our task is to find the normal and shear stresses on a plane that has a normal inclined at an angle θ measured counterclockwise from the X–X axis, as shown in Fig. 17–25(b). The sequence of steps in the construction and interpretation of the

FIGURE 17–25 Axially loaded member and Mohr's circle.

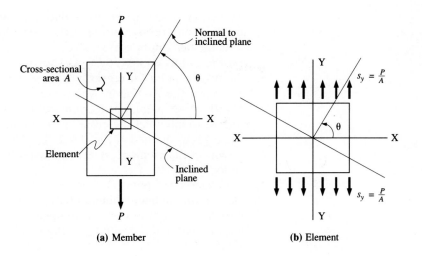

(a) Member   (b) Element

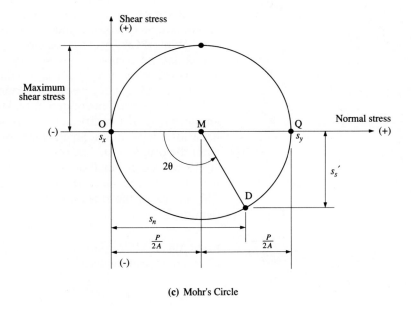

(c) Mohr's Circle

Mohr circle for this application is as follows:

1. Establish a rectangular coordinate system in which normal stresses will be plotted as abscissas (horizontally) and shear stresses will be plotted as ordinates (vertically). (See Fig. 17–25(c).)

2. To some convenient scale, draw a circle with its center on the horizontal axis at point M, and tangent to the vertical axis. For this application, the member is subjected to a tensile load. The circle is drawn to the right of

the vertical axis in the area that represents positive (tensile) normal stress. (Compressive stresses are considered negative.) Point Q represents the maximum normal stress s_y. Therefore, diameter OQ equals P/A and the radius equals $P/2A$, as shown.

3. The origin, point O, represents s_x, which, in this example, is zero. MO represents the X–X axis. MQ represents the Y–Y axis. From the radius MO (the X–X axis) turn a counterclockwise angle of 2θ, draw the radius from M to intersect the circle, and designate the intersection as point D. Angle θ is as defined in Fig. 17–25(b). The counterclockwise direction, as shown on the element, is maintained when turning off the angle on the circle.

4. The abscissa of point D represents the tensile normal stress (s_n) developed on the inclined plane under consideration. The ordinate of D represents the shear stress (s'_s) developed on the inclined plane. Further, note that the maximum ordinate on the circle represents the maximum shear stress.

As may be observed in Fig. 17–25, as θ varies, the shear stress and normal stress on the inclined plane also vary. The circle displays all the combinations of normal and shear stresses.

While Mohr's circle can be a graphical solution, one may also use trigonometric computations in conjunction with the graphical construction. This permits the use of an approximate sketch, not necessarily to scale, while still maintaining acceptable numerical accuracy in the results.

For this brief introduction to Mohr's circle using a uniaxial stress example, signs of the stresses, as shown in Fig. 17–25(c), are based on the usual sign convention, positive stresses being plotted upward and to the right from the origin at O. The ordinate of point D indicates that a negative shear stress was determined on the indicated plane. With reference to Fig. 17–26, we see that this shear stress tends to turn the element counterclockwise. This is one of the rules for Mohr's circle to be discussed in more detail in Section 17–8.

FIGURE 17–26   Shear stress direction on inclined plane.

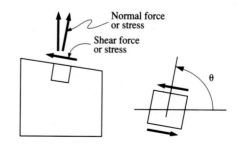

□ **EXAMPLE 17–8**   A 10 in. long metal block, 4 in. by 4 in. in cross section, is subjected to a compressive load of 32,000 lb. Using Mohr's circle, (a) determine the normal (tensile or compressive) stress and shear stress on a plane which has a normal inclined 60° counterclock-

wise from the X–X axis, as shown in Fig. 17–27 and (b) determine the magnitude of
the maximum shear stress and locate the plane on which it acts.

Solution   (a) To construct Mohr's circle, compute the maximum compressive stress on a plane
perpendicular to the longitudinal axis of the member:

$$s_y = -\frac{P}{A} = -\frac{32{,}000}{4(4)} = -2000 \text{ psi (compression)}$$

The radius of Mohr's circle, then, is $P/2A$ or 1000 psi.

FIGURE 17–27   Mohr's circle—
uniaxial stress.

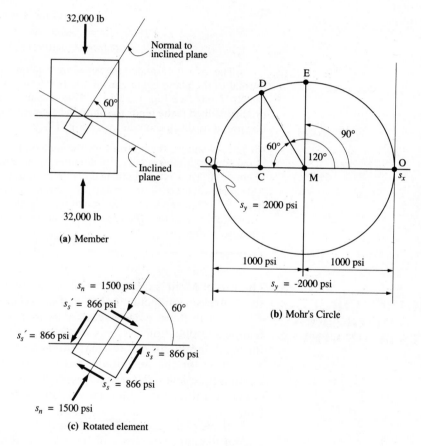

(a) Member

(b) Mohr's Circle

(c) Rotated element

Next draw the circle tangent to the vertical axis using a radius of 1000 psi and
having its center on the horizontal axis at M, as shown in Fig. 17–27. Note that the
circle is drawn on the left side of the origin since s_y is negative. Point Q represents s_y,
point O represents s_x, and MO represents the X–X axis.

From radius MO, turn off a counterclockwise angle of 120° (this is 2θ), draw
the radius from M to intersect the circle, and designate this point as D. The values of
the compressive stress and the shear stress are represented by point D and can be

scaled graphically or can be computed from the triangular relationships shown in the diagram. (Again, the accuracy of the graphical solution will depend on the accuracy with which the circle was constructed.)

Note that the coordinates of point D on Mohr's circle show that the normal stress is a negative value, indicating a compressive stress. Also note that the shear stress is a positive value which, in effect, indicates its direction on the plane under consideration.

Calculating these values using Mohr's circle as the reference, and neglecting signs (since the resulting signs are obvious by inspection),

$$s_n = OC = 1000 + 1000\cos 60°$$
$$= 1500 \text{ psi (compression) (negative)}$$
$$s'_s = DC = 1000\sin 60°$$
$$= 866 \text{ psi (positive)}$$

The two determined stresses are shown on an element rotated so that the normal to the plane being considered forms an angle of 60° (counterclockwise) with the X–X axis (see Fig. 17–27[c]). Note that the direction of the positive shear stress on the inclined plane tends to rotate the element clockwise. This is in agreement with our previous discussion and with Fig. 17–26.

(b) The maximum shear stress for the uniaxially loaded member always occurs on a plane at 45° with the plane perpendicular to the longitudinal axis of the member. If an angle of 90° (this is $2\theta_s$) is turned off on the circle (refer to the manner in which the 120° angle was turned in part [a]), we obtain line ME. By inspection, the ordinate to point E, which represents the maximum shear stress, is equal to the radius of the circle and is equal to +1000 psi. Note that the compressive stress represented by E is equal to −1000 psi.

17–8
MOHR'S CIRCLE— THE GENERAL STATE OF STRESS

The use of Mohr's circle as an alternate method of determining normal and shear stresses on inclined planes of an axially loaded prismatic member was discussed in the previous section. Mohr's circle may also be used in more complex applications. In this section we will study the use of Mohr's circle for the determination of the magnitude and direction of the principal stresses as well as for the magnitude and direction of the maximum shear stress. From a practical design standpoint, these are the values that are usually of concern, rather than the relatively unimportant combined stresses on arbitrary planes.

If we consider an element taken from a machine member subjected to biaxial loading, stresses can be developed as shown in Fig. 17–28(a). The sequence of steps in the construction and interpretation of Mohr's circle for this application is as follows:

1. Make a sketch of the element indicating the known normal and shear stresses with their directions, as shown in Fig. 17–28(a).
2. Establish a rectangular coordinate system with normal stresses to be plotted as abscissas (horizontally) and shear stresses as ordinates (vertically), as shown in Fig. 17–28(b).

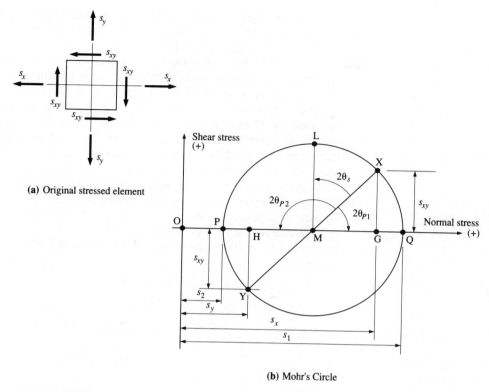

(a) Original stressed element

(b) Mohr's Circle

FIGURE 17–28   Mohr's circle example.

3. With respect to a sign convention, tensile normal stresses are positive and compressive normal stresses are negative. Shear stresses are positive if they tend to rotate the element clockwise and negative if they tend to rotate the element counterclockwise. This is a special rule for shear stress when using Mohr's circle. For this example, the shear stresses on the vertical face are positive and are accompanied by a positive tensile stress s_x and the shear stresses on the horizontal face are negative and are accompanied by a positive tensile stress s_y, as shown in Fig. 17–28(a). Assume, for purposes of this discussion, that $s_x > s_y$ algebraically.

4. With reference to Fig. 17–28(b), plot s_x on the horizontal axis as OG. This is the algebraically larger normal stress. Next plot s_y on the horizontal axis as OH. This is the algebraically smaller normal stress. Both points are plotted to the right of the vertical axis, since both are positive values. (*Note:* a positive (+) stress is algebraically larger than a negative (−) stress; e.g., +100 psi > −500 psi; also, −200 psi > −400 psi.)

5. The shear stress associated with the vertical plane on which s_x acts is $+s_{xy}$. Lay off s_{xy} vertically upward (in a positive direction) from G. Call

this point X. Since the shear stress associated with the horizontal plane on which s_y acts is $-s_{xy}$, lay off s_{xy} vertically downward (in a negative direction) from point H. Call this point Y.

6. Draw the XY line. This is the diameter of Mohr's circle. Line XY intersects the normal stress (horizontal) axis at point M. Draw a circle with its center at point M and passing through points X and Y. This completes the construction of Mohr's circle. Line XY is the reference line that correlates the circle with the particular problem under investigation. It represents the X–Y axes of the original stressed element shown in Fig. 17–28(a): MX represents the X–X axis; MY represents the Y–Y axis. All angles in the circle are doubled; hence, the X–X and Y–Y axes appear to be 180° apart when in reality they are 90° apart.

7. The principal stresses can be determined at points Q and P (dimensions s_1 and s_2) on the circle. Note that these are both positive and, therefore, tensile stresses, and that at these points the shear stresses are zero. From the circle, the equation for the principal stresses can be written as follows:

$$s_{1,2} = \text{OM} \pm \text{circle radius}$$
$$= \text{OM} \pm \text{MX}$$
$$= \frac{s_x + s_y}{2} \pm \sqrt{(\text{MG})^2 + (\text{GX})^2}$$
$$= \frac{s_x + s_y}{2} \pm \sqrt{\frac{(s_x - s_y)^2}{4} + s_{xy}^2}$$

This was previously presented as Eq. (17–8).

8. The orientation of the normals to the principal planes is determined by angle XMQ (and by angle XMP). These angles are measured from the reference line MX clockwise to line MQ and counterclockwise to line MP. Note that angle XMQ is labeled $2\theta_{P1}$ on the circle. This would indicate that the inclination of the maximum principal stress on the stressed element is measured by a clockwise rotation of θ_{P1} from the X–X axis on the stressed element. Similarly, the orientation of the direction of the minimum principal stress is obtained from angle XMP, which is a counterclockwise angle from line MX labeled $2\theta_{P2}$.

9. The orientation of the plane of maximum shear stress can also be determined. Point L represents maximum shear stress $(+s'_s)$. It lies on a plane inclined at θ_s with the X–X axis on the stressed element of Fig. 17–28(a). Angle $2\theta_s$ is indicated on the circle as the counterclockwise angle from line MX to line ML. Note that on the circle it is oriented 90° from the principal planes. Therefore, it is oriented 45° from the principal plane on the stressed element.

10. The results of the analysis are usually displayed graphically on two elements, one rotated to the inclination of the principal planes and the other rotated to the inclination of the planes of maximum shear stress. Appropriate stresses are shown on each element. (See Example 17–9.)

Brief Summary of Mohr's Circle for General Stress

1. Sketch the stressed element showing all applied stresses.
2. Set up the rectangular coordinate system for the circle.
3. Follow the appropriate sign convention: tension—positive; compression—negative; shear—clockwise rotation positive, counterclockwise rotation negative.
4. Plot s_x and s_y on the normal stress (horizontal) axis.
5. Plot the shear stress $\pm s_{xy}$ associated with the normal stress for both the vertical face and the horizontal face. Label points X and Y.
6. Draw the XY line and Mohr's circle.
7. Determine the principal stresses and the maximum shear stress.
8. Determine $2\theta_{P1}$ and/or $2\theta_{P2}$ measuring from the X end of the reference line.
9. Determine $2\theta_s$.
10. Sketch two rotated elements, one for principal stresses and one for maximum shear stresses.

☐ **EXAMPLE 17–9**   An element of a machine member is subjected to the stresses shown in Fig. 17–29. Using Mohr's circle, (a) determine the magnitudes of the principal stresses and the maximum shear stresses, (b) determine the orientations of the principal planes and the planes on which the maximum shear stresses occur, and (c) show the results for the previous two parts on two rotated stressed elements.

FIGURE 17–29   Stressed element.

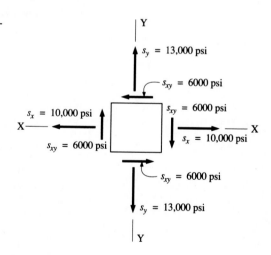

Solution   (a) In accordance with our sign convention, $s_x = +10,000$ psi, $s_y = +13,000$ psi, s_{xy} on the vertical face $= +6000$ psi, and s_{xy} on the horizontal face $= -6000$ psi. Mohr's circle is shown in Fig. 17–30. Since it is not drawn to scale, mathematical relationships will be used along with the graphical construction.

We first plot OG $= +10,000$ psi (this is s_x) and OH $= +13,000$ psi (this is s_y) on the horizontal axis. Since the shear stress s_{xy} associated with s_x is a positive value, we draw a line perpendicular to the horizontal axis (upward) at G and plot point X so that XG $= +6000$ psi. Since s_{xy} associated with s_y is a negative value, we draw a line

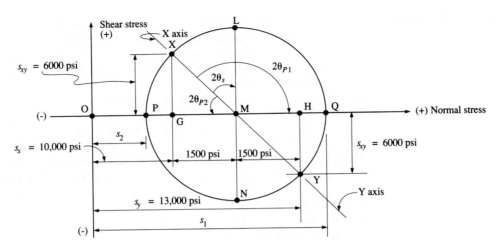

FIGURE 17–30   Mohr's circle.

perpendicular to the horizontal axis (downward) at H and plot point Y so that HY = −6000 psi.

We then draw the diameter XY and locate the point at which it crosses the horizontal axis **M**, which is the center of Mohr's circle. Computing the radius of the circle using triangle MHY,

$$\text{GM} = \text{MH} = \frac{s_y - s_x}{2} = \frac{13{,}000 - 10{,}000}{2} = 1500 \text{ psi}$$

$$\text{MY} = \sqrt{(\text{MH})^2 + (\text{HY})^2} = \sqrt{1500^2 + 6000^2} = 6185 \text{ psi}$$

Distances OQ and OP along the horizontal axis represent the maximum and minimum principal stresses s_1 and s_2:

$$s_1 = \text{OM} + \text{MQ} = (10{,}000 + 1500) + 6185 = +17{,}685 \text{ psi (tension)}$$
$$s_2 = \text{OM} - \text{MP} = (10{,}000 + 1500) - 6185 = +5315 \text{ psi (tension)}$$

The maximum shear stress s'_s is represented by point L (or point N). The magnitude of the maximum shear stress is equal to the radius of the circle, 6185 psi. Note also that the normal stresses acting on the planes of maximum shear stress can be determined at points L and N. (Also, compare with Eq. 17–11.)

$$s_n = \frac{s_x + s_y}{2} = \frac{10{,}000 + 13{,}000}{2} = +11{,}500 \text{ psi (tension)}$$

(b) The inclination of the principal stresses is determined by measuring from the X end of reference axis X–Y to the maximum principal stress at point Q. The clockwise angle of $2\theta_{P1}$ is labeled on the circle. From triangle XMG,

$$2\theta_{P2} = \angle\text{XMG} = \tan^{-1}\frac{\text{XG}}{\text{GM}} = \tan^{-1}\frac{6000}{1500} = 4.0 = 75.96°$$

$$2\theta_{P1} = 180° - 75.96° = 104.04°$$

$$\theta_{P1} = 52.0° \text{ (clockwise)}$$

We see, then, that the maximum principal stress s_1 is inclined at a clockwise angle of 52.0° with the X–X axis of the original stressed element.

The inclination to the plane of maximum positive shear stress is represented by θ_s. Point L on the circle represents this stress and lies clockwise $2\theta_s$ from the X end of the XY reference line. The simplest way to obtain this angle is to subtract 90° from $2\theta_{P1}$:

$$2\theta_s = 2\theta_{P1} - 90°$$

from which

$$\theta_s = \theta_{P1} - 45° = 7.0° \text{ (clockwise)}$$

Since angle $2\theta_s$ is shown as clockwise to the maximum positive shear stress (radius ML), the element is shown rotated clockwise 7° in Fig. 17–31(b). Since the shear stress at point L is positive, it is shown acting downward on the face of the element that corresponds to the right vertical face of the original stressed element. This tends to rotate the element clockwise and is consistent with the previously discussed sign convention. Note that on the stressed element the planes of maximum shear stress are oriented at 45° to the principal planes.

(c) Results from parts (a) and (b) are shown graphically in Fig. 17–31.

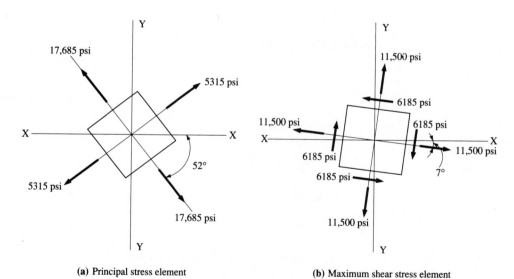

(a) Principal stress element   **(b)** Maximum shear stress element

FIGURE 17–31   Results for Example 17–9.

17–9
SI SYSTEM EXAMPLES

□ **EXAMPLE 17–10**

A steel link in a machine is designed to avoid interference with other moving parts. The link is 150 mm thick. The cross-sectional area of the link is reduced by one-half at section A–A as shown in Fig. 17–32. Compute the maximum tensile stress developed across section A–A. Neglect any stress concentrations.

FIGURE 17–32 Eccentrically
loaded linkage.

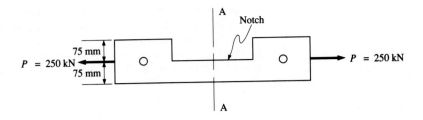

Solution   The load is eccentric with respect to the centroidal axis in the length where the notch
occurs. The eccentric load P results in a force reaction P and a moment reaction Pe
at plane A–A, as shown in the free-body diagram of the link in Fig. 17–33.
The properties of the link at plane A–A are calculated as follows:

$$\text{Area } A = (75 \text{ mm})(150 \text{ mm}) = 11\ 250 \text{ mm}^2$$

$$\text{Section modulus } S = \frac{(150 \text{ mm})(75 \text{ mm})^2}{6} = 140\ 625 \text{ mm}^3$$

$$\text{Eccentricity } e = \frac{75 \text{ mm}}{2} = 37.5 \text{ mm}$$

FIGURE 17–33   Link free body.

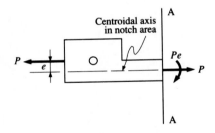

The axial tensile stress due to the load P on plane A–A is

$$s_t = +\frac{P}{A} = \frac{250\ 000 \text{ N}}{11\ 250 \text{ mm}^2} = +22.2 \text{ MPa (tension)}$$

The moment (or couple) Pe acting on plane A–A produces bending stresses of

$$s_b = \pm\frac{Mc}{I} = \pm\frac{Pe}{I/c} = \pm\frac{Pe}{S}$$

$$= \pm\frac{(250\ 000 \text{ N})(37.5 \text{ mm})}{140\ 625 \text{ mm}^3} = \pm66.7 \text{ MPa}$$

The maximum tensile stress is the algebraic sum of the two stresses:

$$s = +22.2 \text{ MPa} + 66.7 \text{ MPa}$$
$$= +88.9 \text{ MPa (tension)}$$

This stress occurs along the top edge of the notch on plane A–A.

□ **EXAMPLE 17–11**   Compute the maximum tensile stress on planes A–A and B–B on the crane boom shown in Fig. 17–34(a). The boom is a hollow steel tubular shape with dimensions as shown in Fig. 17–34(b).

Solution   The reactions and the shear and moment diagrams are shown in Fig. 17–35. You should verify the computed values.

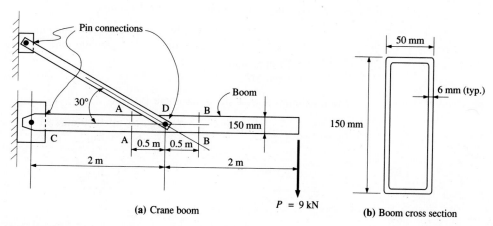

FIGURE 17–34   Crane boom assembly.

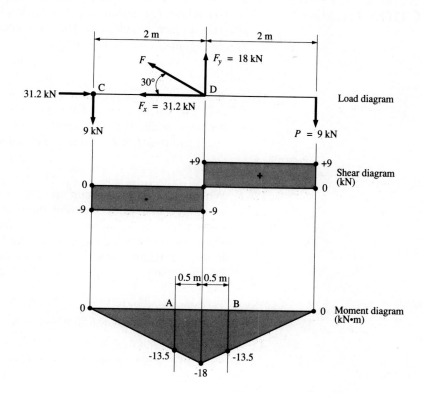

FIGURE 17–35   Load, shear, and moment diagrams.

The area and section modulus of the boom are calculated as follows:

$$A = (50 \text{ mm})(150 \text{ mm}) - (38 \text{ mm})(138 \text{ mm}) = 2256 \text{ mm}^2$$

$$S = \frac{(50 \text{ mm})(150 \text{ mm})^2}{6} - \frac{(38 \text{ mm})(138 \text{ mm})^2}{6} = 66\,890 \text{ mm}^3$$

The maximum tensile stress on plane A–A can now be calculated. Tension is assumed positive and compression is assumed negative.

$$s = -\frac{P}{A} + \frac{M}{S} = -\frac{31\,200 \text{ N}}{2256 \text{ mm}^2} + \frac{(13\,500 \text{ N} \cdot \text{m})(1000 \text{ mm/m})}{66\,890 \text{ mm}^3}$$

$$= +188 \text{ N/mm}^2$$
$$= +188 \text{ MPa}$$

The maximum tensile stress on plane B–B is

$$s = \frac{P}{A} + \frac{M}{S} = 0 + \frac{(13\,500 \text{ N} \cdot \text{m})(1000 \text{ mm/m})}{66\,890 \text{ mm}^3}$$

$$= +202 \text{ N/mm}^2$$
$$= +202 \text{ MPa}$$

SUMMARY—BY SECTION NUMBER

17–2   When a member of relatively short span length is subjected to both transverse loads and direct axial loads, bending stresses and axial stresses are developed. The combined stress at any location is a simple algebraic sum of the two types of stress and is expressed as

$$s = \pm\frac{P}{A} \pm \frac{Mc}{I} \qquad (17\text{–}1)$$

17–3   When a short compression member is subjected to a load that lies on only one of the centroidal axes, bending stresses and axial stresses are developed. If the moment is expressed in terms of the load and eccentricity ($M = Pe$), the combined stress expression is

$$s = -\frac{P}{A} \pm \frac{Pec}{I} \qquad (17\text{–}2)$$

17–4   When a short compression member is subjected to an eccentric load, combined stresses (either tension or compression) will develop at the outer edges of the cross section. If the bending stress is equal to the axial stress, the stress at one outer edge will be zero. The maximum load eccentricity for which this zero stress will develop in a rectangular shape is

$$e = \frac{d}{6} \qquad (17\text{–}3)$$

where d represents the dimension parallel to the eccentricity.

17–5   If a load applied to a short compression member is eccentric to both centroidal axes, the member is subjected to a double eccentricity, and the combined stress expression is

$$s = -\frac{P}{A} \pm \frac{Pe_1c_1}{I_y} \pm \frac{Pe_2c_2}{I_x} \tag{17-4}$$

17-6 The combination of normal and shear stresses results in maximum and minimum normal and shear stresses being developed on planes that are inclined with respect to the planes of stress being combined. The planes on which the normal stresses become maximum or minimum are called *principal planes*. The normal stresses acting on the planes are called *principal stresses* and can be obtained from

$$s_{1,2} = \left(\frac{s_x + s_y}{2}\right) \pm \sqrt{\frac{(s_x - s_y)^2}{4} + s_{xy}^2} \tag{17-8}$$

Shear stresses on the principal planes are equal to zero. The maximum shear stress developed on an inclined plane is

$$s'_{s(max)} = \pm \sqrt{\frac{(s_x - s_y)^2}{4} + s_{xy}^2} \tag{17-10}$$

17-7 and 17-8 Mohr's circle is a graphical (or semigraphical) alternate method of calculating normal stresses and shear stresses acting on any inclined plane at any point in a stressed body. It can be used both for uniaxial and for biaxial loading conditions.

PROBLEMS

Unless noted otherwise, neglect the weights of the members in these problems.

Section 17-2 Combined Axial and Bending Stresses

1. A 1 in. by 4 in. steel bar is subjected to the loads shown in Fig. 17-36. Calculate the combined stresses at points A and B.

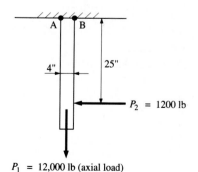

P_1 = 12,000 lb (axial load)

FIGURE 17-36 Problem 1.

2. A W16 × 57 structural steel wide-flange section is used as a horizontal lower chord member of a bridge truss. It is subjected to a tensile load of 300,000 lb. The orientation of the member is such that the Y–Y axis (the weak axis) is horizontal. Calculate the maximum and minimum combined stresses due to the direct axial load and the member's own weight. The panel length (length of the member) is 25 ft.

3. A W12 × 72 structural steel wide-flange section is used as a simply supported beam with a span length of 12 ft. It is subjected to concentrated loads of 20,000 lb at each quarter point and an axial tensile load of 50,000 lb. Calculate the maximum combined tensile and compressive stresses.

4. A solid steel shaft, 3 in. in diameter and 4 ft long, is fixed at one end and free at the other, as shown in Fig. 17-37. The shaft is subjected to an axial tensile load of 50,000 lb. Compute the magnitude of the force P at the free end necessary to reduce the stress at point B to zero. Then compute the maximum combined stress at point A.

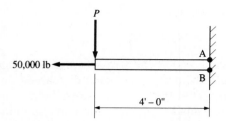

FIGURE 17–37   Problem 4.

Section 17–3   Eccentrically Loaded Members

5. A short compression member is subjected to a compressive load of 300,000 lb at an eccentricity of 1.5 in., as shown in Fig. 17–38. The member is 12 in. by 12 in. in cross section. Calculate the combined stresses at the outer edges AA and BB.

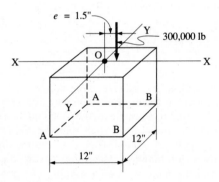

FIGURE 17–38   Problem 5.

6. With reference to Problem 5, calculate the eccentricity that must exist if the combined stress at outer edge AA is to be zero.

7. A section of a 2 in. diameter standard-weight steel pipe is bent into the form shown in Fig. 17–39 and rigidly embedded in a concrete footing. Calculate the stresses at points A and B on the pipe. *P* is 200 lb and *e* is 24 in.

8. For the pipe of Problem 7, compute the maximum load *P* that may be applied if the allowable compressive stress at point B is 12,000 psi.

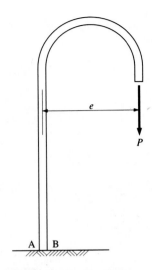

FIGURE 17–39   Problem 7.

Section 17–4   Maximum Eccentricity for Zero Tensile Stress

9. A concrete pedestal is in the shape of a cube and is 6 ft on each side. The pedestal supports a superimposed load of 40,000 lb located on the Y–Y axis, as shown in Fig. 17–40. Calculate the maximum eccentricity *e* for zero tension at the base of the pedestal. Include the weight of the pedestal using a concrete unit weight of 150 pcf.

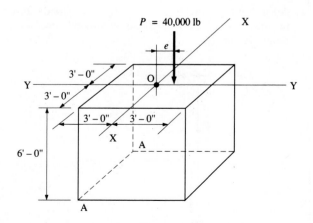

FIGURE 17–40   Problem 9.

FIGURE 17-41 Problem 12.

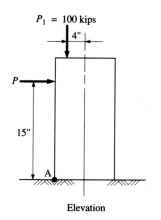

Elevation

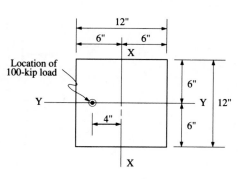

Plan view

10. For the pedestal of Problem 9, assume that the load is placed at the computed eccentricity. Calculate the maximum tensile and compressive stresses developed on a horizontal plane 3 ft below the top of the pedestal.

11. Rework Problem 9, but assume that the applied load is 150,000 lb instead of 40,000 lb.

12. A 12 in. square concrete pedestal is subjected to a 100 kip vertical eccentric load applied on the Y–Y axis, as shown in Fig. 17-41. Determine the required magnitude of the horizontal load P (also in the vertical plane through the Y–Y axis) for a combined stress of zero at point A.

Section 17-5 Eccentric Load Not on Centroidal Axis

13. A short compression member is subjected to a compressive load of 5000 lb at a double eccentricity, as shown in Fig. 17-42. (a) Calculate the combined stress

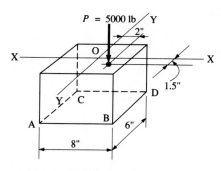

FIGURE 17-42 Problem 13.

at each of the four corners, A, B, C, and D. (b) Locate the line of zero stress.

14. A rectangular concrete footing, 4 ft by 8 ft in plan, is subjected to two column loads, as shown in Fig. 17-43.

FIGURE 17-43 Problem 14.

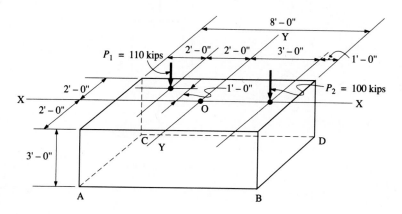

FIGURE 17-44 Problem 15.

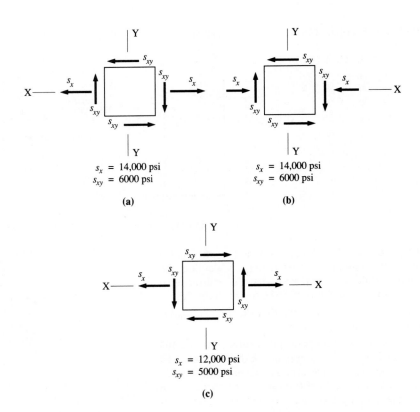

$$s_x = 14{,}000 \text{ psi}$$
$$s_{xy} = 6000 \text{ psi}$$

(a)

$$s_x = 14{,}000 \text{ psi}$$
$$s_{xy} = 6000 \text{ psi}$$

(b)

$$s_x = 12{,}000 \text{ psi}$$
$$s_{xy} = 5000 \text{ psi}$$

(c)

FIGURE 17-45 Problem 16.

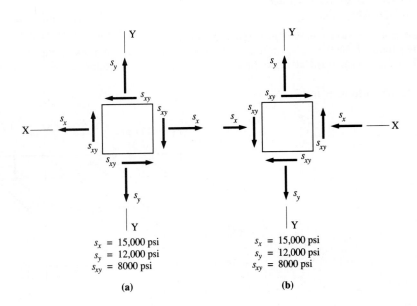

$$s_x = 15{,}000 \text{ psi}$$
$$s_y = 12{,}000 \text{ psi}$$
$$s_{xy} = 8000 \text{ psi}$$

(a)

$$s_x = 15{,}000 \text{ psi}$$
$$s_y = 12{,}000 \text{ psi}$$
$$s_{xy} = 8000 \text{ psi}$$

(b)

Calculate the stress (base pressure) at each corner of the footing. Use a concrete unit weight of 150 pcf. Express the answers in pounds per square foot (psf).

15. The bending and shear stresses developed at a point in a machine part are shown in Fig. 17–44 acting on infinitesimal elements. For each element, calculate (a) the principal stresses and the orientation of the principal planes and (b) the maximum shear stress.

16. Stresses developed at a point in a machine part are shown in Fig. 17–45 acting on infinitesimal elements. For each element, calculate (a) the normal and shear stress intensities on a plane, the normal of which is inclined at an angle of 60° counterclockwise from the X–X axis; (b) the principal stresses and the orientation of the principal planes; and (c) the maximum shear stress.

17. Calculate the principal stresses at points A and B for the bracket in Fig. 17–46. $P = 9000$ lb and $\theta = 30°$. The bracket is 1 in. thick.

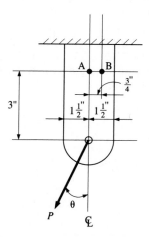

FIGURE 17–46 Problem 17.

18. Rework Problem 17 using $P = 8000$ lb and $\theta = 45°$.

Section 17–7 Mohr's Circle

19. A 1 in. square steel bar is subjected to an axial tensile load of 10,000 lb. Use Mohr's circle to determine the shear stress and normal stress on a plane, which has a normal inclined at 60° counterclockwise from a plane perpendicular to the longitudinal axis of the member.

20. A bar having a cross sectional area of 6 in.^{2} is subjected to an axial tensile load of 39,000 lb. Using Mohr's circle, (a) find the normal and shear stresses on planes with normals inclined at 75°, 65°, and 50° counterclockwise from a plane perpendicular to the longitudinal axis, and (b) find the maximum shear stress.

21. Rework Problem 20, changing the load to a compressive load of 72,000 lb. Also find the normal and shear stresses on a plane, which has a normal inclined at 30° counterclockwise from a plane perpendicular to the longitudinal axis of the bar.

Section 17–8 Mohr's Circle— The General State of Stress

22. Solve Problem 15 using Mohr's circle.

23. For the elements shown in Problem 16, use Mohr's circle to determine (a) the principal stresses and the orientation of the principal planes and (b) the maximum shear stress.

24. Solve Problem 17 using Mohr's circle.

25. In Problem 17, change the load to 8000 lb and the angle to 45° and solve the problem using Mohr's circle.

SI System Problems

26. A timber member 150 mm by 250 mm (S4S) is loaded as shown in Fig. 17–47. Calculate the maximum combined stress acting normal to the plane A–A. Neglect the weight of the member.

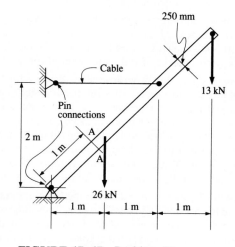

FIGURE 17–47 Problem 26.

27. A short-span cantilever beam is subjected to an inclined load P as shown in Fig. 17–48. The beam is 50 mm by 150 mm in cross section. Calculate the principal stresses, the inclination of the principal planes, and the maximum shear stress at point A. Neglect the weight of the beam.

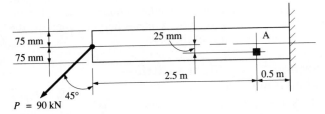

FIGURE 17–48   Problem 27.

28. A 50 mm diameter solid steel shaft is subjected to an axial compressive load of 225 kN and a torque (twisting moment) of 3400 N·m. Calculate the principal stresses and the maximum shear stress on an element on the surface of the shaft.

Computer Problems

For the following computer problems, any appropriate programming language may be used. Input prompts should fully explain what is required of the user (the program should be "user friendly"). The resulting output should be well labeled and self-explanatory.

29. Write a computer program that will allow the user to solve Problem 37. Assume constant dimensions for the wall, but a varying water level. User input, therefore, will be the height of the water behind the wall.

30. Write a computer program that will solve for the normal stress and shear stress on any inclined plane in a uniaxially loaded member, such as the metal block of Example 17–8. User input is to be the cross-sectional area of the member, the axial load, and the inclination θ of a normal to the plane measured counterclockwise from the X–X axis. The input prompts should enable the user to fully understand the required input.

31. Rework Problem 30, but have the program generate a table for normal and shear stresses for values of θ from 0° to 45° in increments of 5°. User input for this program is to be the cross-sectional area of the member and the axial load.

32. Write a program that will allow a user to solve for the principal stresses, the orientation of the principal planes, and the maximum shear stress in a stressed element. Input will be the normal stresses and the associated shear stresses. Data from Problem 15 or Problem 16 may be used to test the program.

Supplemental Problems

33. A simply supported W18 × 50 structural steel wide-flange beam has a span length of 15 ft. It is subjected to a concentrated load of 25,000 lb at midspan and a uniformly distributed load of 350 lb per linear foot. The uniform load includes the weight of the beam. The allowable tensile stress for the beam is 22,000 psi. Calculate the maximum axial tensile force that may be applied to this beam at its ends.

34. An 8 in. square (S4S) vertical timber post is 8 ft long and fixed at its lower end. It supports a vertical axial compressive load of 7000 lb on top and a horizontal load of 1000 lb at a point 3 ft below the top. Compute the maximum and minimum combined stresses at the base. Neglect the weight of the post.

35. Compute the maximum compressive and tensile stresses in the horizontal crane beam in Fig. 17–49. The beam is a W10 × 33 structural steel wide-flange oriented with the strong axis horizontal. Neglect the weight of the beam.

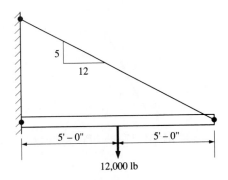

FIGURE 17–49   Problem 35.

36. A short 3 in. square steel bar with a 1 in. diameter axial hole is fixed at its base and loaded at the top as shown in Fig. 17–50. Neglecting the weight of the bar, calculate the value of the force P for which the maximum combined normal stress at the fixed end will not exceed 22 ksi.

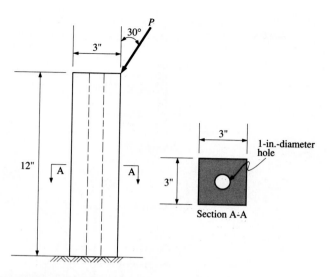

FIGURE 17–50   Problem 36.

37. A concrete wall 8 ft high and 3 ft thick is monolithic with a concrete base. The wall has water behind one face to a height of 6 ft, as shown in Fig. 17–51. Calculate the combined normal stresses at points A and B at the bottom of the wall in pounds per square inch and pounds per square foot. Consider a 1 ft length of wall. Assume the unit weight of concrete to be 150 pcf and the unit weight of water to be 62.4 pcf.

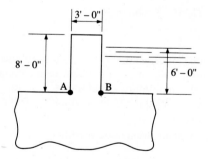

FIGURE 17–51   Problem 37.

38. In Problem 37, assume that the water is full height (8 ft) on one side of the wall and 3 ft high on the other side. Calculate the combined normal stresses at points A and B.

39. A short compression member is subjected to a compressive load of 25,000 lb at an eccentricity of 1.5 in.,

as shown in Fig. 17–52. The member is 6 in. by 8 in. in cross section. Calculate the combined stresses at the outer edges AA and BB.

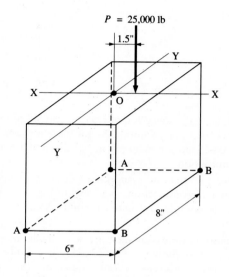

FIGURE 17–52   Problem 39.

40. Calculate the maximum eccentric load that can be applied in Problem 39 if the maximum allowable compressive stress is 1000 psi.

41. A short compression member is subjected to two compressive loads as shown in Fig. 17–53. The member is 6 in. by 12 in. in cross section. Calculate the combined stress at outer edges AA and BB.

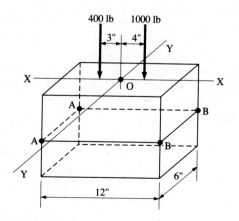

FIGURE 17–53   Problem 41.

FIGURE 17–54   Problem 43.

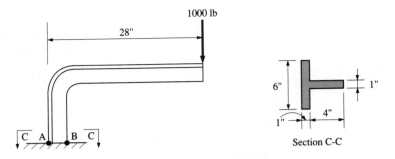

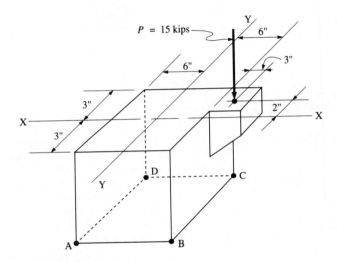

Section C-C

42. Calculate the force P that may be applied to the punch press frame of Example 17–4 if the allowable stresses for cast iron are 6000 psi in tension and 10,000 psi in compression.

43. A load of 1000 lb is supported on a cast-iron bracket with a T cross section, as shown in Fig. 17–54. Calculate the combined stresses at points A and B due to the load.

44. A short compression member is subjected to an eccentric load as shown in Fig. 17–55. The member is 6 in. by 12 in. in cross section. Calculate the combined stress at each of the four corners (A, B, C, and D) and locate the line of zero stress.

45. A W24 × 104 structural steel wide-flange section is used as a short compression strut and subjected to a vertical axial load of 350,000 lb as shown in Fig. 17–56. Calculate the maximum load P that may be applied on the Y–Y axis at the outside face of the flange without exceeding the maximum compressive stress of 20,000 psi.

46. A cast-iron frame for a piece of industrial equipment has the dimensions shown in Fig. 17–57. Calculate the force P that may be applied to this frame if the allowable stresses at section A–A are 6000 psi tension and 10,000 psi compression.

47. The casting in Fig. 17–58 is used in a machine. It supports an axial load P_1 of 8000 lb and an eccentric load

FIGURE 17–55   Problem 44.

P_2 of 8000 lb. The member is a hollow cylinder having a 3 in. outside diameter and a 2 in. inside diameter. Calculate the maximum eccentricity e if the allowable stresses are 8000 psi in tension and 18,000 psi in compression.

48. An element of a machine member is subjected to the

FIGURE 17–56   Problem 45.

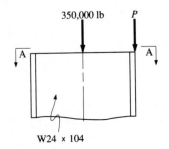

W24 × 104

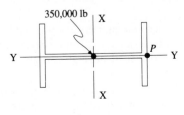

Section A-A

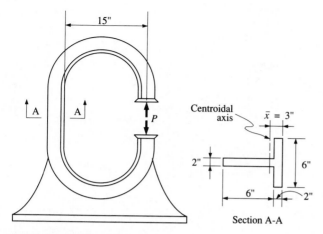

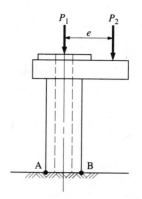

FIGURE 17–58   Problem 47.

FIGURE 17–57   Problem 46.

stresses shown in Fig. 17–59. Calculate the magnitude and direction of the principal stresses and the maximum shear stress. Use the analytical approach.

49. A short-span cantilever built-up beam has the dimensions indicated in Fig. 17–60. Calculate the principal stresses, the inclination of the principal planes, and the maximum shear stress at points A, B, and C. Neglect the weight of the beam and the effect of stress concentrations. Use the analytical approach.

50. Solve Problem 48 using Mohr's circle.

51. A 6 in. diameter solid shaft is subjected to a torque of 40 ft-kips. Using Mohr's circle, (a) determine the magnitude and direction of the principal stresses on the surface of the shaft and (b) determine the stresses acting on the faces of an element rotated 20° clockwise.

52. Rework parts (b) and (c) of Example 17–6 using Mohr's circle.

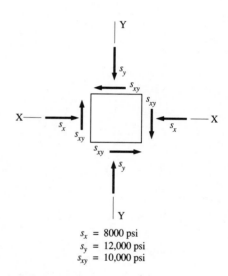

$s_x = 8000$ psi
$s_y = 12,000$ psi
$s_{xy} = 10,000$ psi

FIGURE 17–59   Problem 48.

FIGURE 17–60   Problem 49.

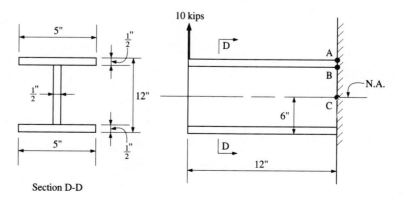

Section D-D

■ ■ ■ ■ 18 Columns

18–1
INTRODUCTION

Members of structures and machines that are subjected to axial compressive loads are called *columns* if their length dimension is significantly larger than their least lateral (or cross-sectional) dimension. When the applied loads are coincident with the longitudinal centroidal axis of the member, the column is said to be axially loaded. This is a special case and one that exists rarely, if at all. Despite this generally accepted fact, a column may still be designed as though it were axially loaded, since an appropriate factor of safety will usually compensate for a minor unintended eccentricity.

When the line of action of the applied load does *not* coincide with the longitudinal centroidal axis, the load is said to be eccentric and the column is said to be eccentrically loaded.

In this chapter, we will use the terms *column* and *compression member* interchangeably. However, while all axially loaded columns are compression members, not all compression members are called columns. Members that carry compressive loads are commonly given names descriptive of their functions, such as pillars, pedestals, shores, props, supports, masts, and piers, to name a few. Truss members in compression, either chords or web members, are other compression members that are not categorized as columns, despite the fact that they are compression members and exhibit a behavior usually described as "column action." Various types of machinery can also have component parts acting as compression members but called something other than columns.

Axially loaded short compression members were discussed in Chapter 9 and eccentrically loaded short compression members were discussed in Chapter 17. In both cases, the use of the direct stress formula ($s = P/A$) was applicable; in addition, in the latter case, the flexure formula ($s = Mc/I$) was employed due to the bending action. Using these generally accepted methods, reasonably reliable solutions were provided for the short compression member. However, as the length of a compression member increases (maintaining a constant cross section) additional factors enter into the problem.

It is convenient to classify columns into three broad categories according to their modes of failure. The three types, short columns, intermediate columns, and long columns, are shown in Fig. 18–1.

The long column (also referred to as a *slender column*), shown in Fig. 18–1(c), will fail by elastic buckling, which occurs at compressive stresses within the elastic range (recall that the upper limit of elastic range is the

FIGURE 18–1 Types of columns.

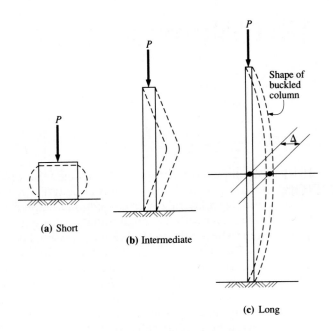

(a) Short

(b) Intermediate

(c) Long

proportional limit or the elastic limit). Buckling of a column can be described as a bending action developed under compressive loads. Failure is said to occur through a lack of stiffness.

A very short and stocky column, as shown in Fig. 18–1(a), will obviously not fail by elastic buckling. It will crush if made of a brittle material (for example, concrete) or yield if made of a ductile material (structural steel) at compressive stresses in the inelastic range. Since crushing and yielding are material failures, the maximum axial load a short column can support is determined solely by the strength of the material. If yielding is the failure criterion, the failure load may be determined as the product of the yield stress of the material and the cross-sectional area of the member.

The intermediate column lies between the two extremes, as shown in Fig. 18–1(b). Most columns or compression members are in the intermediate category. These columns will fail by inelastic buckling initiated when a localized yielding occurs at some point of weakness or crookedness. This failure mode represents a combination of buckling and material failure.

The design and analysis of intermediate columns is much more complex than that of the other columns. They are usually designed and analyzed using empirical formulas based on extensive test results, experience, and judgement. Failure of the intermediate column cannot be predicted using either the elastic buckling criterion of the slender column or the yielding criterion of the short column.

The behavior of compression members can best be understood by beginning with the slender column. This type of column was used originally in the development of the theory of axially loaded elastic column behavior.

**18–2
IDEAL COLUMNS**

The development of the theory of elastic column behavior is attributed to Leonhard Euler (1707–1783), a Swiss mathematician. Euler derived a theoretical equation for the load that would buckle a slender column. The equation is commonly referred to as the *elastic column buckling formula* or, simply, the *Euler formula*. Euler's theory was presented in 1744 and still is the basis for the analysis and design of slender columns.

The buckling of a slender column can be demonstrated by loading an ordinary wooden yardstick in compression. The yardstick will buckle (constituting failure), as shown in Fig. 18–1(c). If the load that first produced this buckling (or bending) remains constant, the lateral deflection Δ will remain constant and the column will support the load. If, however, the load is increased, the column will deflect further and finally collapse. A decrease in load would permit the column to straighten itself. The load that is just sufficient to hold the column in a bent condition is called the *critical load* for the column. The critical load can also be defined as the greatest load that the column will support. Euler's formula gives the buckling load P_e (also called the *critical load* or the *Euler buckling load*) for a theoretically perfect column (also called an *ideal column* or the *Euler column*). The ideal column can be briefly described as a pin-ended homogeneous slender column, initially straight, of an elastic material that is concentrically loaded. The pin-ended column is a column with ends free to rotate but restrained against lateral movement. The Euler buckling load is expressed as

$$P_e = \frac{\pi^2 EI}{L^2} \tag{18–1}$$

where P_e = the critical load; the concentric load that will cause initial buckling (lb) (N)

π = a mathematical constant (3.1416)

E = the modulus of elasticity of the material (psi) (MPa)

I = the least moment of inertia of the cross section (in.⁴) (mm⁴)

L = the length of the column from pin end to pin end (in.) (mm)

Note that the units must be consistent. For instance, it is sometimes convenient to substitute E in terms of ksi, in which case the units of critical load will be kips.

Tests have verified that the Euler formula accurately predicts the buckling load if conditions are such that the buckling stress is less than the proportional limit of the material and adherence to the basic assumptions is maintained. Since the buckling stress must be compared with the proportional limit, Euler's formula is commonly written in terms of stress. This can be derived from the preceding buckling load formula using the relationship $r = \sqrt{I/A}$ or $I = Ar^2$ (see Section 8–5):

$$s_e = \frac{\pi^2 E}{(L/r)^2} \tag{18–2}$$

where s_e = the critical stress; the uniform compressive stress at which initial buckling occurs (same units as E)

r = the least radius of gyration of the cross section (same units as L)

The unitless ratio L/r is called the *slenderness ratio* of the column. Once the critical stress has been determined, the critical load may be found from $P_e = s_e A$ or from Eq. (18–1).

Note that the ideal, or perfect, column is theoretical and does not exist in reality. Even in laboratory conditions, it is impossible to obtain a perfectly straight column, frictionless pinned ends, or a perfectly axial load. Consequently, Euler's formula cannot be used directly as a practical approach to column analysis and design and must be modified.

Note also that the only material property represented in the Euler critical load and critical stress formulas is the modulus of elasticity E. Recall that E is a measure of the stiffness of a material. Material strength, such as the yield stress of steel, does not affect the magnitude of the Euler buckling load.

□ **EXAMPLE 18–1**   A 1 in. diameter steel rod is used as a pin-connected compression member. Calculate the critical load using the Euler formula if the rod is (a) 2 ft long and (b) 4 ft long. The proportional limit for the steel is 34,000 psi and the modulus of elasticity is 30,000,000 psi.

Solution   Referring to Table 8–1 for properties of areas,

$$A = 0.7854 d^2 = 0.7854(1)^2 = 0.7854 \text{ in.}^2$$

$$r = \frac{d}{4} = \frac{1.0}{4} = 0.25 \text{ in.}$$

(a) For a 2 ft length of rod,

$$s_e = \frac{\pi^2 E}{(L/r)^2} = \frac{\pi^2 (30{,}000{,}000)}{[2(12)/0.25]^2}$$
$$= 32{,}128 \text{ psi} < 34{,}000 \text{ psi} \qquad \textbf{OK}$$

The critical load can then be calculated from

$$P_e = s_e A = 32{,}128(0.7854) = 25{,}200 \text{ lb}$$

(b) For a 4 ft length of rod,

$$s_e = \frac{\pi^2 E}{(L/r)^2} = \frac{\pi^2 (30{,}000{,}000)}{[4(12)/0.25]^2}$$
$$= 8032 \text{ psi} < 34{,}000 \text{ psi} \qquad \textbf{OK}$$

The critical load is

$$P_e = s_e A = 8032(0.7854) = 6300 \text{ lb}$$

☐ **EXAMPLE 18-2**   A 1 in. diameter steel shaft is used as an axially loaded pin-connected compression member. The high-strength steel has a proportional limit of 50,000 psi and a modulus of elasticity of 30,000,000 psi. Calculate (a) the shortest length L for which Euler's formula is applicable and (b) the critical load if the length of the shaft is 48 in.

Solution   From Table 8-1,

$$A = 0.7854 d^2 = 0.7854(1)^2 = 0.7854 \text{ in.}^2$$

$$r = \frac{d}{4} = \frac{1.0}{4} = 0.25 \text{ in.}$$

$$I = \frac{\pi d^4}{64} = \frac{\pi(1)^4}{64} = 0.0491 \text{ in.}^4$$

(a) The shortest length L for which the Euler formula applies will be that length for which the critical stress is equal to the proportional limit:

$$s_e = \frac{\pi^2 E}{(L/r)^2}$$

Solving for L,

$$L^2 = \frac{\pi^2 E r^2}{s_e} = \frac{\pi^2(30,000,000)(0.25)^2}{50,000} = 370.1$$

from which

$$L = 19.24 \text{ in.}$$

(b) If $L = 48$ in., Euler's formula is valid (48 in. > 19.24 in.). Therefore, the critical load can be calculated from Eq. (18-1):

$$P_e = \frac{\pi^2 E I}{L^2} = \frac{\pi^2(30,000,000)(0.0491)}{(48)^2} = 6310 \text{ lb}$$

Equation (18-2) established the relationship between modulus of elasticity, slenderness ratio, and critical stress for slender columns. If E is known and if various values of the slenderness ratio are assumed, the corresponding values of s_e can be computed. This value, the Euler critical stress, can then be plotted as a function of the slenderness ratio. The resulting curve is shown in Fig. 18-2. This curve is often referred to as the *Euler curve*, and it clearly depicts the way in which the critical stress at initial buckling decreases, along with the load-carrying capacity of the column, as the slenderness ratio increases.

We see, then, that the slenderness ratio, which is a function of the length and radius of gyration of the column, is a significant factor affecting the load-carrying capacity of the column. The tendency of a member to buckle is a function of the slenderness ratio. A column will always buckle about the axis having the largest slenderness ratio. This is the same as saying that if the unbraced length of the column is the same in all directions, it will buckle about the axis with the least radius of gyration. The concept of

FIGURE 18–2 Euler curve for pinned-end columns.

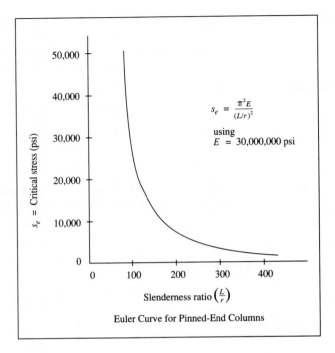

$$s_e = \frac{\pi^2 E}{(L/r)^2}$$

using
$E = 30{,}000{,}000$ psi

Euler Curve for Pinned-End Columns

unbraced length is introduced because some columns are braced at various points. At such a braced point, column buckling (or lateral deflection) is prevented. Further, the bracing may prevent the buckling in a specific direction only. The buckled column must be visualized in order to associate the correct values of radius of gyration with values of length between supports (or braces).

Since Euler's formula is only valid if the resulting critical stress is below the proportional limit, Fig. 18–2 shows that the range in which Euler's formula is applicable will vary with material. As an illustration, assume that a material (steel) has a proportional limit of 34,000 psi. If the 34,000 psi line were shown in Fig. 18–2, it would intersect the Euler curve at a slenderness ratio value of 93.3. Any value of slenderness ratio less than 93.3 would result in a critical stress in excess of 34,000 psi. Therefore, Euler's formula is not applicable for this material if the slenderness ratio is less than 93.3.

**18–3
EFFECTIVE LENGTH**

The Euler formula (Eq. [18–1]) yields the buckling load for an ideal column with pinned ends. A slender column, however, in addition to being less than perfect in other respects, may have end conditions that provide restraint of some magnitude so that the column ends are prevented from rotating freely. In such a case, the use of the Euler formula may be extended to columns having other than pinned ends through the use of an *effective length* in place of the actual unbraced length. The effective length is the distance between points of inflection (contraflexure) on the deflected shape of the column.

These are points of zero bending moment and may be considered analogous to pinned ends, since pinned ends are also points of zero bending moment. For the ideal column, these points occur at the actual column ends. The effect of different end conditions on the deflected shape of a column is shown in Fig. 18–3. As may be seen, the effective length of a column can be quite different for various end conditions, even though the actual length of the column does not vary.

FIGURE 18–3 Effect of end conditions.

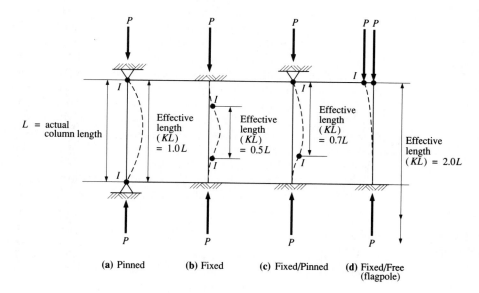

(a) Pinned (b) Fixed (c) Fixed/Pinned (d) Fixed/Free (flagpole)

Note: *I* indicates inflection points

End conditions can be accounted for through the use of an effective length factor K, a dimensionless number, which, when multiplied by the actual length of the column, yields the effective length KL. Therefore, the slenderness ratio is more correctly defined as the ratio of the effective length to the radius of gyration (KL/r.) Euler's formulas can then be rewritten with the inclusion of the effective length:

$$P_e = \frac{\pi^2 EI}{(KL)^2} \tag{18–3}$$

$$s_e = \frac{\pi^2 E}{(KL/r)^2} \tag{18–4}$$

For example, if the ends of a column are fixed against rotation and translation (lateral movement), as shown in Fig. 18–3(b), and the column is then loaded axially until it buckles, points of inflection will develop at the quarter points of the column length. If the middle portion of this fixed-end column is considered separately, it behaves as a pinned-end column having a length $L/2$ or $0.5L$. The length $0.5L$ is the effective length (KL) of the fixed-end column, where the K factor is 0.5. Substituting this into the critical load formula (Eq. [18–3]),

$$P_e = \frac{\pi^2 EI}{(KL)^2} = \frac{\pi^2 EI}{(0.5L)^2} = \frac{4\pi^2 EI}{L^2}$$

This indicates that a slender column with fixed ends that buckles elastically will be four times stronger than the same column with pinned ends.

Note that for the "flagpole" case (Fig. 18–3[d]), one end is fixed and the other end is free with respect to both rotation and translation. In this case there is an inflection point at the top of the column as well as an imaginary point of inflection at a distance L below the column base. The theoretical effective length factor K is 2.0. Note that the curvature shown is similar to the upper half of the pinned-end column of Fig. 18–3(a). The effective length for this "fixed/free" column is $2.0L$. Such a column has only one-quarter the strength of the same column with pinned ends.

Table 18–1 summarizes the K values that are shown in Fig. 18–3. The theoretical K values are for ideal end conditions. In addition, since the idealized end conditions rarely exist, recommended values of K for design and analysis are provided. The latter K values should be used when the idealized end conditions are approximated.

TABLE 18–1 Effective length factors.

Idealized End Conditions	Theoretical K Value	Recommended K Value for Design and Analysis	No. of Times Stronger Than Pinned-End Column*
Pinned	1.0	1.0	1
Fixed	0.5	0.65	4
Fixed/Pinned	0.7	0.80	2
Fixed/Free	2.0	2.10	1/4

* Using the theoretical K value.

18–4
ALLOWABLE AXIAL COMPRESSIVE LOADS

In previous sections we discussed the load-carrying capacity of columns. Among the factors affecting the magnitude of this capacity are end conditions, unbraced length, radius of gyration, eccentricity of load, and column and material imperfections. Although the Euler formula is a theoretical equation and not a practical design formula, it may be modified to make it an expression for an allowable axial compressive load, P_a. This modification is accomplished by introducing a factor of safety (F.S.) to compensate for some of the previously mentioned factors that affect the strength of the column. Rewriting Eq. (18–1),

$$P_a = \frac{\pi^2 EI}{(KL)^2(\text{F.S.})} \tag{18-5}$$

or

$$P_a = \frac{P_e}{\text{F.S.}} = \frac{s_e A}{\text{F.S.}}$$

Equation (18–5) also can be used for design purposes by calculating the required moment of inertia. Following the selection of a member with a moment of inertia in excess of the required moment of inertia, one must then calculate s_e to verify the applicability of the Euler formula. Rewriting Eq. (18–5),

$$\text{Required } I = \frac{P(KL)^2(\text{F.S.})}{\pi^2 E} \tag{18-6}$$

where P in this case represents the applied axial load and all other symbols are as previously defined.

☐ **EXAMPLE 18-3**   A square steel bar 2 in. by 2 in. in cross section is used as a pin-connected axially loaded compression member. The proportional limit for the steel is 34,000 psi and the modulus of elasticity is 30,000,000 psi. Use the modified Euler formula (Eq. (18–5)) with a factor of safety of 3.0 to determine the allowable axial compressive load if the member length is (a) 4.0 ft and (b) 8.0 ft.

Solution   The cross-sectional area is 4 in.^{2}. From Table 8–1,

$$r = \frac{h}{\sqrt{12}} = \frac{2}{\sqrt{12}} = 0.577 \text{ in.}$$

(a) For a length of 4.0 ft,

$$s_e = \frac{\pi^2 E}{(L/r)^2} = \frac{\pi^2(30,000,000)}{[4(12)/0.577]^2}$$
$$= 42,800 \text{ psi} > 34,000 \text{ psi} \qquad \textbf{N.G.}$$

Therefore, the Euler formula is not applicable.

(b) For a length of 8.0 ft,

$$s_e = \frac{\pi^2 E}{(L/r)^2} = \frac{\pi^2(30,000,000)}{[8(12)/0.577]^2}$$
$$= 10,700 \text{ psi} < 34,000 \text{ psi} \qquad \textbf{OK}$$

from which

$$P_a = \frac{s_e A}{\text{F.S.}} = \frac{10,700(4)}{3.0} = 14,270 \text{ lb}$$

☐ **EXAMPLE 18–4**   An ASTM A36 W10 × 45 structural steel wide-flange section is used as an axially loaded column. The proportional limit is 34.0 ksi and the modulus of elasticity is 30,000 ksi. (a) Find the allowable axial compressive load using the Euler formula and a factor of safety of 2.0. The length of the column is 26 ft. The column is fixed at the bottom and pinned at the top. (b) Find the allowable axial compressive load for the same column if the end conditions are fixed/fixed (fixed at each end). (c) Determine the minimum length of the column at which the Euler formula will still be valid if the column is fixed at the bottom and pinned at the top.

Solution   For the W10 × 45, obtain the following properties from Appendix A:

$$A = 13.3 \text{ in.}^2 \quad \text{and} \quad r_y = 2.01 \text{ in.}$$

(a) From Table 18–1, obtain the effective length factor for design and analysis: $K = 0.80$. With this value, calculate the critical stress using Eq. (18–4):

$$s_e = \frac{\pi^2 E}{(KL/r)^2} = \frac{\pi^2(30,000)}{[0.80(26)(12)/2.01]^2}$$

$$= 19.2 \text{ ksi} < 34 \text{ ksi} \qquad \textbf{OK}$$

from which, using Eq. (18–5),

$$P_a = \frac{s_e A}{\text{F.S.}} = \frac{19.2(13.3)}{2.0} = 127.7 \text{ kips}$$

(b) With the column fixed at each end, the K factor is 0.65:

$$s_e = \frac{\pi^2(30,000)}{[0.65(26)(12)/2.01]^2} = 29.1 \text{ ksi} < 34 \text{ ksi} \qquad \textbf{OK}$$

Therefore,

$$P_a = \frac{29.1(13.3)}{2.0} = 193.5 \text{ kips}$$

(c) For the fixed/pinned end conditions, K is 0.80. Find the length at which the critical stress equals the proportional limit. Using Eq. (18–4),

$$s_e = \frac{\pi^2 E}{(KL/r)^2}$$

Solving for L,

$$L = \sqrt{\frac{\pi^2 E}{s_e(K/r)^2}} = \sqrt{\frac{\pi^2(30,000)}{34(0.80/2.01)^2}}$$

$$= 234.5 \text{ in.} = 19.5 \text{ ft}$$

☐ **EXAMPLE 18–5**   A 12 ft long structural steel column must support an axial load of 50 kips. The ends of the column are pin connected. The steel to be used is ASTM A36 with a proportional limit of 34 ksi and a modulus of elasticity of 30,000 ksi. Select the most economical wide-flange (W) shape. Use Euler's modified equation (Eq. (18–6)). Check for applicability. Use a factor of safety of 3.0.

Solution   Equation (18–6) yields

$$\text{Required } I = \frac{P(KL)^2(\text{F.S.})}{\pi^2 E} = \frac{50(1.0 \times 12 \times 12)^2(3.0)}{\pi^2(30{,}000)} = 10.5 \text{ in.}^4$$

From Appendix A we select a W10 × 22, which has the following properties:

$$I_y = 11.4 \text{ in.}^4 \quad \text{and} \quad r_y = 1.33 \text{ in.}$$

We check the applicability of the Euler formula by obtaining the critical stress from Eq. (18–4):

$$s_e = \frac{\pi^2 E}{(KL/r)^2} = \frac{\pi^2(30{,}000)}{[1.0(12)(12)/1.33]^2}$$

$$= 25.3 \text{ ksi} < 34 \text{ ksi} \quad \textbf{OK}$$

Euler's formula is applicable; therefore, use a W10 × 22.

**18–5
ALLOWABLE STRESS
FOR AXIALLY
LOADED STEEL
COLUMNS (AISC)**

In the preceding sections, the Euler formula was used to analyze and design columns. In each case, we checked the applicability of the approach by verifying the fact that s_e was less than the proportional limit. When, indeed, it was, we concluded that the column was slender and that the Euler formula was valid. Practical columns and/or compression members do not always fall into this category, however. Practical analysis/design methods must concern themselves with the entire possible range of slenderness ratios. The Euler formula is not applicable for intermediate and short columns. In fact, there is no single theoretical column formula that can accurately predict the strength for columns in all categories. Many column formulas have been developed over the years based on results of extensive testing and experience.

Various modern design specifications and/or codes stipulate empirical column formulas for use in design. In the field of structural design, with particular reference to buildings, the most commonly accepted and used column formulas are those set forth in the AISC *Specification for Structural Steel Buildings, Allowable Stress Design*, which is available in the AISC *Manual of Steel Construction—Allowable Stress Design*.[1] The AISC column formulas furnish an allowable axial compressive stress, denoted here as $s_{a(\text{all})}$. A plot of the allowable axial compressive stress versus the slenderness ratio is shown in Fig. 18–4. The slenderness ratio KL/r (per AISC) preferably should not exceed 200.

Slenderness ratios in excess of 200 are felt to reflect columns so slender as to be overly sensitive to uncontrollable factors such as the initial straightness of the column. The shape of the elastic buckling portion of the curve is essentially the same as that of the Euler curve, but with a factor of safety

---

[1]   American Institute of Steel Construction, Inc., *Manual of Steel Construction— Allowable Stress Design,* 9th Ed. (Chicago, AISC, 1989).

FIGURE 18–4 Allowable stress vs. slenderness ratio, axially loaded column—ASTM A36 steel, AISC formulas ($K = 1.0$) ($E = 30{,}000{,}000$ psi).

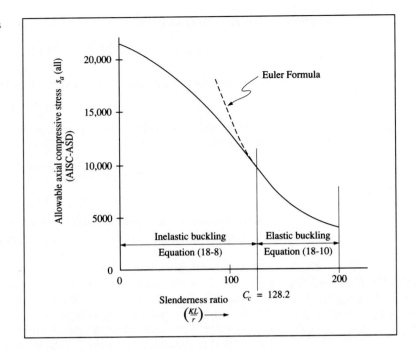

applied. The value of KL/r that separates elastic buckling from inelastic buckling has been arbitrarily established as that value at which the Euler critical stress s_e is equal to $s_Y/2$, where s_Y is the yield stress of the steel. This KL/r value is denoted as C_c. Its value can be calculated from the Euler formula, Eq. (18–4), as follows:

$$s_e = \frac{\pi^2 E}{(KL/r)^2}$$

The substitutions

$$s_e = \frac{s_Y}{2} \quad \text{and} \quad \frac{KL}{r} = C_c$$

will yield

$$\frac{s_Y}{2} = \frac{\pi^2 E}{C_c^2}$$

from which

$$C_c = \sqrt{\frac{2\pi^2 E}{s_Y}} \tag{18–7}$$

For KL/r values less than C_c, the allowable axial compressive stress is determined from

$$s_{a(\text{all})} = \frac{\left(1 - \frac{(KL/r)^2}{2C_c^2}\right)s_Y}{\text{F.S.}} \tag{18-8}$$

where

$$\text{F.S.} = \frac{5}{3} + \frac{3(KL/r)}{8C_c} - \frac{(KL/r)^3}{8C_c^3} \tag{18-9}$$

For KL/r values greater than C_c,

$$s_{a(\text{all})} = \frac{12\pi^2 E}{23(KL/r)^2} \tag{18-10}$$

This is the Euler critical stress formula with a factor of safety of 23/12 or 1.92.

We will now consider applications of the AISC column formulas in Sections 18–6 and 18–7.

**18-6
ANALYSIS OF
AXIALLY LOADED
STEEL COLUMNS
(AISC)**

The application of the AISC column formulas is simplified through the use of the allowable stress tables in the AISC Manual. These tables eliminate the need for solving Eqs. (18–8) and (18–10). Appendix J is similar to the AISC tables; however, for consistency within this text, the tabulated values of $s_{a(\text{all})}$ have been developed using a modulus of elasticity value of 30,000,000 psi. This differs slightly from the value of 29,000,000 psi adopted by the AISC.

Commonly used cross sections for steel compression members include most of the hot-rolled shapes. For larger loads, a built-up cross section is frequently used. In addition to providing increased cross-sectional area, the built-up sections allow a designer to tailor the radius of gyration values about the various axes to meet specific needs. Typical compression members are shown in Fig. 18–5. The dashed lines shown on a cross section represent details such as tie plates or lacing bars that do not contribute to cross-sectional properties but serve to hold components of the cross section in proper relative position and to make the built-up section act as a single unit.

Five examples will now illustrate the analysis method using the AISC column formulas along with the available tables.

□ **EXAMPLE 18-6** Calculate the allowable axial compressive load for a W12 × 50 structural steel column. The material is ASTM A36 steel ($s_Y = 36$ ksi). The column ends are pin connected. Use an unbraced length of (a) 15 ft and (b) 30 ft.

Solution From Appendix A, for the W12 × 50,

$$A = 14.7 \text{ in.}^2 \quad \text{and} \quad r_y = 1.96 \text{ in.}$$

FIGURE 18–5 Structural steel compression member cross sections.

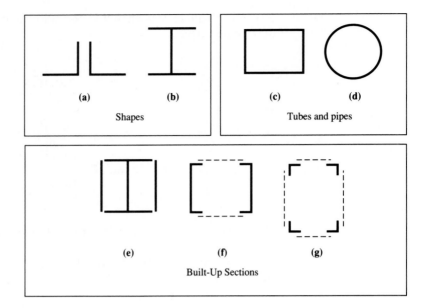

Note that it is only the least radius of gyration that is of interest, since it is this value that will yield the largest slenderness ratio. From Table 18–1, we obtain a K value of 1.0 (pinned ends).

(a) We calculate the KL/r value first:

$$\frac{KL}{r} = \frac{1.0(15)(12)}{1.96} = 92$$

This value has been rounded to the nearest whole number for use in Appendix J. Interpolation is not considered to be warranted.

From Appendix J we obtain a value for allowable axial stress $s_{a(\text{all})}$ of 14.15 ksi. We then calculate the allowable axial compressive load:

$$P_a = s_{a(\text{all})}A = 14.15(14.7) = 208 \text{ kips}$$

(b) Following the same procedure, for $L = 30$ ft,

$$\frac{KL}{r} = \frac{1.0(30)(12)}{1.96} = 184$$

$$s_{a(\text{all})} = 4.56 \text{ ksi}$$

$$P_a = s_{a(\text{all})}A = 4.56(14.7) = 67.0 \text{ kips}$$

☐ **EXAMPLE 18–7**   A W10 × 68 structural steel column of ASTM A36 steel is to carry an axial compressive load of 340 kips. The column has a length of 17 ft. Determine whether or not the column is adequate (a) if the ends are pin connected and (b) if the ends are fixed.

Solution   From Appendix A, for the W10 × 68,

$$A = 20 \text{ in.}^2 \quad \text{and} \quad r_y = 2.59 \text{ in.}$$

(a) From Table 18–1, $K = 1.0$. Hence,

$$\frac{KL}{r} = \frac{1.0(17)(12)}{2.59} = 79$$

From Appendix J, $s_{a(\text{all})} = 15.61$ ksi. The allowable axial compressive load, then, is

$$P_a = s_{a(\text{all})} A = 15.61(20.0) = 312 \text{ kips}$$
$$312 \text{ kips} < 340 \text{ kips} \quad \textbf{N.G.}$$

(b) Following a similar procedure, using $K = 0.65$ (fixed ends),

$$\frac{KL}{r} = \frac{0.65(17)(12)}{2.59} = 51$$
$$s_{a(\text{all})} = 18.34 \text{ ksi}$$
$$P_a = s_{a(\text{all})} A = 18.34(20.0) = 367 \text{ kips}$$
$$367 \text{ kips} > 340 \text{ kips} \quad \textbf{OK}$$

☐ **EXAMPLE 18–8**   Find the allowable axial compressive load for a W8 × 40 structural steel column of ASTM A36 steel. The column has an unbraced length of 24 ft with respect to the X–X axis and 12 ft with respect to the Y–Y axis, as shown in Fig. 18–6. The column is pin connected at the top and fixed at the bottom. (Assume that the column is pin connected at midheight.)

Solution   From Appendix A, for the W8 × 40,

$$A = 11.7 \text{ in.}^2 \quad r_x = 3.53 \text{ in.} \quad r_y = 2.04 \text{ in.}$$

FIGURE 18–6   Unbraced lengths.

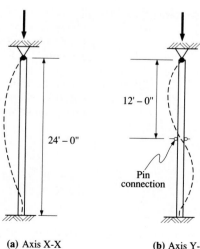

12' – 0"

24' – 0"

Pin connection

(a) Axis X-X   **(b) Axis Y-Y**

With respect to the X–X axis, $L_x = 24$ ft and $K_x = 0.8$. Hence,

$$\frac{K_x L_x}{r_x} = \frac{0.8(24)(12)}{3.53} = 65$$

With respect to the Y–Y axis, $L_y = 12$ ft. $K_y = 1.0$ for the top part of the column. For the bottom part, $K_y = 0.8$. Therefore, with respect to the Y–Y axis, the top part is the more critical (will have the larger slenderness ratio):

$$\frac{K_y L_y}{r_y} = \frac{1.0(12)(12)}{2.04} = 71$$

The controlling slenderness ratio is 71, the larger value. From Appendix J,

$$s_{a(\text{all})} = 16.45 \text{ ksi}$$
$$P_a = s_{a(\text{all})} A = 16.45(11.7) = 192.5 \text{ kips}$$

☐ **EXAMPLE 18–9**   Find the allowable axial compressive load for an 8 in. diameter extra-strong steel pipe column that has an unbraced length of 15 ft. The ends are pin connected and the steel is ASTM A501 ($s_Y = 36$ ksi).

Solution   From Appendix B,

$$A = 12.8 \text{ in.}^2 \quad \text{and} \quad r = 2.88 \text{ in.}$$

Next calculate the slenderness ratio, obtain the allowable axial compressive stress, and determine the allowable axial compressive load.

$$\frac{KL}{r} = \frac{1.0(15)(12)}{2.88} = 63$$
$$s_{a(\text{all})} = 17.24 \text{ ksi}$$
$$P_a = s_{a(\text{all})} A = 17.24(12.8) = 221 \text{ kips}$$

☐ **EXAMPLE 18–10**   Two C12 × 30 structural steel channels of ASTM A36 steel are connected with tie plates, as shown in Fig. 18–7, to form a box-shaped cross section for a column. The back-to-back distance of the channels is to be 12 in. Assume the column ends to be pin connected. The length is 20 ft. Calculate the allowable axial compressive load.

Solution   Since the cross section is built-up, it is not clear about which axis the radius of gyration will be least. Therefore, we will calculate r about both axes. The moment of inertia must be calculated first. For a single C12 × 30, from Appendix C,

$$A = 8.82 \text{ in.}^2 \quad I_x = 162 \text{ in.}^4$$
$$\bar{x} = 0.674 \text{ in.} \quad I_y = 5.14 \text{ in.}^4$$

Using the transfer formula (Eq. (8–4)), we first calculate the moment of inertia about the X–X axis:

$$I_x = \Sigma(I_o + ad^2)$$
$$= 2(162 + 0) = 324 \text{ in.}^4$$

FIGURE 18–7   Column cross section.

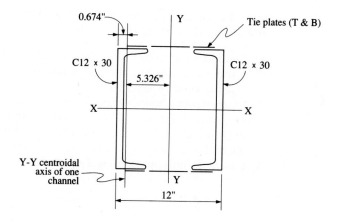

and about the Y–Y axis:

$$I_y = 2[5.14 + 8.82(5.326)^2] = 511 \text{ in.}^4$$

Since I_x is the lesser of the two, it will result in the lower radius of gyration. The X–X axis is the weaker axis. We therefore calculate r_x for use in the remainder of the solution:

$$r_x = \sqrt{\frac{I_x}{A}} = \sqrt{\frac{324}{2(8.82)}} = 4.29 \text{ in.}$$

Completing the solution, we compute the slenderness ratio, obtain $s_{a(\text{all})}$ from Appendix J, and calculate the allowable axial compressive load:

$$\frac{KL}{r_x} = \frac{1.0(20)(12)}{4.29} = 56$$

$$s_{a(\text{all})} = 17.89 \text{ ksi}$$

$$P_a = s_{a(\text{all})}A = 17.89(2)(8.82) = 316 \text{ kips}$$

18–7
DESIGN OF AXIALLY LOADED STEEL COLUMNS (AISC)

Since the allowable axial compressive stress is a function of the slenderness ratio of a column, there is no direct solution for a required area or moment of inertia unless a variation of the Euler formula (such as Eq. (18–10)) governs. Therefore, the design of columns is usually accomplished through a trial-and-error procedure or with the aid of column load tables.

The majority of structural steel compression members are composed of W shapes, steel pipes, and structural tubing. The AISC Manual contains allowable axial load tables for these members, simplifying the design process. Although one may select the proper column section by merely referring to the tables, a designer should understand the application of the formulas through which the tables were developed.

In the absence of allowable column load tables, we recommend the following trial-and-error procedure:

1. Establish the axial column load, the unbraced length of the column with respect to each axis, the end conditions, the type of cross section desired, and the type of steel to be used.
2. Estimate an allowable axial compressive stress $s_{a(all)}$. A rationale for this numerical value could be that most columns will have a slenderness ratio in the range of 30 to 120. Assuming a midpoint slenderness ratio of 75, an allowable axial compressive stress can be determined by formula or table.
3. Calculate an approximate required area:

$$\text{Approx. required } A = \frac{P}{\text{approx. } s_{a(all)}}$$

where P is the applied axial compressive load.
4. Knowing the desired column shape, select a trial section that provides the approximate required area. If the section is to be a W shape, initially consider the W8, W10, W12, and W14 sections that have flange widths approximately equal to the depth of the member.
5. Calculate the slenderness ratio for the trial section and then determine the allowable axial compressive stress by formula or table.
6. Calculate the allowable axial compressive load:

$$P_a = s_{a(all)} A$$

and compare with the applied load. The allowable load must be greater than the applied load. To obtain the most economical member, several trials may have to be performed.

Our selections will be based on least weight. We will assume that this is the most important factor in determining economy. It is the most apparent factor. Overall costs, however, will also depend on such other factors as ease of fabrication, availability, and volume discounts.

☐ **EXAMPLE 18–11**   Select the lightest W shape for a column subjected to an axial compressive load of 250 kips. The unbraced length of the column is 20 ft and the ends are assumed to be pin connected. Use ASTM A36 steel.

Solution   We will assume a slenderness ratio of 75. From Appendix J, we obtain an allowable axial compressive stress $s_{a(all)}$ of 16.04 ksi. The approximate required area can then be determined:

$$\text{Approx. required } A = \frac{P}{\text{approx. } s_{a(all)}} = \frac{250}{16.04} = 15.6 \text{ in.}^2$$

From Appendix A, we select a W10 × 54. The applicable properties are

$$A = 15.8 \text{ in.}^2 \quad \text{and} \quad r_y = 2.56 \text{ in.}$$

The slenderness ratio is computed from

$$\frac{KL}{r} = \frac{1.0(20)(12)}{2.56} = 94$$

From Appendix J,

$$s_{a(all)} = 13.91 \text{ ksi}$$

The allowable axial compressive load is

$$P_a = s_{a(all)}A = 13.91(15.8) = 220 \text{ kips}$$

$$220 \text{ kips} < 250 \text{ kips} \quad \textbf{N.G.}$$

It is apparent that a larger section is required. For our second trial, we will assume that the radius of gyration r will not vary substantially with the next shape to be selected. Therefore, the allowable stress will not vary substantially either. The second trial approximate required area is calculated using $s_{a(all)}$ of 13.91 ksi:

$$\text{Approx. required } A = \frac{250}{13.91} = 18.0 \text{ in.}^2$$

Some possible choices are shown in Table 18–2 along with their resulting allowable axial compressive loads.

TABLE 18–2

Section	A (in.²)	r_{min} (in.)	$\dfrac{KL}{r}$	$s_{a(all)}$ (ksi)	P_a (kips)
W10 × 68	20.0	2.59	93	14.03	281
W12 × 72	21.1	3.04	79	15.61	329
W14 × 74	21.8	2.48	97	13.56	296

The lightest column of the three listed is the W10 × 68. To ensure that we have found the lightest, we will check the next lightest in each depth as shown in Table 18–3.

TABLE 18–3

Section	A (in.²)	r_{min} (in.)	$\dfrac{KL}{r}$	$s_{a(all)}$ (ksi)	P_a (kips)
W10 × 54	15.8	2.56	94	13.91	220
W12 × 58	17.0	2.51	96	13.68	233
W14 × 61	17.9	2.45	98	13.43	240

None of the three preceding sections has sufficient strength.
   Use a W10 × 68.

☐ **EXAMPLE 18–12**   Select the most economical standard-weight steel pipe section to support an axial compressive load of 75 kips. The column is pin connected at each end and has an unbraced length of 12 ft. Use ASTM A501 steel ($s_Y = 36$ ksi).

Solution   Assume $KL/r = 75$, from which $s_{a(all)} = 16.04$ ksi.

$$\text{Approx. required } A = \frac{P}{s_{a(all)}} = \frac{75}{16.04} = 4.68 \text{ in.}^2$$

Properties of the pipe cross sections are found in Appendix B. Possible choices and the resulting allowable axial compressive loads are shown in Table 18–4.

TABLE 18–4

Diameter (in.)	A (in.^{2})	r (in.)	$\dfrac{KL}{r}$	$s_{a(all)}$ (ksi)	P_a (kips)
5	4.30	1.88	77	15.83	68.1
6	5.58	2.25	64	17.15	95.7

The 6 in. diameter pipe has an allowable axial load greater than the applied load of 75 kips.
Use a 6 in. diameter standard-weight steel pipe.

The two preceding AISC column design example problems are not typical of practical column design. The abundance and availability of column load tables and design aids simplify the design process and circumvent the use of a trial-and-error procedure. Refer to the previously referenced AISC Manual and to other sources for a thorough description of a practical approach to steel column design.[2]

**18–8
ANALYSIS AND
DESIGN OF AXIALLY
LOADED STEEL
MACHINE PARTS**

In the design and analysis of machine parts subjected to axial compressive loads, the AISC column formulas are not strictly applicable, despite the fact that they result in solutions that are generally satisfactory. The AISC design specification was developed for the design of steel-framed buildings; it is recognized nationally in that it is incorporated into most state and municipal building codes, thereby making it legal and enforceable. In the area of machine design, however, there is no one nationally accepted or legally adopted design specification that encompasses all machine design applications.

For the design and/or analysis of steel compression members that are a part of some machine, there are some acceptable and appropriate column

---

[2] L. Spiegel and G. F. Limbrunner, *Applied Structural Steel Design*, 2nd ed. (Englewood Cliffs, N.J.: Prentice-Hall, 1993).

equations. that can be used. They are as follows:

1. The member is categorized as slender if

$$\frac{KL}{r} \geq \sqrt{\frac{2\pi^2 E}{s_Y}}$$

The allowable axial compressive stress can then be computed from

$$s_{a(\text{all})} = \frac{\pi^2 E}{(KL/r)^2(\text{F.S.})} \qquad \textbf{(18–11)}$$

Note that the limiting KL/r value that determines whether a member is to be treated as a slender (long) or intermediate column is the same as the value designated as C_c by AISC (see Eq. (18–7)). Also note that Eq. (18–11) is a modified Euler critical stress formula and is similar to the AISC formula (see Eq. (18–10)) except for the factor of safety, which may be varied.

2. The member is categorized as intermediate if

$$\frac{KL}{r} < \sqrt{\frac{2\pi^2 E}{s_Y}}$$

and the allowable axial compressive stress can be determined using the empirical J.B. Johnson formula:

$$s_{a(\text{all})} = \frac{\left[1 - \dfrac{s_Y(KL/r)^2}{4\pi^2 E}\right] s_Y}{\text{F.S.}} \qquad \textbf{(18–12)}$$

where all terms are as were defined in Sections 18–4 and 18–5. This formula is mathematically identical to the AISC column formula for $KL/r < C_c$ (Eq. (18–8)). It is generally considered valid for KL/r values ranging down to zero, as is the case with the AISC formula.

The factor of safety to be used in the machine part applications can be computed from Eq. (18–9) as designated in the AISC design specifications, or it may be selected based on various factors such as type of applied load. Whereas the AISC *maximum* factor of safety for column design is taken as 1.92, a *minimum* factor of safety of 2.0 is generally used in machine design, and larger values are used under conditions of greater uncertainty.

The analysis and design procedure for steel compression member machine parts is similar to that of the AISC column design procedure. Numerous tables and design aids are available from which to determine the allowable axial compressive stress and to simplify analysis. Similarly, column load tables simplify the design process and eliminate the trial-and-error procedure. In the absence of such design aids, we recommend the analysis and design procedures described in the following examples.

☐ **EXAMPLE 18–13**   A square machine member measuring $\frac{1}{2}$ in. by $\frac{1}{2}$ in. has a length of 12 in. The member is to be loaded in axial compression. End conditions are fixed/pinned. The member is made of AISI 1040 hot-rolled steel. Compute the critical load and the allowable load if a factor of safety of 3.0 is used. The modulus of elasticity E is 30,000,000 psi.

Solution   For the $\frac{1}{2}$ in. square cross section (from Table 8–1),

$$A = (0.5)^2 = 0.25 \text{ in.}^2 \quad \text{and} \quad r = \frac{h}{\sqrt{12}} = 0.144 \text{ in.}$$

From Table 18–1, the effective length factor is 0.80. Therefore,

$$\frac{KL}{r} = \frac{0.80(12)}{0.144} = 66.7$$

Next compute the value of the slenderness ratio, which is the lower limit for long columns. The yield stress s_Y is obtained from Appendix G:

$$\sqrt{\frac{2\pi^2 E}{s_Y}} = \sqrt{\frac{2\pi^2(30,000,000)}{42,000}} = 118.7$$

Since $66.7 < 118.7$, the column is intermediate (or short) and the J.B. Johnson formula should be used:

$$s_{a(\text{all})} = \frac{\left[1 - \dfrac{s_Y(KL/r)^2}{4\pi^2 E}\right]s_Y}{\text{F.S.}}$$

$$= \frac{\left[1 - \dfrac{42,000(66.7)^2}{4\pi^2(30,000,000)}\right]42,000}{3.0} = 11,790 \text{ psi}$$

from which, the allowable axial compressive load is

$$P_a = s_{a(\text{all})}A = 11,790(0.25) = 2950 \text{ lb}$$

The critical load can be calculated from

$$P_e = P_a(\text{F.S.}) = 2950(3) = 8850 \text{ lb}$$

☐ **EXAMPLE 18–14**   Determine the required diameter of a round linkage bar of AISI 1040 hot-rolled steel with a length of 15 in. The bar is subjected to an axial compressive load of 3000 lb and is pin connected. A factor of safety of 3.0 is to be used.

Solution   For a round bar, the following properties apply (from Table 8–1):

$$A = 0.7854d^2 \quad \text{and} \quad r = 0.25d$$

From Appendix G, for 1040 hot-rolled steel,

$$s_Y = 42,000 \text{ psi} \quad \text{and} \quad E = 30,000,000 \text{ psi}$$

We will assume that the member is in the category of an intermediate column. Therefore, an allowable axial compressive stress can be computed in terms of the bar diameter using the J.B. Johnson formula. The validity of the assumption will have to be checked before the solution is finalized.

$$s_{a(\text{all})} = \frac{\left[1 - \dfrac{s_Y(KL/r)^2}{4\pi^2 E}\right]s_Y}{\text{F.S.}}$$

$$= \frac{\left[1 - \dfrac{42,000[1(15)/(0.25d)]^2}{4\pi^2(30,000,000)}\right]42,000}{3.0}$$

$$= \frac{\left[1 - \dfrac{0.128}{d^2}\right]42,000}{3.0}$$

$$= 14,000 - \frac{1792}{d^2}$$

Since an allowable axial load of 3000 lb is desired, we substitute P_a/A for $s_{a(\text{all})}$ and the expression becomes

$$\frac{P_a}{A} = 14,000 - \frac{1792}{d^2}$$

Substituting,

$$\frac{3000}{0.7854d^2} = 14,000 - \frac{1792}{d^2}$$

$$\frac{3000}{0.7854} = 14,000d^2 - 1792$$

$$14,000d^2 = 3820 + 1792$$

from which

$$d = 0.633 \text{ in.}$$

Try a $\frac{3}{4}$ in. diameter round bar.

We now check the validity of the J.B. Johnson formula for the bar chosen:

$$\frac{KL}{r} = \frac{1(15)}{0.25(0.75)} = 80$$

Computing the value of the slenderness ratio, which is the lower limit for long columns,

$$\sqrt{\frac{2\pi^2 E}{s_Y}} = \sqrt{\frac{2\pi^2(30,000,000)}{42,000}} = 118.7$$

Since $80 < 118.7$, the use of the J.B. Johnson formula is valid.
Use a $\frac{3}{4}$ in. diameter round bar.

**18–9
ANALYSIS AND
DESIGN OF
AXIALLY LOADED
TIMBER COLUMNS**

The most frequently used wood columns are simple solid pieces of square or rectangular cross section. Simple columns may consist of a single piece of wood or of two or more pieces properly glued together to form a single member. The column of round cross section is a type of simple solid column that is less frequently encountered. There are also other types of timber columns, such as spaced columns and built-up columns, the complexity of which is beyond the scope of this text. Our discussion will be limited to the simple solid wood column of square or rectangular cross section. Generally, axially loaded columns of rectangular cross section have lateral side dimensions that do not differ by more than two inches.

The most widely used column design approach is that furnished in the *National Design Specification for Wood Construction* issued by the National Forest Products Association (NFPA).[3]

Solid columns are classified into three length classes, characterized by the mode of failure at ultimate load. The *short column* is defined as one having an effective unbraced length that is approximately eleven (11) times the least cross-sectional dimension of the column, or less. Failure mode is by crushing.

The *intermediate column* is defined as one with a ratio of effective unbraced length to least cross-sectional dimension (L/d) that lies between 11 and K:

$$11 < \frac{L}{d} < K$$

where K is defined as follows and must not be confused with an effective length factor:

$$K = 0.671 \sqrt{\frac{E}{s_c}} \qquad \textbf{(18–13)}$$

where K = the lowest value of the L/d ratio at which the column can be
           expected to perform as a slender column (an Euler column) (K is
           unitless)
     E = the modulus of elasticity (psi) (Pa, MPa)
     s_c = the allowable stress in compression parallel to the grain (same
           units as E)

The values for both E and s_c can be obtained from Appendix F. Failure mode of an intermediate column is generally a combination of crushing and buckling.

The *long column* (or slender column) is defined as one with an L/d ratio equal to K or greater. The L/d ratio for simple solid timber columns is limited to 50. The failure mode of the slender column is elastic buckling (lateral deflection).

For timber columns, the allowable axial compressive stress parallel to the grain, adjusted for the L/d ratio effects, will be designated $s_{a(\text{all})}$. Note that $s_{a(\text{all})}$ is not to exceed s_c, which is the allowable compressive stress parallel to the grain (not adjusted for L/d) from Appendix F. The value of

---

[3] National Forest Products Association, *National Design Specification for Wood Construction* (Washington, D.C.: NFPA, 1986).

$s_{a(\text{all})}$ can be determined as follows:

1. For short columns ($L/d \le 11$),

$$s_{a(\text{all})} = s_c \qquad\qquad (18\text{–}14)$$

2. For intermediate columns ($11 < L/d < K$),

$$s_{a(\text{all})} = s_c \left[1 - \frac{1}{3} \left(\frac{L/d}{K} \right)^4 \right] \qquad\qquad (18\text{–}15)$$

3. For long columns ($K \le L/d \le 50$),

$$s_{a(\text{all})} = \frac{0.30E}{(L/d)^2} \qquad\qquad (18\text{–}16)$$

Equation (18–16) is the Euler critical stress equation with the radius of gyration r expressed in terms of the least lateral dimension and a factor of safety of 2.74 included.

The preceding formulas can be used for columns of other shapes (round, for instance) by substituting $r\sqrt{12}$ for d, where r is the applicable radius of gyration of the column cross section used. The same substitution should be made when calculating L/d to be used in determining the classification of the column (short, intermediate, or long). Note that these formulas do not include any modifications for abnormal load duration or moisture service conditions. Refer to the NFPA *National Design Specification for Wood Construction* for the criteria governing stress modifications due to abnormal conditions.

As in our previous discussion on end conditions and effective unbraced column lengths, an effective length factor is also used with timber columns. However, it is now designated K_e instead of K as it was in Section 18–3. This K_e designation applies only to timber columns. The effective length factor may still be obtained from Table 18–1. The effective unbraced column length *must* be used in Eqs. (18–15) and (18–16), where

Effective unbraced length $= K_e \times$ (Actual unbraced length)

☐ EXAMPLE 18–15   Find the allowable axial compressive load for a pin-connected 8 in. by 10 in. (S4S) Douglas fir column having an unbraced length of 10 ft. Assume normal moisture conditions and duration of loading.

Solution   From Appendix F, for Douglas fir,

$$s_c = 1050 \text{ psi} \qquad \text{and} \qquad E = 1{,}700{,}000 \text{ psi}$$

From Table 18–1, the effective length factor K_e is 1.0; thus,

$$\frac{L}{d} = \frac{1.0(10)(12)}{7.5} = 16 < 50 \qquad \textbf{OK}$$

$$K = 0.671\sqrt{\frac{E}{s_c}} = 0.671\sqrt{\frac{1{,}700{,}000}{1050}} = 27$$

$$11 < \frac{L}{d} < 27$$

Therefore, the column may be categorized as an intermediate column and the allowable compressive stress can be calculated from

$$s_{a(\text{all})} = s_c \left[1 - \frac{1}{3} \left(\frac{L/d}{K} \right)^4 \right]$$

$$= 1050 \left[1 - \frac{1}{3} \left(\frac{16}{27} \right)^4 \right]$$

$$= 1050(0.959) = 1007 \text{ psi}$$

The allowable axial compressive load is then

$$P_a = s_{a(\text{all})} A = 1007(7.5)(9.5) = 71,700 \text{ lb}$$

☐ **EXAMPLE 18–16**   Find the allowable axial compressive load for an 8 in. by 8 in. (S4S) Hem-fir column having an unbraced length of 22 ft. The column is fixed at the bottom and pin connected at the top. Assume normal moisture conditions and duration of loading.

Solution   For Hem-fir, from Appendix F,

$$s_c = 875 \text{ psi} \quad \text{and} \quad E = 1,400,000 \text{ psi}$$

From Table 18–1, the effective length factor K_e is 0.8; thus,

$$\frac{L}{d} = \frac{0.8(22)(12)}{7.5} = 28.16 < 50 \quad \textbf{OK}$$

$$K = 0.671 \sqrt{\frac{E}{s_c}} = 0.671 \sqrt{\frac{1,400,000}{875}} = 26.8$$

$$\frac{L}{d} > K$$

Therefore, the column may be categorized as a slender column and the allowable compressive stress can be calculated from

$$s_{a(\text{all})} = \frac{0.30E}{(L/d)^2} = \frac{0.30(1,400,000)}{(28.16)^2} = 529.6 \text{ psi}$$

The allowable axial load is then

$$P_a = s_{a(\text{all})} A = 529.6(7.5)(7.5) = 29,790 \text{ lb}$$

The design of axially loaded timber columns is similar to that of steel columns. An abundance of allowable axial load tables for square and rectangular solid timber columns are available. Such tables are usually employed in timber-column design. In their absence, the design process becomes one of trial and error. The following design example will illustrate the trial-and-error procedure.

☐ **EXAMPLE 18–17**   Design a square pin-connected Douglas fir column (S4S) to support an axial compressive load of 45,000 lb. The unbraced length of the column is 14 ft. Assume normal moisture conditions and duration of load.

Solution   For Douglas fir, from Appendix F,

$$s_c = 1050 \text{ psi} \quad \text{and} \quad E = 1,700,000 \text{ psi}$$

From Table 18–1, the effective length factor is 1.0. Assuming $s_{a(all)} = s_c$, one can estimate an approximate required area:

$$\text{Required } A = \frac{P}{s_{a(all)}} = \frac{45,000}{1050} = 42.9 \text{ in.}^2$$

Since the assumed allowable stress is likely to be high, the required area quite possibly will be greater than 42.9 in.^{2}. Therefore, we will try a column with somewhat more area–a nominal 8 in. by 8 in. (S4S) column with an area of 56.25 in.^{2}:

$$\frac{L}{d} = \frac{1.0(14)(12)}{7.5} = 22.4 < 50 \quad \textbf{OK}$$

$$K = 0.671 \sqrt{\frac{E}{s_c}} = 0.671 \sqrt{\frac{1,700,000}{1050}} = 27$$

$$11 < \frac{L}{d} < 27$$

Therefore, the 8 in. by 8 in. column may be categorized as an intermediate column, and the allowable compressive stress can be calculated from Eq. (18–15):

$$s_{a(all)} = s_c \left[1 - \frac{1}{3} \left(\frac{L/d}{K} \right)^4 \right]$$

$$= 1050 \left[1 - \frac{1}{3} \left(\frac{22.4}{27} \right)^4 \right]$$

$$= 1050(0.842) = 884 \text{ psi}$$

The allowable axial compressive load is then

$$P_a = s_{a(all)} A = 884(56.25) = 49,725 \text{ lb}$$

$$49,725 \text{ lb} > 45,000 \text{ lb} \quad \textbf{OK}$$

Use an 8 in by 8 in. (S4S) column.

**18–10
ECCENTRIC LOADS
ON STEEL COLUMNS**

Axially or concentrically loaded compression members, for all practical purposes, are nonexistent in actual structures or machines. All compression members are subjected to some amount of bending moment. The bending moment may be induced by an eccentric load, as shown in Fig. 18–8(a). Such a load produces both a bending effect and an axial load effect. Both must be considered. In many cases, a part of the total load is axial and a part is eccentric, as shown in Fig. 18–8(b). (Refer to Section 17–3 on eccentrically loaded members.)

The bending moment due to the eccentric load is expressed by $M = Pe$; therefore, the bending stress is calculated from

$$s_b = \frac{Mc}{I} = \frac{Pec}{I}$$

FIGURE 18–8   Column loadings.

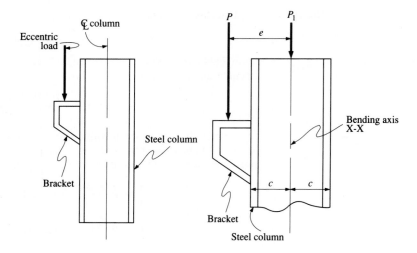

As shown in Fig. 18–8(b), the moment being considered is with respect to the X–X bending axis and I is the moment of inertia of the column cross section about that axis.

The compressive stress due to the axial load effect is developed by summation of the applied loads and is calculated from

$$s_a = \frac{P + P_1}{A}$$

where A is the cross-sectional area of the column.

The sum of the bending stress and the compressive stress does not represent a true value of the induced stresses, since the combination of the two effects generates a secondary bending moment which may be significant. The secondary moment results from a lateral deflection initially caused by the eccentric load and subsequent bending. The product of this deflection and the axial loads causes further bending and creates secondary stresses that are normally not considered in individual beam or column analysis and design.

If the secondary moment is neglected, the maximum compressive stress in an eccentrically loaded column may be approximated as is done for short compression members, where

$$s_{max} = \frac{P}{A} + \frac{Mc}{I} \tag{18–17}$$

Taking this approach, it is an easy problem to compute the approximate maximum stress due to the combined effect. However, the results are of little significance since the allowable bending stress and the allowable axial compressive stress are appreciably different, and an allowable stress for the combined effect has never been established by code.

In an effort to simplify the problem, the previous expression (Eq. (18–17)) can be rewritten as

$$s_{max} = s_a + s_{b_x} \qquad (18\text{–}18)$$

Dividing both sides by s_{max},

$$1.0 = \frac{s_a}{s_{max}} + \frac{s_{b_x}}{s_{max}} \qquad (18\text{–}19)$$

This can be further modified by substituting the applicable allowable stress (maximum values) in place of the s_{max} terms:

$$1.0 = \frac{s_a}{s_{a(all)}} + \frac{s_{b_x}}{s_{b(all)_x}} \qquad (18\text{–}20)$$

where s_a = the computed axial compressive stress
$s_{a(all)}$ = the allowable axial compressive stress
s_{b_x} = the computed maximum compressive bending stress (with respect to the X axis)
$s_{b(all)_x}$ = the allowable compressive bending stress for bending moment alone (with respect to the X axis)

and where all units for stress must be similar.

This formula (Eq (18–20)) shows that if the column is carrying only an axial load and no bending moment, the member is designed as a column based on the allowable axial compressive stress $s_{a(all)}$. If the member carries only moment and no compressive load, the member is designed as a beam based on the allowable bending stress $s_{b(all)}$. Between these two extreme cases, the equation numerically indicates the relative importance of the two effects.

Since the sum of the ratios is shown equal to unity, it is evident that if the sum should result in a value greater than one, the eccentrically loaded column would be considered unsafe. If the sum should result in a value equal to or less than one, the member would be satisfactory. Therefore, Eq. 18–20 may be slightly modified to

$$\frac{s_a}{s_{a(all)}} + \frac{s_{b_x}}{s_{b(all)_x}} \leq 1.0 \qquad (18\text{–}21)$$

This expression was adopted for the AISC Specification and is currently the basis for eccentrically loaded steel column design and analysis. However, it has been modified to reflect the secondary moment effect for columns where the ratio $s_a/s_{a(all)} > 0.15$ by introducing an amplification factor to magnify the actual bending stress s_{b_x}. Refer to footnotes 1 and 2 on pages 595 and 604 for treatment of the analysis and design of eccentrically loaded steel columns using the AISC approach when $s_a/s_{a(all)} > 0.15$.

Note that if bending occurs with respect to both the X and the Y axes, Eq. (18–21) becomes

$$\frac{s_a}{s_{a(all)}} + \frac{s_{b_x}}{s_{b(all)_x}} + \frac{s_{b_y}}{s_{b(all)_y}} \leq 1.0 \qquad (18\text{–}22)$$

This discussion is applicable to any steel member subjected to both bending and axial effects, such as an axially loaded column that supports applied lateral loads.

☐ **EXAMPLE 18–18** A W6 × 25 structural steel column of ASTM A36 steel is subjected to an eccentrically applied load as shown in Fig. 18–9. The column has an unbraced length of 15 ft and may be assumed to have pinned ends. The allowable bending stress is 22 ksi. Determine whether or not the column is adequate.

FIGURE 18–9 Eccentrically loaded column.

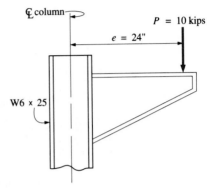

Solution We will use Eq. (18–21), and we will have to determine all the required terms for the substitutions. For the W6 × 25, from Appendix A,

$$A = 7.34 \text{ in.}^2 \qquad r_x = 2.70 \text{ in.}$$
$$S_x = 16.7 \text{ in.}^3 \qquad r_y = 1.52 \text{ in.}$$

We can first replace the eccentric load with a concentric load and a couple (a moment):

$$P = 10 \text{ kips}$$
$$M_x = Pe = 10(24) = 240 \text{ in.-kips}$$

The computed axial compressive stress is

$$s_a = \frac{P}{A} = \frac{10}{7.34} = 1.36 \text{ ksi}$$

We next calculate the slenderness ratio (using an effective length factor of 1.0 from Table 18–1). The slenderness ratio is then used to obtain the allowable axial compressive stress from Appendix J:

$$\frac{KL}{r_y} = \frac{1.0(15)(12)}{1.52} = 118$$
$$s_{a(\text{all})} = 10.85 \text{ ksi}$$

The maximum computed bending stress is

$$s_{b_x} = \frac{M_x c}{I_x} = \frac{M_x}{S_x} = \frac{240}{16.7} = 14.4 \text{ ksi}$$

and the allowable bending stress $s_{b(all)_x}$, which is given, is 22 ksi. We can now check to ensure that Eq. (18–21) is applicable:

$$\frac{s_a}{s_{a(all)}} = \frac{1.36}{10.85} = 0.13 < 0.15 \qquad \textbf{OK}$$

Applying Eq. (18–21),

$$\frac{s_a}{s_{a(all)}} + \frac{s_{b_x}}{s_{b(all)_x}} \le 1.0$$

$$\frac{1.36}{10.85} + \frac{14.4}{22} = 0.78 < 1.0 \qquad \textbf{OK}$$

Therefore, the column is adequate.

18–11
SI SYSTEM EXAMPLES

☐ **EXAMPLE 18–19**

A circular hollow aluminum tube is used as an axially loaded pin-connected column. The tube is 3 m long, has an outside diameter of 50 mm and an inside diameter of 40 mm. The modulus of elasticity E is 70×10^3 MPa and the proportional limit of the material is 220 MPa. (a) Calculate the Euler buckling load. (b) Calculate the mass (kg) that the column can carry if a factor of safety of 2.0 is used.

Solution

From Table 8–1, for properties of areas,

$$A = 0.7854(d^2 - d_i^2) = 0.7854(50^2 - 40^2) = 706.9 \text{ mm}^2$$

$$r = \frac{\sqrt{d^2 + d_i^2}}{4} = \frac{\sqrt{50^2 + 40^2}}{4} = 16.0 \text{ mm}$$

(a) Using Eq. (18–2), the Euler critical stress is

$$s_a = \frac{\pi^2 E}{(L/r)^2} = \frac{\pi^2 (70\ 000 \text{ MPa})}{(3000/16)^2}$$

$$= 19.7 \text{ MPa} < 220 \text{ MPa} \qquad \textbf{OK}$$

The critical load can then be calculated from

$$P_e = s_e A = (19.7 \text{ MPa})(706.9 \text{ mm}^2) = 13\ 926 \text{ N}$$

(Recall that 1 MPa = 1 N/mm².)

(b) Calculating the mass m that the column can carry (at buckling),

$$W = mg$$

$$m = \frac{W}{g} = \frac{13\ 926 \text{ N}}{9.81 \text{ m/s}^2} = 1420 \text{ kg}$$

(Recall that $1 \text{ N} = 1 \text{ kg} \cdot \text{m/s}^2$.) Considering the factor of safety,

$$\text{Mass} = \frac{1420}{\text{F.S.}} = \frac{1420 \text{ kg}}{2.0} = 710 \text{ kg}$$

☐ **EXAMPLE 18–20**   Calculate the allowable axial compressive load for a column having the built-up structural steel section shown in Fig. 18–10. The steel is ASTM A36. The unbraced length is 6 m. Assume that the ends of the column are pin connected. Use the AISC approach.

Solution   From Appendix D, for a single angle,

$$A = 5.45 \times 10^{-3} \text{ m}^2$$
$$I_x = I_y = 11.7 \times 10^{-6} \text{ m}^4$$
$$x = y = 45.2 \text{ mm}$$

Due to symmetry in two directions, the moments of inertia with respect to the X–X and Y–Y axes will be equal. Using the transfer formula (Eq. (8–4)), we will calculate the moment of inertia with respect to the X–X axis:

$$I_x = \Sigma(I_o + ad^2)$$
$$= 4[(11.7 \times 10^{-6} \text{ m}^4) + (5.45 \times 10^{-3} \text{ m}^2)(0.2048 \text{ m})^2]$$
$$= 961.2 \times 10^{-6} \text{ m}^4$$

The total area is

$$A = 4(5.45 \times 10^{-3} \text{ m}^2) = 21.8 \times 10^{-3} \text{ m}^2$$

Calculating the radius of gyration,

$$r_x = r_y = \sqrt{\frac{I_x}{A}} = \sqrt{\frac{961.2 \times 10^{-6} \text{ m}^4}{21.8 \times 10^{-3} \text{ m}^2}} = 0.210 \text{ m}$$

Calculating the slenderness ratio,

$$\frac{KL}{r_x} = \frac{1.0(6 \text{ m})}{0.210 \text{ m}} = 28.6$$

FIGURE 18–10   Built-up column.

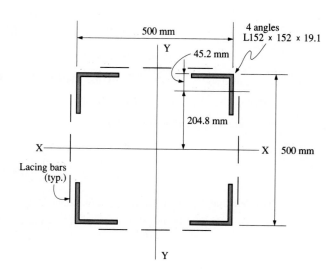

From Appendix J, the allowable axial stress is 20.05 ksi. Converting to the SI system,

$$s_{a(all)} = 20.05(6.895) = 138.2 \text{ MPa} = 138.2 \times 10^6 \text{ Pa}$$

The allowable axial compressive load is then

$$\begin{aligned} P = s_{a(all)}A &= (138.2 \times 10^6 \text{ Pa})(21.8 \times 10^{-3} \text{ m}^2) \\ &= 3010 \times 10^3 \text{ N} \\ &= 3.01 \times 10^6 \text{ N} \\ &= 3.01 \text{ MN} \end{aligned}$$

□ **EXAMPLE 18–21**   Calculate the allowable axial compressive load for a 200 × 200 (S4S) Douglas fir column having an unbraced length of 7 m. Assume that the column is pin connected at both ends. Assume normal moisture conditions and duration of load.

Solution   For Douglas fir, from Appendix F,

$$s_c = 7.24 \text{ MPa} \quad \text{and} \quad E = 12 \times 10^3 \text{ MPa}$$

From Table 18–1, the effective length factor K_e is 1.0; thus,

$$\frac{L}{d} = \frac{1.0(7 \text{ m})(1000 \text{ mm/m})}{191 \text{ mm}} = 36.6 < 50 \quad \text{OK}$$

$$K = 0.671\sqrt{\frac{E}{s_c}} = 0.671\sqrt{\frac{12 \times 10^3 \text{ MPa}}{7.24 \text{ MPa}}} = 27.3$$

$$\frac{L}{d} > K$$

Therefore, the column may be categorized as a slender column with an allowable compressive stress of

$$s_{a(all)} = \frac{0.30E}{(L/d)^2} = \frac{0.30(12 \times 10^3 \text{ MPa})}{(36.6)^2} = 2.69 \text{ MPa}$$
$$= 2.69 \times 10^6 \text{ Pa}$$

The allowable axial load (using an area of 36.5×10^{-3} m² from Appendix E) is then

$$\begin{aligned} P_a = s_{a(all)}A &= (2.69 \times 10^6 \text{ Pa})(36.5 \times 10^{-3} \text{ m}^2) \\ &= 98.2 \times 10^3 \text{ N} \\ &= 98.2 \text{ kN} \end{aligned}$$

SUMMARY—BY SECTION NUMBER

18–1   Columns are usually categorized as short, intermediate, or long (slender). Most real-world columns are in the intermediate category where the failure mode is a combination of buckling of the member and yielding (or crushing) of the material. Buckling is an elastic bending behavior due to an axial compressive load.

18–2   The classic Euler formula for obtaining buckling load or (critical load) for a theoretically perfect (ideal) column is expressed as

$$P_e = \frac{\pi^2 EI}{L^2} \qquad \text{(18–1)}$$

It can also be expressed in terms of a critical stress:

$$s_e = \frac{\pi^2 E}{(L/r)^2} \qquad \text{(18–2)}$$

L/r is called the *slenderness ratio*. The Euler formula is valid only if the resulting critical stress is less than the proportional limit of the material.

18–3   The effective length (KL) of a column is a function of the unbraced length of the column and its end conditions. Values for the effective length factor K are given in Table 18–1.

18–4   The Euler critical load formula is a theoretical equation, not a practical design formula. It can be used as a formula for allowable axial compressive load if an effective length factor and a factor of safety are introduced. The Euler critical load formula then becomes

$$P_a = \frac{\pi^2 EI}{(KL)^2 (\text{F.S.})} \qquad \text{(18–5)}$$

18–5   through **18–7** In the field of steel-frame building design, the AISC recommends the use of Eqs. (18–7) through (18–10) for the determination of allowable compressive stress. The analysis and design of steel columns using the AISC recommendations for allowable stress are simplified through the use of the allowable stress tables furnished in Appendix J.

18–8   The analysis and design procedures for machine parts subjected to axial compressive loads are similar to those outlined for AISC column design. In this text, the formulas used for these procedures are a modified Euler expression for slender columns and the J.B. Johnson formula for other columns.

18–9   The analysis and design procedure for axially loaded timber columns is similar to that used for steel columns except that the allowable stress formulas are those recommended in the National Design Specification of the NFPA (see Eqs. (18–13) through (18–16)).

18–10   Eccentrically loaded columns are those in which the line of action of the applied load is parallel to, but does not coincide with, the longitudinal centroidal axis of the column. Both a bending stress and a direct axial stress are developed. The use of an allowable stress for the combined effect is not practical and has never been established by code. For purposes of design and analysis of eccentrically loaded steel columns in this text, a modified AISC approach is used (see Eqs. (18–17) through (18–22)).

PROBLEMS

In the following problems, unless noted otherwise, assume that the modulus of elasticity of steel is 30,000,000 psi.

Section 18–2 Ideal Columns

1. Calculate the Euler buckling load for an axially loaded pin-connected W14 × 22 structural steel wide-flange shape of ASTM A36 steel. The column length is 12 ft. Assume a proportional limit of 34 ksi.

2. Calculate the Euler buckling load for a pin-connected 8 in. diameter standard-weight steel pipe column for the following lengths: (a) 50 ft, (b) 35 ft, (c) 20 ft, and (d) 15 ft. The proportional limit is 34,000 psi.

3. A pin-connected axially loaded compression member is made of an aluminum alloy. The proportional limit of the material is 32,000 psi and the modulus of elasticity is 10,000,000 psi. Calculate the lowest value of slenderness ratio for which Euler's formula is applicable.

4. A pin-connected axially loaded steel bar is used as a compression member. The bar has a rectangular cross section 1 in. by 2 in. The proportional limit is 30,000 psi. Calculate the minimum length L for Euler's formula to be applicable. Additionally, calculate the critical stress and the critical load if the length of the bar is 5 ft.

Section 18–4 Allowable Axial Compressive Loads

5. A W12 × 22 structural steel wide-flange section of ASTM A36 steel is used as an axially loaded column. Find the allowable axial compressive load using the Euler formula and a factor of safety of 2.0. The column is pin connected at both ends and is 15 ft long. The material has a proportional limit of 34 ksi.

6. Steel columns of ASTM A441 steel are to have end conditions as shown in Fig. 18–3(a–d). Calculate, for each case, the minimum slenderness ratio for the application of the Euler formula. The proportional limit of the material is 40 ksi.

7. Calculate the allowable axial compressive load for a pin-connected hollow tube of aluminum using Euler's formula and a factor of safety of 2.0. The tube is rectangular, 1 in. by 2 in. in cross section, with a wall thickness of 0.1 in. and a length of 48 in. The proportional limit of the material is 30,000 psi.

8. Use the Euler formula and a factor of safety of 2.5 to design a W14 structural steel wide-flange column to support an axial load of 350 kips. The length of the column is 34 ft and its ends are pin connected. The steel is ASTM A36 with a proportional limit of 34 ksi.

Section 18–6 Analysis of Axially Loaded Steel Columns (AISC)

For Problems 9–13, use the AISC column formulas and/or Appendix J. The yield stress of the steel is 36 ksi (ASTM A36 or A501, as appropriate).

9. Calculate the allowable axial compressive load for a W12 × 136 structural steel column having an unbraced length of 16 ft. The ends are pin connected.

10. Calculate the allowable axial compressive load for the column of Problem 9 (a) if the ends are fixed and (b) if they are fixed/pinned.

11. A W10 × 68 structural steel column is to carry an axial load of 300 kips. The unbraced length is 20 ft. Determine whether or not the column is adequate (a) if the ends are pinned and (b) if the ends are fixed/pinned.

12. Find the allowable axial compressive load for a 10 in. diameter standard-weight steel pipe column having an unbraced length of 20 ft. The ends are pin connected.

13. Calculate the allowable axial compressive load for a W12 × 50 structural steel column having a 12 in. by $\frac{1}{2}$ in. plate bolted to each flange. The unbraced length is 30 ft. The ends are pin connected.

Section 18–7 Design of Axially Loaded Steel Columns (AISC)

For Problems 14–18, use the AISC column formulas and/or Appendix J. The yield stress of the steel is 36 ksi (ASTM A36 or A501, as appropriate).

14. Select the lightest standard-weight steel pipe section to support an axial compressive load of 25,000 lb. The column is pin connected at each end and has an unbraced length of 17 ft.

15. Select the lightest standard-weight steel pipe section to support an axial compressive load of 41,000 lb. The column is fixed at one end and pin connected at the other end. The unbraced length is 14 ft.

16. Select the lightest W shape for a column subjected to an axial compressive load of 360 kips. The unbraced length of the column is 24 ft and the ends are pin connected.

17. Select the lightest W shape for a column subjected to an axial compressive load of 100 kips. The unbraced length of the column is 20 ft and the ends are pin connected.

18. Select the lightest W shape for a column subjected to an axial compressive load of 600 kips. The unbraced length of the column is 15 ft and the ends are fixed/pinned.

Section 18–8   Analysis and Design of Axially Loaded Steel Machine Parts

For Problems 19–22, use the Euler-Johnson formulas (Eqs. [18–12] and [18–13]) and a modulus of elasticity of 30,000,000 psi.

19. Calculate the allowable axial compressive load for a bar of rectangular cross section that measures 1 in. by 2 in. and is 20 in. long. The member is made of AISI 1040 hot-rolled steel. Use a factor of safety of 3.5 and assume the member is pin connected.

20. Compute the required diameter of a piston rod of AISI 1040 hot-rolled steel with a length of 20 in. The rod is subjected to an axial compressive load of 6000 lb. Use a factor of safety of 2.5. Assume the rod is pin connected.

21. Compute the required diameter of an air cylinder piston rod of AISI 1040 hot-rolled steel. The rod has a length of 54 in. and is subjected to an axial compressive load of 1900 lb. Assume pinned ends. Use a factor of safety of 3.5.

22. Compute the required diameter of a steel rod of AISI 1050 cold-drawn steel with a yield stress of 85,000 psi and a length of 8 ft. The rod is subjected to an axial compressive load of 18,000 lb. Assume pinned ends. Use a factor of safety of 3.0.

Section 18–9   Analysis and Design of Axially Loaded Timber Columns

For Problems 23–27 assume normal moisture conditions and duration of loading. Refer to Appendix F for values of E and s_c.

23. Find the allowable axial compressive load for an 8 in. by 8 in. (S4S) Douglas fir column. The ends are pin connected. Use an unbraced length of (a) 10 ft and (b) 17 ft.

24. Solve Problem 23 if the column is a 10 in. by 10 in. (S4S) eastern white pine column with fixed/pinned ends.

25. Design a square pin-connected Douglas fir column (S4S) to support an axial compressive load of 90,000 lb. The unbraced length of the column is 18 ft.

26. Design a square or rectangular pin-connected southern pine column (S4S). The unbraced length of the column is 16 ft. The column is to support an axial compressive load of (a) 48,000 lb and (b) 60,000 lb.

27. Solve Problem 26 if the end conditions are fixed/pinned.

Section 18–10   Eccentric Loads on Steel Columns

28. A W12 × 58 structural steel column of ASTM A36 steel supports a vertical load of 40 kips at an eccentricity of 12 in. with respect to the strong (X–X) axis. The column has an unbraced length of 12 ft and is assumed to be pin connected. The allowable bending stress is 22 ksi. Determine whether or not the column is adequate.

29. A W8 × 24 structural steel column of ASTM A36 steel supports a vertical load of 14 kips at an eccentricity of 15 in. with respect to the strong (X–X) axis. The column has an unbraced length of 12 ft and is assumed to be pin connected. The allowable bending stress is 22 ksi. Determine whether or not the column is adequate.

SI System Problems

For Problems 30–34, unless noted otherwise, for steel members use $E = 207 \times 10^3$ MPa, a proportional limit of 234 MPa, and a yield stress of 250 MPa.

30. Calculate the Euler buckling load for a pin connected axially loaded steel rod having a diameter of 25 mm. The length of the rod is (a) 1 m, (b) 2 m, and (c) 4 m.

31. Rework Problem 30 assuming that the material is aluminum with a modulus of elasticity of 70×10^3 MPa and a proportional limit of 220 MPa.

32. Compute the allowable axial compressive load for a W250 × 0.66 structural steel wide-flange section of ASTM A36 steel. Use the AISC column formulas. End conditions and lengths are as follows: (a) pinned, 5 m; (b) pinned, 10 m; (c) pinned/fixed, 10 m; (d) pinned, $L_x = 10$ m and $L_y = 5$ m.

33. A 19 mm diameter steel rod is 350 mm in length and is used as a pin-connected compression member. The rod is AISI 1040 hot-rolled steel. Using the Euler-Johnson formulas with a factor of safety of 2.5, calculate the allowable axial compressive load that the rod can carry.

34. Design a pin-connected southern pine square column (S4S), using the NFPA column formulas. The column is to support an axial compressive load of 450 kN. The unbraced length is 6 m.

Computer Problems

For the following computer problems, any appropriate programming language may be used. Input prompts should fully explain what is required of the user (the program should be "user friendly"). The resulting output should be well labeled and self-explanatory.

35. Write a program that will calculate the Euler buckling load for a given column and check whether the Euler formula is applicable. User input is to be area and least radius of gyration for the cross section, the proportional limit and modulus of elasticity for the material, and the length of the column. Assume pinned ends.

36. Write a program that will calculate the allowable axial compressive load for an S4S square timber column (4 in. by 4 in. minimum) for lengths ranging from 0 to the maximum allowed (in 2 ft increments). Ends are pinned. User input is to be the nominal column size and the values of s_c and E. Output should consist of allowable axial compressive load vs. length in a tabular format.

Supplemental Problems

For Problems 37–57, unless noted otherwise, for steel members use $E = 30,000$ ksi, a proportional limit of 34 ksi, and a yield stress of 36 ksi.

37. Calculate the Euler buckling load for an axially loaded, pin-connected aluminum alloy rod having a diameter of $1\frac{1}{2}$ in. Use a modulus of elasticity of 10,000,000 psi and a proportional limit of 32,000 psi. Calculate the buckling load for the following lengths: (a) 24 in., (b) 60 in., and (c) 96 in.

38. Calculate the Euler buckling load for an axially loaded, pin-connected steel rod $\frac{1}{2}$ in. in diameter and 5 ft long.

39. A W12 × 40 structural steel shape of ASTM A36 steel is used as an axially loaded column. Calculate the minimum length of the column for which Euler's formula may be used. The column is pin connected at each end.

40. For steel columns with end conditions as shown in Fig. 18–3(a–d), determine the lower limit of KL/r for the application of Euler's formula.

41. A built-up steel column is made by welding a 16 in. by 1 in. plate to each flange of a W14 × 48 wide-flange

section. The column is 22 ft long and the ends are pin connected. Calculate the maximum slenderness ratio.

42. A 2 in. diameter standard-weight steel pipe is used as an axially loaded column. Calculate the allowable axial compressive load using the Euler formula and a factor of safety of 4. The pipe column is 84 in. long and is fixed at one end and pin connected at the other end.

43. A structural steel column is 30 ft long and must support an axial compressive load of 20 kips. Using the Euler formula and a factor of safety of 2.0, select the lightest wide-flange section. Assume that the column is pin connected at each end. Check the applicability of the Euler formula.

44. Using the AISC column approach, compute the allowable axial compressive load for a W10 × 54 structural steel wide-flange section of ASTM A36 steel. The yield stress is 36 ksi. End conditions and length are as follows: (a) pinned, 16 ft; (b) pinned, 30 ft; (c) pinned/fixed, 30 ft; (d) pinned, $L_x = 30$ ft and $L_y = 16$ ft.

45. Using the AISC column formulas, compute the allowable axial compressive load for a W14 × 176. The ends are pin connected and the length is 20 ft. The column is of ASTM A441 steel (46 ksi yield stress). (**Hint:** Appendix J is not applicable.)

46. Using the AISC column approach, compute the allowable axial compressive load for the pin-connected built-up ASTM A36 steel column shown in Fig. 18–11. The length of the column is 19 ft.

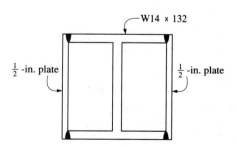

FIGURE 18–11   Problem 46.

47. Using the AISC column approach, select the lightest W shape for a column subjected to an axial compressive load of 280 kips. The unbraced length of the column is 18 ft and the ends are pinned. The steel is ASTM A501.

48. Select the lightest extra-strong steel pipe section to support an axial compressive load of 90 kips. The column is pin connected and has an unbraced length of 16 ft. Use the AISC column approach. The steel is ASTM A501.

49. Compute the required diameter of a steel push-rod subjected to an axial compressive load of 10 kips. The rod is to be made of AISI 1020 cold-drawn steel (yield stress = 50 ksi). The length is 24 in. and the ends are pinned. Use the Euler-Johnson formulas with a factor of safety of 3.0.

50. Compute the allowable axial compressive load for a $\frac{3}{4}$ in. diameter steel linkage rod that is 14 in. long and pin connected. The rod is made of AISI 1020 hot-rolled steel. Use the Euler-Johnson formulas with a factor of safety of 2.0.

51. A pin-connected linkage bar is 16 in. long and subjected to an axial compressive load of 4600 lb. The bar is made of a high-strength steel with a yield stress of 110,000 psi. Compute the required dimensions of the bar if its cross section is to be rectangular with the width being twice the depth. Use the Euler-Johnson formulas with a factor of safety of 3.0.

52. Using the NFPA column formulas, calculate the allowable axial compressive load for a 10 in. by 12 in. (S4S)

Douglas fir column. The ends are fixed/pinned and the length is 16 ft.

53. Using the NFPA column formulas, calculate the allowable axial compressive load for a round southern pine column with a 12 in. diameter. The ends are pin connected and the length is 20 ft.

54. Design a round pin-connected Douglas fir column to support an axial load of 90 kips. The unbraced length of the column is 18 ft. Use the NFPA formulas.

55. A bin weighing 100 tons when full is to be supported by four square (S4S) timber columns, each 20 ft long. Calculate the required size of the columns. The material is to be southern pine. The load is assumed to be equally distributed to each column and the column ends are pin connected.

56. A W10 × 45 structural steel column supports a vertical load of 30 kips at an eccentricity of 3 in. with respect to the weak (Y–Y) axis. The column has an unbraced length of 10 ft and is assumed to be pin connected. The allowable bending stress with respect to the weak axis is 27 ksi. Determine whether or not the member is adequate.

57. For the W12 × 58 structural steel column of Problem 28, calculate the maximum bending moment, with respect to the strong axis, that the member can safely support. The vertical load remains at 40 kips.

19 Connections

**19–1
INTRODUCTION**

In the preceding chapters we considered bending, tension, compression, and torsion and their effects on structural members. Generally, we considered the members to be isolated entities, which served our purpose in studying the behavior of the members. In the real world, however, rather than being isolated, practically all members act in combination with other members to which they are attached. Most machines and structures are actually assemblies composed of members connected together. Each member must be connected in a way that will enable it to transmit its applied loads to another part of the assembly or to a foundation.

In this chapter we will introduce the analysis and design of connections. Since connections serve primarily to transmit load, the design of connections must be based on the theoretical principles discussed in the previous chapters.

Although our emphasis will be on connections for steel, be aware that the same basic principles apply to other metals. For a treatment of connections used in timber construction, refer to publications of the American Institute of Timber Construction (AITC).[1]

Rivets were used for centuries for joining pieces of metal; however, in the recent past, because of their many disadvantages, such as high installation costs and their tendency to loosen under cyclic loads, rivets have become almost obsolete. They have been replaced by bolts and welding in almost all structural steel applications. For our discussion of bolted and welded connections, we will introduce material from the applicable design specifications of the AISC[2] and the AWS.[3]

**19–2
BOLTS AND BOLTED
CONNECTIONS (AISC)**

Several types of bolts can be used for connecting structural steel members. The two types generally used in structural applications are *unfinished bolts* and *high-strength bolts*, which are shown in Fig. 19–1. Proprietary bolts

---

[1] American Institute of Timber Construction, *Timber Construction Manual*, 3d ed. (New York: Wiley, 1985).

[2] American Institute of Steel Construction, Inc., *Specification for Structural Steel Buildings, Allowable Stress Design* (Chicago: AISC, 1989). (Also included in the AISC *Manual of Steel Construction—Allowable Stress Design*.)

[3] American Welding Society, *Welding Handbook* (Miami, FL: AWS, latest edition). (A multivolume series.)

FIGURE 19–1   Structural Bolts. Left; $\frac{7}{8}$ in. diameter, 6 in. long, A325 high-strength bolt; Center: $\frac{3}{4}$ in. diameter A325 high-strength load-indicator bolt, incorporating a splined end to facilitate installation; Right: $\frac{3}{4}$ in. diameter unfinished bolt.

incorporating ribbed shanks, end splines, and slotted ends are also available, but all of these may be considered modifications of the high-strength bolt.

Unfinished bolts are also known as *machine bolts, common bolts, ordinary bolts,* or *rough bolts.* They are designated as ASTM A307 and are threaded bolts of low-carbon steel with rough, unfinished shanks. These bolts range in diameter from $\frac{5}{8}$ in. to $1\frac{1}{2}$ in. inclusive, by $\frac{1}{8}$ in. increments. They are generally inserted in standard holes (circular holes) having a diameter $\frac{1}{16}$ in. larger than the shank of the bolt. Unfinished bolts are relatively inexpensive and can be tightened with a hand wrench. However, since permissible loads on these bolts are significantly less than those for high-strength bolts of similar size, their application is limited.

High-strength bolts are undoubtedly the most commonly used mechanical fastener for structural steel. In fact, bolting with high-strength bolts has become the primary means of connecting steel members in the shop as well as in the field. These fasteners offer advantages in speed of installation, strength, simplicity, and safety.

The two basic types of high-strength bolts are the ASTM A325 high-strength carbon steel bolt and the ASTM A490 heat-treated high-strength bolt.[4,5] The A490 bolt has the higher material strength. Both are threaded structural bolts used with heavy hex nuts. In general, the A325 and A490 bolts are available in diameters ranging from $\frac{1}{2}$ in. to $1\frac{1}{2}$ in. inclusive, by $\frac{1}{8}$ in. increments, with $\frac{3}{4}$ in. and $\frac{7}{8}$ in. diameters being the most commonly used sizes in structural applications. They are inserted in holes having a diameter $\frac{1}{16}$ in. larger than the diameter of the shank of the bolt.

The performance of a high-strength bolt depends on the proper installation of the bolt. Until recently, the installation of all high-strength bolts required that the bolts be tightened in such a way that the tension induced into the bolt be equal to, or greater than, 70% of the specified minimum tensile strength for that steel, as prescribed by the AISC Specification for Structural Joints.[6] This requirement now applies only to slip-resistant connections which, as will be discussed in Section 19–4, are designated as *slip-critical* connections. Other types of bolted connections, known as *bearing-type* connections, may be installed by tightening to a "snug tight" condition. *Snug tight* is defined as the tightness that exists when all plies in a joint are in firm contact. The tension induced into the bolt based on the snug-tight concept is somewhat less than the tension induced based on the 70% concept. This in effect permits some slippage in the joint.

When the bolt is tightened in accordance with the AISC requirements, a significant clamping force is developed between the connected parts. As a result of this clamping force, the contact surfaces in the slip-critical connection are capable of transmitting loads entirely by friction. In the bearing-type connection, frictional resistance between the connected parts is neglected and the loads are assumed to be transmitted into and out of the bolts by direct bearing.

Bolted connections serve primarily to transmit loads between members or elements of members; hence, the design of connections must be based on the same structural principles used in the design of the members. This involves creating a connection that is structurally adequate, economical, and practical.

The simplest form of bolted connection is the lap joint shown in Fig. 19–2(a). Some joints in structures are of this general type, but it is not a

---

4   American Society for Testing and Materials, *Standard Specification for High-Strength Bolts for Structural Steel Joints* (Philadelphia, PA: ASTM, latest edition).

5   American Society for Testing and Materials, *Standard Specification for Heat Treated Steel Structural Bolts* (Philadelphia, PA: ASTM, latest edition).

6   American Institute of Steel Construction, Inc., *Specification for Structural Joints Using ASTM A325 or A490 Bolts* (Chicago: AISC, 1985). (With commentary.)

FIGURE 19–2   Typical lap joint.

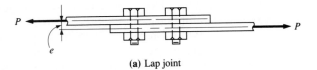

(a) Lap joint

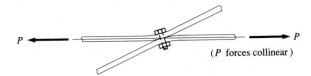

(*P* forces collinear)

(b) Bending in lap joint

commonly used detail since, as shown in Fig. 19–2(b), the connected members have a tendency to bend as a result of the applied tensile loads not being collinear. The joint will deform so that the loads *will* be collinear. A more desirable connection that eliminates this tendency of the connected members to bend is the butt joint shown in Fig. 19–3. The connected members lie in the same plane and splice plates are used. The splice plates, in effect, replace the member at the point where it is cut. A few other commonly used bolted connections are illustrated in Fig. 19–4.

FIGURE 19–3   Butt joint.

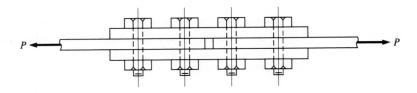

19–3
MODES OF FAILURE
OF A BOLTED
CONNECTION

A simple bolted lap joint subjected to an axial tensile load is shown in Fig. 19–5. The stresses developed in such a joint are quite complex. For analysis and design purposes, the following failure modes are usually cons dered: (a) shear in the bolts, (b) bearing of the bolts on the connected material, (c) tension in the connected members, (d) end tear-out, and (e) block shear.

(a) *Failure in shear* is a failure in the bolts. It is usually assumed that each bolt will carry its proportional share of the total load on the joint. In the lap joint shown in Fig. 19–5, the bolts have a tendency to shear off along the single contact plane of the two plates. Since the bolt is resisting the tendency of the plates to slide past one another along the contact surface and is being sheared on a single plane, the bolt is said to be in *single shear*. It is a single cross-sectional area in each bolt that resists the applied load on that bolt.

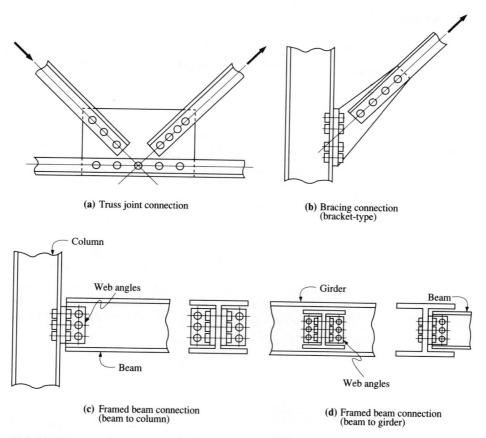

(a) Truss joint connection

(b) Bracing connection
(bracket-type)

(c) Framed beam connection
(beam to column)

(d) Framed beam connection
(beam to girder)

FIGURE 19–4   Common types of bolted connections.

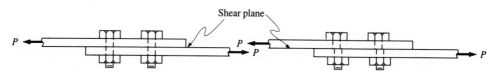

FIGURE 19–5   Lap joint—bolts in single shear.

In a butt joint, such as that shown in Fig. 19–6, there are two contact planes. Therefore, each bolt has a tendency to shear off along two contact planes and is said to be in *double shear*. Hence, two cross-sectional areas in each bolt resist the applied load on that bolt.

(b) *Failure in bearing* is a failure in the connected material. Bolts may be adequate in shear, but the connection may fail if the material joined is not capable of transmitting the load into the bolts. This is a crushing or

FIGURE 19–6 Butt joint—bolts in double shear.

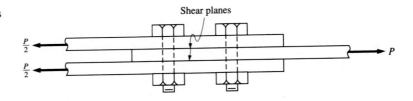

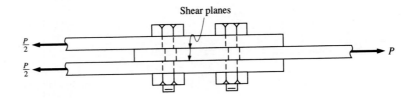

compressive-type behavior whereby the bolts and connected plates bear against each other, as shown in Fig. 19–7. The bolt may crush the material of the plate against which it bears, resulting in an elongation of the hole and an associated deformation of the member. This is the usual case. In addition, the bolt, itself, may be deformed by the plate acting on it.

FIGURE 19–7 Bearing pressures.

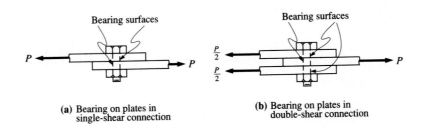

(a) Bearing on plates in single-shear connection

(b) Bearing on plates in double-shear connection

(c) *Failure in tension* is a failure in the connected material, *not* a tension failure of the bolts. The connected members of a bolted connection, such as the lap joint shown in Fig. 19–8, may fail in tension in the plane of the bolt holes. This is the plane having the least resisting area—a likely area for rupture to occur. This area may be termed the *net area*. A tensile failure could also occur on a plane where no holes exist; however, this type of failure is one of excessive deformation due to yielding as opposed to one in which rupture occurs. Both tensile failure modes are recognized by the AISC Specification and will be discussed further in Section 19–4.

(d) *End tear-out* is shown in Fig. 19–9. It could occur where the distance parallel to the applied load, between the edge of the plate and the bolt hole, is insufficient. This failure mode is easily avoided by providing

FIGURE 19–8 Lap joint—failure in tension.

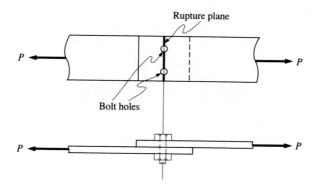

FIGURE 19–9 Lap joint—end tear-out.

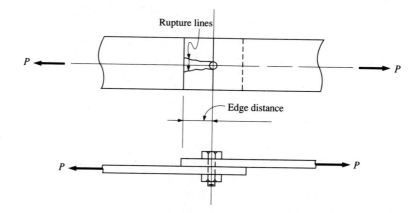

sufficient edge distance, as recommended in the AISC Specification; hence, it will not be investigated in this text.

(e) *Block shear* is another type of failure of structural connections which must be considered for both tension member connections and beam connections and is depicted in Fig. 19–10.

Block shear is a tearing failure that can occur at end connections along the perimeter of a group of bolt holes. Depending on the end connection, the failure could occur in either the member itself or in the member to which it is attached (e.g., a tension member connected to a gusset plate). It is characterized by a combination of shear failure along a plane through the bolt holes and a simultaneous tension failure along a perpendicular plane.

Block shear may be avoided by providing proper connection geometry using adequate bolt spacings and edge distances. Therefore, it will not be considered in connection analysis and design in this text. For further discussion of block shear, along with applications, refer to

FIGURE 19–10 Block shear in end connections of tension members.

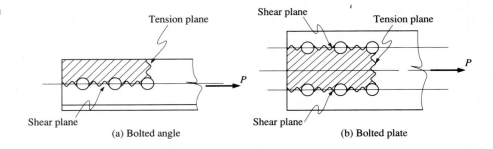

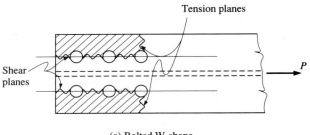

Applied Structural Steel Design[7] and to the reference in footnote 2 on page 623.

19–4 HIGH-STRENGTH BOLTED CONNECTIONS

The strength of connections composed of high-strength bolts is analyzed by considering shear, bearing, and tension separately. The allowable tensile load, or tensile capacity, for the connection will then be the smallest of the computed values.

Shear Strength

The shear strength for the connection, based on bolt shear, is the product of the cross-sectional area of the shank, its allowable shear stress, and the number of bolts in the connection. The allowable load is determined from

$$P_s = A_B s_{s(\text{all})} N \qquad (19\text{–}1)$$

where P_s = the allowable load for the connection, based on bolt shear (lb, kips) (N)

A_B = the cross-sectional area of one bolt (in.^{2}) (mm^2)

$s_{s(\text{all})}$ = the allowable shear stress in the bolt material (psi, ksi) (MPa)

N = the number of bolts contained in the connection being considered

---

7   L. Spiegel and G. F. Limbrunner, *Applied Structural Steel Design*, 2nd Edition (Englewood Cliffs, NJ, Prentice Hall, 1993).

In a lap joint, such as shown in Fig. 19–5, each bolt is in single shear. When a bolt is subjected to more than one plane of shear, such as the double shear in the butt joint of Fig. 19–6, the shear strength of each bolt must be multiplied by 2, the number of shear planes (or shear areas) in the bolt. Thus, Eq. (19–1) can be rewritten as

$$P_s = n A_B s_{s(\text{all})} N \qquad (19\text{--}2)$$

where n is the number of shear planes per bolt.

The allowable shear stress is a function of the type of high-strength bolt (A325 or A490) and the type of connection (to be discussed later in this section). The type of bolt hole in the connected material also affects the allowable shear stress in the bolt. For purposes of this text, the bolt holes will be assumed to be of standard size and of circular shape (as opposed to oversized or slotted holes). Allowable stress values taken from the AISC Specification are presented in Table 19–1.

TABLE 19–1 Allowable shear stress on steel fasteners.

Description of Fastener	Allowable Shear Stress ($s_{s(\text{all})}$) [ksi (MPa)]	
	Slip-Critical Connection	Bearing-Type Connection
A307 low-carbon bolts	—	10.0 (68.9)
A325 bolts—threads in shear plane	17.0 (117)	21.0 (145)
A325 bolts—threads excluded from shear plane	17.0 (117)	30.0 (207)
A490 bolts—threads in shear plane	21.0 (145)	28.0 (193)
A490 bolts—threads excluded from shear plane	21.0 (145)	40.0 (276)
A502 Grade 1 hot-driven rivets	—	17.5 (121)
A502 Grade 2 hot-driven rivets	—	22.0 (152)

Note: U.S. Customary System values are from the AISC Specification. SI values are converted from the U.S. Customary System values.

Bearing Strength   The bearing strength of a connection is a function of the bearing (crushing) strength of the connected material and the resisting contact area. The true distribution of the bearing pressure on the material around the perimeter of a hole is unknown. However, satisfactory results have been obtained by assuming a uniform bearing pressure acting on the projection of the contact area. This projected area, a rectangular area, is obtained as the product of the nominal diameter of the bolt and the thickness of the connected material. We obtain the strength of the connection, based on bearing on the connected material, as the product of the resisting contact area, an allowable bearing stress, and the number of bolts in the connection:

$$P_p = dts_{p(all)}N \qquad (19\text{--}3)$$

where P_p = the allowable load for the connection, based on bearing on the connected material (lb, kips) (N)

d = the nominal bolt diameter (in.) (mm)

t = the thickness of the connected part (in.) (mm)

$s_{p(all)}$ = the allowable bearing stress on the connected material (psi, ksi) (MPa)

N = the number of bolts contained in the connection being considered

The allowable bearing stress on the connected material may be taken as a maximum of $1.5 \times s_{t(ult)}$, where $s_{t(ult)}$ represents the lowest specified ultimate tensile strength of the connected material (psi, ksi, or MPa). Some values for $s_{p(all)}$ and $s_{t(ult)}$ for steels are given in Table 19–2. The values of $s_{p(all)}$ are based on the assumption that minimum bolt spacings and edge distances satisfy the requirements of the AISC Specification.

TABLE 19–2   Allowable stresses [ksi (MPa)].

| | | Allowable Bearing Stress | | | Allowable Tensile Stress | |
| | | | | | (Gross) | (Net) |
Structural steels	$s_{t(ult)}$	$s_{p(all)} = (1.5)s_{t(ult)}$	s_Y	$s_{t(all)} = (0.60)s_Y$	$s_{t(all)} = (0.50)s_{t(ult)}$
A36 carbon	58* (400)	87* (600)	36 (250)	22 (152)	29* (200)
A441 high-strength low-alloy	60 (414)	90 (620)	40 (276)	24 (165)	30 (207)
	63 (434)	95 (655)	42 (290)	25 (172)	32 (221)
	67 (462)	101 (696)	46 (317)	28 (193)	34 (234)
	70 (483)	105 (724)	50 (345)	30 (207)	35 (241)
A572 high-strength low-alloy					
Grade 42	60 (414)	90 (620)	42 (290)	25 (172)	30 (207)
Grade 50	65 (448)	98 (676)	50 (345)	30 (207)	33 (228)
Grade 60	75 (517)	113 (779)	60 (414)	36 (248)	38 (262)
Grade 65	80 (552)	120 (827)	65 (448)	39 (269)	40 (276)
A588 corrosion-resistant high-strength low-alloy	70 (483)	105 (724)	50 (345)	30 (207)	35 (241)

* Minimum values.

Note: U.S. Customary System values are from the AISC Specification. SI values are converted from the U.S. Customary System values.

Tensile Strength The tensile strength, or the allowable tensile load, of a connection subjected to an axial tensile load is a function of a resisting area and an allowable tensile stress $s_{t(\text{all})}$. In calculating the allowable tensile load, the AISC Specification requires that two conditions be checked. These two conditions involve either a net area A_n or a gross area A_g as the resisting area. Net area and gross area are depicted in Fig. 19–11. When the gross area is used, the allowable tensile load is calculated from

$$P_g = A_g s_{t(\text{all})} = A_g(0.60)s_Y \qquad (19\text{–}4)$$

where P_g = the allowable tensile load on the gross area of the member (lb, kips) (N)

A_g = the gross cross-sectional area (in.²) (mm²)

s_Y = the yield stress of the material (psi, ksi) (MPa)

FIGURE 19–11 Lap joint—plates in tension.

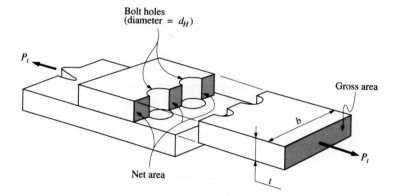

For the plate shown in Fig. 19–11, the gross area is the product of b and t. Note in Eq. (19–4) that the allowable tensile stress is $0.60 \times s_Y$. Also note that Eq. (19–4) ignores the reduction in cross-sectional area due to the bolt holes and is based on a yielding mode of failure.

When the net area is used, it can be calculated as follows:

$$A_n = A_g - A_H \qquad (19\text{–}5)$$

which can be rewritten as

$$A_n = bt - N_F d_H t$$

And the allowable tensile load is calculated from

$$P_n = A_n s_{t(\text{all})} = A_n(0.50)s_{t(\text{ult})} \qquad (19\text{–}6)$$

where A_n = the net area of the cross section (in.²) (mm²)

A_H = the projection ($d_H \times t$) of the area of the holes (in.²) (mm²)

b = the width of the plate (in.) (mm)

t = the thickness of the plate (in.) (mm)

N_F = the number of holes in the fracture plane

d_H = the diameter of the holes (in.) (mm)

P_n = the allowable tensile load on the net area of the member (lb) (kips) (N)

and the other terms are as previously defined. Note in this equation that the allowable tensile stress is equal to $0.50 \times s_{t(ult)}$. Note also that the hole diameter, for analysis and design purposes, is taken as the bolt diameter plus $\frac{1}{8}$ in. according to the AISC Specification. This provides for both the clearance around the bolt and an allowance for misshape of the holes, etc. (For selected values for ultimate tensile stress $s_{t(ult)}$, yield stress s_Y, and allowable tensile stress $s_{t(all)}$, see Table 19–2.)

The smaller of the two calculated allowable tensile loads (P_n or P_g) is then used as the controlling allowable tensile load in the determination of the strength of the connection.

As mentioned in Section 19–2, high-strength bolted connections may be of two types, either *slip-critical* or *bearing-type*. In the slip critical connection, it is assumed that the applied load is transmitted from one connected part to another entirely by the friction that results from the high tension induced in the bolt upon installation. The theory behind the slip-critical connection is that no slippage occurs between the connected parts, and that the bolts are *not* actually loaded in shear or bearing. However, for the design and analysis calculations, it is assumed that the bolts are in shear and bearing and, therefore, allowable stress values are furnished. Even though no slippage is expected to occur, the bearing and shear considerations will result in an acceptable design should the remotely possible slippage take place. (For allowable shear stresses for a slip-critical connection with standard holes and with contact surfaces free of oil, paint, lacquer, or other coatings, see Table 19–1.)

In the bearing-type connection, it is assumed that the connected parts do slip and come into bearing contact with the bolts, and that the load is transmitted by the shear resistance of the bolt as well as by bearing of the connected parts on the bolt. The frictional resistance between the connected parts is neglected and is of no concern.

Bearing-type connections can be designed with the bolt threads in the shear plane or out of the shear plane. The allowable shear stress will reflect the particular condition. As a convenience, the AISC Specification uses the following bolt designations:

Bolt Designation	Type of Application
A325-SC, A490-SC	Slip-critical connection
A325-N, A490-N	Bearing-type connection with threads in the shear plane
A325-X, A490-X	Bearing-type connection with threads excluded from the shear plane

The difference between the two types of connections lies in the allowable stresses used in the analysis or design. A slip-critical connection will generally contain more bolts than a bearing-type connection. For both types of connections, the contact surfaces must be brought into solid contact. No loose mill scale, burrs, dirt, or other foreign material is permitted.

☐ **EXAMPLE 19–1** Compute the allowable tensile load P for the single-shear lap joint shown in Fig. 19–12. The plates are ASTM A36 steel and the high-strength bolts are $\frac{3}{4}$ in. diameter A325-SC (slip-critical connection) in standard holes.

FIGURE 19–12 Single-shear lap joint.

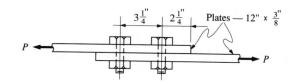

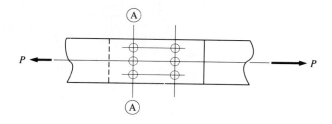

Solution First we check bolt shear. From Table 19–1, the allowable shear stress $s_{s(all)}$ is 17.0 ksi. The allowable load for the connection, based on bolt shear, is calculated from Eq. (19–2):

$$P_s = n A_B s_{s(all)} N$$
$$= 1.0(0.7854)(0.75)^2(17.0)(6)$$
$$= 45.1 \text{ kips}$$

Next we check bearing on the $\frac{3}{8}$ in. thick plate. From Table 19–2, the allowable bearing stress $s_{p(all)}$ is 87 ksi. The allowable load for the connection, based on bearing on the connected material, is calculated from Eq. (19–3):

$$P_p = dt s_{p(all)} N$$
$$= 0.75(0.375)(87)(6)$$
$$= 147 \text{ kips}$$

Lastly, we check the tensile capacity of the plates. Using the allowable tensile stresses from Table 19–2, the allowable tensile load, based on gross area, is calculated from Eq. (19–4):

$$P_g = A_g s_{t(all)} = 12(0.375)(22) = 99.0 \text{ kips}$$

And, based on net area, from Eqs. (19–5) and (19–6),

$$A_n = bt - N_F d_H t = 12(0.375) - 3(0.875)(0.375) = 3.52 \text{ in.}^2$$
$$P_n = A_n s_{t(\text{all})} = 3.52(29) = 102 \text{ kips}$$

Therefore, bolt shear governs the strength of the lap joint. The joint (or connection) has an allowable tensile load of 45.1 kips.

□ **EXAMPLE 19–2**   Compute the allowable tensile load for the double-shear butt joint shown in Fig. 19–13. The plates are ASTM A441 steel with an ultimate tensile stress $s_{t(\text{ult})}$ of 70 ksi. The high-strength bolts are $\frac{3}{4}$ in. diameter A325 in standard holes. Assume that the connection is (a) slip-critical (A325-SC) and (b) bearing-type, with threads in the shear plane (A325-N).

FIGURE 19–13   Double-shear butt joint.

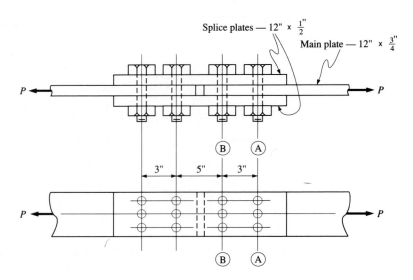

Splice plates — 12" × $\frac{1}{2}$"

Main plate — 12" × $\frac{3}{4}$"

Solution   Note that the bolts are in double shear.

(a) For A325-SC bolts, the allowable shear stress $s_{s(\text{all})}$ from Table 19–1 is 17.0 ksi. The allowable load, based on bolt shear, is calculated from Eq. (19–2):

$$P_s = n A_B s_{s(\text{all})} N$$
$$= 2(0.7854)(0.75)^2(17.0)(6)$$
$$= 90.1 \text{ kips}$$

We now check bearing on the $\frac{3}{4}$ in. main plate. (Note that bearing on the splice plates need not be checked since the total splice-plate thickness exceeds that of the main plate.) Using an allowable bearing stress of 105 ksi from Table 19–2, Eq. (19–3) yields

$$P_p = dt s_{p(\text{all})} N$$
$$= 0.75(0.75)(105)(6)$$
$$= 354 \text{ kips}$$

Next we check tension in the $\frac{3}{4}$ in. main plate on bolt line AA. This is the critical section in the main plate, since the full load is in the main plate at this section. At bolt line BB only one-half of the load is in the main plate, since the other half has been transferred (by the bolts on line AA) into the splice plates. Calculating the gross area of the plate and the area of the holes on line AA,

$$A_g = 12(\tfrac{3}{4}) = 9.0 \text{ in.}^2$$
$$A_H = (\tfrac{3}{4} + \tfrac{1}{8})(0.75)(3) = 1.97 \text{ in.}^2$$

The net area of the plate can be calculated from Eq. (19–5):

$$A_n = A_g - A_H = 9.0 - 1.97 = 7.03 \text{ in.}^2$$

Tensile capacities based on gross area and net area can now be determined using Eqs. (19–4) and (19–6). Allowable tensile stresses are obtained from Table 19–2:

$$P_g = A_g s_{t(\text{all})} = 9.0(30) = 270 \text{ kips}$$
$$P_n = A_n s_{t(\text{all})} = 7.03(35) = 246 \text{ kips}$$

We see, then, that bolt shear governs the strength of the butt joint. The allowable tensile load for the joint is 90.1 kips.

(b) For a bearing-type connection (A325-N), the allowable shear stress $s_{s(\text{all})}$ from Table 19–1 is 21.0 ksi. From Eq. (19–2),

$$
\begin{aligned}
P_s &= n A_B s_{s(\text{all})} N \\
&= 2(0.7854)(0.75)^2(21.0)(6) \\
&= 111 \text{ kips}
\end{aligned}
$$

We consider bearing and tension next; these values will be unchanged from part (a). Therefore,

$$P_p = 354 \text{ kips}$$
$$P_n = 246 \text{ kips}$$

Bolt shear again governs, and the allowable tensile load for the joint is 111 kips.

___

☐ **EXAMPLE 19–3**   A butt splice is to be designed for the plates shown in Fig. 19–14. Use $\frac{3}{4}$ in. diameter A325 high-strength bolts in standard holes. Assume a slip-critical connection (A325-SC). The plates are ASTM A36 steel and the axial tensile load is 90 kips. Compute the number of bolts required on each side of the splice and indicate the bolt arrangement by means of a sketch (AISC Specification.)

Solution   Two $\frac{1}{4}$ in. thick splice plates are selected, since the sum of the splice-plate thicknesses should be at least equal to the thickness of the plates being spliced.
Considering bolt shear first, the allowable shear stress $s_{s(\text{all})}$ from Table 19–1 is 17.0 ksi. We then solve Eq. (19–2) for the required number of bolts N. Note that the bolts are in double shear.

$$\text{Required } N = \frac{P}{n A_B s_{s(\text{all})}} = \frac{90}{2(0.7854)(0.75)^2(17.0)} = 5.99 \text{ bolts}$$

FIGURE 19–14 Double-shear butt joint.

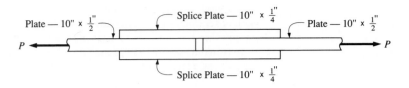

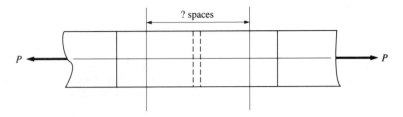

Next we consider bearing on the connected material. Since there are two $\frac{1}{4}$ in. thick splice plates and one $\frac{1}{2}$ in. thick main plate, we see that the splice plates and the main plate are equally critical and we base the design on a $\frac{1}{2}$ in. material thickness. From Table 19–2 we obtain the allowable bearing stress $s_{p(\text{all})}$ of 87 ksi. Solving Eq. (19–3) for the required number of bolts,

$$\text{Required } N = \frac{P}{dts_{p(\text{all})}} = \frac{90}{0.75(0.5)(87)} = 2.8 \text{ bolts}$$

Of the two preceding considerations, shear in the bolts is the more critical. We select a 6 bolt pattern for each side of the splice as shown in Fig. 19–15. We then check the allowable tensile load based on gross area and net area of the main plate, using allowable stresses from Table 19–2. From Eq. (19–4),

$$P_g = A_g s_{t(\text{all})} = 10(0.5)(22) = 110 \text{ kips} > 90 \text{ kips} \quad \textbf{OK}$$

From Eqs. (19–5) and (19–6),

$$A_n = bt - N_F d_H t$$
$$= 10(0.5) - 3(0.875)(0.5) = 3.69 \text{ in.}^2$$
$$P_n = 3.69(29) = 107 \text{ kips} > 90 \text{ kips} \quad \textbf{OK}$$

Therefore, the bolt arrangement shown in Fig. 19–15 is satisfactory.

FIGURE 19–15 Bolt pattern.

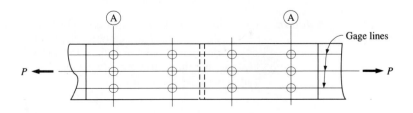

19–5
CONNECTIONS USING RIVETS AND COMMON BOLTS (AISC)

The modes of failure discussed in Section 19–3 are also applicable to structural steel connections in which rivets and common (or unfinished) bolts are used.

In connections using these two types of fasteners, it is assumed that the connected materials are not clamped together tightly enough to cause the development of any frictional resistance capable of transmitting a load. Therefore, structural steel connections using unfinished bolts (ASTM A307) or rivets (hot-driven ASTM A502) are designed and analyzed as bearing-type connections.

As was the case with the bearing-type connections using high-strength bolts, one must consider shear, bearing, and tension in joint analysis and design. The allowable shear stresses for rivets and unfinished bolts are indicated in Table 19–1. Note that the unfinished bolt allowable stresses are appreciably less than those permitted for rivets or for high-strength bolts. The allowable stresses for bearing and tension are functions of the connected materials; therefore, they are the same as for connections using high-strength bolts.

☐ **EXAMPLE 19–4**

Rework Example 19–2 assuming that the fasteners are $\frac{3}{4}$ in. diameter ASTM A307 unfinished bolts.

Solution

For unfinished bolts, the allowable shear stress $s_{s(all)}$ from Table 19–1 is 10 ksi. The bolts are in double shear. From Eq. (19–2),

$$P_s = nA_B s_{s(all)} N$$
$$= 2(0.7854)(0.75)^2(10.0)(6)$$
$$= 53.0 \text{ kips}$$

Considering bearing and tension next, we note that these values will be identical with those determined in Example 19–2. Therefore,

$$P_p = 354 \text{ kips}$$
$$P_n = 246 \text{ kips}$$

Bolt shear again governs, and the allowable tensile load for the joint is 53.0 kips.

The analysis and design of riveted connections, as well as connections using unfinished bolts, involve procedures similar to those used in the analysis and design of the bearing-type high-strength bolted connections.

19–6
INTRODUCTION TO WELDING

The function of welds in connections is similar to that of bolts and rivets. The weld is the fastening medium that provides the path by which the applied loads are safely and predictably transferred from one element to another.

Welding may be defined as a process in which two or more pieces of metal are fused together by heat to form a joint. In most welding processes, this is accomplished through the addition of filler metal from an electrode. The surfaces to be connected or fused are subjected to heat by means of an electric arc, a gas flame, or a combination of electric resistance and pressure. The welding processes used in the design of steel structures, pressure vessels, boilers, tanks, and machines are almost exclusively all of the electric-arc type.

In the arc-welding process, an electric arc is formed between the end of a metal electrode and the steel components to be welded. The heat from the arc (approximately 6500°F) raises the temperature of the electrode and the base metal (the material to be connected) in the immediate area of the arc to the point where they melt together to form a localized molten pool of steel on the surface of the member. The electrode, with its electric arc, is moved along the joint to be welded, with the molten pool solidifying rapidly as the temperature of the pool behind it drops below the melting point.

The chemical and mechanical properties of the added weld metal from the electrode should be as similar as possible to the properties of the base metal. A variety of electrodes are available to satisfy the requirements of the various steels. Electrodes are designated by a numbering system with an E prefix (indicating electrode). In structural welding, the prefix is followed by two or three digits (e.g., E60, E70, E100). The number denotes the minimum ultimate tensile strength (in ksi) of the weld metal in the electrode. Thus, an E70 electrode would have an ultimate tensile strength of 70,000 psi. Two other digits are added to the electrode designation to denote special factors relative to the welding process other than strength properties. Proper electrode-base metal combinations have been established by the AISC. These result in weld metal that is always stronger than the base metal.

The two types of load-carrying welds most commonly used are the *fillet weld* and the *groove weld*. Other load-carrying welds are the *plug weld* and the *slot weld*, which generally are used only under special circumstances (e.g., where space is limited or where the fillet weld lacks adequate load-carrying capacity). The four weld types are shown in Fig. 19–16. Our discussion will be limited to the commonly used load-carrying fillet and groove welds.

FIGURE 19–16   Weld types.

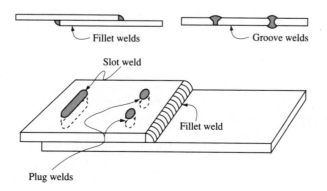

In structural applications, the fillet weld finds most frequent use, whereas in pressure vessels, boilers, tanks, and ship building applications, the groove weld predominates. Groove welds generally require extensive edge preparation as well as special fabrication. As a result, they are more costly.

In any given joint, the adjoining members to be connected may be oriented with respect to each other in several ways. The joints are usually categorized as *butt, tee, lap, corner,* and *edge,* as shown in Fig. 19–17.

FIGURE 19–17 Joint types.

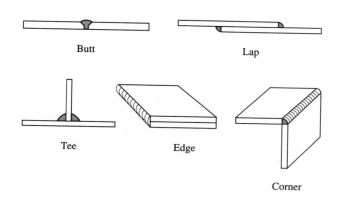

Butt

Lap

Tee

Edge

Corner

**19–7
STRENGTH AND
BEHAVIOR
OF WELDED
CONNECTIONS (AISC)**

Fillet welds are welds of theoretically triangular cross section joining two surfaces that are at approximate right angles to each other, such as in a lap joint, or a tee joint. The cross section of a typical fillet weld is a right triangle with equal legs, as shown in Fig. 19–18. The size of the weld is designated by the leg size. The *root* is the vertex of the triangle or the point at which the legs intersect. The *face* of the weld is the hypotenuse of the weld triangle and is a theoretical plane since in reality weld faces will form as either convex or concave, as shown in Fig. 19–19. The convex fillet weld is the more desirable of the two since it has less of a tendency to crack as a result of shrinkage while cooling. The distance from the theoretical face of a weld to the root is called the *throat*.

FIGURE 19–18 Typical fillet weld.

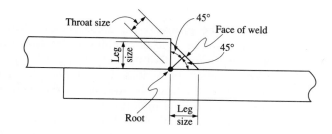

Throat size

45°

Face of weld

45°

Leg size

Root

Leg size

FIGURE 19–19   Fillet welds.

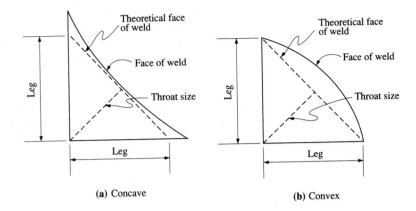

(a) Concave   (b) Convex

Groove welds are welds made in a groove between adjacent ends or surfaces of two parts to be joined, as in a butt, corner, or tee joint. (A tee joint may be fillet welded or groove welded.) The edge preparation for a welded butt joint can be made in any one of a wide variety of configurations, a few of which are shown in Fig. 19–20. With the exception of the square-groove weld, some edge preparation is required for either one or both of the members to be connected. Because of their high cost as compared to fillet welds, groove welds should be used only when specifically required.

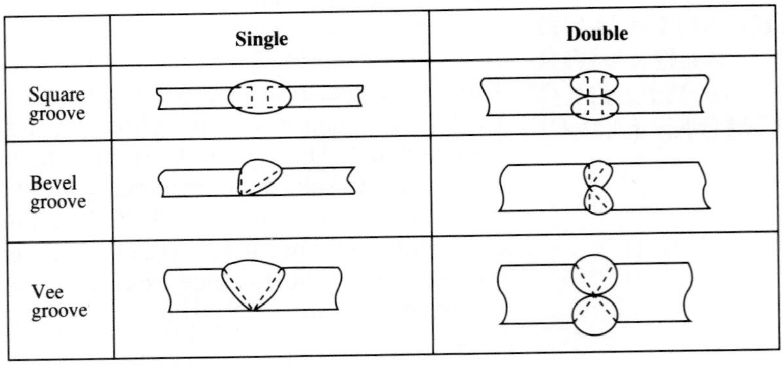

FIGURE 19–20   Groove welds.

	Single	**Double**
Square groove		
Bevel groove		
Vee groove		

In the groove-weld connections, there is no need to calculate the stresses in the weld or attempt to determine its size, since all groove welds are full-strength; that is, their strength is equal to or greater than that of the steel members being joined. Assuming the proper electrode is used with the base metals, allowable stresses in the weld may be taken as the same as those for the base materials.

With respect to the design of fillet-welded connections, it is necessary to determine the size and length of the fillets to avoid overwelding or underwelding. Tests have shown that fillet welds will fail in shear. There-fore, the strength of a fillet weld is based on the shear strength of the effec-tive throat area of the weld. If we consider a one-inch length of weld, the effective throat area is the product of the throat dimension and the one-inch length of the weld, and the strength is the product of the area and an allow-able shear stress. Hence, a value for shear strength is usually expressed in units of kips/in. or lb/in.

In a fillet with equal leg sizes, and where a cross-sectional shape of the weld is theoretically a 45° right triangle, the effective throat size is

$$\sin 45° \times \text{leg size} = 0.707 \times \text{leg size}$$

If weld metal exists outside the theoretical right triangle, this additional weld metal is considered to be reinforcement and is assumed to contribute no additional strength.

The allowable shear stress for the weld metal, as specified by the AISC, can be taken as

$$s_{s(\text{all})} = 0.3s_{t(\text{ult})} \tag{19–7}$$

where $s_{t(\text{ult})}$ is the specified minimum ultimate tensile strength of the elec-trode weld metal. Therefore, the shear strength of a fillet weld per lineal inch length of weld is

$$P_s = s_{s(\text{all})}(0.707)(\text{leg size})$$
$$= 0.3s_{t(\text{ult})}(0.707)(\text{leg size})$$

from which

$$P_s = 0.212s_{t(\text{ult})}(\text{leg size}) \tag{19–8}$$

Using Eq. (19–8), the strength per linear inch (or unit strength) for a fillet weld having a leg size of $\frac{1}{16}$ in. can be calculated. This is a hypothetical value, since, according to the AISC, the minimum allowable weld size in structural welding is $\frac{1}{8}$ in. However, this will be a convenient reference value for calculations. The unit strength of other size fillet welds can be obtained by multiplying by the number of sixteenths in the leg size. For an E70 electrode ($s_{t(\text{ult})} = 70$ ksi),

$$P = 0.212(70)(\tfrac{1}{16}) = 0.928 \text{ kips/in.}$$

This value is generally rounded off to 0.925 kips/in. Using 0.925 kips/in. as a basic value, the unit strength of the other sizes of fillet welds can be com-puted and tabulated (see Table 19–3).

For example, the unit strength of a $\frac{3}{16}$ in. fillet weld would be

$$(0.925)(3) = 2.78 \text{ kips/in.}$$

A similar approach could be used for other electrodes, such as the E60 electrode, where $s_{t(\text{ult})} = 60$ ksi.

The strength of a fillet weld depends on the direction of the applied load, which may be parallel or perpendicular to the direction of the weld. In

TABLE 19–3 Unit strengths for fillet welds (kips/in.).

Weld Size (in.)	E70 Electrode	E60 Electrode
$\frac{1}{16}$	0.925	0.795
$\frac{1}{8}$	1.85	1.59
$\frac{3}{16}$	2.78	2.39
$\frac{1}{4}$	3.70	3.18
$\frac{5}{16}$	4.63	3.98
$\frac{3}{8}$	5.55	4.77
$\frac{7}{16}$	6.48	5.57
$\frac{1}{2}$	7.40	6.36
$\frac{9}{16}$	8.33	7.16
$\frac{5}{8}$	9.25	7.95
$\frac{11}{16}$	10.18	8.75
$\frac{3}{4}$	11.10	9.54
$\frac{13}{16}$	12.03	10.34
$\frac{7}{8}$	12.95	11.13

Note: Values are based on the shielded metal-arc welding process.

both cases, the weld fails in shear, but the plane of rupture is not the same. If the load is parallel to the weld, failure will occur along the 45° plane of the throat area. If the load is perpendicular to the weld, failure will occur on a plane oriented at approximately 67.5° from one side of the triangular cross section.

Tests have shown that a fillet weld loaded in a direction perpendicular to the weld is approximately one-third stronger than one loaded in a parallel direction (see Fig. 19–21). However, the AISC Specification does not allow

FIGURE 19–21 Direction of applied load.

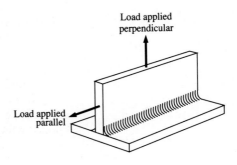

Load applied perpendicular

Load applied parallel

TABLE 19–4   Fillet weld sizes.

Thickness of Material	Minimum Weld Size*	Maximum Weld Size
$\frac{1}{8}$	$\frac{1}{8}$	$\frac{1}{8}$
$\frac{3}{16}$	$\frac{1}{8}$	$\frac{3}{16}$
$\frac{1}{4}$	$\frac{1}{8}$	$\frac{3}{16}$
$\frac{5}{16}$	$\frac{3}{16}$	$\frac{1}{4}$
$\frac{3}{8}$	$\frac{3}{16}$	$\frac{5}{16}$
$\frac{7}{16}$	$\frac{3}{16}$	$\frac{3}{8}$
$\frac{1}{2}$	$\frac{3}{16}$	$\frac{7}{16}$
$\frac{9}{16}$	$\frac{1}{4}$	$\frac{1}{2}$
$\frac{5}{8}$	$\frac{1}{4}$	$\frac{9}{16}$
$\frac{11}{16}$	$\frac{1}{4}$	$\frac{5}{8}$
$\frac{3}{4}$	$\frac{1}{4}$	$\frac{11}{16}$
$\frac{13}{16}$	$\frac{5}{16}$	$\frac{3}{4}$
$\frac{7}{8}$	$\frac{5}{16}$	$\frac{13}{16}$

*   Weld size is determined by the thicker of the two parts joined, except that the weld size need not exceed the thickness of the thinner part joined.

for this consideration in weld design. The strengths of all fillets are based on the values calculated for loads applied in a parallel direction.

In addition to the strength criteria, the AISC Specification includes design requirements with respect to minimum and maximum sizes of fillet welds. These are summarized in Table 19–4.

Additional AISC requirements for fillet-welded connections are summarized as follows:

1. The minimum effective length of a fillet weld must not be less than four times the nominal size of the weld. If a weld is intermittent, the minimum length of each weld is $1\frac{1}{2}$ in.
2. Side or end fillet welds should be returned continuously around the corners for a distance not less than twice the nominal size of the weld. (This is called an *end return*.)
3. If fillet welds parallel to the applied load are used alone (without perpendicular welds other than end returns) in end connections of flat-plate tension members, the length of each fillet weld cannot be less than the perpendicular distance between them.
4. Where lap joints are used, the minimum amount of lap should be five times the thickness of the thinner part joined, but not less than 1 in.

□ **EXAMPLE 19–5**   Calculate the allowable axial tensile load that can be applied to the connection in Fig. 19–22. The plates are ASTM A36 steel. The weld is a $\frac{7}{16}$ in. fillet weld and is made using an E70 electrode.

Solution   The total length of the weld is 20 in. From Table 19–3, the unit strength of a $\frac{7}{16}$ in. weld is 6.48 kips/in. Therefore,

$$\text{Weld capacity} = 6.48(20) = 129.6 \text{ kips}$$

From Table 19–2, the allowable axial tensile stress is 22.0 ksi. The tensile capacity of the smaller plate is then calculated from

$$P = A_g s_{t(\text{all})} = 8(0.75)(22) = 132 \text{ kips}$$

Therefore, the allowable axial tensile load is 129.6 kips.

FIGURE 19–22   Welded lap joint.

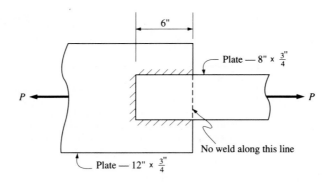

□ **EXAMPLE 19–6**   Calculate the allowable axial tensile load that can be applied to the welded connection in Fig. 19–23. The plates are ASTM A36 steel. The weld is a $\frac{5}{16}$ in. fillet weld made using an E70 electrode.

Solution   The total length of the weld is 20 in. The unit strength of a $\frac{5}{16}$ in. fillet weld, from Table 19–3, is 4.63 kips/in. Therefore,

$$\text{Weld capacity} = 4.63(20) = 92.6 \text{ kips}$$

FIGURE 19–23   Transverse-welded lap joint.

From Table 19–2, the allowable axial tensile stress is 22.0 ksi. The tensile capacity of the plate is then calculated from

$$P = A_g s_{t(all)} = 10(0.5)(22) = 110 \text{ kips}$$

Therefore, the allowable axial tensile load is 92.6 kips.

□ **EXAMPLE 19–7**   Design the fillet welds parallel to the applied load to develop the full tensile capacity of the 8 in. by $\frac{1}{2}$ in. plate in Fig. 19–24. The plates are ASTM A36 steel and the electrode is an E70. Standard end returns may be used.

FIGURE 19–24   Parallel-loaded lap joint.

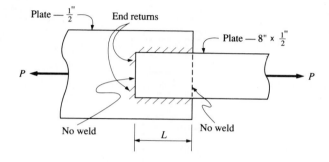

Solution   From Table 19–2, the allowable axial tensile stress is 22.0 ksi. The tensile capacity for the 8 in. by $\frac{1}{2}$ in. steel plate is then calculated from

$$P = A_g s_{t(all)} = 8(0.5)(22) = 88 \text{ kips}$$

The maximum weld size, from Table 19–4, is a $\frac{7}{16}$ in. weld. However, for economy reasons, we use a $\frac{5}{16}$ in. weld, which is the largest weld that can be made manually in a single pass. From Table 19–3, the unit strength of a $\frac{5}{16}$ in. weld is 4.63 kips/in. The length of weld required is then

$$\text{Required length} = \frac{88}{4.63} = 19 \text{ in.}$$

AISC requirements call for end returns with a minimum length of two times the leg size:

$$2 \times (\tfrac{5}{16}) = (\tfrac{10}{16})$$

Use 1 in. end returns.

The required length for each side weld is then

$$\text{Required } L = \frac{19 - 2(1)}{2} = 8.5 \text{ in.}$$

The minimum length of the side fillet welds must not be less than the perpendicular distance between them. Therefore, the minimum length required is 8 in. Since 8.5 in. > 8 in., the design is OK.

<table>
<tr><td>

19–8
SI SYSTEM
EXAMPLES

</td></tr>
</table>

☐ **EXAMPLE 19–8**

Compute the allowable tensile load P for the double-shear butt joint in Fig. 19–25. The plates are ASTM A36 steel and the fasteners are 20 mm diameter A325 high-strength bolts in standard holes. Assume that the connection is (a) slip-critical (A325-SC) and (b) bearing-type, with the threads in the shear plane (A325-N).

FIGURE 19–25 Double-shear butt joint.

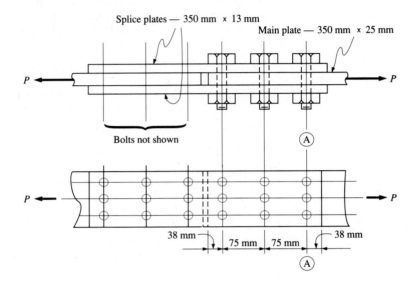

Solution   (a) For A325-SC bolts, the allowable shear stress $s_{s(\text{all})}$ from Table 19–1 is 117 MPa. The allowable load, based on bolt shear, is calculated from Eq. (19–2):

$$
\begin{aligned}
P_s &= n A_B s_{s(\text{all})} N \\
&= 2(0.7854)(20 \text{ mm})^2(117 \text{ MPa})(9) \\
&= 662\,000 \text{ N} \\
&= 662 \text{ kN}
\end{aligned}
$$

The allowable load, based on bearing, on the 25 mm main plate is calculated from Eq. (19–3). From Table 19–2, the allowable bearing stress $s_{p(\text{all})}$ is 600 MPa:

$$
\begin{aligned}
P_p &= dts_{p(\text{all})} N \\
&= (20 \text{ mm})(25 \text{ mm})(600 \text{ MPa})(9) \\
&= 2\,700\,000 \text{ N} \\
&= 2700 \text{ kN}
\end{aligned}
$$

Tension in the 25 mm main plate is considered next. Tension on both the gross area and the net area must be considered. The full load acts on bolt line AA in the main plate. The net area is calculated at that location. In the calculation of net area, 3 mm is added to the bolt diameter to determine the hole diameter for analysis purposes. For the gross area,

$$
A_g = (350)(25) = 8750 \text{ mm}^2
$$

From Eq. (19–5), the net area is

$$A_n = A_g - A_H$$
$$= 8750 - 3(20 + 3)(25) = 7025 \text{ mm}^2$$

Using allowable tensile stresses from Table 19–2, tensile capacities based on gross area and net area can be calculated. From Eq. (19–4),

$$P_g = A_g s_{t(\text{all})} = A_g(0.60 s_Y)$$
$$= (8750 \text{ mm}^2)(152 \text{ MPa})$$
$$= 1\ 330\ 000 \text{ N}$$
$$= 1330 \text{ kN}$$

From Eq. (19–6),

$$P_n = A_n s_{t(\text{all})} = A_n(0.50)(s_{t(\text{ult})})$$
$$= (7025 \text{ mm}^2)(200 \text{ MPa})$$
$$= 1\ 405\ 000 \text{ N}$$
$$= 1405 \text{ kN}$$

Therefore, bolt shear governs the strength of the butt joint. The allowable tensile load for the joint is 662 kN.

(b) For A325-N bolts, the allowable shear stress $s_{s(\text{all})}$ is obtained from Table 19–1. The allowable load based on shear is calculated from Eq. (19–2):

$$P_s = n A_B s_{s(\text{all})} N$$
$$= 2(0.7854)(20 \text{ mm})^2(145 \text{ MPa})(9)$$
$$= 820\ 000 \text{ N}$$
$$= 820 \text{ kN}$$

The allowable loads based on bearing and tension are unchanged from part (a). Therefore,

$$P_p = 2700 \text{ kN}$$
$$P_g = 1330 \text{ kN}$$

Bolt shear again governs, and the allowable tensile load for the joint is 820 kN.

SUMMARY—BY SECTION NUMBER

19–1 The connection of structural steel members or elements of members is accomplished almost exclusively through the use of bolts or welding.

19–2 The most frequently used bolts in structural steel connections are unfinished bolts (ASTM A307) and high-strength bolts (ASTM A325 or A490.)

19–3 The possible failure modes for bolted joints are (a) shear in the bolts, (b) bearing of the bolts on the connected material, (c) tension in the connected parts, (d) end tear-out and (e) block shear.

19–4 The allowable load for a high-strength bolted connection is determined by considering shear, bearing, and tension separately. The shear strength is obtained from

$$P_s = n A_B s_{s(\text{all})} N \qquad \textbf{(19–2)}$$

The bearing strength is obtained from

$$P_p = dt s_{p(all)} N \qquad (19\text{–}3)$$

The tensile strength is obtained from

$$P_g = A_g s_{t(all)} = A_g(0.60)s_Y \qquad (19\text{–}4)$$

or from

$$P_n = A_n s_{t(all)} = A_n(0.50)s_{t(ult)} \qquad (19\text{–}6)$$

Allowable, yield, and ultimate tensile stresses are listed in Tables 19–1 and 19–2. High-strength bolted connections are categorized as either *slip-critical* or *bearing-type*. Various allowable stresses apply.

19–5 Failure modes for connections using rivets (ASTM A502) and common bolts (ASTM A307) are similar to those of the high-strength bolted connections. These connections, however, are categorized only as bearing-type connections.

19–6 Welding is a process in which two or more pieces of metal are fused together by heat to form a joint. The most common welding process is electric-arc welding. Fillet welds and groove welds are the two types of load-carrying welds most commonly used.

19–7 The cross section of a typical fillet weld is a right triangle with equal leg sizes. A fillet weld will fail in shear, with its shear strength (per linear inch of weld) equal to

$$P_s = 0.212 s_{t(ult)}(\text{leg size}) \qquad (19\text{–}8)$$

This value can also be obtained from Table 19–3. Groove welds are full-strength welds requiring no strength calculations if the proper electrode is used.

PROBLEMS

Where applicable, refer to Tables 19–1 and 19–2 for allowable stresses. Assume standard holes for all bolted connections.

Section 19–4   High-Strength Bolted Connections

1. Compute the allowable tensile load for the single-shear lap joint in Fig. 19–26. The plates are ASTM A36 steel and the high-strength bolts are $\frac{7}{8}$ in. diameter A325 bolts in standard holes. Assume a bearing-type connection with threads excluded from the shear plane (A325-X).

2. Rework Problem 1 assuming a slip-critical connection (A325-SC).

3. Rework Problem 1 assuming a bearing-type connection with bolt threads in the shear plane (A325-N).

4. Compute the allowable tensile load for the double-shear butt joint in Fig. 19–27. The plates are ASTM A441 steel with $s_{t(ult)}$ of 60 ksi. The bolts are $\frac{7}{8}$ in. diameter A325 high-strength bolts in standard holes. Assume a slip-critical connection (A325-SC).

5. Rework Problem 4 assuming a bearing-type connection (a) with bolt threads in the shear plane (A325-N) and (b) with bolt threads excluded from the shear plane (A325-X).

6. Rework Problem 4 assuming that the bolts are $\frac{3}{4}$ in. diameter A490 high-strength bolts in standard holes.

7. Select the number and arrangement of $\frac{3}{4}$ in. diameter A325 high-strength bolts required to resist an axial tensile load of 150 kips for the butt joint in Fig. 19–28.

FIGURE 19–26   Problem 1.

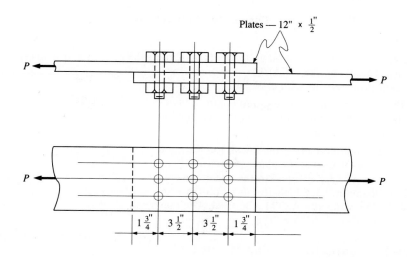

FIGURE 19–27   Problem 4.

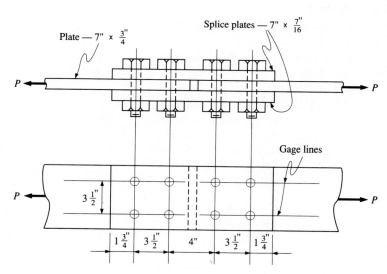

FIGURE 19–28   Problem 7.

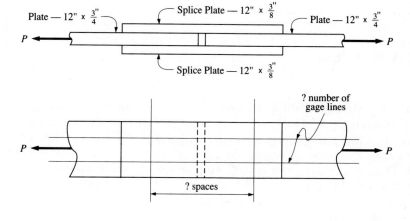

Assume a slip-critical connection (A325-SC) and standard holes. The plates are all ASTM A36 steel. Indicate the bolt arrangement with a sketch.

8. For Problem 2, determine the required width of plates so that the strength of the plates in tension is equal to that of the bolts in shear.

9. Calculate the minimum plate thickness for the joint shown in Fig. 19–26 so that the plates, which are 12 in. wide, will have a capacity equal to the shear capacity of the $\frac{7}{8}$ in. diameter A325 high-strength bolts in standard holes. The plates are ASTM A36 steel. The connection is slip-critical (A325-SC).

Section 19–5   Connections Using Rivets and Common Bolts (AISC)

10. Rework Problem 1 assuming the fasteners used are $\frac{7}{8}$ in. diameter ASTM A502 Grade 1 hot-driven rivets.

11. Rework Problem 1 assuming the fasteners used are $\frac{7}{8}$ in. diameter ASTM A307 unfinished bolts.

12. Rework Problem 4 assuming the fasteners used are $\frac{7}{8}$ in. diameter ASTM A502 Grade 2 hot-driven rivets.

13. Two plates are connected by the lap joint in Fig. 19–29. Using $\frac{3}{4}$ in. diameter ASTM A502 Grade 2 hot-driven rivets in standard holes, calculate the allowable axial tensile load P. The plates are of ASTM A36 steel.

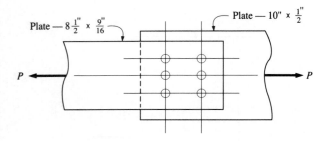

FIGURE 19–29   Problem 13.

Section 19–7   Strength and Behavior of Welded Connections (AISC)

14. Calculate the allowable tensile load for the connection in Fig. 19–30. The plates are ASTM A36 steel and the weld is a $\frac{3}{8}$ in. fillet weld, which is made using an E70 electrode.

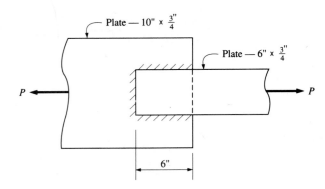

FIGURE 19–30   Problem 14.

15. In the connection in Fig. 19–31, $\frac{1}{4}$ in. side and end fillet welds are used to connect the 3 in. by 1 in. tension member to the plate. The applied load is 60,000 lb. Find the required dimension L. The steel is ASTM A36 and the electrode used is an E70.

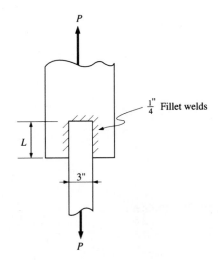

FIGURE 19–31   Problem 15.

16. Design the fillet welds parallel to the applied load to develop the full allowable tensile load of the 6 in. by $\frac{3}{8}$ in. ASTM A36 steel plate in Fig. 19–32. The electrode is an E70. Minimum end returns may be used.

17. A fillet weld between two steel plates intersecting at right angles was made with one $\frac{1}{4}$ in. leg and one $\frac{9}{16}$ in.

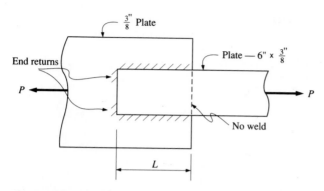

FIGURE 19–32   Problem 16.

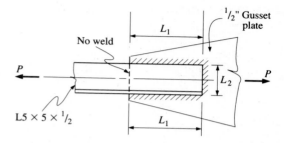

FIGURE 19–33   Problem 18.

leg using an E70 electrode. Determine the strength of this weld in kips/in.

18. Design an end connection using longitudinal welds and an end transverse weld to develop the full tensile capacity of the angle shown in Fig. 19–33. Use A36 steel and E70 electrodes. *Note:* the AISC Specification utilizes a reduction coefficient for the computation of the net area for a tension member that does not have all of its cross-sectional elements connected to the supporting member. This factor is neglected in this text.

SI System Problems

19. Calculate the allowable tensile load for the lap joint in Fig. 19–34. The fasteners are 25 mm diameter A325

high-strength bolts in a slip-critical connection. The plates are ASTM A36 steel.

20. Calculate the allowable tensile load for the butt joint in Fig. 19–35. The fasteners are 20 mm diameter A325-SC high-strength bolts and the plates are ASTM A36 steel.

21. Rework Problem 20 changing the bolts to A490-SC and the plates to ASTM A441 steel (ultimate tensile stress = 414 MPa).

22. Rework Problem 16 assuming that both plates are 12 mm thick and that the smaller plate is 150 mm wide. Use a 10 mm fillet weld.

23. Rework Problem 18 assuming that the angle is an L127 × 89 × 12.7 and that the gusset plate is 13 mm thick. Assume a 10 mm fillet weld. Assume that the long leg of the angle is connected to the gusset plate.

FIGURE 19–34   Problem 19.

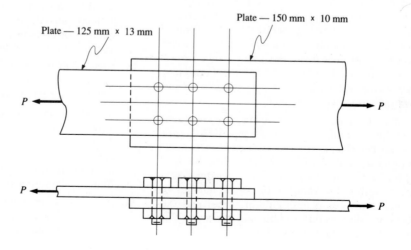

FIGURE 19–35 Problem 20.

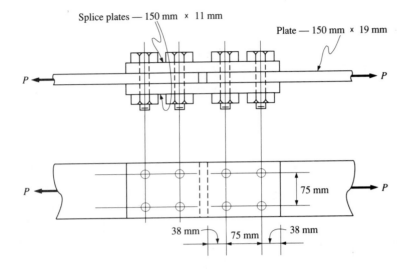

Computer Problems

For the following computer problems, any appropriate programming language may be used. Input prompts should fully explain what is required of the user (the program should be "user friendly"). The resulting output should be well labeled and self-explanatory.

24. The load that can be transmitted by a single fastener in shear is sometimes called the *shear value*. The shear value is the product of the cross-sectional area of the shank of the fastener and an allowable shear stress. Write a program that will generate a table of shear values for fasteners having the allowable stresses given in Table 19–1. Fastener diameters are to range from $\frac{5}{8}$ in. to $1\frac{1}{2}$ in. (in $\frac{1}{8}$ in. increments). The tabular format should be such that the diameters vary by column and the allowable stress varies by row.

25. A single-shear lap joint is to have fasteners arranged on three gage lines, similar to the joint shown in Figure 19–12. The steel plates are to be of ASTM A36 steel. Write a program that will calculate the allowable tensile load for such a joint assuming $\frac{3}{4}$ in. diameter A325-SC high-strength bolts. Input is to be the number of bolts, which must be in multiples of three, and the cross-sectional dimensions (width and thickness) of the main plates. Assume standard holes.

Supplemental Problems

26. Calculate the allowable tensile load for the butt joint in Fig. 19–36. The fasteners are $\frac{7}{8}$ in. diameter A325 high-strength bolts in a slip-critical connection (A325-SC). The plates are of ASTM A36 steel.

27. Rework Problem 26 using $\frac{3}{4}$ in. diameter A502 Grade 2 hot-driven rivets.

28. Two ASTM A36 steel plates, each 12 in. by $\frac{1}{2}$ in., are connected by a lap joint. The fasteners are $\frac{3}{4}$ in. diameter A325 high-strength bolts. The connection is subjected to an axial tensile load of 100 kips. Calculate the number of bolts required for a slip-critical connection and sketch the bolt layout pattern.

29. Rework Problem 28 changing the fasteners to (a) $\frac{3}{4}$ in. diameter A307 unfinished bolts and (b) $\frac{3}{4}$ in. diameter A502 Grade 1 hot-driven rivets.

30. Calculate the minimum main plate thickness for the joint shown in Fig. 19–27 so that the plates, which are 7 in. wide, will have a capacity equal to the shear capacity of the $\frac{7}{8}$ in. diameter A325 high-strength bolts in standard holes. The plates are ASTM A441 steel. The connection is slip-critical. Assume $s_{t(ult)}$ of 70 ksi.

31. A roof truss tension member is made up of 2L6 $\times$ 4 $\times$ $\frac{1}{2}$ (long legs back to back) of A36 steel as shown in Fig. 19–37. The bolts are $\frac{3}{4}$ in. diameter A325 high-strength

FIGURE 19–36 Problem 26.

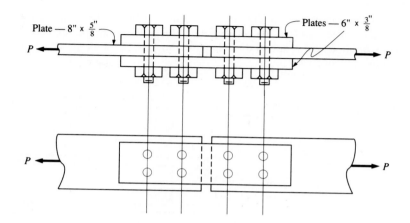

Plate — 8" × $\frac{5}{8}$"   Plates — 6" × $\frac{3}{8}$"

FIGURE 19–37 Problem 31.

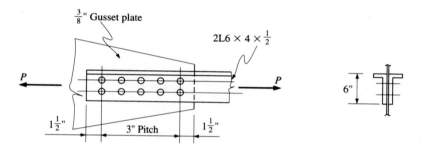

$\frac{3}{8}$" Gusset plate

2L6 × 4 × $\frac{1}{2}$

6"

$1\frac{1}{2}$"   3" Pitch $1\frac{1}{2}$"

bolts in standard holes. Compute the tensile capacity P for the connection shown if the connection is slip-critical (A325-SC). The special note for angle tension connections of Problem 18 applies.

32. Rework Problem 31 changing the fasteners to (a) $\frac{3}{4}$ in. diameter A307 unfinished bolts and (b) $\frac{3}{4}$ in diameter A502 Grade 1 hot-driven rivets. Assume a bearing-type connection.

33. Determine the allowable tensile load that can be applied to the connection shown in Fig. 19–38. The plates are of ASTM A36 steel and the weld is made using an E70 electrode.

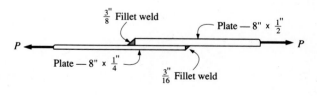

$\frac{3}{8}$" Fillet weld   Plate — 8" × $\frac{1}{2}$"

Plate — 8" × $\frac{1}{4}$" $\frac{3}{16}$" Fillet weld

8"

FIGURE 19–38 Problem 33.

34. The welded connection shown in Fig. 19–39 is subjected to an axial tensile load P of 100,000 lb. Calculate the minimum size fillet weld required if the length L is 7 in. Assume an E70 electrode.

35. In Problem 34, use a $\frac{3}{8}$ in. fillet weld, change L to 9 in., and find the allowable tensile load as governed by the welds.

FIGURE 19–39 Problem 34.

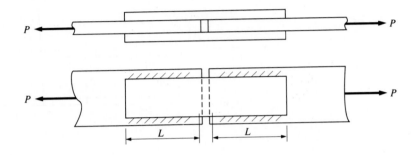

20 Pressure Vessels

20–1
INTRODUCTION

Pressure vessels may be described as leak-proof containers. They are found in various shapes and sizes. Boilers, fire extinguishers, shaving cream cans, and pipes are common examples. The sophisticated vessels used in the space, nuclear, and chemical industries are also examples of pressure vessels.

Pressure vessels commonly have the form of spheres, cylinders, ellipsoids, or some composite of these with the primary purpose of containing liquids and/or gases under pressure. In practice, vessels are usually composed of a complete pressure-containing shell together with flange rings and fastening devices for connecting and securing mating parts.

In this chapter, we are primarily concerned with the stresses developed in the walls of spheres and cylinders. Stress analysis can be performed by various means. Our limited discussion assumes a simple vessel with no consideration of shape discontinuities. If the vessel exhibits abrupt cross-sectional changes or discontinuities such as penetrations of the shell wall for the attachment of pipes or nozzles, irregularities in the stress distribution will occur. These, in general, are developed in only localized portions of the vessel and are called *localized stresses* or *stress concentrations*. In these cases, the problem becomes too complex for an analytical solution. A computer solution or an experimental solution will then be required. Our discussion is limited to the analytical approach.

The stress analysis is only one phase of a total design. One significant consideration in design is the selection of the material to be used and its relationship to the environment to which it will be subjected. Based on the safety demands of nuclear reactors, space vehicles, and deep-diving submersibles, considerable research has been directed toward new and improved materials and their behavior in specific environments. No one perfect pressure vessel material is suitable for all situations. Material selection must be consistent with the application and the environment. This is particularly true for nuclear plant pressure vessels.

Our discussion in this chapter will consider only the stresses in thin-walled pressure vessels at sections other than the discontinuity locations. Thin-walled pressure vessels are defined as having a wall thickness not in excess of one-tenth (0.10) of the internal radius of the vessel. In most cylindrical and spherical vessels, the internal pressures are low so that the wall thickness is relatively small compared to the other dimensions of the vessel.

For a thin-walled vessel, assume that the tensile stress across the wall thickness is constant; the value can be obtained using the basic equations of equilibrium. This results in an average stress with an error less than 4 or 5 percent. For thick-walled vessels ($r_i/t < 10$) as in the cases of gun barrels or high-pressure hydraulic presses, the variation in the stress from the inner surface to the outer surface becomes appreciable, a higher stress existing at the inner surface. This problem is more complex and the thin-walled formulas cannot be used.

**20–2
STRESSES IN
THIN-WALLED
PRESSURE VESSELS**

A *thin-walled pressure vessel* has been defined as one with a wall thickness not in excess of one-tenth (0.1) of the internal radius of the vessel. It is usually made up of curved plates fastened together by continuous joints or seams. In a cylindrical pressure vessel, the plates are joined using two types of joints: a longitudinal, or lengthwise, joint and a circumferential joint. Both types are shown in Fig. 20–1. A spherical pressure vessel, by virtue of its complete symmetry, has only the circumferential-type joint.

FIGURE 20–1   Cylindrical pressure vessel.

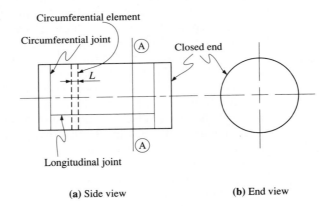

(a) Side view   (b) End view

Cylindrical and spherical thin-walled pressure vessels are generally subjected to some magnitude of internal gas and/or liquid pressure. As a result of the internal pressure, tensile stresses are developed in the pressure vessel walls. As with other structural shapes, the induced tensile stresses may not exceed specified allowable tensile stresses. The internal pressure tends to rupture the pressure vessel along either a longitudinal or a circumferential joint. The walls of a thin-walled pressure vessel are assumed to act as a membrane in that no bending of the walls takes place.

We begin our analysis of pressure vessels by considering a cylindrical pressure vessel. Figure 20–2(a) shows section A–A, obtained by passing plane A–A through the pressure vessel of Fig. 20–1. This is a typical cross section of a cylindrical thin-walled pressure vessel subjected to an internal pressure p. The internal pressure at any point acts equally in all directions and is always perpendicular to any surface on which it acts. Therefore, in Fig. 20–2(a), the pressure is seen to act radially outward.

FIGURE 20–2 Cylindrical pressure vessel.

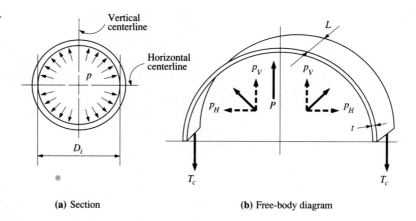

(a) Section

(b) Free-body diagram

The radially acting internal pressure tends to rupture the longitudinal joints. To resist this tendency, tensile stresses are developed in the walls of the pressure vessel. Those stresses are called *circumferential stresses* (s_{t_c}) and they act on the longitudinal joint. In order to establish an expression for the circumferential stress, let us consider a circumferential element of length L taken from the cylindrical pressure vessel of Fig. 20–1. The free-body diagram of half of this element is shown in Fig. 20–2(b), which also shows the forces and pressures acting on the free body. Note, as shown in Fig. 20–2(a), that the internal pressure p acts radially and is uniformly distributed over the curved surface. For the free body shown in part (b), the horizontal components p_H of the radial pressure act in opposite directions, thus cancelling each other by virtue of symmetry about the vertical centerline. However, the vertical components p_V act in the same direction and must be resisted in order for a state of equilibrium to exist. The total upward acting force on the curved surface is the sum of all the vertical components of the forces due to the internal pressure and can be represented by a single resultant force P. It can be shown that this resultant vertical force P is obtained as the product of the pressure p and the area of the curved wall surface projected on a horizontal plane. Since the projected area is equal to LD_i, the resultant vertical force can be calculated from

$$P = pLD_i$$

The resistance to this force is furnished by the walls of the pressure vessel. A resisting tensile force T_c is developed in each cut wall as shown in the free body of Fig. 20–2(b). A summation of vertical forces shows that $P = 2T_c$, from which $T_c = P/2$. Solving for the circumferential tensile stress (also called the *hoop stress*) in the wall of the pressure vessel,

$$s_{t_c} = \frac{T_c}{tL} = \frac{P/2}{tL} = \frac{pLD_i}{2tL} = \frac{pD_i}{2t} \qquad \textbf{(20–1)}$$

where s_{t_c} = the circumferential tensile stress in the wall of the pressure vessel (psi, ksi) (Pa)

  p = the internal pressure (psi, ksi) (Pa)
  D_i = the inside diameter (in.) (mm)
  t = the wall thickness of the pressure vessel (in.) (mm)

Equation (20–1) can also be used for design purposes, in order to calculate a required wall thickness to resist a given internal pressure while not exceeding an allowable tensile stress $s_{t_{c(all)}}$. The equation can be rewritten as follows:

$$\text{Required } t = \frac{pD_i}{2s_{t_{c(all)}}} \tag{20–2}$$

In addition to the tendency of a closed-end cylindrical pressure vessel to rupture along a longitudinal joint, the pressure simultaneously tends to push out the ends and pull the vessel apart on a circumferential joint. To resist this tendency, tensile stresses are developed in the walls of the pressure vessel which act on a circumferential joint. These are called *longitudinal stresses* and are perpendicular to the circumferential stresses.

In order to establish an expression for the longitudinal stresses, let us consider a free body of the closed-end portion of a cylindrical pressure vessel subjected to an internal pressure, as shown in Fig. 20–3.

FIGURE 20–3 Longitudinal stresses in cylindrical pressure vessel.

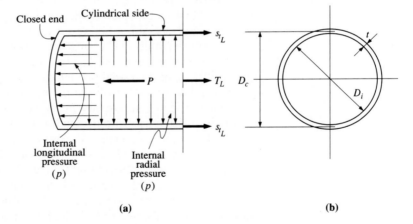

(a)   (b)

Since the end of the pressure vessel is closed, the end area on which the pressure acts is circular in shape. The total resultant force tending to push out the end of the vessel can be calculated as the product of pressure and area:

$$P = p\left(\frac{\pi D_i^2}{4}\right)$$

In order for a state of equilibrium to exist, force P must equal the total resistance T_L offered by the walls of the vessel around the entire circumferential joint. Assuming the longitudinal stress s_{t_L} in the walls of the vessel to be uniform throughout the wall thickness, we can calculate the stress as

$$s_{t_L} = \frac{T_L}{\pi D_c t} = \frac{P}{\pi D_c t}$$

where D_c is the vessel diameter center-to-center of walls. Recognizing that $D_i \approx D_c$ (very close), we make this substitution as well as the substitution for P from the preceding expression:

$$s_{t_L} = \frac{p(\pi D_i^2/4)}{\pi D_i t} = \frac{pD_i}{4t} \qquad \textbf{(20–3)}$$

where s_{t_L} is the longitudinal stress in the walls of the pressure vessel (psi, ksi) (Pa) and the other terms are as previously defined for Eq. (20–1).

Note that the longitudinal stress developed is one-half the circumferential stress. Thus, if a fluid in a closed vessel (a tank or pipe) freezes, the vessel will rupture along a longitudinal joint. Note that the two expressions for stresses are not accurate in the immediate vicinity of a closed end.

One often encounters a thin-walled spherical pressure vessel subjected to an internal pressure. Although the spherical shape is more efficient than the cylindrical shape, it is not as commonly used and is generally more expensive to manufacture. Whereas the cylindrical shape is subjected to two different magnitudes of stress, as previously discussed, every point in the spherical shape is subjected to the same maximum tensile stress. This tensile stress is the same as that for the longitudinal stress in a thin-walled cylindrical pressure vessel and can be evaluated using Eq. (20–3).

Note that our discussion in this section, including the equations developed, applies only to thin-walled pressure vessels and the case of internal pressure. Since we are neglecting all discontinuities and localized stresses, the state of stress for an element of a thin-walled cylindrical-shaped pressure vessel may, for all practical purposes, be considered biaxial, as shown in Fig. 20–4. Recall from previous discussion that the circumferential stress s_{t_c} is twice that of the longitudinal stress s_{t_L}.

If an additional tensile force P is applied to the closed ends as shown in Fig. 20–5, the longitudinal stress s_{t_L} is increased by the amount of

$$s_{t_p} = \frac{P}{A} = \frac{P}{\pi D_i t}$$

Therefore, the total longitudinal stress can be written:

$$\text{Total } s_{t_L} = s_{t_L} + s_{t_p}$$

$$= \frac{pD_i}{4t} + \frac{P}{\pi D_i t} \qquad \textbf{(20–4)}$$

FIGURE 20–4   State of biaxial stress.

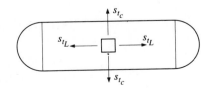

FIGURE 20–5 Modified state of biaxial stress.

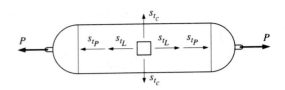

☐ **EXAMPLE 20–1**   Calculate the circumferential and longitudinal stresses developed in the walls of a cylindrical gas storage tank. The tank is of steel, has a 48 in. inside diameter, and a wall thickness of $\frac{3}{8}$ in. The internal gage pressure is 150 psi. Show that the thin-wall theory is applicable to this problem.

Solution   The internal pressure is gage pressure. This means that it is an internal pressure in excess of an external pressure. Usually the external pressure is atmospheric. The thin-wall theory (along with its equations) is applicable if $t \leq 0.1 \times$ (radius):

$$0.1(48/2) = 2.4 \text{ in.}$$

$$0.375 \text{ in.} < 2.4 \text{ in.} \qquad \textbf{OK}$$

The circumferential stress developed in the tank walls can be found from Eq. (20–1):

$$s_{t_c} = \frac{pD_i}{2t} = \frac{150(48)}{2(0.375)} = 9600 \text{ psi}$$

The longitudinal stress developed in the tank walls can be found from Eq. (20–3):

$$s_{t_L} = \frac{pD_i}{4t} = \frac{150(48)}{4(0.375)} = 4800 \text{ psi}$$

☐ **EXAMPLE 20–2**   Calculate the maximum allowable internal gage pressure that can be contained by a closed cylindrical steel tank having an internal diameter of 36 in. and a wall thickness of $\frac{1}{4}$ in. The steel has an allowable tensile stress of 15,000 psi.

Solution   First, check the applicability of the thin-wall theory:

$$0.1(36/2) = 1.8 \text{ in.}$$

$$0.25 \text{ in.} < 1.8 \text{ in.} \qquad \textbf{OK}$$

Since the circumferential stress developed will be twice the longitudinal stress developed, the allowable internal pressure must be based on Eq. (20–1), where

$$s_{t_c} = \frac{pD_i}{2t}$$

Rewriting and solving for p,

$$p = \frac{2ts_{t_{c(\text{all})}}}{D_i} = \frac{2(0.25)(15,000)}{36} = 209 \text{ psi}$$

☐ **EXAMPLE 20–3**

Calculate the stress developed in the walls of a spherical steel tank having a 12 ft inside diameter and a wall thickness of $\frac{3}{4}$ in. The tank is subjected to an internal gage pressure of 300 psi.

Solution

First, check the applicability of the thin-wall theory:

$$0.1(144/2) = 7.2 \text{ in.}$$

$$0.75 \text{ in.} < 7.2 \text{ in.} \qquad \textbf{OK}$$

The tensile stress developed in the walls of a spherical pressure vessel can be found from Eq. (20–3):

$$s_{t_L} = \frac{pD_i}{4t} = \frac{300(12)(12)}{4(0.75)} = 14{,}400 \text{ psi}$$

☐ **EXAMPLE 20–4**

A cylindrical pressure vessel made of steel has a 24 in. inside diameter and a wall thickness of $\frac{3}{8}$ in. The vessel is subjected to an internal gage pressure of 200 psi. The vessel is simultaneously subjected to an axial tensile force of 90,000 lb. Calculate the circumferential and longitudinal stresses developed in the walls.

Solution

First, check the applicability of the thin-wall theory:

$$0.1\left(\frac{24}{2}\right) = 1.2 \text{ in.}$$

$$0.375 \text{ in.} < 1.2 \text{ in.} \qquad \textbf{OK}$$

The circumferential stress developed in the walls of the vessel is due to the internal pressure only (Eq. (20–1)):

$$s_{t_c} = \frac{pD_i}{2t} = \frac{200(24)}{2(0.375)} = 6400 \text{ psi}$$

The longitudinal stress developed in the walls is the sum of the stresses due to the internal pressure and the axial tensile load (Eq. (20–4)):

$$s_{t_L} = \frac{pD_i}{4t} + \frac{P}{\pi D_i t}$$

$$= \frac{200(24)}{4(0.375)} + \frac{90{,}000}{\pi(24)(0.375)} = 6380 \text{ psi}$$

**20–3
JOINTS IN
THIN-WALLED
PRESSURE VESSELS**

The design of joints (frequently called *boiler joints*) in thin-walled pressure vessels, as well as the design of the pressure vessels, themselves, is standardized by the American Society of Mechanical Engineers (ASME) Boiler and Pressure Vessel Code.[1] The ASME Code constitutes an international standard providing recommended design and manufacturing criteria for pressure-containing structures.

---

[1] American Society of Mechanical Engineers, *Boiler and Pressure Vessel Code, Section VIII, Division 1* (New York: ASME, 1980).

As stated previously, thin-walled pressure vessels are generally manufactured from curved plates fastened together by designed joints that must remain sealed under internal pressure. The most common type of joint for the pressure vessel is the continuous welded butt joint for which the plates are groove welded using some form of edge preparation as discussed in Section 19–7. For all practical purposes, the welded boiler joints have replaced the once commonly used riveted boiler joints.

The welds are generally a full-penetration-type weld. The thickness of the weld is equal to the thickness of the plates. The allowable tensile stress for the connected plates, based on the ASME Boiler Code, is generally adopted as either one-fourth of the ultimate tensile strength of the steel or two-thirds of the yield strength of the steel, whichever is less. The allowable tensile stress for the weld may be assumed to be equal to, or some percentage of, the allowable tensile stress of the connected plates (75 percent, for example). The percentage is often called a *percentage factor*. The factor may vary greatly and has the effect of reducing the internal pressure capacity of a vessel. The percentage factor is also referred to as the *efficiency of the joint*. The maximum efficiency of a welded butt joint may be taken as 100 percent. This implies a full-strength weld as well as proper and comprehensive inspection during fabrication. Using a radiographic inspection process, a 100% joint efficiency may be assumed if the full continuous welded butt joint is X-rayed. With a partial (or spot) X ray, a joint efficiency of 85% should be used. If the joint is not X-rayed, a joint efficiency of 70% is recommended.

☐ **EXAMPLE 20–5**   A cylindrical steel gas storage tank has an inside diameter of 18 in. and a wall thickness of $\frac{3}{8}$ in. All joints are full-strength groove-welded butt joints. Assume a joint efficiency (percentage factor) of 70%. Calculate the safe internal gage pressure if the allowable tensile stress for the steel is 12,000 psi. Check the applicability of the thin-wall theory.

Solution   First, check the applicability of the thin-wall theory:

$$0.1(18/2) = 0.9 \text{ in.}$$

$$0.375 \text{ in.} < 0.9 \text{ in.} \quad \textbf{OK}$$

Since the circumferential stress developed will be twice the longitudinal stress developed, the safe (allowable) internal pressure must be based on Eq. (20–1) modified by the joint efficiency (J.E.), where

$$s_{t_c} = \frac{pD_i}{2t(\text{J.E.})}$$

Rewriting and solving for p,

$$p = \frac{2t(\text{J.E.})s_{t_c(\text{all})}}{D_i} = \frac{2(0.375)(0.70)(12,000)}{18} = 350 \text{ psi}$$

Note that, in effect, the joint efficiency reduces the allowable tensile stress and, therefore, the allowable internal gage pressure.

□ **EXAMPLE 20–6**   A spherical tank used for gas storage is 38 ft in diameter and is subjected to an internal gage pressure of 60 psi. Determine and select the required thickness of the tank wall if it is constructed of steel plate with an allowable tensile stress of 20,000 psi. Assume butt-welded joints with 85% efficiency. Check the applicability of the thin-wall theory.

Solution   The tensile stress developed in the welded joints of a spherical pressure vessel can be found from Eq. (20–3), where

$$s_{tt} = \frac{pD_i}{4t}$$

Rewriting, introducing the 85% joint efficiency, and solving for t,

$$\text{Required } t = \frac{pD_i}{4s_{tt(\text{all})}(\text{J.E.})} = \frac{60(38)(12)}{4(20,000)(0.85)} = 0.402 \text{ in.}$$

Use a $\frac{7}{16}$ in. thick plate.

The thin-wall theory applies if $t \leq 0.1 \times$ (radius):

$$0.1(38/2)(12) = 22.8 \text{ in.}$$

$$0.4375 \text{ in.} < 22.8 \text{ in.} \qquad \textbf{OK}$$

In liquid-filled tanks and pipes that are open to the air at the top, the internal pressure at any point is equal to the weight of a column of the liquid of the same height as the vertical distance from the point to the surface of the liquid. This vertical distance is called the *head*. If the head, in feet, is designated h, the pressure in psi for water, which weighs 62.5 pounds per cubic foot, is equal to

$$p = \frac{62.5(h)}{144} = 0.434(h)$$

Note that h is in feet and p is in psi.

□ **EXAMPLE 20–7**   Calculate and select the required wall thickness of the bottom section of a 15 ft diameter cylindrical steel standpipe subjected to an 80 ft head of water. Assume that the top of the standpipe is open. The allowable tensile stress for the steel plate wall of the standpipe is 12,600 psi. The plates are connected by a full-strength butt weld of 100% efficiency. Add approximately $\frac{1}{8}$ in. to the computed wall thickness to compensate for corrosion. Localized stresses at the junction of the vertical walls and the bottom may be neglected.

Solution   Since the contained liquid is water, the pressure p in any direction at the base of the standpipe is calculated as

$$p = 0.434(h) = 0.434(80) = 34.7 \text{ psi}$$

The pressure is equal in all directions and it acts radially against the vertical standpipe wall, as shown in Fig. 20–6. It varies from zero at the top of the standpipe to a maximum at the bottom. Since the top of the standpipe is open, there is no

FIGURE 20–6   Section through standpipe.

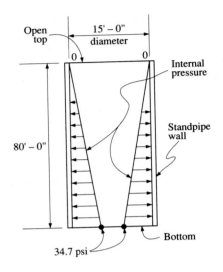

longitudinal stress developed. The required wall thickness is based on an allowable circumferential tensile stress and is obtained from Eq. (20–2):

$$\text{Required } t = \frac{pD_i}{2s_{tc(all)}}$$

The point that must be considered for design is the bottom of the standpipe where the radial pressure is at a maximum. Substituting,

$$\text{Required } t = \frac{pD_i}{2s_{tc(all)}} = \frac{34.7(15)(12)}{2(12,600)} = 0.248 \text{ in.}$$

Adding the corrosion allowance of $\frac{1}{8}$ in. yields

$$\text{Required } t = 0.248 + 0.125 = 0.373 \text{ in.}$$

Use a $\frac{3}{8}$ in. thick plate.

**20–4
DESIGN AND
FABRICATION
CONSIDERATIONS**

The basic expressions developed in Section 20–2 are based on the assumption that there is continuous elastic behavior throughout the pressure vessel and that the tensile stresses developed are uniformly distributed over the wall cross section. Discontinuities that cause stress concentrations may invalidate this assumption.

The stress concentrations may be of little significance if the vessel is subjected to a steady internal pressure and is fabricated using a ductile material such as a mild steel. The steel should be capable of yielding at highly stressed locations, thereby allowing a transfer of stress from overstressed areas to adjacent understressed ones. When the internal pressure is repetitive, localized stresses become significant because of fatigue, even though the material is ductile and has a large measure of static reserve strength. For this condition, the average stress formulas of Section 20–2 must be modified by a stress concentration factor in order to obtain a realistic maximum stress value.

Stress concentrations cannot be avoided in pressure vessels. All vessels must have supports and openings for attachments as necessary operating features. These stress concentrations invariably result in geometric discontinuities and cause localized stresses that must be considered in vessel design.

The majority of pressure vessels are made from various parts that have been fabricated, by welding, into shapes such as cylinders and hemispheres to form the base vessel. To this base vessel are also attached, by welding, the necessary appurtenances such as pipes, access openings, and support attachments. Only those closures that must be frequently removed for service or maintenance are attached by bolts, studs, or other mechanical closure devices. This contributes to a leak-proof vessel since the number of mechanical joints is kept to a minimum. In addition, the welding process, assuming acceptable quality control for the welding procedures, will contribute toward eliminating or controlling the effect of stress concentrations. Actually, almost all failures of pressure vessels occur as a result of fatigue in the areas of high stress concentrations. Therefore, the elimination or reduction of these localized stresses provides a partial solution to safe pressure vessel construction.

20–5
SI SYSTEM
EXAMPLE

□ **EXAMPLE 20–8**

A steel cylinder with a 300 mm diameter contains a gas at a gage pressure of 1.5 MPa. Using a steel having a yield stress of 350 MPa and a factor of safety (based on yielding) of 3.0, compute the required wall thickness of the cylinder. Check the applicability of the thin-wall theory.

Solution

Since the circumferential stress developed will be twice the longitudinal stress developed, the required wall thickness will be determined using Eq. (20–2):

$$\text{Required } t = \frac{pD_i}{2s_{t_{c(\text{all})}}} = \frac{(1.5 \text{ MPa})(300 \text{ mm})}{2(350/3 \text{ MPa})} = 1.93 \text{ mm}$$

Next, check the applicability of the thin-wall theory:

$$0.1 \left(\frac{300}{2}\right) = 15 \text{ mm}$$

$$1.93 \text{ mm} < 15 \text{ mm} \quad \textbf{OK}$$

SUMMARY—BY
SECTION NUMBER

20–1 Pressure vessels (in this text) are leakproof containers with no consideration of shape discontinuities or structural discontinuities.

20–2 Thin-walled pressure vessels are generally cylindrical or spherical. They are made up of curved plates fastened together by continuous longitudinal and/or circumferential welded joints. Pressure vessels are generally subjected to an internal gas or liquid pressure, thereby developing tensile stresses in the walls of the vessel. The circumferential tensile stress (developed on a longitudinal joint) is

$$s_{t_c} = \frac{pD_i}{2t} \tag{20-1}$$

The longitudinal tensile stress (developed on a circumferential joint) is

$$s_{t_L} = \frac{pD_i}{4t} \tag{20-3}$$

20-3   The most common type of joint used for pressure vessels is a continuous, full penetration, welded butt joint. In this type of a joint, the edges of the plates are carefully shaped prior to being groove welded.

20-4   Localized stresses cannot be eliminated in pressure vessels. Almost all failures occur at such locations. Vessel design (including material selection) and fabrication must be such as to minimize the effect of stress concentrations.

PROBLEMS

For each of the following problems, check the applicability of the thin-wall theory. Unless otherwise noted, the stated diameter represents an inside diameter.

Section 20-2   Stresses in Thin-Walled Pressure Vessels

1. A welded water pipe has a diameter of 9 ft and a wall of steel plate $\frac{13}{16}$ in. thick. After fabrication, this pipe was tested under an internal pressure of 225 psi. Calculate the circumferential stress developed in the walls of the pipe.

2. Calculate the tensile stresses developed (circumferential and longitudinal) in the walls of a cylindrical boiler 6 ft in diameter with a wall thickness of $\frac{1}{2}$ in. The boiler is subjected to an internal gage pressure of 150 psi.

3. Calculate the internal water pressure that will burst a 16 in. diameter cast-iron water pipe if the wall thickness is $\frac{5}{8}$ in. Use an ultimate tensile strength of 60,000 psi for the pipe.

4. Calculate the wall thickness required for a 4 ft diameter cylindrical steel tank containing gas at an internal gage pressure of 500 psi. The allowable tensile stress for the steel is 15,000 psi.

5. A spherical gas container, 52 ft in diameter, is to hold gas at a pressure of 50 psi. Calculate the thickness of the steel wall required. The allowable tensile stress is 20,000 psi.

6. Calculate the tensile stresses (circumferential and longitudinal) developed in the walls of a cylindrical pressure vessel. The inside diameter is 18 in. The wall thickness is $\frac{1}{4}$ in. The vessel is subjected to an internal gage pressure of 300 psi and a simultaneous external axial tensile load of 50,000 lb.

Section 20-3   Joints in Thin-Walled Pressure Vessels

7. A gas is to be stored in a spherical steel tank 20 ft in diameter. The wall thickness is $\frac{7}{8}$ in. Assume all joints to be groove-welded butt joints with a 90% joint efficiency. Calculate the maximum safe internal gage pressure if the allowable tensile stress for the steel is 15,000 psi.

8. The longitudinal joint in an 8 ft diameter closed cylindrical steel boiler is butt welded with a joint efficiency of 85%. The boiler plate is $\frac{3}{4}$ in. thick with an allowable tensile stress of 11,000 psi. Calculate the maximum safe internal pressure.

9. A spherical steel tank, 52 ft in diameter, is to hold gas at a pressure of 40 psi. The joints are butt welded with an efficiency of 75%. The allowable tensile stress for the steel plate is 18,000 psi. Calculate the required thickness of the walls.

10. A closed cylindrical tank for an air compressor is 24 in. in diameter and is subjected to an internal pressure of

450 psi. The tank walls are of steel and have an allowable tensile stress of 11,000 psi. Joint efficiency is 100%. Calculate the required wall thickness.

11. Calculate the wall thickness required for a 36 in. diameter steel penstock with an allowable tensile stress of 11,000 psi. The penstock, which is a pipe for conveying water to a hydroelectric turbine, operates under a head of 400 ft. Assume welded joints of 70% efficiency.

SI System Problems

12. Compute the stress developed in the walls of a spherical steel tank having an outside diameter of 500 mm and an inside diameter of 490 mm. The tank is subjected to an internal gage pressure of 14.2 MPa.

13. A closed cylindrical air-compressor tank is 600 mm in diameter and is subjected to an internal pressure of 5 MPa. The tank walls are of steel, which has an allowable tensile stress of 110 MPa. Calculate the required wall thickness.

14. Calculate the tensile stresses (circumferential and longitudinal) developed in the walls of a cylindrical pressure vessel. The inside diameter is 0.50 m. The wall thickness is 7 mm. The vessel is subjected to an internal gage pressure of 2.50 MPa and a simultaneous external axial tensile load of 200 kN.

Computer Problems

For the following computer problems, any appropriate language may be used. Input prompts should fully explain what is required of the user (the program should be "user friendly"). The resulting output should be well labeled and should be self-explanatory.

15. Write a program that will compute the tensile stresses in the wall of a thin-walled pressure vessel. User input should be type of vessel (cylindrical or spherical), internal diameter, wall thickness, and internal pressure. The program should also verify that the pressure vessel is thin-walled. The program can be used to check some of the preceding problems.

16. Rework the program of Problem 15 giving the user the option of specifying the application of an external axial tensile load.

17. Viking Vessel Company manufactures cylindrical gas storage tanks. The company foresees a market for nominal 100 ft³ tanks of varying proportions. The tanks are to be rated at 60 psi. Viking uses steel with an allowable tensile stress of 15,000 psi. Welded joints are

fully inspected and a 100% joint efficiency may be assumed. Write a program that will generate a table of required wall thicknesses for tanks which have length-to-diameter ratios ranging (approximately) from 1.0 to 4.0. Diameters are to be in 2 in. increments. Lengths should be rounded up to the next 2 in. increment, so that the volume of the tank is 100 ft³, minimum. The output should be a list showing diameter, length, volume, and required wall thickness. (The thickness may be rounded up to 0.01 in.)

Supplemental Problems

18. Calculate the circumferential and longitudinal tensile stresses developed in the walls of a cylindrical steel plate boiler. The diameter of the boiler is 8 ft and the wall thickness is 1 in. The boiler is subjected to an internal gage pressure of 230 psi.

19. A welded aluminum alloy cylindrical pressure vessel has a diameter of 8 in. and a wall thickness of $\frac{3}{16}$ in. The ultimate tensile strength of the alloy is 65,000 psi. Calculate the internal pressure that will cause rupture.

20. A spherical tank is to hold 400 ft³ of gas at a pressure of 150 psi. The steel to be used has an ultimate tensile strength of 65 ksi and a yield strength of 42 ksi. Assume a 70% joint efficiency. Determine the required wall thickness.

21. A cast-iron pipe, 3 ft in diameter, is subjected to an internal gage pressure of 100 psi. Calculate the wall thickness required if the allowable tensile stress is 11,000 psi.

22. A steel pipe located in a hydroelectric power plant is 24 in. in diameter and has a wall thickness of $\frac{3}{8}$ in. The system operates under a maximum head of 750 ft. Assuming an ultimate tensile strength of 55,000 psi, calculate the factor of safety for the pipe.

23. A 5 ft diameter steel pipe is made of $\frac{3}{8}$ in. plate with groove-welded longitudinal joints. Assuming an allowable tensile strength of 13,000 psi and a joint efficiency of 100%, compute the safe head of water on the pipe.

24. Rework Problem 23 changing the joint efficiency to 80%.

25. A piece of 10 in. diameter pipe of 0.10 in. wall thickness was closed off at the ends. This assembly was placed in a testing machine and subjected simultaneously to an axial tensile load P and an internal gage pressure of 240 psi. If the longitudinal tensile stress in the pipe walls is not to exceed 12,000 psi, calculate the maximum value of the applied load P.

21 Statically Indeterminate Beams

21–1
INTRODUCTION

The beams considered thus far have been statically determinate; that is, the equations of static equilibrium were sufficient to compute the external reactions. This permitted us to analyze and design the beams based on stresses and deflections. The types of beams introduced in Section 13–1 and categorized as statically determinate were the simple, cantilever, and overhanging beams (see Fig. 13–2). Their reactions can be determined by using the three basic laws of equilibrium: $\Sigma F_V = 0$, $\Sigma F_H = 0$, and $\Sigma M = 0$.

In statically indeterminate beams, there are more unknown reactions than there are equations of equilibrium. In this chapter, three types of statically indeterminate beams will be considered. The common designations for these types of beams are *fixed beam, propped cantilever beam,* and *continuous beam.* These beams are shown in Fig. 13–2. Since their reactions cannot be determined by the three laws of equilibrium alone, additional equations based on deflection or end rotation of the beam must be introduced.

Although statically determinate beams are more common than statically indeterminate beams, statically indeterminate beams frequently provide advantages in economy, function, or aesthetics.

In this chapter, only horizontal beams will be considered. Additionally, for any given beam, cross-sectional properties (area and moment of inertia) and material properties (modulus of elasticity) will be assumed to be constant.

21–2
RESTRAINED
BEAMS

Restrained beams include the propped cantilever beam and the fixed beam. The propped cantilever is fixed at one end and simply supported at either the other end or at a point near the other end. The fixed beam, as the name implies, is fixed at both ends. Both types are shown in Fig. 21–1.

The upper diagram for each beam shows the conventional representation of the beam including the shape of the elastic curve. The lower diagram for each beam shows the beam as a free body with applied loads and resisting shears and moments that, in effect, hold the free body in equilibrium.

For purposes of analysis, the important fact about a fixed end is that the tangent to the elastic curve at that point is horizontal (the slope of the tangent line is zero). The fixed end, or point of restraint, is considered to be at the face of the wall. This assumes that the restraint is furnished by anchoring or embedding the end of the beam in a wall (other means of restraint may be furnished). Therefore, the beam span extends to the face of the wall (or walls, if the beam is fixed at both ends).

FIGURE 21–1 Restrained beams.

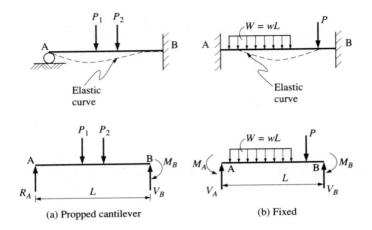

(a) Propped cantilever   (b) Fixed

With respect to the propped cantilever, the free-body diagram shows that in addition to the vertical forces R_A and V_B the beam is acted on by a moment of unknown magnitude at B. Note that this constitutes three unknowns. However, only two equations of static equilibrium exist for the determination of the forces acting on the beam (there are no horizontal forces). The two equations are not sufficient to determine the three unknowns. In fact, there are many combinations of values of M_B, R_A, and V_B that will satisfy the conditions of equilibrium. Therefore, some condition in addition to the two furnished is required to establish which combination of values is the correct one. Usually, for the propped cantilever, this additional condition is that the deflection at point A is zero.

When a beam is fixed at both ends, as shown in Fig. 21–1(b), there are four unknown reactive elements. As with the propped cantilever, the two equations of static equilibrium are not sufficient to determine the four unknowns. Therefore, two conditions in addition to the two furnished by the equilibrium equations must be introduced. The two conditions are that the slopes of the ends of the beam must remain unchanged. Consider the restrained fixed beam to be equivalent to a simple beam acted on by the given loading as well as end moments M_A and M_B. The end moments must be sufficient to rotate the ends of the beam until the slopes of the tangent at the ends correspond to the slopes at the ends of the restrained beam.

21–3
PROPPED
CANTILEVER BEAMS

There are several methods of analysis for this type beam. The method of superposition offers the simplest solution by utilizing deflection equations for cantilever beams as furnished in Appendix H.

The method of superposition, as used in this application, is based on the principle that the deflection at any point in the propped cantilever beam is the algebraic sum of two deflections:

1. The deflection at that point due to the given loads with the reaction R_A from the simple support removed (This, in effect, is the deflection of the cantilever.)

2. The deflection at that point in the same cantilever beam (with the applied loads removed) caused by the reaction R_A that was previously removed.

Generally, the desired final deflection at the simple support of the loaded propped cantilever beam is zero. This means that each support will be on the same level and that the reaction R_A has such a value that it produces an upward deflection at the support equal to the downward deflection at the same point due to the given loads.

This relationship can be expressed mathematically with the solution yielding a value for R_A. Figure 21–2(a) depicts a propped cantilever beam with both supports on the same level. If the support at A is removed, the beam becomes a cantilever with a maximum downward deflection at the free end as shown in Fig. 21–2(b).

FIGURE 21–2 Propped cantilever beam.

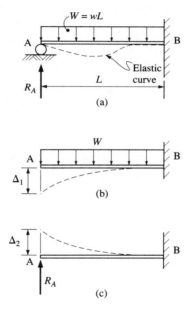

From Appendix H, the maximum deflection at the free end is calculated from

$$\Delta_1 = \frac{wL^4}{8EI} = \frac{WL^3}{8EI}$$

Since it is generally the case that zero deflection is desired at the propped end, the reaction R_A must be of such a magnitude so as to eliminate Δ_1. Since R_A occurs at the end of the beam, as shown in Fig. 21–2(c), and the uniformly distributed load is removed, the upward deflection due to R_A is calculated from

$$\Delta_2 = \frac{R_A L^3}{3EI}$$

Since the final deflection equals zero, Δ_1 must equal Δ_2. Substitution then yields

$$\frac{WL^3}{8EI} = \frac{R_A L^3}{3EI}$$

from which

$$R_A = \frac{3W}{8}$$

Note that no sign convention has been used.

Since one unknown has now been computed, the other two unknowns can be determined using the two applicable laws of static equilibrium ($\Sigma M = 0$ and $\Sigma F_y = 0$).

☐ **EXAMPLE 21–1**   Select the most economical (lightest) structural steel wide-flange section (W shape) for the propped cantilever beam which supports the loads shown in Fig. 21–3. Consider moment and shear. The allowable bending stress is 24 ksi and the allowable shear stress is 14.5 ksi. Neglect the weight of the beam. The vertical deflection at support A is zero.

Solution   Removing the support at A as shown in Fig. 21–3(b) results in a cantilever beam loaded with a concentrated load P. From Appendix H,

$$\Delta_1 = \frac{Pb^2}{6EI}(3L - b) = \frac{10(20)^2(12)^2}{6EI}[3(25) - 20](12)$$

$$= \frac{63,360,000}{EI}$$

FIGURE 21–3   Propped cantilever beam for Example 21–1.

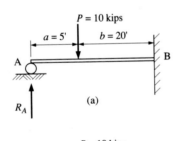

(a)

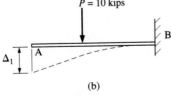

(b)

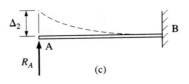

(c)

With the loading removed and reaction R_A applied as shown in Fig. 21–3(c),

$$\Delta_2 = \frac{R_A L^3}{3EI} = \frac{R_A(25)^3(12)^3}{3EI} = \frac{R_A(9,000,000)}{EI}$$

Since the final deflection must equal zero, Δ_1 must equal Δ_2:

$$\Delta_1 = \Delta_2$$

$$\frac{63,360,000}{EI} = \frac{R_A(9,000,000)}{EI}$$

$$R_A = 7.04 \text{ kips}$$

Solving for the shear V_B at the fixed support and referring to Fig. 21–4(a),

$$\Sigma F_y = 0$$

$$R_A + V_B - 10 = 0$$

$$7.04 + V_B - 10 = 0$$

from which $\qquad\qquad V_B = 2.96 \text{ kips}$

The complete shear and moment diagrams can then be drawn (see Fig. 21–4). Next, select the shape based on the required section modulus:

$$\text{Required } S_x = \frac{M}{s_{b(\text{all})}} = \frac{35.2(12)}{24} = 17.6 \text{ in.}^3$$

From Appendix I, select a W10 × 22 which provides a section modulus of 23.2 in.^{3}.

FIGURE 21–4 Shear and moment diagrams.

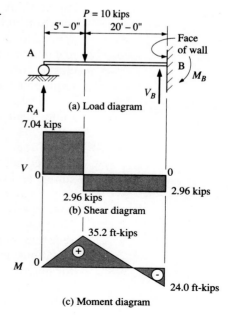

(a) Load diagram

(b) Shear diagram

(c) Moment diagram

Next, check shear (using properties from Appendix A):

$$s_s = \frac{V}{dt_w} = \frac{7.04}{10.17(0.240)} = 2.88 \text{ ksi}$$

$$2.88 \text{ ksi} < 14.5 \text{ ksi} \qquad \textbf{OK}$$

The W10 × 22 is satisfactory for both shear and moment.
  Use a W10 × 22.

Note that both supports of the beam in Example 21–1 are on the same level. If the simple support (left end) is raised above the level of the tangent to the elastic curve at the fixed end, the reaction R_A will increase. If the simple support settles below the level of the tangent line at the fixed end, the reaction will decrease. The reaction becomes zero when the simple support is lowered until the beam carries the load as a cantilever (if it is able to do so). The solution of such problems is similar to that when both supports are on the same level.

For instance, in Example 21–1, assume that the support at A settles 1.0 inch. The equation then used to determine R_A would be written

$$\Delta_1 - \Delta_2 = 1.0 \text{ in.}$$

where Δ_1 = downward deflection due to load
  Δ_2 = upward deflection due to the reaction R_A

21–4
FIXED BEAMS

Fixed beams are generally designated *fixed-end beams*. It is also generally assumed that the ends have complete fixity, which implies that the tangent to the elastic curve has a slope of zero at each support. It is sometimes convenient to think of a fixed beam as a beam simply supported at each end and acted on not only by a load, or a system of loads, but also by moments applied to the beam at its ends. If these moments are of the proper sense and magnitude, they will rotate the ends of the beam until the tangents to the elastic curve are horizontal at those points. The beam then meets all the requirements of a beam "fixed at both ends."

The method of superposition can also be used for this type beam, but rather than use deflection equations to supplement the static equilibrium equations, slope equations for the elastic curve will be used to evaluate two of the four unknown reactive elements. Tables 21–1 and 21–2 furnish derived equations for the slope of the elastic curve at designated locations for simple and cantilever beams under various types of loads.

Figure 21–5(a) depicts a horizontal beam fixed at both ends with a concentrated load at any point. The weight of the beam has been neglected. Note that at fixed ends A and B there are four unknown reactive elements: resisting moments M_A and M_B and the sheares V_A and V_B. Therefore, two equations, in addition to the two equations of static equilibrium, are required for determining the reactions.

Two equations can be obtained by considering the beam in Fig. 21–5(a) to be made up of the beams in parts (b), (c), and (d) of Fig. 21–5. Note that in

FIGURE 21–5   Fixed beam.

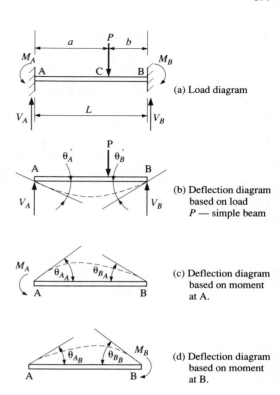

(a) Load diagram

(b) Deflection diagram based on load P — simple beam

(c) Deflection diagram based on moment at A.

(d) Deflection diagram based on moment at B.

order for the slopes at A and B to be equal to zero, it is necessary that

$$\theta'_A = \theta_{A_A} + \theta_{A_B}$$

and

$$\theta'_B = \theta_{B_A} + \theta_{B_B}$$

where θ_{A_A} is the slope at A due to a moment at A, and θ_{A_B} is the slope at A due to the moment at B.

In other words, the beam in Fig. 21–5(a) is equivalent to the superposition of the beams in parts (b), (c), and (d) of Fig. 21–5. Using Table 21–1, the equations for the slopes can be substituted in the above relationships and equations for M_A and M_B can be established as follows:

$$\theta'_A = \theta_{A_A} + \theta_{A_B}$$

$$\frac{Pb(L^2 - b^2)}{6LEI} = \frac{M_A L}{3EI} + \frac{M_B L}{6EI} \qquad \textbf{(Eq. 1)}$$

also,

$$\theta'_B = \theta_{B_A} + \theta_{B_B}$$

$$\frac{Pab(2L - b)}{6LEI} = \frac{M_A L}{6EI} + \frac{M_B L}{3EI} \qquad \textbf{(Eq. 2)}$$

TABLE 21–1 Slopes of beams on two supports.

Case	Simple Beam and Loading	Slope at Supports
1		$\theta_A = \theta_B = \dfrac{PL^2}{16EI}$
2		$\theta_A = \dfrac{Pb\,(L^2 - b^2)}{6LEI}$ $\theta_B = \dfrac{Pab\,(2L - b)}{6LEI}$
3		$\theta_A = \theta_B = \dfrac{wL^3}{24EI}$
4		$\theta_{A_A} = \dfrac{M_A L}{3EI}$ $\theta_{B_A} = \dfrac{M_A L}{6EI}$
5		$\theta_{A_B} = \dfrac{M_B L}{6EI}$ $\theta_{B_B} = \dfrac{M_B L}{3EI}$
6		$\theta_A = \theta_B = \dfrac{5WL^2}{96EI}$

TABLE 21–2 Slopes of cantilever beams.

Case	Cantilever Beam and Loading	Slope at Free End
1		$\theta = \dfrac{PL^2}{2EI}$
2		$\theta = \dfrac{Pa^2}{2EI}$
3		$\theta = \dfrac{wL^3}{6EI}$
4		$\theta = \dfrac{wa^3}{6EI}$
5		$\theta = \dfrac{ML}{EI}$
6		$\theta = \dfrac{WL^2}{12EI}$

Solving Eq. 1 and Eq. 2 simultaneously yields

$$M_A = \frac{Pab^2}{L^2}$$

$$M_B = \frac{Pa^2b}{L^2}$$

Since two unknowns are now determined, the other two unknowns can be calculated using the two applicable laws of static equilibrium ($\Sigma M = 0$ and $\Sigma F_y = 0$).

Note that no sign convention has been used with respect to upward or downward rotation.

☐ **EXAMPLE 21–2**   In the construction of walls, it is necessary to provide beams over openings such as doors or windows. This is especially the case where the walls are constructed of brick or masonry units. The beams are called *lintels*, and they must support the weight of the wall and any other loads above the opening. Due to an arching action that generally develops above the opening, the loading on the lintel will be of a triangular shape. Assuming complete fixity at each end, design a W-shape structural steel lintel beam for the span and loading shown in Fig. 21–6. Consider moment and shear. Use an allowable bending stress of 22 ksi and an allowable shear stress of 14.5 ksi. Neglect the weight of the beam. Due to the wall thickness, the flange width must be between 8 and 12 inches.

FIGURE 21–6   Fixed lintel beam.

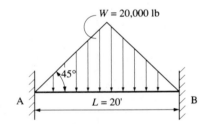

$W = 20,000$ lb

45°

A $L = 20'$ B

Solution   Using the method of superposition, the beam shown in Fig. 21–7(a) is equivalent to the superposition of the beams in parts (b), (c), and (d) of Fig. 21–7.

In order for the slopes at A and B to be equal to zero, it is necessary that

$$\theta'_A = \theta_{A_A} + \theta_{A_B}$$

and

$$\theta'_B = \theta_{B_A} + \theta_{B_B}$$

Substituting in both relationships from Table 21–1,

$$\theta'_A = \theta_{A_A} + \theta_{A_B}$$

$$\frac{5WL^2}{96EI} = \frac{M_AL}{3EI} + \frac{M_BL}{6EI}$$

FIGURE 21–7 Fixed lintel beam.

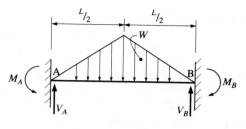

(a) Load diagram

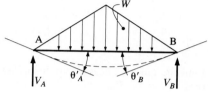

(b) Deflection diagram based on load W—simple beam.

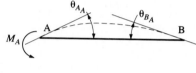

(c) Deflection diagram for end-moment at A.

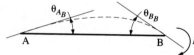

(d) Deflection diagram for end-moment at B.

This can be simplified to

$$M_B = \frac{5}{16}WL - 2M_A \qquad \textbf{(Eq. 1)}$$

Also,

$$\theta'_B = \theta_{B_A} + \theta_{B_B}$$

$$\frac{5WL^2}{96EI} = \frac{M_AL}{6EI} + \frac{M_BL}{3EI}$$

This can be simplified to

$$\frac{5WL}{16} = M_A + 2M_B \qquad \textbf{(Eq. 2)}$$

Substituting Eq. 1 into Eq. 2,

$$\frac{5WL}{16} = M_A + 2\left(\frac{5}{16}WL - 2M_A\right)$$

$$M_A = \frac{5WL}{48}$$

Since the loading is symmetrical,

$$M_A = M_B = \frac{5WL}{48} = \frac{5(20,000)(20)}{48} = 41,670 \text{ ft-lb}$$

Using the laws of static equilibrium,

$$\Sigma F_y = W - V_A - V_B = 0$$

Since the loading is symmetrical ($V_A = V_B$),

$$W = 2V_A$$

$$V_A = \frac{20,000}{2} = 10,000 \text{ lb} = V_B$$

Find the moment at midspan using laws of static equilibrium and a free-body diagram as shown in Fig. 21–8.

FIGURE 21–8   Free body.

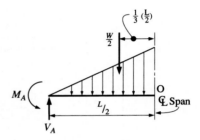

Taking moments at midspan (point O) and assuming moments causing compression in the top of the beam to be positive,

$$\Sigma M_O = +V_A\left(\frac{L}{2}\right) - M_A - \frac{W}{2}\left(\frac{1}{3}\right)\left(\frac{L}{2}\right)$$

$$= +10,000(10) - 41,670 - \frac{20,000}{2}\left(\frac{1}{3}\right)\left(\frac{20}{2}\right)$$

$$= 25,000 \text{ ft-lb}$$

Therefore, moment is maximum (41,670 ft-lb) at the fixed ends and the required section modulus can be calculated:

$$\text{Required } S_x = \frac{M}{S_{b(\text{all})}} = \frac{41,670(12)}{22,000} = 22.7 \text{ in.}^3$$

From Appendix I, select a W10 × 33 with a section modulus of 35.0 in.³. Note that there are lighter W-shapes that will satisfy the S_x requirements. However, the desired minimum flange width is 8 inches and the flange width for the W10 × 33 is 7.96 inches (say OK).

Check shear (using the properties of Appendix A):

$$s_s = \frac{V}{t_w d} = \frac{10,000}{0.29(9.73)} = 3540 \text{ psi} < 14,500 \text{ psi} \qquad \textbf{OK}$$

Use a W10 × 33.

A continuous beam is one that rests on more than two supports as shown in Fig. 21–9. This type of member is frequently used in modern structures and generally offers economy (in weight of steel required) when compared with a series of simple beams over the same spans. The effect of the beam continuity is to reduce the maximum bending moment from that of the simple beam, thereby reducing the required beam size. Continuous beams are statically indeterminate; and, therefore, the external reactions cannot be found by the equations of static equilibrium alone.

In general, the method of superposition used for the one-span indeterminate beams cannot be conveniently used for continuous beams. An exception is a two-span continuous beam such as shown in Fig. 21–10. The solution is similar to those of the one-span indeterminate beams.

FIGURE 21–9   Continuous beam.

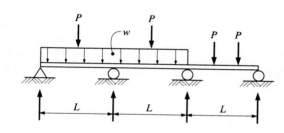

FIGURE 21–10   Two-span continuous beam.

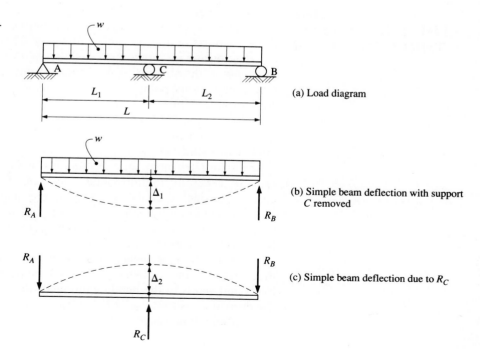

(a) Load diagram

(b) Simple beam deflection with support C removed

(c) Simple beam deflection due to R_C

With reference to Fig. 21–10, if the support at C is removed, beam AB becomes a simple beam with a span length equal to the sum of the two spans and will deflect downward a distance Δ_1 due to the applied uniformly distributed load w.

However, a support does exist at C and the task becomes one of determining the magnitude of the support reaction R_C that will deflect the beam AB upward the same distance that it deflected downward under the load w. Note in Fig. 21–10(c) that the uniformly distributed load is removed when applying R_C. Since the final deflection must be zero,

$$\Delta_1 = \Delta_2$$

$$\frac{5wL^4}{384EI} = \frac{R_C L^3}{48EI}$$

$$R_C = \frac{5}{8}\, wL$$

With R_C computed, the other reactions R_A and R_B can be found and subsequently shears and moments can be computed using the equations of static equilibrium. Note that the reaction R_C and the beam moments will change if the interior support is not at the same level as the two exterior supports.

This method of solution is not typical for continuous beams. However, it is applicable for the case discussed.

21–6
THE THEOREM OF
THREE MOMENTS

A very convenient method of finding bending moments in continuous beams is by means of a relationship that exists between the bending moments at the three supports of any two adjacent spans. This relationship is expressed as an equation and is commonly called the *theorem of three moments*. After the moments at the supports have been computed, shears, reactions, and moments at various points can be determined using the equations of static equilibrium.

In Section 21–4 we showed that a loaded beam fixed at each end is equivalent to a simple beam having the same span and load, with moments applied at each end so as to make the tangents to the elastic curve at the ends of the span horizontal. In a continuous beam, the tangents at the end of a span are generally not horizontal. Any span of a continuous beam, however, can be considered equivalent to a simple beam having the same span and load and acted on by end moments of sufficient magnitude to give the tangents at the ends of the beam the slope of the elastic curve of the continuous beam at the supports in question.

Figure 21–11 depicts a continuous beam subjected to uniformly distributed loads w_1 and w_2. An equation will be derived that will relate the given loads and spans to the bending moments at points A, B, and C. The key to the derivation is that $\theta'_{B_1} = \theta'_{B_2}$. That is, the beam is continuous so there is one and only one tangent at point B.

FIGURE 21–11 Continuous beam.

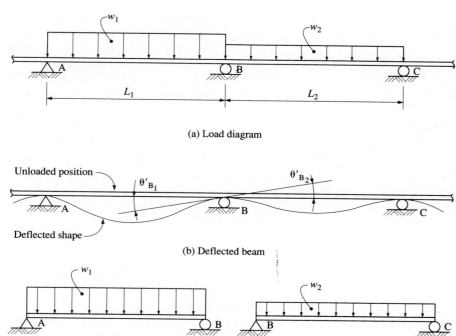

(a) Load diagram

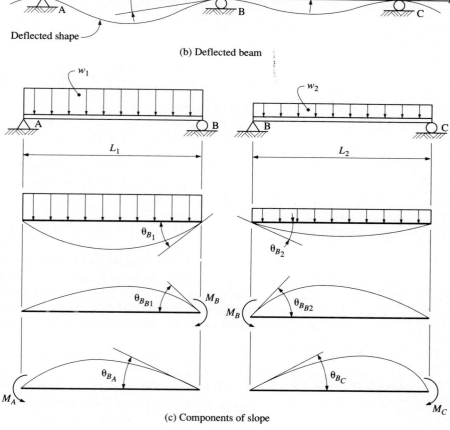

(b) Deflected beam

(c) Components of slope

The slope angle in each span is composed of three parts (Fig. 21–11(c)), one part produced by the applied load and the other two parts produced by the (unknown) end moments. The slope angle in each span will be computed and then the two will be equated.

Negative moments will be assumed (tension in the top of the beam) and will produce slope components as shown. Therefore,

$$\theta'_{B_1} = \theta_{B_1} + \theta_{B_{B_1}} + \theta_{B_A}$$
$$\theta'_{B_2} = \theta_{B_2} + \theta_{B_{B_2}} + \theta_{B_C}$$

Recognizing that slopes θ'_{B_1} and θ'_{B_2} must be of equal magnitude, but of opposite sign,

$$\theta'_{B_1} = -\theta'_{B_2}$$

Substituting,

$$\theta_{B_1} + \theta_{B_{B_1}} + \theta_{B_A} = -(\theta_{B_2} + \theta_{B_{B_2}} + \theta_{B_C})$$

$$\frac{w_1 L_1^3}{24EI} + \frac{M_B L_1}{3EI} + \frac{M_A L_1}{6EI} = -\frac{w_2 L_2^3}{24EI} - \frac{M_B L_2}{3EI} - \frac{M_C L_2}{6EI}$$

From which

$$M_A L_1 + 2M_B(L_1 + L_2) + M_C L_2 = -\frac{w_1 L_1^3}{4} - \frac{w_2 L_2^3}{4} \qquad \textbf{(21–1)}$$

This is a special form of the theorem of three moments. A general form, which can be derived using the moment-area principles of Chapter 16, can be written:

$$M_A L_1 + 2M_B(L_1 + L_2) + M_C L_2 = -\frac{6A_1 x_1}{L_1} - \frac{6A_2 x_2}{L_2} \qquad \textbf{(21–2)}$$

where the A and x terms on the right side result from the area and centroid locations of the moment diagrams. These terms vary with the type loading. Table 21–3 shows two other types of loading and the appropriate terms for substitution in the right side of the equation. For the case of equal spans and equal uniformly distributed load on each span, the equation becomes

$$M_A + 4M_B + M_C = -\frac{wL^2}{2} \qquad \textbf{(21–3)}$$

For a combination of uniformly distributed loads and concentrated loads, the right side of the equation becomes a combination of the terms shown. This will be illustrated in Example 21–4.

The three-moment equation is applicable to continuous beams with any number of spans since the equations give the relationship between the moments at any three successive supports. An equation is written for each two successive spans. Assuming that the ends of the continuous beam are not fixed, there will then be two fewer equations than the number of unknown moments. If the support is at the end of the beam and not fixed, the moment at the support is zero. If the beam overhangs the end support, the moment at

TABLE 21–3 Substitution terms for theorem of three moments.

Loading	$\dfrac{6A_1X_1}{L_1}$	$\dfrac{6A_2X_2}{L_2}$
P_1, P_2 loading	$\dfrac{P_1a}{L_1}(L_1^2-a^2)$	$\dfrac{P_2b}{L_2}(L_2^2-b^2)$
W_1, W_2 triangular loading	$\dfrac{4W_1L_1^2}{15}$	$\dfrac{4W_2L_2^2}{15}$

the support is computed by static conditions, the overhanging part being used as a free body. Note that if the beam is convex upward at a support, as is the usual case, the moment is negative and should be used as such in the equations.

After the moments are computed, the first span at either end, including the overhang, if any, may be taken as a free body to determine the end reactions. The first and second span together may next be taken as a free body to determine the second reaction, and so on across the span, until finally the entire beam may be taken as a free body.

After the reactions are computed, the complete shear and moment diagrams can be drawn. The entire process will be illustrated in three examples.

□ **EXAMPLE 21–3** A three-span continuous beam carries the uniformly distributed loads shown in Fig. 21–12. The loads include an assumed beam weight. (a) Compute the moments at each support. (b) Compute the four reactions.

Solution (a) As indicated previously, the special form of the three-moment equation (Eq. (21–1)) for a continuous beam supporting uniformly distributed loads is

$$M_AL_1 + 2M_B(L_1 + L_2) + M_CL_2 = -\frac{w_1L_1^3}{4} - \frac{w_2L_2^3}{4}$$

FIGURE 21–12 Beam for Example 21–3.

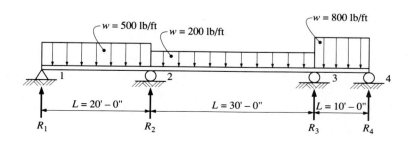

For the first two spans (points 1, 2 and 3), note that $L_1 = 20$ ft, $L_2 = 30$ ft, $M_B = M_2$, $M_C = M_3$, and $M_A = M_1 = 0$ (since the support at point 1 is a simple support). Substituting into Eq. (21–1) (with units being ft and lb),

$$0(20) + 2M_2(20 + 30) + M_3(30) = -\frac{500(20)^3}{4} - \frac{200(30)^3}{4}$$

$$100M_2 + 30M_3 = -1,000,000 - 1,350,000$$

Dividing by 30 yields

$$3.333M_2 + M_3 = -78,333 \qquad \textbf{(Eq. 1)}$$

For the next two spans (points 2, 3 and 4), note that $L_1 = 30$ ft, $L_2 = 10$ ft, $M_A = M_2$, $M_B = M_3$, and $M_C = M_4 = 0$ (since the support at point 4 is a simple support). Substituting into Eq. (21–1),

$$M_2(30) + 2M_3(30 + 10) + M_4(10) = -\frac{200(30)^3}{4} - \frac{800(10)^3}{4}$$

$$30M_2 + 80M_3 + 0(10) = -1,350,000 - 200,000$$

Dividing by 80 yields

$$0.375M_2 + M_3 = -19,375 \qquad \textbf{(Eq. 2)}$$

Solving Eq. 1 and Eq. 2 simultaneously for M_2 yields

$$M_2 = -19,930 \text{ ft-lb}$$

Substituting the value of M_2 into either Eq. 1 or Eq. 2 yields

$$M_3 = -11,900 \text{ ft-lb}$$

(b) Compute the four reactions. Note that the support moments are negative, indicating that at the support the top fibers of the beam are in tension and the bottom fibers are in compression. Also note that in the following calculations for ΣM, counterclockwise is assumed as positive.
For reaction R_1, take $\Sigma M = 0$ at point 2. Refer to Fig. 21–13.

$$-R_1(20) + \frac{500(20)^2}{2} - 19,930 = 0$$

$$R_1 = 4003 \text{ lb}$$

For reaction R_2, take $\Sigma M = 0$ at point 3. Refer to Fig. 21–14.

$$-4003(50) - R_2(30) - 11,900 + 500(20)(40) + 200(30)(15) = 0$$

$$R_2 = 9265 \text{ lb}$$

FIGURE 21–13 Free body—Span 1–2.

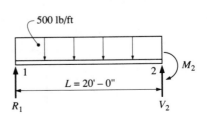

FIGURE 21–14 Free body—
Spans 1–2 and 2–3.

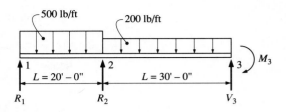

For reaction R_3, take $\Sigma M = 0$ at point 4. Refer to Fig. 21–12.

$$-4003(60) - 9265(40) - R_3(10) + 500(20)(50) + 200(30)(25) + 800(10)(5) = 0$$
$$R_3 = 7922 \text{ lb}$$

For reaction R_4, consider the entire beam as a free body and take a summation of vertical forces ($\Sigma F_y = 0$). Up is taken as positive.

$$+4003 + 9265 + 7922 + R_4 - 500(20) - 200(30) - 800(10) = 0$$
$$R_4 = 2810 \text{ lb}$$

□ **EXAMPLE 21–4**   For the continuous beam shown in Fig. 21–15, calculate the moments at the supports. Neglect the weight of the beam.

FIGURE 21–15 Beam for Example 21–4.

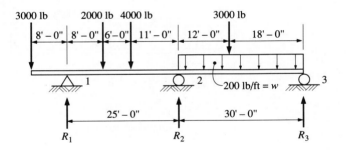

Solution   Due to the overhang at the left end, the moment at support 1 can be computed using a statics equation:

$$M_1 = -3000(8) = -24,000 \text{ ft-lb}$$

The end support at point 3 is a simple support; therefore, $M_3 = 0$. The only unknown moment is at support 2 (M_2).

A combined form of the three-moment equation is

$$M_A L_1 + 2M_B(L_1 + L_2) + M_C L_2 = -\Sigma\left(\frac{P_1 a}{L_1}(L_1^2 - a^2) - \frac{w_1 L_1^3}{4}\right)$$
$$- \Sigma\left(\frac{P_2 b}{L_2}(L_2^2 - b^2) - \frac{w_2 L_2^3}{4}\right)$$

There are only two spans to consider, span 1–2 and span 2–3. Note that $L_1 = 25$ ft, $L_2 = 30$ ft, $M_A = M_1$, $M_B = M_2$, and $M_C = M_3 = 0$. Distances a and b are defined in Table 21–3.

Substituting in the combined form of the three-moment equation,

$$-24{,}000(25) + 2M_2(25 + 30) + 0(30) = -\frac{2000(8)}{25}(25^2 - 8^2)$$

$$-\frac{4000(14)}{25}(25^2 - 14^2) - 0 - \frac{3000(18)}{30}(30^2 - 18^2) - \frac{200(30)^3}{4}$$

Solving for M_2 yields

$$M_2 = -28{,}240 \text{ ft-lb}$$

If a continuous beam is fixed at both ends, the foregoing approach must be modified. The moments at the supports can be computed if at each fixed end an auxiliary span, the length of which is zero, is assumed. With this assumption, the three-moment equation will furnish enough equations to compute the moments at the fixed ends and at the interior supports. The approach is illustrated in Example 21–5.

☐ **EXAMPLE 21–5**   A three-span continuous beam is fixed at both ends and is supported and loaded as shown in Fig. 21–16. (a) Calculate the moments at the fixed ends and at the interior supports. (b) Calculate the shears at the fixed ends and the interior support reactions.

FIGURE 21–16   Beam for Example 21–5.

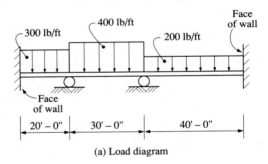

(a) Load diagram

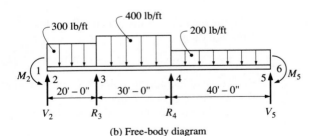

(b) Free-body diagram

Solution   (a) At the left of point 2 and at the right of point 5, auxiliary spans of length zero are assumed. The three-moment equation can then be written for points 1, 2, and 3 and then for points 2, 3, and 4, and so on. As indicated previously, the special form (Eq. 21–1) of the equation for a continuous beam supporting uniformly distributed loads is

$$M_A L_1 + 2M_B(L_1 + L_2) + M_C L_2 = -\frac{w_1 L_1^3}{4} - \frac{w_2 L_2^3}{4}$$

For points 1, 2, and 3, where $w_1 = 0$ and $L_1 = 0$,

$$M_1(0) + 2M_2(0 + 20) + M_3(20) = -0 - \frac{300(20)^3}{4}$$

$$40M_2 + 20M_3 = -600,000 \qquad \textbf{(Eq. 1)}$$

For points 2, 3, and 4,

$$M_2(20) + 2M_3(20 + 30) + M_4(30) = -\frac{300(20)^3}{4} - \frac{400(30)^3}{4}$$

$$20M_2 + 100M_3 + 30M_4 = -3,300,000 \qquad \textbf{(Eq. 2)}$$

For points 3, 4, and 5,

$$M_3(30) + 2M_4(30 + 40) + M_5(40) = -\frac{400(30)^3}{4} - \frac{200(40)^3}{4}$$

$$30M_3 + 140M_4 + 40M_5 = -5,900,000 \qquad \textbf{(Eq. 3)}$$

For points 4, 5, and 6,

$$M_4(40) + 2M_5(40 + 0) + M_6(0) = -\frac{200(40)^3}{4}$$

$$40M_4 + 80M_5 + 0 = -3,200,000 \qquad \textbf{(Eq. 4)}$$

Solving Eqs. 1, 2, 3, and 4 simultaneously yields

$$M_2 = -3333 \text{ ft-lb}$$
$$M_3 = -23,333 \text{ ft-lb}$$
$$M_4 = -30,000 \text{ ft-lb}$$
$$M_5 = -25,000 \text{ ft-lb}$$

(b) Solve for shear at points 2 and 5 and reactions at points 3 and 4. With reference to Fig. 21–17, take $\Sigma M = 0$ at point 3 to determine V_2 (note that counterclockwise is positive):

$$-V_2(20) + 3333 - 23,333 + 300(20)(10) = 0$$
$$V_2 = 2000 \text{ lb}$$

Next, determine R_3 by taking $\Sigma M = 0$ at point 4 (Fig. 21–18):

$$-2000(50) - 30R_3 - 30,000 + 3333 + 300(20)(40) + 400(30)(15) = 0$$
$$R_3 = 9778 \text{ lb}$$

FIGURE 21–17 Free body— Span 2–3.

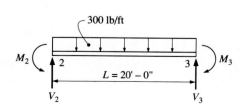

FIGURE 21–18   Free body—
Spans 2–3 and 3–4.

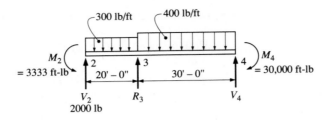

Then, using the free-body diagram shown in Fig. 21–16(b), taking $\Sigma M = 0$ at point 5, solve for R_4:

$$-2000(90) - 9778(70) - R_4(40) + 3333 - 25{,}000 + 300(20)(80)$$
$$+ 400(30)(55) + 200(40)(20) = 0$$
$$R_4 = 10{,}347 \text{ lb}$$

Lastly, using the free-body diagram shown in Fig. 21–19, taking $\Sigma M = 0$ at point 4, solve for V_5:

$$+40V_5 + 30{,}000 - 25{,}000 - 200(40)(20) = 0$$
$$V_5 = 3875 \text{ lb}$$

FIGURE 21–19   Free body—
Span 4–5.

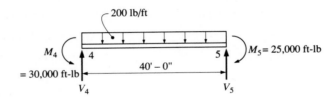

Note the shears at points 2 and 5 are equivalent to vertical reactions at those points. As a check, the sum of all the vertical forces on the beam must equal zero. Taking ΣF_y (up is positive) yields

$$+2000 + 9778 + 10{,}347 + 3875 - 300(20) - 400(30) - 200(40) = 0 \qquad \textbf{OK}$$

**SUMMARY—BY
SECTION NUMBER**

21–1   A statically indeterminate beam is one that has more unknown reactions than there are equations of equilibrium. To determine the unknown reactions, additional equations based on deflection or end rotation must be used.

21–2   Restrained beams include the propped cantilever and the fixed beam. A fixed support keeps the end of the beam from rotating (and translating).

21–3   A propped cantilever beam can be conveniently analyzed using the method of superposition. The important fact for this analysis is that the supports are on the same level and the deflection at the simply supported end is to be zero. The reaction at the simply supported end must cause a deflection equal in magnitude and opposite in sense to that caused by the load.

21–4   A fixed beam or a fixed-end beam can be analyzed using the method of superposition. In this case, it is convenient to consider slopes at the supports. Since the slope of the elastic curve at the supports is zero, the unknown moments must cause slopes equal in magnitude and opposite in sense to those caused by the loads. Tables of derived slope equations are furnished in this section.

21–5   A continuous beam is one which rests on more than two supports. In general, the method of superposition used for the one-span indeterminate beams cannot be conveniently used for continuous beams. An exception is a two-span symmetrically loaded continuous beam.

21–6   The theorem of three moments is a convenient method of finding bending moments in continuous beams. The analysis utilizes the relationship that exists between the bending moments at the three supports of any two adjacent spans. Equations (21–1) and (21–2) allow the determination of moments at the supports, after which reactions and shears can be found. Complete shear and moment diagrams can then be drawn.

PROBLEMS

All beams in these problems are assumed to have constant E and I values.

Section 21–3   Propped Cantilever Beams

1. Rework Example 21–1 changing dimension *a* to 10 ft, dimension *b* to 15 ft, and the point load to 15 kips.

2. Use the method of superposition to determine the reactions for the propped cantilever beam shown in Fig. 21–2. Span length is 20 ft and the uniformly distributed load *w* is 2.5 kips/ft. Neglect beam weight.

3. Use the method of superposition to determine the reactions for the propped cantilever beams shown in Fig. 21–20. Draw complete shear and moment diagrams. Neglect the beam weight.

Section 21–4   Fixed Beams

4. through 6. Draw complete shear and moment diagrams for the fixed beams shown in Fig. 21–21, Fig. 21–22, and Fig. 21–23. Select the lightest structural steel W shape. Neglect the beam weight. Assume that the allowable bending stress is 24 ksi and the allowable shear stress is 14.5 ksi.

FIGURE 21–20   Problem 3.

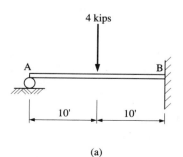

(a)

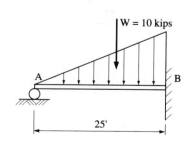

(b)

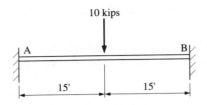

FIGURE 21–21   Problem 4.

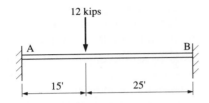

FIGURE 21–22   Problem 5.

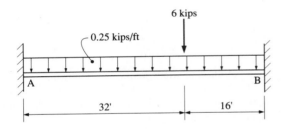

FIGURE 21–23   Problem 6.

Section 21–5   Continuous Beams—Superposition

7. Determine the reactions for the beam shown in Fig. 21–24.

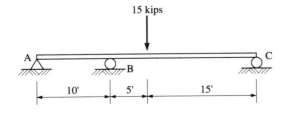

FIGURE 21–24   Problem 7.

8. Determine the reactions for the beam shown in Fig. 21–25.

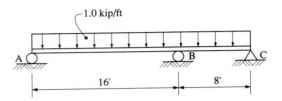

FIGURE 21–25   Problem 8.

9. Select a Southern pine timber beam (S4S) shown in Fig. 21–26. Consider shear and moment. See Appendix F for design values. Neglect the weight of the beam.

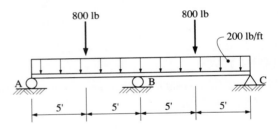

FIGURE 21–26   Problem 9.

Section 21–6   The Theorem of Three Moments

10. through **12.** For the continuous beams shown in Fig. 21–27, Fig. 21–28, and Fig. 21–29, find moments at the supports and the reactions. Draw complete shear and moment diagrams.

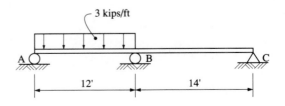

FIGURE 21–27   Problem 10.

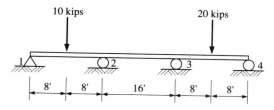

FIGURE 21–28   Problem 11.

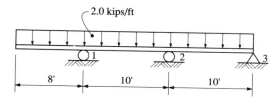

FIGURE 21–29   Problem 12.

Supplemental Problems

For these problems, use any appropriate method of analysis. Unless noted otherwise, neglect the beam weight in all problems.

13. A structural steel W12 × 30 has a span length of 24 ft. The beam is fixed at one end and simply supported at the other. Both supports are on the same level. The beam is subjected to a uniformly distributed load of 800 lb/ft which includes its own weight. Draw the shear and moment diagrams and calculate the maximum bending stress. Label the simple support *A* and the fixed support *B*.

14. Rework Problem 13 assuming that the simple support settles 0.6 in.

15. Rework Problem 13 assuming that the simple support is raised 0.6 in. above the level of the fixed end.

16. A structural steel W12 × 40 has a span length of 20 ft. The beam is fixed at one end and simply supported at the other end. Both supports are on the same level. The beam is loaded with a concentrated load of 20,000 lb at a point 8 ft from the simply supported end. Compute the maximum bending stress in the beam.

17. Rework Problem 16 assuming that the simply supported end is raised one inch above the level of the fixed end.

18. Draw complete shear and moment diagrams for the beam shown in Fig. 21–30.

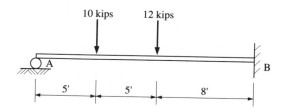

FIGURE 21–30   Problem 18.

19. Rework Problem 18 adding a uniformly distributed load of 1 kip/ft for the full span.

20. Select the most economical (lightest) structural steel W-shape beam to carry the superimposed loads shown in Fig. 21–31. Draw the shear and moment diagrams. Use an allowable bending stress of 22,000 psi and an allowable shear stress of 14,500 psi. Assume complete fixity at each end with supports on the same level.

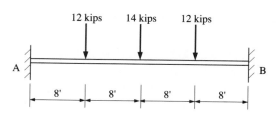

FIGURE 21–31   Problem 20.

21. Draw complete shear and moment diagrams for the beam shown in Fig. 21–32.

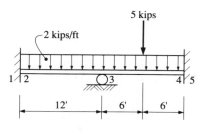

FIGURE 21–32   Problem 21.

22. Draw complete shear and moment diagrams for the beam shown in Fig. 21–33.

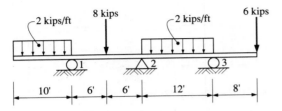

FIGURE 21–33   Problem 22.

23. A structural steel W18 × 50 supports loads as shown in Fig. 21–34. (a) Find the reactions and draw complete shear and moment diagrams. (b) Assume that the center support settles 0.3 in. Find the reactions and draw complete shear and moment diagrams.

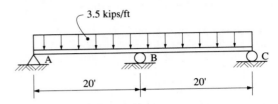

FIGURE 21–34   Problem 23.

Appendixes

Tables of SI usage and conversions are printed on the inside of the covers.

APPENDIXES NOTES   The following appendixes contain helpful data in both U.S. Customary units and SI units. Data on material properties and design values (Appendixes F and G) are average values presented expressly for use with the problems and solutions contained in this book. These values could vary significantly with specific materials; therefore, for actual applications, the reader should verify the data by referring to publications from the proper technical associations, such as the Aluminum Association, the American Society for Testing Materials, the National Forest Products Association, and so forth. The SI portions of these appendixes represent a "soft" conversion from U.S. Customary units (using the conversion factors printed inside the back cover of this book).

The following detailed notes are applicable:

1. Appendixes A–D. The U.S. Customary portion of this data was reproduced with permission of the American Institute of Steel Construction (AISC). For a more complete listing of available products, refer to the AISC Manual and/or manufacturers' literature. The SI nominal depths of wide-flange shapes and channels have been rounded to the nearest 10 mm. Other SI dimensions and properties for the most part reflect the numerical accuracy of the AISC Manual.

2. **Appendix E.** For the SI portion of this table, U.S. Customary dressed sizes were converted to millimeter sizes, and these were used to calculate the other properties. Nominal sizes have been rounded to the nearest 10 mm; dressed sizes and properties are shown to three significant digits.

3. **Appendix H.** This appendix was reproduced with permission of the American Institute of Steel Construction.

4. **Appendix J.** The AISC column formulas provided the basis for this appendix. However, for consistency within the text, the modulus of elasticity of steel used was 30,000,000 psi. This varies slightly from the AISC value of 29,000,000 psi. Also, no associated SI values of allowable axial compressive stress are given. To obtain the allowable axial compressive stress in MPa, multiply the tabulated value (in ksi) by 6.895.

Selected W Shapes—Dimensions and Properties

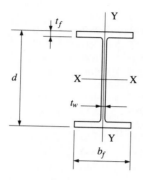

TABLE A–1 U.S. Customary

Designation	A in.^{2}	d in.	t_w in.	b_f in.	t_f in.	I_x in.^{4}	S_x in.^{3}	r_x in.	I_y in.^{4}	S_y in.^{3}	r_y in.
W44 × 285	83.8	44.02	1.024	11.811	1.772	24600	1120	17.1	490	83.0	2.42
W44 × 224	65.8	43.31	0.787	11.811	1.416	19200	889	17.1	391	66.0	2.44
W40 × 655	192.0	43.62	1.970	16.870	3.540	56500	2590	17.2	2860	339	3.86
W40 × 531	156.0	42.34	1.610	16.510	2.910	44300	2090	16.9	2200	266	3.75
W40 × 436	128.0	41.34	1.34	16.240	2.400	35400	1710	16.6	1720	212	3.67
W40 × 328	96.4	40.00	0.910	17.910	1.730	26800	1340	16.7	1660	185	4.15
W40 × 268	78.8	39.37	0.750	17.750	1.415	21500	1090	16.5	1320	149	4.09
W40 × 244	71.7	39.06	0.710	17.710	1.260	19200	983	16.4	1170	132	4.04
W40 × 183	53.7	38.98	0.650	11.810	1.220	13300	682	15.7	336	56.9	2.50

TABLE A–1   U.S. Customary (*continued*)

Designation	A in.^{2}	d in.	t_w in.	b_f in.	t_f in.	I_x in.^{4}	S_x in.^{3}	r_x in.	I_y in.^{4}	S_y in.^{3}	r_y in.
W36 × 848	249.0	42.45	2.520	18.130	4.530	67400	3170	16.4	4550	501	4.27
W36 × 650	190.0	40.47	1.970	17.575	3.540	48900	2420	16.0	3230	367	4.12
W36 × 527	154.0	39.21	1.610	17.220	2.910	38300	1950	15.8	2490	289	4.02
W36 × 439	128.0	38.26	1.360	16.965	2.440	31000	1620	15.6	1990	235	3.95
W36 × 359	105.0	37.40	1.120	16.730	2.010	24800	1320	15.4	1570	188	3.87
W36 × 300	88.3	36.74	0.945	16.655	1.680	20300	1110	15.2	1300	156	3.83
W36 × 260	76.5	36.26	0.840	16.550	1.440	17300	953	15.0	1090	132	3.78
W36 × 230	67.6	35.90	0.760	16.470	1.260	15000	837	14.9	940	114	3.73
W36 × 194	57.0	36.49	0.765	12.115	1.260	12100	664	14.6	375	61.9	2.56
W36 × 170	50.0	36.17	0.680	12.030	1.100	10500	580	14.5	320	53.2	2.53
W36 × 150	44.2	35.85	0.625	11.975	0.940	9040	504	14.3	270	45.1	2.47
W33 × 424	124.0	36.34	1.380	16.315	2.480	26900	1480	14.7	1800	221	3.81
W33 × 318	93.5	35.16	1.040	15.985	1.890	19500	1110	14.4	1290	161	3.71
W33 × 241	70.9	34.18	0.830	15.860	1.400	14200	829	14.1	932	118	3.63
W33 × 201	59.1	33.68	0.715	15.745	1.150	11500	684	14.0	749	95.2	3.56
W33 × 141	41.6	33.30	0.605	11.535	0.960	7450	448	13.4	246	42.7	2.43
W33 × 118	34.7	32.86	0.550	11.480	0.740	5900	359	13.0	187	32.6	2.32
W30 × 391	114.0	33.19	1.360	15.590	2.440	20700	1250	13.5	1550	198	3.68
W30 × 292	85.7	32.01	1.020	15.255	1.850	14900	928	13.2	1100	144	3.58
W30 × 191	56.1	30.68	0.710	15.040	1.185	9170	598	12.8	673	89.5	3.46
W30 × 132	38.9	30.31	0.615	10.545	1.000	5770	380	12.2	196	37.2	2.25
W30 × 116	34.2	30.01	0.565	10.495	0.850	4930	329	12.0	164	31.3	2.19
W30 × 108	31.7	29.83	0.545	10.475	0.760	4470	299	11.9	146	27.9	2.15
W30 × 99	29.1	29.65	0.520	10.450	0.670	3990	269	11.7	128	24.5	2.10
W27 × 307	90.2	29.61	1.160	14.445	2.090	13100	884	12.0	1050	146	3.42
W27 × 235	69.1	28.66	0.910	14.190	1.610	9660	674	11.8	768	108	3.33
W27 × 161	47.4	27.59	0.660	14.020	1.080	6280	455	11.5	497	70.9	3.24
W27 × 114	33.5	27.29	0.570	10.070	0.930	4090	299	11.0	159	31.5	2.18
W27 × 94	27.7	26.92	0.490	9.990	0.745	3270	243	10.9	124	24.8	2.12
W24 × 450	132.0	29.09	1.810	13.955	3.270	17100	1170	11.4	1490	214	3.36
W24 × 306	89.8	27.13	1.260	13.405	2.280	10700	789	10.9	919	137	3.20
W24 × 207	60.7	25.71	0.870	13.010	1.570	6820	531	10.6	578	88.8	3.08
W24 × 162	47.7	25.00	0.705	12.955	1.220	5170	414	10.4	443	68.4	3.05
W24 × 131	38.5	24.48	0.605	12.855	0.960	4020	329	10.2	340	53.0	2.97
W24 × 104	30.6	24.06	0.500	12.750	0.750	3100	258	10.1	259	40.7	2.91
W24 × 84	24.7	24.10	0.470	9.020	0.770	2370	196	9.79	94.4	20.9	1.95
W24 × 76	22.4	23.92	0.440	8.990	0.680	2100	176	9.69	82.5	18.4	1.92
W24 × 62	18.2	23.74	0.430	7.040	0.590	1550	131	9.23	34.5	9.80	1.38

TABLE A-1 U.S. Customary (*continued*)

Designation	A in.^{2}	d in.	t_w in.	b_f in.	t_f in.	I_x in.^{4}	S_x in.^{3}	r_x in.	I_y in.^{4}	S_y in.^{3}	r_y in.
W21 × 147	43.2	22.06	0.720	12.510	1.150	3630	329	9.17	376	60.1	2.95
W21 × 122	35.9	21.68	0.600	12.390	0.960	2960	273	9.09	305	49.2	2.92
W21 × 101	29.8	21.36	0.500	12.290	0.800	2420	227	9.02	248	40.3	2.89
W21 × 83	24.3	21.43	0.515	8.355	0.835	1830	171	8.67	81.4	19.5	1.83
W21 × 73	21.5	21.24	0.455	8.295	0.740	1600	151	8.64	70.6	17.0	1.81
W21 × 62	18.3	20.99	0.400	8.240	0.615	1330	127	8.54	57.5	13.9	1.77
W21 × 50	14.7	20.83	0.380	6.530	0.535	984	94.5	8.18	24.9	7.64	1.30
W18 × 119	35.1	18.97	0.655	11.265	1.060	2190	231	7.90	253	44.9	2.69
W18 × 97	28.5	18.59	0.535	11.145	0.870	1750	188	7.82	201	36.1	2.65
W18 × 76	22.3	18.21	0.425	11.035	0.680	1330	146	7.73	152	27.6	2.61
W18 × 71	20.8	18.47	0.495	7.635	0.810	1170	127	7.50	60.3	15.8	1.70
W18 × 60	17.6	18.24	0.415	7.555	0.695	984	108	7.47	50.1	13.3	1.69
W18 × 50	14.7	17.99	0.355	7.495	0.570	800	88.9	7.38	40.1	10.7	1.65
W18 × 40	11.8	17.90	0.315	6.015	0.525	612	68.4	7.21	19.1	6.35	1.27
W16 × 100	29.4	16.97	0.585	10.425	0.985	1490	175	7.10	186	35.7	2.51
W16 × 77	22.6	16.52	0.455	10.295	0.760	1110	134	7.00	138	26.9	2.47
W16 × 57	16.8	16.43	0.430	7.120	0.715	758	92.2	6.72	43.1	12.1	1.60
W16 × 45	13.3	16.13	0.345	7.035	0.565	586	72.7	6.65	32.8	9.34	1.57
W16 × 36	10.6	15.86	0.295	6.985	0.430	448	56.5	6.51	24.5	7.00	1.52
W16 × 26	7.68	15.69	0.250	5.500	0.345	301	38.4	6.26	9.59	3.49	1.12
W14 × 730	215.0	22.42	3.070	17.890	4.910	14300	1280	8.17	4720	527	4.69
W14 × 605	178.0	20.92	2.595	17.415	4.160	10800	1040	7.80	3680	423	4.55
W14 × 500	147.0	19.60	2.190	17.010	3.500	8210	838	7.48	2880	339	4.43
W14 × 398	117.0	18.29	1.770	16.590	2.845	6000	656	7.16	2170	262	4.31
W14 × 311	91.4	17.12	1.410	16.230	2.260	4330	506	6.88	1610	199	4.20
W14 × 233	68.5	16.04	1.070	15.890	1.720	3010	375	6.63	1150	145	4.10
W14 × 176	51.8	15.22	0.830	15.650	1.310	2140	281	6.43	838	107	4.02
W14 × 132	38.8	14.66	0.645	14.725	1.030	1530	209	6.28	548	74.5	3.76
W14 × 109	32.0	14.32	0.525	14.605	0.860	1240	173	6.22	447	61.2	3.73
W14 × 90	26.5	14.02	0.440	14.520	0.710	999	143	6.14	362	49.9	3.70
W14 × 74	21.8	14.17	0.450	10.070	0.785	796	112	6.04	134	26.6	2.48
W14 × 61	17.9	13.89	0.375	9.995	0.645	640	92.2	5.98	107	21.5	2.45
W14 × 48	14.1	13.79	0.340	8.030	0.595	485	70.3	5.85	51.4	12.8	1.91
W14 × 38	11.2	14.10	0.310	6.770	0.515	385	54.6	5.87	26.7	7.88	1.55
W14 × 34	10.0	13.98	0.285	6.745	0.455	340	48.6	5.83	23.3	6.91	1.53
W14 × 30	8.85	13.84	0.270	6.730	0.385	291	42.0	5.73	19.6	5.82	1.49
W14 × 22	6.49	13.74	0.230	5.000	0.335	199	29.0	5.54	7.00	2.80	1.04

TABLE A–1 U.S. Customary (*continued*)

Designation	A in.^{2}	d in.	t_w in.	b_f in.	t_f in.	I_x in.^{4}	S_x in.^{3}	r_x in.	I_y in.^{4}	S_y in.^{3}	r_y in.
W12 × 336	98.8	16.82	1.775	13.385	2.955	4060	483	6.41	1190	177	3.47
W12 × 279	81.9	15.85	1.530	13.140	2.470	3110	393	6.16	937	143	3.38
W12 × 230	67.7	15.05	1.285	12.895	2.070	2420	321	5.97	742	115	3.31
W12 × 170	50.0	14.03	0.960	12.570	1.560	1650	235	5.74	517	82.3	3.22
W12 × 136	39.9	13.41	0.790	12.400	1.250	1240	186	5.58	398	64.2	3.16
W12 × 96	28.2	12.71	0.550	12.160	0.900	833	131	5.44	270	44.4	3.09
W12 × 72	21.1	12.25	0.430	12.040	0.670	597	97.4	5.31	195	32.4	3.04
W12 × 58	17.0	12.19	0.360	10.010	0.640	475	78.0	5.28	107	21.4	2.51
W12 × 50	14.7	12.19	0.370	8.080	0.640	394	64.7	5.18	56.3	13.9	1.96
W12 × 40	11.8	11.94	0.295	8.005	0.515	310	51.9	5.13	44.1	11.0	1.93
W12 × 30	8.79	12.34	0.260	6.520	0.440	238	38.6	5.21	20.3	6.24	1.52
W12 × 22	6.48	12.31	0.260	4.030	0.425	156	25.4	4.91	4.66	2.31	0.847
W12 × 16	4.71	11.99	0.220	3.990	0.265	103	17.1	4.67	2.82	1.41	0.773
W10 × 112	32.9	11.36	0.755	10.415	1.250	716	126	4.66	236	45.3	2.68
W10 × 88	25.9	10.84	0.605	10.265	0.990	534	98.5	4.54	179	34.8	2.63
W10 × 68	20.0	10.40	0.470	10.130	0.770	394	75.7	4.44	134	26.4	2.59
W10 × 54	15.8	10.09	0.370	10.030	0.615	303	60.0	4.37	103	20.6	2.56
W10 × 45	13.3	10.10	0.350	8.020	0.620	248	49.1	4.32	53.4	13.3	2.01
W10 × 33	9.71	9.73	0.290	7.960	0.435	170	35.0	4.19	36.6	9.20	1.94
W10 × 22	6.49	10.17	0.240	5.750	0.360	118	23.2	4.27	11.4	3.97	1.33
W10 × 17	4.99	10.11	0.240	4.010	0.330	81.9	16.2	4.05	3.56	1.78	0.844
W10 × 12	3.54	9.87	0.190	3.960	0.210	53.8	10.9	3.90	2.18	1.10	0.785
W8 × 58	17.1	8.75	0.510	8.220	0.810	228	52.0	3.65	75.1	18.3	2.10
W8 × 40	11.7	8.25	0.360	8.070	0.560	146	35.5	3.53	49.1	12.2	2.04
W8 × 31	9.13	8.00	0.285	7.995	0.435	110	27.5	3.47	37.1	9.27	2.02
W8 × 24	7.08	7.93	0.245	6.495	0.400	82.8	20.9	3.42	18.3	5.63	1.61
W8 × 18	5.26	8.14	0.230	5.250	0.330	61.9	15.2	3.43	7.97	3.04	1.23
W8 × 13	3.84	7.99	0.230	4.000	0.255	39.6	9.91	3.21	2.73	1.37	0.843
W6 × 25	7.34	6.38	0.320	6.080	0.455	53.4	16.7	2.70	17.1	5.61	1.52
W6 × 15	4.43	5.99	0.230	5.990	0.260	29.1	9.72	2.56	9.32	3.11	1.46
W6 × 12	3.55	6.03	0.230	4.000	0.280	22.1	7.31	2.49	2.99	1.50	0.918
W5 × 19	5.54	5.15	0.270	5.030	0.430	26.2	10.2	2.17	9.13	3.63	1.28
W4 × 13	3.83	4.16	0.280	4.060	0.345	11.3	5.46	1.72	3.86	1.90	1.00

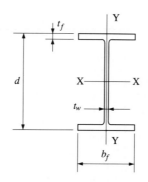

TABLE A–2   SI

Designation mm × kN/m	A m² × 10⁻³	d mm	t_w mm	b_f mm	t_f mm	I_x m⁴ × 10⁻⁶	S_x m³ × 10⁻³	r_x mm	I_y m⁴ × 10⁻⁶	S_y m³ × 10⁻³	r_y mm
W1120 × 4.16	54.1	1118	26.01	300	45.01	10240	18.4	434	204	1.36	61.5
W1120 × 3.27	42.5	1100	19.99	300	35.97	7990	14.6	434	163	1.08	62.0
W1020 × 9.56	123.9	1108	50.04	428	89.92	23500	42.4	437	1190	5.56	98.0
W1020 × 7.75	100.7	1075	40.89	419	73.91	18440	34.2	429	916	4.36	95.3
W1020 × 6.36	82.6	1050	34.04	412	60.96	14730	28.0	422	716	3.47	93.2
W1020 × 4.79	62.2	1016	23.11	455	43.94	11160	22.0	424	691	3.03	105
W1020 × 3.91	50.8	1000	19.05	451	35.94	8950	17.9	419	549	2.44	104
W1020 × 3.56	46.3	992	18.03	450	32.00	7990	16.1	417	487	2.16	103
W1020 × 2.67	34.6	990	16.51	300	30.99	5540	11.2	399	140	0.932	63.5
W910 × 12.4	160.7	1078	64.01	461	115.1	28100	51.9	417	1894	8.21	108
W910 × 9.49	122.6	1028	50.04	446	82.92	20400	39.7	406	1344	6.01	105
W910 × 7.69	99.4	996	40.89	437	73.91	15940	32.0	401	1036	4.74	102
W910 × 6.41	82.6	972	34.54	431	61.98	12900	26.5	396	828	3.85	100
W910 × 5.24	67.7	950	28.45	425	51.05	10320	21.6	391	653	3.08	98.3
W910 × 4.38	57.0	933	24.0	423	42.7	8450	18.2	386	541	2.56	97.3
W910 × 3.79	49.4	921	21.3	420	36.6	7200	15.6	381	454	2.16	96.0
W910 × 3.36	43.6	912	19.3	418	32.0	6240	13.7	378	391	1.87	94.7
W910 × 2.83	36.8	927	19.4	308	32.0	5040	10.9	371	156	1.01	65.0
W910 × 2.48	32.3	919	17.3	306	27.9	4370	9.50	368	133	0.872	64.3
W910 × 2.19	28.5	911	15.9	304	23.9	3770	8.26	363	112	0.739	62.7
W840 × 6.19	80.0	923	35.1	414	63.0	11200	24.3	373	749	3.62	96.8
W840 × 4.64	60.3	893	26.4	406	48.0	8120	18.2	366	537	2.64	94.2
W840 × 3.52	45.7	868	21.1	403	35.6	5910	13.6	358	388	1.93	92.2
W840 × 2.93	38.1	855	18.2	400	29.2	4790	11.2	356	312	1.56	90.4
W840 × 2.06	26.8	846	15.4	293	24.4	3100	7.34	340	102	0.700	61.7
W840 × 1.72	22.4	835	14.0	292	18.8	2460	5.88	330	77.8	0.534	58.9

TABLE A–2 SI (*continued*)

Designation mm × kN/m	A m² × 10⁻³	d mm	t_w mm	b_f mm	t_f mm	I_x m⁴ × 10⁻⁶	S_x m³ × 10⁻³	r_x mm	I_y m⁴ × 10⁻⁶	S_y m³ × 10⁻³	r_y mm
W760 × 5.71	73.6	843	34.5	396	62.0	8620	20.5	343	645	3.24	93.5
W760 × 4.26	55.3	813	25.9	387	47.0	6200	15.2	335	458	2.36	90.9
W760 × 2.79	36.2	779	18.0	382	30.1	3820	9.80	325	280	1.47	87.9
W760 × 1.93	25.1	770	15.6	268	25.4	2400	6.23	310	81.6	0.610	57.2
W760 × 1.69	22.1	762	14.4	267	21.6	2050	5.39	305	68.3	0.513	55.6
W760 × 1.58	20.5	758	13.8	266	19.3	1860	4.90	302	60.8	0.457	54.6
W760 × 1.44	18.8	753	13.2	265	17.0	1660	4.41	297	53.3	0.401	53.3
W690 × 4.48	58.2	752	29.5	367	53.09	5450	14.5	305	437	2.39	86.9
W690 × 3.43	44.6	728	23.1	360	40.89	4020	11.0	300	320	1.77	84.6
W690 × 2.35	30.6	701	16.8	356	27.4	2610	7.46	292	207	1.16	82.3
W690 × 1.66	21.6	693	14.5	256	23.6	1700	4.90	279	66.2	0.516	55.4
W690 × 1.37	17.9	684	12.4	254	18.9	1360	3.98	277	51.6	0.406	53.8
W610 × 6.57	85.2	739	46.0	354	83.0	7120	19.2	290	620	3.51	85.3
W610 × 4.47	57.9	689	32.0	340	57.9	4450	12.9	277	383	2.25	81.3
W610 × 3.02	39.2	653	22.1	330	39.9	2840	8.70	269	241	1.46	78.2
W610 × 2.36	30.8	635	17.9	329	31.0	2150	6.78	264	184	1.12	77.5
W610 × 1.91	24.8	622	15.4	327	24.4	1670	5.39	259	142	0.869	75.4
W610 × 1.52	19.7	611	12.7	324	19.0	1290	4.23	257	108	0.667	73.9
W610 × 1.23	15.9	612	11.9	229	19.6	986	3.21	249	39.3	0.342	49.5
W610 × 1.11	14.5	608	11.2	228	17.3	874	2.88	246	34.3	0.302	48.8
W610 × 0.905	11.7	603	10.9	179	15.0	645	2.15	234	14.4	0.161	35.1
W530 × 2.15	27.8	560	18.3	318	29.2	1510	5.39	233	157	0.985	74.9
W530 × 1.78	23.2	551	15.2	315	24.4	1230	4.47	231	127	0.806	74.2
W530 × 1.47	19.2	543	12.7	312	20.3	1010	3.72	229	103	0.660	73.4
W530 × 1.21	15.7	544	13.1	212	21.2	762	2.80	220	33.9	0.320	46.5
W530 × 1.07	13.9	539	11.6	211	18.8	666	2.47	219	29.4	0.279	46.0
W530 × 0.90	11.8	533	10.2	209	15.6	554	2.08	217	23.9	0.228	45.0
W530 × 0.73	9.48	529	9.7	166	13.6	410	1.55	208	10.4	0.125	33.0
W460 × 1.74	22.6	482	16.6	286	26.9	912	3.79	201	105	0.736	68.3
W460 × 1.42	18.4	472	13.6	283	22.1	728	3.08	199	83.7	0.592	67.3
W460 × 1.11	14.4	463	10.8	280	17.3	554	2.39	196	63.3	0.452	66.3
W460 × 1.04	13.4	469	12.6	194	20.6	487	2.08	191	25.1	0.259	43.2
W460 × 0.88	11.4	463	10.5	192	17.7	410	1.77	190	20.9	0.218	42.9
W460 × 0.73	9.48	457	9.0	190	14.5	333	1.46	187	16.7	0.175	41.9
W460 × 0.58	7.61	455	8.0	153	13.3	255	1.12	183	7.95	0.104	32.3

TABLE A–2 SI (*continued*)

Designation mm × kN/m	A m² × 10⁻³	d mm	t_w mm	b_f mm	t_f mm	I_x m⁴ × 10⁻⁶	S_x m³ × 10⁻³	r_x mm	I_y m⁴ × 10⁻⁶	S_y m³ × 10⁻³	r_y mm
W410 × 1.46	19.0	431	14.9	265	25.0	620	2.87	180	77.4	0.585	63.8
W410 × 1.12	14.6	420	11.6	261	19.3	462	2.20	178	57.4	0.441	62.7
W410 × 0.83	10.8	417	10.9	181	18.2	316	1.51	171	17.9	0.198	40.6
W410 × 0.66	8.58	410	8.8	179	14.4	244	1.19	169	13.7	0.153	39.9
W410 × 0.53	6.84	403	7.5	177	10.9	186	0.926	165	10.2	0.115	38.6
W410 × 0.38	4.95	399	6.4	140	8.8	125	0.629	159	3.99	0.0572	28.4
W360 × 10.65	139	569	78.0	454	124.7	5950	21.0	208	1960	8.64	119
W360 × 8.83	115	531	65.9	442	105.7	4500	17.0	198	1530	6.93	116
W360 × 7.30	94.8	498	55.6	432	88.9	3420	13.7	190	1200	5.56	113
W360 × 5.81	75.5	465	45.0	421	72.3	2500	10.7	182	903	4.29	109
W360 × 4.54	59.0	435	35.8	412	57.4	1800	8.29	175	670	3.26	107
W360 × 3.40	44.2	407	27.2	404	43.7	1250	6.15	168	479	2.38	104
W360 × 2.57	33.4	387	21.1	398	33.3	891	4.60	163	349	1.75	102
W360 × 1.93	25.1	372	16.4	374	26.2	637	3.42	160	228	1.22	95.5
W360 × 1.59	20.6	364	13.3	371	21.8	516	2.83	158	186	1.00	94.7
W360 × 1.31	17.1	356	11.2	369	18.0	416	2.34	156	151	0.818	94.0
W360 × 1.08	14.1	360	11.4	256	19.9	331	1.84	153	55.8	0.436	63.0
W360 × 0.89	11.5	353	9.5	254	16.4	266	1.51	152	44.5	0.352	62.2
W360 × 0.70	9.10	350	8.6	204	15.1	202	1.15	149	21.4	0.210	48.5
W360 × 0.55	7.23	358	7.9	172	13.1	160	0.895	149	11.1	0.129	39.4
W360 × 0.50	6.45	355	7.2	171	11.6	142	0.796	148	9.70	0.113	38.9
W360 × 0.44	5.71	352	6.9	171	9.8	121	0.688	146	8.16	0.0954	37.8
W360 × 0.32	4.19	349	5.8	127	8.5	82.8	0.475	141	2.91	0.0459	26.4
W300 × 4.90	63.7	427	45.1	340	75.1	1690	7.91	163	495	2.90	88.1
W300 × 4.07	52.8	403	38.9	334	62.7	1290	6.44	156	390	2.34	85.9
W300 × 3.36	43.7	382	32.6	328	52.6	1010	5.26	152	309	1.88	84.1
W300 × 2.48	32.3	356	24.4	319	39.6	690	3.85	146	215	1.35	81.8
W300 × 1.98	25.7	341	20.1	315	31.7	516	3.05	142	166	1.05	80.3
W300 × 1.40	18.2	323	14.0	309	22.9	347	2.15	138	112	0.728	78.5
W300 × 1.05	13.6	311	10.9	306	17.0	249	1.60	135	81.2	0.531	77.2
W300 × 0.85	11.0	310	9.1	254	16.3	198	1.28	134	44.5	0.351	63.8
W300 × 0.73	9.48	310	9.4	205	16.3	164	1.06	132	23.4	0.228	49.8
W300 × 0.58	7.61	303	7.5	203	13.1	129	0.850	130	18.4	0.180	49.0
W300 × 0.44	5.67	313	6.6	166	11.2	99.1	0.633	132	8.45	0.102	38.6
W300 × 0.32	4.18	313	6.6	102	10.8	64.9	0.416	125	1.94	0.0379	21.5
W300 × 0.23	3.04	305	5.6	101	6.7	42.9	0.280	119	1.17	0.0231	19.6

TABLE A–2 SI (*continued*)

Designation mm × kN/m	A m² × 10⁻³	d mm	t_w mm	b_f mm	t_f mm	I_x m⁴ × 10⁻⁶	S_x m³ × 10⁻³	r_x mm	I_y m⁴ × 10⁻⁶	S_y m³ × 10⁻³	r_y mm
W250 × 1.63	21.2	289	19.2	265	31.8	298	2.06	118	98.2	0.742	68.1
W250 × 1.28	16.7	275	15.4	261	25.1	222	1.61	115	74.5	0.570	66.8
W250 × 0.99	12.9	264	11.9	257	19.6	164	1.24	113	55.8	0.433	65.8
W250 × 0.79	10.2	256	9.4	255	15.6	126	0.983	111	42.9	0.338	65.0
W250 × 0.66	8.58	257	8.9	204	15.7	103	0.805	110	22.2	0.218	51.1
W250 × 0.48	6.26	247	7.4	202	11.0	70.8	0.574	106	15.2	0.151	49.3
W250 × 0.32	4.19	258	6.1	146	9.1	49.1	0.380	108	4.75	0.0651	33.8
W250 × 0.25	3.22	257	6.1	102	8.4	34.1	0.266	103	1.48	0.0292	21.5
W250 × 0.18	2.28	251	4.8	101	5.3	22.4	0.179	99.1	0.907	0.0180	19.9
W200 × 0.85	11.0	222	13.0	209	20.6	94.9	0.852	92.7	31.3	0.300	53.3
W200 × 0.58	7.55	210	9.1	205	14.2	60.8	0.582	89.7	20.4	0.200	51.8
W200 × 0.45	5.89	203	7.2	203	11.0	45.8	0.451	88.1	15.4	0.152	51.3
W200 × 0.35	4.57	201	6.2	165	10.2	34.5	0.342	86.9	7.62	0.0923	40.9
W200 × 0.26	3.39	207	5.8	133	8.4	25.8	0.249	87.1	3.32	0.0498	31.2
W200 × 0.19	2.48	203	5.8	102	6.5	16.5	0.162	81.5	1.14	0.0225	21.4
W150 × 0.36	4.47	162	8.1	154	11.6	22.2	0.274	68.6	7.12	0.0919	38.6
W150 × 0.22	2.86	152	5.8	152	6.6	12.1	0.159	65.0	3.88	0.0510	37.1
W150 × 0.18	2.29	153	5.8	102	7.1	9.2	0.120	63.2	1.24	0.0246	23.3
W130 × 0.28	3.57	131	6.9	128	10.9	10.9	0.167	55.1	3.80	0.0595	32.5
W100 × 0.19	2.47	106	7.1	103	8.76	4.7	0.089	43.7	1.61	0.0311	25.4

APPENDIX B

Pipe—Dimensions and Properties

TABLE B–1 U.S. Customary

Nominal Diameter in.	Outside Diameter in.	Inside Diameter in.	Wall Thickness in.	Weight lb/ft	A in.^{2}	I in.^{4}	S in.^{3}	r in.
Standard Weight								
2	2.375	2.067	0.154	3.65	1.07	0.666	0.561	0.787
2½	2.875	2.469	0.203	5.79	1.70	1.53	1.06	0.947
3	3.500	3.068	0.216	7.58	2.23	3.02	1.72	1.16
3½	4.000	3.548	0.226	9.11	2.68	4.79	2.39	1.34
4	4.500	4.026	0.237	10.79	3.17	7.23	3.21	1.51
5	5.563	5.047	0.258	14.62	4.30	15.2	5.45	1.88
6	6.625	6.065	0.280	18.97	5.58	28.1	8.50	2.25
8	8.625	7.981	0.322	28.55	8.40	72.5	16.8	2.94
10	10.750	10.020	0.365	40.48	11.9	161	29.9	3.67
12	12.750	12.000	0.375	49.56	14.6	279	43.8	4.38
Extra Strong								
2	2.375	1.939	0.218	5.02	1.48	0.868	0.731	0.766
2½	2.875	2.323	0.276	7.66	2.25	1.92	1.34	0.924
3	3.500	2.900	0.300	10.25	3.02	3.89	2.23	1.14
3½	4.000	3.364	0.318	12.50	3.68	6.28	3.14	1.31
4	4.500	3.826	0.337	14.98	4.41	9.61	4.27	1.48
5	5.563	4.813	0.375	20.78	6.11	20.7	7.43	1.84
6	6.625	5.761	0.432	28.57	8.40	40.5	12.2	2.19
8	8.625	7.625	0.500	43.39	12.8	106	24.5	2.88
10	10.750	9.750	0.500	54.74	16.1	212	39.4	3.63
12	12.750	11.750	0.500	65.42	19.2	362	56.7	4.33

TABLE B–2   SI

Nominal Diameter mm	Outside Diameter mm	Inside Diameter mm	Wall Thickness mm	Weight kN/m $\times 10^{-3}$	A m^2 $\times 10^{-3}$	I m^4 $\times 10^{-6}$	S m^3 $\times 10^{-6}$	r mm
Standard Weight								
50	60.33	52.50	3.91	53.3	0.690	0.277	9.19	20.0
65	73.03	62.71	5.16	84.5	1.097	0.637	17.37	24.1
75	88.90	77.93	5.49	110.6	1.439	1.257	28.2	29.5
90	101.6	90.12	5.74	132.9	1.729	1.994	39.2	34.0
100	114.3	102.3	6.02	157.5	2.05	3.01	52.6	38.4
125	141.3	128.2	6.55	213	2.77	6.33	89.3	47.8
150	168.3	154.1	7.11	277	3.60	11.70	139	57.2
205	219.1	202.7	8.18	417	5.42	30.2	275	74.5
255	273.1	254.5	9.27	591	7.68	67.0	490	93.2
305	323.9	304.8	9.53	723	9.42	116.1	718	111
Extra Strong								
50	60.33	49.25	5.54	73.3	0.955	0.361	12.0	19.5
65	73.03	59.00	7.01	111.8	1.452	0.799	22.0	23.5
75	88.90	73.66	7.62	149.6	1.948	1.619	36.5	29.0
90	101.6	85.45	8.08	182.4	2.37	2.61	51.5	33.3
100	114.3	97.18	8.56	219	2.85	4.00	70.0	37.6
125	141.3	122.3	9.53	303	3.94	8.62	122	46.7
150	168.3	146.3	10.97	417	5.42	16.86	200	55.6
205	219.1	193.7	12.70	633	8.26	44.1	401	73.2
255	273.1	247.7	12.70	799	10.39	88.2	646	92.2
305	323.9	298.5	12.70	955	12.39	150.7	929	110

Selected Channels—Dimensions and Properties

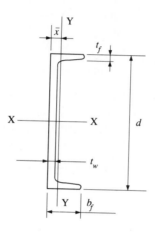

TABLE C–1 U.S. Customary

Designation	A in.²	d in.	t_w in.	b_f in.	Avg. t_f in.	$\bar{x}$ in.	I_x in.⁴	S_x in.³	r_x in.	I_y in.⁴	S_y in.³	r_y in.
C15 × 50	14.7	15.00	0.716	3.716	0.650	0.798	404	53.8	5.24	11.0	3.78	0.867
C15 × 40	11.8	15.00	0.520	3.520	0.650	0.777	349	46.5	5.44	9.23	3.37	0.886
C15 × 33.9	9.96	15.00	0.400	3.400	0.650	0.787	315	42.0	5.62	8.13	3.11	0.904
C12 × 30	8.82	12.00	0.510	3.170	0.501	0.674	162	27.0	4.29	5.14	2.06	0.763
C12 × 25	7.35	12.00	0.387	3.047	0.501	0.674	144	24.1	4.43	4.47	1.88	0.780
C12 × 20.7	6.09	12.00	0.282	2.942	0.501	0.698	129	21.5	4.61	3.88	1.73	0.799

TABLE C–1   U.S. Customary (*continued*)

Designation	A in.²	d in.	t_w in.	b_f in.	Avg. t_f in.	$\bar{x}$ in.	I_x in.⁴	S_x in.³	r_x in.	I_y in.⁴	S_y in.³	r_y in.
C10 × 30	8.82	10.00	0.673	3.033	0.436	0.649	103	20.7	3.42	3.94	1.65	0.669
C10 × 25	7.35	10.00	0.526	2.886	0.436	0.617	91.2	18.2	3.52	3.36	1.48	0.676
C10 × 20	5.88	10.00	0.379	2.739	0.436	0.606	78.9	15.8	3.66	2.81	1.32	0.692
C10 × 15.3	4.49	10.00	0.240	2.600	0.436	0.634	67.4	13.5	3.87	2.28	1.16	0.713
C9 × 20	5.88	9.00	0.448	2.648	0.413	0.583	60.9	13.5	3.22	2.42	1.17	0.642
C9 × 15	4.41	9.00	0.285	2.485	0.413	0.586	51.0	11.3	3.40	1.93	1.01	0.661
C9 × 13.4	3.94	9.00	0.233	2.433	0.413	0.601	47.9	10.6	3.48	1.76	0.962	0.669
C8 × 18.75	5.51	8.00	0.487	2.527	0.390	0.565	44.0	11.0	2.82	1.98	1.01	0.599
C8 × 13.75	4.04	8.00	0.303	2.343	0.390	0.553	36.1	9.03	2.99	1.53	0.854	0.615
C8 × 11.5	3.38	8.00	0.220	2.260	0.390	0.571	32.6	8.14	3.11	1.32	0.781	0.625
C7 × 14.75	4.33	7.00	0.419	2.299	0.366	0.532	27.2	7.78	2.51	1.38	0.779	0.564
C7 × 12.25	3.60	7.00	0.314	2.194	0.366	0.525	24.2	6.93	2.60	1.17	0.703	0.571
C7 × 9.8	2.87	7.00	0.210	2.090	0.366	0.540	21.3	6.08	2.72	0.968	0.625	0.581
C6 × 13	3.83	6.00	0.437	2.157	0.343	0.514	17.4	5.80	2.13	1.05	0.642	0.525
C6 × 10.5	3.09	6.00	0.314	2.034	0.343	0.499	15.2	5.06	2.22	0.866	0.564	0.529
C6 × 8.2	2.40	6.00	0.200	1.920	0.343	0.511	13.1	4.38	2.34	0.693	0.492	0.537

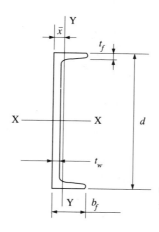

TABLE C–2 SI

Designation mm × kN/m	A m^2 $\times 10^{-3}$	d mm	t_w mm	b_f mm	Avg. t_f mm	$\bar{x}$ mm	I_x m^4 $\times 10^{-6}$	S_x m^3 $\times 10^{-6}$	r_x mm	I_y m^4 $\times 10^{-6}$	S_y m^3 $\times 10^{-6}$	r_y mm
C380 × 0.730	9.48	381.0	18.2	94.4	16.5	20.3	168	882	133	4.58	61.9	22.0
C380 × 0.584	7.61	381.0	13.2	89.4	16.5	19.7	145	762	138	3.84	55.2	22.5
C380 × 0.495	6.43	381.0	10.2	86.4	16.5	20.0	131	688	143	3.38	51.0	23.0
C300 × 0.438	5.69	304.8	13.0	80.5	12.7	17.1	67.4	442	109	2.14	33.8	19.4
C300 × 0.365	4.74	304.8	9.83	77.4	12.7	17.1	59.9	395	113	1.86	30.8	19.8
C300 × 0.302	3.93	304.8	7.16	74.7	12.7	17.7	53.7	352	117	1.61	28.3	20.3
C250 × 0.438	5.69	254.0	17.1	77.0	11.1	16.5	42.9	339	86.9	1.64	27.0	17.0
C250 × 0.365	4.74	254.0	13.4	73.3	11.1	15.7	38.0	298	89.4	1.40	24.3	17.2
C250 × 0.292	3.79	254.0	9.63	69.6	11.1	15.4	32.8	259	93.0	1.17	21.6	17.6
C250 × 0.223	2.90	254.0	6.10	66.0	11.1	16.1	28.1	221	98.3	0.949	19.0	18.1
C230 × 0.292	3.79	228.6	11.4	67.3	10.5	14.8	25.3	221	81.8	1.01	19.2	16.3
C230 × 0.219	2.85	228.6	7.24	63.1	10.5	14.9	21.2	185	86.4	0.803	16.6	16.8
C230 × 0.196	2.54	228.6	5.92	61.8	10.5	15.3	19.9	174	88.4	0.733	15.8	17.0
C200 × 0.274	3.55	203.2	12.4	64.2	9.91	14.4	18.3	180	71.6	0.824	16.6	15.2
C200 × 0.201	2.61	203.2	7.70	59.5	9.91	14.0	15.0	148	75.9	0.637	14.0	15.6
C200 × 0.168	2.18	203.2	5.59	57.4	9.91	14.5	13.6	133	79.0	0.549	12.8	15.9
C180 × 0.215	2.79	177.8	10.6	58.4	9.30	13.5	11.3	127	63.8	0.574	12.8	14.3
C180 × 0.179	2.32	177.8	7.98	55.7	9.30	13.3	10.1	114	66.0	0.487	11.5	14.5
C180 × 0.143	1.85	177.8	5.33	53.1	9.30	13.7	8.87	99.6	69.1	0.403	10.2	14.8
C150 × 0.190	2.47	152.4	11.1	54.8	8.71	13.1	7.24	95.0	54.1	0.437	10.5	13.3
C150 × 0.153	1.99	152.4	7.98	51.7	8.71	12.7	6.33	82.9	56.4	0.360	9.24	13.4
C150 × 0.120	1.55	152.4	5.08	48.8	8.71	13.0	5.45	71.8	59.4	0.288	8.06	13.6

Angles—Properties for Designing

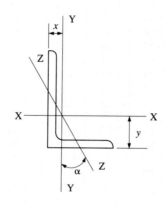

TABLE D–1 U.S. Customary

Designation	Wt. lb/ft	A in.²	I_x in.⁴	S_x in.³	r_x in.	y in.	I_y in.⁴	S_y in.³	r_y in.	x in.	r_z in.	Tan α
L8 × 8 × 1⅛	56.9	16.7	98.0	17.5	2.42	2.41	98.0	17.5	2.42	2.41	1.56	1.000
L8 × 8 × 3/4	38.9	11.4	69.7	12.2	2.47	2.28	69.7	12.2	2.47	2.28	1.58	1.000
L8 × 8 × 1/2	26.4	7.75	48.6	8.36	2.50	2.19	48.6	8.36	2.50	2.19	1.59	1.000
L8 × 6 × 1	44.2	13.0	80.8	15.1	2.49	2.65	38.8	8.92	1.73	1.65	1.28	0.543
L8 × 6 × 3/4	33.8	9.94	63.4	11.7	2.53	2.56	30.7	6.92	1.76	1.56	1.29	0.551
L8 × 6 × 1/2	23.0	6.75	44.3	8.02	2.56	2.47	21.7	4.79	1.79	1.47	1.30	0.558
L7 × 4 × 3/4	26.2	7.69	37.8	8.42	2.22	2.51	9.05	3.03	1.09	1.01	0.860	0.324
L7 × 4 × 1/2	17.9	5.25	26.7	5.81	2.25	2.42	6.53	2.12	1.11	0.917	0.872	0.335
L7 × 4 × 3/8	13.6	3.98	20.6	4.44	2.27	2.37	5.10	1.63	1.13	0.870	0.880	0.340
L6 × 6 × 1	37.4	11.0	35.5	8.57	1.80	1.86	35.5	8.57	1.80	1.86	1.17	1.000
L6 × 6 × 3/4	28.7	8.44	28.2	6.66	1.83	1.78	28.2	6.66	1.83	1.78	1.17	1.000
L6 × 6 × 5/8	24.2	7.11	24.2	5.66	1.84	1.73	24.2	5.66	1.84	1.73	1.18	1.000

TABLE D–1   U.S. Customary (*continued*)

Designation	Wt. lb/ft	A in.^{2}	I_x in.^{4}	S_x in.^{3}	r_x in.	$\bar{y}$ in.	I_y in.^{4}	S_y in.^{3}	r_y in.	$\bar{x}$ in.	r_z in.	Tan α
L6 × 6 × 1/2	19.6	5.75	19.9	4.61	1.86	1.68	19.9	4.61	1.86	1.68	1.18	1.000
L6 × 4 × 3/4	23.6	6.94	24.5	6.25	1.88	2.08	8.68	2.97	1.12	1.08	0.860	0.428
L6 × 4 × 1/2	16.2	4.75	17.4	4.33	1.91	1.99	6.27	2.08	1.15	0.987	0.870	0.440
L6 × 4 × 3/8	12.3	3.61	13.5	3.32	1.93	1.94	4.90	1.60	1.17	0.941	0.877	0.446
L5 × 5 × 7/8	27.2	7.98	17.8	5.17	1.49	1.57	17.8	5.17	1.49	1.57	0.973	1.000
L5 × 5 × 1/2	16.2	4.75	11.3	3.16	1.54	1.43	11.3	3.16	1.54	1.43	0.983	1.000
L5 × 5 × 5/16	10.3	3.03	7.42	2.04	1.57	1.37	7.42	2.04	1.57	1.37	0.994	1.000
L5 × 3½ × 3/4	19.8	5.81	13.9	4.28	1.55	1.75	5.55	2.22	0.977	0.996	0.748	0.464
L5 × 3½ × 1/2	13.6	4.00	9.99	2.99	1.58	1.66	4.05	1.56	1.01	0.906	0.755	0.479
L5 × 3½ × 5/16	8.7	2.56	6.60	1.94	1.61	1.59	2.72	1.02	1.03	0.838	0.766	0.489
L5 × 3 × 3/8	9.8	2.86	7.37	2.24	1.61	1.70	2.04	0.888	0.845	0.704	0.654	0.364
L5 × 3 × 1/4	6.6	1.94	5.11	1.53	1.62	1.66	1.44	0.614	0.861	0.657	0.663	0.371
L4 × 4 × 3/4	18.5	5.44	7.67	2.81	1.19	1.27	7.67	2.81	1.19	1.27	0.778	1.000
L4 × 4 × 3/8	9.8	2.86	4.36	1.52	1.23	1.14	4.36	1.52	1.23	1.14	0.788	1.000
L4 × 4 × 1/4	6.6	1.94	3.04	1.05	1.25	1.09	3.04	1.05	1.25	1.09	0.795	1.000
L4 × 3½ × 1/2	11.9	3.50	5.32	1.94	1.23	1.25	3.79	1.52	1.04	1.00	0.722	0.750
L4 × 3½ × 5/16	7.7	2.25	3.56	1.26	1.26	1.18	2.55	0.994	1.07	0.932	0.730	0.757
L4 × 3 × 1/2	11.1	3.25	5.05	1.89	1.25	1.33	2.42	1.12	0.864	0.827	0.639	0.543
L4 × 3 × 3/8	8.5	2.48	3.96	1.46	1.26	1.28	1.92	0.866	0.879	0.782	0.644	0.551
L4 × 3 × 1/4	5.8	1.69	2.77	1.00	1.28	1.24	1.36	0.599	0.896	0.736	0.651	0.558
L3½ × 3 × 3/8	7.9	2.30	2.72	1.13	1.09	1.08	1.85	0.851	0.897	0.830	0.625	0.721
L3½ × 3 × 1/4	5.4	1.56	1.91	0.776	1.11	1.04	1.30	0.589	0.914	0.785	0.631	0.727
L3 × 3 × 1/2	9.4	2.75	2.22	1.07	0.898	0.932	2.22	1.07	0.898	0.932	0.584	1.000
L3 × 3 × 5/16	6.1	1.78	1.51	0.707	0.922	0.865	1.51	0.707	0.922	0.865	0.589	1.000
L3 × 3 × 3/16	3.71	1.09	0.962	0.441	0.939	0.820	0.962	0.441	0.939	0.820	0.596	1.000
L3 × 2½ × 3/8	6.6	1.92	1.66	0.810	0.928	0.956	1.04	0.581	0.736	0.706	0.522	0.676
L3 × 2½ × 3/16	3.39	0.996	0.907	0.430	0.954	0.888	0.577	0.310	0.761	0.638	0.533	0.688
L3 × 2 × 3/8	5.9	1.73	1.53	0.781	0.940	1.04	0.543	0.371	0.559	0.539	0.430	0.428
L3 × 2 × 1/4	4.1	1.19	1.09	0.542	0.957	0.993	0.392	0.260	0.574	0.493	0.435	0.440
L2½ × 2½ × 1/4	4.1	1.19	0.703	0.394	0.769	0.717	0.703	0.394	0.769	0.717	0.491	1.000
L2½ × 2 × 3/8	5.3	1.55	0.912	0.547	0.768	0.831	0.514	0.363	0.577	0.581	0.420	0.614
L2 × 2 × 3/8	4.7	1.36	0.479	0.351	0.594	0.636	0.479	0.351	0.594	0.636	0.389	1.000
L2 × 2 × 1/4	3.19	0.938	0.348	0.247	0.609	0.592	0.348	0.247	0.609	0.592	0.391	1.000
L2 × 2 × 1/8	1.65	0.484	0.190	0.131	0.626	0.546	0.190	0.131	0.626	0.546	0.398	1.000

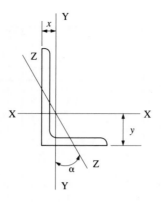

TABLE D–2   SI

Designation	Wt. kN/m × 10⁻³	A m² × 10⁻³	I_x m⁴ × 10⁻⁶	S_x m³ × 10⁻⁶	r_x mm	y mm	I_y m⁴ × 10⁻⁶	S_y m³ × 10⁻⁶	r_y mm	x mm	r_z mm	Tan α
L203 × 203 × 28.6	830	10.8	40.8	287	61.5	61.2	40.8	287	61.5	61.2	39.6	1.000
L203 × 203 × 19.1	568	7.35	29.0	200	62.7	57.9	29.0	200	62.7	57.9	40.1	1.000
L203 × 203 × 12.7	385	5.00	20.2	137	63.5	55.6	20.2	137	63.5	55.6	40.4	1.000
L203 × 152 × 25.4	645	8.39	33.6	247	63.2	67.3	16.2	146	43.9	41.9	32.5	0.543
L203 × 152 × 19.1	493	6.41	26.4	192	64.3	65.0	12.8	113	44.7	39.6	32.8	0.551
L203 × 152 × 12.7	336	4.35	18.4	131	65.0	62.7	9.03	78.5	45.5	37.3	33.0	0.558
L178 × 102 × 19.1	382	4.96	15.7	138	56.4	63.8	3.77	49.7	27.7	25.7	21.8	0.324
L178 × 102 × 12.7	261	3.39	11.1	95.2	57.2	61.5	2.72	34.7	28.2	23.3	22.1	0.335
L178 × 102 × 9.5	198	2.57	8.57	72.8	57.7	60.2	2.12	26.7	28.7	22.1	22.4	0.340
L152 × 152 × 25.4	546	7.10	14.8	140	45.7	47.2	14.8	140	45.7	47.2	29.7	1.000
L152 × 152 × 19.1	419	5.45	11.7	109	46.5	45.2	11.7	109	46.5	45.2	29.7	1.000
L152 × 152 × 15.9	353	4.59	10.1	92.8	46.7	43.9	10.1	92.8	46.7	43.9	30.0	1.000
L152 × 152 × 12.7	286	3.71	8.28	75.5	47.2	42.7	8.28	75.5	47.2	42.7	30.0	1.000
L152 × 102 × 19.1	344	4.48	10.2	102	47.8	52.8	3.61	48.7	28.4	27.4	21.8	0.428
L152 × 102 × 12.7	236	3.06	7.24	71.0	48.5	50.5	2.61	34.1	29.2	25.1	22.1	0.440
L152 × 102 × 9.5	179	2.33	5.62	54.4	49.0	49.3	2.04	26.2	29.7	23.9	22.3	0.446
L127 × 127 × 22.2	397	5.15	7.41	84.7	37.8	39.9	7.41	84.7	37.8	39.9	24.7	1.000
L127 × 127 × 12.7	236	3.06	4.70	51.8	39.1	36.3	4.70	51.8	39.1	36.3	25.0	1.000
L127 × 127 × 7.9	150	1.95	3.09	33.4	39.9	34.8	3.09	33.4	39.9	34.8	25.2	1.000
L127 × 89 × 19.1	289	3.75	5.79	70.1	39.4	44.5	2.31	36.4	24.8	25.3	19.0	0.464
L127 × 89 × 12.7	198	2.58	4.16	49.0	40.1	42.2	1.69	25.6	25.7	23.0	19.2	0.479
L127 × 89 × 7.9	127	1.65	2.75	31.8	40.9	40.4	1.13	16.7	26.2	21.3	19.5	0.489
L127 × 76 × 9.5	143	1.85	3.07	36.7	40.9	43.2	0.849	14.6	21.5	17.9	16.6	0.364
L127 × 76 × 6.4	96	1.25	2.13	25.1	41.1	42.2	0.599	10.1	21.9	16.7	16.8	0.371
L102 × 102 × 19.1	270	3.51	3.19	46.0	30.2	32.3	3.19	46.0	30.2	32.3	19.8	1.000
L102 × 102 × 9.5	143	1.85	1.81	24.9	31.2	29.0	1.81	24.9	31.2	29.0	20.0	1.000
L102 × 102 × 6.4	96	1.25	1.27	17.2	31.8	27.7	1.27	17.2	31.8	27.7	20.2	1.000
L102 × 89 × 12.7	174	2.26	2.21	31.8	31.2	31.8	1.58	24.9	26.4	25.4	18.3	0.750
L102 × 89 × 7.9	112	1.45	1.48	20.6	32.0	30.0	1.06	16.3	27.2	23.7	18.5	0.757
L102 × 76 × 12.7	162	2.10	2.10	31.0	31.8	33.8	1.01	18.4	21.9	21.0	16.2	0.543

TABLE D–2   SI (*continued*)

Designation	Wt. kN/m $\times 10^{-3}$	A m^2 $\times 10^{-3}$	I_x m^4 $\times 10^{-6}$	S_x m^3 $\times 10^{-6}$	r_x mm	y mm	I_y m^4 $\times 10^{-6}$	S_y m^3 $\times 10^{-6}$	r_y mm	x mm	r_z mm	Tan α
L102 × 76 × 9.5	124	1.60	1.65	23.9	32.0	32.5	0.799	14.2	22.3	19.9	16.4	0.551
L102 × 76 × 6.4	85	1.09	1.15	16.4	32.5	31.5	0.566	9.82	22.8	18.7	16.5	0.558
L89 × 76 × 9.5	115	1.48	1.13	18.5	27.7	27.4	0.770	13.9	22.9	21.1	15.9	0.721
L89 × 76 × 6.4	79	1.01	0.795	12.7	28.2	26.4	0.541	9.65	23.2	19.9	16.0	0.727
L76 × 76 × 12.7	137	1.77	0.924	17.5	22.8	23.7	0.924	17.5	22.8	23.7	14.8	1.000
L76 × 76 × 7.9	89	1.15	0.629	11.6	23.4	22.0	0.629	11.6	23.4	22.0	15.0	1.000
L76 × 76 × 4.8	54	0.703	0.400	7.23	23.9	20.8	0.400	7.23	23.9	20.8	15.1	1.000
L76 × 64 × 9.5	96	1.24	0.691	13.3	23.6	24.3	0.433	9.52	18.7	17.9	13.3	0.676
L76 × 64 × 4.8	49	0.643	0.378	7.05	24.2	22.6	0.240	5.08	19.3	16.2	13.5	0.688
L76 × 51 × 9.5	86	1.12	0.637	12.8	23.9	26.4	0.226	6.08	14.2	13.7	10.9	0.428
L76 × 51 × 6.4	60	0.768	0.454	8.88	24.3	25.2	0.163	4.26	14.6	12.5	11.0	0.440
L64 × 64 × 6.4	60	0.768	0.293	6.46	19.5	18.2	0.293	6.46	19.5	18.2	12.5	1.000
L64 × 51 × 9.5	77	1.00	0.380	8.96	19.5	21.1	0.214	5.95	14.7	14.8	10.7	0.614
L51 × 51 × 9.5	69	0.877	0.199	5.75	15.1	16.2	0.199	5.75	15.1	16.2	9.88	1.000
L51 × 51 × 6.4	47	0.605	0.145	4.05	15.5	15.0	0.145	4.05	15.5	15.0	9.93	1.000
L51 × 51 × 3.2	24	0.312	0.079	2.15	15.9	13.9	0.079	2.15	15.9	13.9	10.1	1.000

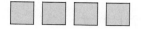

Properties of Structural Timber

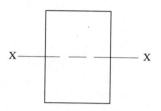

TABLE E–1 U.S. Customary

Nominal Size in.	Dressed Size in.	A in.^{2}	Wt. lb/ft	I_x in.^{4}	S_x in.^{3}
2 × 4	$1\frac{1}{2} \times 3\frac{1}{2}$	5.25	1.46	5.36	3.06
× 6	× $5\frac{1}{2}$	8.25	2.29	20.8	7.56
× 8	× $7\frac{1}{4}$	10.9	3.02	47.6	13.1
× 10	× $9\frac{1}{4}$	13.9	3.85	98.9	21.4
× 12	× $11\frac{1}{4}$	16.9	4.68	178	31.6
× 14	× $13\frac{1}{4}$	19.9	5.52	291	43.9
× 16	× $15\frac{1}{4}$	22.9	6.35	443	58.2
× 18	× $17\frac{1}{4}$	25.9	7.17	642	74.5
3 × 4	$2\frac{1}{2} \times 3\frac{1}{2}$	8.75	2.42	8.93	5.10
× 6	× $5\frac{1}{2}$	13.8	3.82	34.7	12.6
× 8	× $7\frac{1}{4}$	18.1	5.04	79.4	21.9
× 10	× $9\frac{1}{4}$	23.1	6.42	165	35.6
× 12	× $11\frac{1}{4}$	28.1	7.81	297	52.7
× 14	× $13\frac{1}{4}$	33.1	9.20	485	73.2
× 16	× $15\frac{1}{4}$	38.1	10.6	739	96.9
× 18	× $17\frac{1}{4}$	43.1	12.0	1070	124

TABLE E–1 U.S. Customary (*continued*)

Nominal Size in.	Dressed Size in.	A in.^{2}	Wt. lb/ft	I_x in.^{4}	S_x in.^{3}
4 × 4	3½ × 3½	12.3	3.40	12.5	7.15
× 6	× 5½	19.3	5.35	48.5	17.6
× 8	× 7¼	25.4	7.05	111	30.7
× 10	× 9¼	32.4	8.93	231	49.9
× 12	× 11¼	39.4	10.9	415	73.8
× 14	× 13¼	46.4	12.9	678	102
× 16	× 15¼	53.4	14.9	1030	136
× 18	× 17¼	60.4	16.8	1500	174
6 × 6	5½ × 5½	30.3	8.40	76.3	27.7
× 8	× 7½	41.3	11.4	193	51.6
× 10	× 9½	52.3	14.5	393	82.7
× 12	× 11½	63.3	17.5	697	121
× 14	× 13½	74.3	20.6	1130	167
× 16	× 15½	85.3	23.6	1710	220
× 18	× 17½	96.3	26.7	2460	281
× 20	× 19½	108	29.8	3400	349
8 × 8	7½ × 7½	56.3	15.6	264	70.3
× 10	× 9½	71.3	19.8	536	113
× 12	× 11½	86.3	23.9	951	165
× 14	× 13½	101	28.0	1540	228
× 16	× 15½	116	32.0	2330	300
× 18	× 17½	131	36.4	3350	383
× 20	× 19½	146	40.6	4630	475
× 22	× 21½	161	44.8	6210	578
10 × 10	9½ × 9½	90.3	25.0	679	143
× 12	× 11½	109	30.3	1200	209
× 14	× 13½	128	35.6	1950	289
× 16	× 15½	147	40.9	2950	380
× 18	× 17½	166	46.1	4240	485
× 20	× 19½	185	51.4	5870	602
× 22	× 21½	204	56.7	7870	732
× 24	× 23½	223	62.0	10300	874
12 × 12	11½ × 11½	132	36.7	1460	253
× 14	× 13½	155	43.1	2360	349
× 16	× 15½	178	49.5	3570	460
× 18	× 17½	201	55.9	5140	587
× 20	× 19½	224	62.3	7110	729
× 22	× 21½	247	68.7	9520	886
× 24	× 23½	270	75.0	12400	1060

Notes: Properties and weights are for dressed sizes. Assumed unit weight of timber is 40 pcf. Moment of inertia and section modulus are about the strong axis.

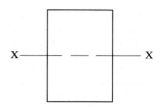

TABLE E–2 SI

Nominal Size mm	Dressed Size mm	A m^2 $\times\,10^{-3}$	Wt. kN/m $\times\,10^{-3}$	I_x m^4 $\times\,10^{-6}$	S_x m^3 $\times\,10^{-3}$
50×100	38.1×88.9	3.39	21.3	2.23	0.0502
$\times 150$	$\times 140$	5.33	33.5	8.71	0.124
$\times 200$	$\times 184$	7.01	44.0	19.8	0.215
$\times 250$	$\times 235$	8.95	56.3	41.2	0.351
$\times 300$	$\times 286$	10.9	68.5	74.3	0.519
$\times 360$	$\times 337$	12.8	80.7	122	0.721
$\times 410$	$\times 387$	14.7	92.6	184	0.951
$\times 460$	$\times 438$	16.7	105	267	1.22
80×100	63.5×88.9	5.65	35.5	3.72	0.0836
$\times 150$	$\times 140$	8.89	55.9	14.5	0.207
$\times 200$	$\times 184$	11.7	73.4	33.0	0.358
$\times 250$	$\times 235$	14.9	93.8	68.7	0.584
$\times 300$	$\times 286$	18.2	114	124	0.866
$\times 360$	$\times 337$	21.4	134	203	1.20
$\times 410$	$\times 388$	24.6	155	309	1.59
$\times 460$	$\times 438$	27.8	175	445	2.03
100×100	88.9×88.9	7.90	49.7	5.21	0.117
$\times 150$	$\times 140$	12.4	78.2	20.3	0.290
$\times 200$	$\times 184$	16.4	103	46.2	0.502
$\times 250$	$\times 235$	20.9	131	96.1	0.818
$\times 300$	$\times 286$	25.4	160	173	1.21
$\times 360$	$\times 337$	30.0	188	284	1.68
$\times 410$	$\times 388$	34.5	217	433	2.23
$\times 460$	$\times 438$	38.9	245	623	2.84
150×150	140×140	19.6	123	32.0	0.457
$\times 200$	$\times 191$	26.7	168	81.3	0.851
$\times 250$	$\times 241$	33.7	212	163	1.36
$\times 300$	$\times 292$	40.9	257	290	1.99
$\times 360$	$\times 343$	48.0	302	471	2.75
$\times 410$	$\times 394$	55.2	347	714	3.63
$\times 460$	$\times 445$	62.3	391	1030	4.62
$\times 510$	$\times 496$	69.4	436	1420	5.74

TABLE E–2 SI (*continued*)

Nominal Size mm	Dressed Size mm	A m^2 $\times 10^{-3}$	Wt. kN/m $\times 10^{-3}$	I_x m^4 $\times 10^{-6}$	S_x m^3 $\times 10^{-3}$
200 × 200	191 × 191	36.5	229	111	1.16
× 250	× 241	46.0	289	223	1.85
× 300	× 292	55.8	350	396	2.71
× 360	× 343	65.5	412	642	3.75
× 410	× 394	75.3	473	974	4.94
× 460	× 445	85.0	534	1400	6.30
× 510	× 495	94.5	594	1930	7.80
× 560	× 546	104	655	2590	9.49
250 × 250	241 × 241	58.1	365	281	2.33
× 300	× 292	70.4	442	500	3.42
× 360	× 343	82.7	519	810	4.73
× 410	× 394	95.0	597	1230	6.24
× 460	× 445	107	674	1770	7.95
× 510	× 495	119	750	2440	9.84
× 560	× 546	132	827	3270	12.0
× 610	× 597	144	904	4270	14.3
300 × 300	292 × 292	85.3	536	606	4.15
× 360	× 343	100	629	982	5.73
× 410	× 394	115	723	1490	7.55
× 460	× 445	130	816	2140	9.64
× 510	× 495	145	908	2950	11.9
× 560	× 546	159	1000	3960	14.5
× 610	× 597	174	1100	5180	17.3

Notes: Properties and weights are for dressed sizes. Assumed unit weight of timber is 6.28 kilonewtons per cubic meter. Moment of inertia and section modulus are about the strong axis.

Design Values for Timber Construction

Note: Values shown are approximate and are provided only for use in solving problems in this text.

TABLE F–1 U.S. Customary

Species	Allowable Stress* (psi)					Modulus of Elasticity E (ksi)
	s_c	s_{c_p}	s_t	s_b	s_s	
Douglas fir	1050	385	625	1450	95	1700
Southern pine	1250	410	825	1600	90	1700
Hem-fir	875	245	500	1000	75	1400
Eastern white pine	725	220	400	975	65	1100
California redwood	1050	270	650	1350	100	1100

* Allowable stresses:

s_c—compression parallel to grain
s_{c_p}—compression perpendicular to grain
s_t—tension parallel to grain
s_b—bending
s_s—horizontal shear

TABLE F–2   SI

Species	Allowable Stress* (MPa)					Modulus of Elasticity E (MPa × 10³)
	s_c	s_{c_p}	s_t	s_b	s_s	
Douglas fir	7.24	2.65	4.31	10.0	0.65	12
Southern pine	8.62	2.83	5.69	11.0	0.62	12
Hem-fir	6.03	1.69	3.45	6.89	0.52	9.7
Eastern white pine	5.00	1.52	2.76	6.72	0.45	7.6
California redwood	7.24	1.86	4.48	9.31	0.69	7.6

* Allowable stresses:

     s_c—compression parallel to grain
     s_{c_p}—compression perpendicular to grain
     s_t—tension parallel to grain
     s_b—bending
     s_s—horizontal shear

APPENDIX G

Typical Average Properties of Some Common Materials

Note: Values shown are approximate and are provided only for use in solving problems in this text.

TABLE G–1 U.S. Customary

Material	Weight pcf	Mod. of Elasticity E ksi	Mod. of Rigidity G ksi	Tensile Yield Strength ksi	Ultimate Strength ksi			Coeff. of Thermal Expansion α in./in./F° $\times 10^{-6}$	Poisson's Ratio μ
					Tens.	Comp.	Shear		
Steel (carbon) ASTM A36 or A501	490	30,000	12,000	36	70			6.5	0.25
Steel (alloy) ASTM A441	490	30,000	12,000	45	65			6.5	0.25
Steel AISI 1020 hot-rolled	490	30,000	11,500	30	55			6.5	0.25
Steel AISI 1040 hot-rolled	490	30,000	11,500	42	76			6.5	0.25
Stainless steel (annealed)	490	29,000	11,600	40	85		60	6.5	0.25
Cast iron (gray)	450	15,000	6000		20	80	32	5.9	0.26
Cast iron (malleable)	450	25,000	12,500	45	65	220	48	6.6	0.27

ASTM - AMERICAN SOCIETY FOR TESTING OF MATERIALS
AISI - AMER. INST FOR STEEL

TABLE G–1 U.S. Customary (*continued*)

Material	Weight pcf	Mod. of Elasticity E ksi	Mod. of Rigidity G ksi	Tensile Yield Strength ksi	Ultimate Strength ksi			Coeff. of Thermal Expansion α in./in./F° $\times 10^{-6}$	Poisson's Ratio μ
					Tens.	Comp.	Shear		
Wrought iron	480	28,000	11,000	28	48	48	38	6.7	0.27
Aluminum alloy 6061-T6	165	10,000	4000	35	42		27	12.8	0.33
Titanium alloy	275	16,500	6500	150	170		100	6.0	
Magnesium alloy	112	6500	2400	20	40		20	14.5	0.34
Brass (rolled)	535	14,000	6000	50	60		50	10.4	0.34
Bronze (cast)	535	12,000	5000	25	33	56		10.1	0.35
Copper (hard drawn)	550	15,000	6000	40	55		38	9.3	0.35
Concrete	150	3120				3.0		5.5	0.20
Concrete	150	3605				4.0		5.5	0.20

TABLE G–2 SI

Material	Weight kN/m³	Mod. of Elasticity E MPa × 10³	Mod. of Rigidity G MPa × 10³	Tensile Yield Strength MPa	Ultimate Strength MPa			Coeff. of Thermal Expansion α m/m/C° × 10⁻⁶	Poisson's Ratio μ
					Tens.	Comp.	Shear		
Steel (carbon) ASTM A36 or A501	77.0	207	83	250	480			11.7	0.25
Steel (alloy) ASTM A441	77.0	207	83	310	450			11.7	0.25
Steel AISI 1020 hot-rolled	77.0	207	79.3	210	380			11.7	0.25
Steel AISI 1040 hot-rolled	77.0	207	79.3	290	520			11.7	0.25
Stainless steel (annealed)	77.0	200	80	280	580		410	11.7	0.25
Cast iron (gray)	70.7	100	41		140	550	220	10.6	0.26
Cast iron (malleable)	70.7	172	86	310	450	1510	330	11.9	0.27
Wrought iron	75.4	190	76	190	330	330	260	12.1	0.27
Aluminum alloy 6061-T6	25.9	70	28	240	290		190	23.0	0.33
Magnesium alloy	17.6	45	17	210	280		140	26.1	0.34
Titanium alloy	43.2	114	45	1030	1170		690	10.8	
Brass (rolled)	84.0	97	41	340	410		340	18.7	0.34
Bronze (cast)	84.0	83	34	170	230	390		18.2	0.35
Copper (hard drawn)	86.4	103	41	280	380		260	16.7	0.35
Concrete	23.6	21.5				21		9.9	0.20
Concrete	23.6	24.9				28		9.9	0.20

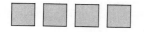

Beam Diagrams and Formulas

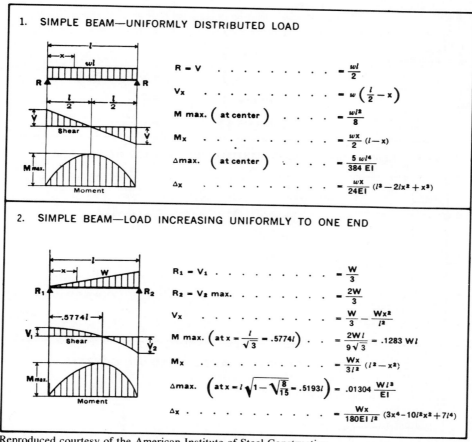

1. SIMPLE BEAM—UNIFORMLY DISTRIBUTED LOAD

$$R = V \quad \ldots \ldots \ldots \ldots = \frac{wl}{2}$$

$$V_x \quad \ldots \ldots \ldots \ldots = w\left(\frac{l}{2} - x\right)$$

$$M \text{ max.} \left(\text{at center}\right) \quad \ldots \ldots = \frac{wl^2}{8}$$

$$M_x \quad \ldots \ldots \ldots \ldots = \frac{wx}{2}(l - x)$$

$$\Delta \text{max.} \left(\text{at center}\right) \quad \ldots \ldots = \frac{5\,wl^4}{384\,EI}$$

$$\Delta_x \quad \ldots \ldots \ldots \ldots = \frac{wx}{24EI}(l^3 - 2lx^2 + x^3)$$

2. SIMPLE BEAM—LOAD INCREASING UNIFORMLY TO ONE END

$$R_1 = V_1 \quad \ldots \ldots \ldots = \frac{W}{3}$$

$$R_2 = V_2 \text{ max.} \quad \ldots \ldots = \frac{2W}{3}$$

$$V_x \quad \ldots \ldots \ldots = \frac{W}{3} - \frac{Wx^2}{l^2}$$

$$M \text{ max.} \left(\text{at } x = \frac{l}{\sqrt{3}} = .5774l\right) \ldots = \frac{2Wl}{9\sqrt{3}} = .1283\,Wl$$

$$M_x \quad \ldots \ldots \ldots = \frac{Wx}{3l^2}(l^2 - x^2)$$

$$\Delta \text{max.} \left(\text{at } x = l\sqrt{1 - \sqrt{\tfrac{8}{15}}} = .5193l\right) = .01304\,\frac{Wl^3}{EI}$$

$$\Delta_x \quad \ldots \ldots \ldots = \frac{Wx}{180EI\,l^2}(3x^4 - 10l^2x^2 + 7l^4)$$

Reproduced courtesy of the American Institute of Steel Construction

BEAM DIAGRAMS AND FORMULAS
For various static loading conditions

3. SIMPLE BEAM—LOAD INCREASING UNIFORMLY TO CENTER

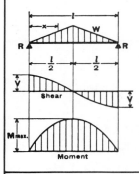

$$R = V \quad \ldots \ldots \ldots \ldots = \frac{W}{2}$$

$$V_x \quad \left(\text{when } x < \frac{l}{2}\right) \ldots \ldots = \frac{W}{2l^2}(l^2 - 4x^2)$$

$$M \text{ max.} \left(\text{at center}\right) \ldots \ldots = \frac{Wl}{6}$$

$$M_x \quad \left(\text{when } x < \frac{l}{2}\right) \ldots \ldots = Wx\left(\frac{1}{2} - \frac{2x^2}{3l^2}\right)$$

$$\Delta \text{max.} \left(\text{at center}\right) \ldots \ldots = \frac{Wl^3}{60EI}$$

$$\Delta_x \quad \left(\text{when } x < \frac{l}{2}\right) \ldots \ldots = \frac{Wx}{480\,EI\,l^2}(5l^2 - 4x^2)^2$$

4. SIMPLE BEAM—UNIFORM LOAD PARTIALLY DISTRIBUTED AT ONE END

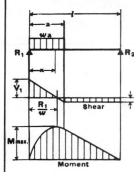

$$R_1 = V_1 \text{ max.} \ldots \ldots \ldots = \frac{wa}{2l}(2l - a)$$

$$R_2 = V_2 \ldots \ldots \ldots \ldots = \frac{wa^2}{2l}$$

$$V_x \quad \left(\text{when } x < a\right) \ldots \ldots = R_1 - wx$$

$$M \text{ max.} \left(\text{at } x = \frac{R_1}{w}\right) \ldots \ldots = \frac{R_1^2}{2w}$$

$$M_x \quad \left(\text{when } x < a\right) \ldots \ldots = R_1 x - \frac{wx^2}{2}$$

$$M_x \quad \left(\text{when } x > a\right) \ldots \ldots = R_2(l - x)$$

$$\Delta_x \quad \left(\text{when } x < a\right) \ldots \ldots = \frac{wx}{24EIl}\left(a^2(2l-a)^2 - 2ax^2(2l-a) + lx^3\right)$$

$$\Delta_x \quad \left(\text{when } x > a\right) \ldots \ldots = \frac{wa^2(l-x)}{24EIl}(4xl - 2x^2 - a^2)$$

5. SIMPLE BEAM—CONCENTRATED LOAD AT CENTER

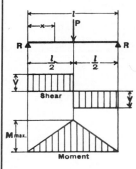

$$R = V \quad \ldots \ldots \ldots \ldots = \frac{P}{2}$$

$$M \text{ max.} \left(\text{at point of load}\right) \ldots \ldots = \frac{Pl}{4}$$

$$M_x \quad \left(\text{when } x < \frac{l}{2}\right) \ldots \ldots = \frac{Px}{2}$$

$$\Delta \text{max.} \left(\text{at point of load}\right) \ldots \ldots = \frac{Pl^3}{48EI}$$

$$\Delta_x \quad \left(\text{when } x < \frac{l}{2}\right) \ldots \ldots = \frac{Px}{48EI}(3l^2 - 4x^2)$$

BEAM DIAGRAMS AND FORMULAS
For various static loading conditions

6. SIMPLE BEAM—CONCENTRATED LOAD AT ANY POINT

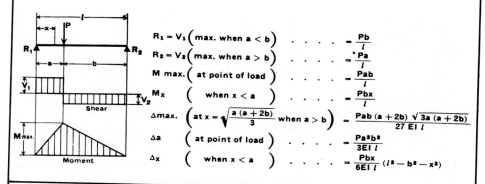

$$R_1 = V_1 \left(\text{max. when } a < b \right) \quad \cdots \quad = \frac{Pb}{l}$$

$$R_2 = V_2 \left(\text{max. when } a > b \right) \quad \cdots \quad = \frac{Pa}{l}$$

$$M \text{ max.} \left(\text{at point of load} \right) \quad \cdots \quad = \frac{Pab}{l}$$

$$M_x \quad \left(\text{when } x < a \right) \quad \cdots \quad = \frac{Pbx}{l}$$

$$\Delta \text{max.} \left(\text{at } x = \sqrt{\frac{a(a+2b)}{3}} \text{ when } a > b \right) = \frac{Pab(a+2b)\sqrt{3a(a+2b)}}{27 \, EI \, l}$$

$$\Delta a \quad \left(\text{at point of load} \right) \quad \cdots \quad = \frac{Pa^2 b^2}{3EI \, l}$$

$$\Delta x \quad \left(\text{when } x < a \right) \quad \cdots \quad = \frac{Pbx}{6EI \, l} (l^2 - b^2 - x^2)$$

7. SIMPLE BEAM—TWO EQUAL CONCENTRATED LOADS SYMMETRICALLY PLACED

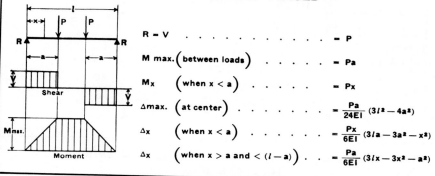

$$R = V \quad \cdots \cdots \quad = P$$

$$M \text{ max.} \left(\text{between loads} \right) \quad \cdots \quad = Pa$$

$$M_x \quad \left(\text{when } x < a \right) \quad \cdots \quad = Px$$

$$\Delta \text{max.} \left(\text{at center} \right) \quad \cdots \quad = \frac{Pa}{24EI} (3l^2 - 4a^2)$$

$$\Delta x \quad \left(\text{when } x < a \right) \quad \cdots \quad = \frac{Px}{6EI} (3la - 3a^2 - x^2)$$

$$\Delta x \quad \left(\text{when } x > a \text{ and } < (l-a) \right) \quad \cdots \quad = \frac{Pa}{6EI} (3lx - 3x^2 - a^2)$$

8. BEAM FIXED AT ONE END, SUPPORTED AT OTHER— UNIFORMLY DISTRIBUTED LOAD

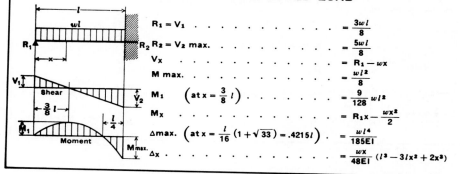

$$R_1 = V_1 \quad \cdots \cdots \quad = \frac{3wl}{8}$$

$$R_2 = V_2 \text{ max.} \quad \cdots \cdots \quad = \frac{5wl}{8}$$

$$V_x \quad \cdots \cdots \quad = R_1 - wx$$

$$M \text{ max.} \quad \cdots \cdots \quad = \frac{wl^2}{8}$$

$$M_1 \quad \left(\text{at } x = \frac{3}{8} l \right) \quad \cdots \quad = \frac{9}{128} wl^2$$

$$M_x \quad \cdots \cdots \quad = R_1 x - \frac{wx^2}{2}$$

$$\Delta \text{max.} \left(\text{at } x = \frac{l}{16}(1 + \sqrt{33}) = .4215l \right) = \frac{wl^4}{185EI}$$

$$\Delta x \quad \cdots \cdots \quad = \frac{wx}{48EI} (l^3 - 3lx^2 + 2x^3)$$

BEAM DIAGRAMS AND FORMULAS
For various static loading conditions

9.   BEAM FIXED AT BOTH ENDS—UNIFORMLY DISTRIBUTED LOADS

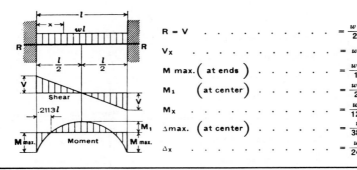

$$R = V = \frac{wl}{2}$$

$$V_x = w\left(\frac{l}{2} - x\right)$$

$$M \text{ max.} \left(\text{at ends}\right) = \frac{wl^2}{12}$$

$$M_1 \left(\text{at center}\right) = \frac{wl^2}{24}$$

$$M_x = \frac{w}{12}(6lx - l^2 - 6x^2)$$

$$\Delta \text{max.} \left(\text{at center}\right) = \frac{wl^4}{384EI}$$

$$\Delta_x = \frac{wx^2}{24EI}(l - x)^2$$

10.   BEAM FIXED AT BOTH ENDS—CONCENTRATED LOAD AT CENTER

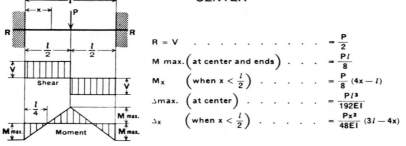

$$R = V = \frac{P}{2}$$

$$M \text{ max.} \left(\text{at center and ends}\right) = \frac{Pl}{8}$$

$$M_x \left(\text{when } x < \frac{l}{2}\right) = \frac{P}{8}(4x - l)$$

$$\Delta \text{max.} \left(\text{at center}\right) = \frac{Pl^3}{192EI}$$

$$\Delta_x \left(\text{when } x < \frac{l}{2}\right) = \frac{Px^2}{48EI}(3l - 4x)$$

11.   CANTILEVER BEAM—LOAD INCREASING UNIFORMLY TO FIXED END

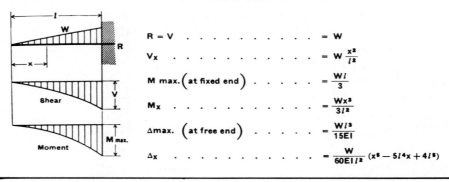

$$R = V = W$$

$$V_x = W\frac{x^2}{l^2}$$

$$M \text{ max.} \left(\text{at fixed end}\right) = \frac{Wl}{3}$$

$$M_x = \frac{Wx^3}{3l^2}$$

$$\Delta \text{max.} \left(\text{at free end}\right) = \frac{Wl^3}{15EI}$$

$$\Delta_x = \frac{W}{60EIl^2}(x^5 - 5l^4x + 4l^5)$$

BEAM DIAGRAMS AND FORMULAS
For various static loading conditions

12. CANTILEVER BEAM—UNIFORMLY DISTRIBUTED LOAD

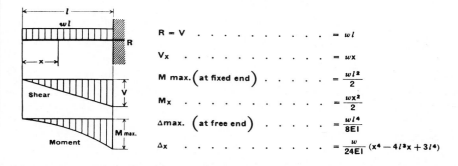

$R = V$ . . . . $= wl$

V_x . . . . $= wx$

M max. $\left(\text{at fixed end}\right)$ . . . . . $= \dfrac{wl^2}{2}$

M_x . . . . $= \dfrac{wx^2}{2}$

Δ max. $\left(\text{at free end}\right)$ . . . . . $= \dfrac{wl^4}{8EI}$

Δ_x . . . . $= \dfrac{w}{24EI}\,(x^4 - 4l^3x + 3l^4)$

13. CANTILEVER BEAM—CONCENTRATED LOAD AT ANY POINT

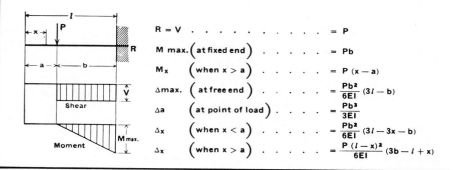

$R = V$ . . . . $= P$

M max. $\left(\text{at fixed end}\right)$ . . . . . $= Pb$

M_x $\left(\text{when } x > a\right)$ . . . . . $= P\,(x - a)$

Δ max. $\left(\text{at free end}\right)$ . . . . . $= \dfrac{Pb^2}{6EI}\,(3l - b)$

Δ_a $\left(\text{at point of load}\right)$ . . . . $= \dfrac{Pb^3}{3EI}$

Δ_x $\left(\text{when } x < a\right)$ . . . . . $= \dfrac{Pb^2}{6EI}\,(3l - 3x - b)$

Δ_x $\left(\text{when } x > a\right)$ . . . . . $= \dfrac{P\,(l - x)^2}{6EI}\,(3b - l + x)$

14. CANTILEVER BEAM—CONCENTRATED LOAD AT FREE END

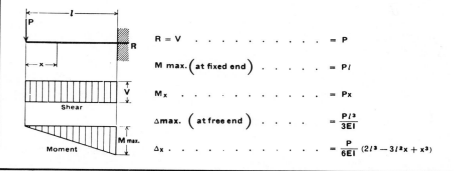

$R = V$ . . . . $= P$

M max. $\left(\text{at fixed end}\right)$ . . . . . $= Pl$

M_x . . . . $= Px$

Δ max. $\left(\text{at free end}\right)$ . . . . $= \dfrac{Pl^3}{3EI}$

Δ_x . . . . $= \dfrac{P}{6EI}\,(2l^3 - 3l^2x + x^3)$

APPENDIX I

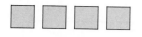

Beam Selection Table (Elastic Design)

TABLE I–1 U.S. Customary

S_x in.^{3}	Shape	S_x in.^{3}	Shape	S_x in.^{3}	Shape	S_x in.^{3}	Shape
3170	**W36 × 848**	**684**	**W33 × 201**	**176**	**W24 × 76**	**56.5**	**W16 × 36**
				175	W16 × 100	54.6	W14 × 38
2590	**W40 × 655**	**682**	**W40 × 183**	173	W14 × 109	51.9	W12 × 40
		674	W27 × 235	171	W21 × 83	49.1	W10 × 45
2420	**W36 × 650**	664	W36 × 194				
		598	W30 × 191	**151**	**W21 × 73**	**48.6**	**W14 × 34**
2090	**W40 × 531**			146	W16 × 76		
		580	**W36 × 170**	143	W14 × 90	**42.0**	**W14 × 30**
1950	**W36 × 527**	531	W24 × 207	134	W16 × 77		
						38.6	**W12 × 30**
1710	**W40 × 436**	**504**	**W36 × 150**				
1620	W36 × 439	455	W27 × 161	**131**	**W24 × 62**		
						38.4	**W16 × 26**
1480	**W33 × 424**	**448**	**W33 × 141**	**127**	**W21 × 62**	35.0	W10 × 33
		414	W24 × 162	127	W18 × 71		
1340	**W40 × 328**			112	W14 × 74	**29.0**	**W14 × 22**
1320	W36 × 359	**380**	**W30 × 132**			27.5	W8 × 31
1250	W30 × 391						
1170	W24 × 450	**359**	**W33 × 118**	**108**	**W18 × 60**	**25.4**	**W12 × 22**
				97.4	W12 × 72		
1120	**W44 × 285**	**329**	**W30 × 116**			**23.2**	**W10 × 22**
1110	W36 × 300	329	W24 × 131	**94.5**	**W21 × 50**	20.9	W8 × 24
1110	W33 × 318	329	W21 × 147	92.2	W16 × 57		
				92.2	W14 × 61	**17.1**	**W12 × 16**
1090	**W40 × 268**	**299**	**W30 × 108**			16.7	W6 × 25
		299	W27 × 114			16.2	W10 × 17
983	**W40 × 244**	273	W21 × 122	**88.9**	**W18 × 50**	15.2	W8 × 18
953	W36 × 260			78.0	W12 × 58		
928	W30 × 292	**269**	**W30 × 99**				
		258	W24 × 104	**72.9**	**W16 × 45**	**10.9**	**W10 × 12**
889	**W44 × 224**	**243**	**W27 × 94**	70.3	W14 × 48	10.2	W5 × 19
884	W27 × 307	231	W18 × 119			9.91	W8 × 13
		227	W21 × 101			9.72	W6 × 15
837	**W36 × 230**			**68.4**	**W18 × 40**		
829	W33 × 241	**196**	**W24 × 84**	64.7	W12 × 50	**7.31**	**W6 × 12**
789	W24 × 306	188	W18 × 97	60.0	W10 × 54	5.46	W4 × 13

TABLE I–2   SI

S_x m^3 $\times 10^{-3}$	Shape mm × kN/m	S_x m^3 $\times 10^{-3}$	Shape mm × kN/m	S_x m^3 $\times 10^{-3}$	Shape mm × kN/m	S_x m^3 $\times 10^{-3}$	Shape mm × kN/m
51.9	**W910 × 12.4**	**11.2**	**W840 × 2.93**	**2.88**	**W610 × 1.11**	**0.918**	**W410 × 0.53**
				2.87	W410 × 1.46	0.895	W360 × 0.55
42.4	**W1020 × 9.56**	**11.2**	**W1020 × 2.67**	2.83	W360 × 1.59	0.850	W300 × 0.58
		11.0	W690 × 3.43	2.80	W530 × 1.21	0.805	W250 × 0.66
39.7	**W910 × 9.49**	10.9	W910 × 2.83				
		9.80	W760 × 2.79	**2.47**	**W530 × 1.07**	**0.796**	**W360 × 0.50**
34.2	**W1020 × 7.75**			2.39	W410 × 1.11		
		9.50	**W910 × 2.48**	2.34	W360 × 1.31	**0.688**	**W360 × 0.44**
32.0	**W910 × 7.69**	8.70	W610 × 3.02	2.20	W410 × 1.12		
						0.633	**W300 × 0.44**
28.0	**W1020 × 6.36**	**8.26**	**W910 × 2.19**				
26.5	W910 × 6.41	7.46	W690 × 2.35	**2.15**	**W610 × 0.90**		
		7.34	W840 × 2.06			**0.629**	**W410 × 0.38**
24.3	**W840 × 6.19**	6.78	W610 × 2.36	**2.08**	**W530 × 0.90**	0.574	W250 × 0.48
				2.08	W460 × 1.04		
22.0	**W1020 × 4.79**	**6.23**	**W760 × 1.93**	1.84	W360 × 1.08	**0.475**	**W360 × 0.32**
21.6	W910 × 5.24					0.451	W200 × 0.45
20.5	W760 × 5.71	**5.88**	**W840 × 1.72**				
19.2	W610 × 6.57			**1.77**	**W460 × 0.88**	0.416	W300 × 0.32
		5.39	**W760 × 1.69**	1.60	W300 × 1.05		
18.4	**W1120 × 4.16**	5.39	W610 × 1.91			**0.380**	**W250 × 0.32**
18.2	W910 × 4.38	5.39	W530 × 2.15	**1.55**	**W530 × 0.73**	0.342	W200 × 0.35
18.2	W840 × 4.64			1.51	W410 × 0.83		
		4.90	**W760 × 1.58**	1.51	W360 × 0.89	**0.280**	**W300 × 0.23**
17.9	**W1020 × 3.91**	4.90	W690 × 1.66			0.274	W150 × 0.36
		4.47	W530 × 1.78	**1.46**	**W460 × 0.73**	0.265	W250 × 0.25
16.1	**W1020 × 3.56**			1.28	W300 × 0.85	0.249	W200 × 0.26
15.6	W910 × 3.79	**4.41**	**W760 × 1.44**				
15.2	W760 × 4.26	4.23	W610 × 1.52	**1.19**	**W410 × 0.66**	**0.179**	**W250 × 0.18**
		3.98	W690 × 1.37	1.15	W360 × 0.70	0.167	W130 × 0.28
14.6	**W1120 × 3.27**	3.79	W460 × 1.74			0.162	W200 × 0.19
14.5	W690 × 4.48	3.72	W530 × 1.47			0.159	W150 × 0.22
				1.12	**W460 × 0.58**		
13.7	**W910 × 3.36**	**3.21**	**W610 × 1.23**	1.06	W300 × 0.73	**0.120**	**W150 × 0.18**
13.7	W840 × 3.52	3.08	W460 × 1.42	0.983	W250 × 0.79	0.089	W100 × 0.19
12.9	W610 × 4.47						

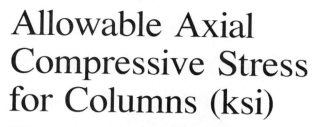

Allowable Axial Compressive Stress for Columns (ksi)

$\frac{KL}{r}$	$S_{a(all)}$	$\frac{KL}{r}$	$S_{a(all)}$	$\frac{KL}{r}$	$S_{a(all)}$	$\frac{KL}{r}$	$S_{a(all)}$	$\frac{KL}{r}$	$S_{a(all)}$	$\frac{KL}{r}$	$S_{a(all)}$
1	21.56	35	19.62	69	16.65	102	12.94	135	8.48	168	5.47
2	21.52	36	19.55	70	16.55	103	12.82	136	8.35	169	5.41
3	21.48	37	19.47	71	16.45	104	12.69	137	8.23	170	5.35
4	21.44	38	19.40	72	16.35	105	12.56	138	8.11	171	5.28
5	21.40	39	19.32	73	16.25	106	12.44	139	8.00	172	5.22
6	21.35	40	19.24	74	16.14	107	12.31	140	7.88	173	5.16
7	21.31	41	19.16	75	16.04	108	12.18	141	7.77	174	5.10
8	21.26	42	19.09	76	15.93	109	12.05	142	7.66	175	5.04
9	21.21	43	19.01	77	15.83	110	11.92	143	7.55	176	4.99
10	21.16	44	18.93	78	15.72	111	11.79	144	7.45	177	4.93
11	21.11	45	18.84	79	15.61	112	11.66	145	7.35	178	4.88
12	21.06	46	18.76	80	15.50	113	11.52	146	7.25	179	4.82
13	21.01	47	18.68	81	15.40	114	11.39	147	7.15	180	4.77
14	20.96	48	18.59	82	15.29	115	11.25	148	7.05	181	4.72
15	20.90	49	18.51	83	15.18	116	11.12	149	6.96	182	4.66
16	20.85	50	18.42	84	15.06	117	10.98	150	6.87	183	4.61
17	20.79	51	18.34	85	14.95	118	10.85	151	6.78	184	4.56
18	20.74	52	18.25	86	14.84	119	10.71	152	6.69	185	4.51
19	20.68	53	18.16	87	14.73	120	10.57	153	6.60	186	4.47
20	20.62	54	18.07	88	14.61	121	10.43	154	6.51	187	4.42
21	20.56	55	17.98	89	14.50	122	10.29	155	6.43	188	4.37
22	20.50	56	17.89	90	14.38	123	10.15	156	6.35	189	4.32
23	20.44	57	17.80	91	14.27	124	10.01	157	6.27	190	4.28
24	20.37	58	17.71	92	14.15	125	9.86	158	6.19	191	4.23
25	20.31	59	17.62	93	14.03	126	9.72	159	6.11	192	4.19
26	20.25	60	17.53	94	13.91	127	9.57	160	6.03	193	4.15
27	20.18	61	17.43	95	13.80	128	9.43	161	5.96	194	4.10
28	20.11	62	17.34	96	13.68	129	9.28	162	5.89	195	4.06
29	20.05	63	17.24	97	13.56	130	9.14	163	5.81	196	4.02
30	19.98	64	17.15	98	13.43	131	9.00	164	5.74	197	3.98
31	19.91	65	17.05	99	13.31	132	8.87	165	5.67	198	3.94
32	19.84	66	16.95	100	13.19	133	8.73	166	5.61	199	3.90
33	19.77	67	16.85	101	13.07	134	8.60	167	5.54	200	3.86
34	19.69	68	16.75								

Notes:
1. Values are for compression members of 36 ksi (250 MPa) yield stress steel.
2. To obtain allowable stress in MPa, multiply tabulated value by 6.895.
3. For this table, $E = 30{,}000{,}000$ psi.

APPENDIX K

Centroids of Areas by Integration

In Section 7–3, we saw that by applying Varignon's theorem, equations for $\bar{x}$ and $\bar{y}$, which locate the centroid of an area with respect to given reference axes, could be written as

$$\bar{x} = \frac{\Sigma ax}{A} \quad \text{or} \quad \frac{\Sigma ax}{\Sigma a} \tag{7–3}$$

$$\bar{y} = \frac{\Sigma ay}{A} \quad \text{or} \quad \frac{\Sigma ay}{\Sigma a} \tag{7–4}$$

Recall that in using Eqs. (7–3) and (7–4), the total area in question was divided into component areas of known geometric shapes. For each component area, the centroid location and the area were known.

Alternatively, if an area is bounded by lines and/or curves that can be defined mathematically, the location of its centroid can be obtained using integral calculus. Since integration is the process of summing up infinitesimal quantities, the total area is divided into infinitesimally small component areas (commonly termed *differential areas* or *differential elements*) and the integration process is performed. The equations, with respect to the X and Y axes, then become

$$\bar{x} = \frac{\int x\, dA}{\int dA} \tag{7–3/K}$$

$$\bar{y} = \frac{\int y\, dA}{\int dA} \tag{7–4/K}$$

When using integration, note that we must still be able to determine the magnitude of the area of the typical differential element, as well as the location of its centroid. In all cases, limits of integration must be established so that all the differential elements are included in the integration.

□ **EXAMPLE K–1** A triangle is shown in Fig. K–1. Using integration, find the vertical distance from the base of the triangle (the reference axis) to its centroid.

Solution A typical differential area dA (where $dA = L\,dy$) is shown. The length L is determined by similar triangles:

$$\frac{L}{b} = \frac{h - y}{h}$$

$$L = \frac{b(h - y)}{h}$$

FIGURE K–1   Centroid of a triangle by integration.

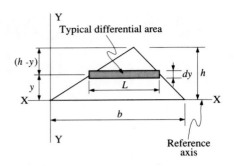

From Eq. (7–4/K), and integrating from 0 to h:

$$\bar{y} = \frac{\int y \, dA}{\int dA} = \frac{\int_0^h y \, L dy}{\int_0^h L dy} = \frac{\int_0^h y \, \dfrac{b(h-y)}{h} dy}{\int_0^h \dfrac{b(h-y)}{h} dy}$$

$$= \frac{\int_0^h y(h-y)dy}{\int_0^h (h-y)dy} = \frac{\int_0^h hy\,dy - \int_0^h y^2 dy}{\int_0^h h\,dy - \int_0^h y\,dy}$$

$$= \frac{\left[\dfrac{hy^2}{2}\right]_0^h - \left[\dfrac{y^3}{3}\right]_0^h}{[hy]_0^h - \left[\dfrac{y^2}{2}\right]_0^h} = \frac{\dfrac{h^3}{2} - \dfrac{h^3}{3}}{h^2 - \dfrac{h^2}{2}} = \frac{h}{3}$$

□ **EXAMPLE K–2**   A semicircle of radius r is shown in Fig. K–2. Using integration, find the distance from the base (reference axis) of the semicircle to its centroid.

Solution   This problem could be solved using rectangular coordinates, but polar coordinates are more convenient. The typical differential area shown is a triangle of height r and base $r d\theta$ (where θ is measured in radians). Therefore, the area of the differential area is

$$dA = \frac{1}{2}(rd\theta)r = \frac{r^2 \, d\theta}{2}$$

FIGURE K–2   Centroid of a semicircle by integration.

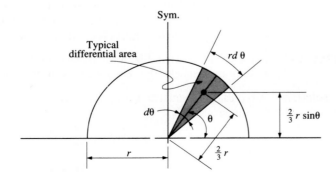

From Example K–1, we know that the centroid of a triangle lies $\frac{2}{3}$ of its height from the vertex. The distance from the reference axis to the centroid of the differential area is then $\frac{2}{3}r \sin \theta$. From Eq. (7–4/K), and integrating from 0 to π:

$$\bar{y} = \frac{\int y \, dA}{\int dA} = \frac{\int_0^\pi \left(\frac{2}{3} r \sin \theta\right) \left(\frac{r^2 \, d\theta}{2}\right)}{\int_0^\pi \frac{r^2 \, d\theta}{2}}$$

$$= \frac{\frac{r^3}{3} \int_0^\pi \sin \theta \, d\theta}{\frac{r^2}{2} \int_0^\pi d\theta} = \frac{\frac{r^3}{3} [-\cos \theta]_0^\pi}{\frac{r^2}{2} [\theta]_0^\pi}$$

$$= \frac{\frac{r^3}{3} [-(-1) - (-1)]}{\frac{r^2}{2} [\pi - 0]} = \frac{4r}{3\pi}$$

APPENDIX L

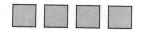

Area Moments of Inertia by Integration

In Chapter 8, the moment of inertia of a plane area A was defined as

$$I_x = \Sigma ay^2 \qquad \text{(8–1)}$$
$$I_y = \Sigma ax^2 \qquad \text{(8–2)}$$

The term a represented a small area, which was a component of area A, and x or y represented the distance from the small component area a to the axis being considered. Equations (8–1) and (8–2) furnished an approximate moment of inertia, the accuracy being a function of the size of the small area chosen.

If the area A can be defined mathematically by lines and/or curves, we may use integral calculus to perform the summation. The integration process is accomplished by dividing the plane area into an infinite number of differential areas (each designated dA) and then summing the moments of inertia of all the differential areas. The result is an exact moment of inertia. With reference to Fig. 8–1, the moment of inertia with respect to axes X–X and Y–Y may then be expressed as

$$I_x = \int y^2 dA \qquad \text{(8–1/L)}$$
$$I_y = \int x^2 dA \qquad \text{(8–2/L)}$$

It is the integration process that is used to derive the theoretical and exact moment of inertia formulas of the simple geometric shapes presented in Table 8–1. Examples illustrating this process follow.

□ **EXAMPLE L–1**　Determine the moment of inertia for the rectangular area shown in Fig. L–1 with respect to (a) the centroidal axis parallel to the base and (b) an axis coinciding with the base.

Solution　(a) The typical differential area dA (where $dA = bdy$) is shown. Selecting limits of integration as $h/2$ and $-h/2$, the moment of inertia about axis X_o–X_o is calculated from

$$I_{x_o} = \int y^2 dA$$
$$= \int_{-h/2}^{h/2} y^2 bdy = b \int_{-h/2}^{h/2} y^2 dy$$
$$= \frac{b}{3} [y^3]_{-h/2}^{h/2} = \frac{bh^3}{12}$$

Appendix L

FIGURE L–1 Moment of inertia of a rectangular area by integration.

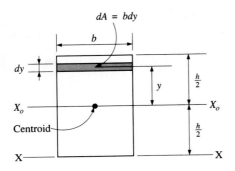

(b) Selecting limits of integration of 0 and h, the moment of inertia about axis X–X is calculated from

$$I_x = \int_0^h y^2 b\,dy = b \int_0^h y^2\,dy$$

$$= \frac{b}{3}\,[y^3]_0^h = \frac{bh^3}{3}$$

☐ **EXAMPLE L–2**   Determine the moment of inertia with respect to the base for the triangular area shown in Fig. L–2.

Solution   The typical differential area dA (where $dA = x\,dy$) is shown. The length x must be written in terms of y. From similar triangles,

$$\frac{x}{b} = \frac{h - y}{h}$$

from which

$$x = \frac{b}{h}\,(h - y)$$

FIGURE L–2 Moment of inertia of a triangular area by integration.

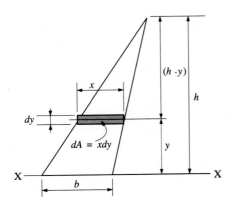

Selecting limits of integration of 0 and h, the moment of inertia with respect to axis X–X is calculated:

$$I_x = \int y^2 dA = \int y^2 x \, dy$$

$$= \int_0^h y^2 \frac{b}{h}(h - y) \, dy$$

$$= \frac{b}{h} \left[\int_0^h hy^2 \, dy - \int_0^h y^3 \, dy \right]$$

$$= \frac{b}{h} \left[\frac{hy^3}{3} - \frac{y^4}{4} \right]_0^h = \frac{bh^3}{12}$$

Notation

Symbols

A = total cross-sectional area, cross-sectional area over which stress develops, required cross-sectional area

A_M = area of moment diagram

C = compression, compressive force

C_c = arbitrary value of effective slenderness ratio that separates elastic and inelastic buckling

D = diameter

E = modulus of elasticity (Young's modulus)

F = force, load, frictional force

G = modulus of elasticity in shear, modulus of rigidity

I = area moment of inertia

J = area polar moment of inertia

K = effective length factor

L = length, span length of beam, lead of screw

M = moment of a force

M_p = plastic moment

M_R = allowable moment

M_y = yield moment

N = force or reaction acting perpendicular (normal) to a surface, number of bolts

P = force, load, axial load capacity

P_e = critical load for column buckling

Q = force, statical moment of area

R = reaction, resultant force or load, radius of curvature of the elastic curve

S = section modulus

T = tension, tensile force, torque

V = shear force

W = weight of a body, load, total distributed load, total weight

Z = plastic section modulus

a = component area, infinitesimal area

b = width of cross-section

c = radial distance to outer fiber

d = perpendicular distance between moment center and a force or between forces of a couple, diameter or linear dimension of a geometric shape, transfer distance for moment of inertia calculation

e = base of natural logarithms (2.718), eccentricity

k = stress concentration factor

g = acceleration of gravity (9.81 m/sec^2)

h = linear dimension for a geometric shape, height of cross-section

m = mass

n = modular ratio, ratio of modulus of elasticity values

n_r = number of revolutions per minute

p = pitch of thread, internal pressure

r = radius, mean radius of a screw, radius of gyration, radial distance

s = average computed stress, unit stress

t = thickness

w = uniformly distributed load intensity, weight of component element

$\bar{x}$ = distance from a resultant force to a reference point or plane, distance to centroidal Y–Y axis from reference line

y = vertical displacement

$\bar{y}$ = distance to centroidal X–X axis from reference line

Abbreviations

AISI = American Iron and Steel Institute

AITC = American Institute of Timber Construction

ASD = Allowable stress design

ASME = American Society of Mechanical Engineers

ASTM = American Society for Testing and Materials

AWS = American Welding Society

$C°$ = celsius degrees

CG = center of gravity

$F°$ = farenheit degrees

FS = factor of safety

HP = horsepower

LRFD = Load and Resistance Factor Design

N = newton

NFPA = National Forest Products Association

Pa = pascal (one newton per square meter (N/m^2))

W = watt

ft = foot

in. = inches

kip = kilopound (1000 lb)

ksi = kips per square inch

kW = kilowatt

lb = pound

m = meter

psf = pounds per square foot

psi = pounds per square inch

rpm = revolutions per minute

r/s = radian per second

sym = symmetrical

Greek Letter Symbols

α (lower case alpha) = angular value, linear coefficient of thermal expansion

β (lower case beta) = angle of contact for belt friction, angle of wrap

Δ (upper case delta) = deflection

ΔT = change in temperature

δ (lower case delta) = total deformation (total change in length)

ϵ (lower case epsilon) = strain, unit strain

θ (lower case theta) = angular value, lead angle, angle of twist, angle between two tangents

μ (lower case mu) = coefficient of friction, Poisson's ratio

ϕ (lower case phi) = angular value, angle of friction, resistance factor

Answers to Selected Problems

Chapter 1

1. $h = 34.2$ ft

3. $AB = 20$ ft, $BC = 34.18$ ft, $\angle A = 36.87°$, $\angle C = 20.56°$

5. (a) $c = 14.22$ ft, $\angle A = 43.83°$, $\angle B = 56.17°$
(b) $a = 94.72$ ft, $\angle B = 51.55°$, $\angle C = 56.45°$

7. Pads $= 3.20$ lb, shock $= 8.70$ lb

9. Vertical ht. $= 500$ ft, slant ht. $= 625$ ft, base width $= 750$ ft

11. (a) 8800 yds   **(b)** 26,400 ft

13. 6000 psf, 41.7 psi

15. 18.44 m, 27.43 m, 125.9 m

17. (a) 5670 mm   **(b)** 2.692 m   **(c)** 0.6985 m

19. (a) 0.4045 m^2 **(b)** 11.71×10^6 mm^2
(c) 51.1 kN/m   **(d)** 58 400 kPa   **(e)** 83.3 kPa
(f) 134 kPa   **(g)** 38.8 kN/m^3

21. $w = 117.6$ ft

23. 835.6 mi

25. (a) $c = 45.68$, $\angle A = 4.875°$, $\angle B = 160.1°$
(b) $c = 10.35$, $\angle A = \angle B = 75°$
(c) $c = 19.742$, $\angle A = 29.14°$, $\angle B = 76.86°$

27. 2,244,300 gal/min, 1.071 cu mi/yr

29. $DG = 575.4$ ft, $CB = 813.8$ ft

31. $AB = 318.2$ ft

Chapter 2

1. (a) $F_x = +520$ lb, $F_y = -300$ lb
(b) $F_x = -3.54$ kips, $F_y = -3.54$ kips
(c) $F_x = -600$ lb, $F_y = +1039$ lb

3. (a) $P_x = +300.7$ lb, $P_y = -109.4$ lb
(b) $P_x = +554.3$ lb, $P_y = -320$ lb
(c) $P_x = +245.1$ lb, $P_y = -205.7$ lb
(d) $P_x = +11.17$ lb, $P_y = -319.8$ lb

5. (a) $P_x = +104$ kN, $P_y = -60.0$ kN
(b) $P_x = +31.1$ kN, $P_y = -115.9$ kN
(c) $P_x = -31.1$ kN, $P_y = -115.9$ kN

7. $F_x = 107.3$ N, $F_y = 53.67$ N

9. $F = 142.8$ N

13. (a) $F_y = -611$ lb, $F_x = +222$ lb
(b) $F_y = -115$ lb, $F_x = +277$ lb

15. $F_x = +28.2$ lb

17. $\theta = 26.6°$

19. (a) $F = 361$ lb, $\theta_x = 33.7°$
(b) $F = 583$ lb, $\theta_x = 31.0°$
(c) $F = 433$ lb, $\theta_x = 56.3°$
(d) $F = 524$ lb, $\theta_x = 61.5°$

21. $T_V = 1879$ lb, $T_H = 684$ lb

23. 6 lb force: $F_y = 5.65$ lb, $F_x = 2.02$ lb
9 lb force: $P_y = 8.07$ lb, $P_x = 3.99$ lb

Chapter 3

1. $R = 64.0$ lb, $\theta_x = 27.41°$

3. $R = 65.32$ lb, $\theta_x = 76.62°$

5. $R = 13.60$ kips, $\theta_x = 68.07°$

7. $F = 87.94$ lb, $\theta_y = 3.87°$

9. $R = 124.2$ lb, $\theta_x = 65.41°$

11. $R = 388.4$ lb, $\theta_x = 38.0°$

13. $R = 1330$ lb, $\theta_x = 12.17°$

15. (a) $M_O = +300, +116.5, 0$ ft-lb
(b) $M_O = +416.5$ ft-lb c.c.
(c) $R = 84.07$ lb, $\theta_x = 55.65°$
(d) $M = +416.5$ ft-lb (checks)

17. $R = 14.018$ lb, $\theta_x = 61.37°$, $M_O = -49.2$ ft-lb

19. $M_A = +102,720$ ft-lb

21. $M_O = -1429$ ft-lb (both points)

23. $M_A = +2150$ ft-lb

25. $R = +10$ kips, $\bar{x} = 6.7$ ft

27. $R = -3$ kips, $\bar{x} = 5.67$ ft

29. $F_1 = 80$ lb $\downarrow$, $a = 11.75$ ft

31. $R = -9600$ lb, $\bar{x} = 7.83$ ft

33. $M = -1100$ in.-lb (clockwise)

35. $F = 64$ lb

37. $R = 565.3$ lb, $\theta_x = 15.39°$, $\bar{x} = 26.4$ in.

39. $R = 71.6$ kips, $\theta_x = 77.9°$, $\bar{x} = 4.43$ ft

41. $R = 59.4$ N, $\theta_x = 47.8°$

43. $R = 126.3 \times 10^3$ N, $\theta_x = 87.2°$, $\bar{x} = 1.53$ m

49. $F_2 = 579$ lb, $F_1 = 579$ lb

51. $R = 47.5$ lb, $\theta_x = 84.9°$

53. $F_2 = 240.3$ lb, $F_1 = 161.4$ lb

55. $M_A = -55.0$ ft-kips

57. $R = -2820$ lb, $\bar{x} = 22.6$ ft

59. (a) $M_A = -19{,}100$ ft-lb

(b) $R = 1700$ lb, 23.88 ft above point A

61. $R = -1$ kip, $\bar{x} = 27$ ft

63. $F_2 = 105$ lb, $F_1 = 145$ lb

Chapter 4

7. $F_A = 53.2$ lb, $F_B = 68.4$ lb

9. $C_R = 346$ lb, $F = 239$ lb

11. $F = 57.8$ lb

13. $R_B = +12.55$ kips, $R_{A_V} = +12.45$ kips

15. $R_B = +16.75$ kips, $R_{A_V} = +12.25$ kips

17. $x = 4.4$ ft

19. $R_B = +706$ lb, $R_{A_V} = +560$ lb, $R_{A_H} = 200$ lb

21. $R_{B_H} = +5328$ lb, $R_{A_H} = +5706$ lb, $R_{A_V} = +2089$ lb

23. $T = +431$ lb, $R_{A_H} = +305$ lb, $R_{A_V} = +195$ lb

25. $R_B = 5226$ lb, $R_{C_x} = +4311$ lb, $R_{C_y} = +4304$ lb

27. $CA = 18.07 \times 10^3$ N, $B_H = +12.78 \times 10^3$ N, $B_V = -2.77 \times 10^3$ N

29. $R_{A_V} = +166.8 \times 10^3$ N, $R_{B_V} = +33.4 \times 10^3$ N

37. $W = 1723$ lb

39. $R_{B_V} = 62.3$ kips, $R_A = 47.7$ kips

41. $R_B = 547$ lb, $R_{A_V} = 1353$ lb

43. $L = 10$ ft and 8.5 ft

45. $T = 7.01$ lb, $\theta = 44.5°$

47. $R_{B_H} = +138.6$ lb, $R_{A_H} = +138.6$ lb, $R_{A_V} = +330$ lb

49. $R_{B_H} = +3640$ lb, $R_{A_H} = +3640$ lb, $R_{A_V} = +4300$ lb

51. $R_{B_V} = +7.87$ kips, $R_{A_V} = +12.79$ kips, $R_{B_H} = +8$ kips

53. $R_D = 2711$ lb (@45°), $R_C = 2249$ lb, $\theta_x = 27.9°$

Chapter 5

1. $AB = 7.07$ kips (C), $AC = 5.0$ kips (T)

3. (lb): $AB = 427$ (C), $BC = 774$ (C), $AD = 670$ (T), $BD = 600$ (T), $DC = 670$ (T)

5. (kips): $AB = 0$, $AC = 30$ (C), $BC = 14.14$ (C), $BD = 20$ (C), $CD = 20$ (C), $CE = 70$ (C), $CF = 42.43$ (T), $DF = 20$ (C), $EF = 0$

7. $BC = 2697$ lb (T), $BE = 0$, $FE = 2342$ lb (C)

9. $BC = 6$ kips (T), $BG = 13.41$ kips (T), $FG = 12$ kips (C)

11. Pin reactions (lb)   A: 84.4, B: 178.9, C: 96.0, E: 468.2, F: 517.0

13. Pin reactions (lb)   A: 9798, B: 8200, D: 7017

15. $AB = 100$ lb (C), $C_V = 100$ lb, $C_H = 115.5$ lb, Reaction at $C = 152.8$ lb

17. (kN) $AB = 10$ (C), $AC = 100$ (C), $BC = 16.7$ (T), $BD = 113.3$ (C), $CD = 30$ (C), $CE = 86.7$ (C), $DE = 50$ (T), $EF = 0$, $DF = 153.3$ (C)

19. $BD = 36.5$ kN (C), $BE = 16.7$ kN (C)

21. (lbs) $AB = 14{,}170$ (C), $AD = 11{,}330$ (T), $BD = 6000$ (T), $BC = 5000$ (C), $BE = 9160$ (C), $DE = 11{,}330$ (T), $CE = 3000$ (T)

23. (lbs) $AB = 9900$ (T), $AD = 3000$ (T), $BD = 4240$ (C), $BC = 5660$ (T), $BE = 0$, $DE = 4000$ (C), $EC = 4000$ (C)

25. (kips) $AB = 10$ (C), $AD = 0$, $BC = 22.5$ (T), $BE = 43.3$ (C), $CE = 37.5$ (C), $DE = 22.5$ (C), $DF = 56.7$ (T), $FG = 10$ (T), $DG = 16.7$ (C), $EG = 73.3$ (C), $BD = 54.1$ (T)

27. (kips) $AB = 13$ (C), $AE = 0.29$ (C), $BE = 5.0$ (T), $EF = 0.29$ (C), $BF = 5.47$ (T), $BC = 13.79$ (C), $CF = 3.37$ (T), $FD = 3.79$ (T), $CD = 5.08$ (C)

29. $CD = 14.5$ kips (T), $CK = 13.3$ kips (C), $EM = 16$ kips (T)

31. $AF = 45.6$ kips (T), $BG = 26$ kips (T), $FG = 16.7$ kips (C)

33. $BE = 1.33$ kips (T), $CE = 1.703$ kips (C), $CF = 0$

35. $BC = 10{,}182$ lb (C), $CD = 8050$ lb (T), $AD = 10{,}800$ lb (T)

37. Pin reactions (lbs)   A: 313, B: 313, C: 280, D: 140, E: 140

39. Pin reactions (lbs)   A: 1127, B: 656, C: 1947

41. $F = 12.80$ lb

43. (a) $F = 150$ lb   (b) $F = 171$ lb

Chapter 6

1. 34.6 lb $<$ 39 lb (Block will not slide)

3. (a) $P = 92$ lb   (b) $P = 28$ lb

5. $P = 96.7$ lb

7. $P = 57.5$ lb

9. $P = 280$ lb

11. $P = 325.3$ lb

13. $T_L = 12,200$ lb

15. $\beta = 132.4°$

17. 1.56 turns

19. $Q = 51.7$ lb

21. $C = 4884$ lb

23. $P = 522$ N

25. $\theta = 26.6°$

27. $P = 3573$ N

29. $T_L = 296.4$ N, $T_S = 26.3$ N

33. $\mu_s = 0.40$

35. (a) $P = 700$ lb   (b) $P = 800$ lb

37. (a) $F = 17.1$ lb   (b) $P = 40.6$ lb
     (c) Zero force required

39. Body will remain at rest.

41. $P = 280$ lb

43. $P = 56$ lb, $h = 57.1$ in.

45. $\theta = 67.2°$

47. $\theta = 71.0°$

49. $W_1 = 244$ lb

51. $P = 1076$ lb

53. $T_L = 633$ lb

55. 2.41 turns

57. (a) $W = 12,280$ lb
     (b) $W = 9834$ lb (Decrease $= 2446$ lb)

Chapter 7

1. $\bar{x} = 2.68$ ft

3. $\bar{x} = 25.3$ in.

5. $l = 35.1$ in.

7. (a) $\bar{x} = 1.94$ in., $\bar{y} = 4.66$ in.
     (b) $\bar{x} = 2.81$ in., $\bar{y} = 2.30$ in
     (c) $\bar{x} = 4.91$ in., $\bar{y} = 3.80$ in.
     (d) $\bar{x} = 8.54$ in., $\bar{y} = 9.44$ in.

9. (a) $\bar{x} = 7.77$ in., $\bar{y} = 3.62$ in.
     (b) $\bar{x} = 5.5$ in., $\bar{y} = 6.5$ in.

11. $\bar{x} = 8.72$ m

13. $\bar{y} = 0.321$ m

19. $\bar{y} = 2.96$ in.

21. (a) $\bar{x} = 2.33$ in., $\bar{y} = 4.33$ in.
     (b) $\bar{x} = 9.34$ in., $\bar{y} = 3.27$ in.
     (c) $\bar{x} = 4.60$ in., $\bar{y} = 6.65$ in.

23. (a) $\bar{x} = 18.75$ in., $\bar{y} = 10.25$ in.   (b) $\bar{y} = 3.56$ in.

Chapter 8

1. (a) $I_x = 13,623$ in.^{4} (b) $I_x = 527$ in.^{4}
     (c) $I_x = 1643$ in.^{4}

3. $I_x = 1733$ in.^{4}

5. (a) $I_{x(\text{base})} = 3402$ in.^{4} (b) $I_{x(\text{base})} = 3432$ in.^{4}

7. (a) $I_x = 628$ in.^{4}, $I_y = 2979$ in.^{4}
     (b) $I_x = 246$ in.^{4}, $I_y = 61.5$ in.^{4}
     (c) $I_x = 596$ in.^{4}, $I_y = 464$ in.^{4}
     (d) $I_x = 1528$ in.^{4}, $I_y = 8327$ in.^{4}

9. $I_x = 1506$ in.^{4}, $I_y = 3476$ in.^{4}

11. $s = 7.366$ in.

13. $r_x = 3.87$ in., $r_y = 2.09$ in.

15. $r_x = 0.69$ in., $r_y = 0.67$ in.

17. $r_x = 4.36$ in., $r_y = 4.57$ in.

19. $J = 4.12$ in.^{4}

21. $I_x = 5178 \times 10^{-6}$ m^4

23. (a) $r_y = 88.8 \times 10^{-3}$ m   (b) $r_y = 115.1 \times 10^{-3}$ m

25. (a) $J = 1273 \times 10^{-6}$ m^4 (b) $J = 1.700 \times 10^{-3}$ m^4

29. $I_x = 63.4$ in.^{4}, $I_y = 30.7$ in.^{4}

31. $I_x = 1328.5$ in.^{4}

33. $I_x = 2218$ in.^{4}

35. (a) $I_x = 690.2$ in.^{4}, $I_y = 98.2$ in.^{4}
     (b) $r_x = 5.43$ in., $r_y = 2.05$ in.

Chapter 9

3. $s_{c_A} = 8.13$ ksi, $s_{c_B} = 5.5$ ksi

5. Required $d = 1.49$ in.   Use $1\frac{1}{2}$ in. diam. rod.

7. $P = 65{,}980$ lb

9. $s_s = 50{,}930$ psi

11. Required $w = 0.24$ in.

13. (a) $\epsilon = 0.0010$ in./in.   (b) $L = 1666.7$ ft
(c) $\delta = 0.42$ in.

15. $\epsilon = 0.001333$ in./in.

17. (a) $s_t = 19{,}100$ psi   (b) $\epsilon = 0.000637$
(c) $\delta = 0.1529$ in.

19. $\delta = 0.0735$ in.

21. (a) $s_t = 56$ MPa   (b) $s_t = 3.96$ MPa
(c) $s_t = 142.6$ MPa

23. $P = 350$ kN

25. Required $d = 25.9$ mm

27. $P = 10.16 \times 10^3$ N, $s_t = 517.3$ MPa

29. $\delta = 0.316$ mm

31. $s_s = 122.2$ MPa

37. Column: 5.08 ksi, Base plate: 0.306 ksi,
Pedestal: 0.133 ksi, Footing: 3.75 ksf

39. (a) $s_t = 1361.6$ psi   (b) $P = 624.8$ lb

41. $x = 2.67$ ft

43. (a) $\delta = 0.01$ ft   (b) $s_t = 3000$ psi

45. $P = 16{,}490$ lb

47. (a) $w = 3.2$ in.   (b) $s_t = 20{,}833$ psi

49. $\delta = 0.7308$ in.

51. $\delta = 0.0832$ in.

53. $y = 0.163$ in.

55. $s_t = 5250$ psi

57. (a) $s_s = 509$ psi   (b) $s_p = 848$ psi

Chapter 10

1. (a) $s_t = 26{,}160$ psi   (b) $\epsilon = 0.000894$
(c) $E = 29{,}261{,}000$ psi

3. Pt. 1: $E = 29{,}155{,}000$ psi; Pt. 2: $E = 29{,}138{,}000$ psi;
Pt. 3: N.G. (Stress > P.L.)

5. Required $d = 1.07$ in.

7. (a) $P = 189{,}640$ lb   (b) $s_t = 17{,}400$ psi

9. $\theta_{max} = 135.2°$

11. $E = 20.39 \times 10^3$ MPa

13. (a) F.S. = 1.67   (b) F.S. = 3.94

15. $P = 258.7$ kN

17. $L = 4$ m

21. 6 plates required; $s_t = 23{,}700$ psi

23. Required length = 6 in.

25. $P = 6000$ lb

27. (a) Required $A = 10.53$ in.^{2}
(b) Required $A = 8.0$ in.^{2}

29. (a) $\delta_{DC} = 0.009$ in., $\delta_{AB} = 0.004$ in.
(b) Required ratio of areas = 4.5/1.0

Chapter 11

1. (a) $\epsilon_{axial} = 0.0040$ (b) $\mu = 0.30$

3. (a) $G = 11{,}720{,}000$ psi   (b) $\mu = 0.333$

5. At 32°F, $s_t = 10{,}608$ psi (T); at 90°F,
$s_t = 3184$ psi (T)

7. $s = 11{,}700$ psi

9. $s_{CU} = 13{,}360$ psi, $s_{ST} = 26{,}720$ psi

11. (a) $P = 49.1$ kips   (b) $s_{ST} = 12.28$ ksi

13. (a) $s_t = 8000$ psi   (b) $s_t = 10{,}700$ psi
(c) $s_{t(max)} = 18{,}190$ psi

15. (a) $s_s' = 30{,}400$ psi   (b) $s_{n(max)} = 61{,}100$ psi
(c) $s_n = 27{,}400$ psi

17. $s_s' = 3460$ psi, $s_n = 2000$ psi

19. (a) $\theta = 26.57°$ (b) $\Sigma F = 0$ (check)

21. $\delta_z = 0.269$ mm (increase),
$\delta_y = 0.0472$ mm (decrease),
$\delta_x = 1.053 \times 10^{-3}$ mm (decrease)

23. $s = 51.8$ MPa

25. $\mu = 0.258$

27. $P_{BR} = 95.75$ kN, $P_{ST} = 42.13$ kN

29. $s_{avg} = 32.14$ MPa, $s_{max} = 75.5$ MPa

31. (a) $s_s' = 70.6$ MPa, $s_n = 122.2$ MPa
(b) $s_{s(max)}' = 81.5$ MPa, $s_{n(max)} = 163.0$ MPa

33. $P_{all} = 96$ kN

39. $\delta_{axial} = 0.0256$ in. (increase),
$\delta_{trans} = 0.000266$ in. (decrease)

41. $E = 29{,}530{,}000$ psi; $\mu = 0.2496$

43. Dimensions: $x = 2.997$ in., $y = 12.016$ in., $z = 0.9994$ in.

45. $T = 61.6°F$, $s = 9.44$ ksi

47. Corrected dist. = 2209.19 ft; at 20°F, dist. = 2207.87 ft

49. Temp. rises: $s = 6880$ psi (T); temp. falls: $s = 22,480$ psi (T)

51. $s_{ST} = 22,300$ psi, $s_{CU} = 11,150$ psi

53. $196.2°F$

55. $s_{ST} = 13.18$ ksi, $s_{CI} = 6.59$ ksi, $s_{CON} = 1.37$ ksi

57. (a) $s_{BR} = 6100$ psi, $s_{ST} = 13,070$ psi
(b) $P_{BR} = 23,975$ lb, $P_{ST} = 41,040$ lb
(c) $\delta = 0.00523$ in.

59. (a) $s_{BR} = 6802$ psi, $s_{ST} = 17,006$ psi
(b) $\delta = 0.136$ in.

61. $W = 108,100$ lb

63. $r_{min} = 0.75$ in.

65. $P_{max} = 282,700$ lb

67. $s'_s = 866$ psi, $s_n = 1500$ psi

69. $s_{n(max)} = 9200$ psi (T) & (C)

Chapter 12

1. A: 4 in.-kips, B: 13 in.-kips, C: 31 in.-kips

3. $s_s = 7130$ psi

5. $T_R = 77,300$ in.-lb, $s_s = 6750$ psi

7. $s_{s(outer)} = 8800$ psi, $s_{s(inner)} = 4400$ psi

9. (a) $P_3 = 414.7$ lb   **(b)** $s_{s(max)} = 1736$ psi

11. $\theta = 1.824°$

13. $\theta = 2.407°$

15. $\theta = 3.438°$

17. $T = 1621$ in.-lb

19. Required $d = 0.634$ in.; use an $\frac{11}{16}$ in. diam. shaft.

21. (a) $s_s = 5016$ psi (O.K.)   **(b)** $\theta = 0.862°$ (O.K.)

23. $s_s = 4760$ psi

25. $s_{s(max)} = 37.1$ MPa

27. $W = 17.35$ kN

29. $s_{s(max)} = 88.4$ MPa

35. O.D. = 2 in., I.D. = 1 in.

37. Required $d = 1.822$ in.; use a $1\frac{7}{8}$ in. diam. shaft.

39. $G = 12,028,000$ psi

41. O.D. = 9.5 in., I.D. = 6.25 in., $\theta = 1.25°$

43. (a) $s_s = 11,150$ psi   **(b)** $s_s = 2970$ psi

45. Required $d = 1.528$ in.

47. $s_{s(solid)} = 4013$ psi, $s_{s(hollow)} = 4093$ psi

Chapter 13

1. (a) $R_A = R_B = 16$ kips
(b) $R_A = 11.11$ kips, $R_B = 8.89$ kips

3. (a) $R_A = 19.2$ kips, $R_B = 4.80$ kips
(b) $R_A = 26.36$ kips, $R_B = 9.64$ kips

5. (a) $R_A = 42.6$ kips, $R_B = 42.4$ kips
(b) $R_A = 6.2$ kips, $R_B = 20.8$ kips

7. (a) $V = -8.3$ kips, $M = +250$ ft-kips
(b) $V_{left} = +9$ kips, $V_{right} = -9$ kips, $M = +240$ ft-kips

9. (a) $V_{max} = +19.8$ kips   **(b)** $V_{max} = \pm16$ kips

11. (a) $V_{max} = +26.43$ kips   **(b)** $V_{max} = +21.67$ kips

13. (a) $V_{max} = \pm5$ kips, $M_{max} = +25$ ft-kips
(b) $V_{max} = +11.78$ kips, $M_{max} = +74.02$ ft-kips

15. (a) $V_{max} = 71.15$ kips, $M_{max} = +446.8$ ft-kips
(b) $V_{max} = +15.62$ kips, $M_{max} = +126.4$ ft-kips

17. $V_{max} = +15.33$ kips, $+M_{max} = +22.2$ ft-kips, $-M_{max} = -32.0$ ft-kips

19. Abs. $V_{max} = +33.3$ kips, Abs. $M_{max} = +208.4$ ft-kips

21. Abs. $V_{max} = +43.9$ kips, Abs. $M_{max} = +471$ ft-kips

23. (a) $R_A = 230$ kN, $R_B = 30$ kN
(b) $R_A = 153.4$ kN, $R_B = 86.6$ kN

25. (a) $V_{10} = -15.6$ kN, $M_{10} = +1094$ kN·m, $V_{16(left)} = -90.6$ kN, $V_{16(right)} = -165.6$ kN, $M_{16} = +663$ kN·m
(b) $V_{10} = +5.90$ kN, $M_{10} = +145.4$ kN·m, $V_{16(left)} = -46.6$ kN, $V_{16(right)} = +40$ kN, $M_{16} = -79.6$ kN·m

27. $V_{max} = +150$ kN, $M_{max} = -240$ kN·m

29. $V_{max} = -106$ kN, $M_{max} = -384$ kN·m

35. (a) $R_A = 140$ lb, $R_B = 460$ lb
(b) $R_A = 4200$ lb, $R_B = 5800$ lb
(c) $R_A = 7140$ lb, $R_B = 11,860$ lb
(d) $R_A = 21$ kips, $R_B = 45$ kips

37. (a) $V_4 = +140$ lb, $M_4 = +560$ ft-lb, $V_{10(left)} = -260$ lb, $V_{10(right)} = +200$ lb, $M_{10} = -600$ ft-lb
(b) $V_4 = +200$ lb, $M_4 = -7600$ ft-lb, $V_{10} = -1000$ lb, $M_{10} = -7600$ ft-lb

39. $V_{max} = -2700$ lb, $M_{max} = -12,750$ ft-lb

41. $V_{max} = +9.6$ kips, $M_{max} = +46.1$ ft-kips

43. $V_{max} = +26.4$ kips, $M_{max} = +174.2$ ft-kips

45. $V_{max} = -7860$ lb, $M_{max} = +23,400$ ft-lb

47. $V_{max} = -5000$ lb, $M_{max} = -15,000$ ft-lb

49. $V_{max} = +1920$ lb, $M_{max} = +3840$ ft-lb

51. Abs. $V_{max} = 64$ kips, Abs. $M_{max} = +384$ ft-kips

53. $R_{A_V} = 1.897$ kips, $R_{A_H} = 0.662$ kips, $R_B = 1.286$ kips

Chapter 14

1. $S_x = 82.7$ in.^{3}
3. $s_{b(max)} = 752$ psi
5. $s_{b(max)} = 26.4$ ksi
7. **(a)** $M_R = 256$ in.-kips **(b)** $M_R = 220$ in.-kips
9. **(a)** $s_s = 12.7$ ksi **(b)** $s_s = 11.53$ ksi
11. $s_s = 7550$ psi
13. $w = 922.1$ lb/ft
15. $w = 17,742$ lb/ft
17. **(a)** $S_x = 25.15$ in.^{3}, $Z_x = 45$ in.^{3}, S/F = 1.79
    (b) $S_x = 108.7$ in.^{3}, $Z_x = 138$ in.^{3}, S/F = 1.27
19. $P = 55$ kips
21. $s_b = 50.3$ MPa
23. **(a)** $s_b = 10.4$ MPa **(b)** $s_b = 5.20$ MPa
25. **(a)** $M_R = 3.23$ MN·m **(b)** $M_R = 447.1$ kN·m
27. **(a)** $s_s = 81.3$ MPa **(b)** $s_s = 72.1$ MPa
29. $P_{max} = 14.74$ kN
33. **(a)** $S_x = 255.2$ in.^{3}
    (b) $S_{x(bot)} = 191.5$ in.^{3}, $S_{x(top)} = 129.7$ in.^{3}
35. $s_{b(max)} = 23,472$ psi
37. $s_{b(max)} = 15,840$ psi
39. $s_{s(max)} = 107.0$ psi
41. s_s (ksi) at: N.A.: 0.511; web-flange junction (in web): 0.486; web-flange junction (in flange): 9.27; 2 in. from tip of flange: 5.17
43. $s_b = 21.1$ ksi, $s_s = 14.7$ ksi
45. $s_{b(top)} = 9580$ psi (C), $s_{b(bot)} = 3710$ psi (T), $s_{s(max)} = 1630$ psi (at N.A.)
47. **(a)** $P_{max} = 1080$ lb **(b)** $s_{b(max)} = 1620$ psi
49. Max. screw spacing = 4.76 in.
51. $P_{max} = 6845$ lb
53. $M_y = 266.7$ ft-kips, $M_p = 296.1$ ft-kips, $w_{max} = 2.63$ kips/ft

Chapter 15

1. W18 × 50
3. W18 × 40
5. W21 × 73
7. Required $h = 1\frac{1}{2}$ in., required $w = \frac{1}{2}$ in.
9. 6 × 18 (S4S)
11. 6 × 18 (S4S)

13. **(a)** $\phi_b M_n = 780$ ft-kips **(b)** $\phi_b M_n = 212$ ft-kips
    (c) $\phi_b M_n = 54.3$ ft-kips
15. W27 × 94
17. W27 × 94
19. W610 × 1.11
21. 50 × 360 (S4S)
25. W27 × 94
27. W14 × 22
29. Beams: W16 × 26, Girders: W21 × 62
31. Required $d = 1.765$ in., use $1\frac{7}{8}$ in. diam. rod.
33. 12 × 16 (S4S)
35. 10 × 20 (S4S)
37. 2 × 14 (S4S)
39. 3 × 14 (S4S)
41. 10 × 18 (S4S)
43. Use 4 planks.
45. W27 × 94

Chapter 16

1. $s_{b(max)} = 20.0$ ksi
3. Min. $D = 333.3d$
5. $R = 971.7$ ft
7. $\Delta = 0.283$ in. < 0.60 in. (O.K.)
9. $P_{max} = 3030$ lb
11. Required $I = 332$ in.^{4}
13. $\theta = 0.392°$
15. $\theta = 3.06°$
19. $P = 6.74$ kips
21. $\Delta = 0.1661$ in.
23. $M_{max} = +60$ ft-kips
25. $M_{max} = +4800$ ft-lb
27. $\Delta_{max} = 0.9124$ in.; at load, $\Delta = 0.747$ in.
29. $E = 1,609,000$ psi
31. Between supports, $\Delta_{max} = 0.655$ in.; at free-end, $\Delta = 0.292$ in. (upward)
33. $\Delta_{max} = 0.823$ in.
35. $\theta = 0.45°$, $\Delta_{max} = 1.511$ in.
37. $\Delta = 0.680$ in. (to the right)
39. $\Delta = \dfrac{23PL^3}{648EI}$

41. $\Delta = 1.968$ mm

43. Required $d = 65.1$ mm, $s_b = 25.8$ MPa < 165 MPa (O.K.), $s_s = 1.201$ MPa < 100 MPa (O.K.)

45. $\Delta_{\text{centerline}} = 73.2$ mm; at concentrated load, $\Delta = 52.6$ mm

51. $M = 844$ in.-lb

53. (a) $\Delta_{\text{max}} = 0.482$ in.   (b) $s_{b(\text{max})} = 21.2$ ksi

55. $R_C = 6350$ lb

57. $\Delta = 9.56$ in.

59. $\Delta_{\text{max}} = 1.207$ in.

63. $\theta = 0.291°$, $\Delta_A = 0.857$ in., $\Delta_C = 0.282$ in.

65. $L = 8.23$ ft

67. $P_{\text{max}} = 9.747$ kips

69. W24 $\times$ 76

71. $$\Delta = \frac{5PL^3}{9EI}$$

Chapter 17

1. $s_A = -8250$ psi (C), $s_B = +14,250$ psi (T)

3. Bottom: $s_{\text{max}} = +17,154$ psi (T), top: $s_{\text{min}} = -12,414$ psi (C)

5. $s_{AA} = -520$ psi (C), $s_{BB} = -3646$ (C)

7. $s_A = +8369$ psi (T), $s_B = -8743$ psi (C)

9. $e = 21.71$ in.

11. $e = 14.59$ in.

13. $s_A = -104.2$ psi (C), $s_B = -417$ psi (C), $s_C = +208$ psi (T), $s_D = -104.2$ psi (C)

15. (a) $s_1 = +16,220$ psi (T), $s_2 = -2220$ psi (C), $\theta_P = 69.7°$, $s_s' = \pm9220$ psi
(b) $s_1 = +2220$ psi (T), $s_2 = -16,220$ psi (C), $\theta_P = 20.3°$, $s_s' = \pm9220$ psi
(c) $s_1 = +13,810$ psi (T), $s_2 = -1810$ psi (C), $\theta_P = 19.9°$, $s_s' = \pm7810$ psi

17. Point A: $s_1 = +3900$ psi (T), $s_2 = -1300$ psi (C);
Point B: $s_1 = +7480$ psi (T), $s_2 = -381$ psi (C)

19. $s_n = +7500$ psi (T), $s_s' = -4330$ psi

21. For $\theta = 75°$: $s_n = -11,200$ psi (C), $s_s' = +3000$ psi.
For $\theta = 65°$: $s_n = -9860$ psi (C), $s_s' = +4600$ psi.
For $\theta = 50°$: $s_n = -7040$ psi (C), $s_s' = +5910$ psi.
For $\theta = 30°$: $s_n = -3000$ psi (C), $s_s' = +5200$ psi.
$s_{s(\text{max})}' = +6000$ psi

23. (a) $s_1 = +21,640$ psi (T), $s_2 = +5360$ psi (T), $\theta_{P_1} = 39.69°$ (clockwise), $s_{s(\text{max})}' = \pm8140$ psi

(b) $s_1 = +14,190$ psi (T), $s_2 = -17,190$ psi (C), $\theta_{P_2} = 15.33°$ (clockwise), $s_{s(\text{max})}' = \pm15,690$ psi

25. Point A: $s_1 = +3925$ psi (T), $s_2 = -2039$ psi (C);
Point B: $s_1 = +8100$ psi (T), $s_2 = -556$ psi (C)

27. $s_1 = +0.470$ MPa (T), $s_2 = -274.8$ MPa (C), $\theta_P = 87.64°$, $s_{s(\text{max})}' = \pm137.62$ MPa

33. $P_{\text{max}} = 117,845$ lb

35. $s_{\text{bot}} = +8803$ psi (T), $s_{\text{top}} = -11,769$ psi (C)

37. $s_A = -2697.6$ psf $= -18.73$ psi (C), $s_B = 297.6$ psf $= +2.06$ psi (T)

39. $s_{AA} = +260.4$ psi (T), $s_{BB} = -1302$ psi (C)

41. $s_{AA} = 0$, $s_{BB} = -38.9$ psi (C)

43. $s_A = +1808$ psi (T), $s_B = -4553$ psi (C)

45. $P_{\text{max}} = 108,000$ lb

47. $e = 3.21$ in.

49. Point A: $s_1 = -3.26$ ksi (C), $s_{s(\text{max})}' = \pm1.63$ ksi on 45° plane;
Point B (in web): $s_1 = +0.48$ ksi (T), $s_2 = -3.48$ ksi (C), $\theta_P = 20.50°$, $s_{s(\text{max})}' = \pm1.98$ ksi;
Point C: $s_1 = s_2 = \pm1.99$ ksi, $\theta_P = 45°$

51. (a) $s_1 = +11.318$ ksi (T), $s_2 = 11.318$ ksi (C), $\theta_P = 45°$ (clockwise)
(b) $s_n = \pm7.275$ ksi, $s_s' = \pm8.67$ ksi

Chapter 18

1. $P_e = 100.2$ kips

3. Min. $L/r = 55.5$

5. $P_a = 21.2$ kips

7. $P_a = 1925$ lb

9. $P_a = 695$ kips

11. (a) $P_a = 281$ kips < 300 kips (N.G.)
(b) $P_a = 323$ kips > 300 kips (O.K.)

13. $P_a = 240$ kips

15. Use a 4 in. diam. std.-weight pipe.

17. W12 $\times$ 40

19. $P_a = 19.91$ kips

21. Required $d = 1.075$ in., use a $1\frac{1}{8}$ in. diam. rod.

23. (a) $P_a = 56.7$ kips   (b) $P_a = 38.8$ kips

25. 12 $\times$ 12 (S4S)

27. (a) 8 $\times$ 8 (S4S)   (b) 8 $\times$ 10 (S4S)

29. $0.593 < 1.0$ (Col. is adequate.)

31. (a) $P_e = 13.25$ kN   **(b)** $P_e = 3.31$ kN
   (c) $P_e = 828$ kN

33. $P_a = 26.6$ kN

37. (a) $P_e = 42.6$ kips   **(b)** $P_e = 6.81$ kips
   (c) $P_e = 2.66$ kips

39. $L = 15.01$ ft

41. Max. $KL/r = 66.2$

43. W8 $\times$ 24

45. $P_a = 1114$ kips

47. W10 $\times$ 68

49. $1\frac{1}{8}$ in. diam. rod

51. Required dimensions: 0.517 in. $\times$ 1.035 in.

53. $P_a = 105.5$ kips

55. 10 $\times$ 10 (S4S)

57. $M_{max} = 124.2$ ft-kips

Chapter 19

1. $P_n = 130.5$ kips

3. $P_s = 113.6$ kips

5. (a) $P_s = 101$ kips   **(b)** $P_n = 112.5$ kips

7. 10 bolts each side of splice, 2 gage lines (5 per line).

9. Required $t_{min} = 0.352$ in.

11. $P_s = 54.1$ kips

13. $P_s = 58.3$ kips

15. Required $L = 6.6$ in., use 7 in.

17. $P_s = 4.8$ kips/in.

19. $P_n = 179.4$ kN

21. $P_s = 364$ kN

23. End weld $L = 127$ mm, side welds $L = 126$ mm each side.

27. $P_s = 77.8$ kips

29. (a) Use 24 bolts, 3 gage lines (8 per line).
   (b) Use 14 rivets, 2 gage lines (7 per line).

31. $P_s = 150.2$ kips

33. All. tens. load = 44 kips

35. All. tens. load = 199.8 kips

Chapter 20

1. $s_{t_c} = 14,950$ psi

3. $p = 4690$ psi

5. Required $t = 0.39$ in.

7. $p_{max} = 196.9$ psi

9. Required $t = 0.462$ in.

11. Required $t = 0.406$ in.

13. Required $t = 13.64$ mm

19. $p = 3050$ psi

21. Required $t = 0.164$ in.

23. Safe head = 374.4 ft

25. $P_{max} = 18,850$ lb

Chapter 21

1. W16 $\times$ 26

3. (a) $R_A = 1.25$ kips, $V_B = 2.75$ kips, $M_B = 15$ ft-kips
   (b) $R_A = 2.0$ kips, $V_B = 8.0$ kips, $M_B = 33.3$ ft-kips

5. W16 $\times$ 26

7. $R_A = -3.28$ kips (uplift), $R_B = 16.17$ kips, $R_C = 2.11$ kips

9. 4 $\times$ 10 (S4S)

11. $M_1 = 0$, $M_2 = -8$ ft-kips, $M_3 = -28$ ft-kips, $M_4 = 0$, $R_1 = 4.5$ kips, $R_2 = 4.25$ ft-kips, $R_3 = 13$ kips, $R_4 = 8.25$ kips

13. $R_A = 7.2$ kips, $V_B = 12$ kips, $M_B = 57.6$ ft-kips, $s_b = 17.91$ ksi

15. $R_A = 7.74$ kips, $V_B = 11.46$ kips, $M_B = 44.7$ ft-kips, $s_b = 13.90$ ksi

17. $s_b = 19.74$ ksi

19. $R_A = 15.72$ kips, $V_B = 24.28$ kips, $M_{max} = -105.04$ ft-kips

21. $M_2 = -22.13$ ft-kips, $M_3 = -27.75$ ft-kips, $M_4 = -33.38$ ft-kips

23. (a) $R_A = R_C = 26.25$ kips, $R_B = 87.5$ kips
   (b) $R_A = R_C = 27.8$ kips, $R_B = 84.4$ kips

Index

ISBN 0-02-414961-6

9 780024 149619

90000>

TABLE 1–3 *SI Prefixes*

Prefix	SI Symbol	Factor	Example
giga	G	$1\ 000\ 000\ 000 = 10^9$	Gm = 1 000 000 000 meters
mega	M	$1\ 000\ 000 = 10^6$	Mm = 1 000 000 meters
kilo	k	$1000 = 10^3$	km = 1000 meters
milli	m	$0.001 = 10^{-3}$	mm = 0.001 meters
micro	μ	$0.000\ 001 = 10^{-6}$	μm = 0.000 001 meters